AUTODESK® REVIT Basics Training Manual

AUTODESK® REVIT Basics Training Manual

Brian W. Clayton

Industrial Press, Inc.

Industrial Press, Inc.
32 Haviland Street, Suite 3
South Norwalk, CT 06854
Phone: 203-956-5593 Toll-Free in USA: 888-528-7852 Fax: 203-354-9391
Email: info@industrialpress.com

Author: Brian W. Clayton

Title: Autodesk® REVIT Basics Training Manual, First Edition

Library of Congress Control Number: 2017947360

Printed in the United States of America.

ISBN (print): 978-0-8311-3621-5

ISBN (ePDF): 978-0-8311-9458-1

ISBN (ePUB): 978-0-8311-9459-8

ISBN (eMobi): 978-0-8311-9460-4

Editorial Director: Judy Bass

Cover and Interior Designer: Janet Romano-Murray

industrialpress.com ebooks.industrialpress.com

Table of Contents

Acknowledgments

I would like to give thanks to the people who have assisted me along the way in getting my business started, the ones who have helped guide me in my training for my knowledge of the software to pass along to others, and an individual person who advised me along the way in my curriculum advancements and publishing my manual.

Personal thanks to my family, Robert Hughes, Daniel John Stine, CSI, CDT and Thomas R. Gauthier, Ed.D, as well as my college instructors.

Many thanks to Industrial Press, Inc. for making *all* this possible!

About the Author

Brian W. Clayton is the owner of Clayton Consulting and Design, LLC. located in Greensboro, NC., with over 20 years of AutoCAD experience and 8 years of Autodesk Revit experience. He has an A.A.S. Mechanical Engineering Degree, A.A.S. Mechanical Engineering Degree in CAD Support, and an A.A.S. Architectural Technology Degree, all from Guilford Technical Community College. He currently works as a commercial building space planner/designer, and provides commercial space measuring and upfits for property management companies, along with residential home designs and renovations. He is also an AutoCAD and Revit specialist for architectural and engineering firms, providing corporate training on the latest Revit /BIM and AutoCAD software, along with providing companies with informative ways to improve productivity in the use of the software.

Mr. Clayton is a faculty member at various higher education institutions, specializing in the facilitation of the course work to students enrolled in the Architectural Technology degree, Graphic Communication Systems and Continuing Education programs. His concentration is in technical training in various Revit/BIM and AutoCAD courses for the students and professionals in the surrounding city of Greensboro, NC, who want to continue their education and keep up to date with the latest versions of the software. The various higher education institutions Mr. Clayton works with include Guilford Technical Community College, Forsyth Technical Community College and North Carolina A&T State University. Mr. Clayton currently teaches courses in AutoCAD Fundamentals, Advanced AutoCAD, Revit Fundamentals, Advanced Revit, Revit MEP, Revit Interior Design, Revit Structure and Autodesk Inventor.

Please visit Brian Clayton's website for a list of his clientele and photos of projects he has completed: www.claytonconsultinganddesign.com.

Foreword

Brian Clayton's *Autodesk® Revit Basics Training Manual* is a contemporary and relevant text for all those looking to learn or advance their design capabilities into the future of building information modeling (BIM). Today, the building design and construction industry has been and continues to be influenced by modern technological gains. Like several other industries, building design and construction requires a deep commitment to understanding BIM and its application. Professional educators and consultants facilitating design and construction curricula need to adopt a learning culture and create professional development opportunities in order to provide the industry with a contemporary skill set. The practice of designing a structure by hand is nearly obsolete, and while computer aided drafting software has been used for several years, BIM is a relatively new concept for the design and construction industry.

Currently there is very little literature available that can be used as an instructional manual for the application of the Revit software. *Autodesk Revit Basics Training Manual* presents an important contribution to the building design and construction industry and provides colleges and universities with a step-by-step instructional manual for the application of Revit. While this manual can be beneficial to numerous design and construction disciplines and trades, the text is focused on the architectural and interior design application. However, the author does briefly introduce the reader to other concepts such as plumbing for bathroom fixtures, and electrical for lighting and ceiling fans. The author also discusses Revit's mechanical application used for HVAC design.

Brian Clayton has developed an easy-to-read, systematic Revit user's manual. This work provides students and professionals alike with the skills required to be competitive in today's design and construction market, and also offers educators a supplement to their career and technical program pedagogy as a result of collaborations between educators and employers[1]. Likewise, students will be able to use this manual as a career-long reference once they are employed in the design and construction industry. In addition, if students or design professionals have prior Revit experience, this manual can offer useful tips and supplement their prior knowledge, improving their design productivity by offering a more efficient way to apply Revit in their discipline.

Revit can be used by large architectural firms with various teams within a firm working on a model and it can be used by small, one-person design firms to help facilitate the design and construction process. Revit projects can range in size from large, complex designs to small, relatively simple designs. Since Revit is BIM software, it is able to jump back and forth between different disciplines. Therefore, it is considered a multifunctional software that supports collaboration efforts among the various trades and disciplines involved in large or small design projects. In today's LEED[2]-based design economy, Revit software offers the support required to accurately respond to LEED design inquiries. Because of the prevalence of LEED, many design firms are required to use BIM software to complete their contract draw-

[1] *Hora, M.T., Benbow, R.J., & Oleson, A.K. (2016). Beyond the skills gap Cambridge, MA: Harvard Education Press*

[2] *LEED stands for Leadership in Energy and Environmental Design, a rating by the USGBC—the United States Green Building Council—to evaluate the environmental performance of a building, and encourage market transformation towards sustainable design.*

ings. While the LEED application of the Revit software is outside the scope of this manual, the author provides learners with a solid foundation in Revit upon which they can build.

Brian Clayton has written this manual to provide those professionals interested in learning Revit with the objective of increasing design production and to offer career and technical pedagogy with an informed, easy-to-read, step-by-step user's manual. While this manual offers a streamlined facilitation of the software, Revit instruction in general is an advanced pedagogy. Therefore, this manual is intended to target those professionals and students who are already familiar with Autodesk AutoCAD software. Indeed, because Revit is a more complex design software, the ability for a first-time Revit user to complete a small design project using this manual as a reference is favorable.

This Revit manual helps to bridge the gap between the lack of Revit instruction literature and the demand for the software to be used in the industry. Brian Clayton's *Autodesk Revit Basics Training Manual* is a valuable tool for students and professionals.

Thomas R. Gauthier, EdD.
Associate Professor
Guilford Technical Community College

Preface

This manual is written for the basics of the Revit software. It focuses on Revit architecture and interior design. It does briefly go into the other Revit disciplines, such as plumbing for bathroom fixtures, electrical for lighting and ceiling fans, as well as the mechanical for the air conditioning and furnace locations.

The manual is intended for the student who is already familiar with Autodesk® AutoCAD software and the experience of navigating through the interface ribbon. The purpose of this work is to provide a more efficient way of learning the Revit software that is straight forward and to the point, without trying to make it more complicated than it needs to be.

For students or professionals with prior Revit experience, this manual can provide useful tips and improve their productivity, and may even provide a more efficient way of using Revit in their disciplines or normal day-to-day Revit use.

Revit is also known as a *BIM* software, which stands for *Building Information Modeling* (or Model). There are many other types of BIM software out there, such as Bentley and Tekla. The most common is Revit, produced by Autodesk, Inc.

In this manual we focus on the architectural side for the examples used to navigate through the ribbon and the functions. There are chapters that will briefly touch on interior design, interior renderings, and 3-point perspective layouts. One important thing to understand about the Revit software is that once you understand how Revit works and the fundamentals behind the software, you can begin to cut your design time down by 25–65% with much more accuracy.

Revit can be used not only by large architectural firms, but by small, one-person design firms, MEP engineering firms, and both large and small interior design firms. The Revit projects can range from very large to projects that are just over 2000–5000 square feet. Because Revit is able to transfer between different disciplines, it becomes a *multifunctional* software—don't let it intimidate you. Rest assured that you can quickly and easily learn to use Revit no matter the size of your design task.

Revit has its own family of custom doors, windows and other features that come with the software that will assist in making your designs easier, simpler and more efficient. This manual will show you how to get to these functions and how and when to utilize them to your advantage.

In order to make following the steps a bit easier, I've included icons throughout the text. These are:

ESSENTIAL REVIT FACT

Provides information that only applies to the Revit software.

EXAMPLE

Shows a sample of how the instructions look after they are implemented.

Tells readers where they might get additional information about a topic.

Gives readers a bit more insight into the task at hand.

REMINDER

Repeats vital instructions about certain tasks that were discussed earlier in the manual.

EXAMPLE

Offers insider information of how I tend to organize my designs.

WARNING

Cautions you about doing or not doing something that could interfere in your design.

Enjoy the Revit/BIM/3D software and this manual, and see how your drawings will change the way you and your clients look at designs.

CHAPTER 1

Revit Templates

GETTING STARTED

When beginning with a Revit project, you need to understand the different types of templates that are available to you. Revit has default templates, as well as certain templates you have to browse and select for your application.

When beginning a project, the opening Revit screen looks like the following screen:

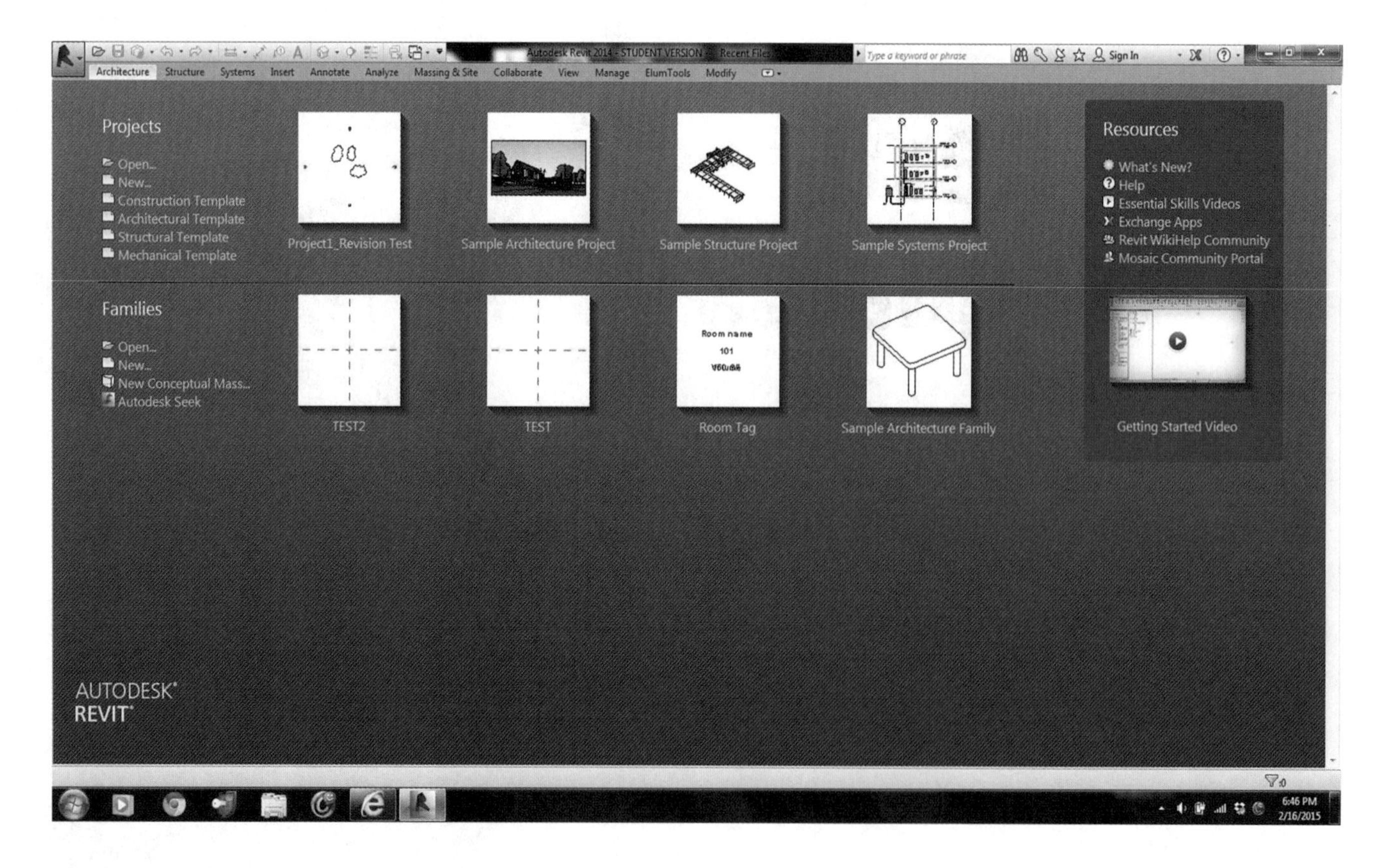

Let's take a look at the templates that are available as soon as you open the Revit software.

These templates will quickly take you to a drawing to begin your project.

NOTES

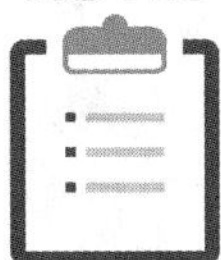

As you select a template to use, you need to pay close attention to which Revit "families" are applied to your project.

EXAMPLE

For example, if you click on the ARCHITECTURAL ***template, and you begin your drawing, you will notice it defaults to a commercial wall with metal studs for a commercial design.***

The structural template is mainly for structural concrete and steel design. It doesn't have any floor plan details for commercial or residential designs.

The mechanical template is mainly used for heating, ventilation, and air conditioning system design.

NOTES

This template will *not* ***have structural layout designs or floor plans for commercial and residential designs.***

These various templates are separated for their specific designs and application in different disciplines.

If you want to get to a specific template, such as a residential template, click on the **blue R** in the Application Menu.

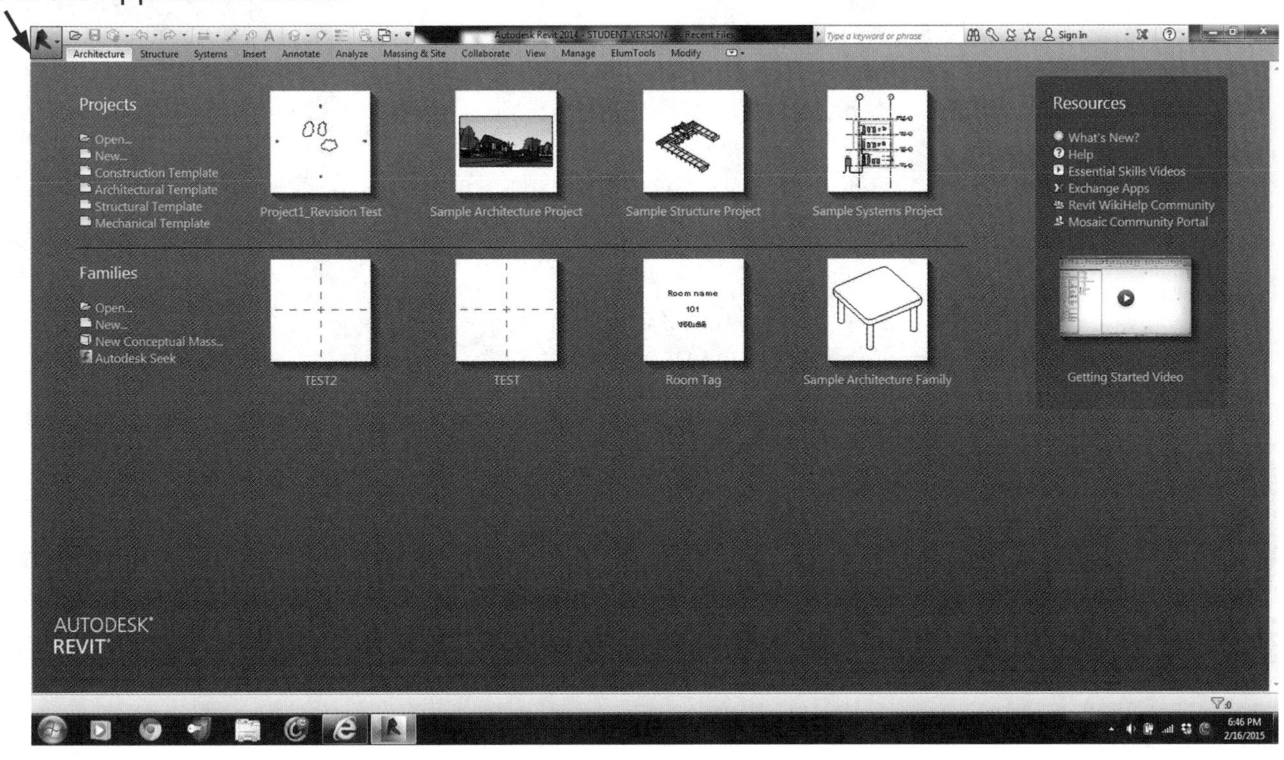

The following drop-down window will open:

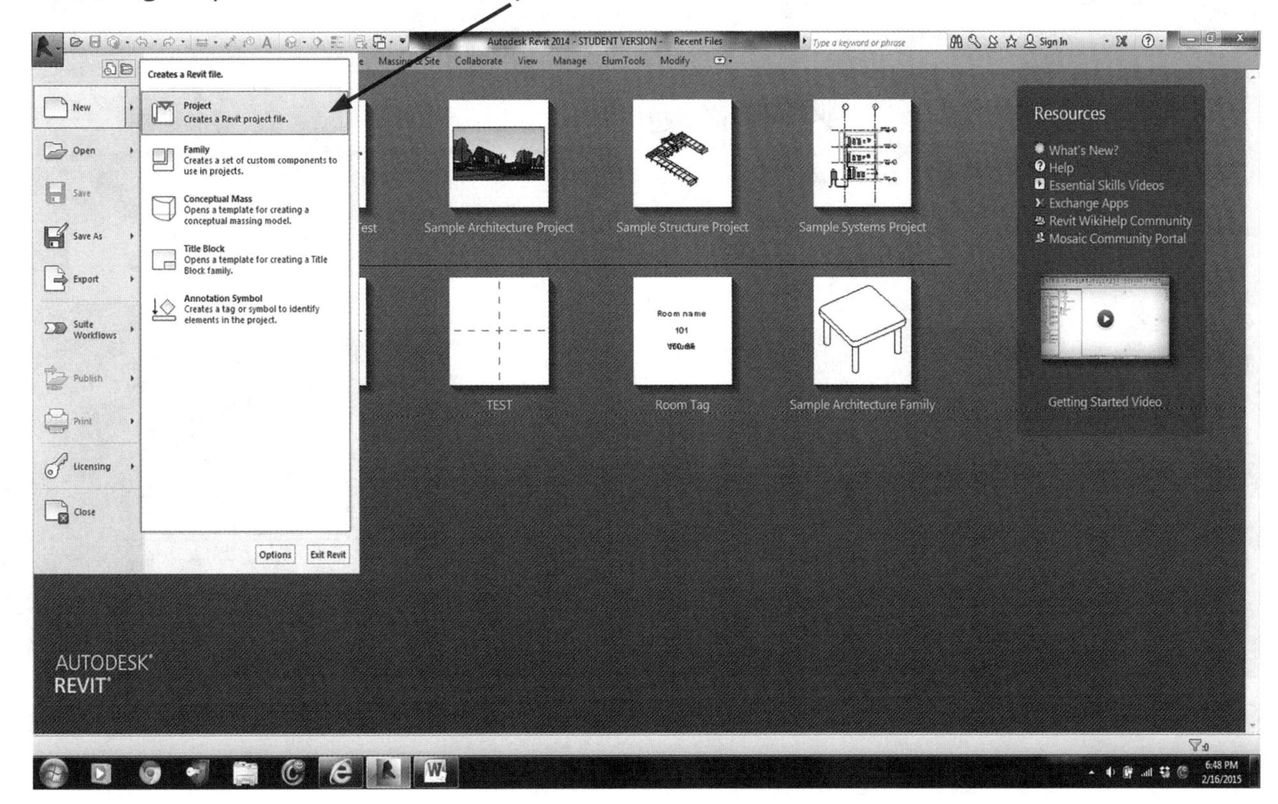

Next, click on **NEW**. Then go to **PROJECT**.

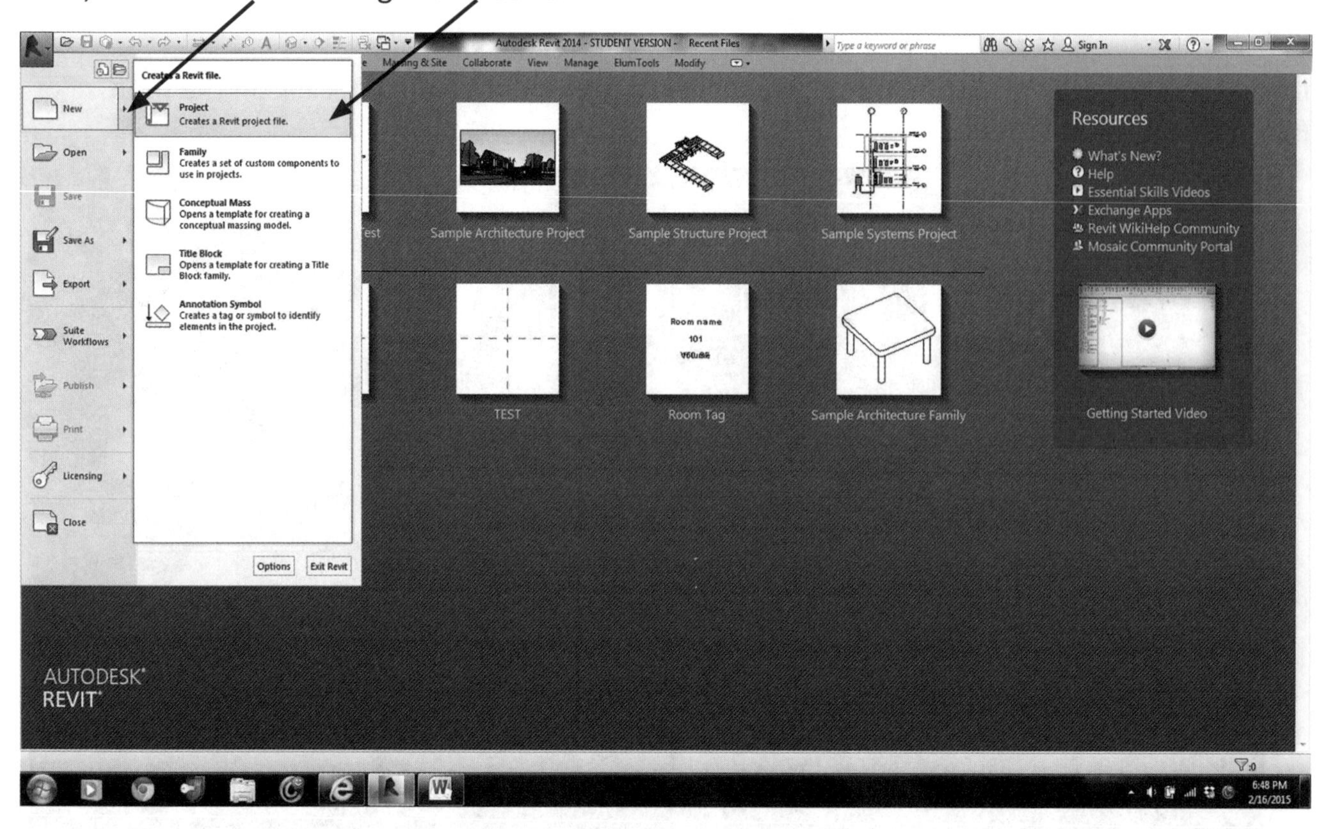

A new Project Dialog Box will appear on the screen:

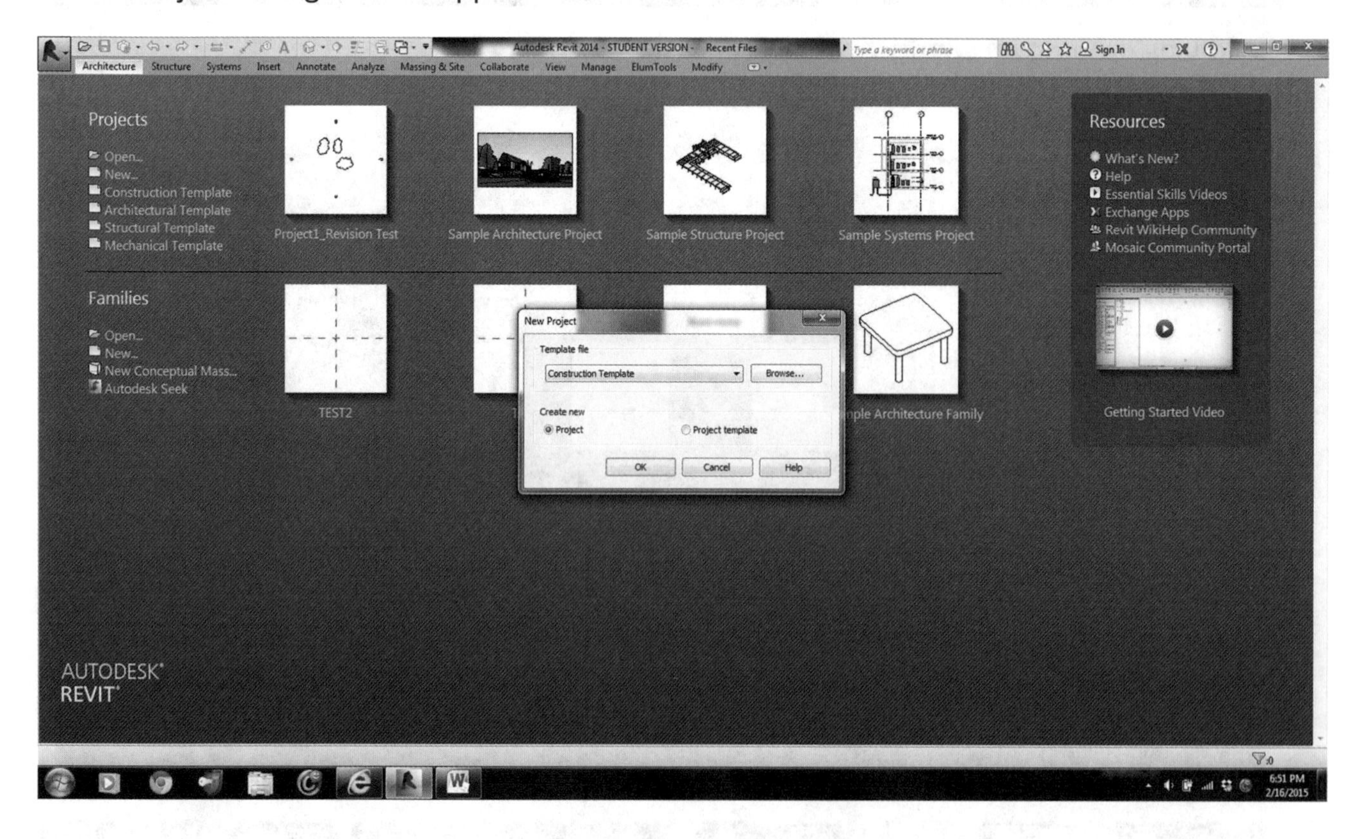

Click on the **BROWSE** button and a new dialog box will open with more templates to select from to assist you in what you will be designing.

WARNING

Make sure the PROJECT *button is selected, and do not select the* PROJECT TEMPLATE. *The reason you want to select the* PROJECT *button is so your drawing will be a Revit project, and can be opened and shared by others and other companies as such.*

NOTES

If you select the PROJECT TEMPLATE*, you cannot share any Revit file or project with others. It becomes a* **Default Template** *and part of the template list, just like the ones we discussed earlier*

You will see that the **RESIDENTIAL- DEFAULT** and **COMMERCIAL-DEFAULT** options appear.

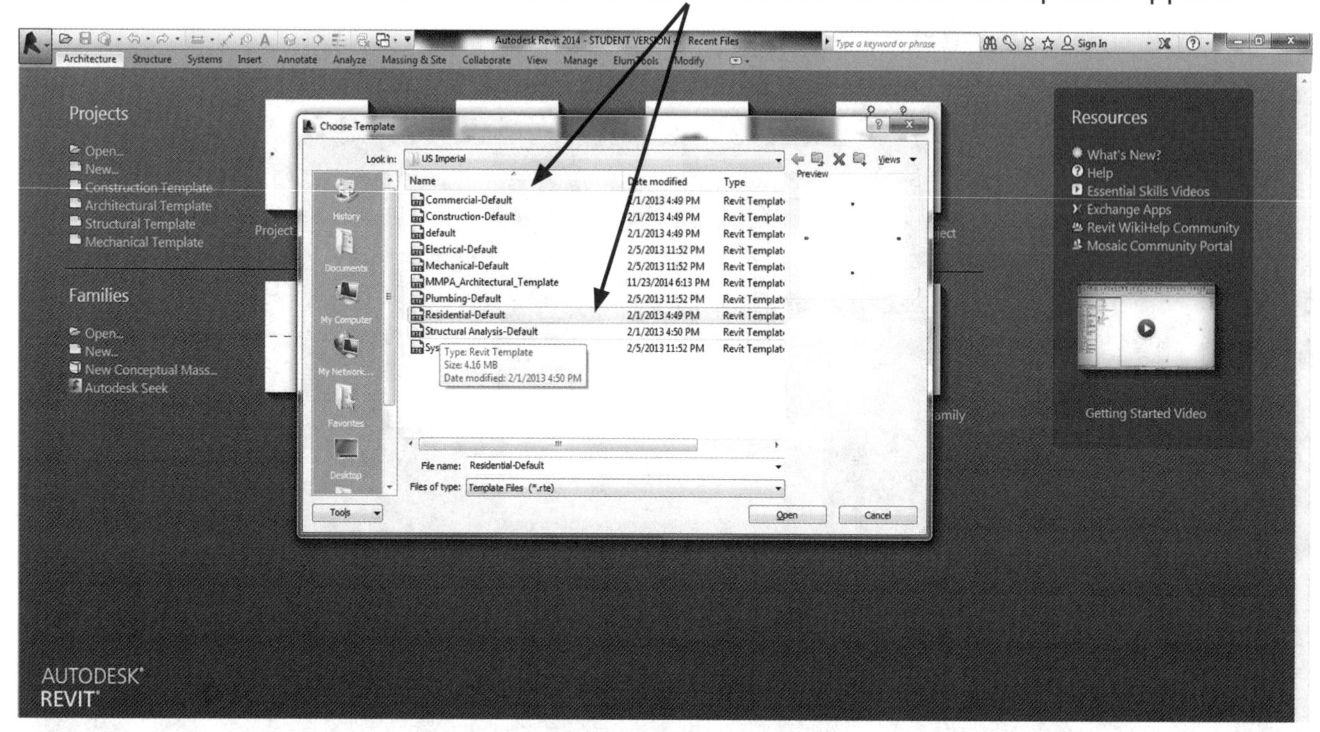

After selecting your **RESIDENTIAL** or **COMMERCIAL** template, click **OPEN**. As discussed earlier, there are different templates and different functions within each template.

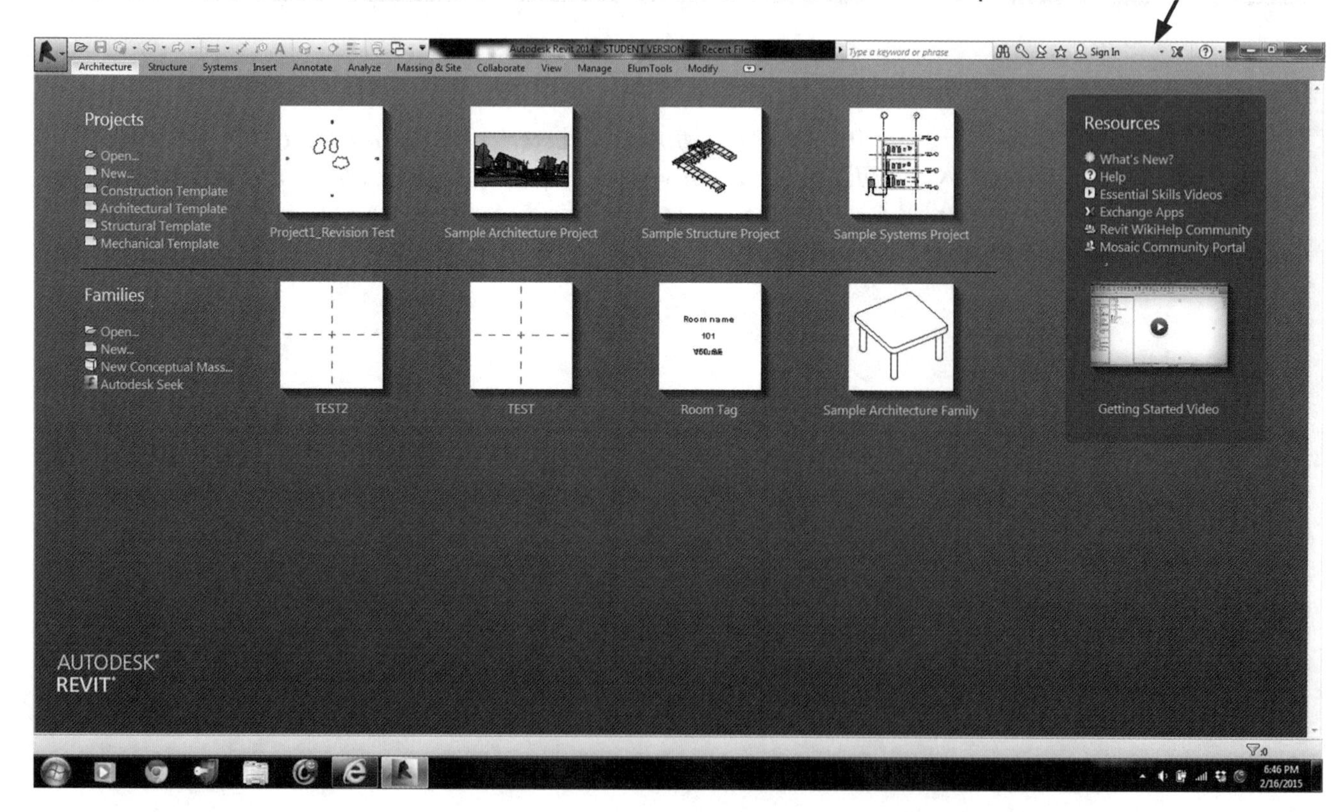

Chapter 13 will explain how to modify walls to suit your type of application.

HOW TO LOCATE A SAVED PROJECT ON YOUR COMPUTER, FLASH DRIVE, REMOVABLE OR EXTERNAL DRIVE

When beginning a project, the opening Revit screen looks like this:

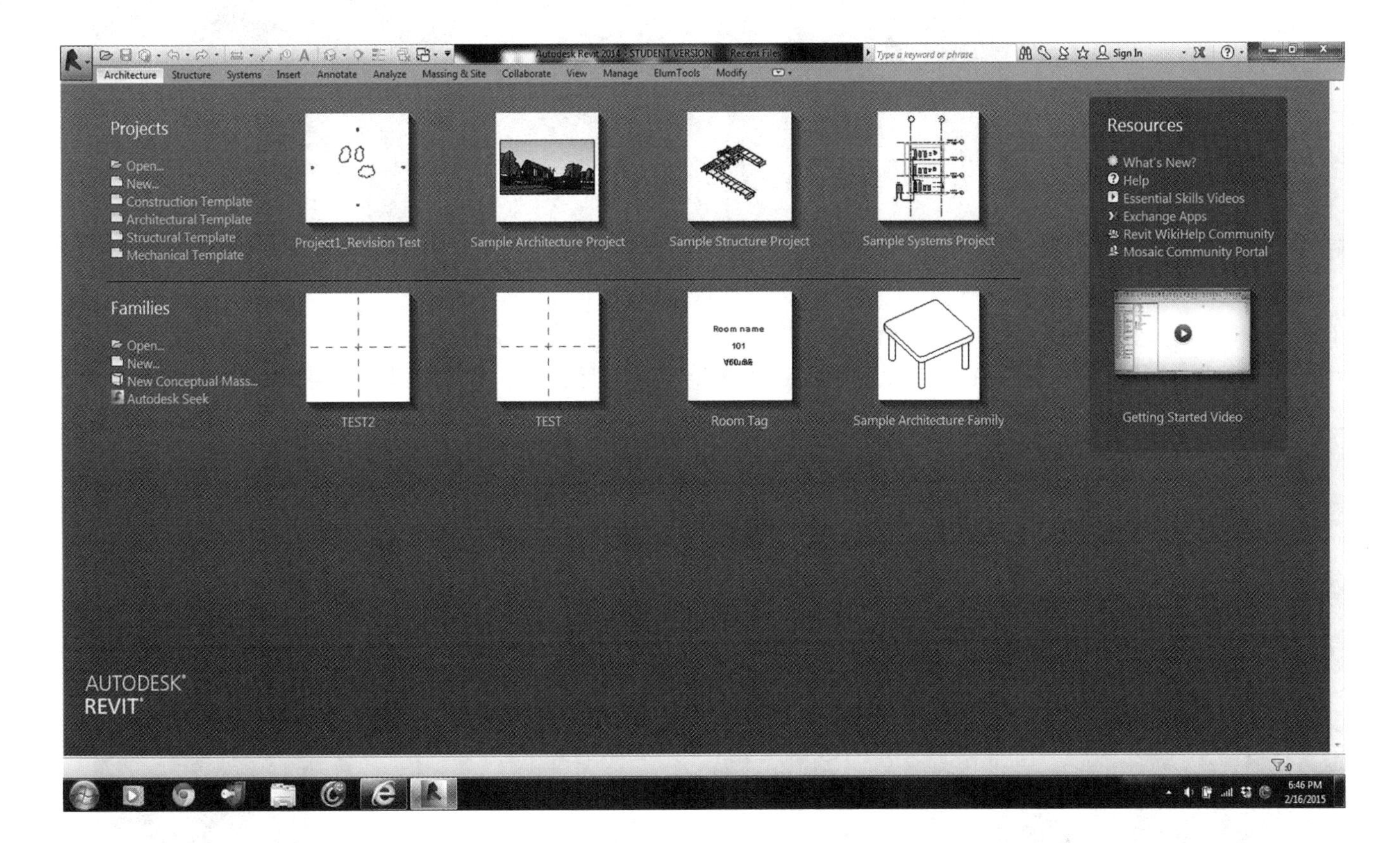

ESSENTIAL REVIT FACT

If you already have a project that was started and saved, you can follow these instructions to retrieve it.

Click on the **R** (called the Application Menu), then go to **OPEN** and **PROJECT**.

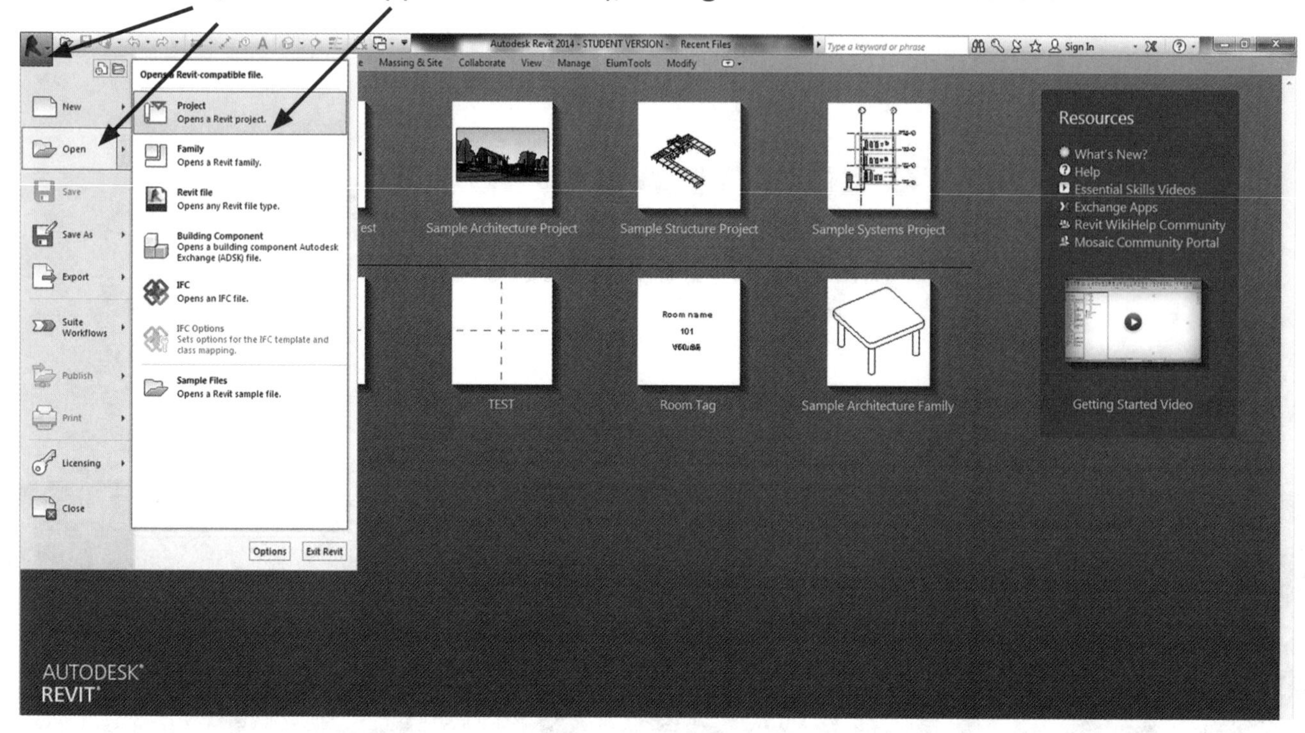

The next step is where you will go to the location where you saved your drawing either on your desktop, your flash drive, or on your computer in your C: drive under the client's name.

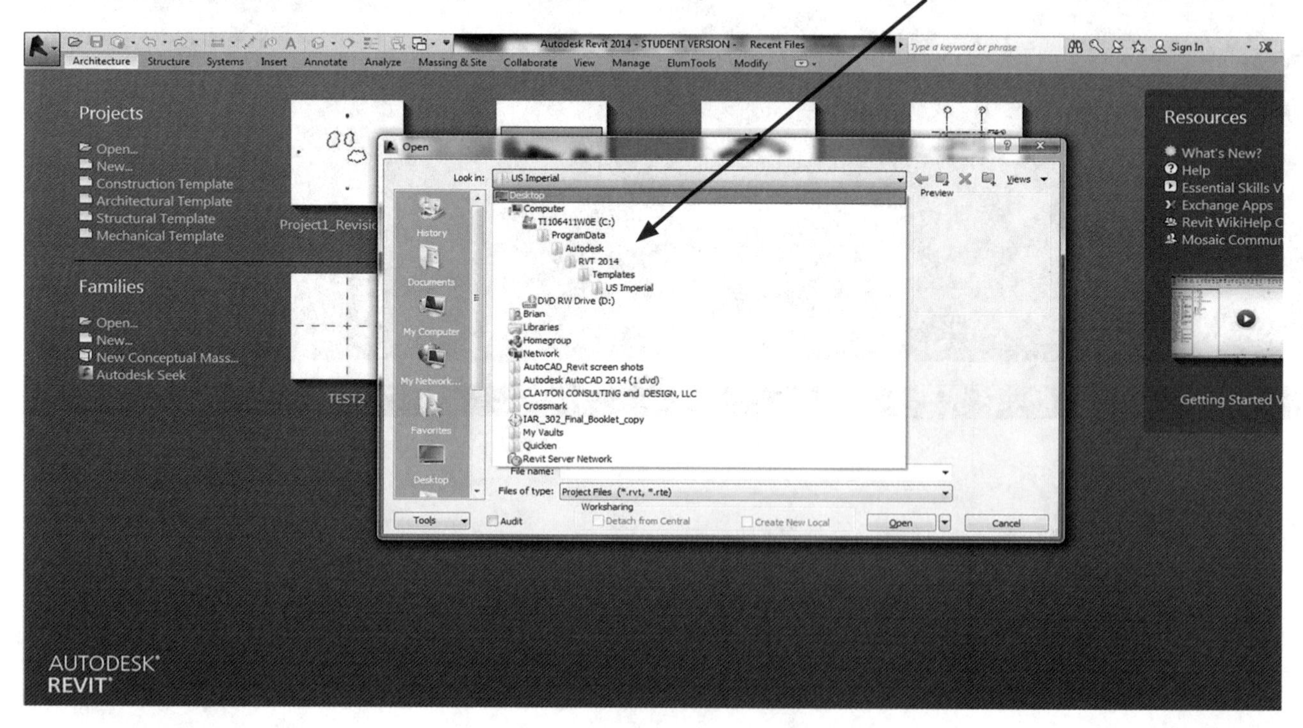

REMINDER

I would recommend creating a folder with each client's name, just to stay organized.

Once you have located the file, click **OPEN**. You can start drawing where you left off.

As we have noticed in this chapter, we can see how Revit has made it efficient to begin your project by having various Revit templates to assist in the beginning stages of your project, depending on the project's complexity, and whether it is a residential, commercial, or even a structural project.

Now let's look at Chapter 2, *Dimensioning in Revit*, and how Revit makes this process easier by having the dimension styles preset for the user to begin dimensioning a project.

← Locate Saved Project

Start New Project

File

Project

Project   Project template

Browse...

Residential or Commercial

Open

CHAPTER 2

Dimensioning in Revit

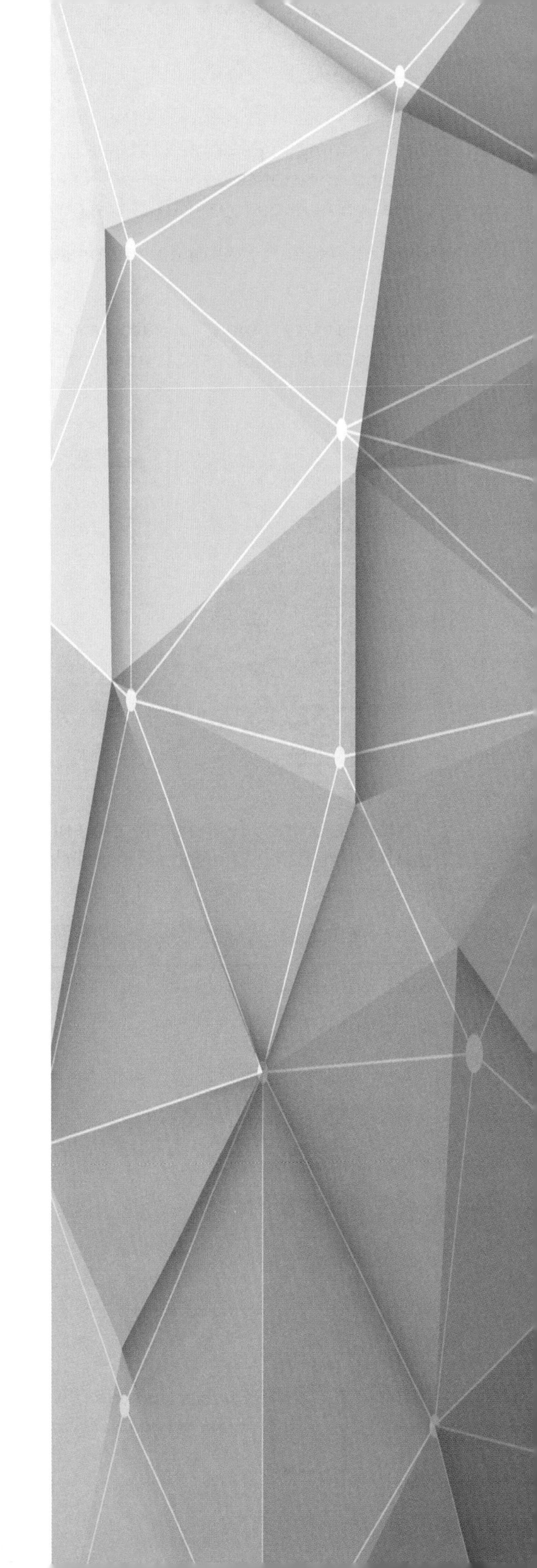

Dimensioning in Revit means to provide a clear and complete description of the object, house, building or distance between objects. A complete set of dimensions will permit only one interpretation needed to construct the object, house or building.

Revit offers two types of dimension styles. There are temporary dimensions and permanent dimensions.

The temporary dimension automatically appears when you draw a wall or place any object as the default. It appears as a faint dimension line, which allows you to type in your specific dimensions.

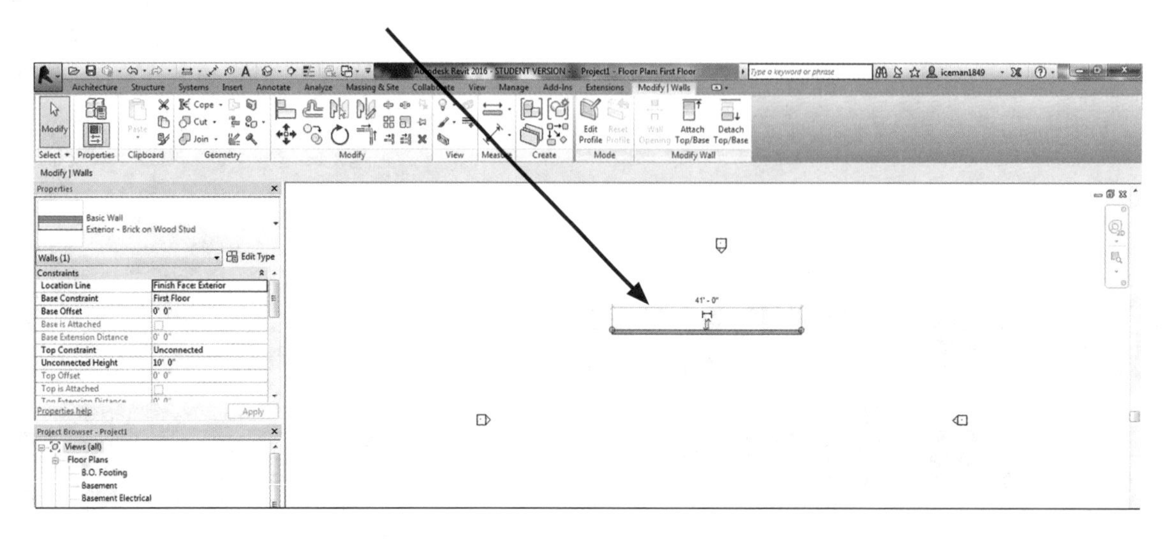

To change or modify the dimension number (numeric), first zoom into the number or location on the dimension line and simply click on the number. This will activate the **MODIFY** Box.

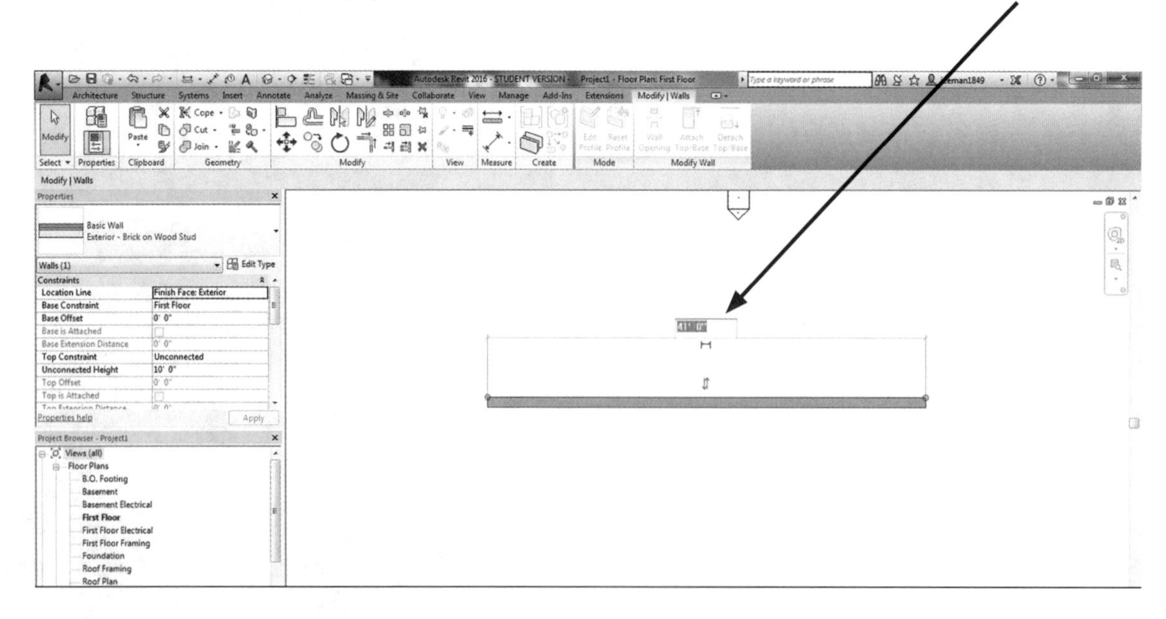

NOTES

You can only modify the dimension number when it appears faintly. Once the number is bolder, you cannot change it.

Then type in the new, desired dimension number or length, and press the **ENTER** key, or click anywhere in the screen.

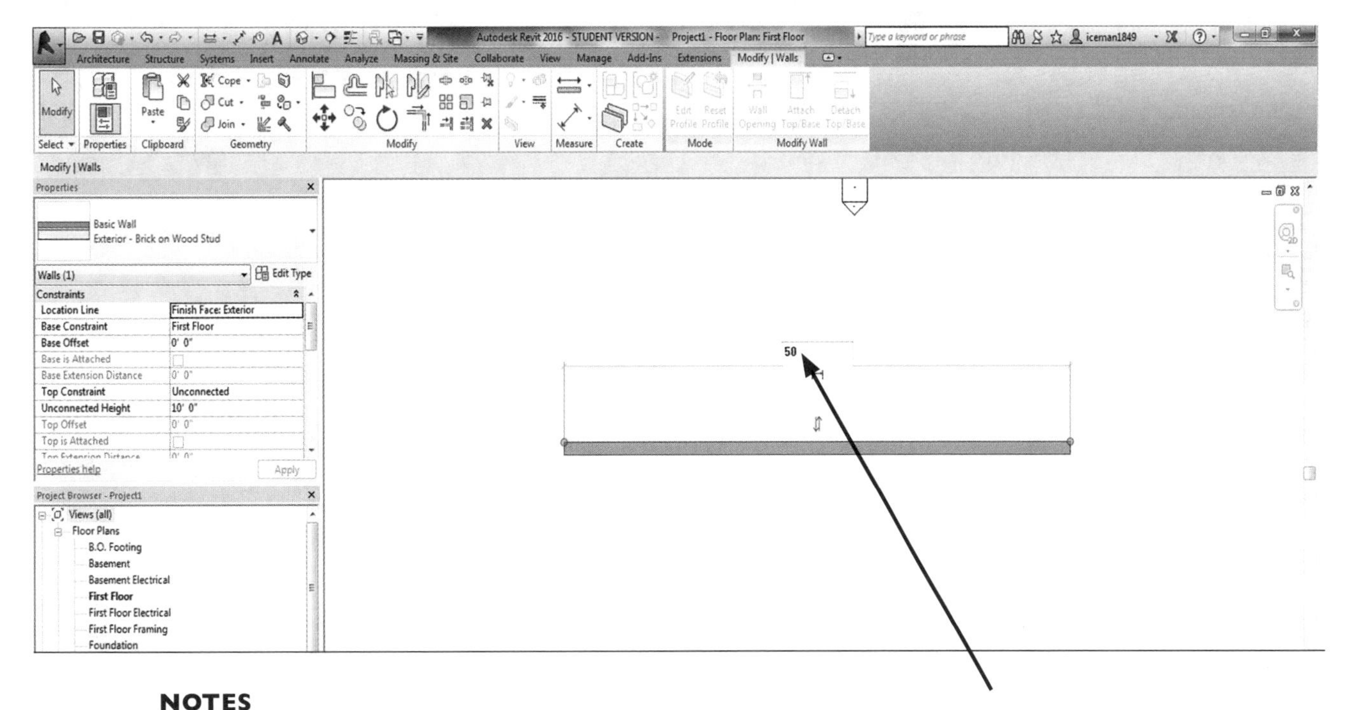

NOTES

With Revit, you don't have to use the foot mark (') to designate feet. Revit automatically defaults to the foot measurement. However, you do have to use the inches symbol (") to denote inches.

After the dimensions have been changed, your temporary dimension will look like the following image.

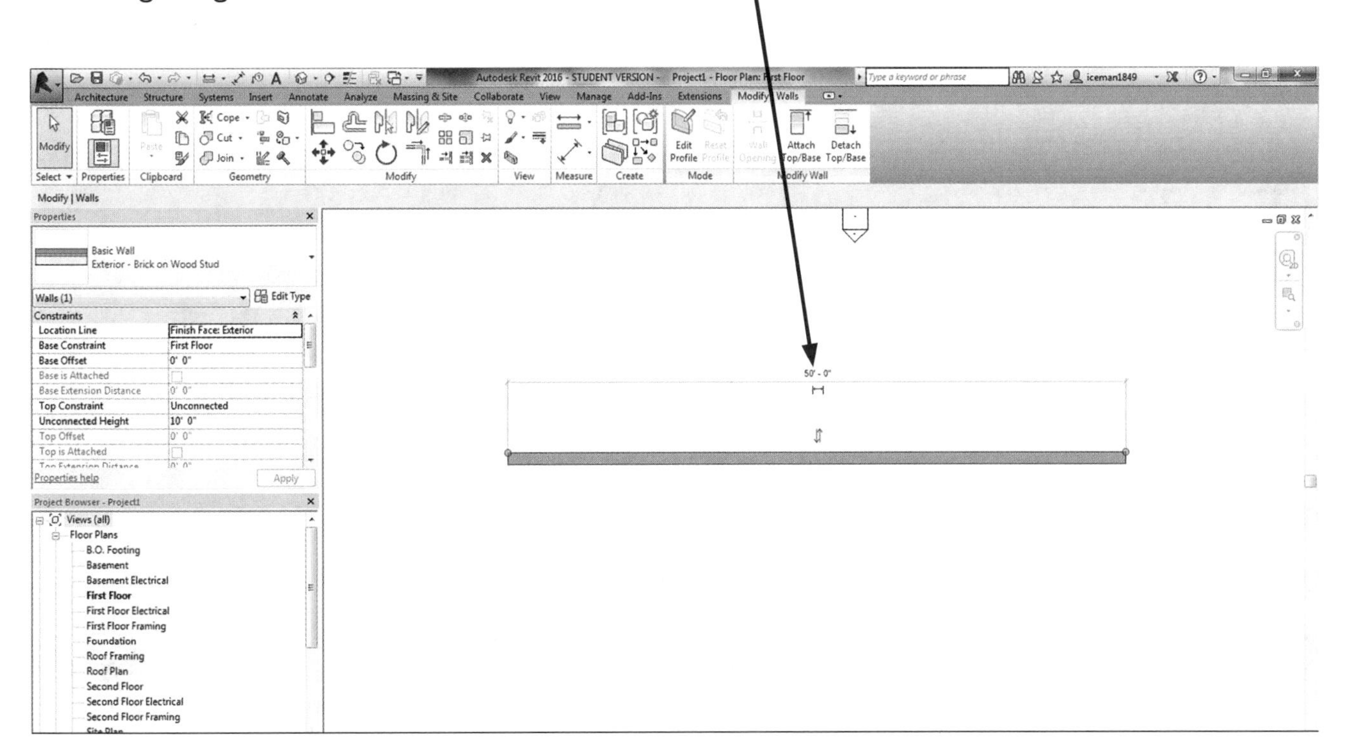

To finish the command, click **ENTER** or anywhere on the white screen in the drawing window.

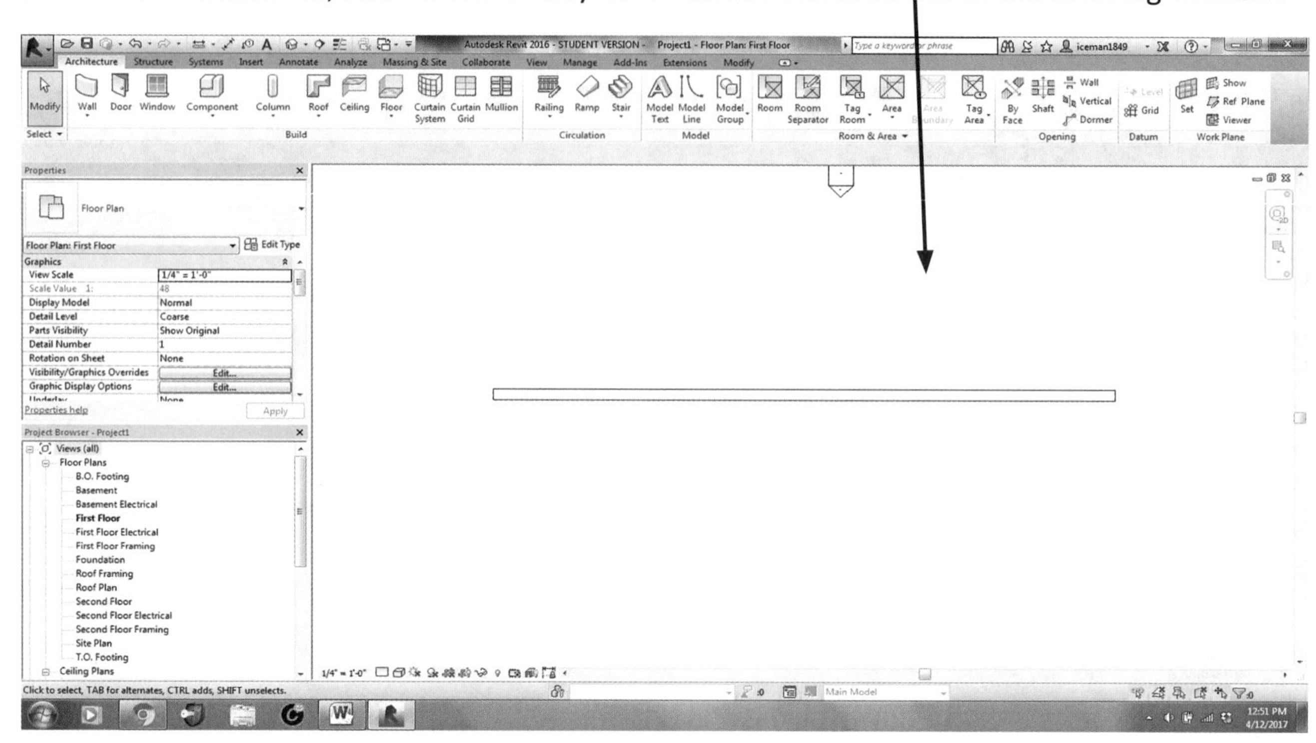

To change the temporary dimension to a permanent dimension, first click on the wall or object and the temporary dimension lines will appear. Activate the temporary dimensions.

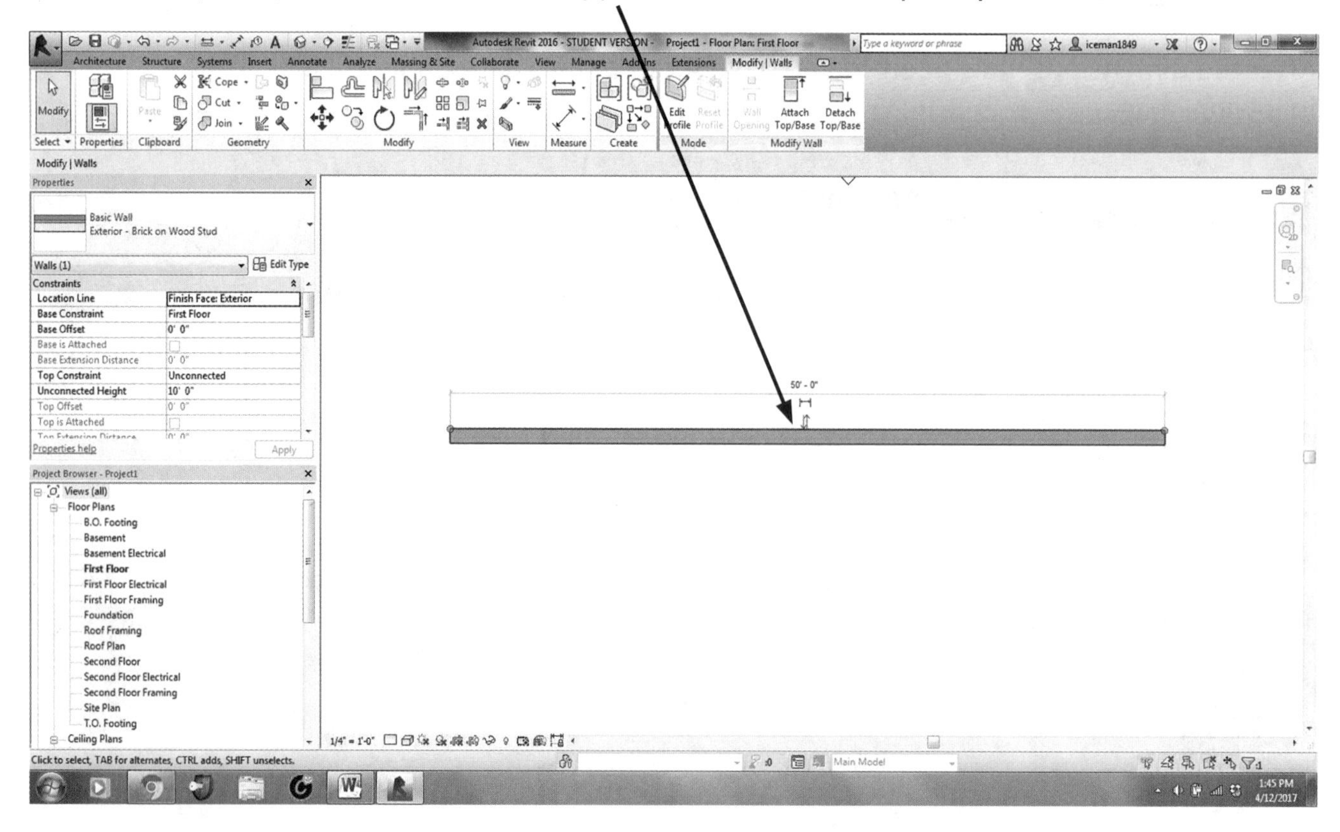

Click on the **DIMENSIONS** symbol beneath the dimensions line. This will turn the dimension line black, and grips will appear.

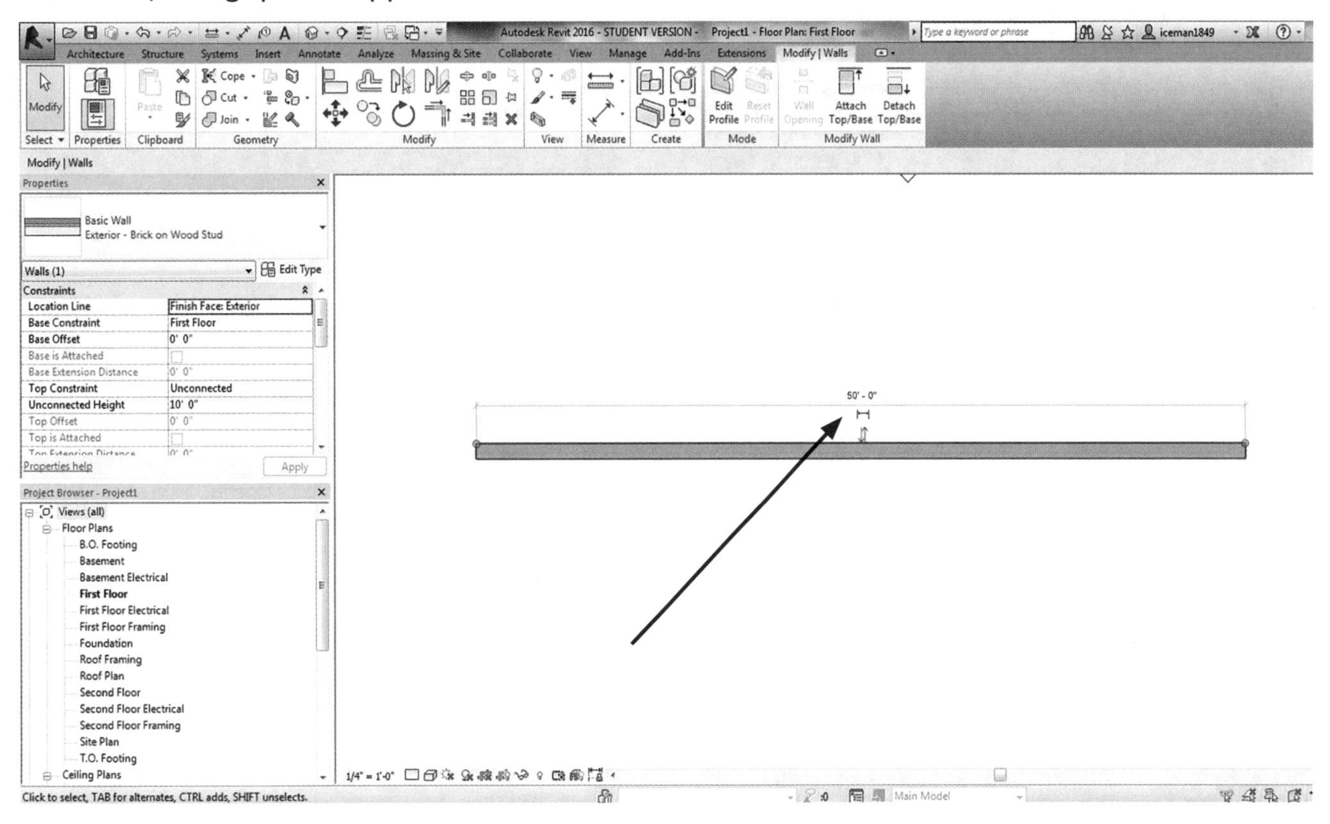

Then click anywhere on the white screen in the drawing window to complete the function and the numbers will turn black (bold). Everything will be permanent.

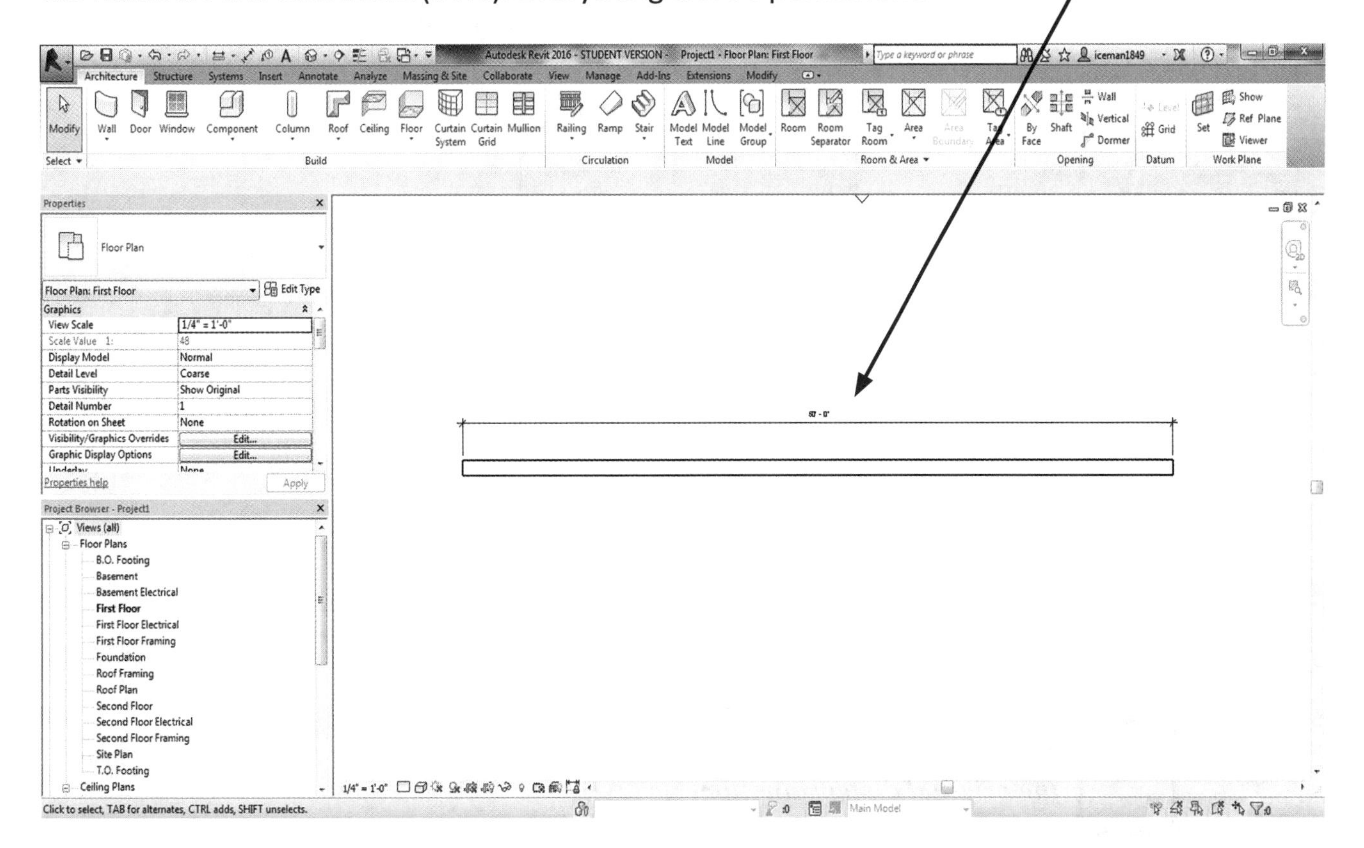

WARNING

From here you** cannot **change or override the dimension numeric text. If you double-click on the numeric text, a dialog box will appear for you to** only **add or type in text.

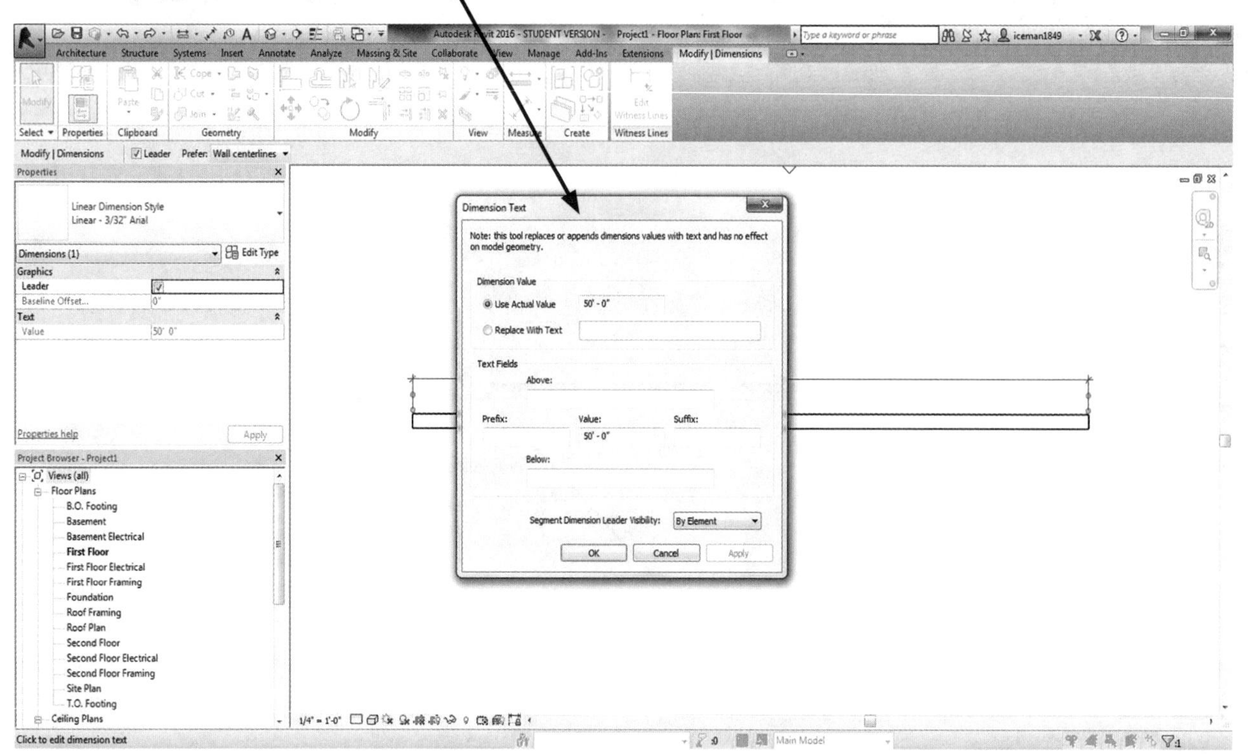

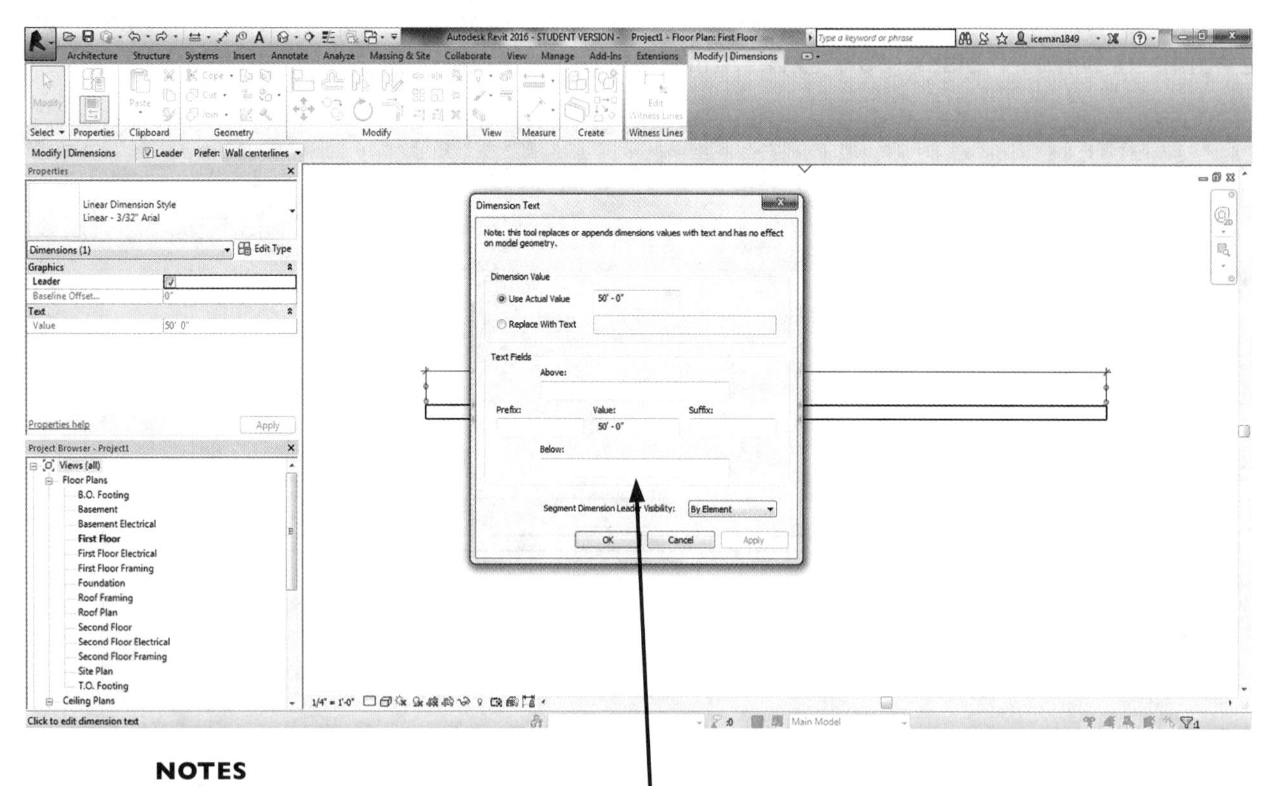

NOTES

The text can be placed above or below the dimension line, but** cannot **replace the original dimension numeric text.

The following image is an example of text applied below the dimension line.

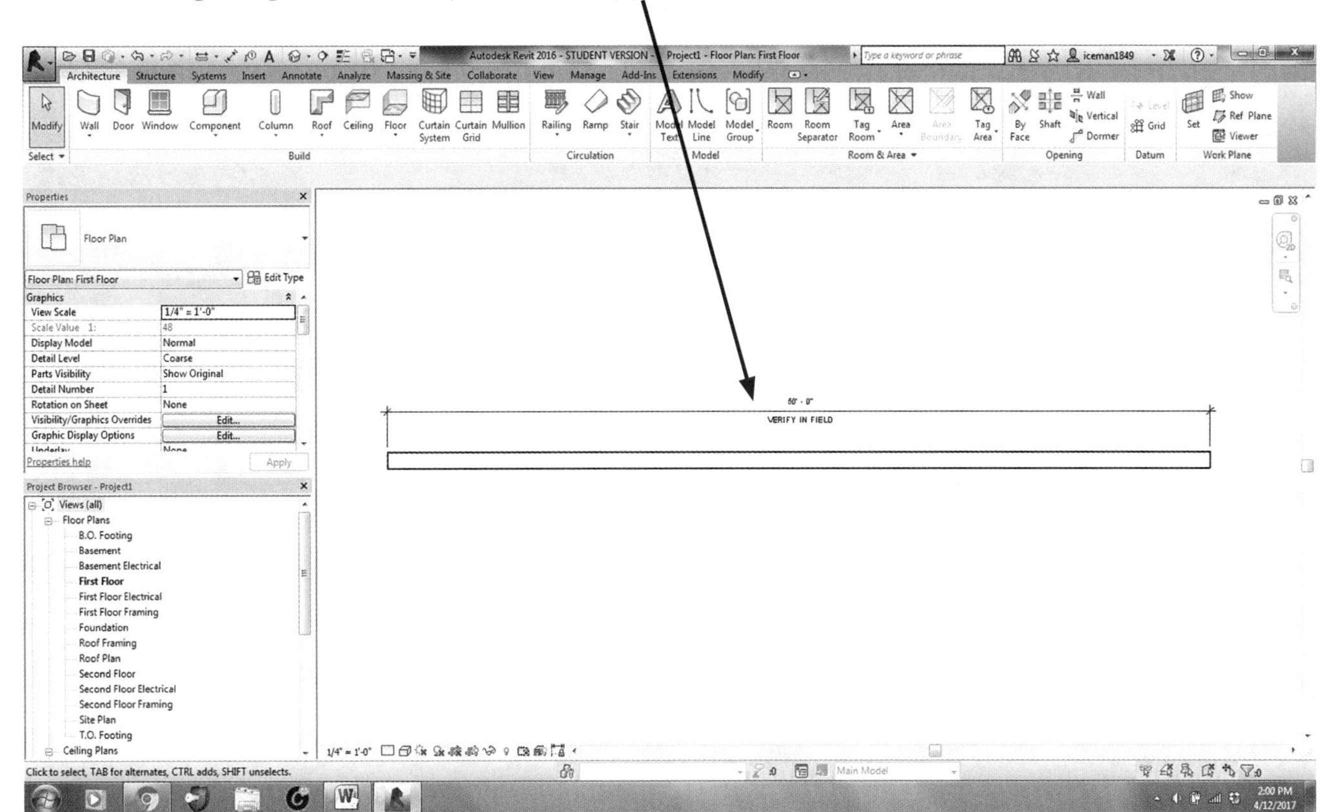

Revit makes dimensioning much easier and more efficient, which speeds up the design process. As we have seen, Revit allows you to not only draw in the wall and place a dimension, but you can modify the dimension and move the wall before anything becomes permanent. Then you can switch over, and make all the temporary dimensions permanent with a simple click of the mouse on the dimension symbol.

The next chapter is Chapter 3, *Beginning a Drawing*. Here we start harnessing the power of the Revit software, and the fun really begins.

CHAPTER

3

Beginning a Drawing

To begin drawing a design, you must draw in a clockwise direction, and try to center your drawing between the elevation markers.

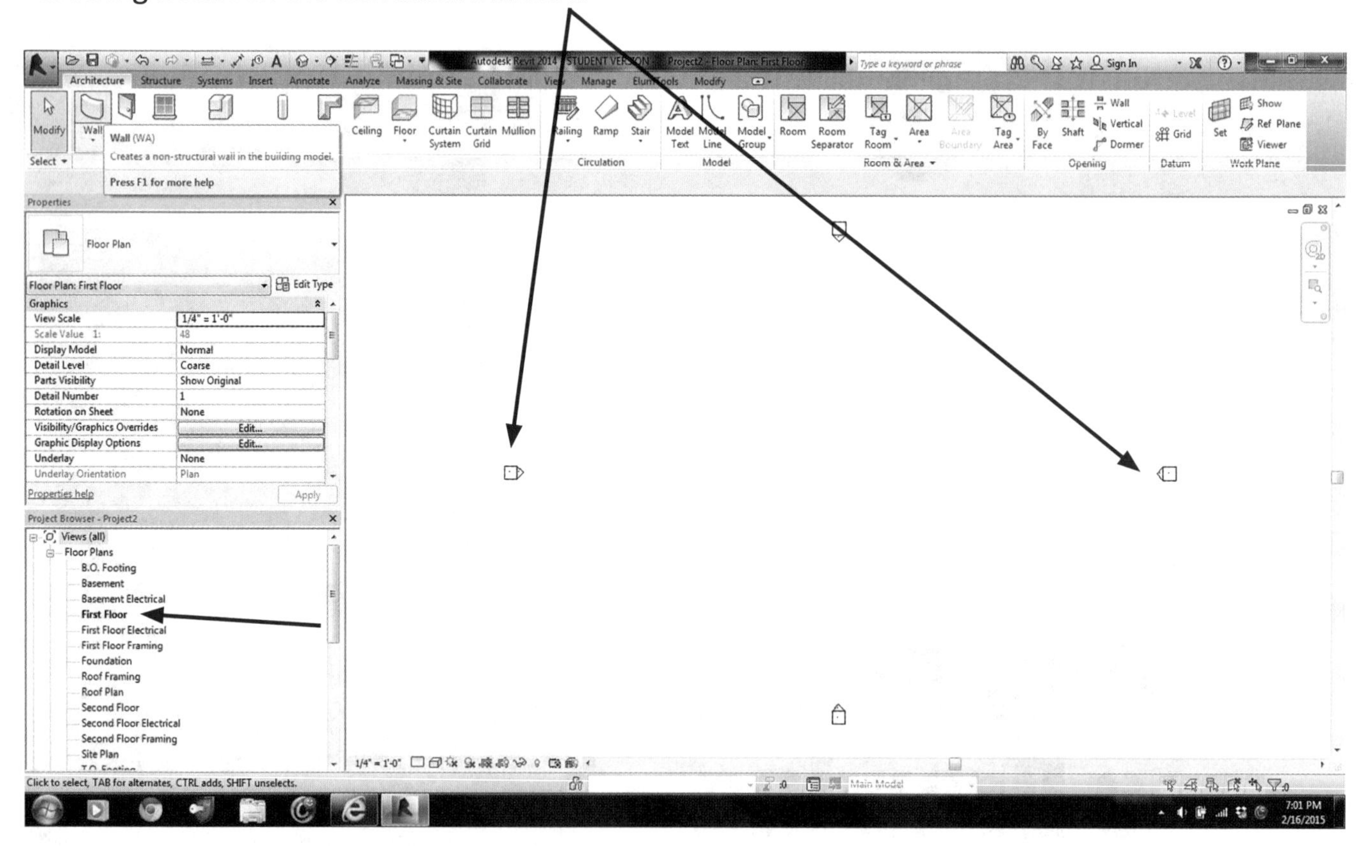

For this project and demonstration, we will be designing a single-story house/building with 10'-0" high exterior walls.

NOTES

Notice that when you open the drawing, it automatically defaults to the first floor.

First, be sure you are on the Architecture Tab, then select the **WALL** command.

If you click on the top section of the tile, it will quickly take you to the Architectural Wall default.

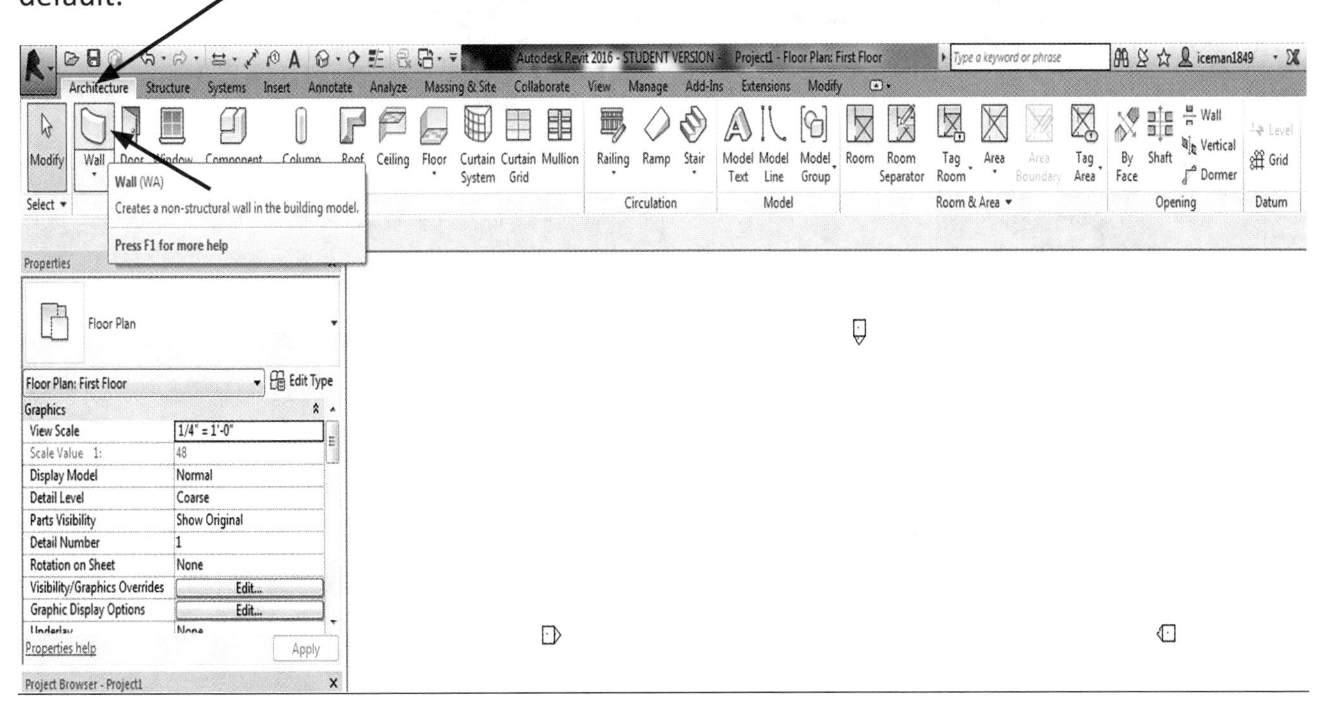

NOTES

Notice that the Wall tile is split into two sections. The top section defaults directly to the Architectural Wall command, and the bottom section of the tile allows you to be more selective on the type of walls you pick from.

You'll see that the Architectural Wall default in the Properties Box has opened.

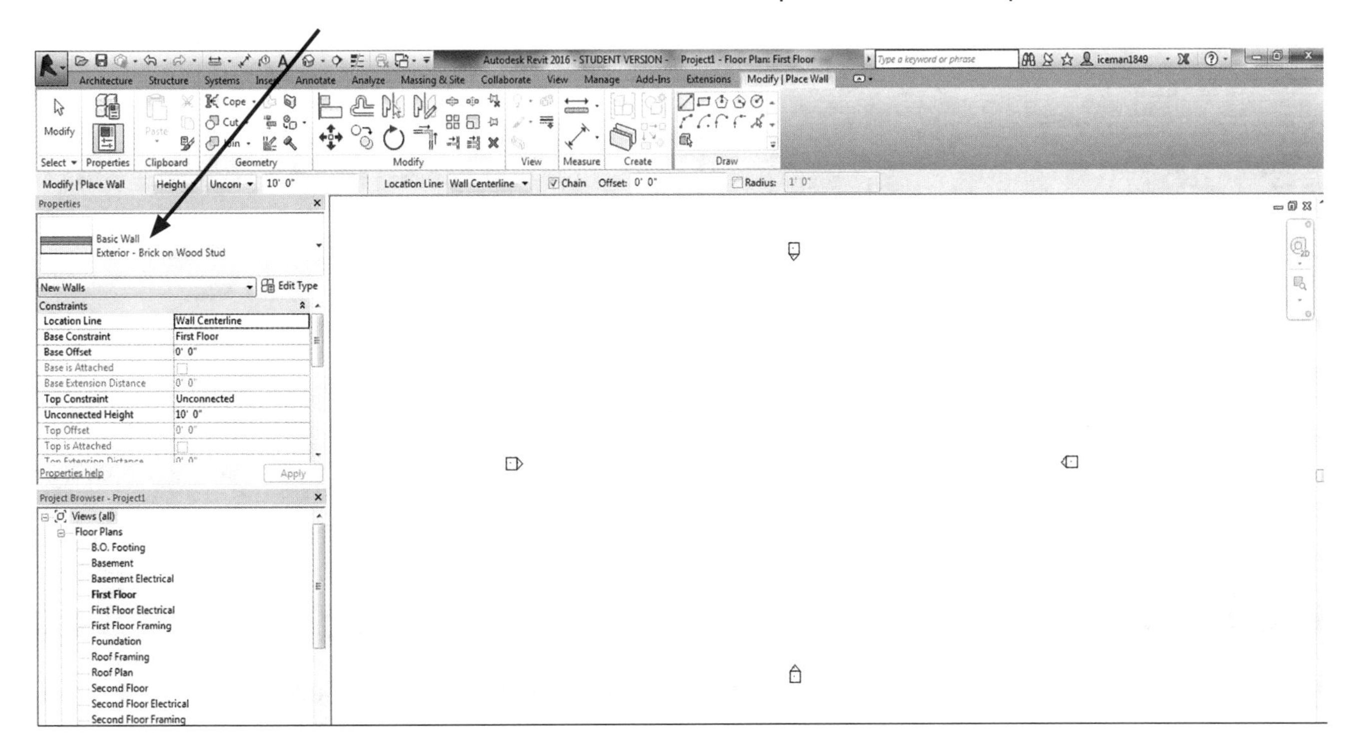

If you would like to select additional walls to use at this time, you can select the drop-down arrow of the Wall Tile and the wall selections will appear.

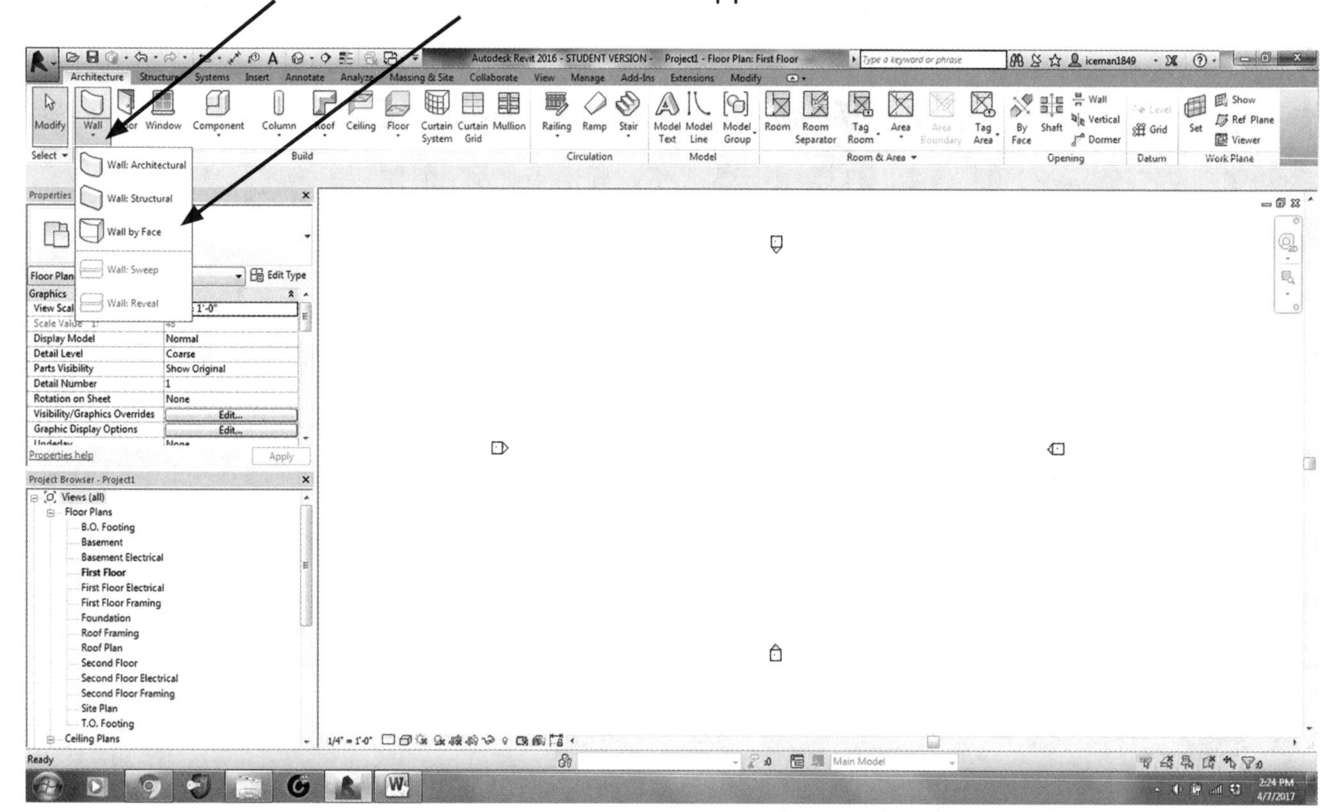

Once you select the wall type you want to use, the Properties Box will open.

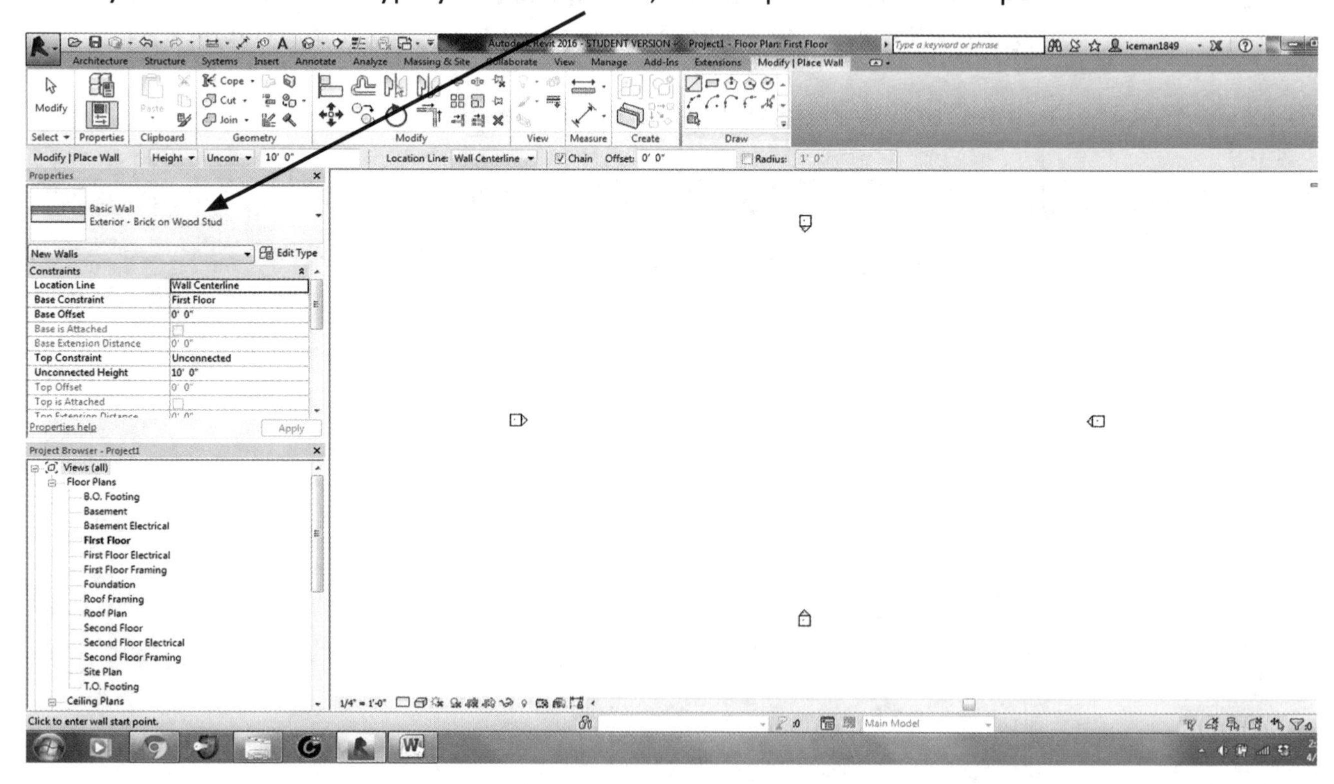

It just depends which steps you feel comfortable using and are more efficient.

After clicking on the **WALL** command, you will then select the type of exterior wall you want to use.

Next you need to adjust the wall height. Since we are doing a single story, you should change the wall height to 10'-0". After that, you should determine how the walls will connect to each other.

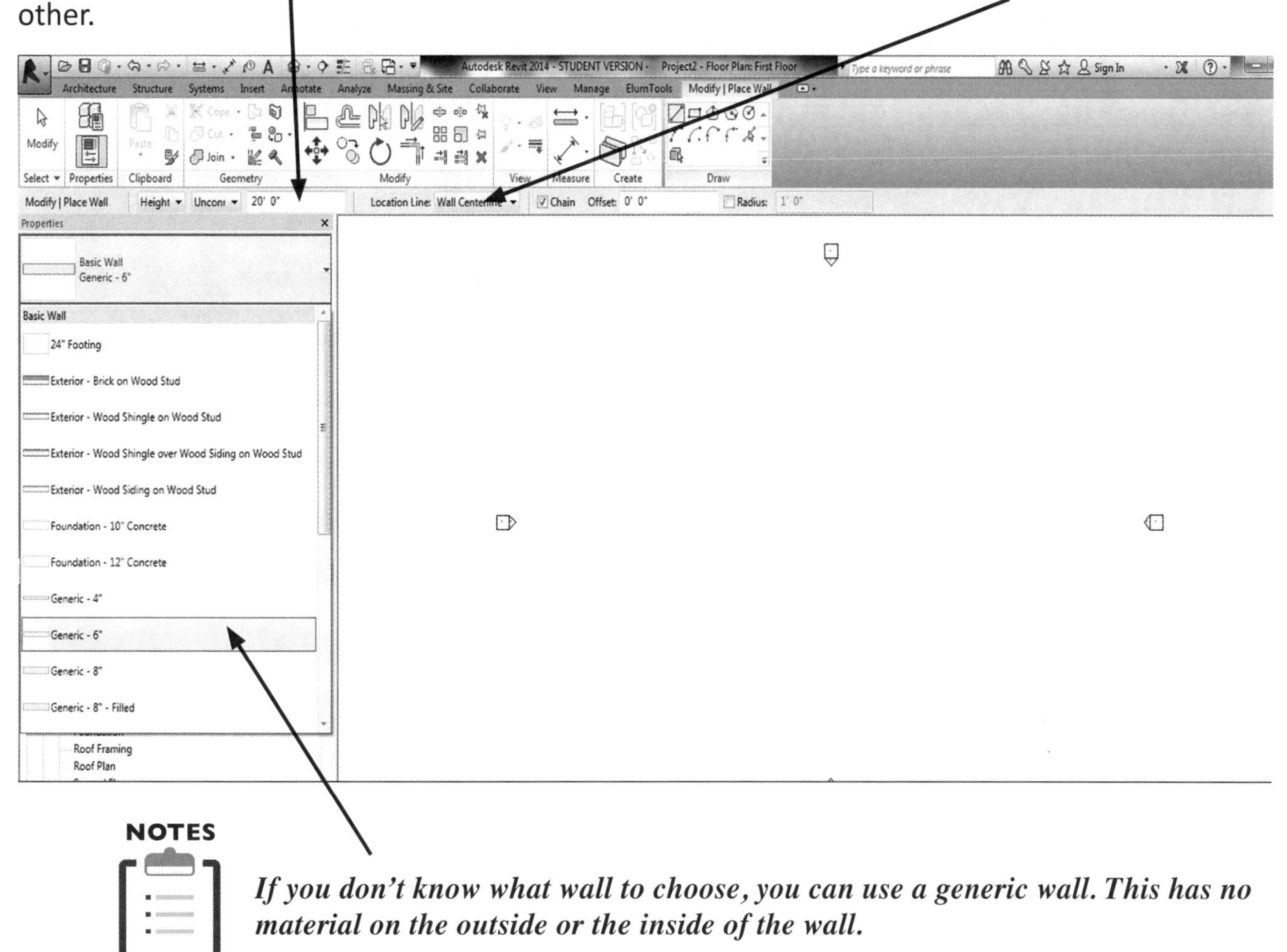

NOTES

If you don't know what wall to choose, you can use a generic wall. This has no material on the outside or the inside of the wall.

Once you have selected all the important factors of the wall, the next step is to begin drawing the wall.

ESSENTIAL REVIT FACT

Revit only likes the walls to be drawn in clockwise order, so when you do select a wall with material on the exterior of the wall, the exterior material hatch is on the *outside* ***of the wall, not the inside.***

NOTES

For this exercise, we will be using a 50 foot by 50 foot square space.

After you have clicked on your wall and have selected either **GENERIC** or **BRICK ON WOOD STUD**, click the starting of your wall close to the top left and inside the elevation markers.

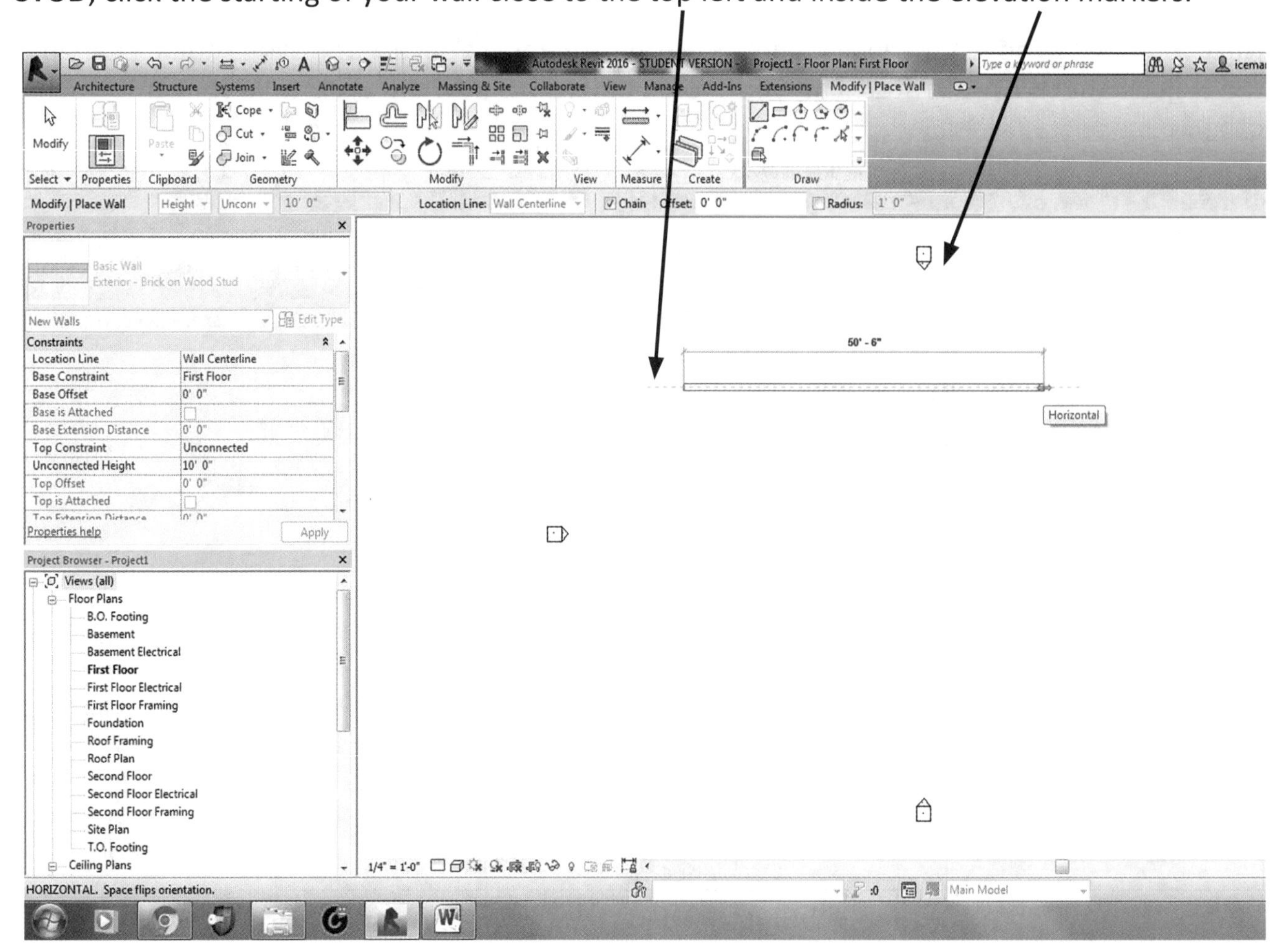

As you draw your wall, notice the measurement numbers are in 6” increments. You can either drag the length of the wall to 50 feet long, or you can just begin drawing the wall and type in the distance of 50 feet (or whatever distance you need the walls to be). If you zoom in close to the wall, you will see the wall measurement numbers will change to increments of one inch.

Continue to draw the walls all around and make them 50 feet on each side for this exercise. Your drawing should look like the following screenshot.

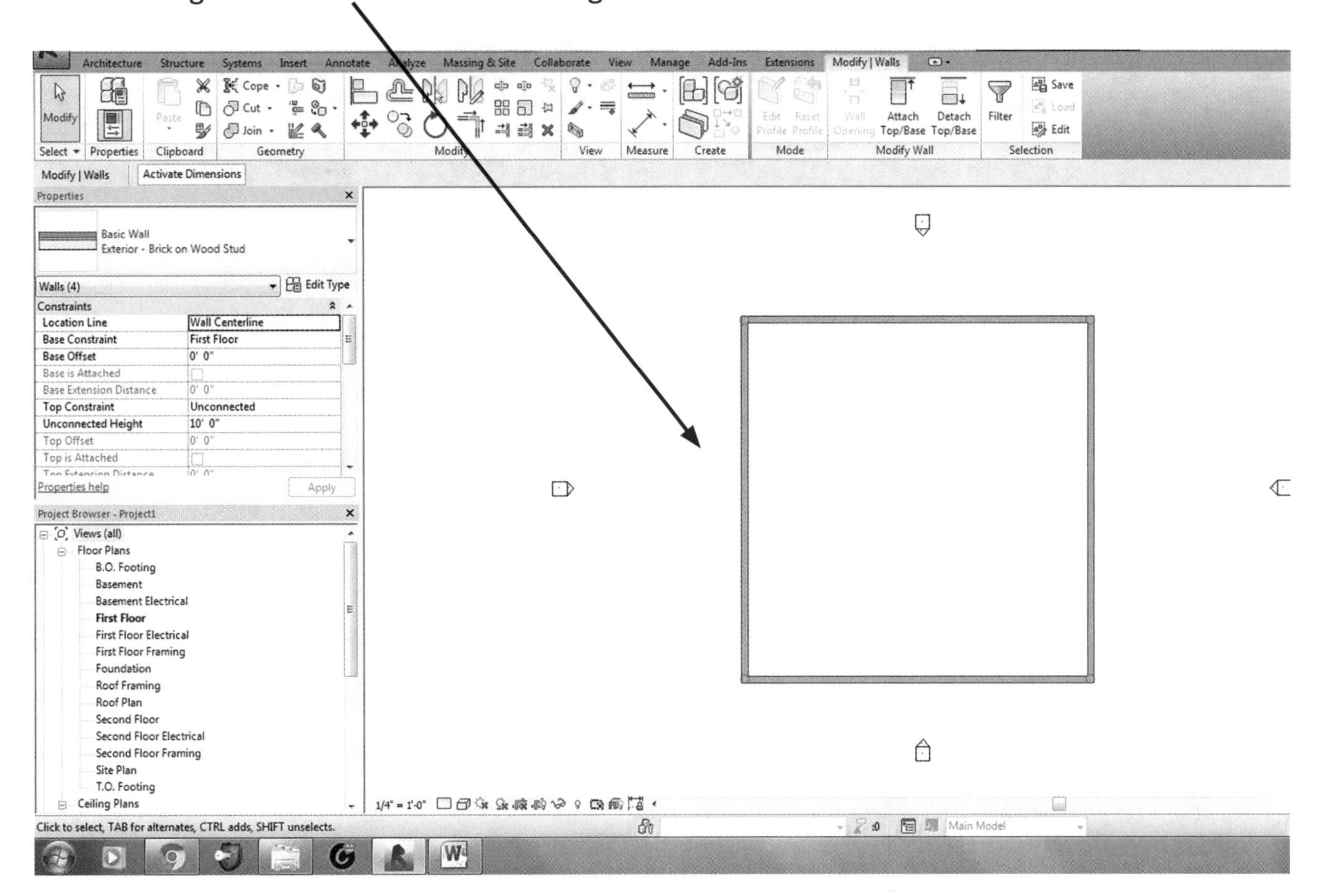

NOTES

In this example, the wall that was selected was the BASIC WALL: EXTERIOR – BRICK ON WOOD STUD.

Once the walls are all placed, click the **MODIFY** Arrow.

WARNING

Please do not click the ESC Key, as this will cancel out other functions in later steps in Revit which you may not desire. If you click the ESC Key in some instances it is the equivalent to hitting CRTL + Z, or the UNDO command numerous times, and clearing out your entire project or entire work. If you do this, you might have to start over again.

If you click on the **DETAIL LEVEL** Button at the bottom left, change the detail level to **FINE** and then you can see the hatches symbols in your material on the wall detail.

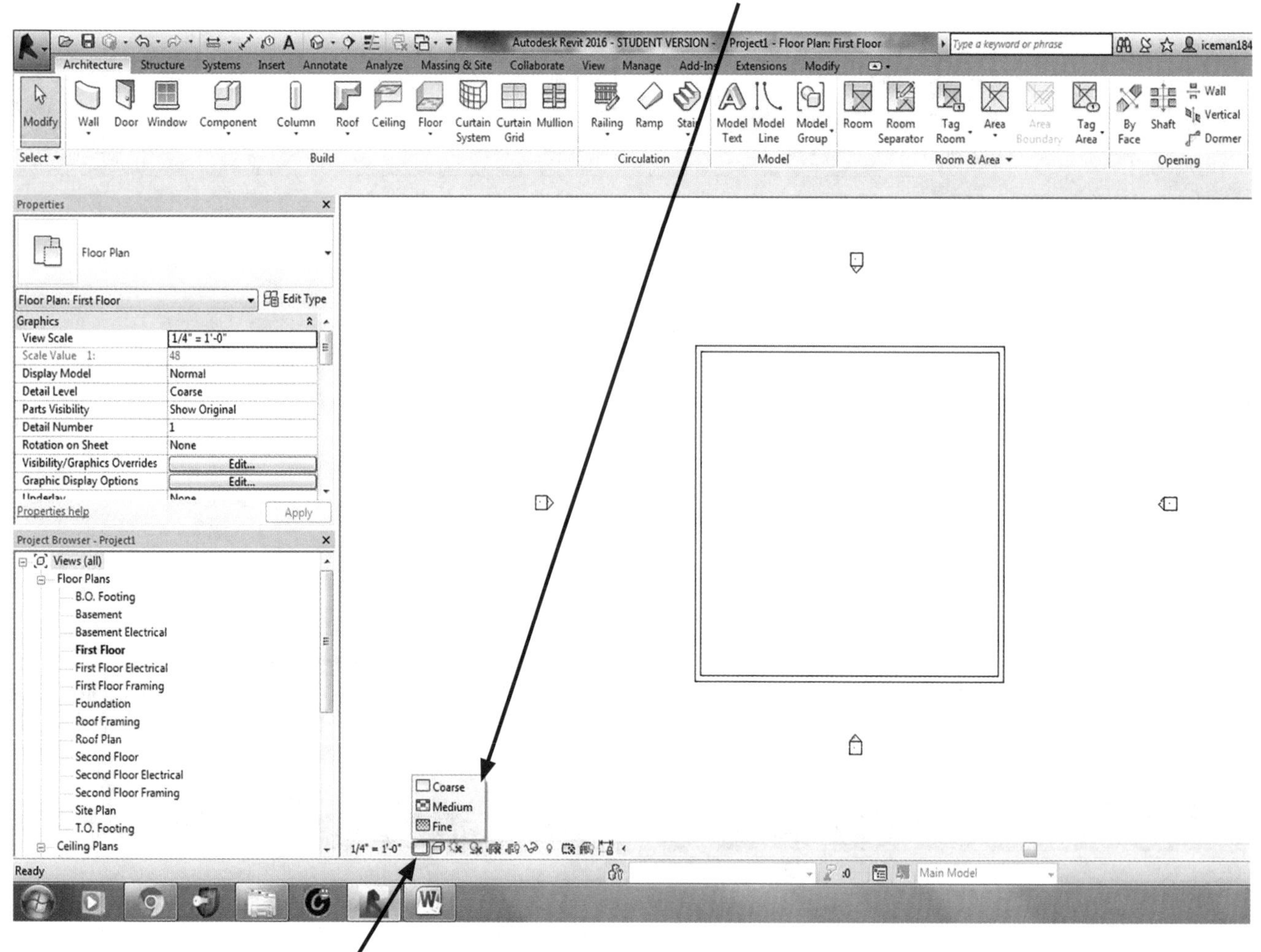

Use the **DETAIL LEVEL** Button to show the hatch detail for walls and other objects.

REMINDER

You won't see the hatches if you used a generic wall, since it doesn't have material on the exterior or interior.

The next step will be placing interior walls in the space. Follow the same steps as before to select a wall, but select an Interior **4-3/4" Partition** (that is a basic interior wall thickness). Once selected, it will appear in the view window.

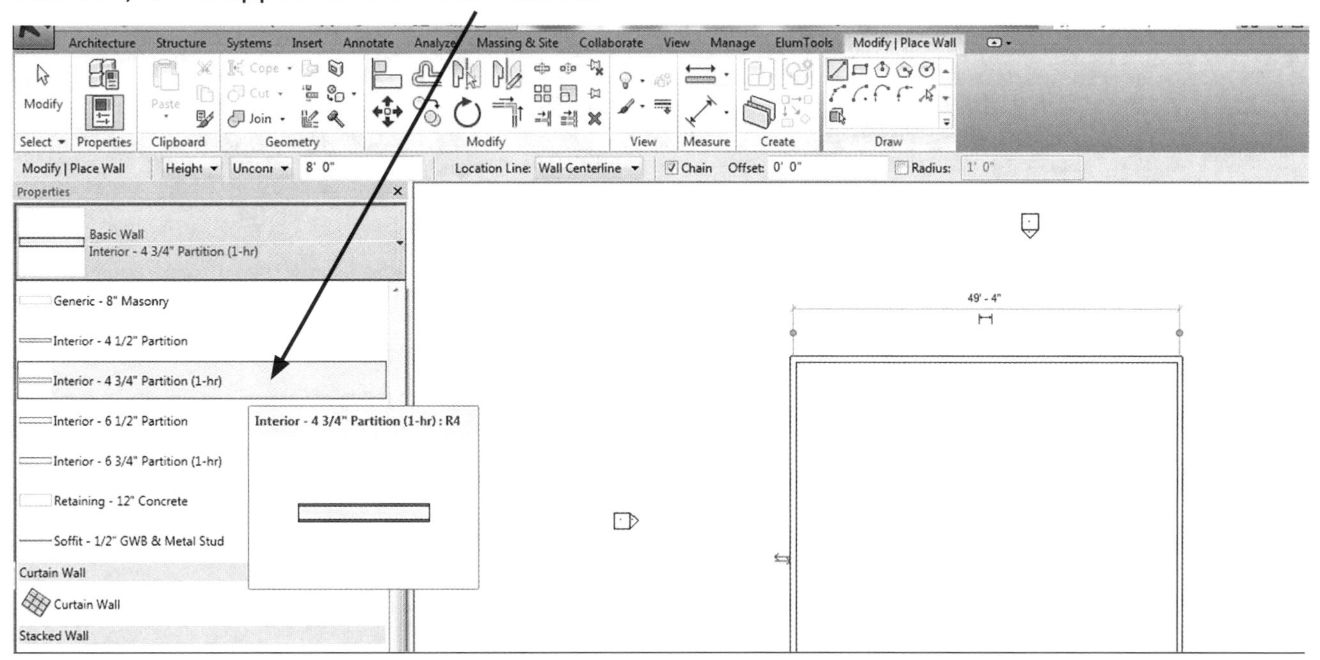

In this example, we will be creating a room, 17'-0" x 17'-4", in the bottom left corner. Begin by clicking the mouse on the exterior wall to place the new interior wall you selected in the bottom left corner and (release the mouse button) drag the wall up the screen 17'-0" to make the wall and click the mouse button to stop the wall.

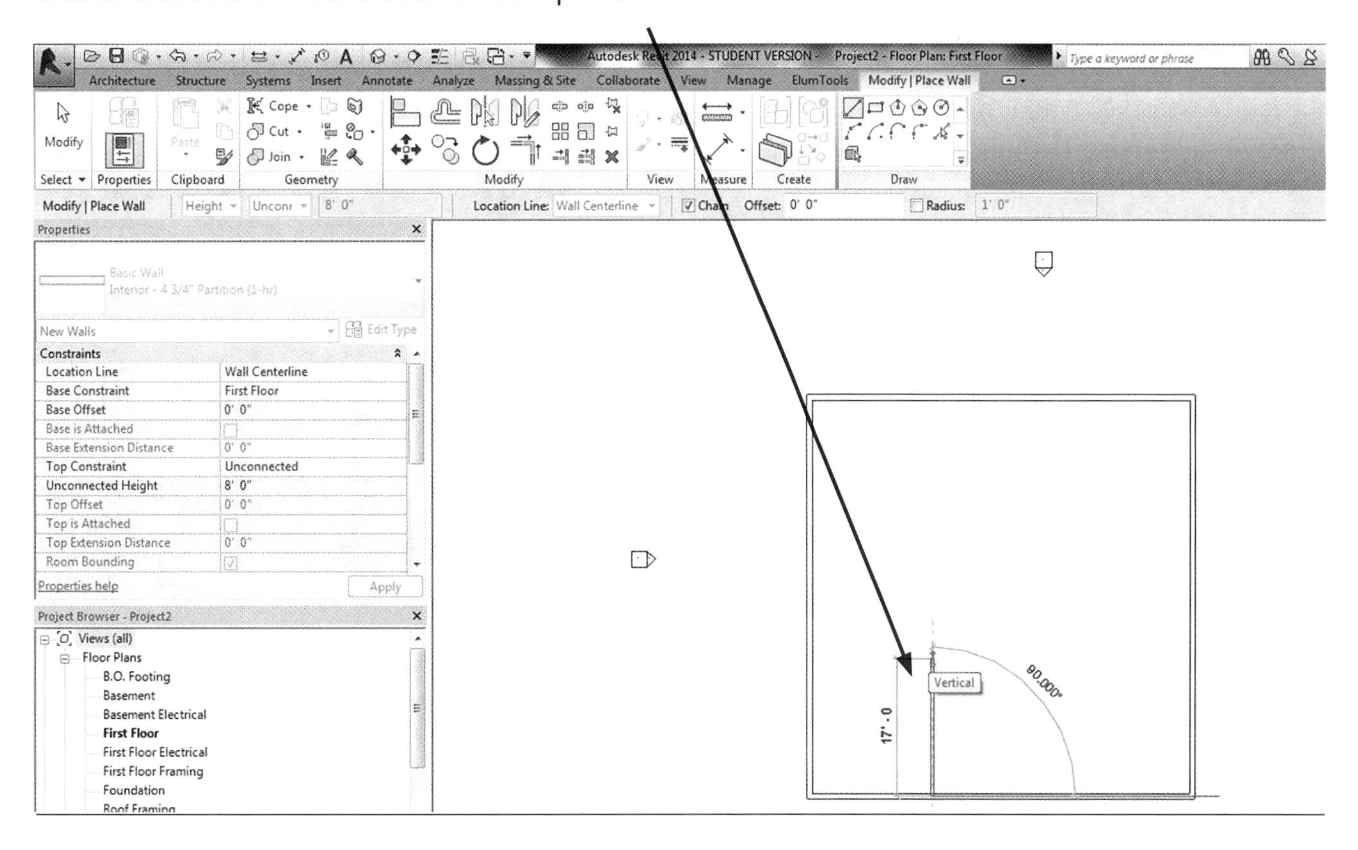

Next, we will complete the room by drawing the other wall. If you are still in the Modify/Place Wall command on the ribbon, then you can continue to extend the wall over to the left or whichever wall you are connecting to.

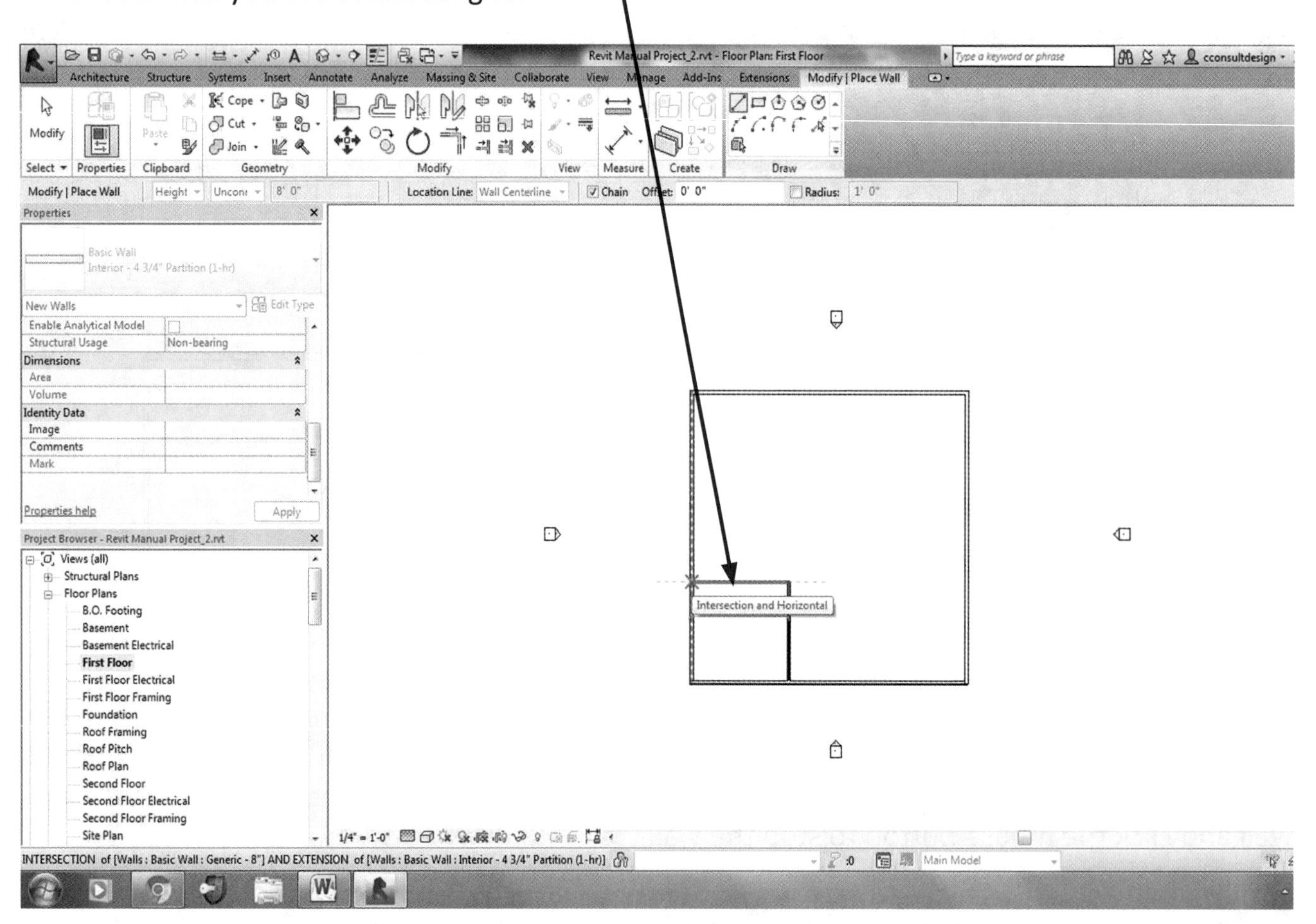

Click anywhere on the white screen in the drawing window and the command will end and the project will have a completed room.

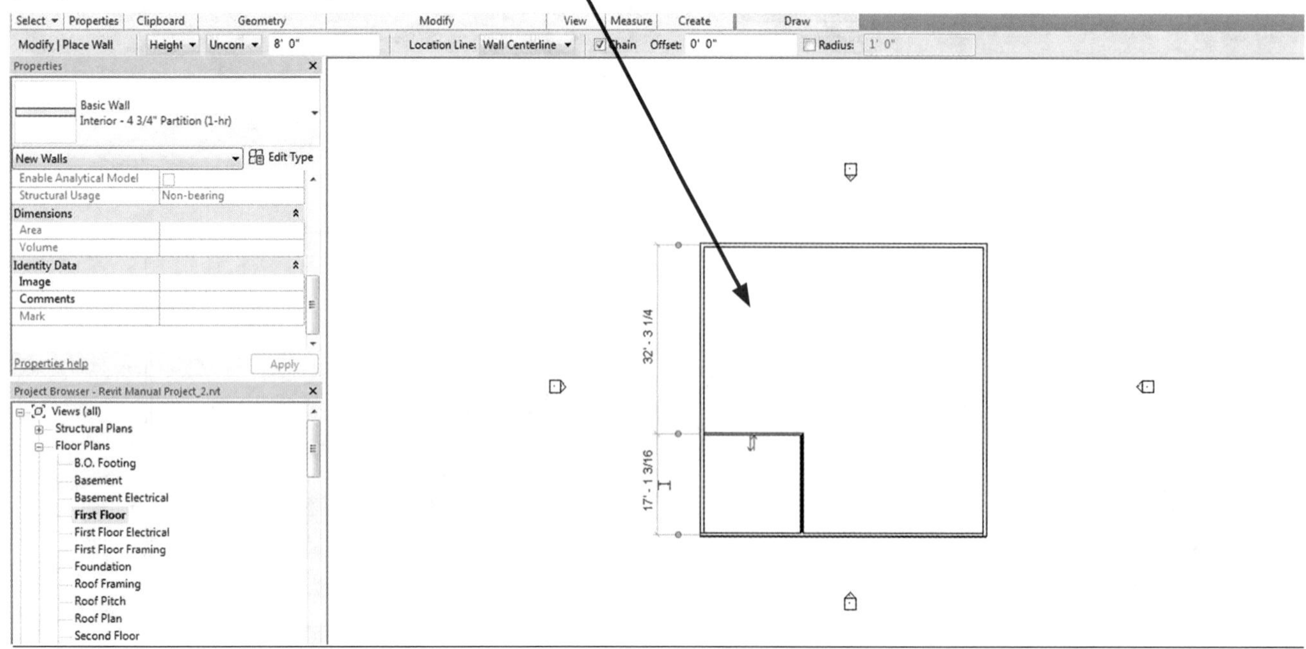

If by chance you had drawn the wall and gotten out of the wall command before extending the other wall, as seen in the following screen shot, don't worry. There is an easy fix for this.

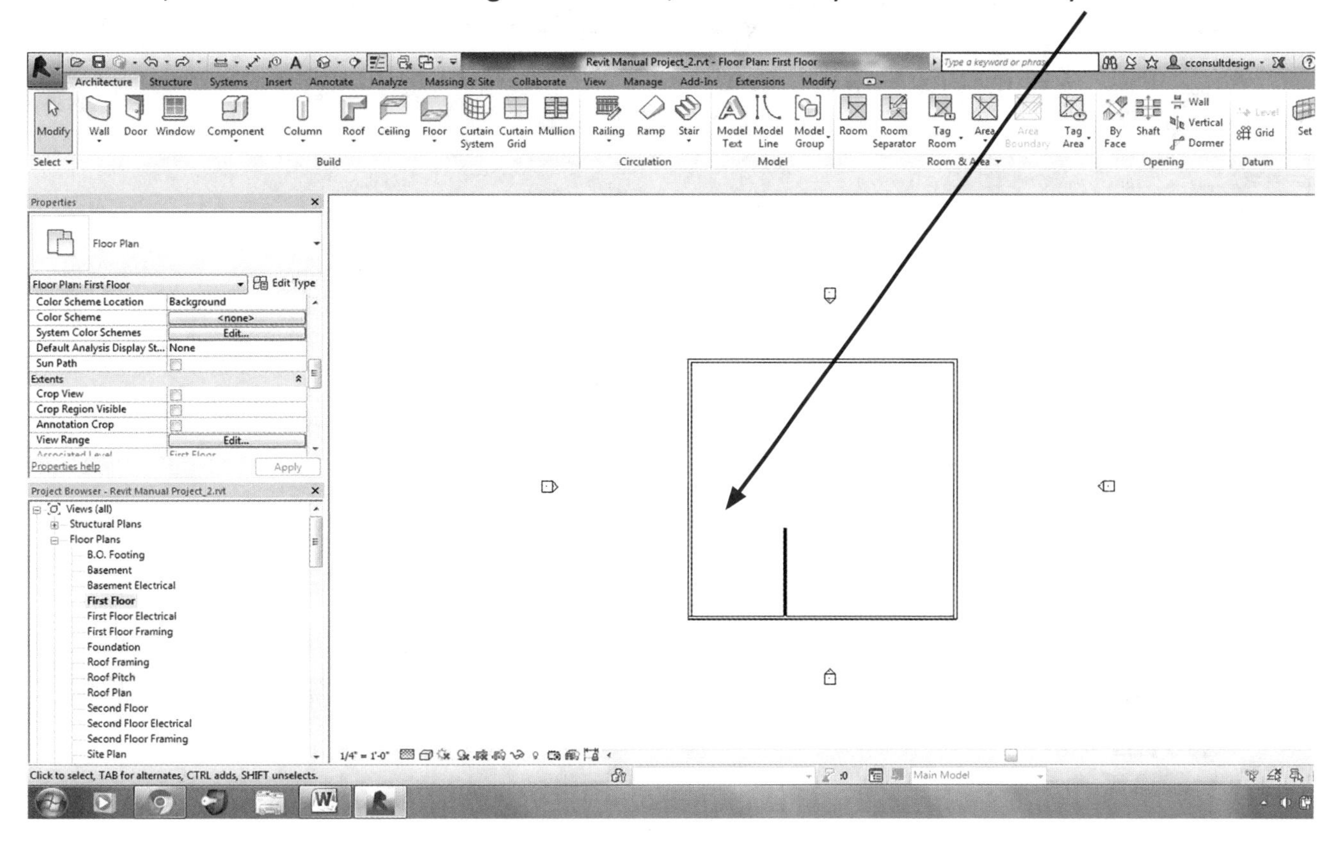

Simply click on the wall command in the Architectural Ribbon, and select the **INTERIOR** 4 ¾" **PARTITION** wall again.

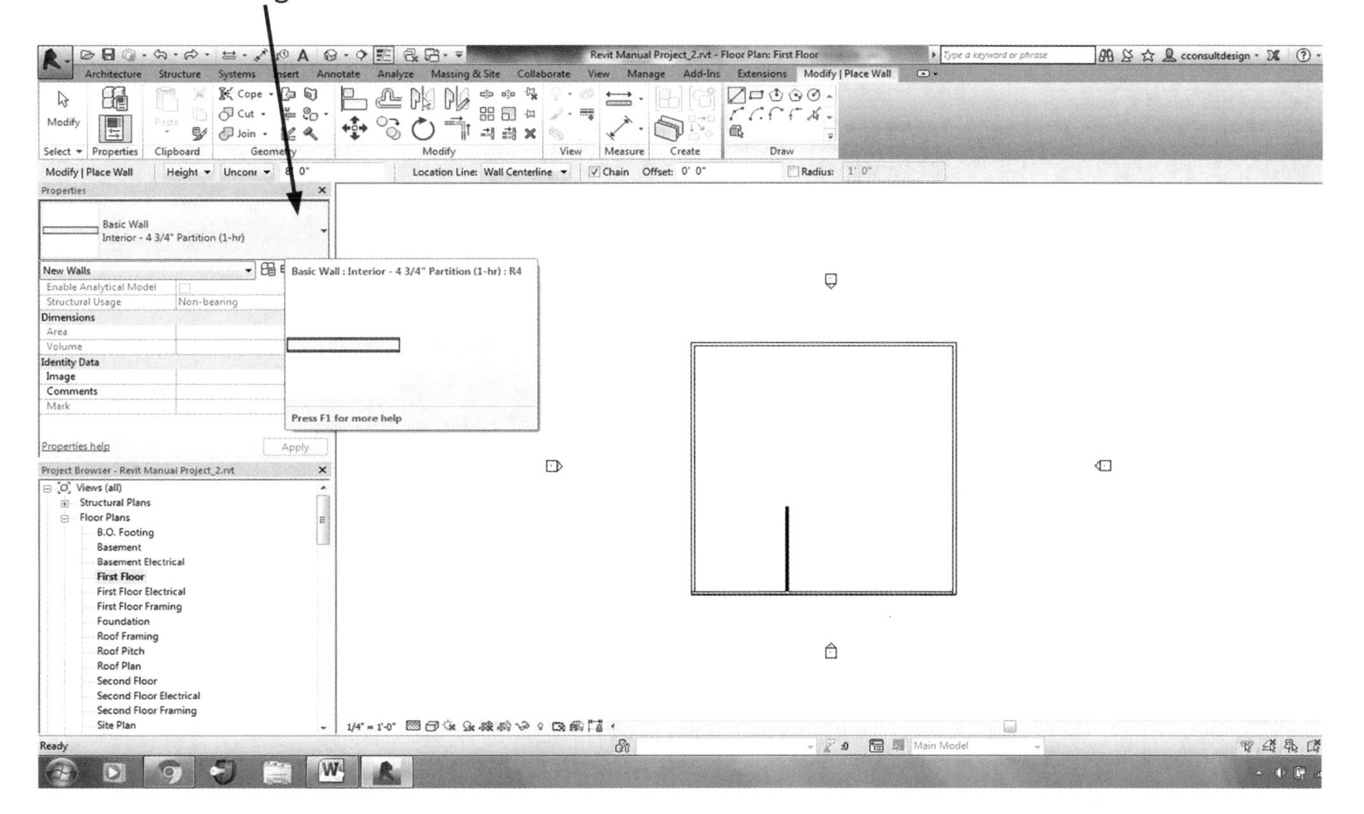

Connect the wall to the existing wall in your drawing.

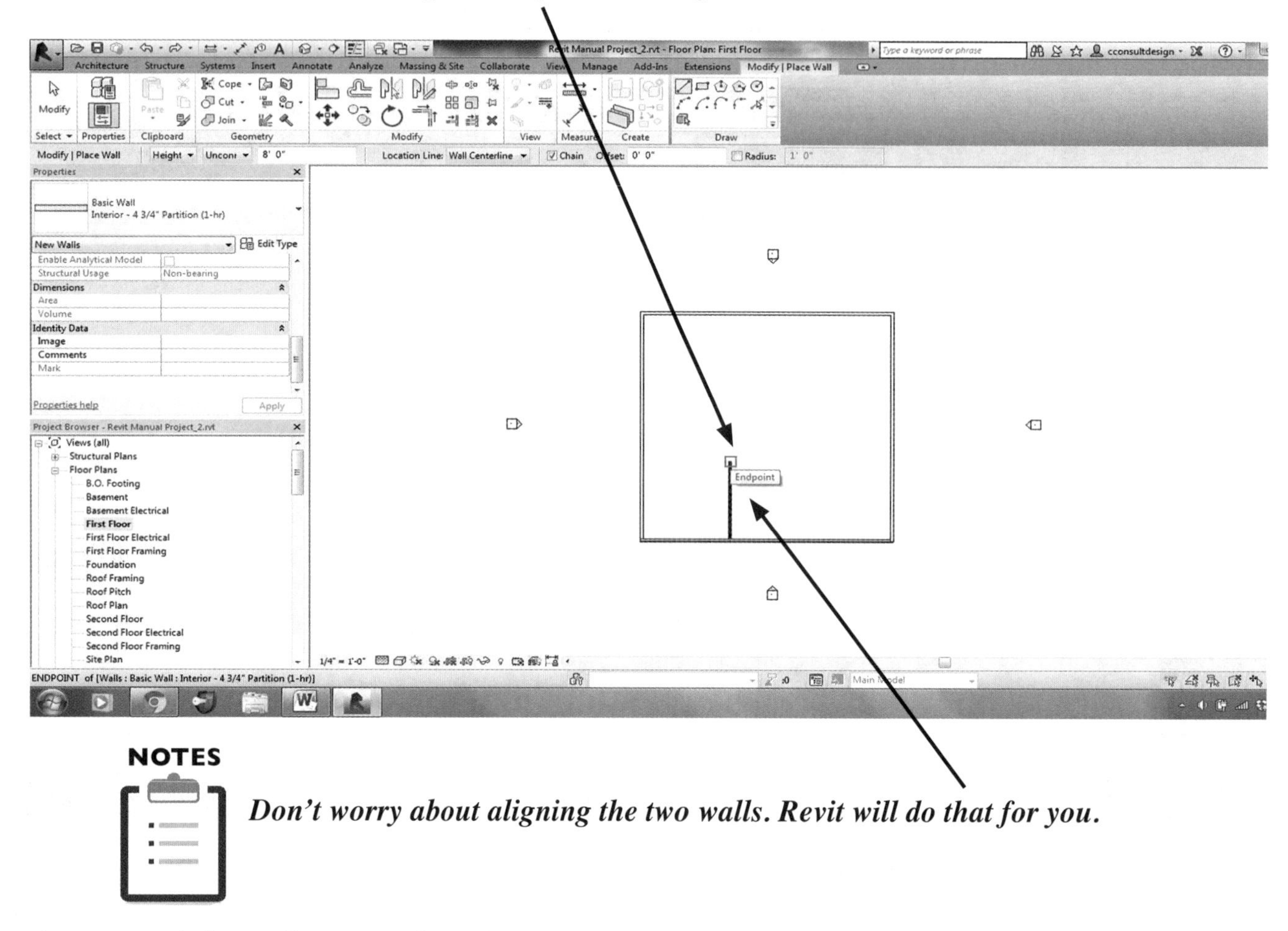

NOTES

Don't worry about aligning the two walls. Revit will do that for you.

Then extend the wall over to the other side or to another wall you want to connect with.

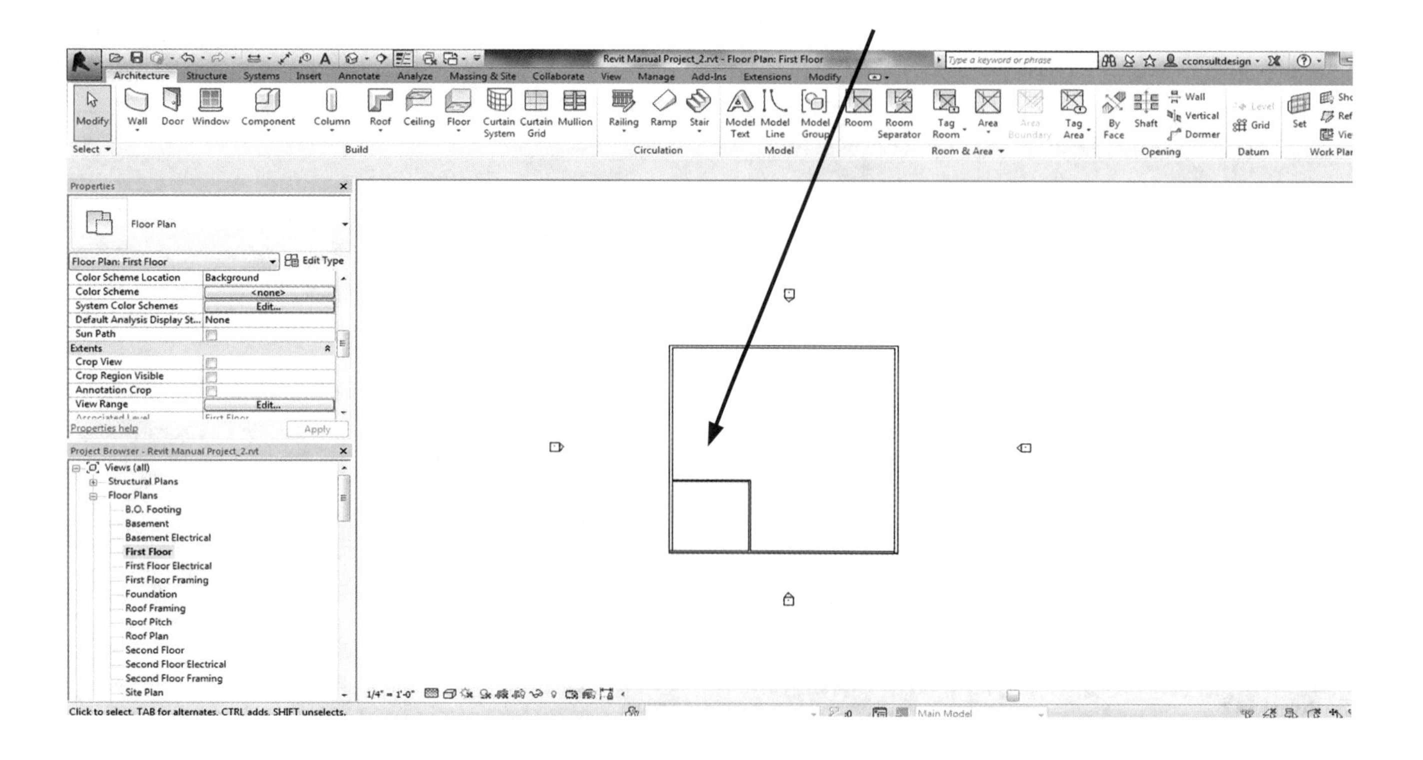

The completed room layout should look like this:

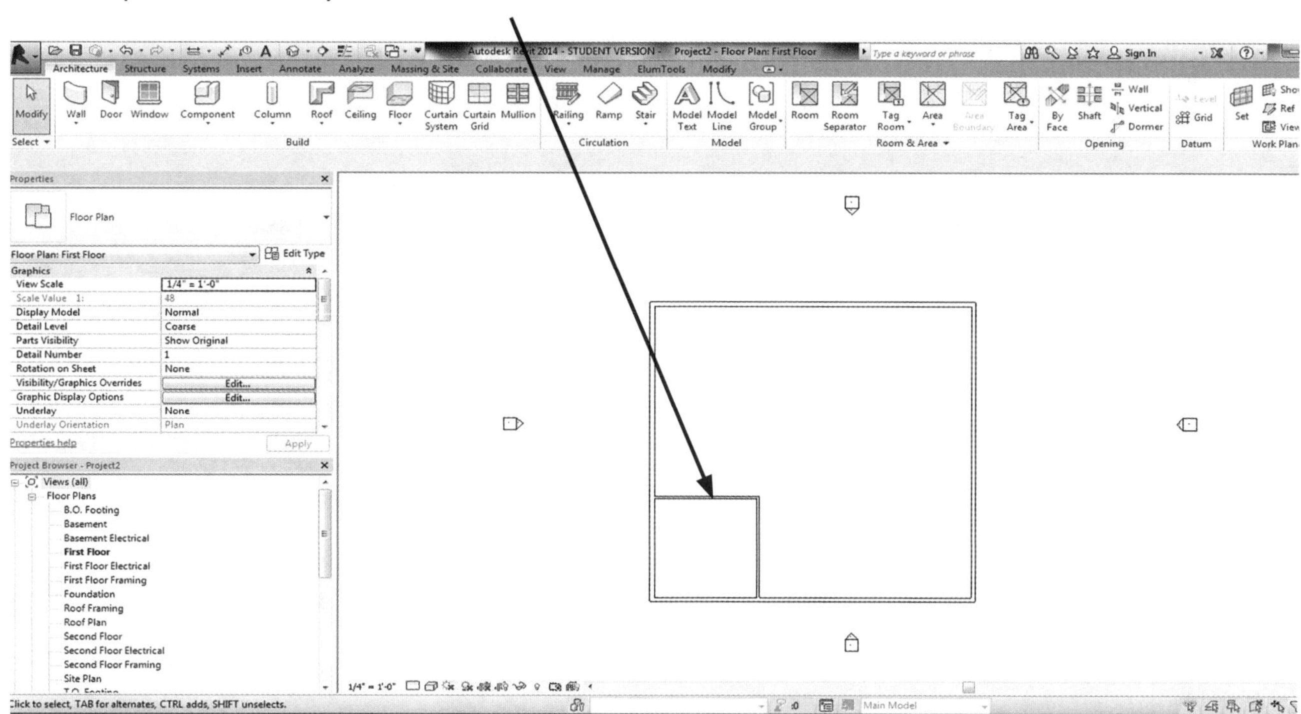

Revit has allowed you to design a complete exterior and interior of a home or commercial building with a few clicks of a mouse and an extensive library of walls. These walls are already pre-made for you to use and modify at a later point. This will allow you to become very precise in your designs.

If you notice your room doesn't match the dimensions shown, refer back to *Chapter 2, Dimensioning in Revit*, to adjust the wall length. Simply click on the wall and adjust the temporary dimension length, and the wall will move to your desired position.

Now we can move on to making your project inviting. Let's take a look at Chapter 4, *Placing Doors and Windows.*

CHAPTER 4

Placing Doors and Windows

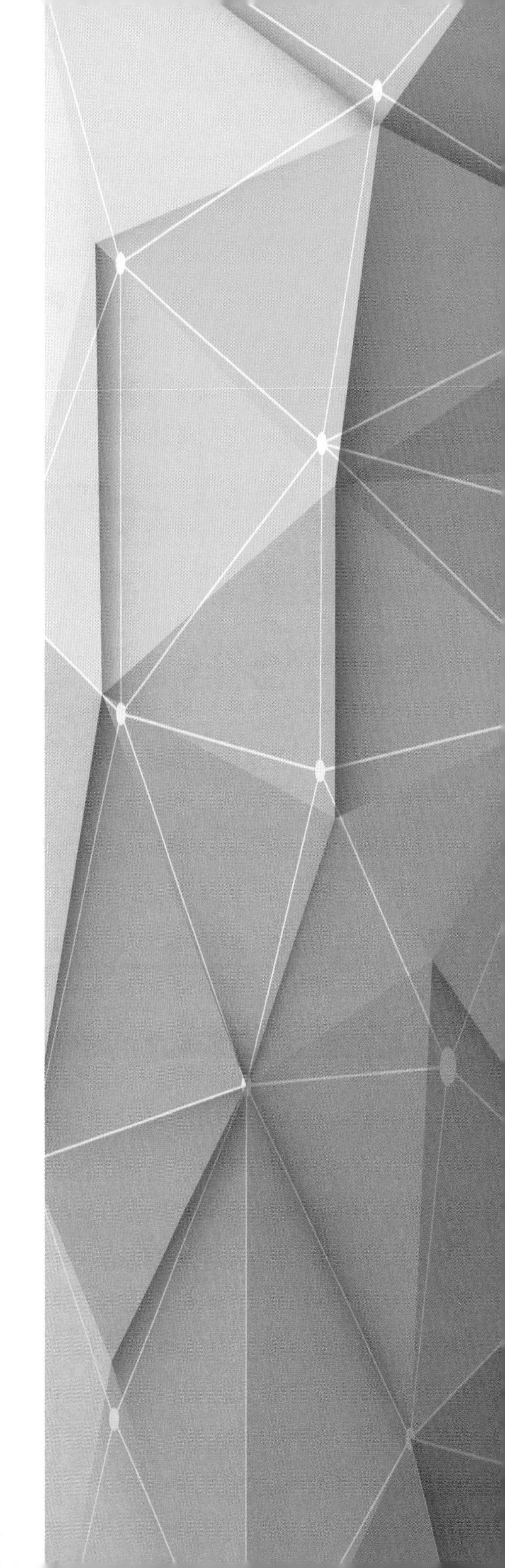

PLACING DOORS

To begin placing a door in your drawing, go to the Architecture Tab, and select the **DOOR** tile to activate a list of doors in Revit.

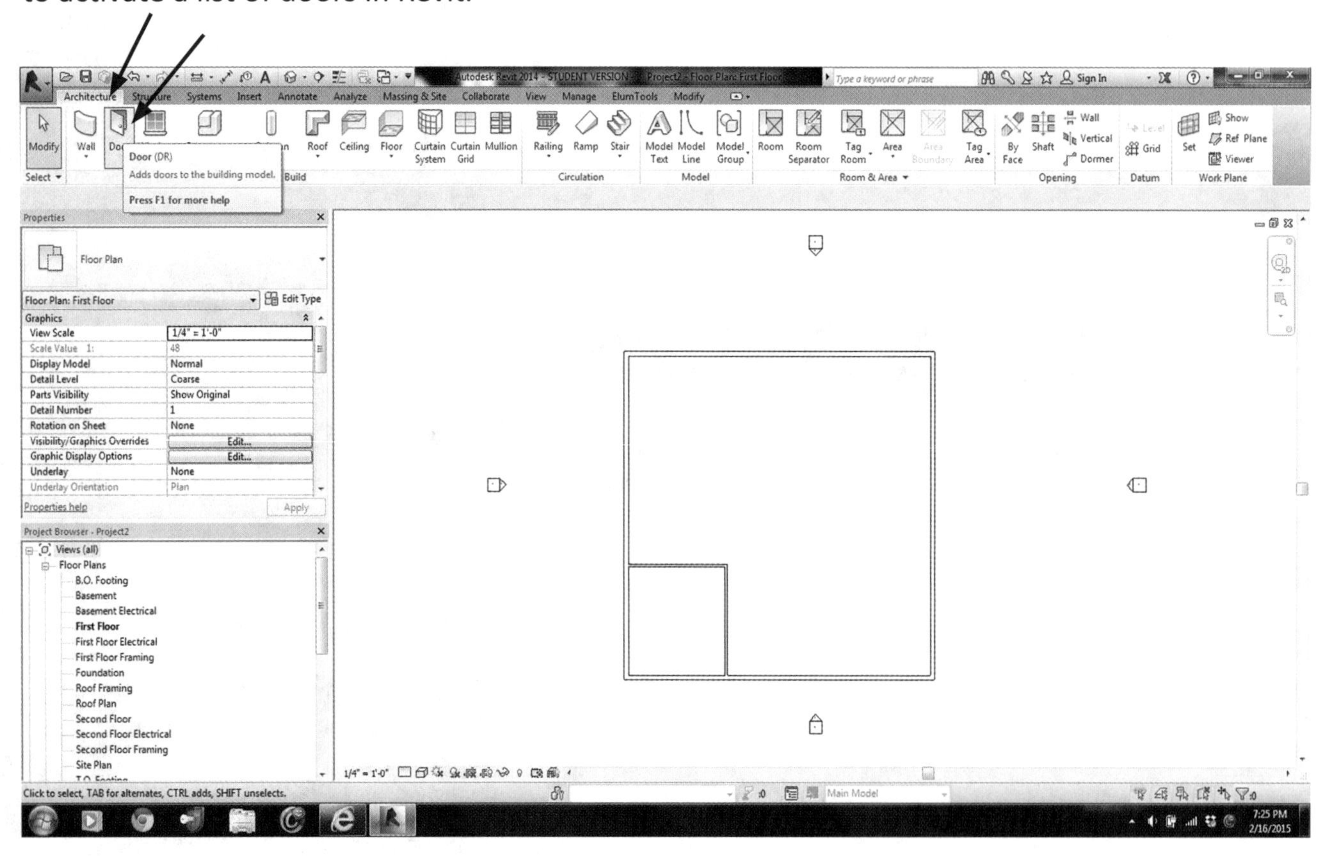

This command will open up a quick door selection of what is already available, so you can choose which door you want to use in your design. You can then drag the door of your choice to your design, and place it in the specified location in the wall.

NOTES

If you want the door to have a mark or door tag identifier such as 101 or 101A, then make sure* TAG ON PLACEMENT *is highlighted.

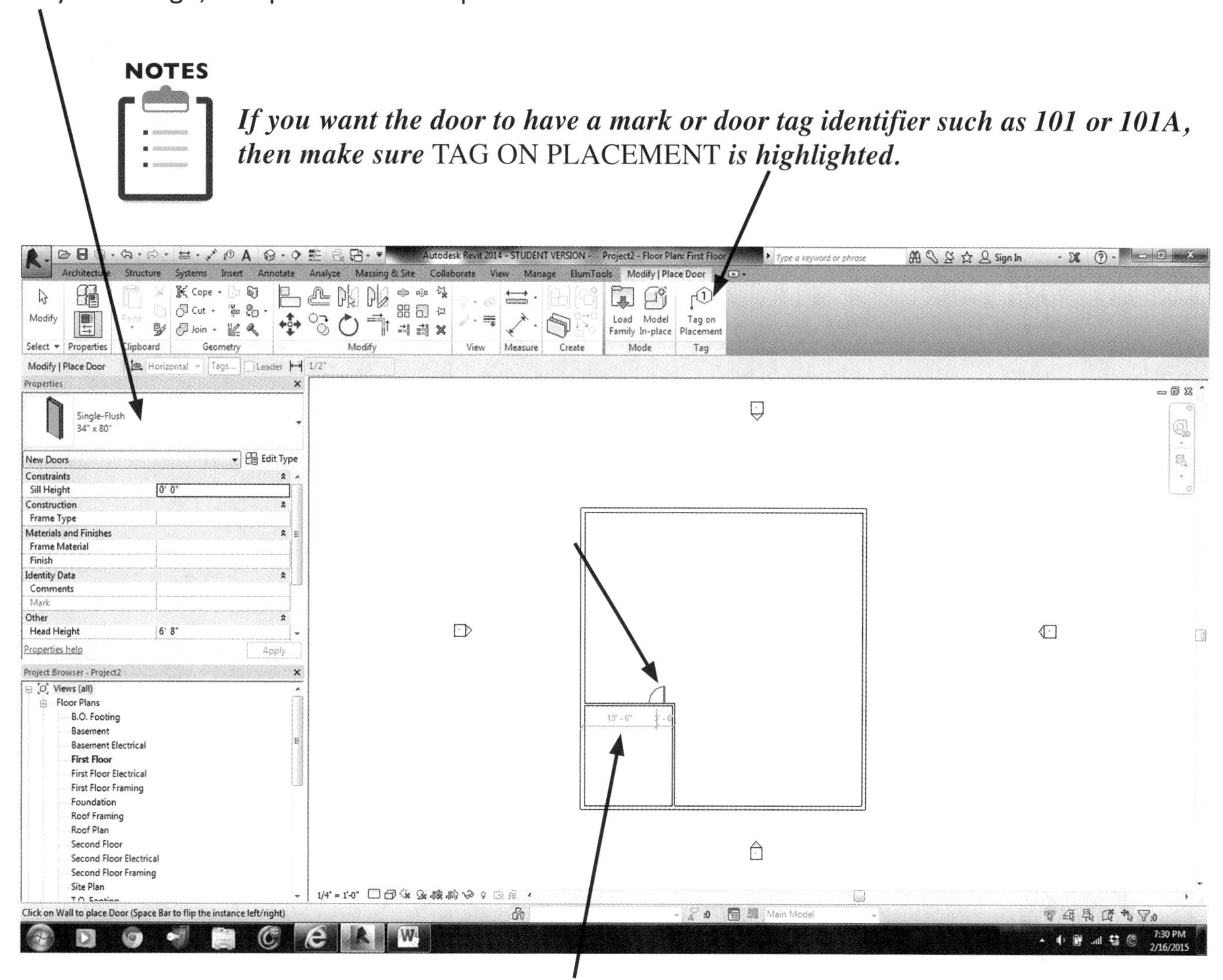

When placing a door, notice that the temporary dimensions appear just like we discussed in Chapter 2, *Dimensioning in Revit.* So, if you need to place a door in a specific location, you can click on the dimension and type in the specific distance, and the door will move to your new desired location. For example, if you want the door to be placed 6 inches from the edge of a wall (like in your home), then you can specify that distance.

Here is another example of placing a front door following the same steps as before. Click on the door drop-down category and drag the door into position.

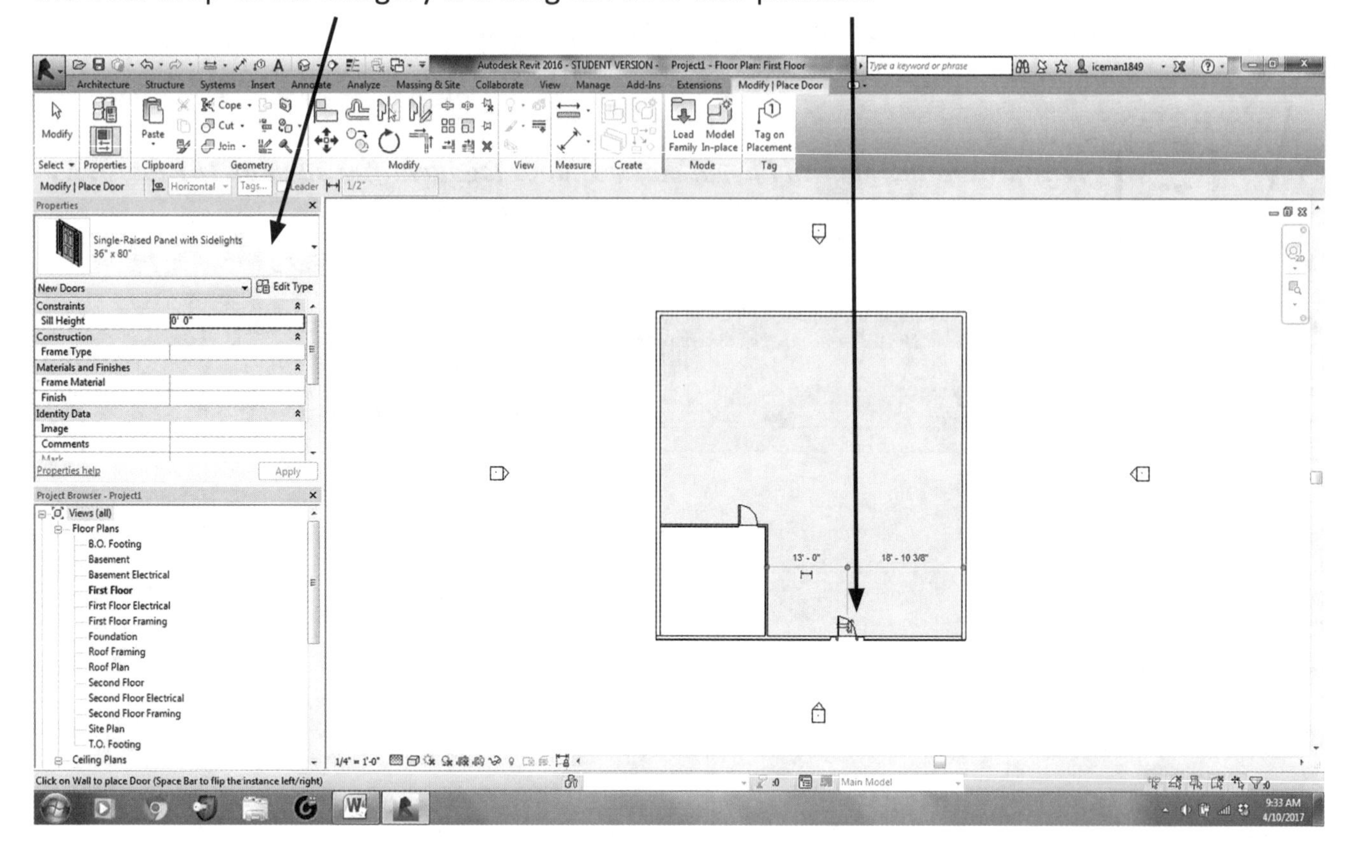

ESSENTIAL REVIT FACT

When placing doors in a project, doors are "host-based," which means that the doors have to be placed in a wall. They cannot be placed out in the middle of nowhere. If you attempt to put a door in without walls, you will notice that you get a circle with a line through it. This means that you can't put the door in that position.

Once you have selected the door and placed it in your house or building, click the **MODIFY** Arrow to end the command. Please don't click the **ESC** Key.

REMINDER

As explained earlier, you do not want to hit the ESC *Key, as it can undo everything you've just done, and you may have to start over.*

NOTES

If the doors that are in the quick drop-down box in Properties are not what you are looking for, then you can go to the Revit Library inside the Revit software to find more options.

Click on the **INSERT** tab, and click on **LOAD FAMILY**.

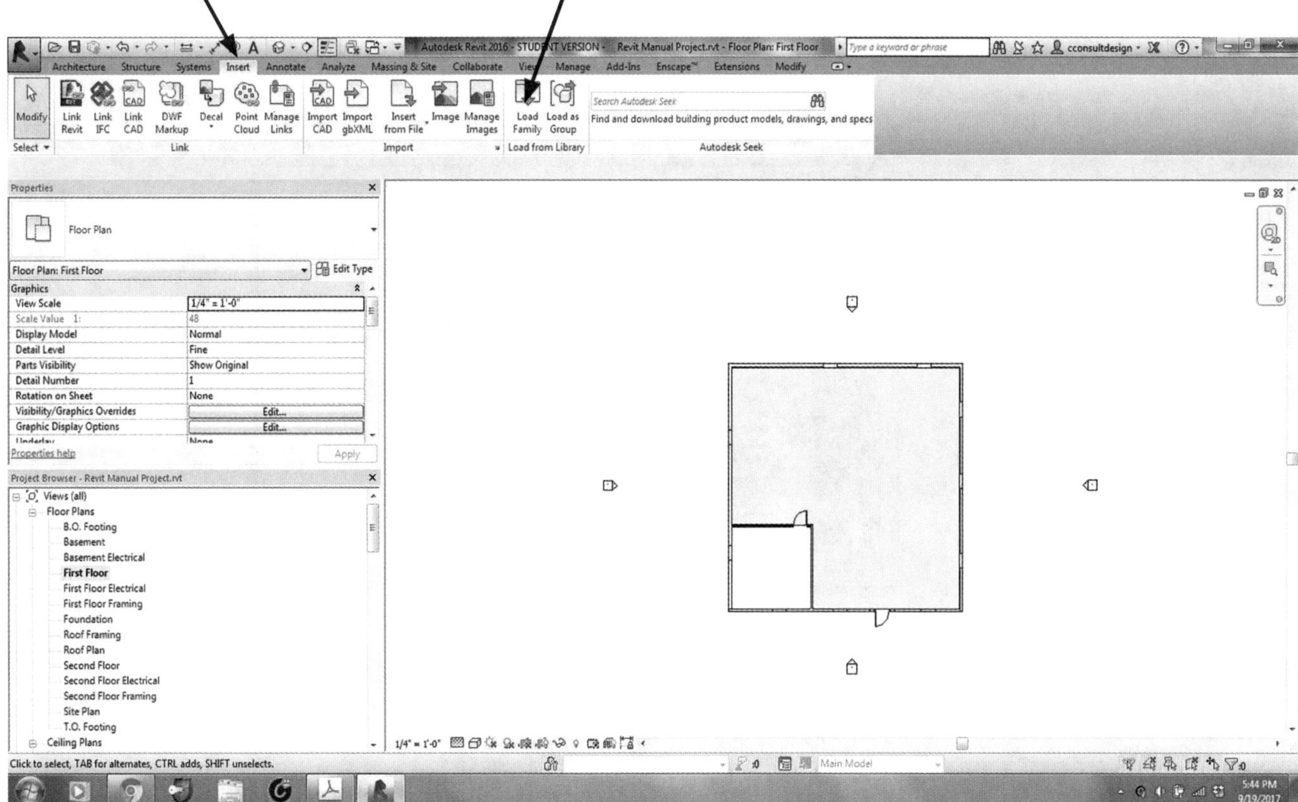

You will get a dialog box with **LOAD FAMILY** template folders that are stored in Revit.

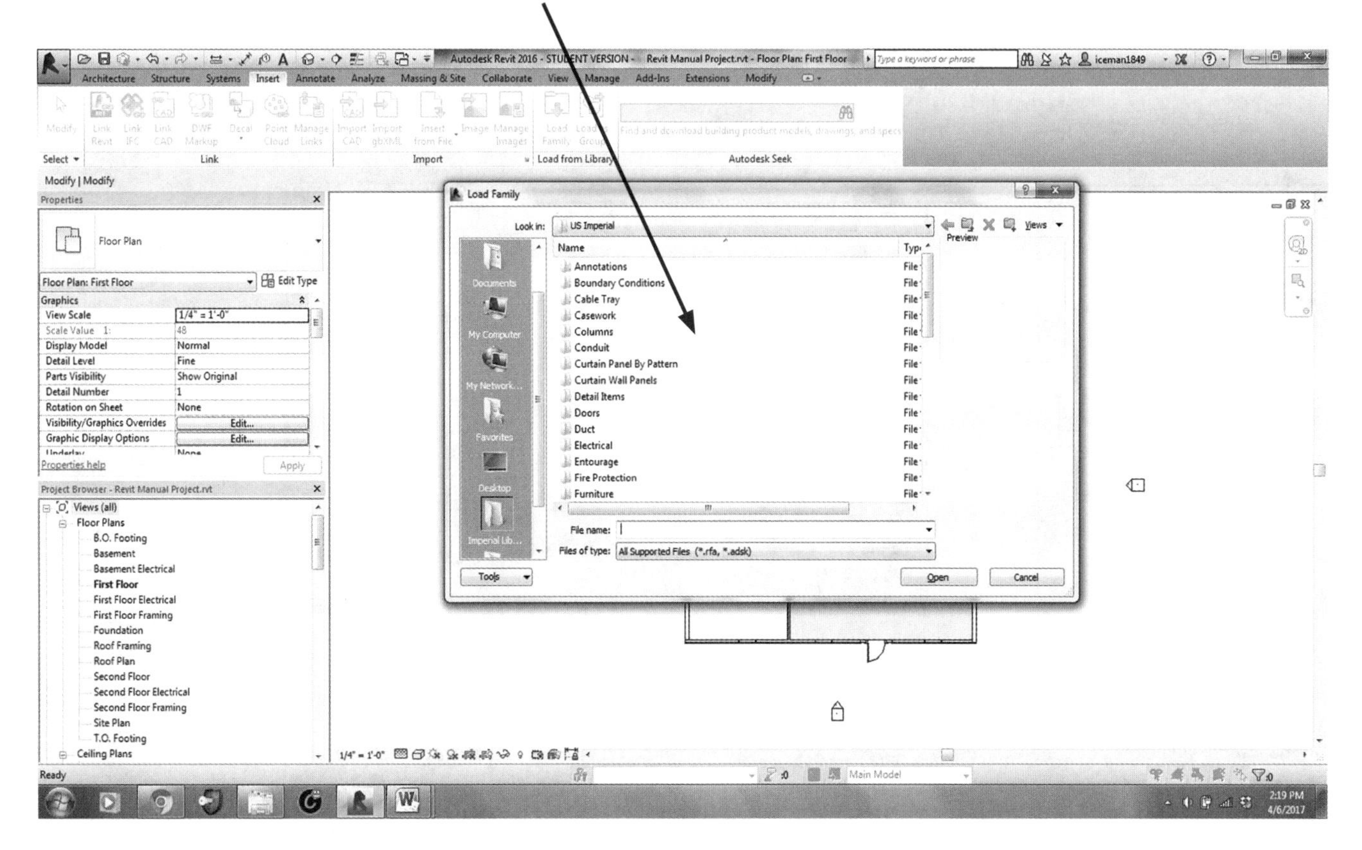

Next, click on **DOORS** and open that folder.

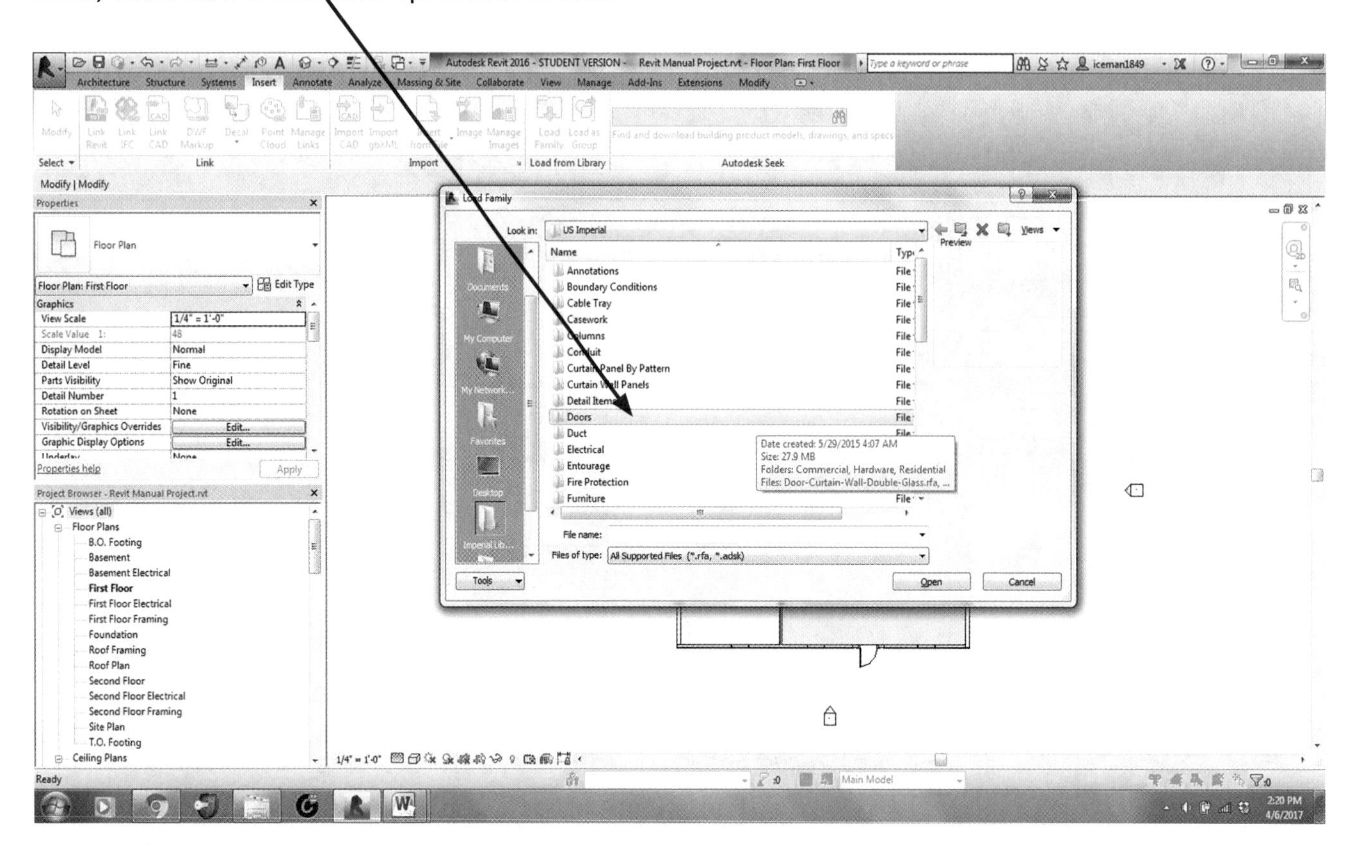

Click inside the **DOORS** folder.

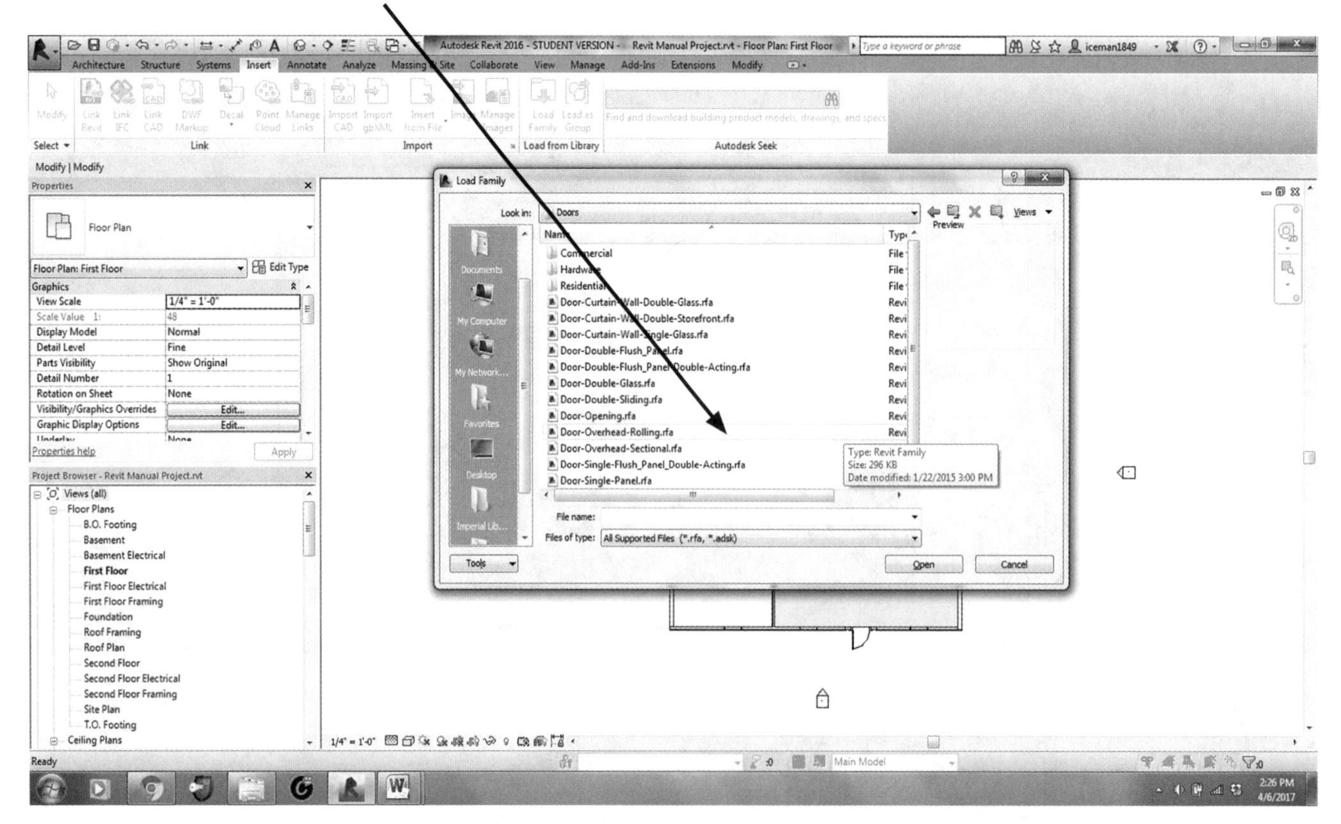

Here you can select from many more doors for your design. You will even get a picture in the top right, so you can see exactly what you are selecting.

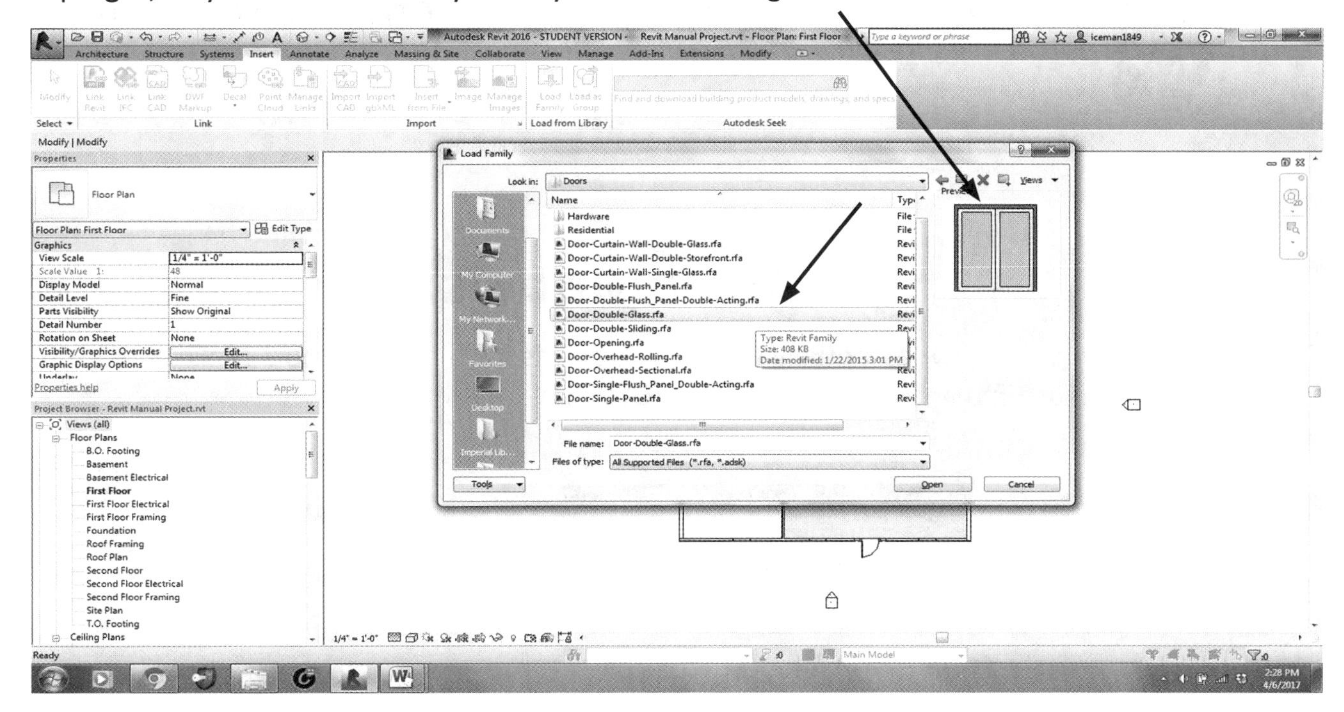

Once you have selected the door you want for your design, click **OPEN**, and the door will be loaded into your project in the Project Browser.

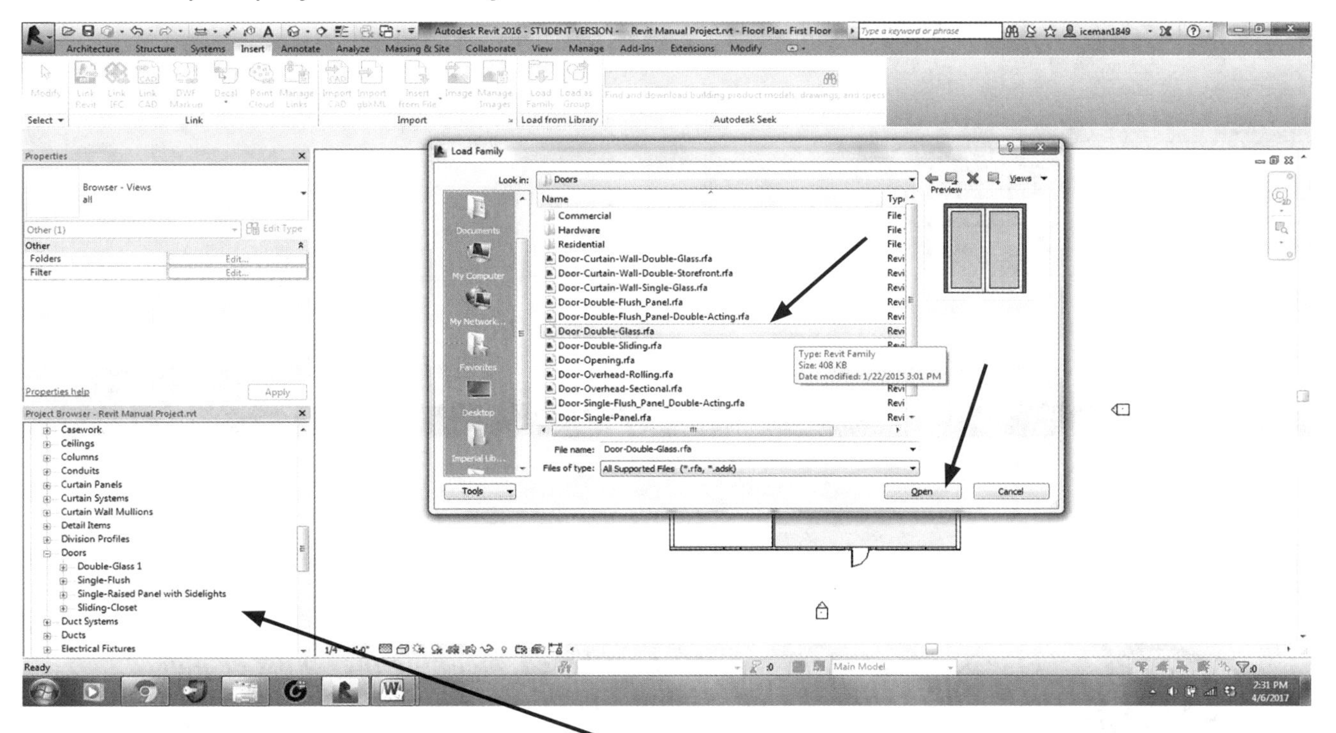

Once the door has been loaded into the Project Browser, follow the steps of expanding the tree, selecting the door size you desire for your house or building, and dragging the door into your design.

REMINDER

Remember that your door must be hosted by a wall, or you will get that circle with a line through it.

PLACING WINDOWS

To place a window, you follow essentially the same steps as for placing a door. Make sure you are in the Architecture Tab and click on the **WINDOW** Tab.

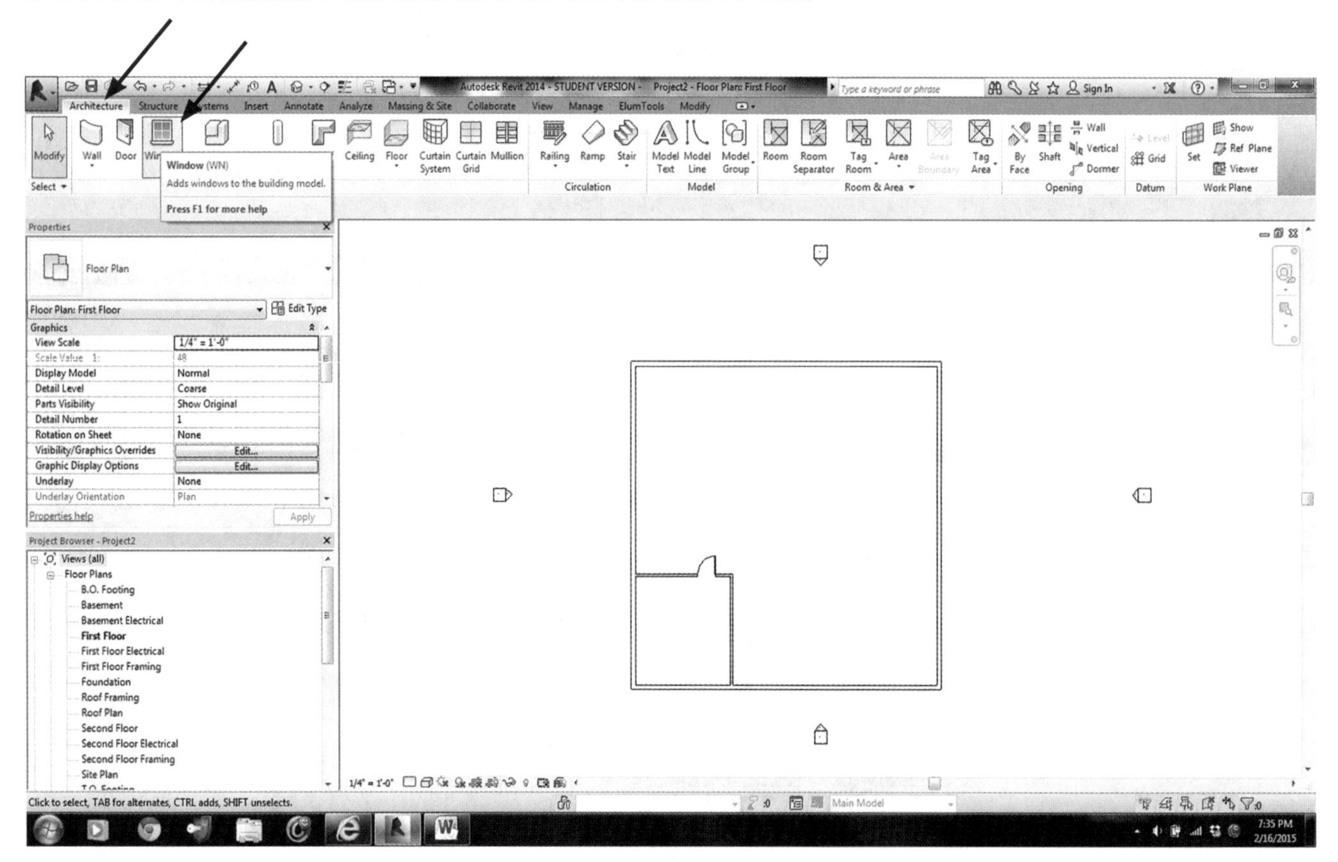

ESSENTIAL REVIT FACT

Like doors, windows are also host-based, meaning they must be placed in a wall. They cannot be placed in the middle of nowhere.

Select the window you want to use in your drawing, then drag the window to your design and place it in the specified location. Place the window just as you did the door, select it, then left-click with your mouse and drag the window into the desired location on your wall.

ESSENTIAL REVIT FACT

Be sure your cursor is on the outside of the wall for the window to be placed correctly.

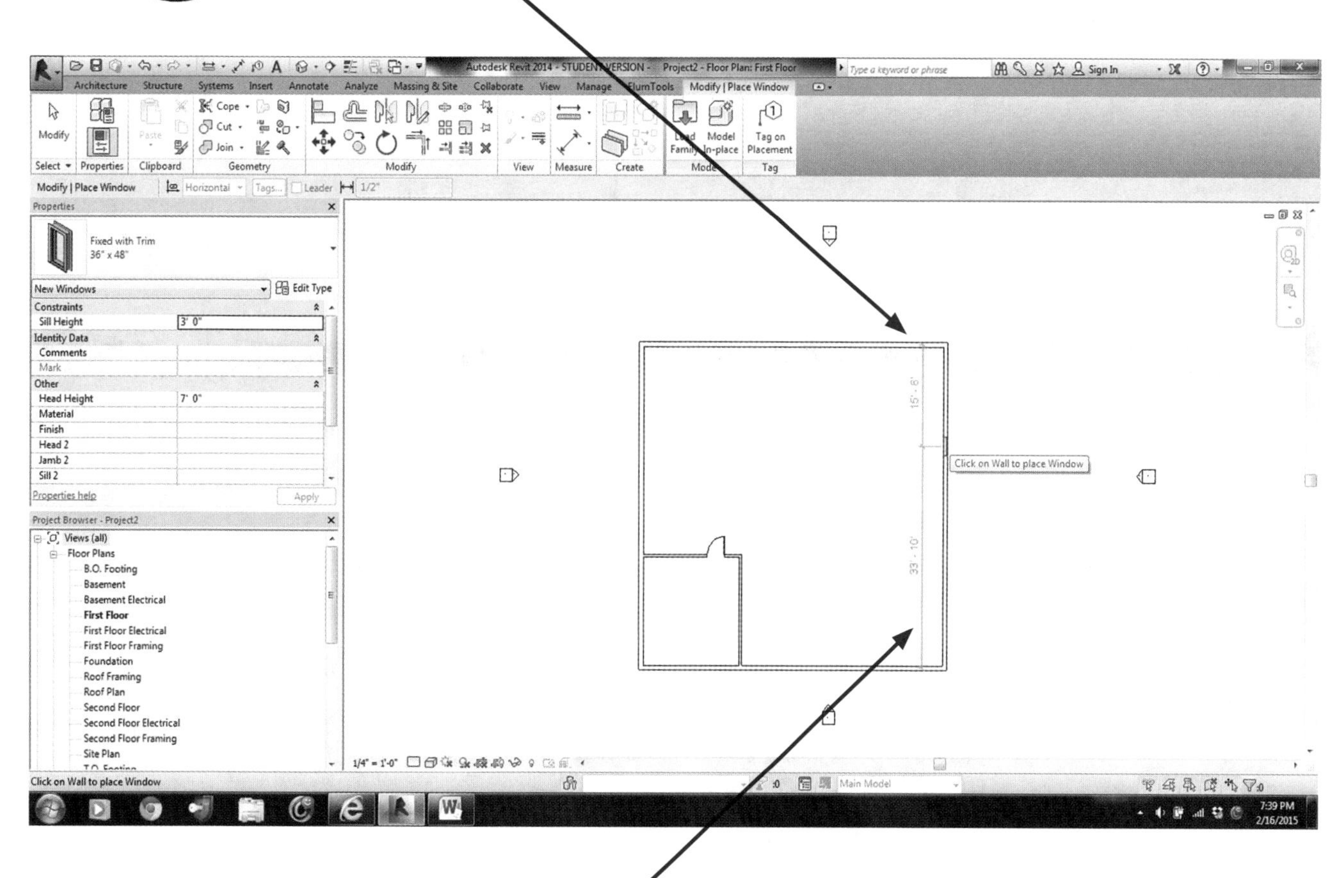

When placing a window, notice that the temporary dimensions appear just like we discussed in the doors section, as well as in Chapter 2, *Dimensioning in Revit.* If you need to place a window in a specific location, you can click on the dimension and type in the specific distance, and the window will move to your new desired location. This is particularly helpful if you are placing a window a certain distance from the edge of the wall to the center of the window.

NOTES

Please note that the window that you have highlighted or activated will be the one that moves to the new or specified location.

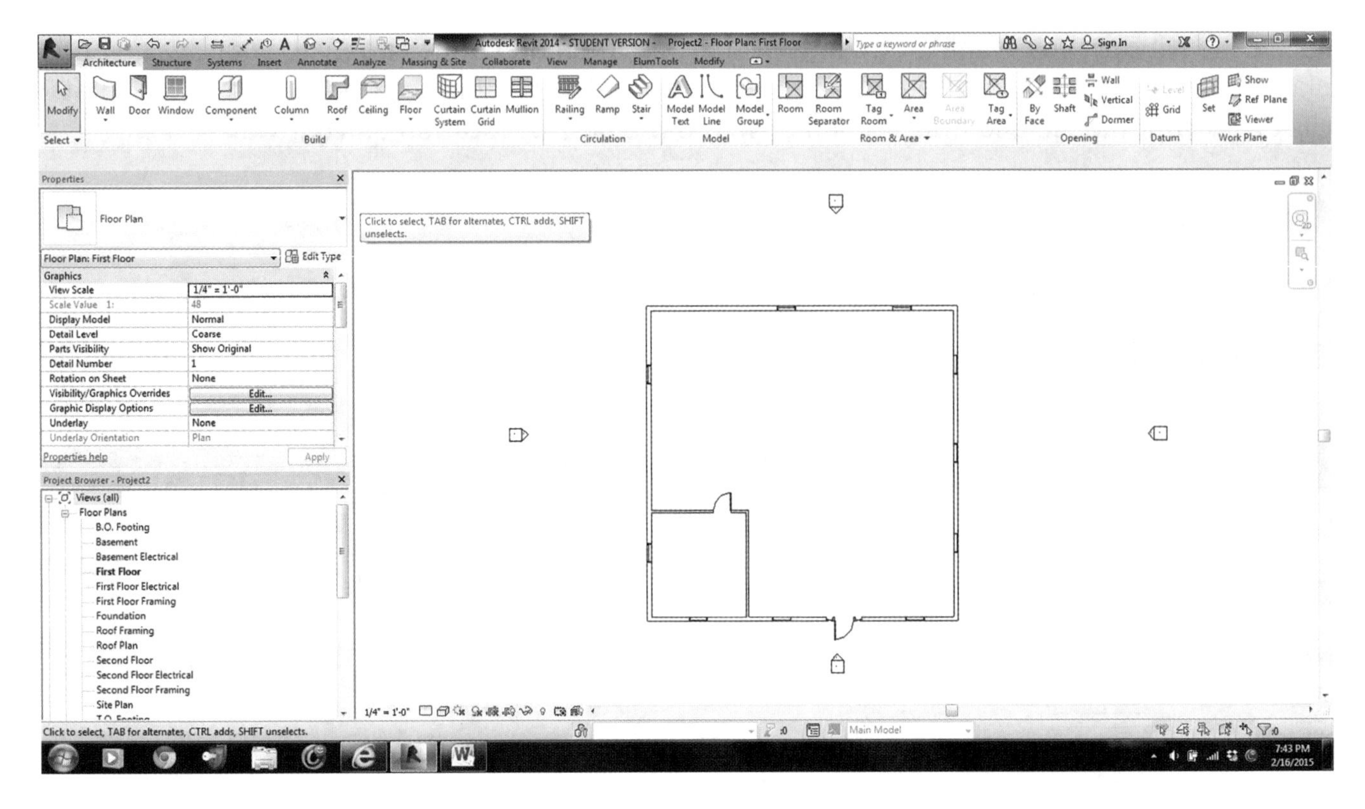

Just like with doors, if you don't see the window size or style you want, you can go into the Load Family category of Revit, and choose additional options.

Here you will get a dialog box with Load Family template folders that are stored in the Revit software.

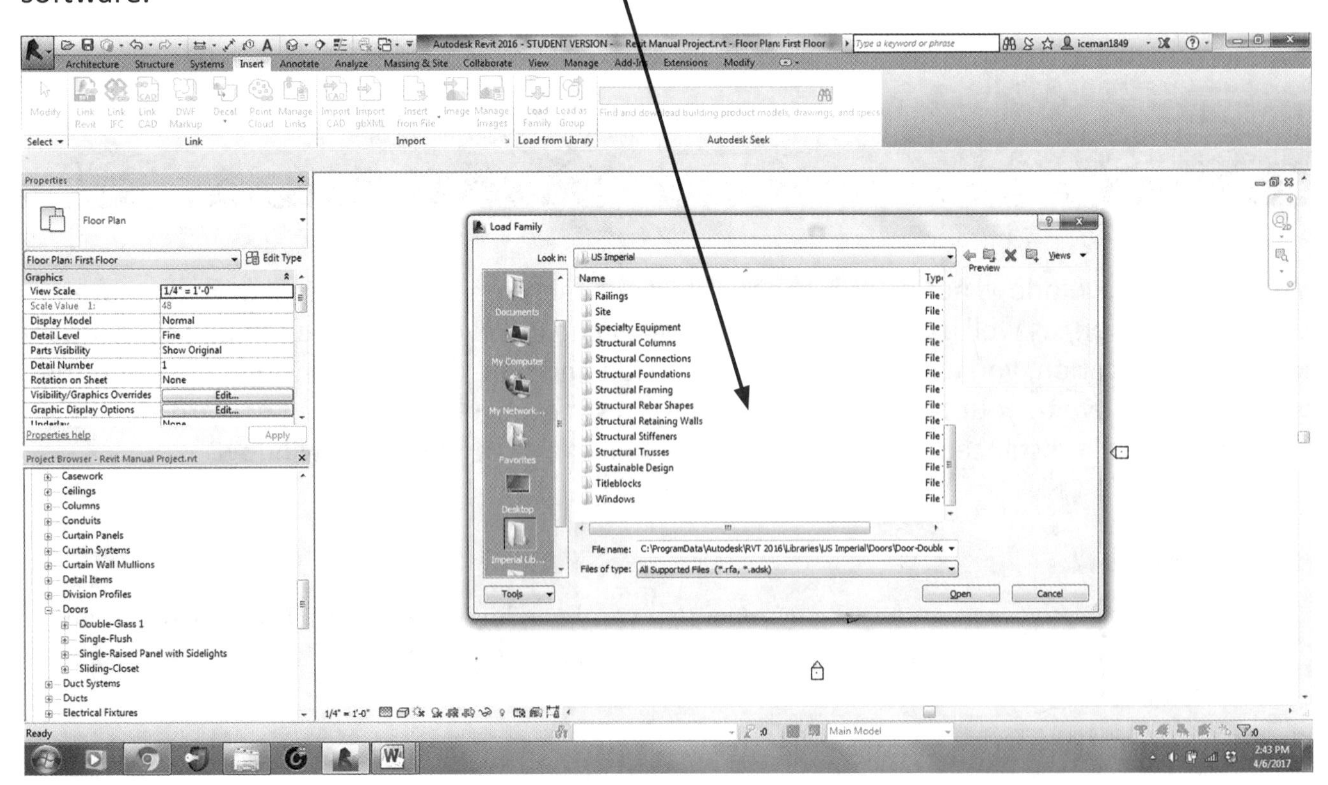

Next click on **WINDOWS**, open that folder, and click on the **WINDOWS** folder.

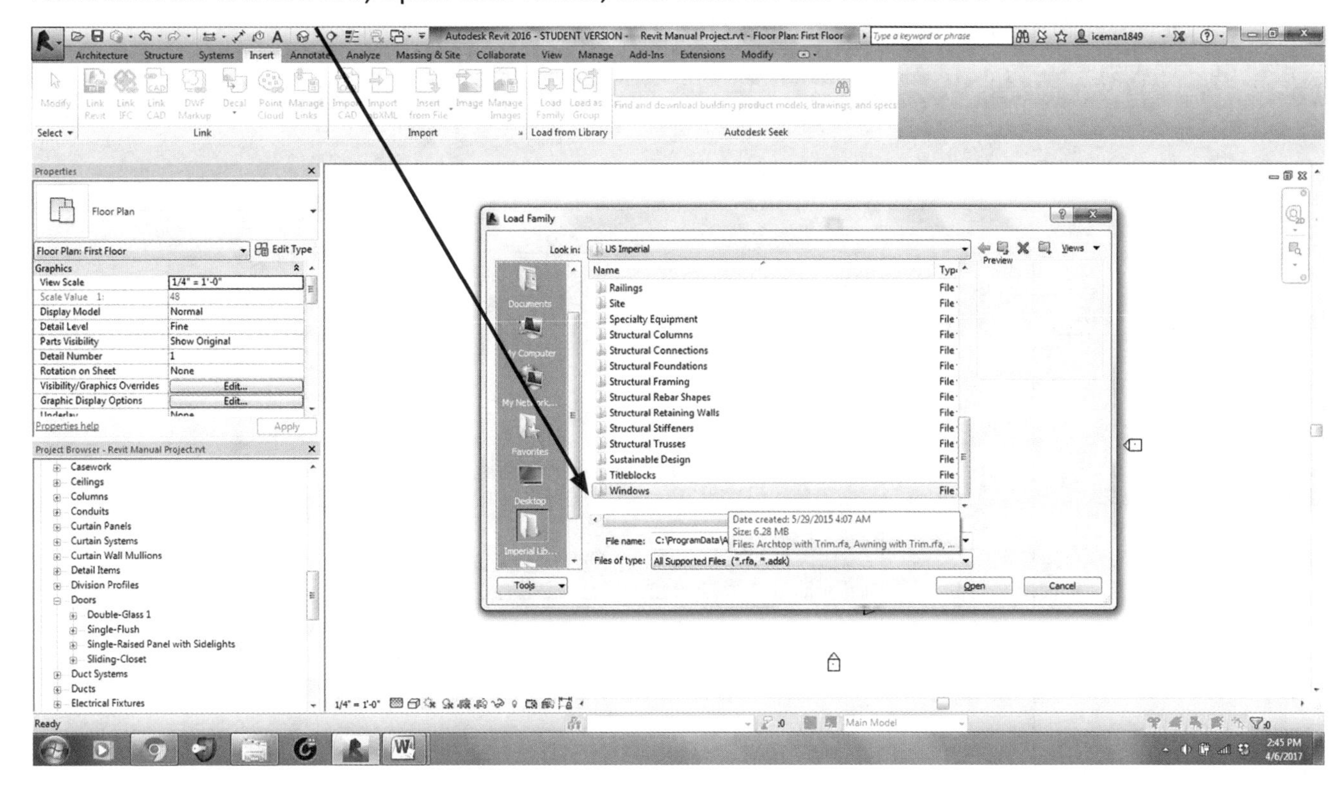

Then click inside the **WINDOWS** folder.

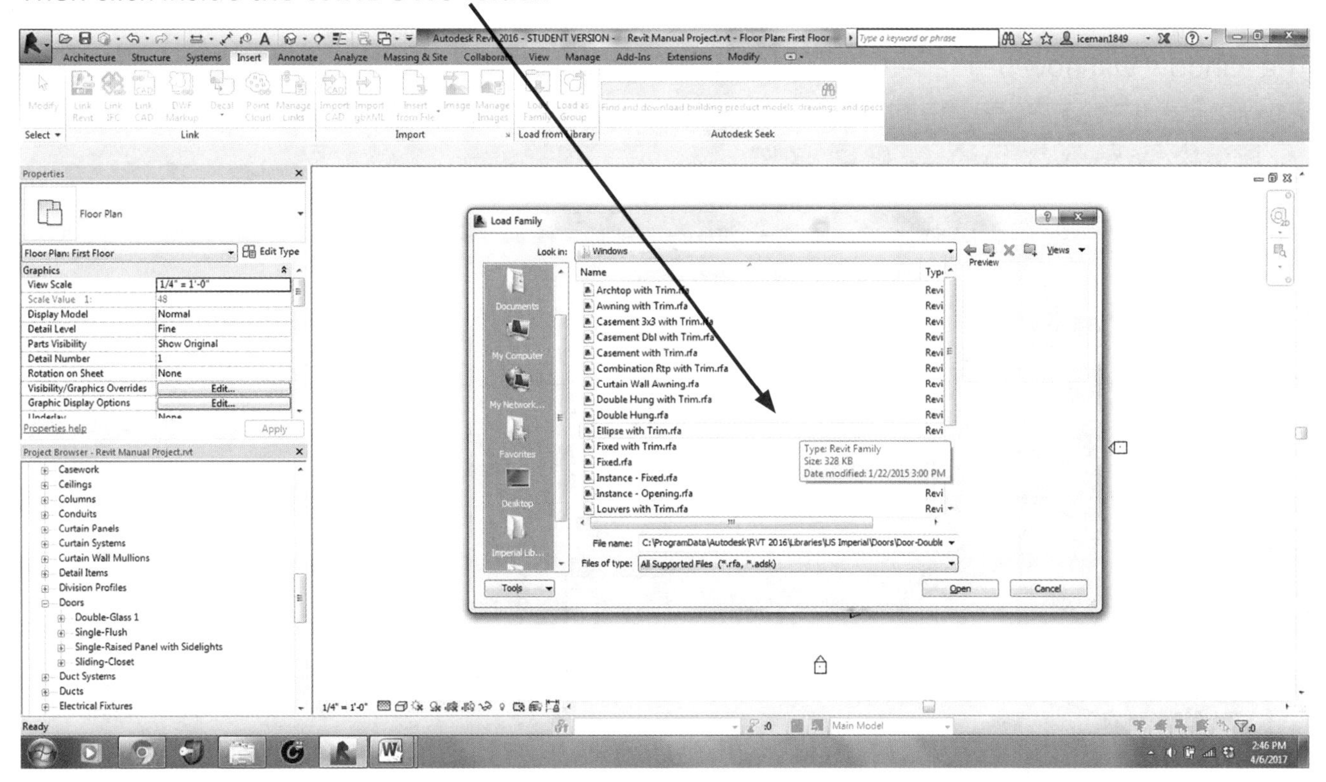

After selecting the window that you want for your project, click on the window name and you will get a picture in the top right. Then click **OPEN**.

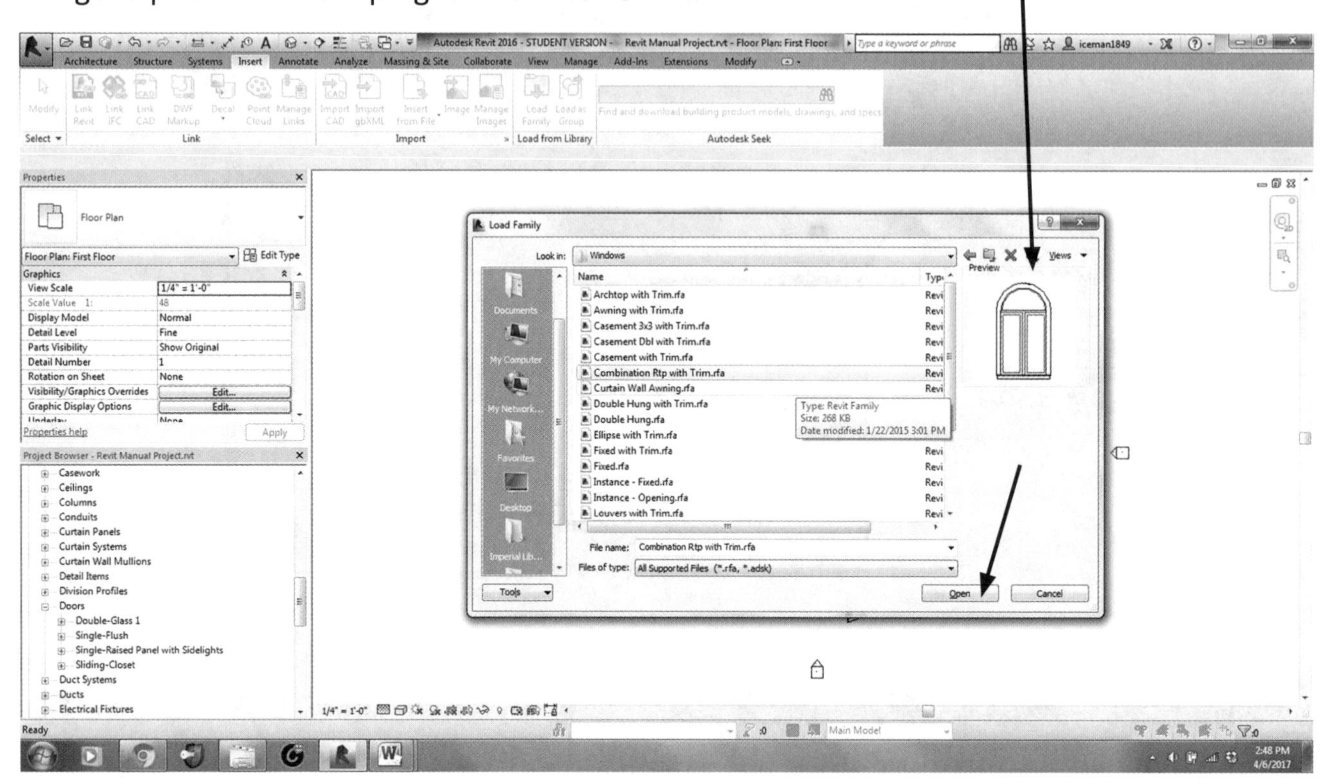

The window will now be loaded into the Project Browser of your project for your drawing.

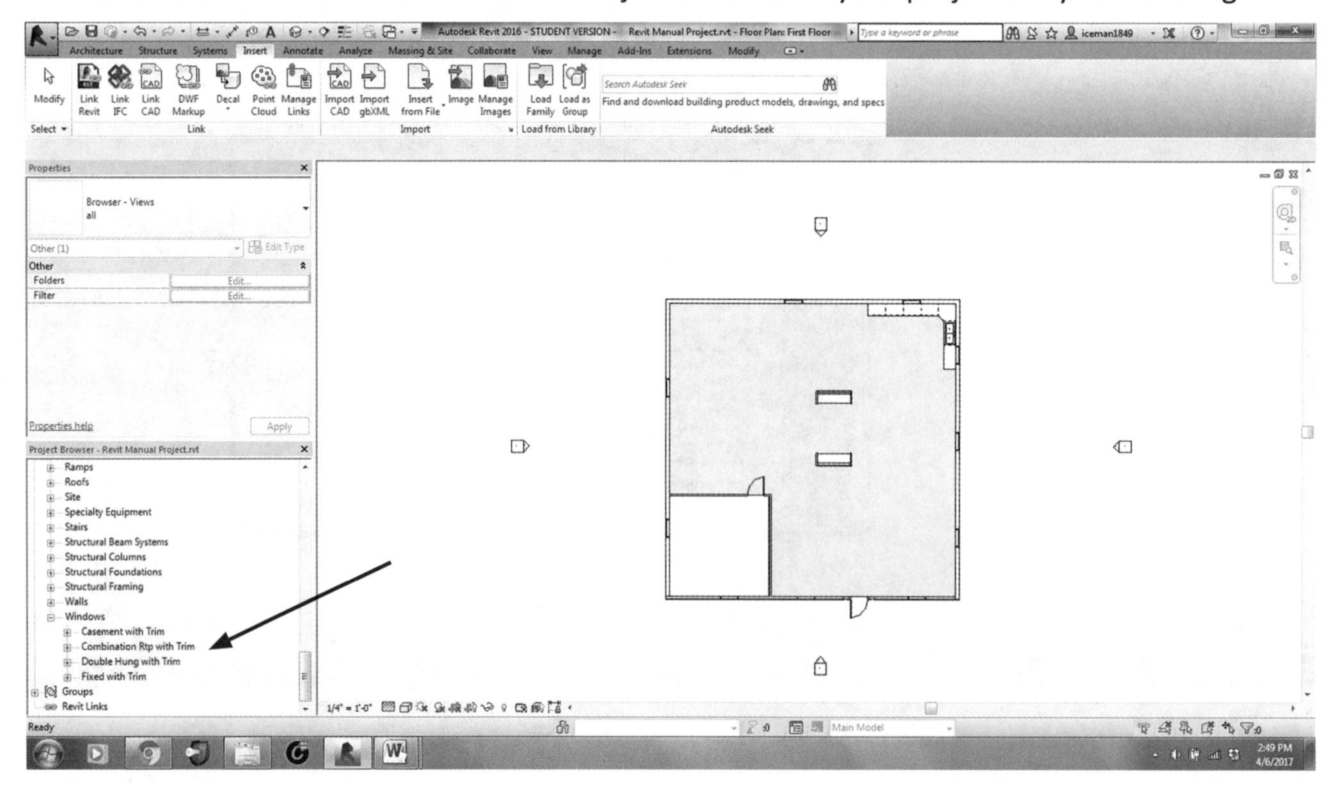

Placing doors and windows in your project design using the Revit software has made it so much easier since you don't need to determine which direction the doors should go, or where the windows need to be placed. The doors and windows will always snap to a wall. You, the user, just have to decide where to place them and what size you desire.

Now let's go further with Revit to the next Chapter 5, *Creating Floors*.

CHAPTER

5

Creating Floors

When creating a floor, click on the **ARCHITECTURE** Tab, then click on the **FLOOR** Tab.

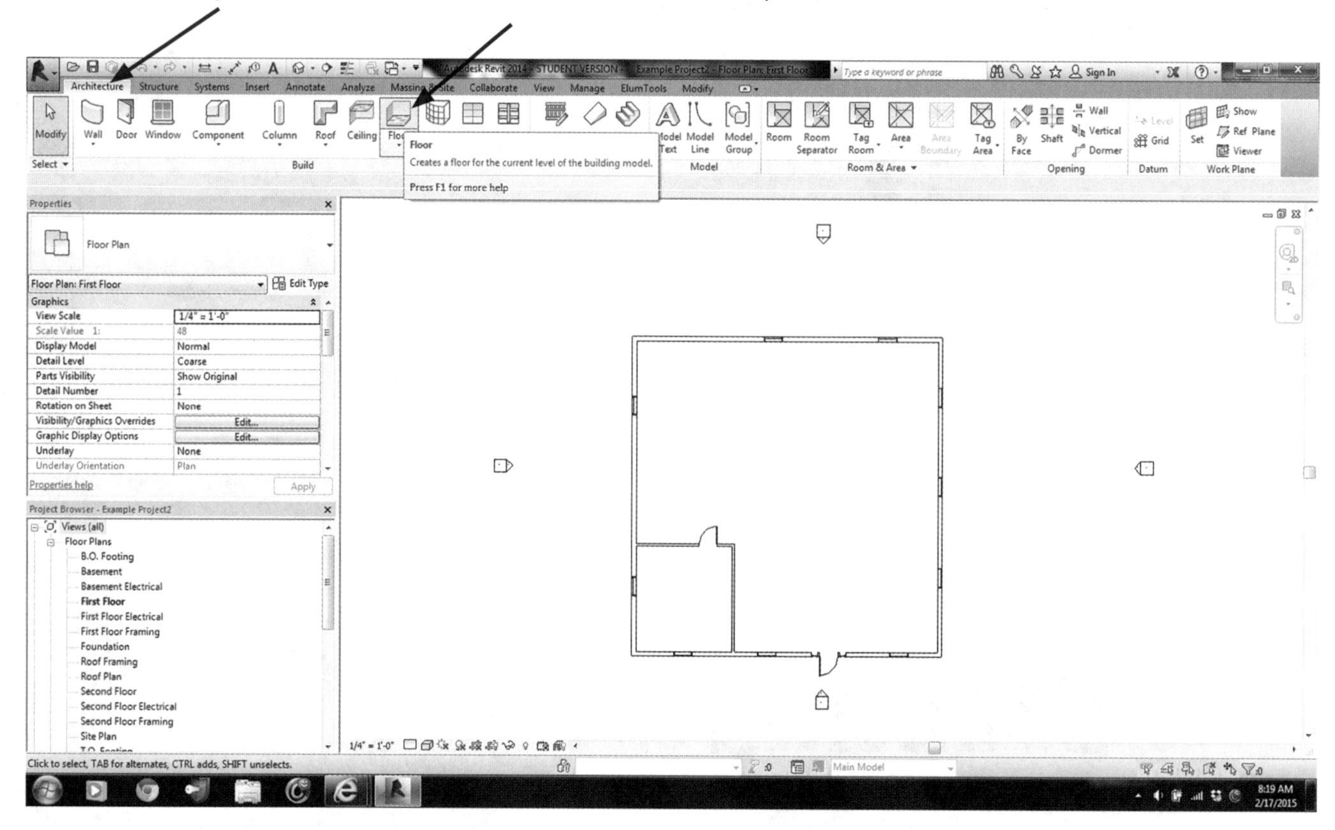

The ribbon will change to **MODIFY/CREATE FLOOR BOUNDARY.** Here you can sketch the floor boundary you want for your design.

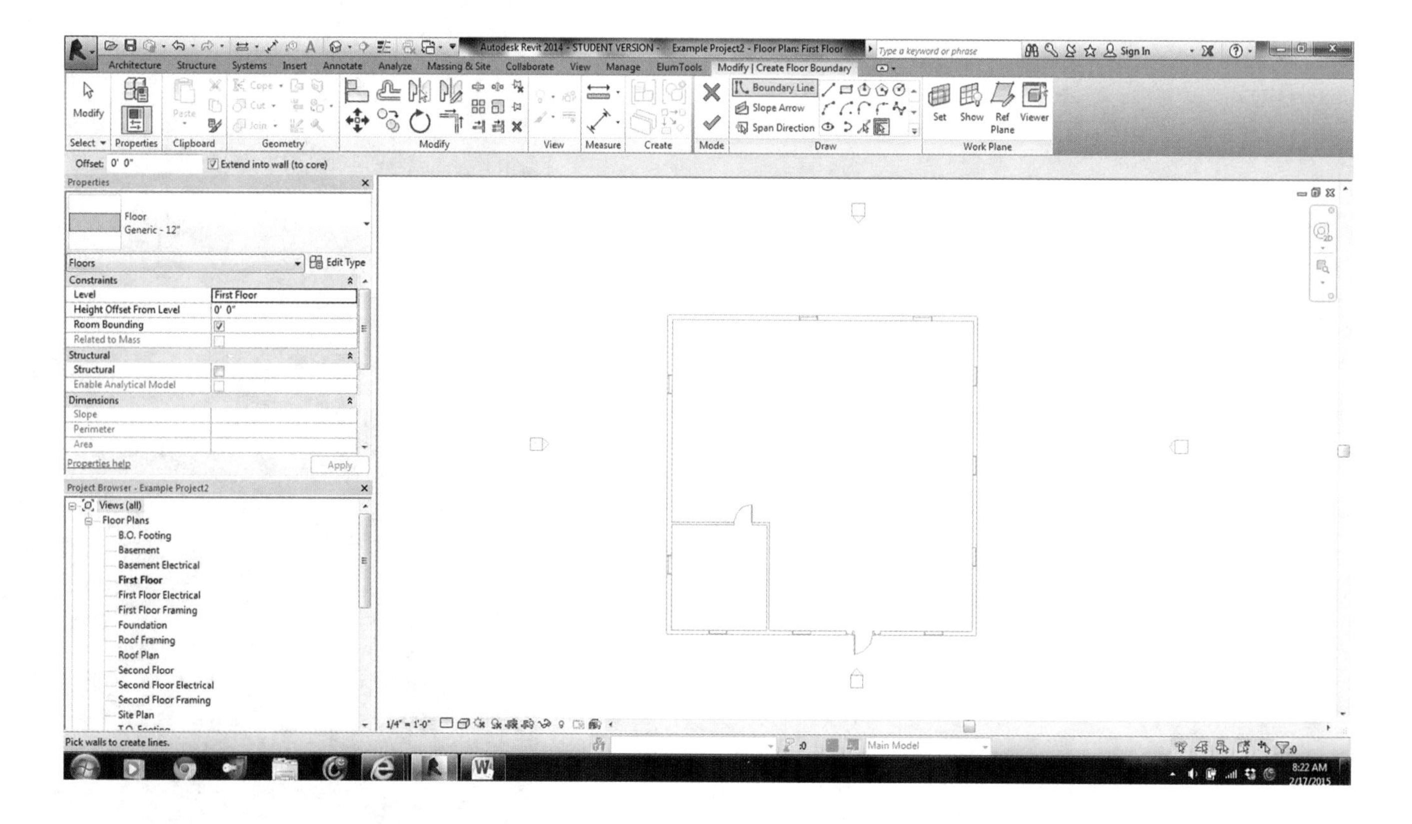

Click on the **PROPERTIES BOX** and then click the drop-down box to select the type of floor you want to use for your design.

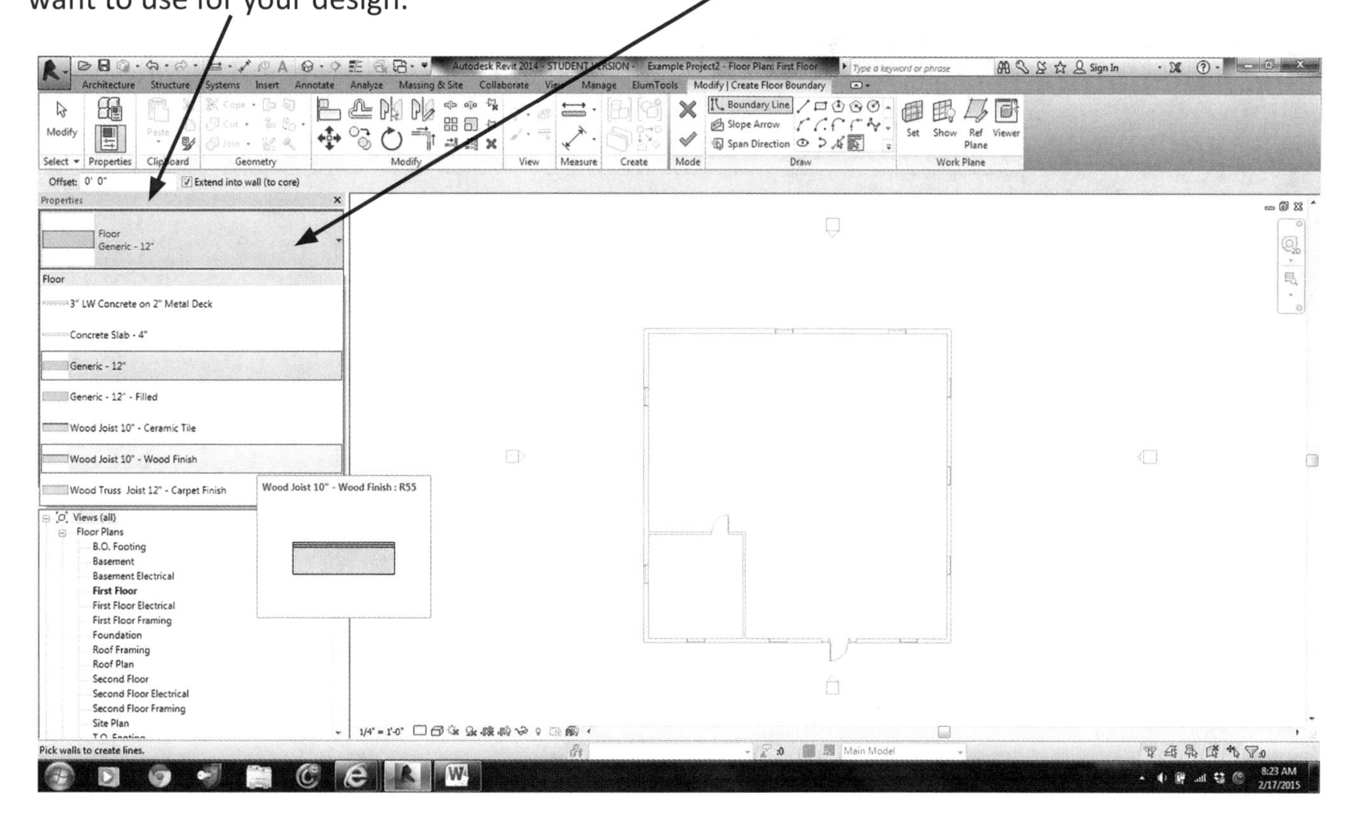

After picking the floor in the draw box, the **PICK WALLS** Button comes up automatically, so you can simply select the walls and automatically trace the complete design.

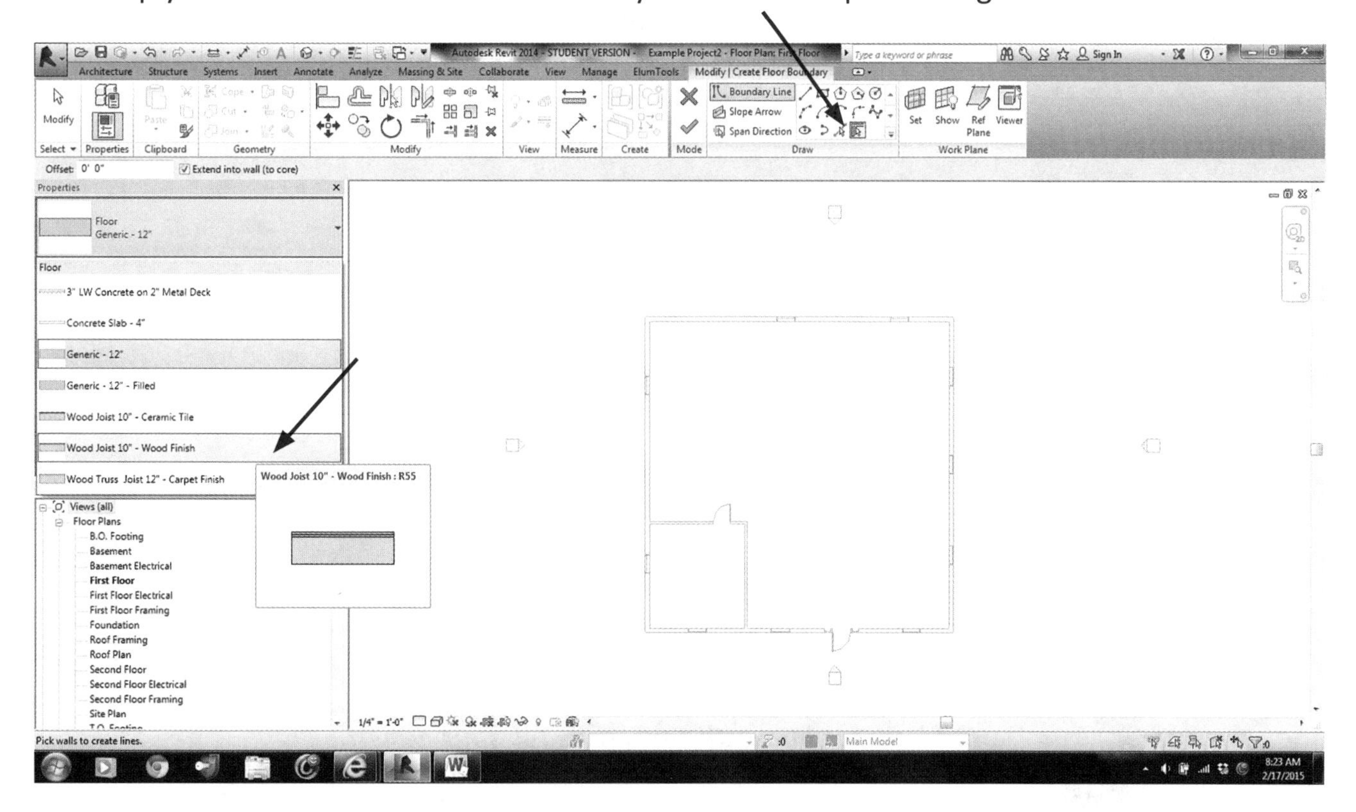

Just begin picking the walls to trace your design (click the inside of the building or the inside of the walls).

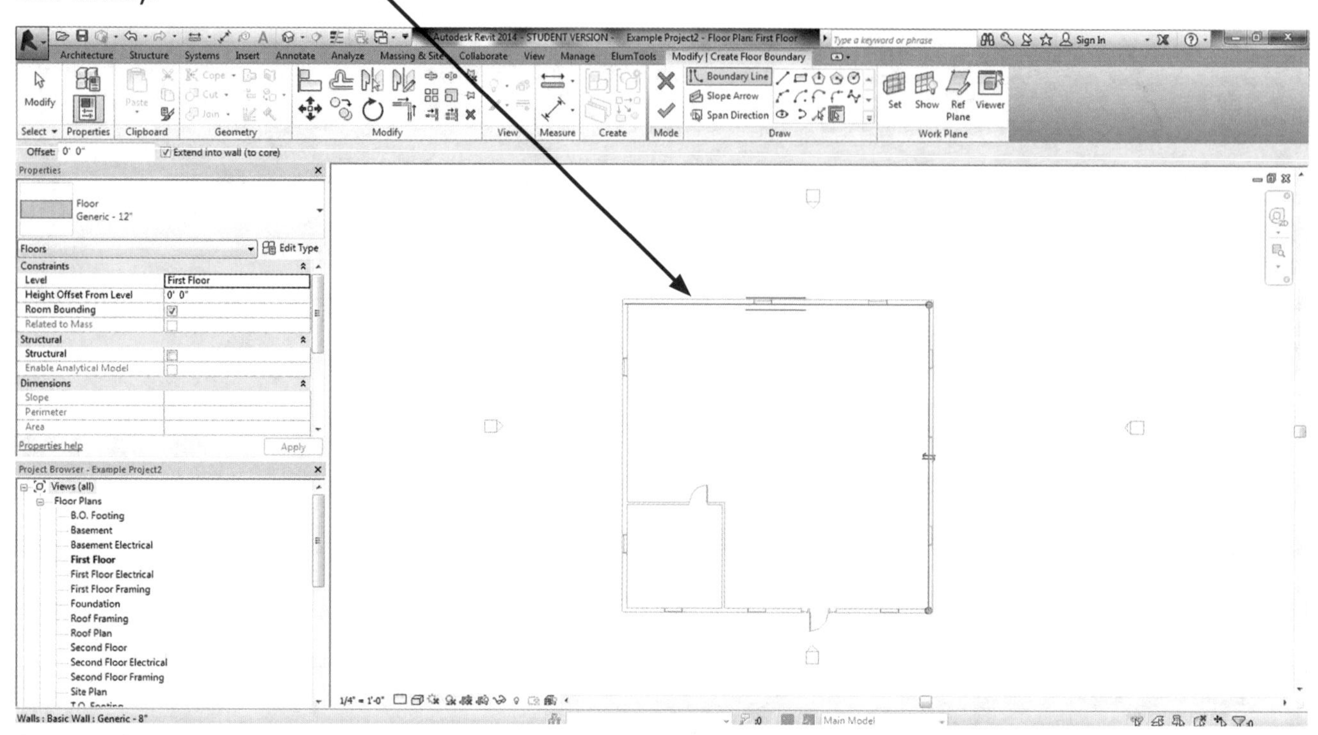

Once you have completed tracing your design, click on the green **CHECK MARK** to accept or complete the floor sketch.

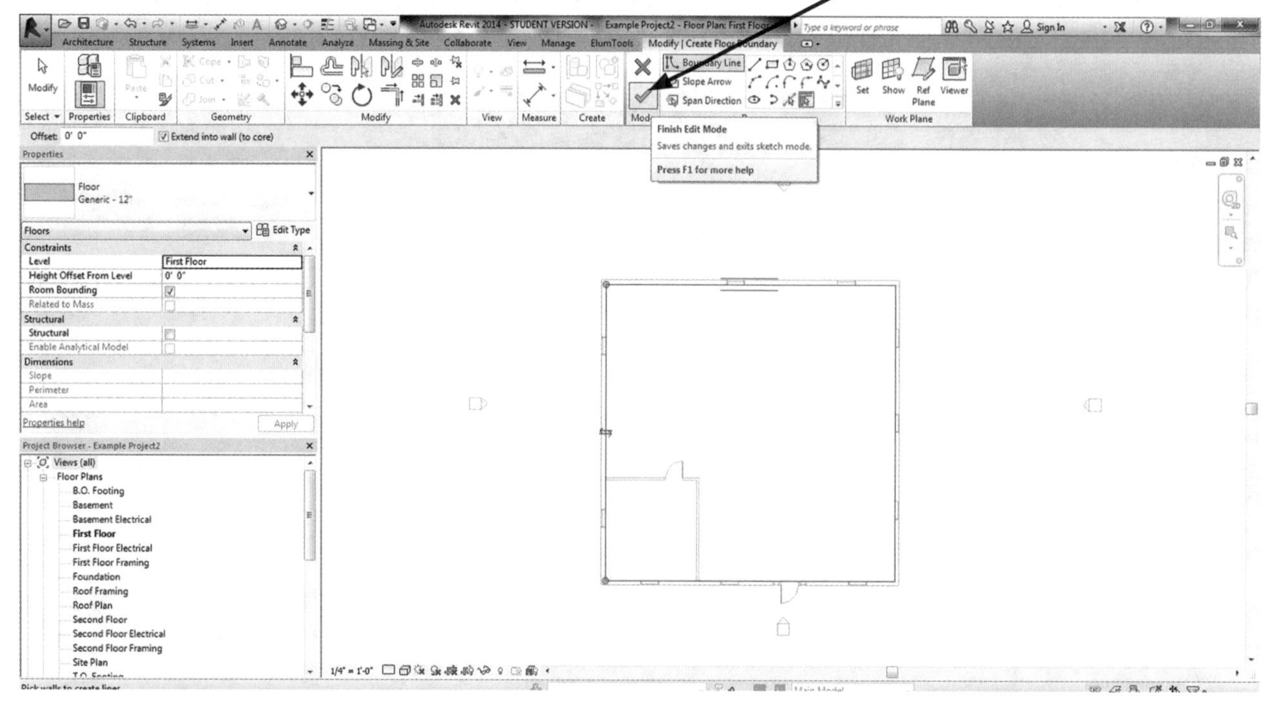

REMINDER

Don't forget to choose all four walls for your design.

If you want to place two different types of flooring in your designs, you would use the same steps using the **PICK WALLS** Button around your whole design.

EXAMPLE

You may want to use wood floors on most of the floor, and ceramic tiles for your bathroom space.

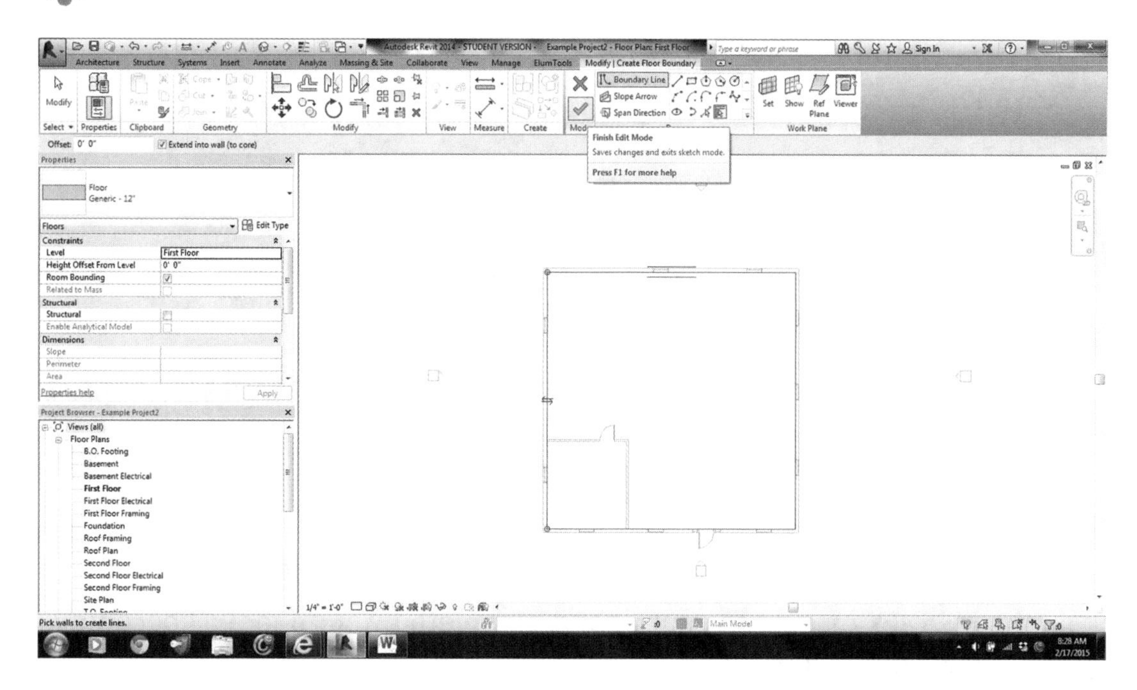

Then go to the draw box, select the **LINE TOOL** option, and sketch around the area you don't want to include in your design. We will trace a different floor in the room at the bottom left.

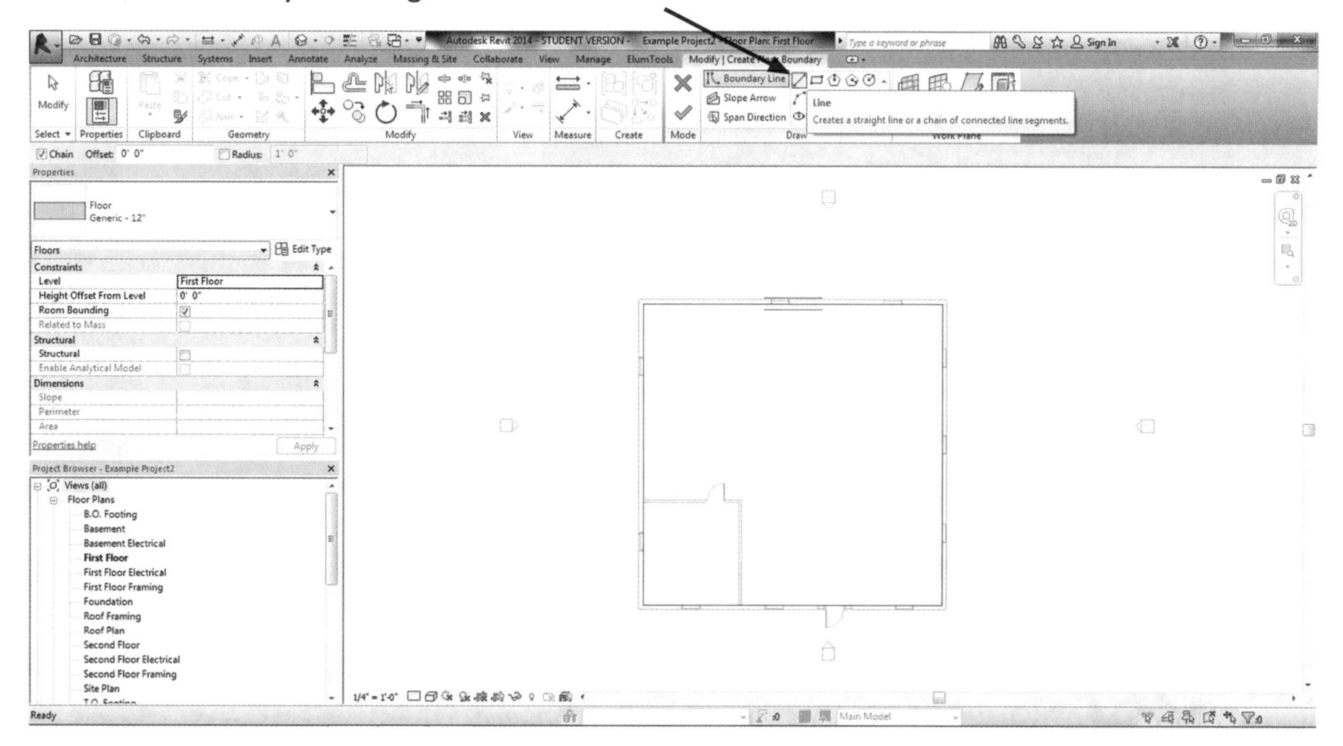

Once you trace the area you want to exclude from your design, click in the **MODIFY** Box and select the **TRIM** command. Click on the lines you want to keep.

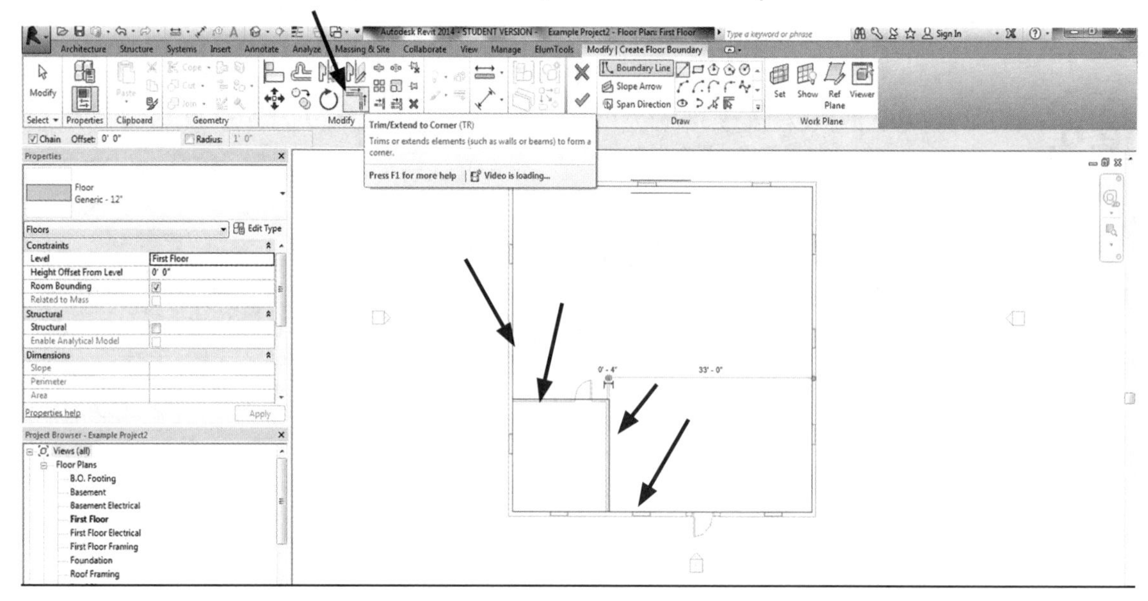

LOOK AHEAD

The Trim Command is explained more fully in Chapter 20, Revit Modify Commands.

Then click on the **MODIFY** Arrow and your floor is completed. Your floor will become highlighted to let you know what the floor boundary looks like.

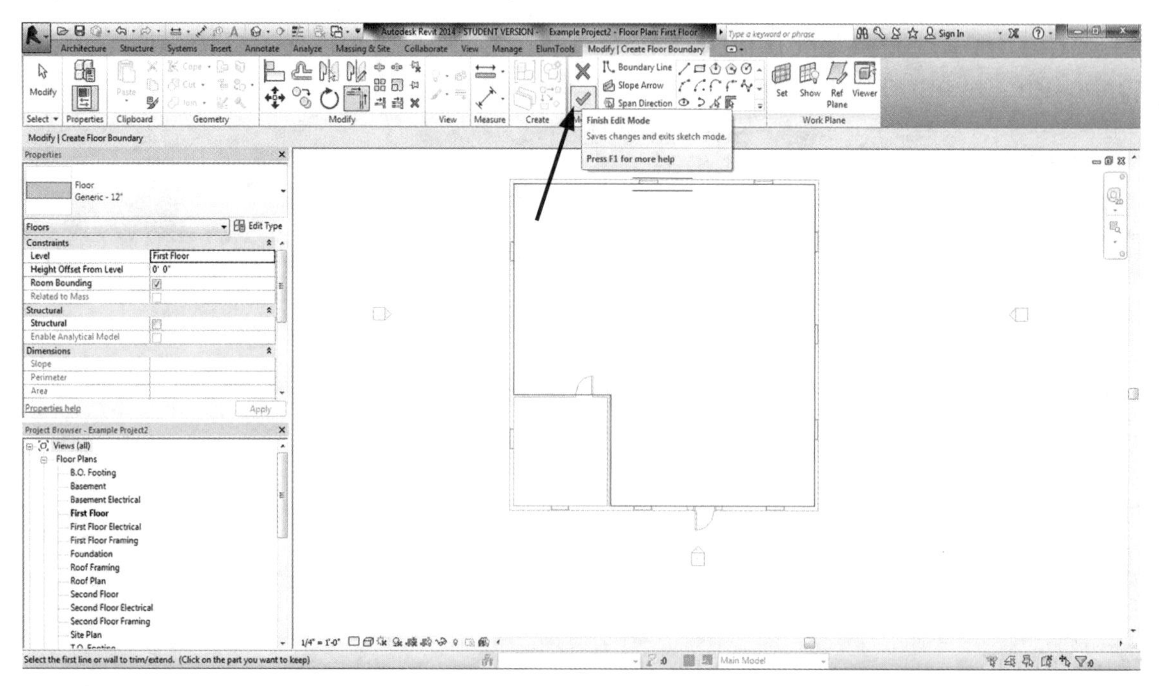

WARNING

Do not click on the lines you want to exclude for that particular design. Click on the green CHECK MARK ***to finish the command. Then click on the*** MODIFY ***Arrow, and your floor is completed.***

To place a different type of flooring in the vacant area, follow the same steps and change the floor type in the Properties Box.

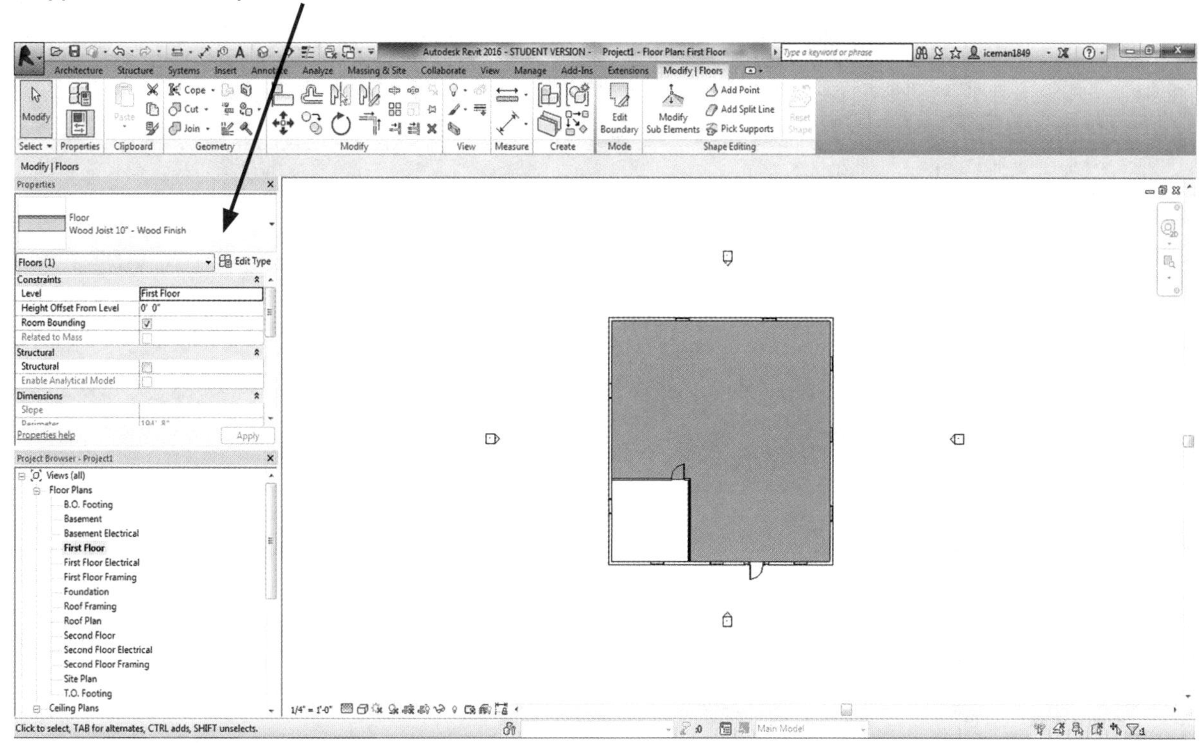

Here you can change the floor type to either carpet or ceramic tile.

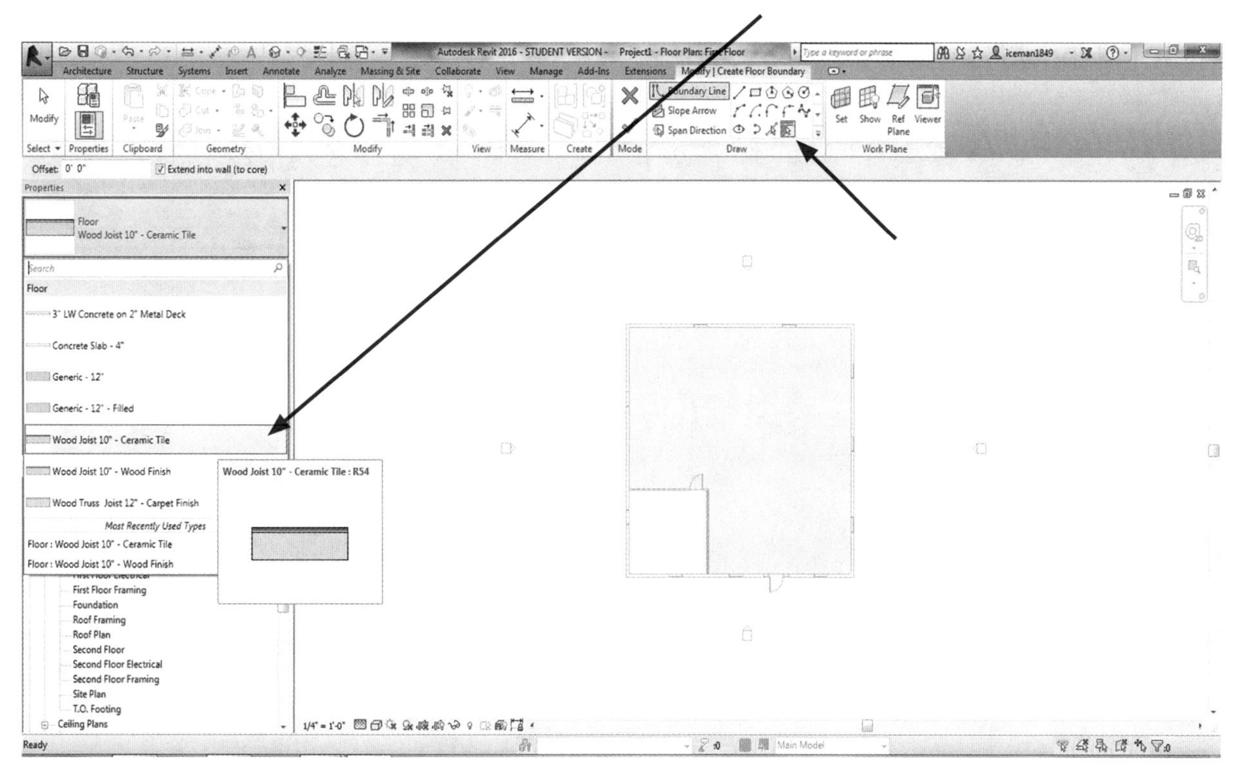

Then use the **PICK WALLS** Command to trace out the vacant space. The end result is that you'll have two different floor types.

After you have placed the new ceramic tile floor, click the green **CHECK MARK** to finish the command. You will have placed the ceramic tile in another room of the house or building.

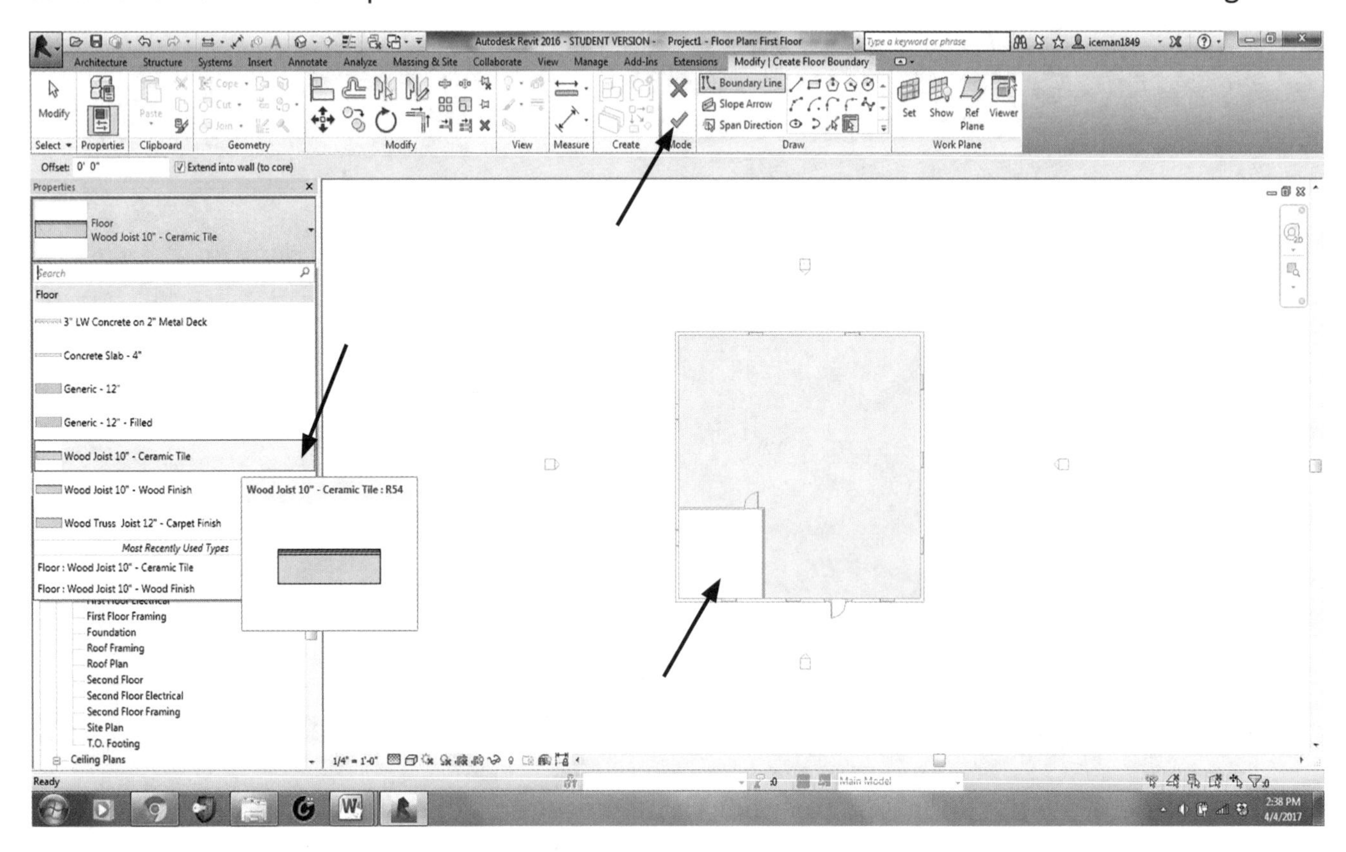

Your drawing should look like the following screen:

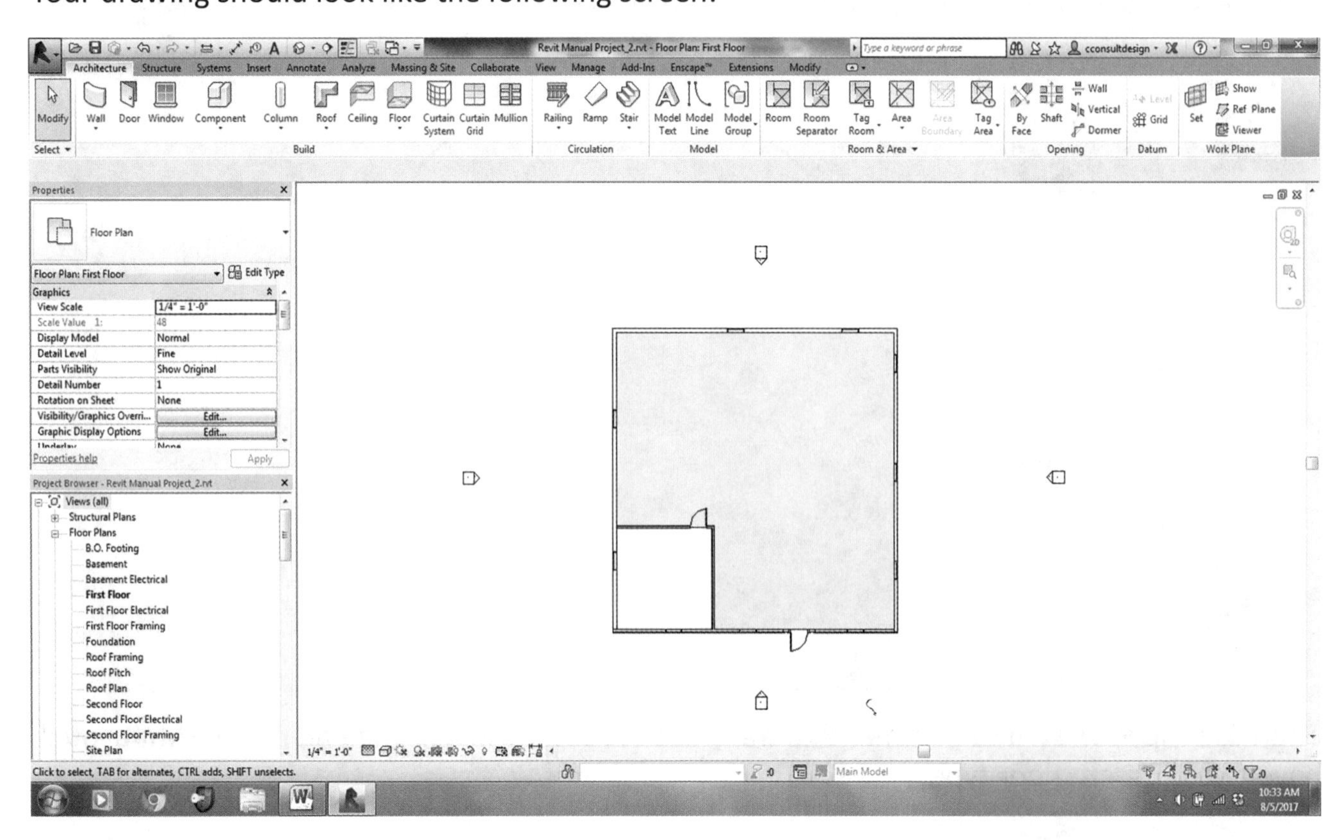

NOTES *If your tile pattern doesn't show up in the view, you can make the pattern visible. Take your mouse pointer and hover over the edge of the inside of the ceramic floor near the wall, and click the* TAB *Button on the keyboard until you see the floor highlight with a border.*

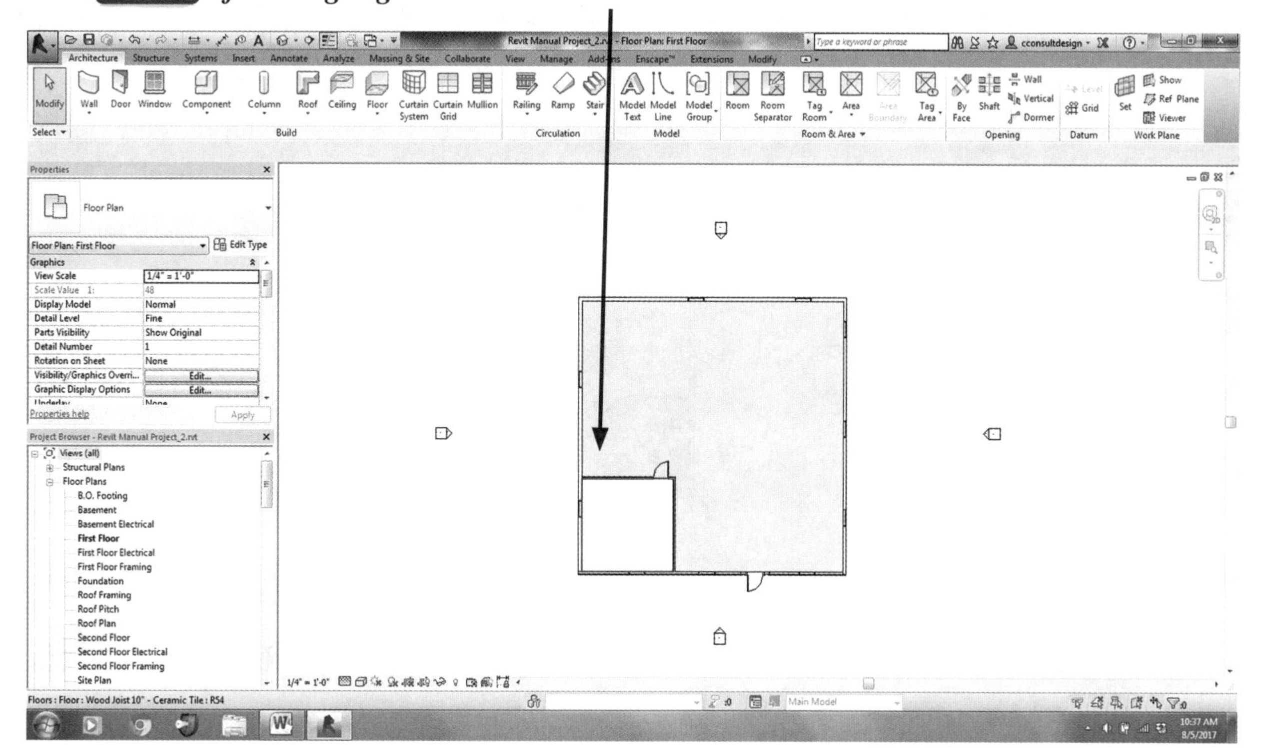

Then click inside that floor boundary, and you will see the floor highlight, meaning you selected the floor for editing. This has changed the screen to the **MODIFY FLOORS** command.

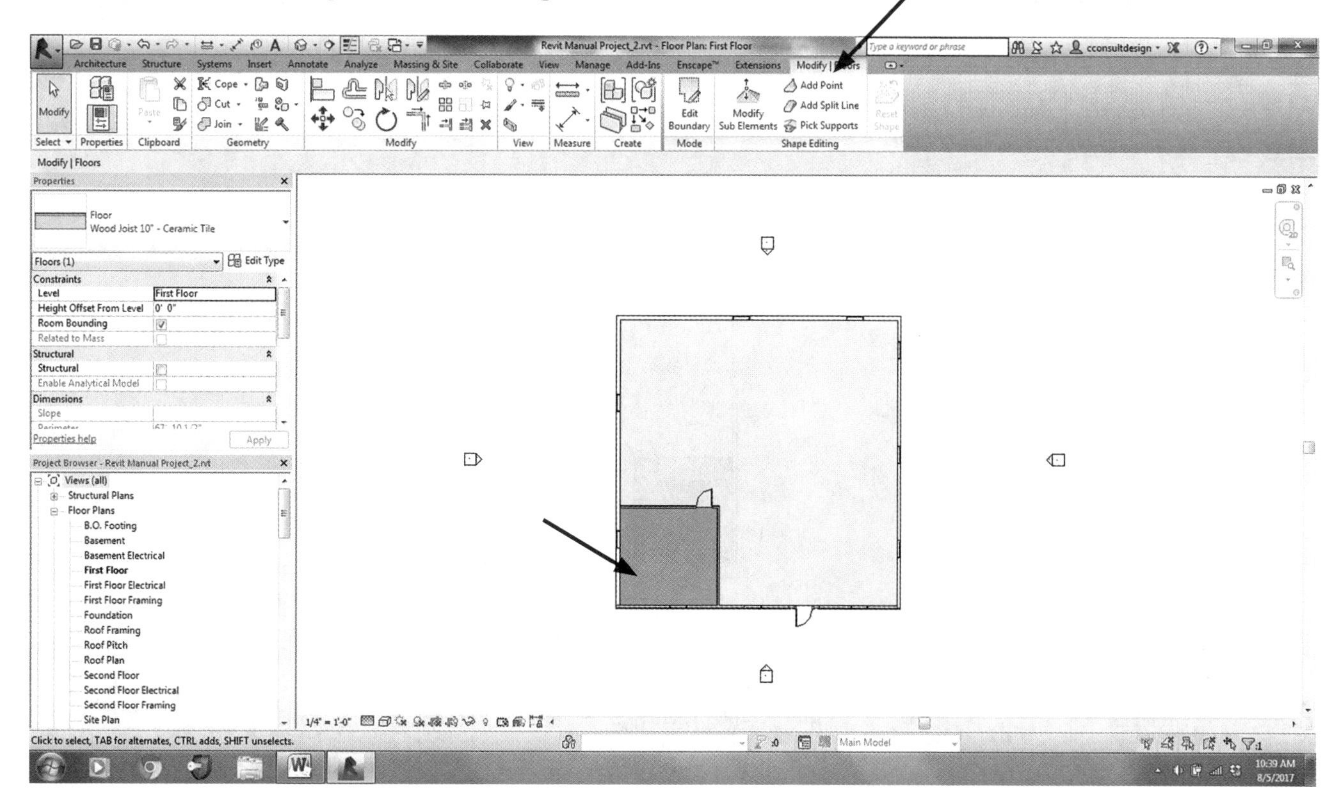

Notice the Properties Box has changed to edit the floor properties.

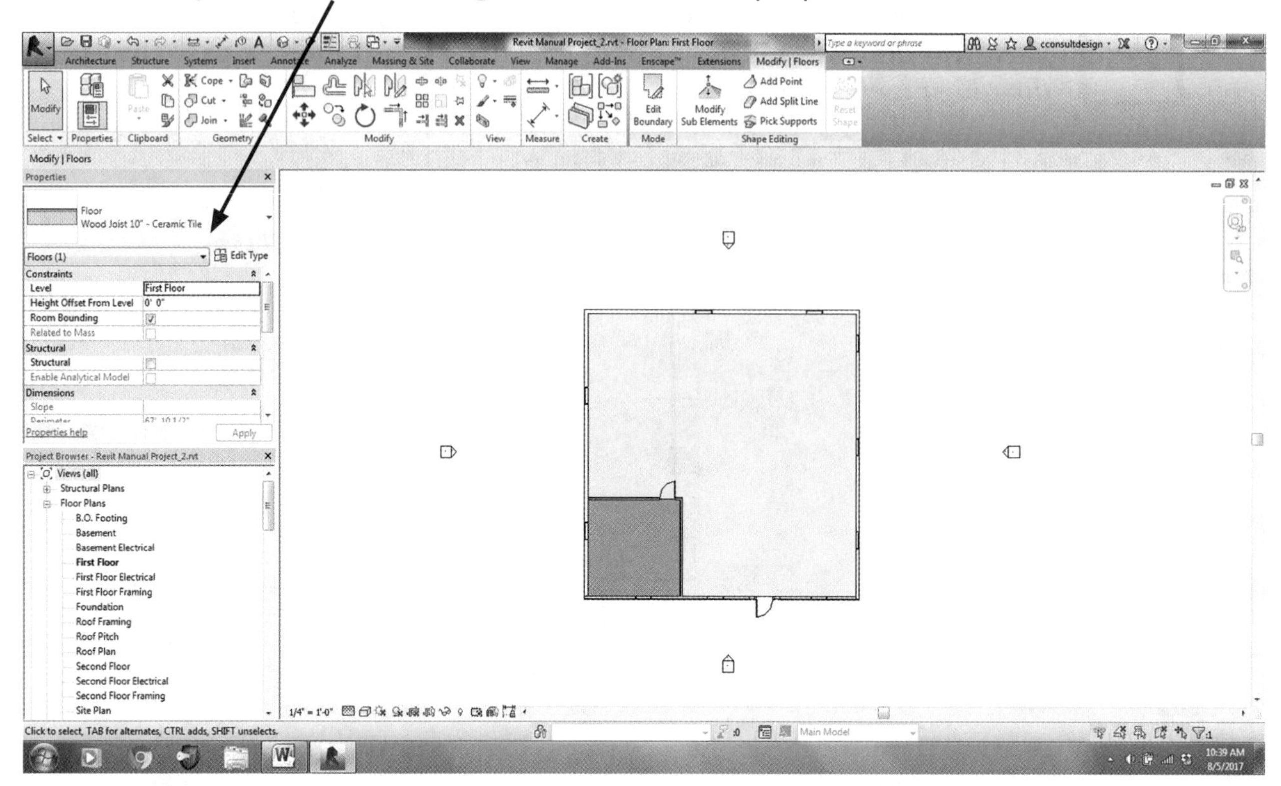

In the Properties Box, click on the **EDIT TYPE** Button and the Type Properties Dialog Box appears.

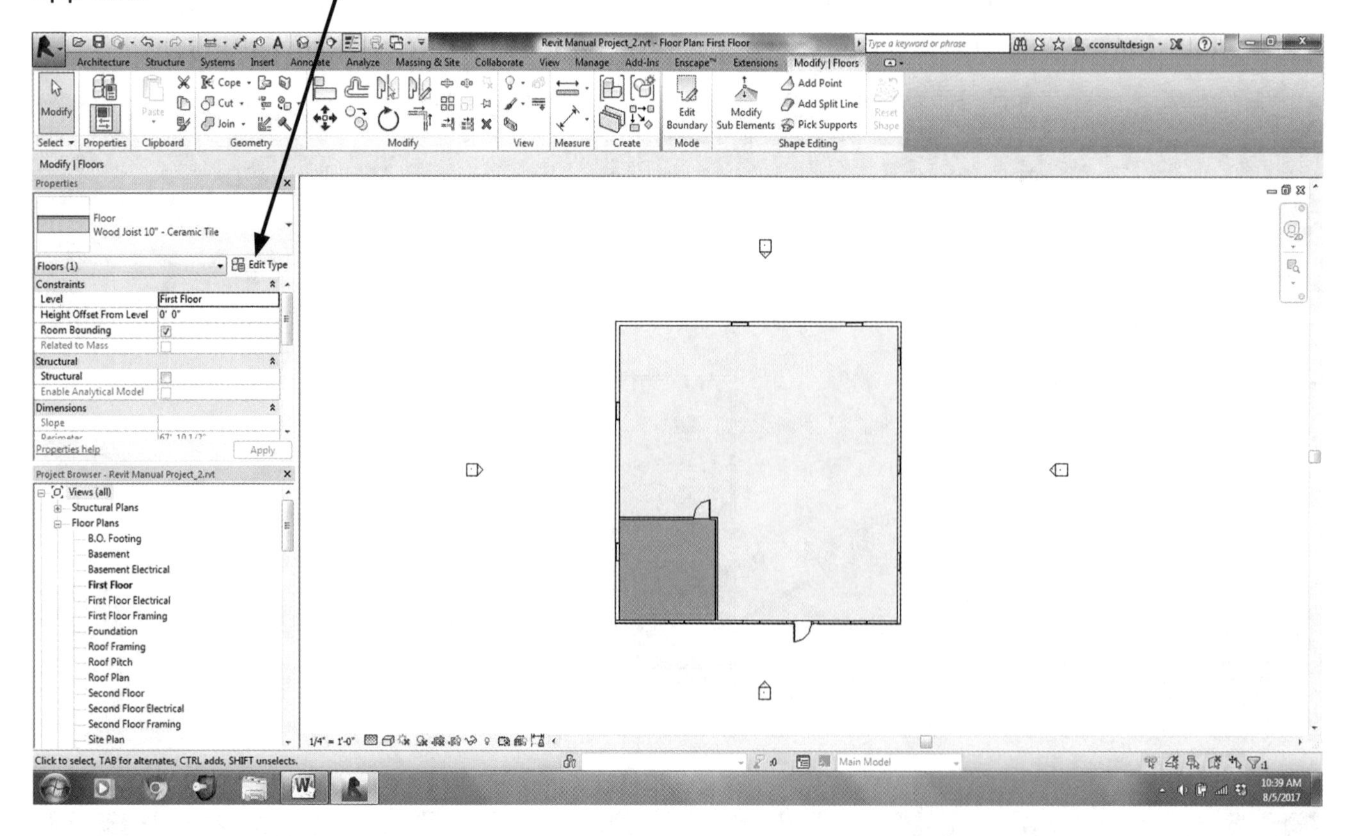

Once the Type Properties Box appears, you can modify the visibility of the floor pattern.

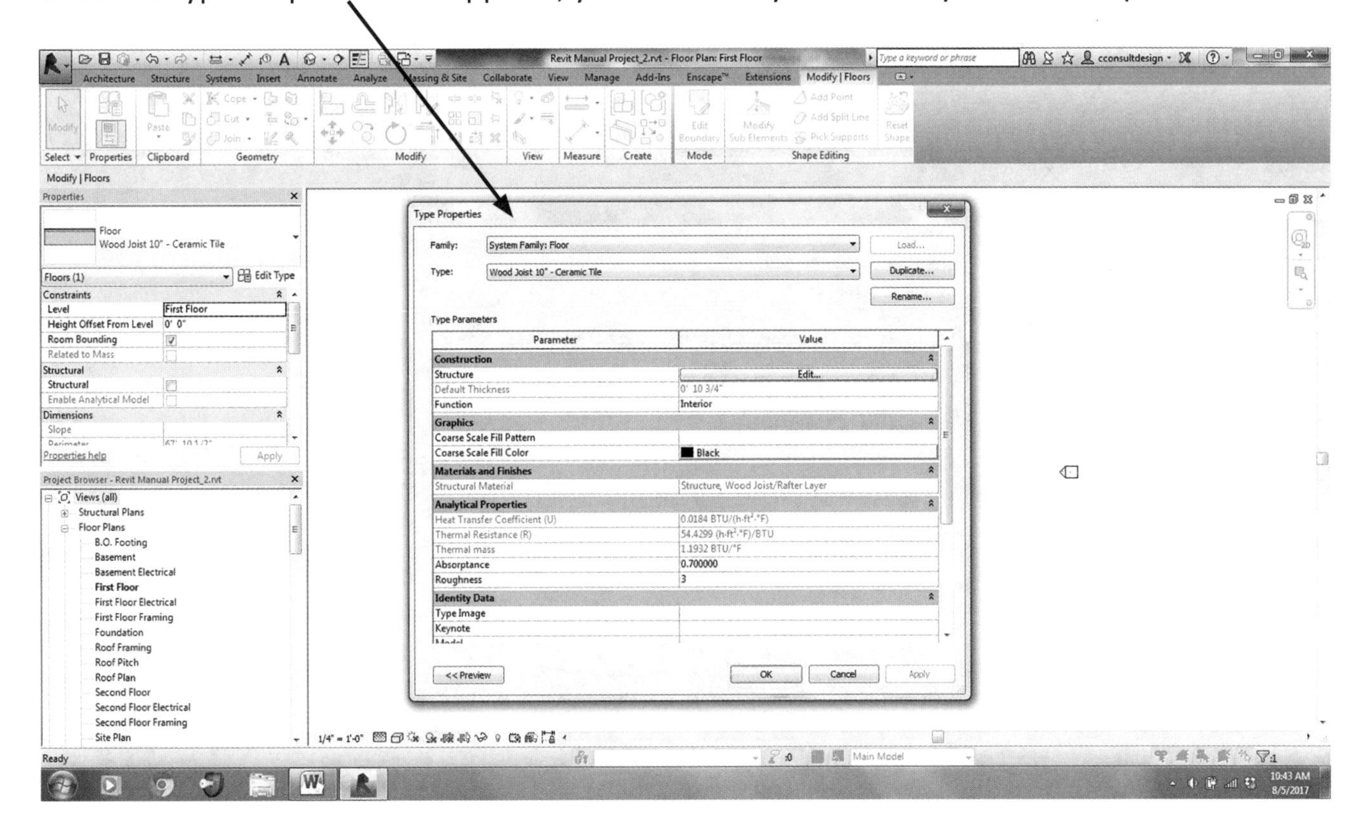

Click on the **EDIT** Tab in the dialog box.

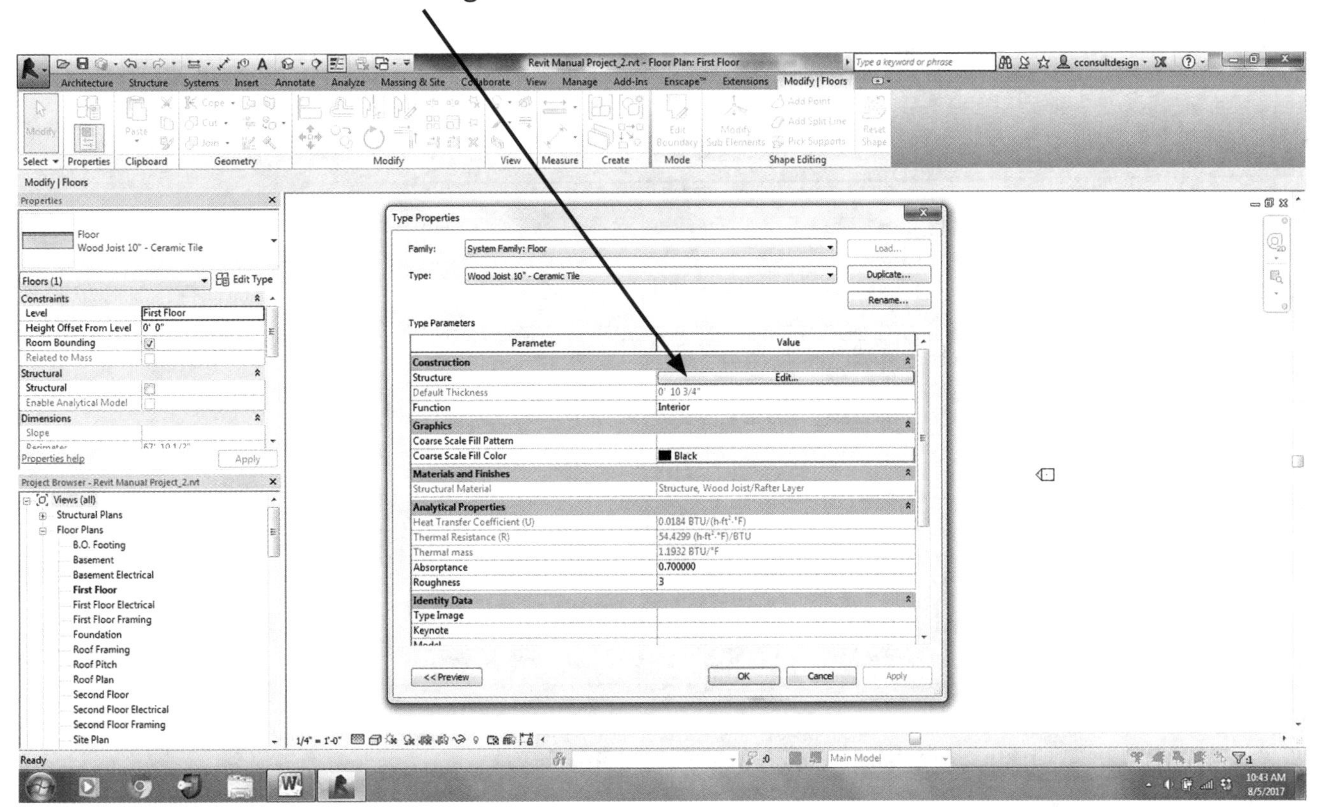

The Edit Assembly Dialog Box appears for the visibility pattern.

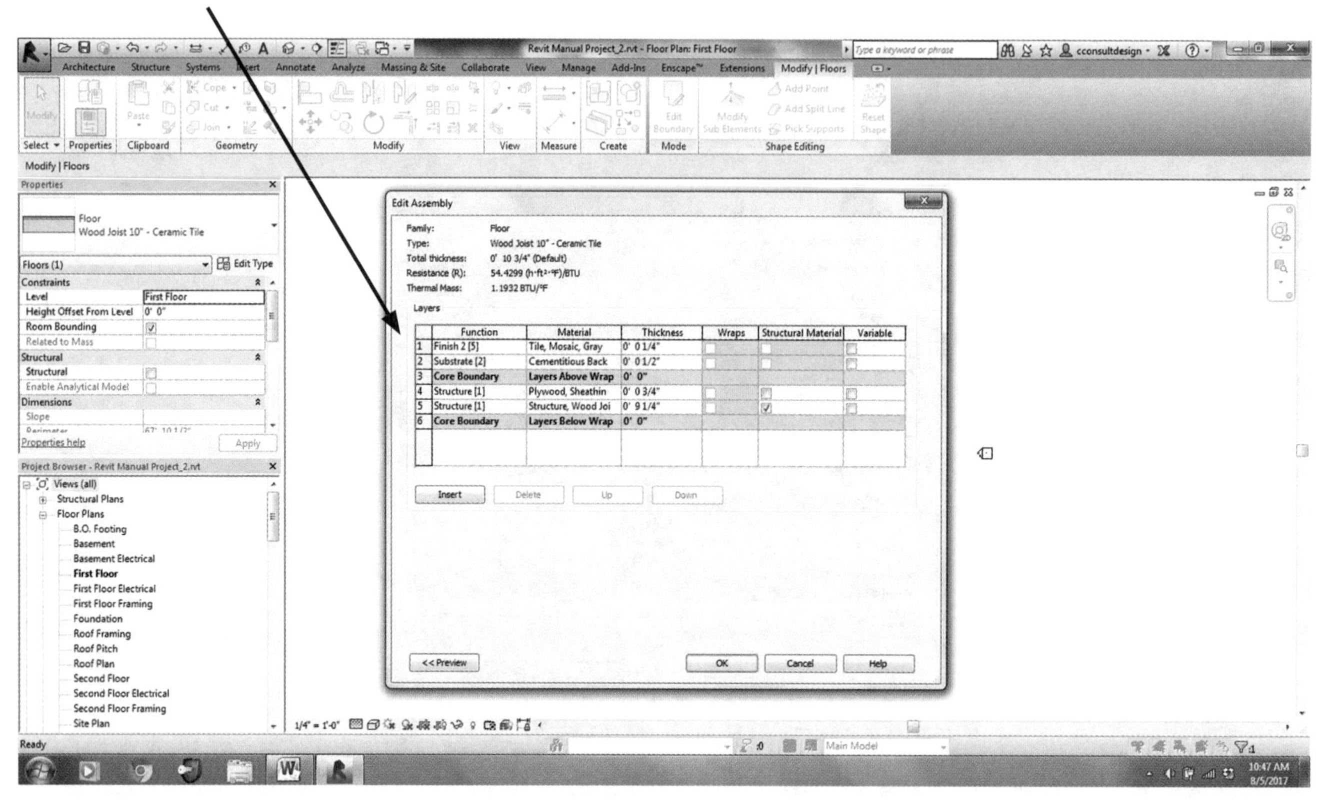

In row 1, you'll see "Tile, Mosaic, Gray," in the Material Box. Click on the material, and a small dialog button appears on the right. Click on that small button.

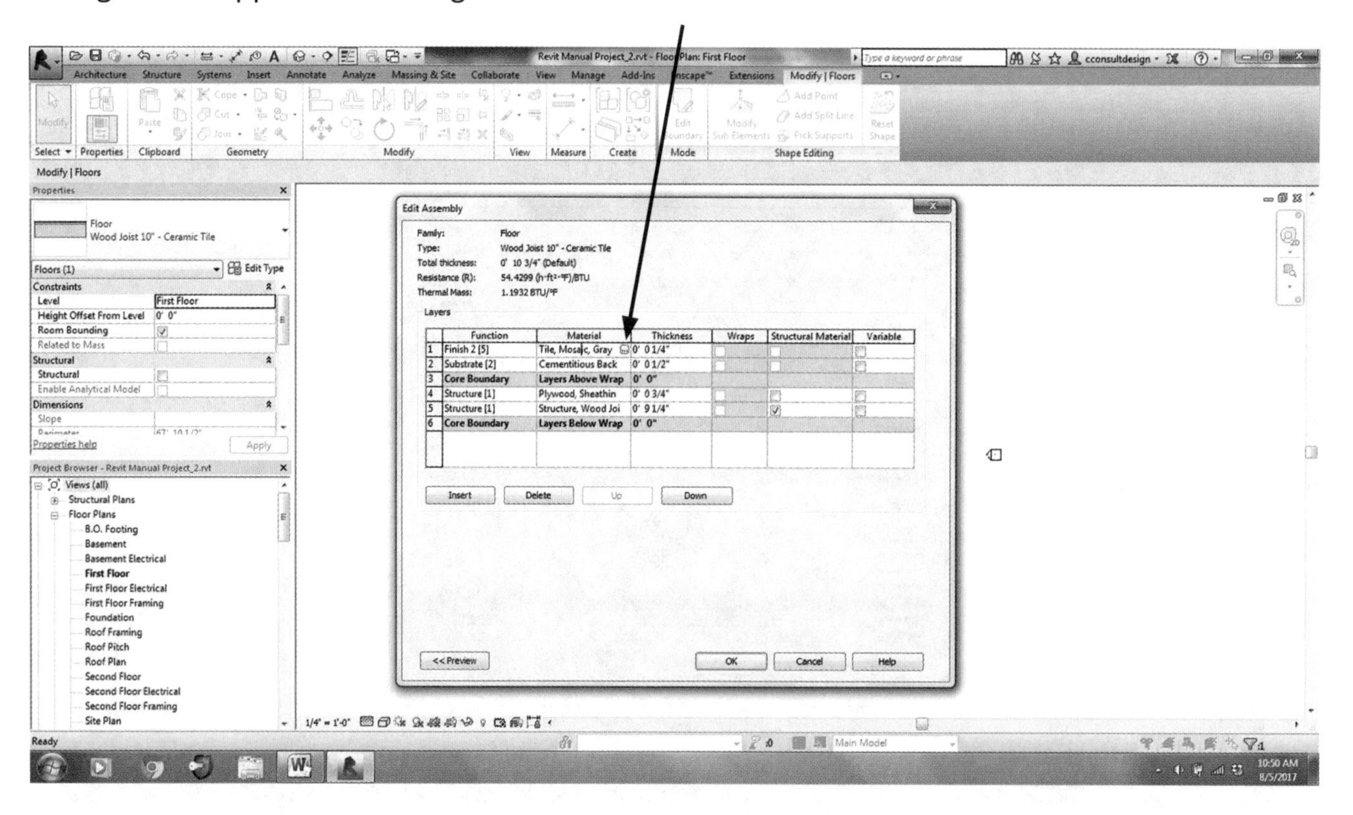

After clicking on the small information button, the Material Browser Dialog Box appears.

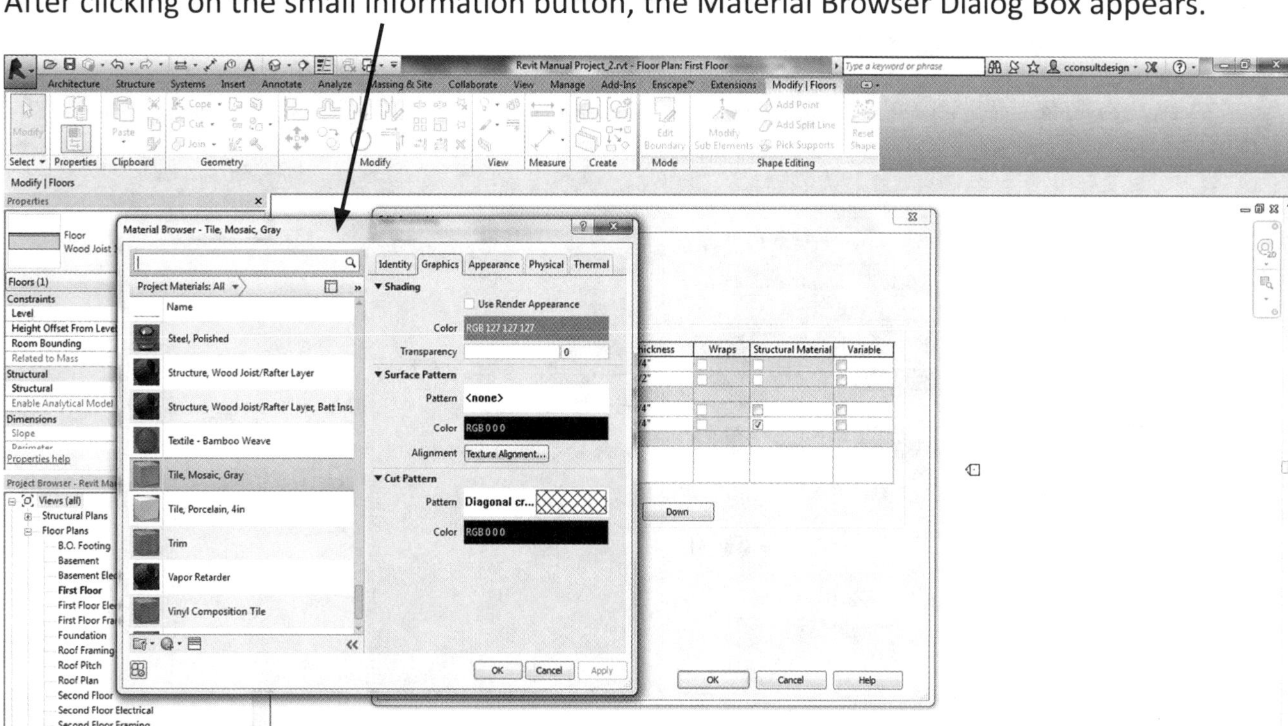

Notice that the Tile, Mosaic, Gray material is automatically highlighted. This indicates that this material is ready for editing.

On the right of the Materials Browser, you will see a section named Surface Pattern.

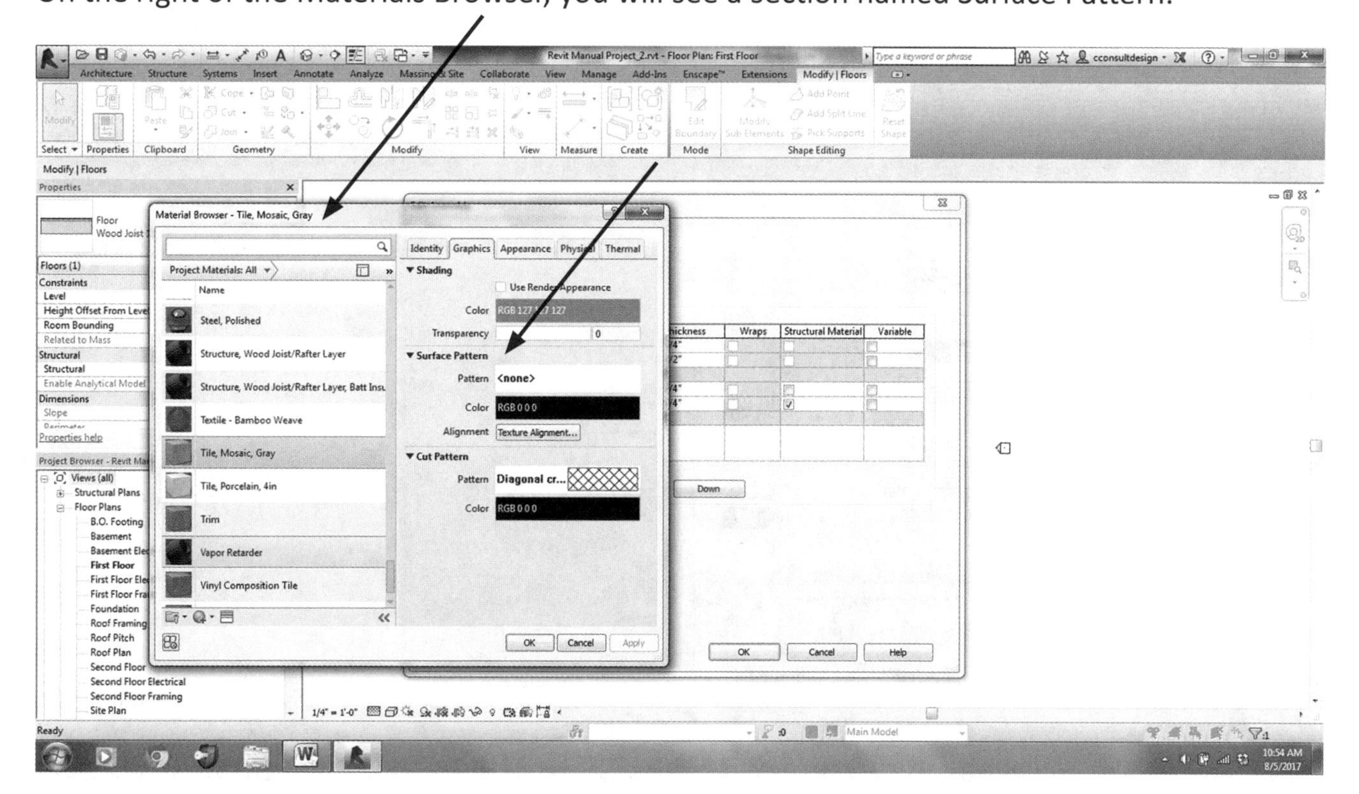

Click on the surface pattern that reads **NONE** and a Fill Patterns Dialog will appear with a list of line patterns from which you can select.

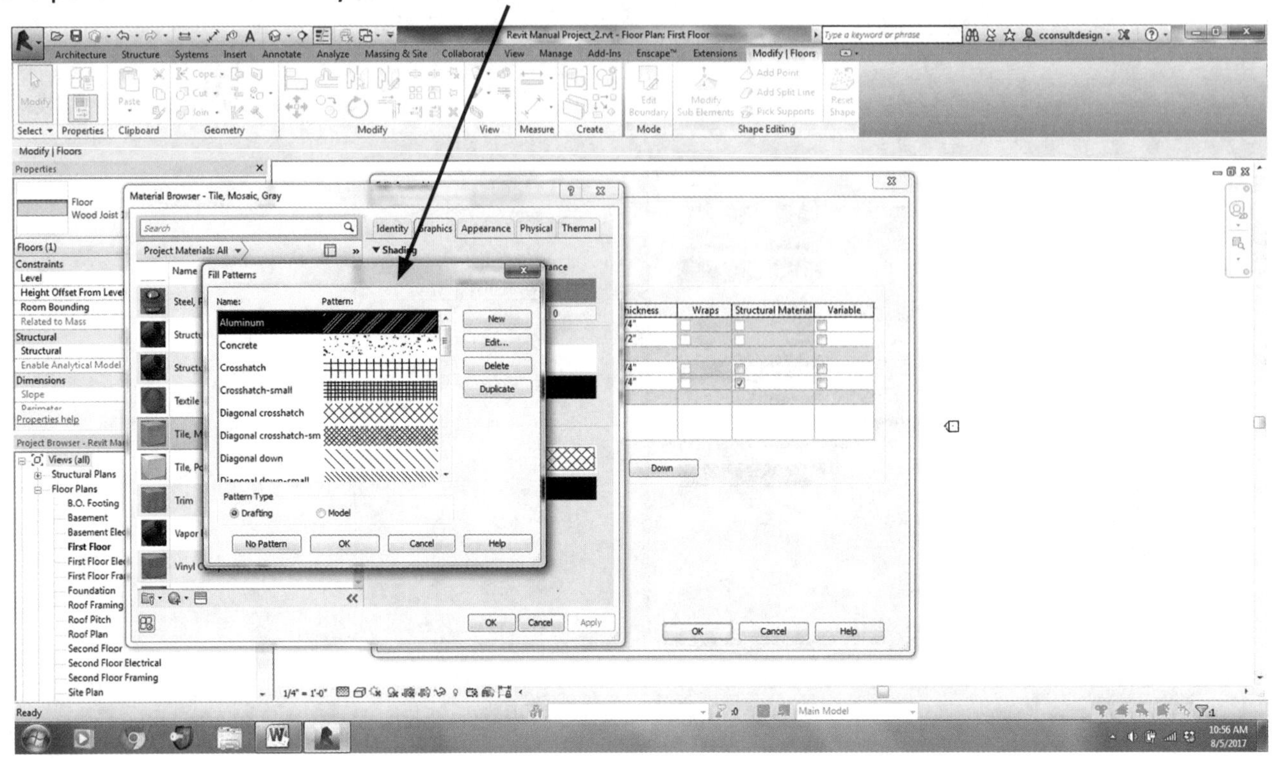

At the bottom, you will see Pattern Type. One is set for Drafting and one for Model. Since we are modeling a house, we need to switch the Pattern Type to **MODEL** so we can see a tile pattern.

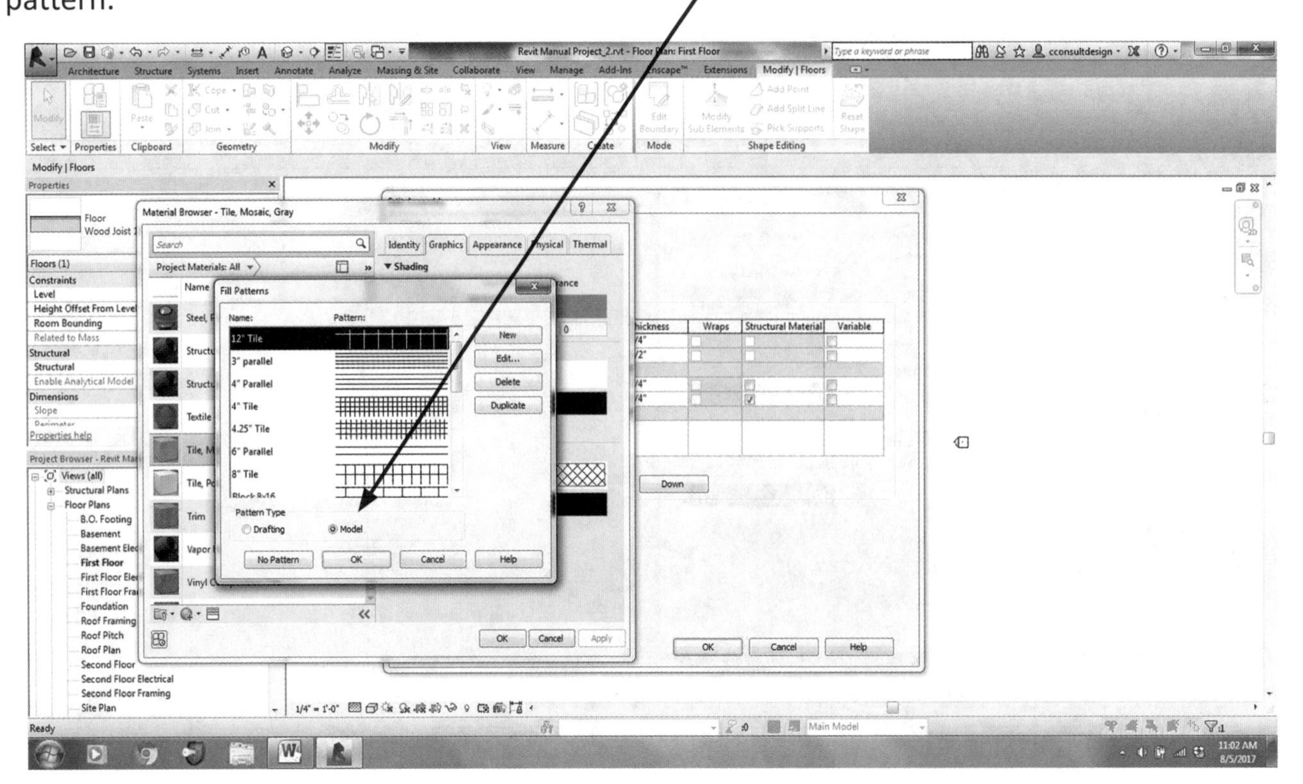

After selecting the **MODEL** type in the Pattern, scroll down until you locate 12" tile.

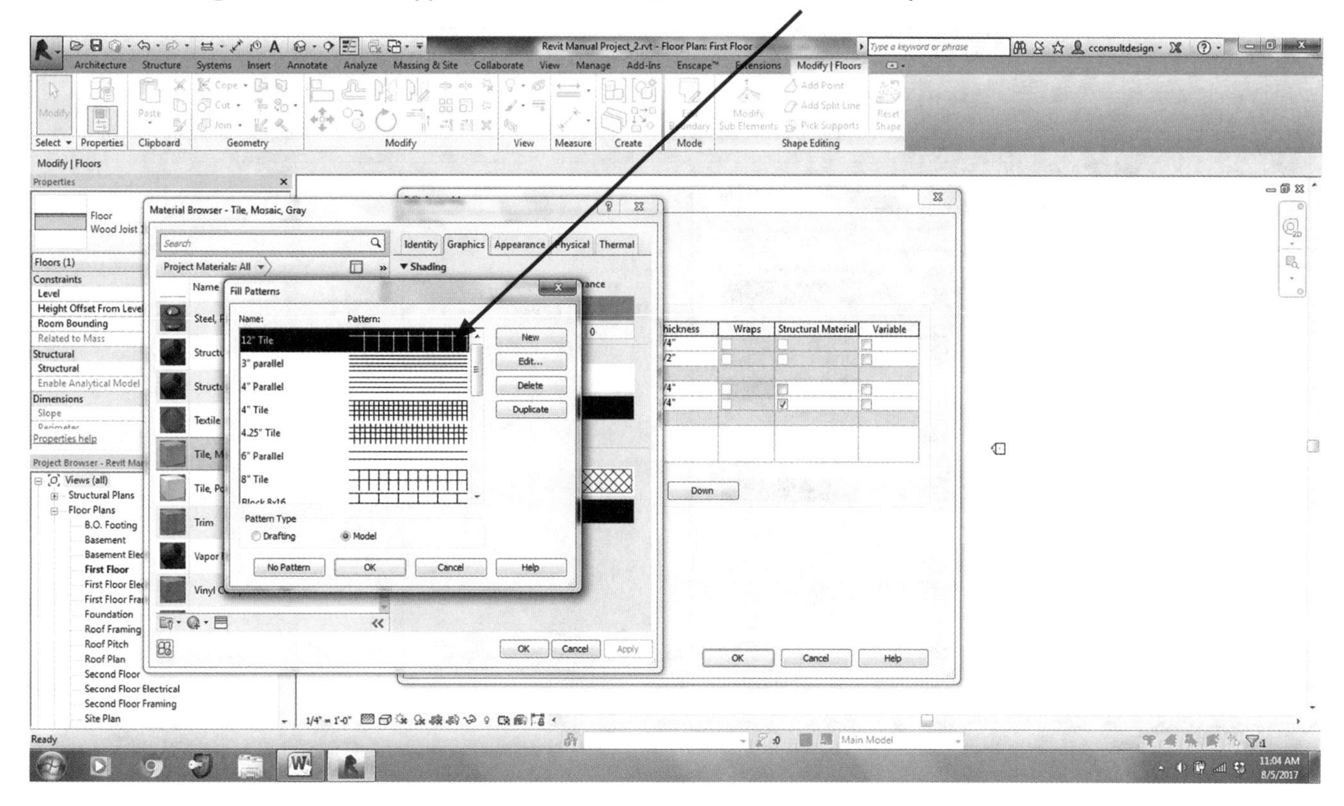

If you decide you want other sizes, you can select from 4", 6", and so on. Once you have selected the tile pattern you want, click the **OK** button.

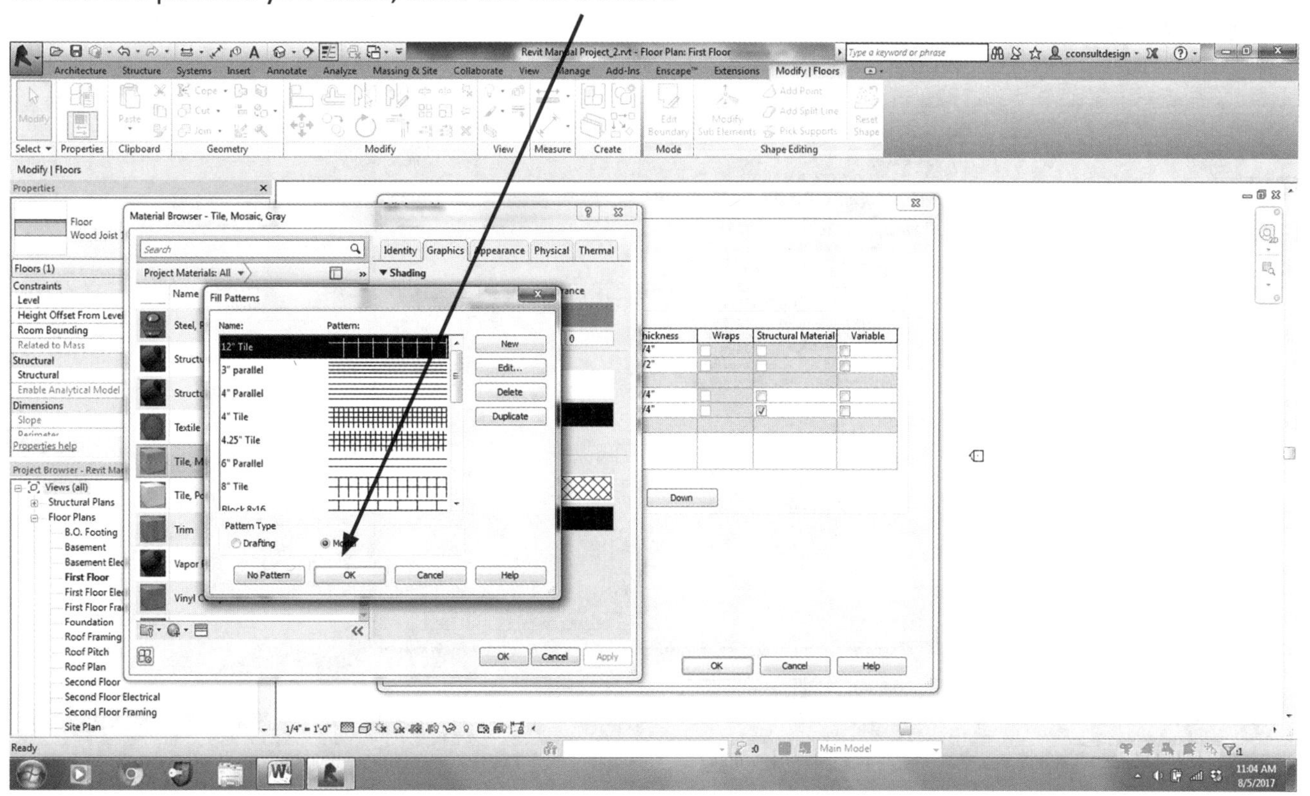

Then click the **OK** Button on the Material Browser Dialog Box.

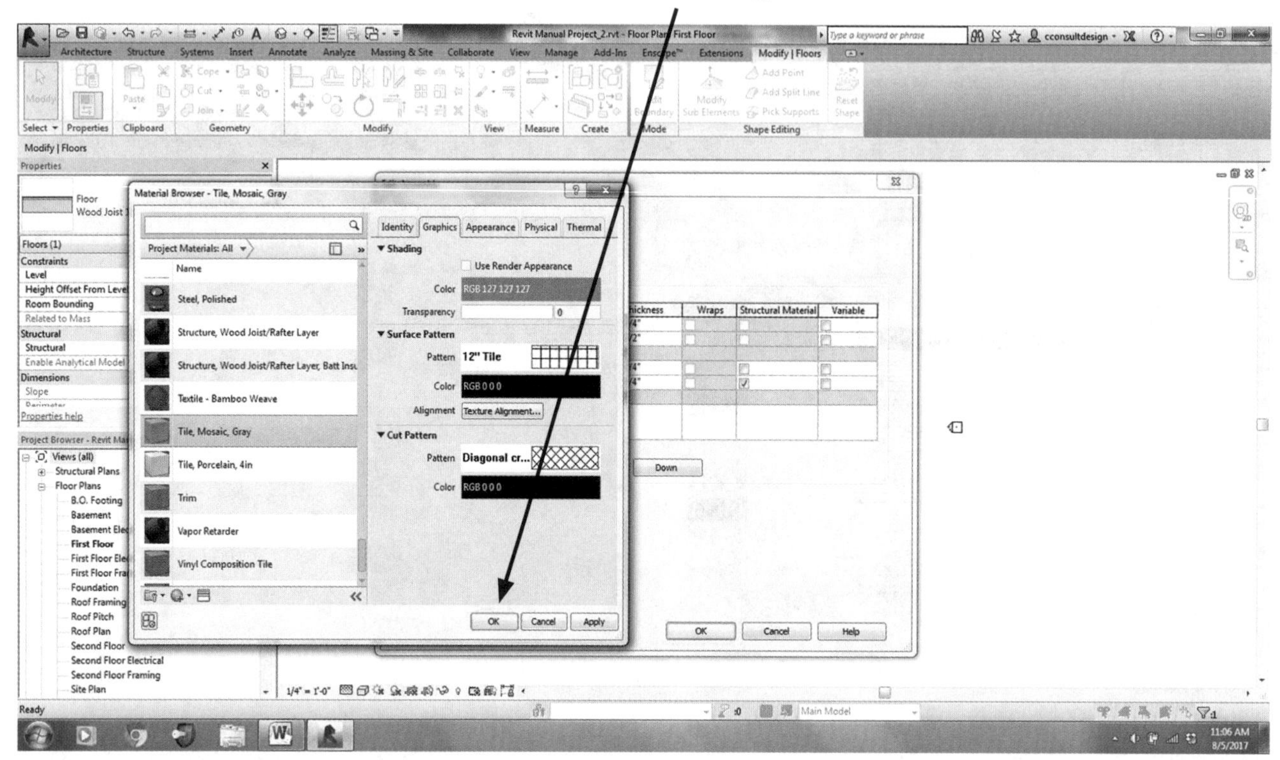

Next, click **OK** on the Edit Assembly Button.

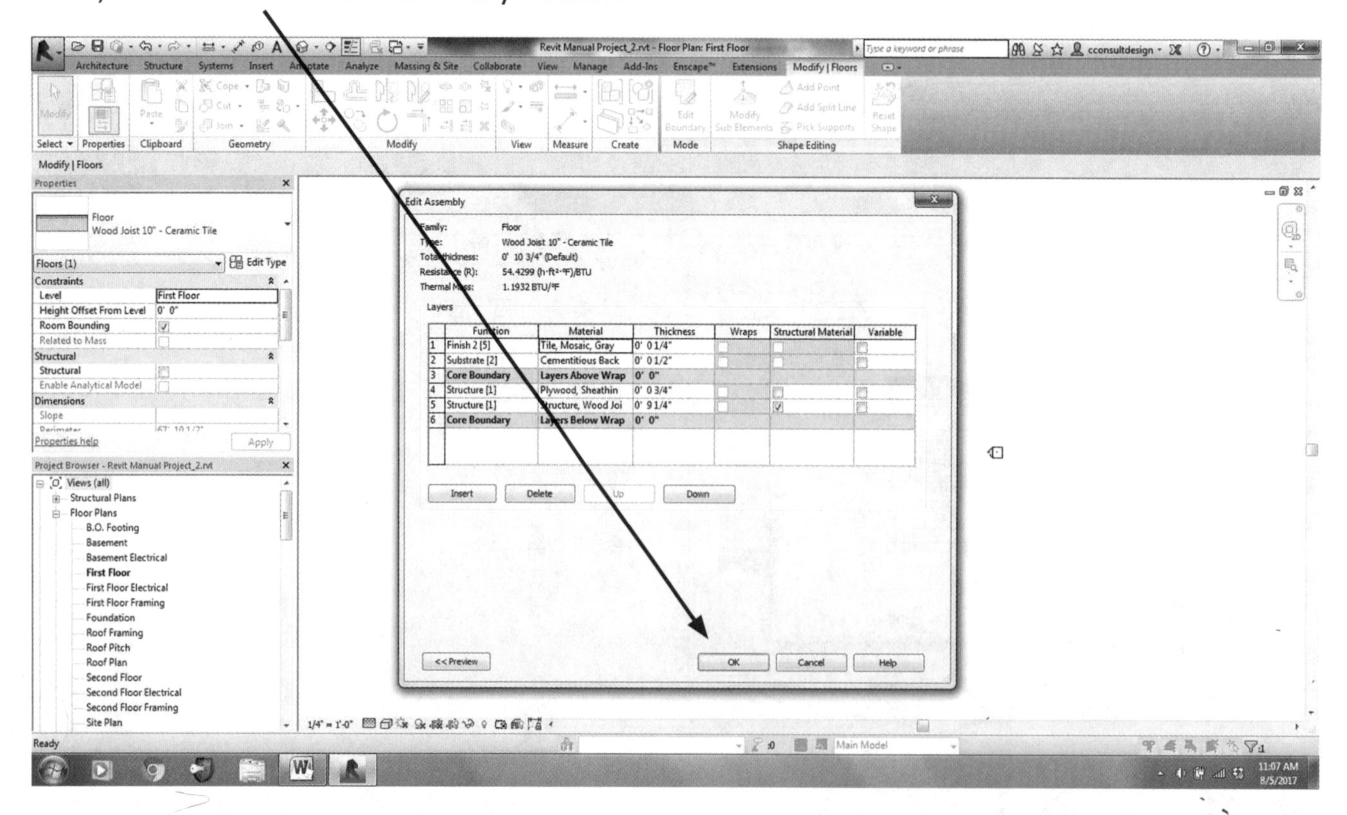

Then click **OK** on the Type Properties Dialog Box.

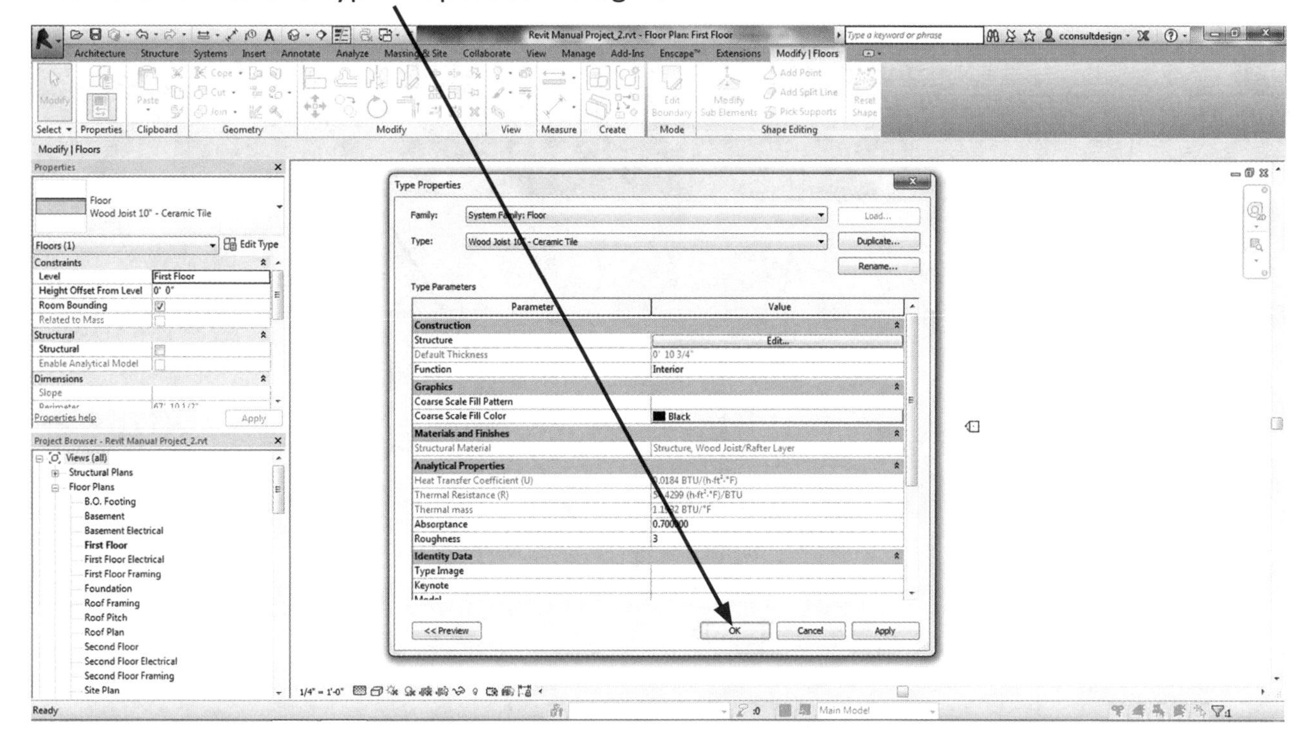

Notice that once you have selected the pattern you want for the floor, the tile pattern automatically appears in the floor area.

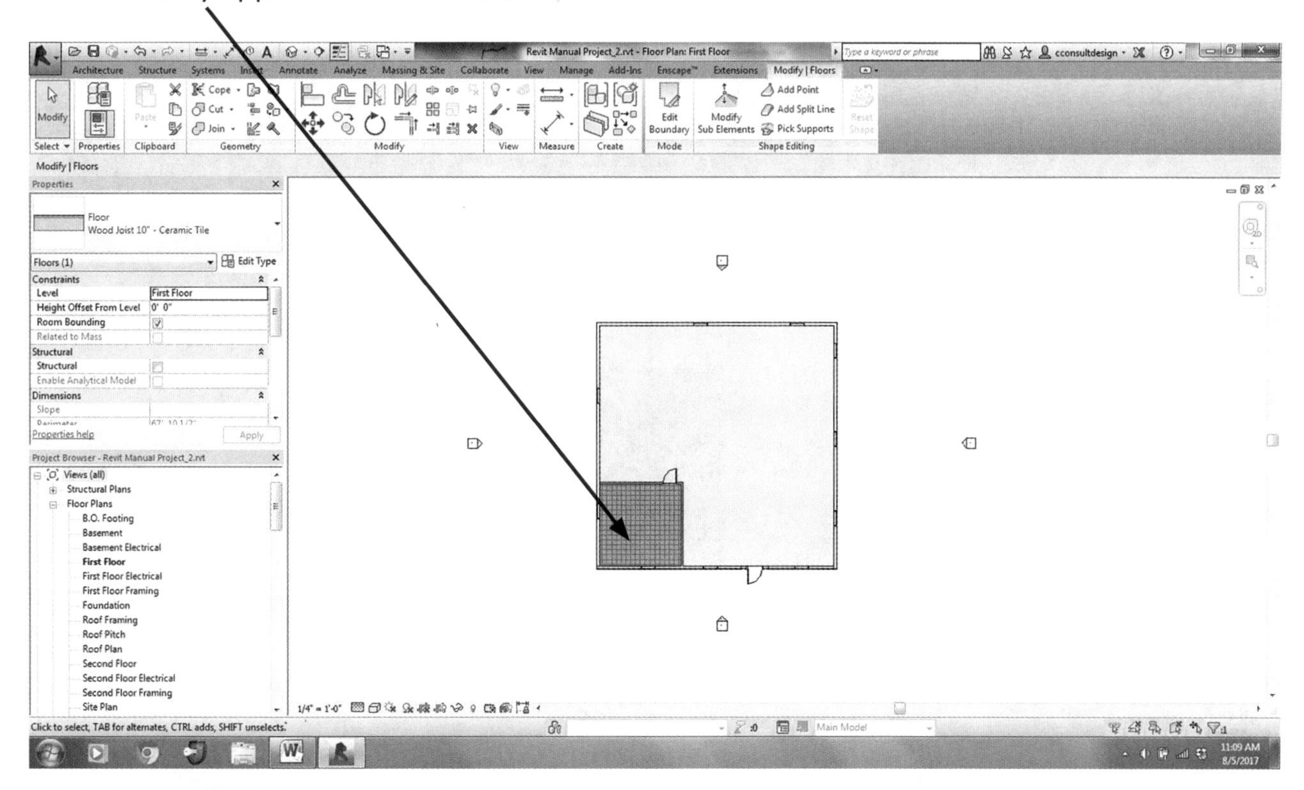

Click the **MODIFY** Arrow and the blue floor will return to normal. Now let's assume you want to add carpet to the main floor in your house or building. You can select the main floor and edit that boundary.

Take your cursor and hover over the edge of the main flooring area until the floor boundary highlights.

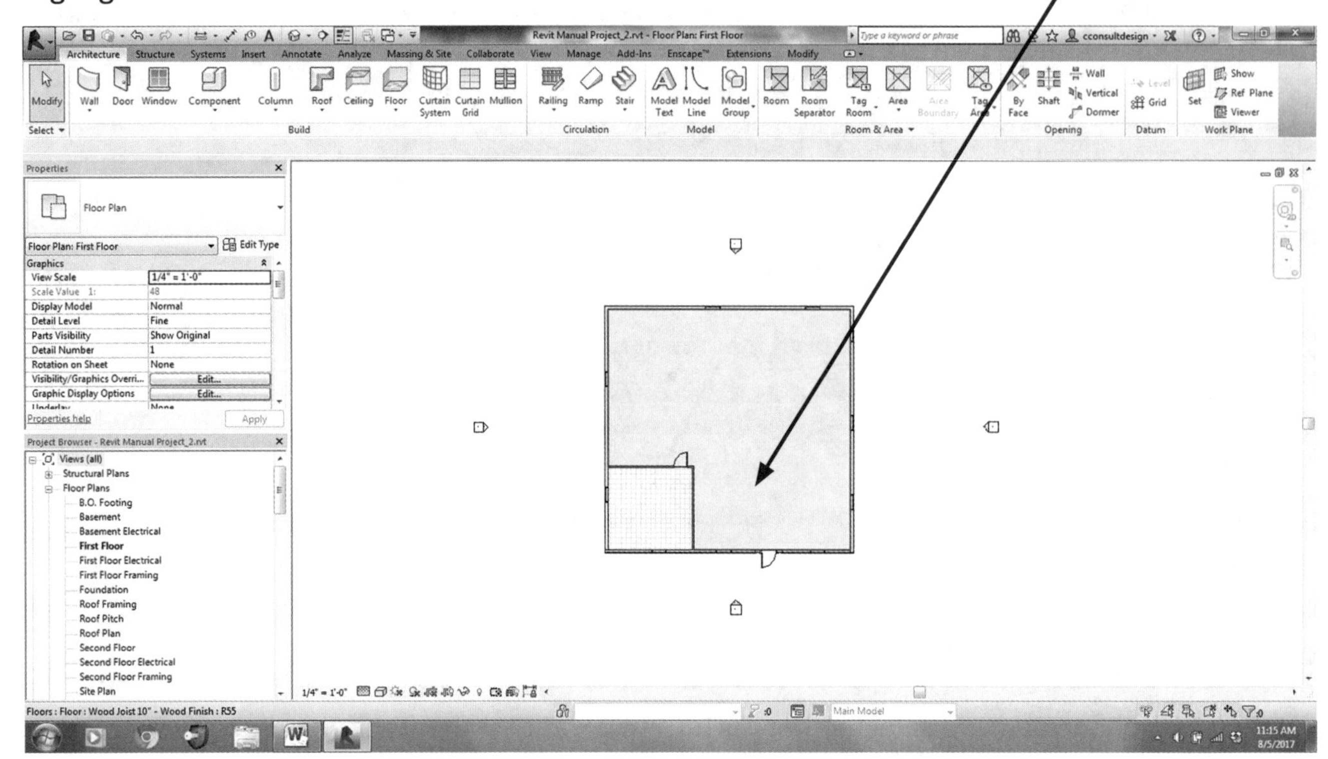

Then click on the floor boundary and the floor will turn solid.

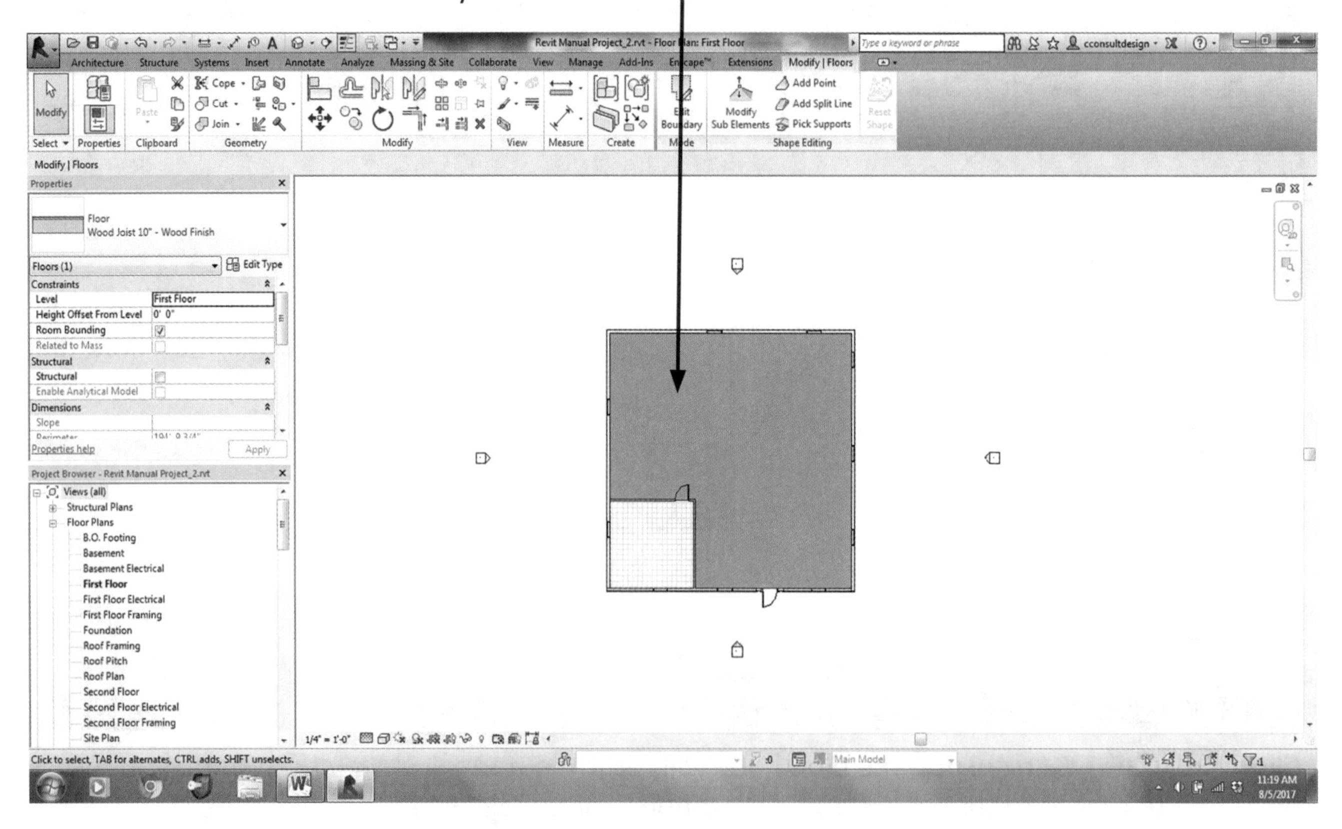

You are now in the Modify Floor command, and can begin editing the floor. Click on the **EDIT BOUNDARY** tile in the Ribbon.

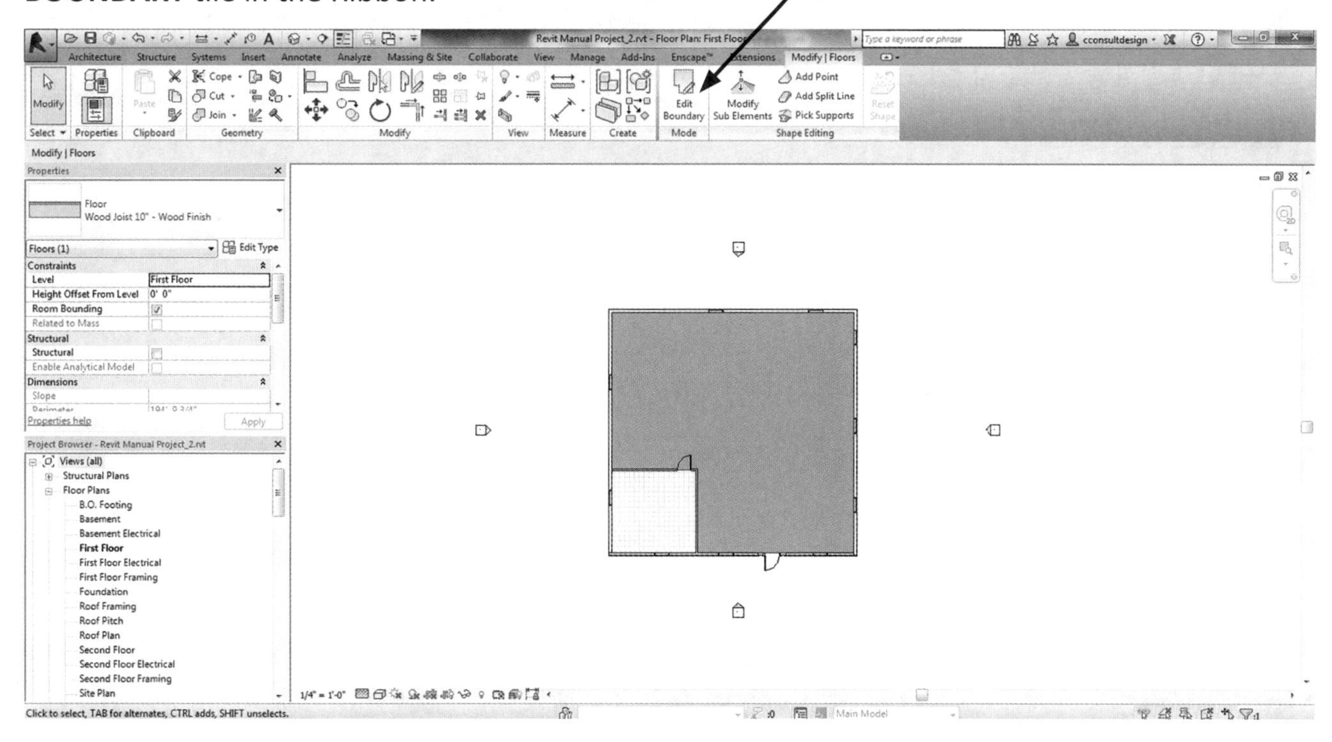

The Edit Boundary lines will activate, a magenta colored boundary will appear, and you can begin to edit the floor layout.

Click on the **LINE** command and begin to draw your new layout to separate the existing hardwood floor to where you want to place carpet or other material.

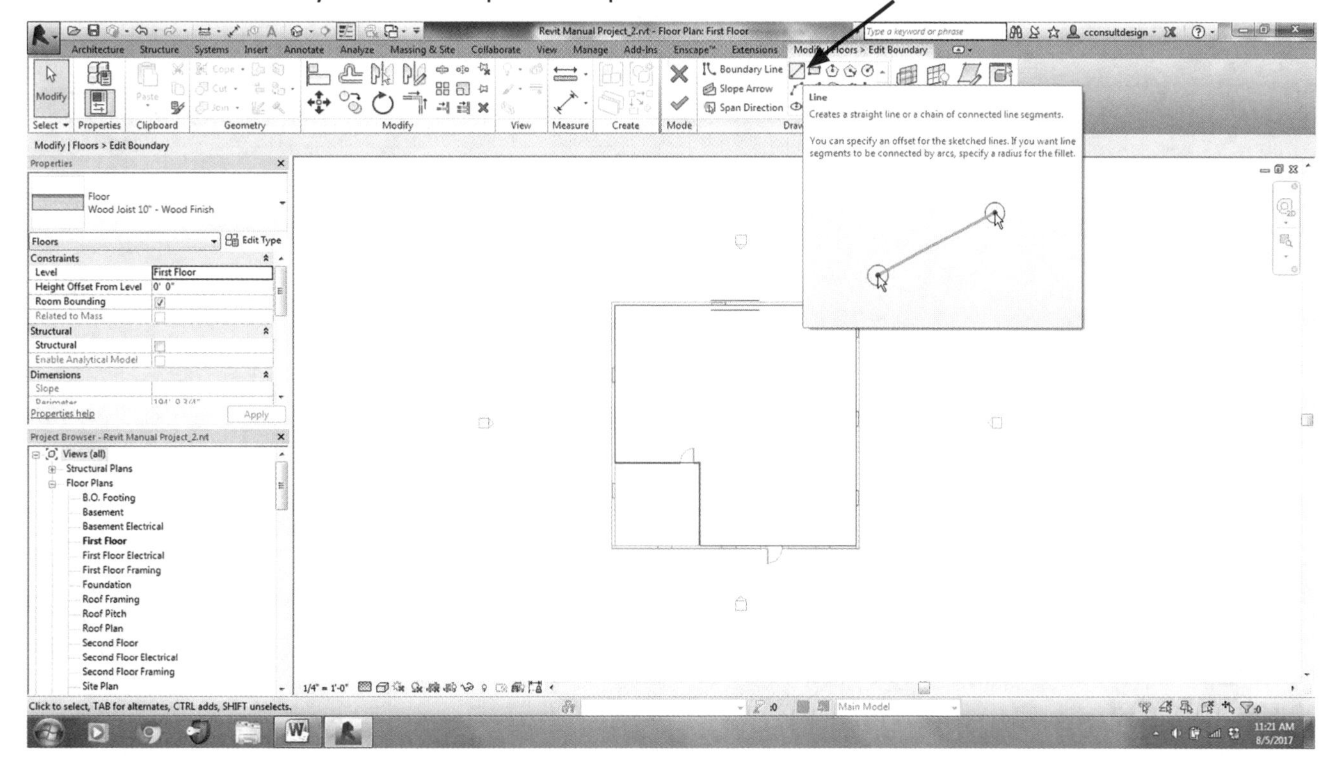

Once you select the **LINE** tool, begin drawing your new layout.

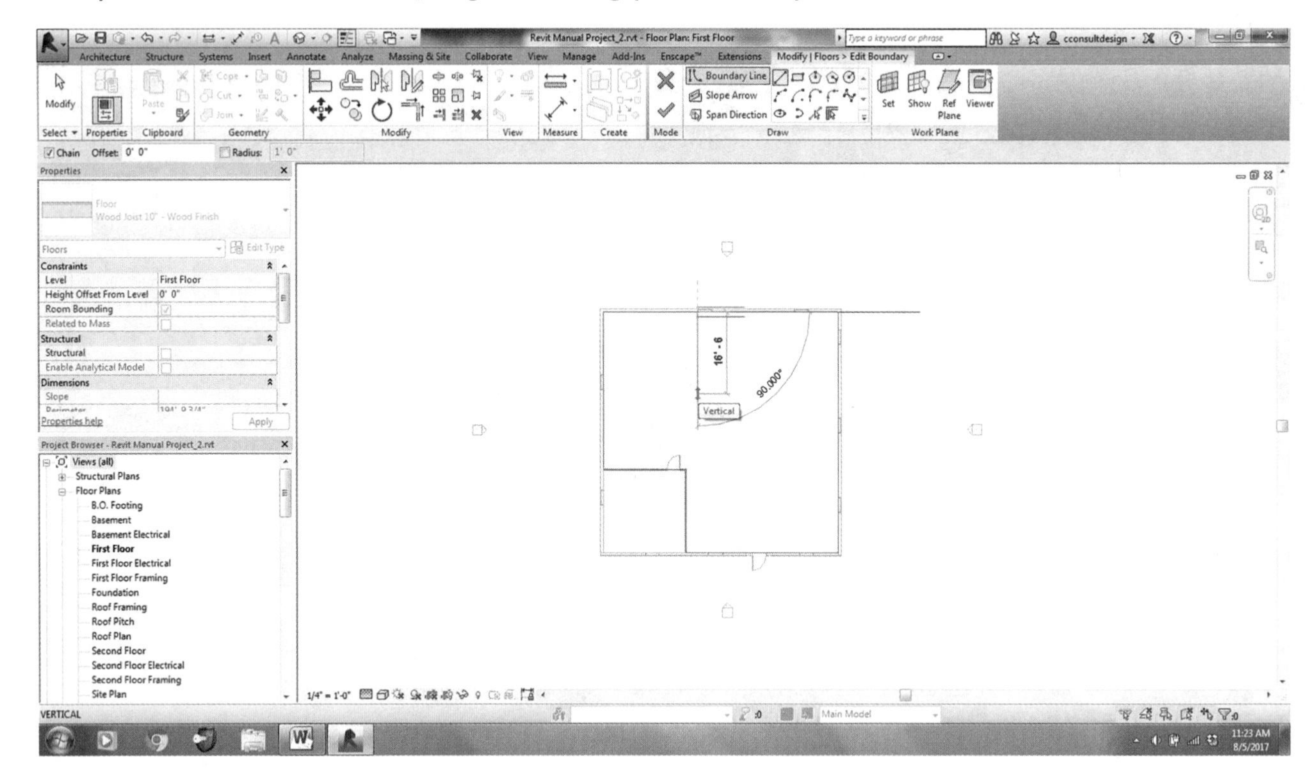

You can pick any location to begin, then draw to your next location. Be sure your lines are fully connected.

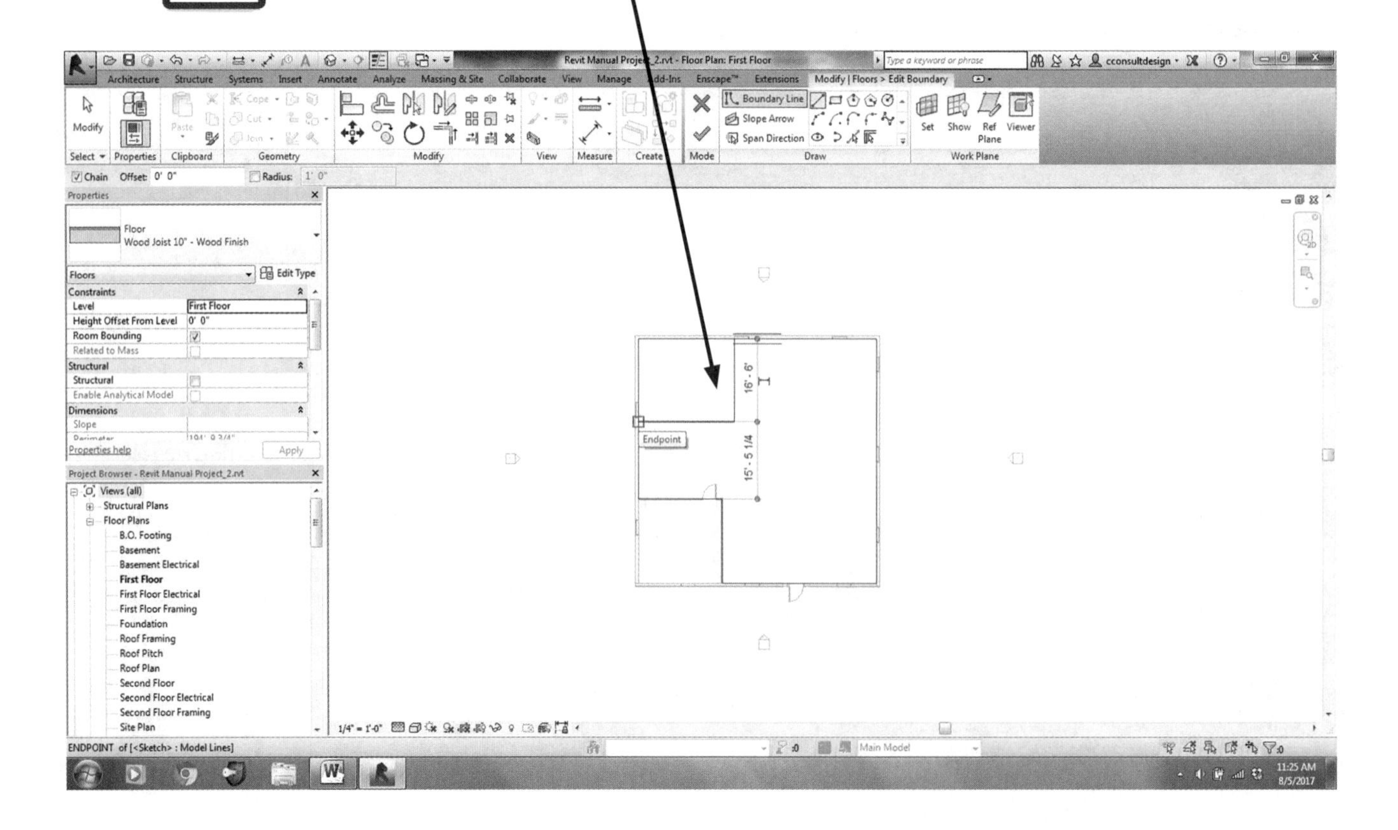

Before clicking the green **CHECK MARK**, you must first trim the floor boundary lines. Click the **MODIFY** Arrow.

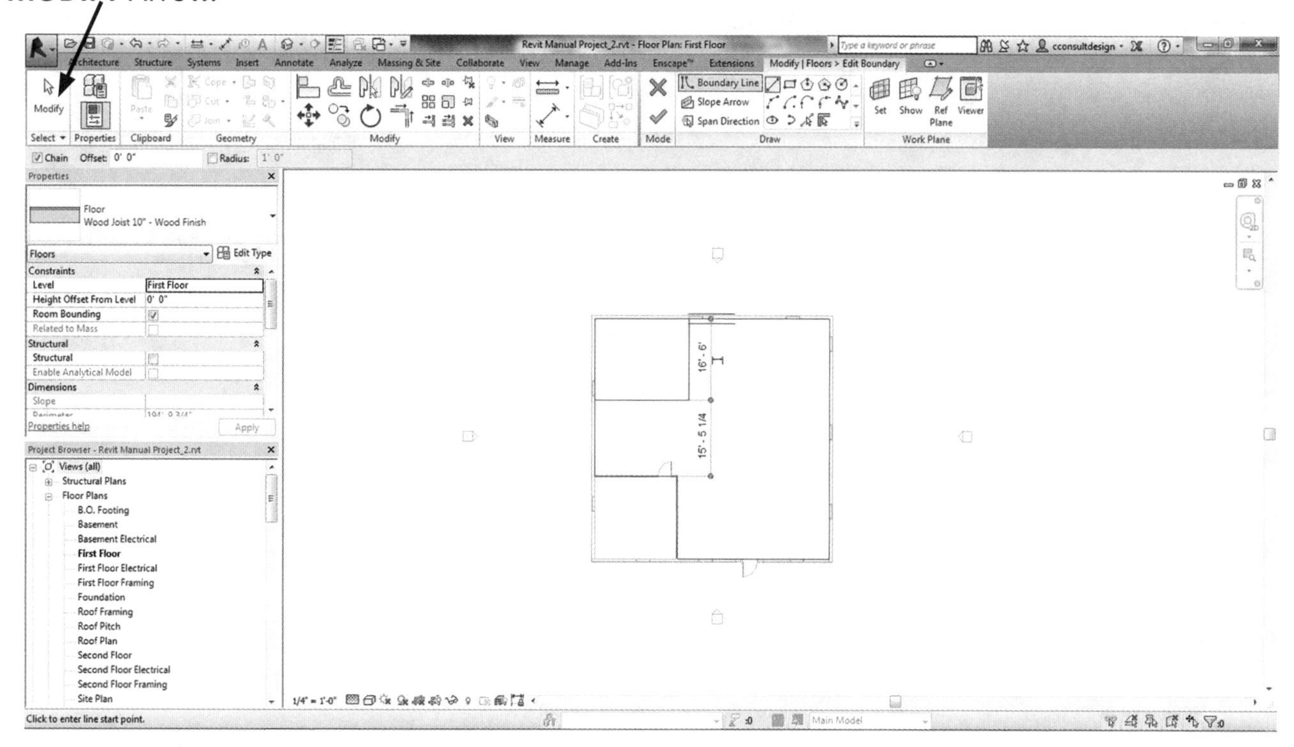

Now you need to trim out the section of the floor you want to remove. To trim the floor, click on the lines you want to *keep*, *not* the line you want to remove.

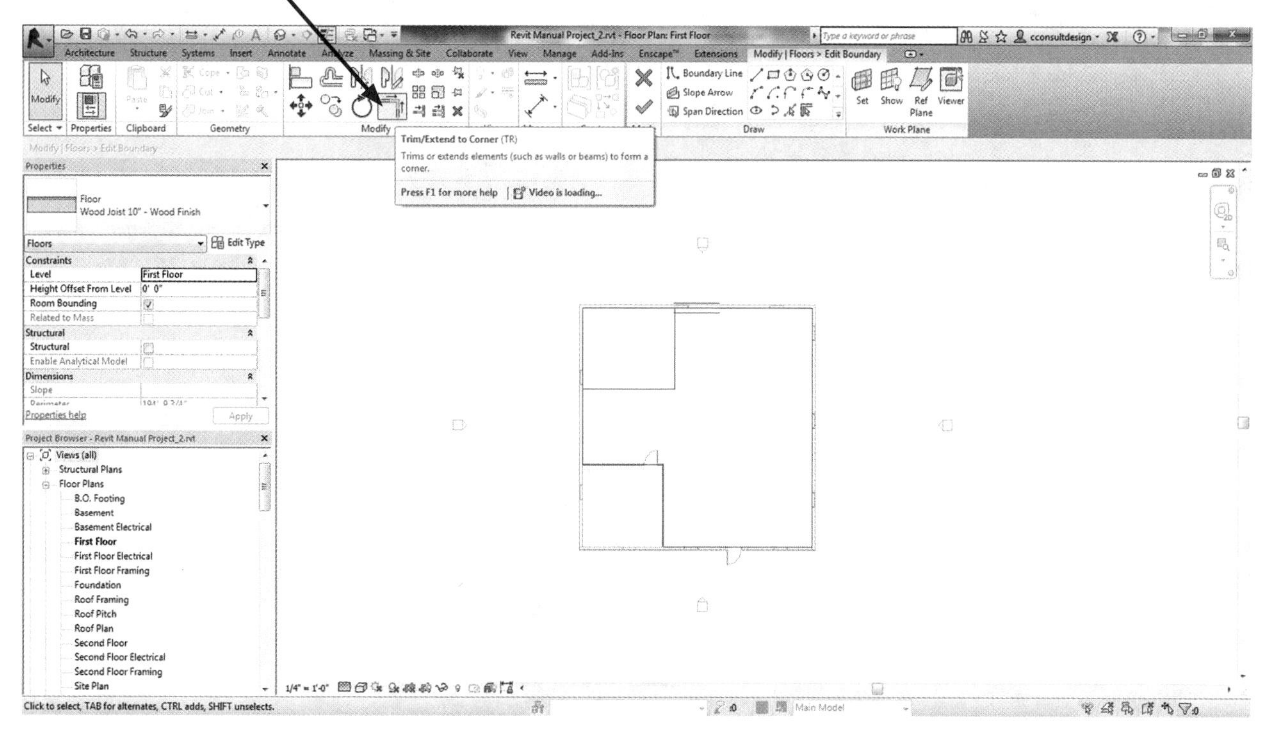

LOOK AHEAD

Trim command instructions are covered more in-depth in Chapter 20, Revit Modify Commands.

Click on the *inside* lines, *not* the outside lines, and the outside boundary lines will be erased.

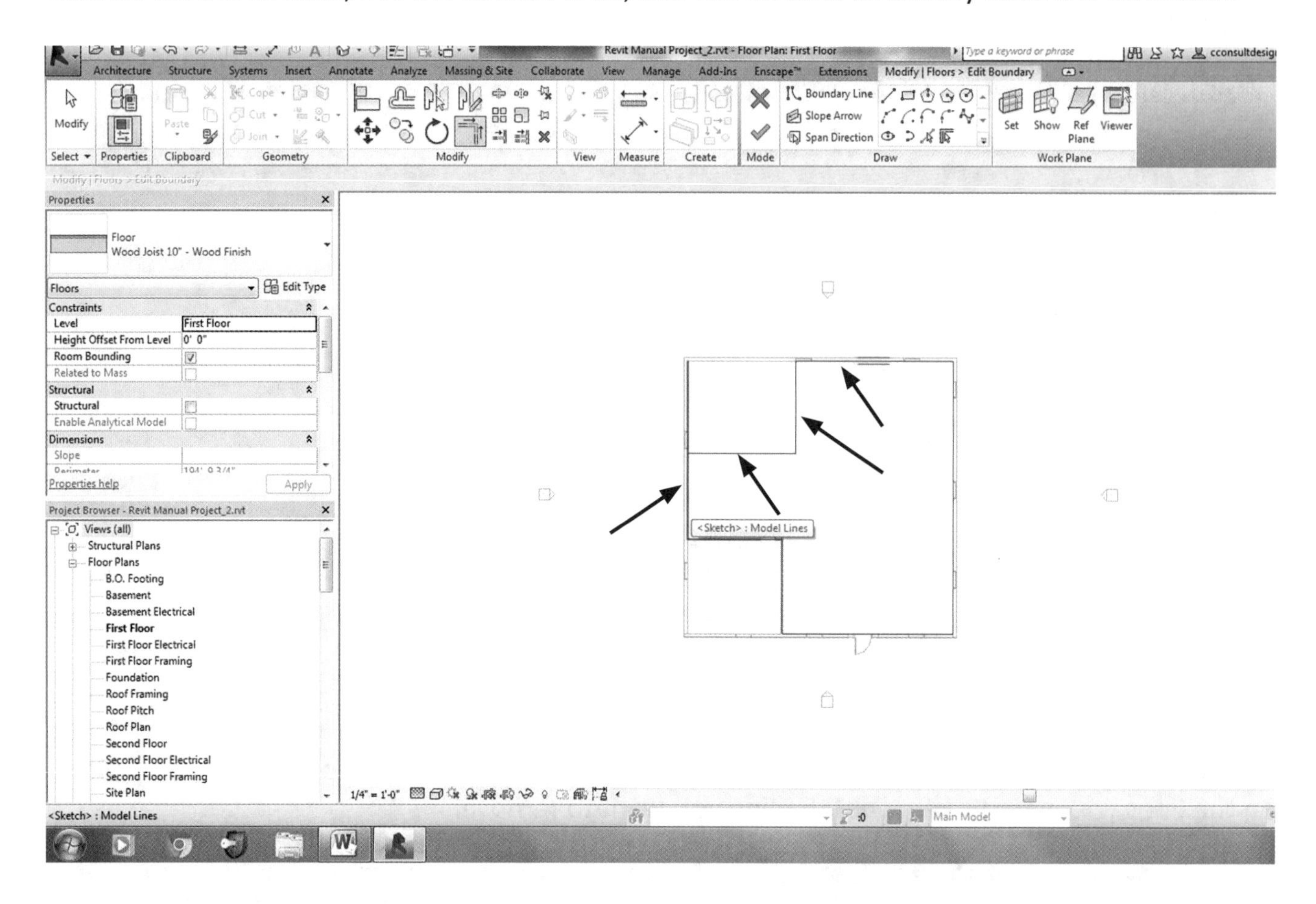

Then click the **green CHECK MARK** to complete the command.

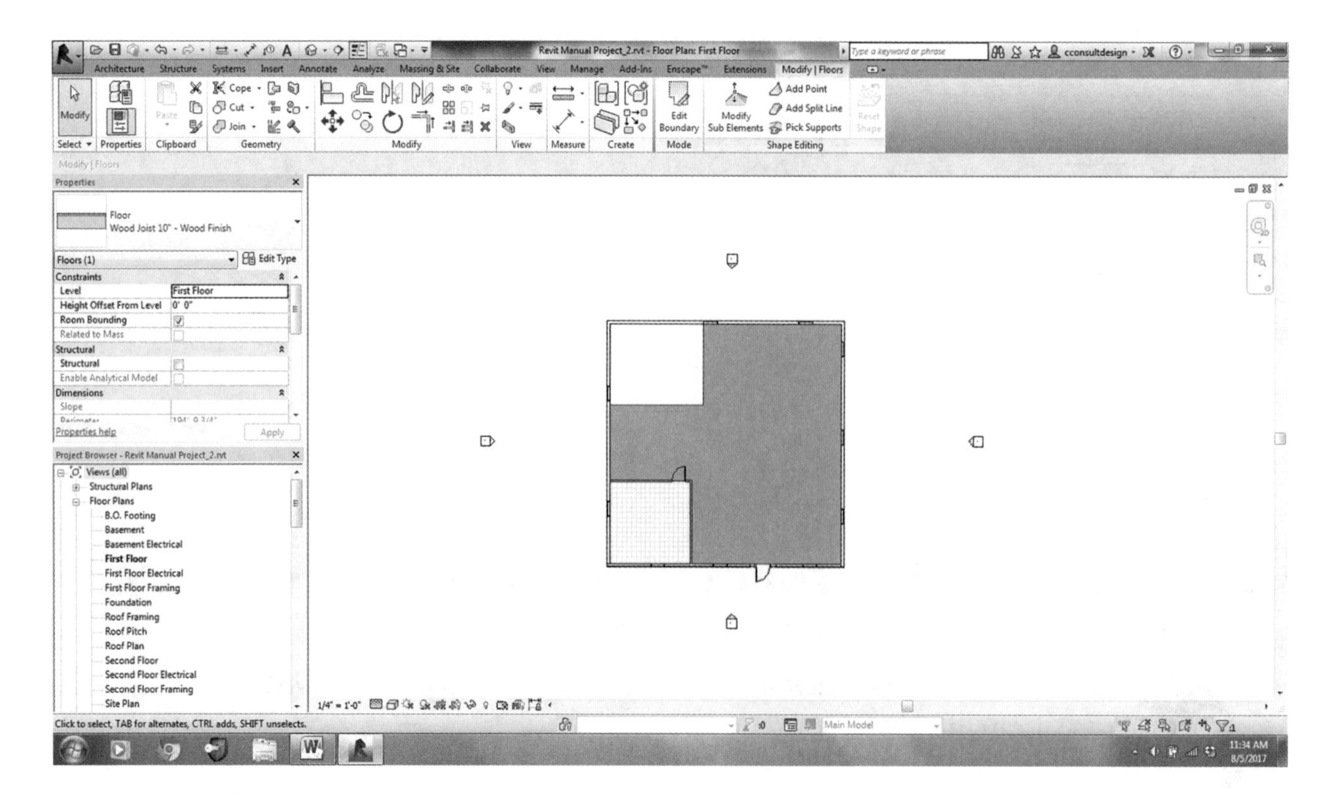

Click the **MODIFY** Arrow, and the highlighted floor will clear away. To add a new floor type in the vacant space that was created, click on the **FLOOR** Tile in the ribbon.

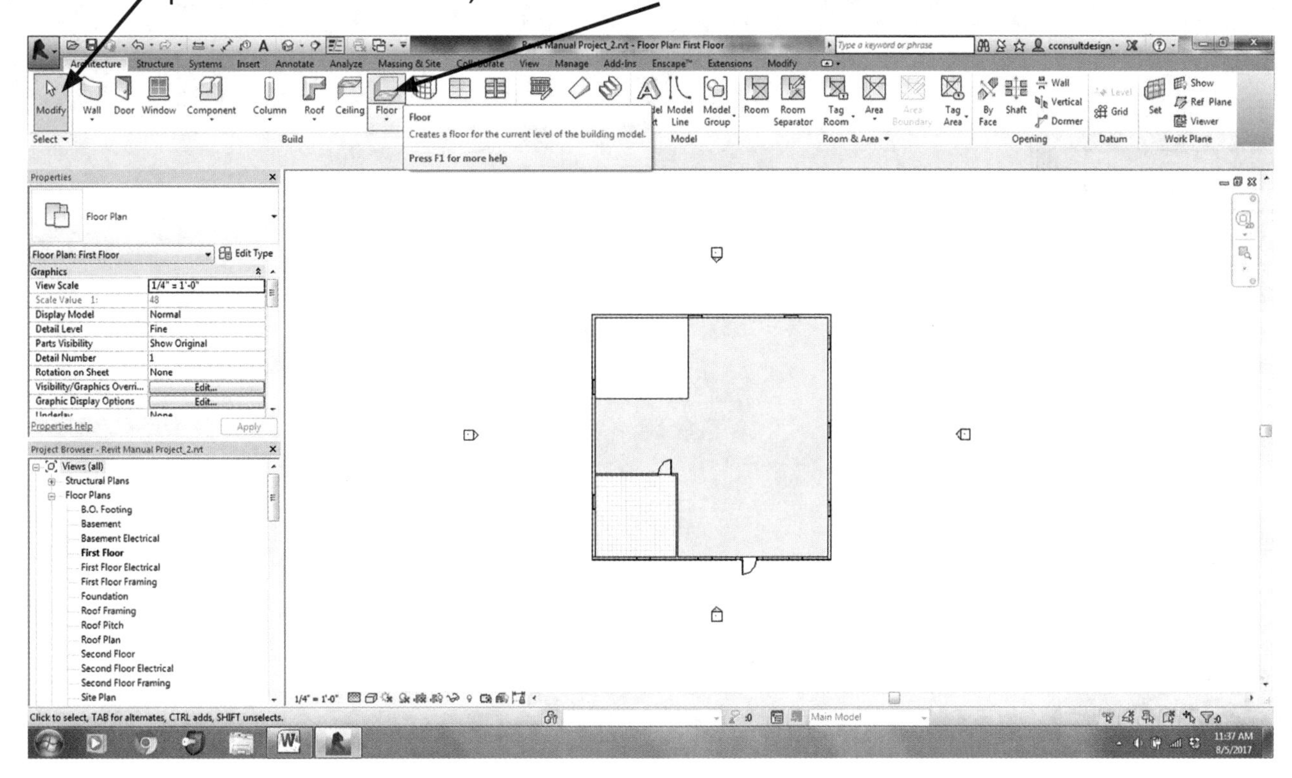

The Properties Box changes, and you can select the type of flooring you want to place in the vacant space that was created.

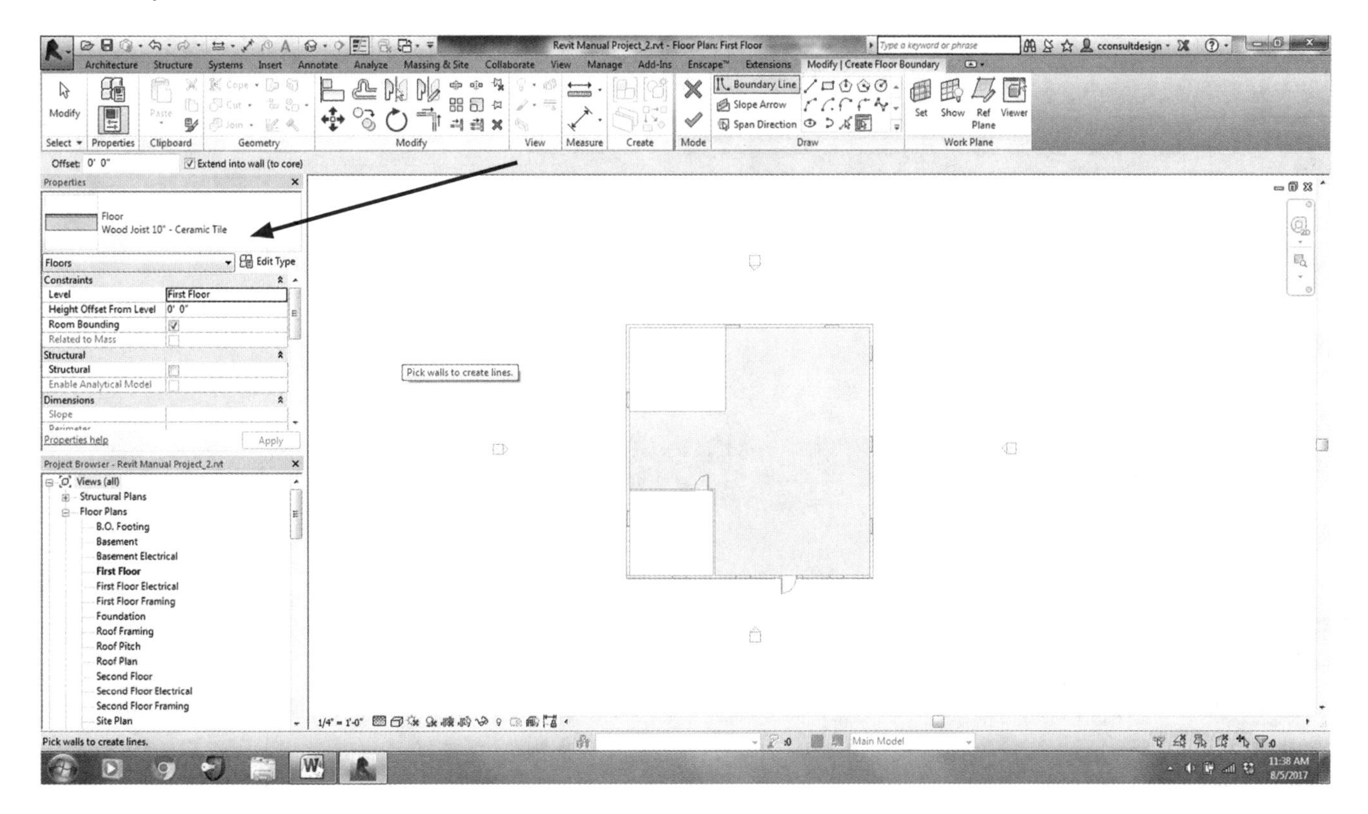

Scroll down, and select the carpet finish you desire.

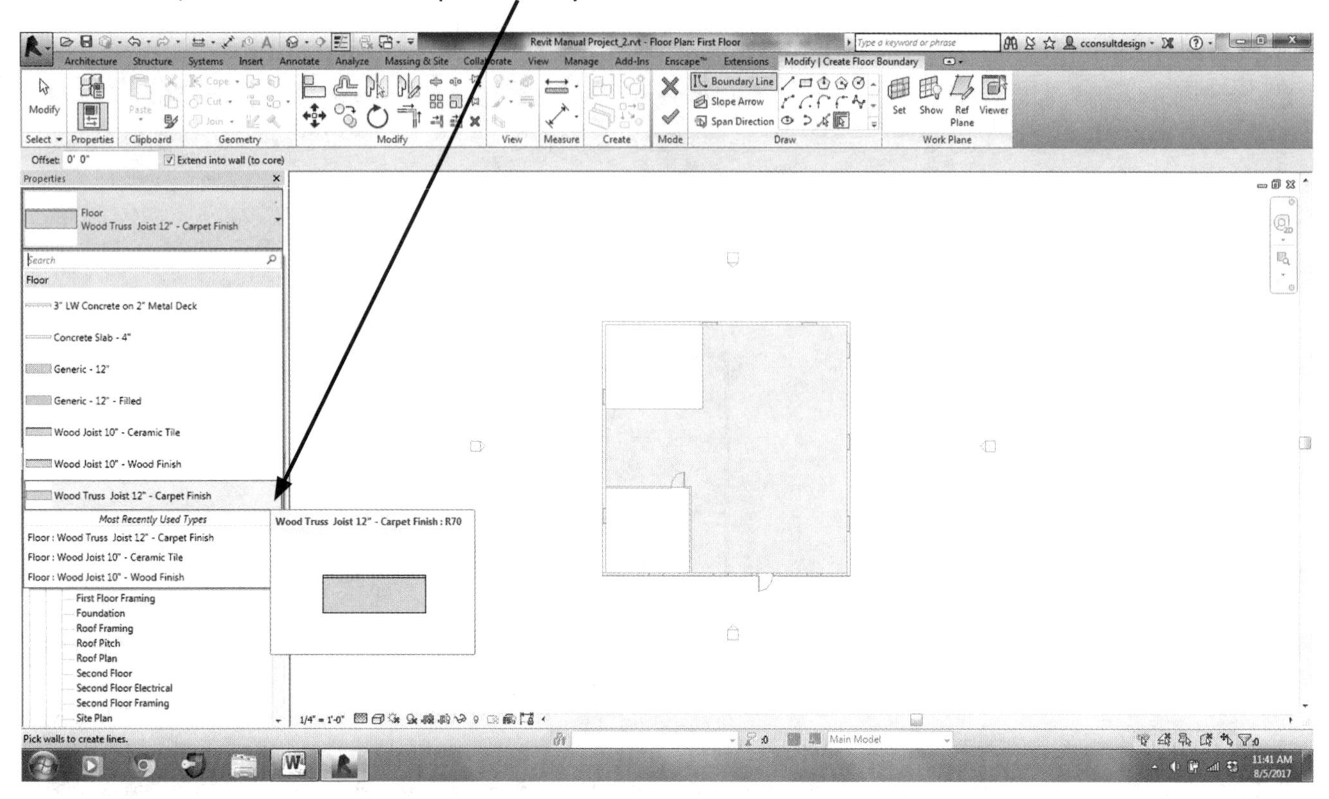

After selecting the type of floor you want to replace, you can use the **RECTANGLE** Tool to quickly draw a rectangle on the opening, or line tool to trace the opening.

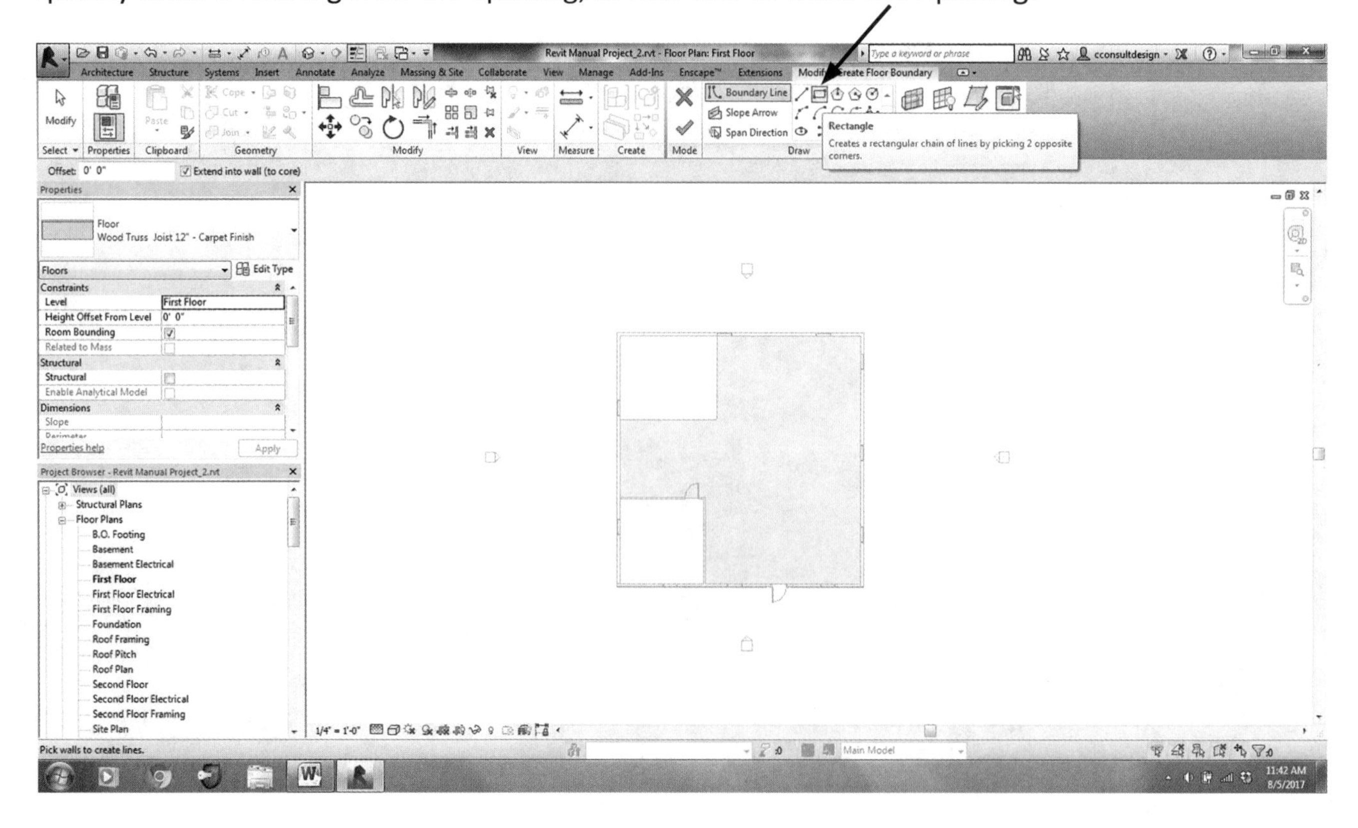

The magenta lines will trace your opening, and that will place your new flooring type in the vacant space.

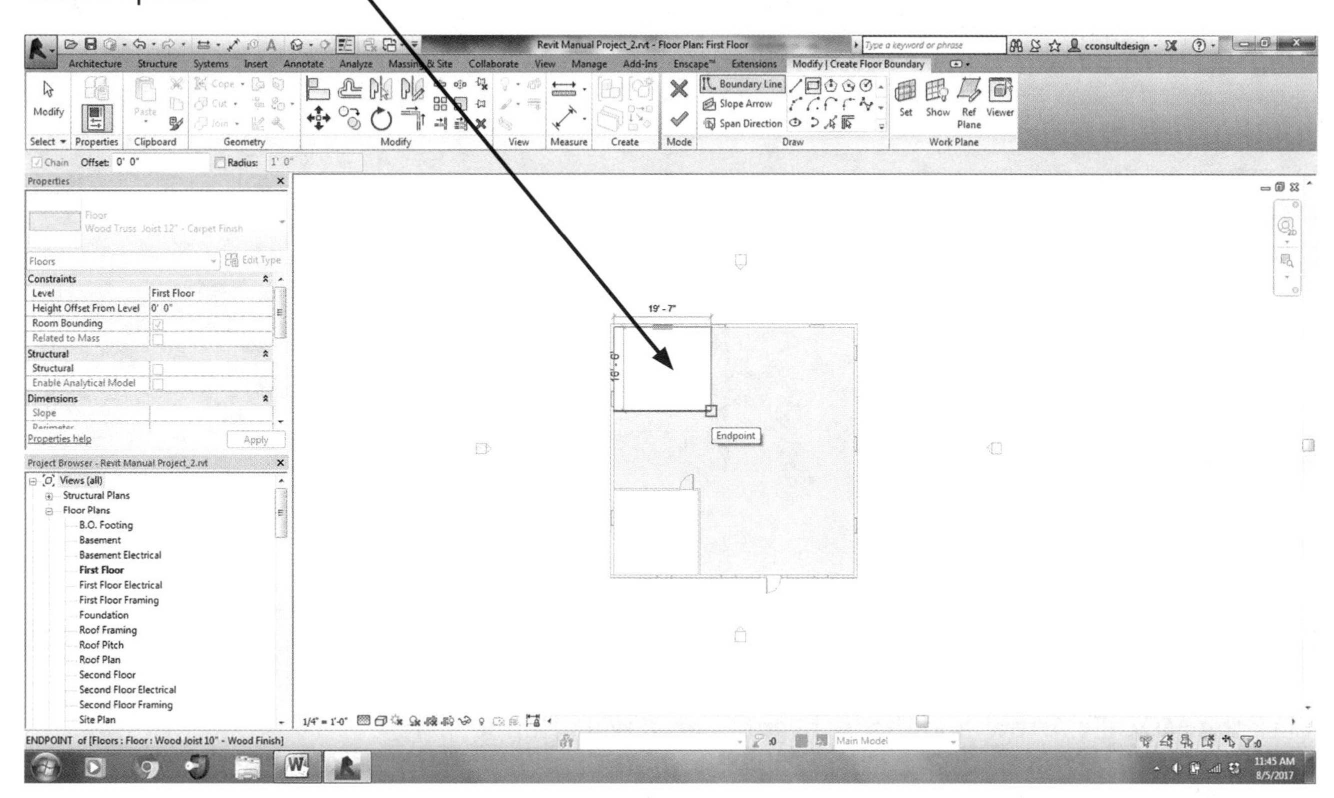

Once the new boundary lines are placed, click the green **CHECK MARK** and you will have completed the floor command.

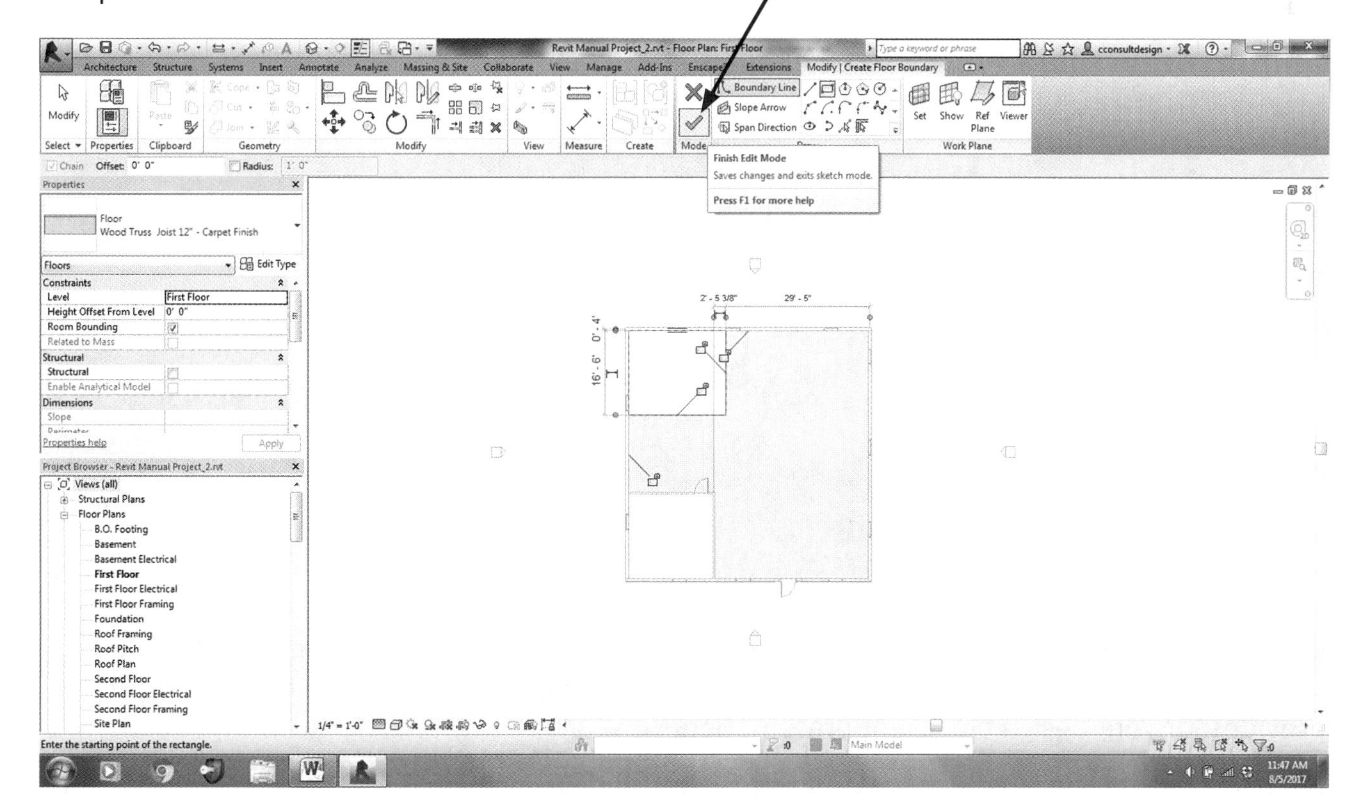

The floor boundary will be highlighted to show this is the area you modified.

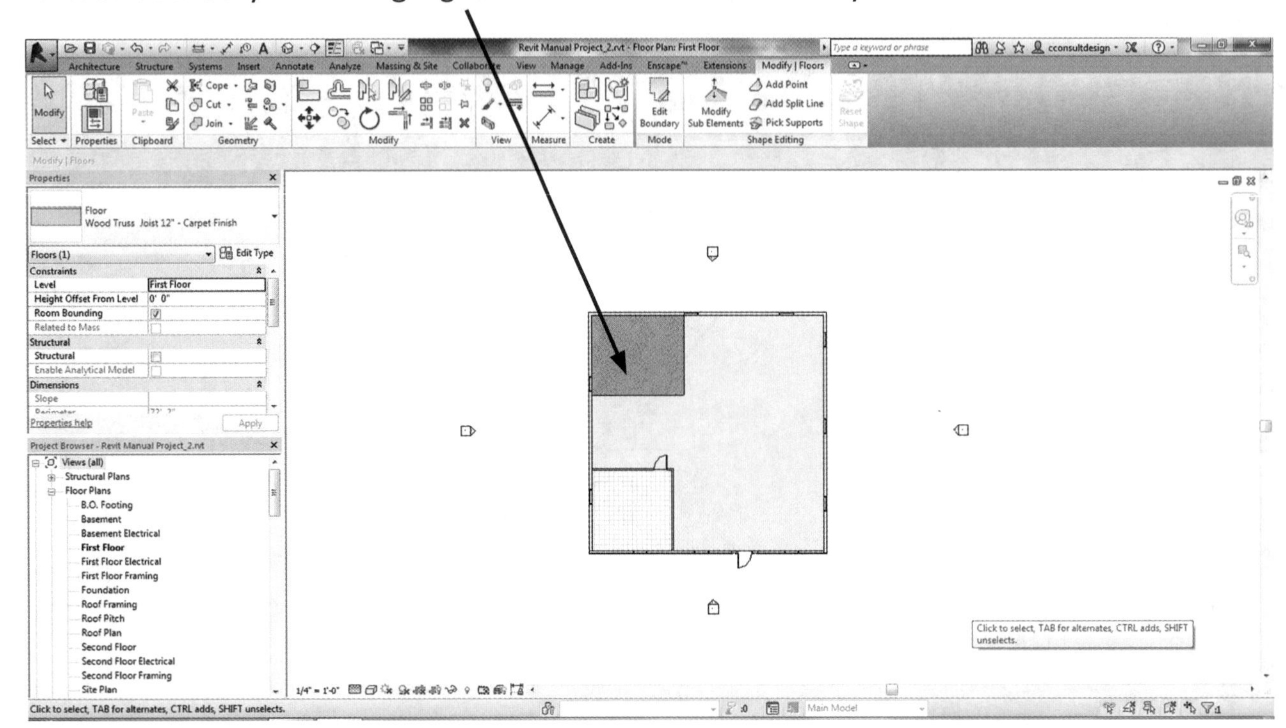

Click the **MODIFY** Arrow, and the highlighted area will disappear.

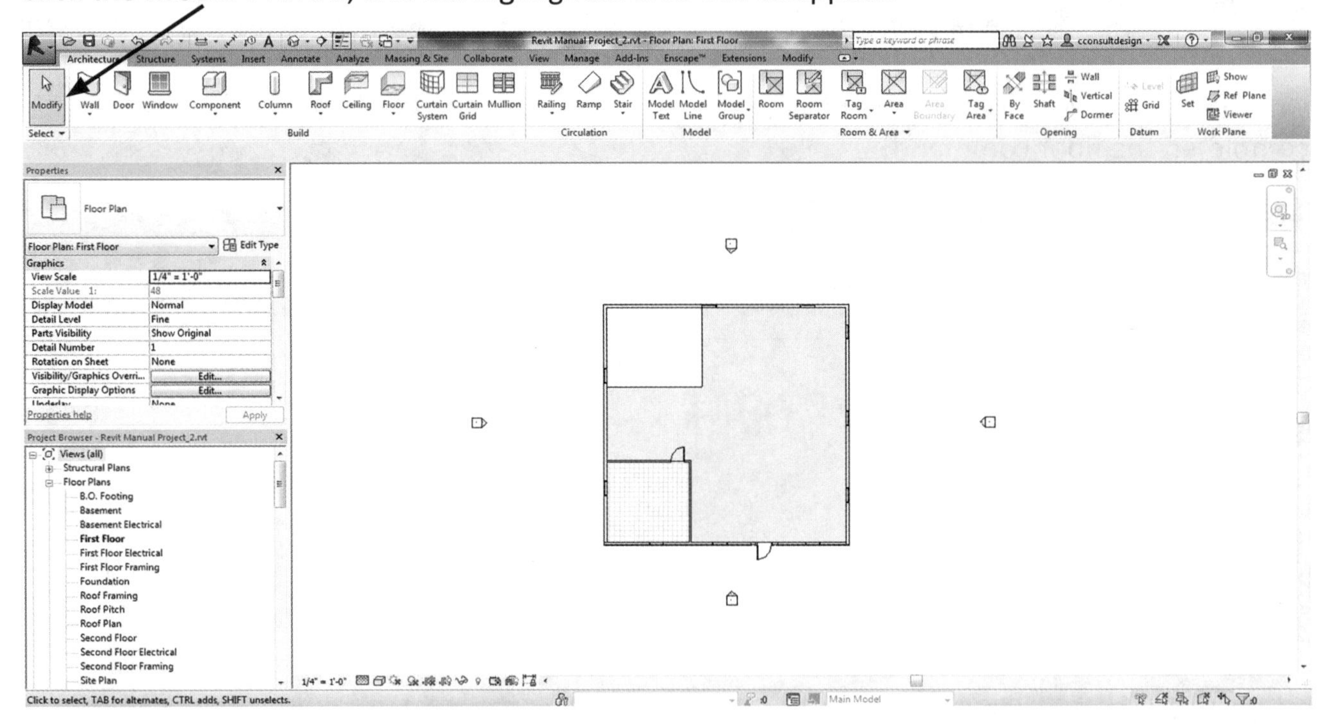

In this chapter, we have seen different ways to add various types of flooring in different sections of the house or building in Revit. With the aid of the materials browser, you can select the type of flooring material and patterns you are able to add, and the level of detail Revit provides for your design and client presentation.

In Chapter 6, we will look at adding ceilings to your house.

CHAPTER 6

Creating Ceilings

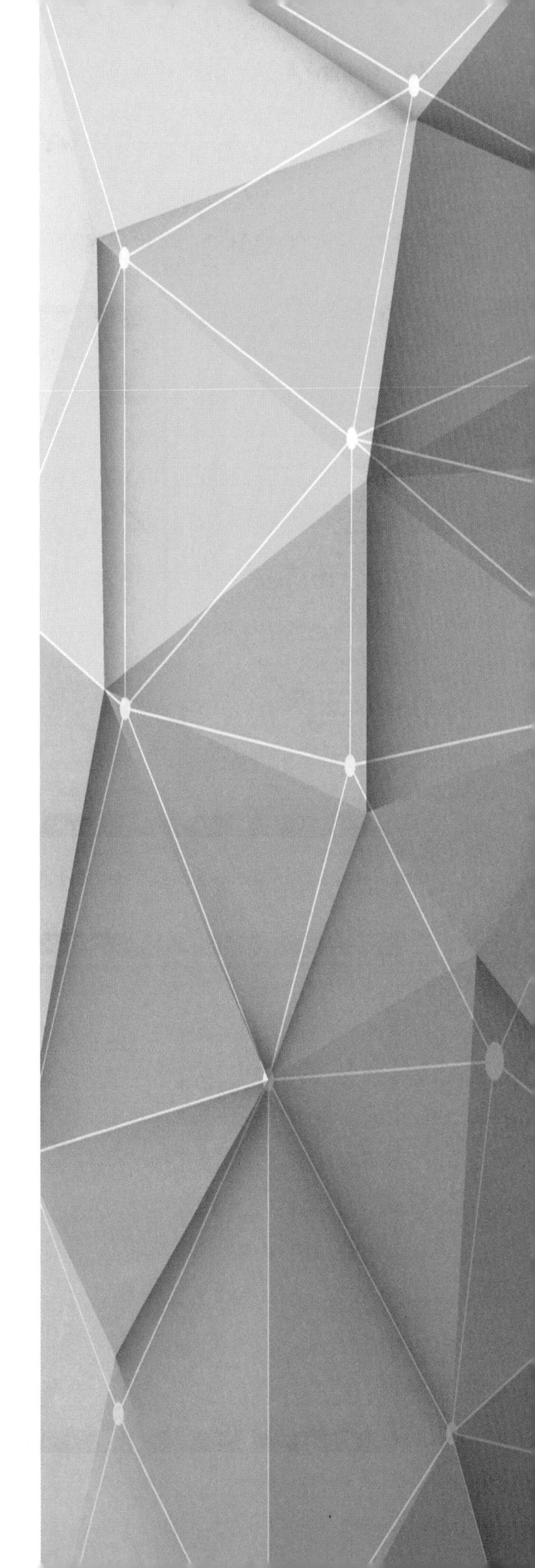

CREATING A CEILING

When creating a ceiling make sure you are in the Ceiling Plan view in the Project Browser, and on the proper selected floor plan.

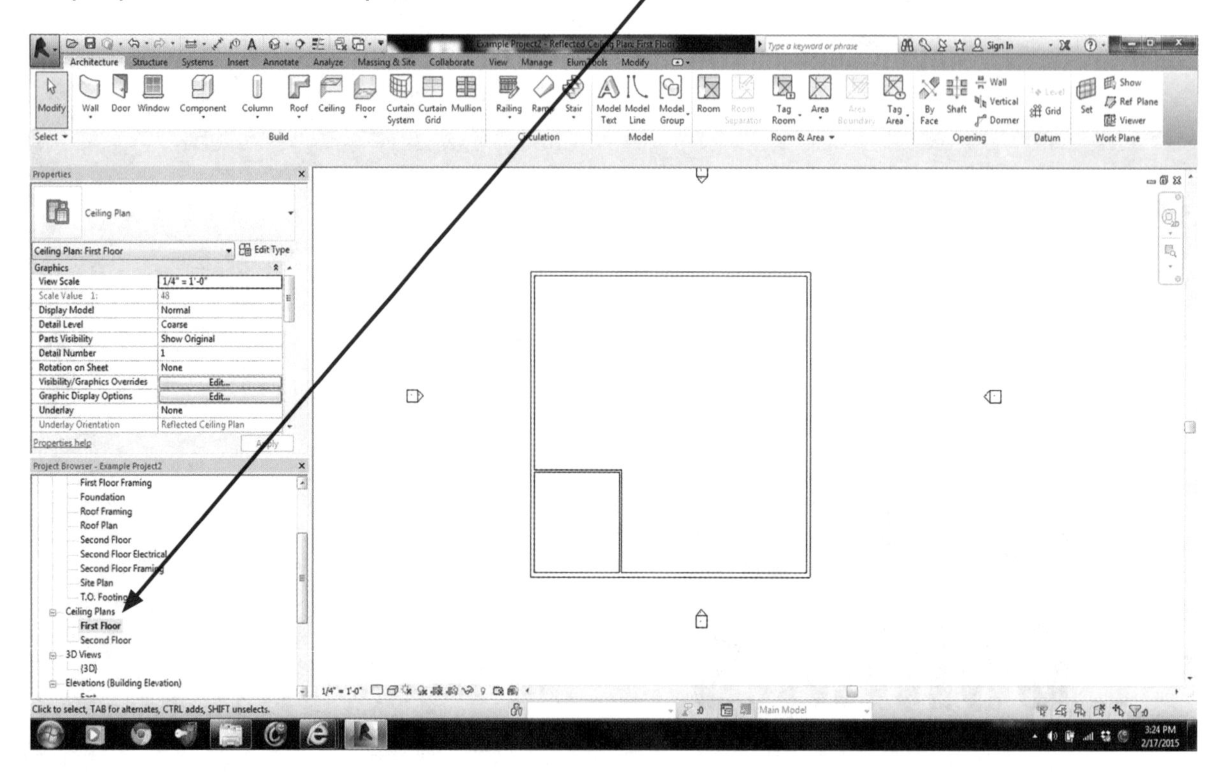

Once you are in the Ceiling Plan view in the Architecture Tab, click on the **CEILING** Tab**.**

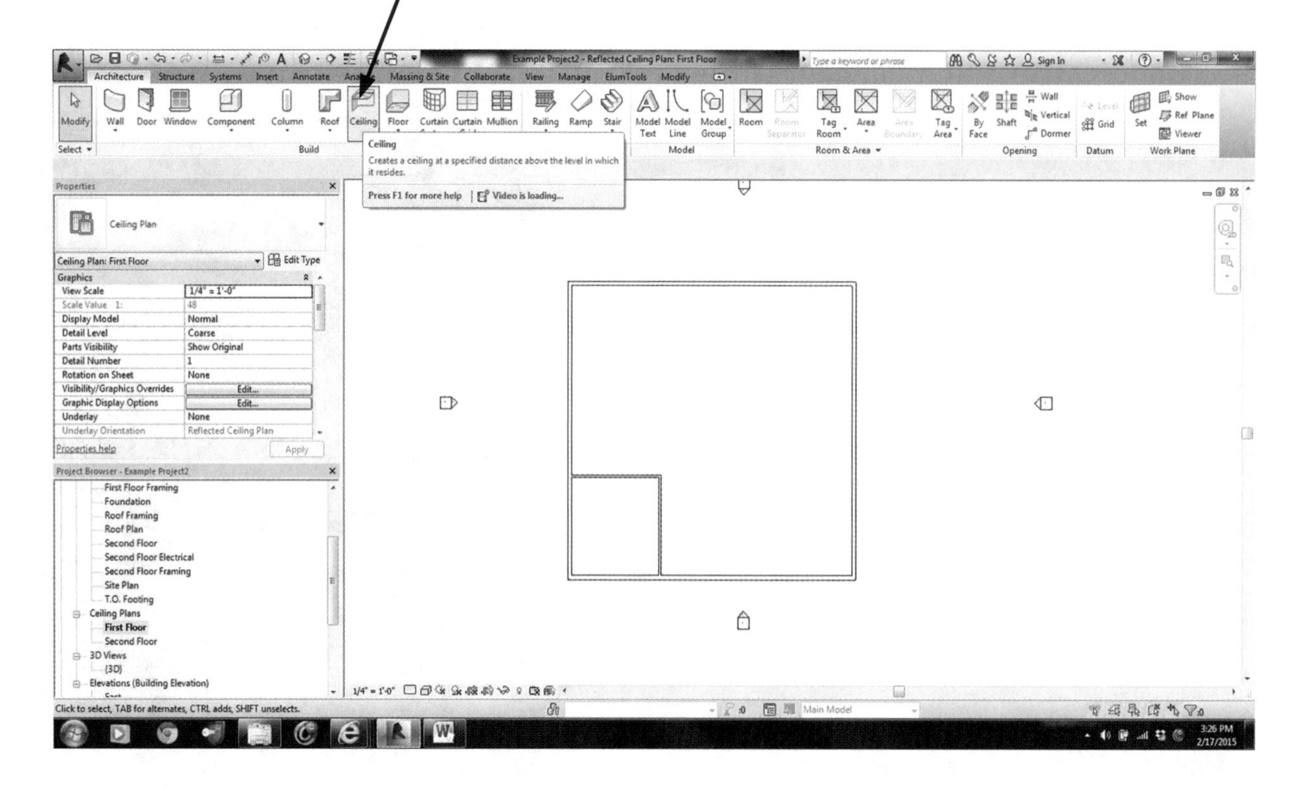

Using the same steps as you did when creating a floor plan (in Chapter 5), you will select the type of ceiling you want.

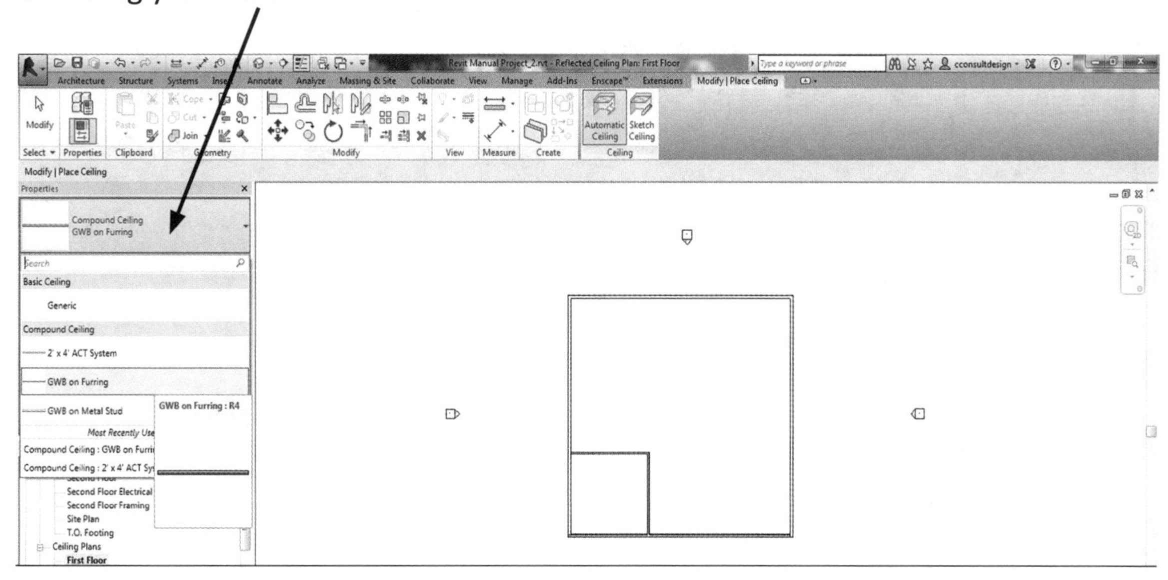

Once you select the ceiling type, you will get 2 options, either Automatic Ceiling or Sketch Ceiling. Automatic Ceiling allows you to pick the entire space and place a ceiling all at once. Sketch Ceiling allows you to go to the Draw Box and draw lines around the space where you want a ceiling.

REMINDER

The Sketch Ceiling option allows you to create a boundary just like you did when creating a floor layout.

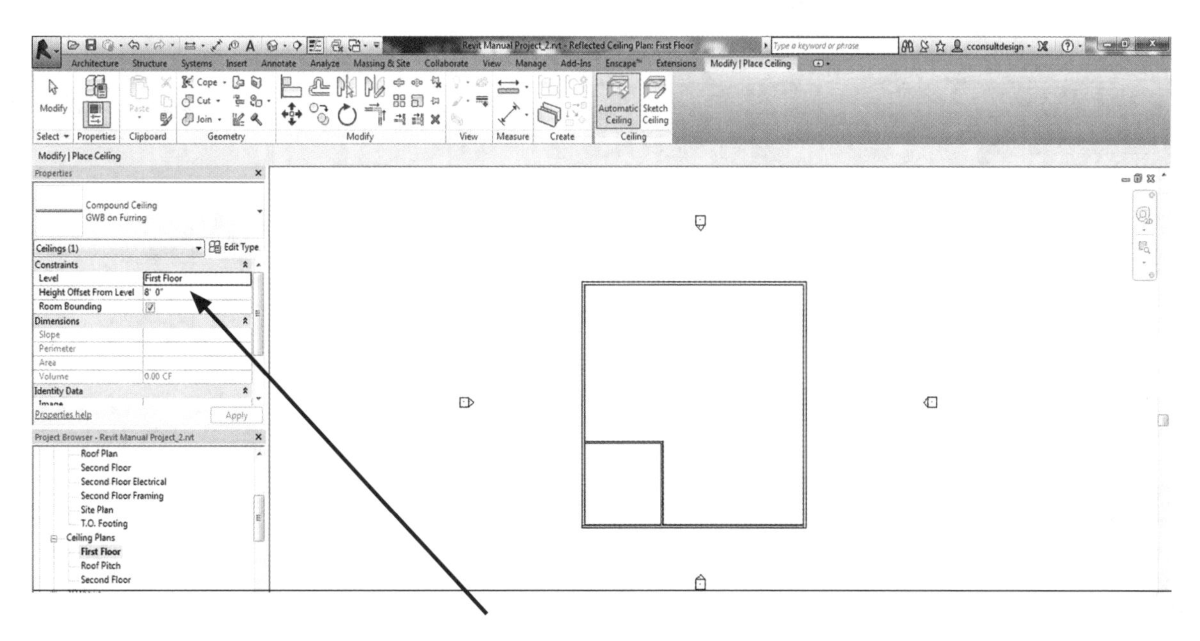

This is the Automatic Ceiling option. To change the ceiling height, simply hover over the space, and the ceiling boundary will appear. When the boundary appears you can change the ceiling height in the Properties Box on the left under "Height Offset From Level."

In this Sketch Ceiling option, use the **PICK WALLS** tool or **LINES** tool and trace the ceiling boundary, then change the ceiling height.

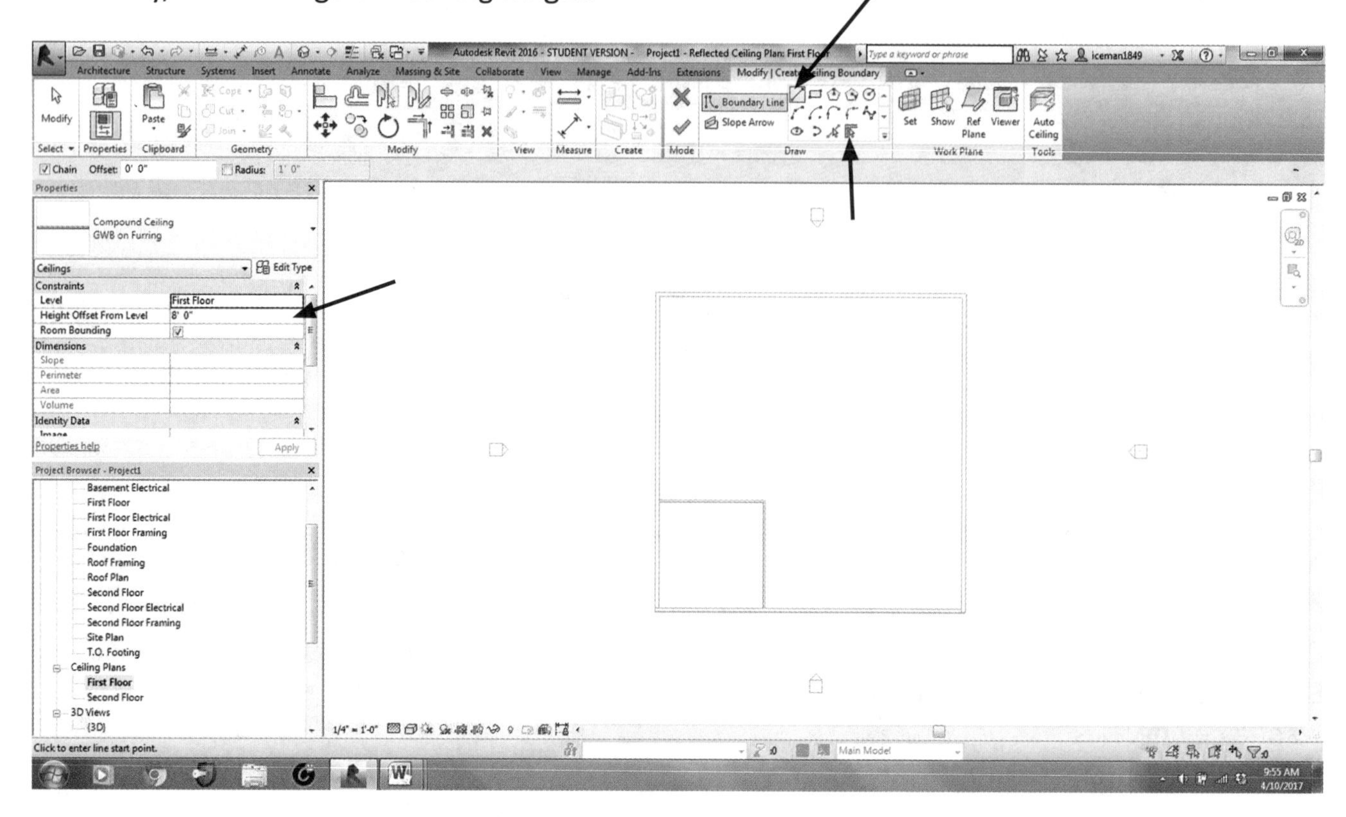

Once the ceiling is in place (after sketching and changing the ceiling height), click on the green **CHECK MARK** to complete to ceiling.

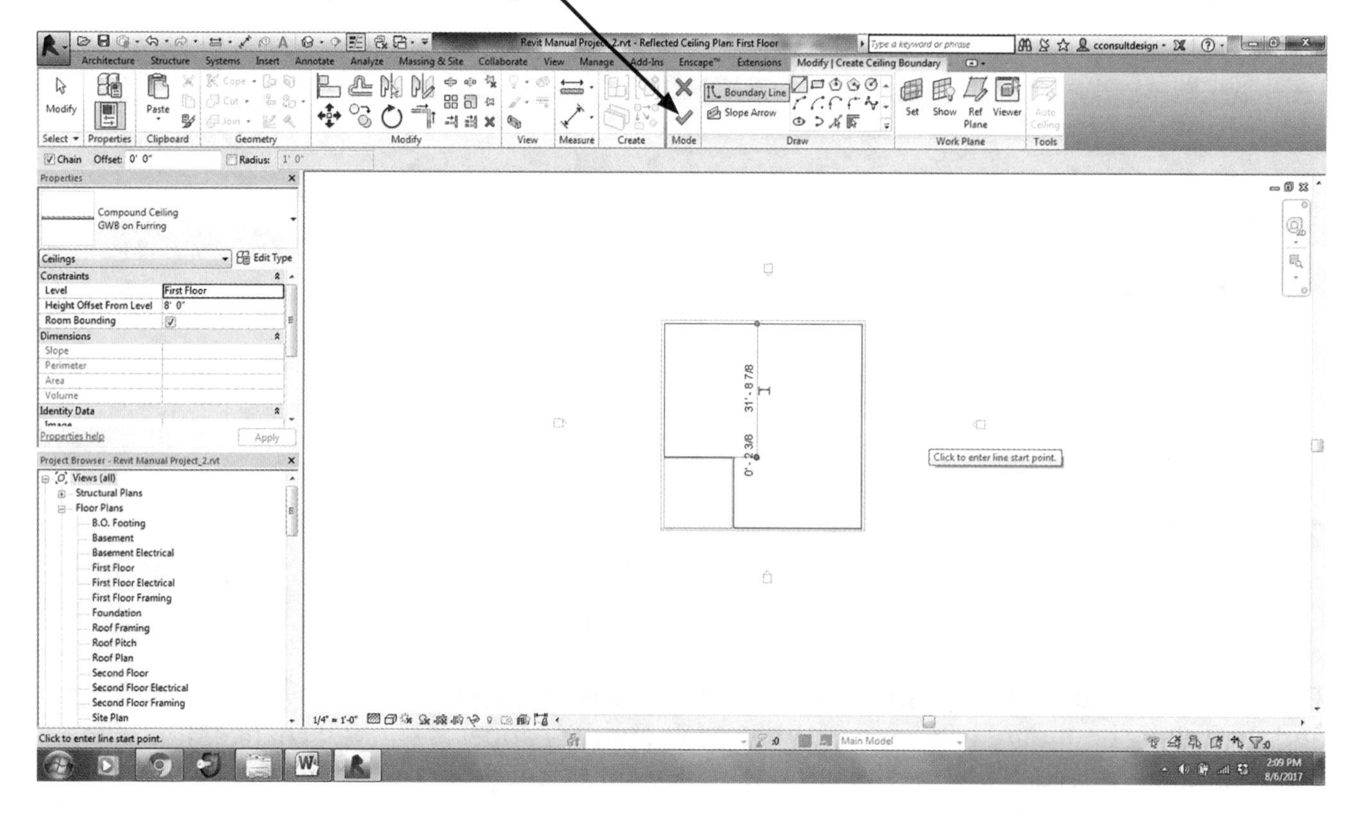

Then click the **MODIFY** Arrow to complete the task.

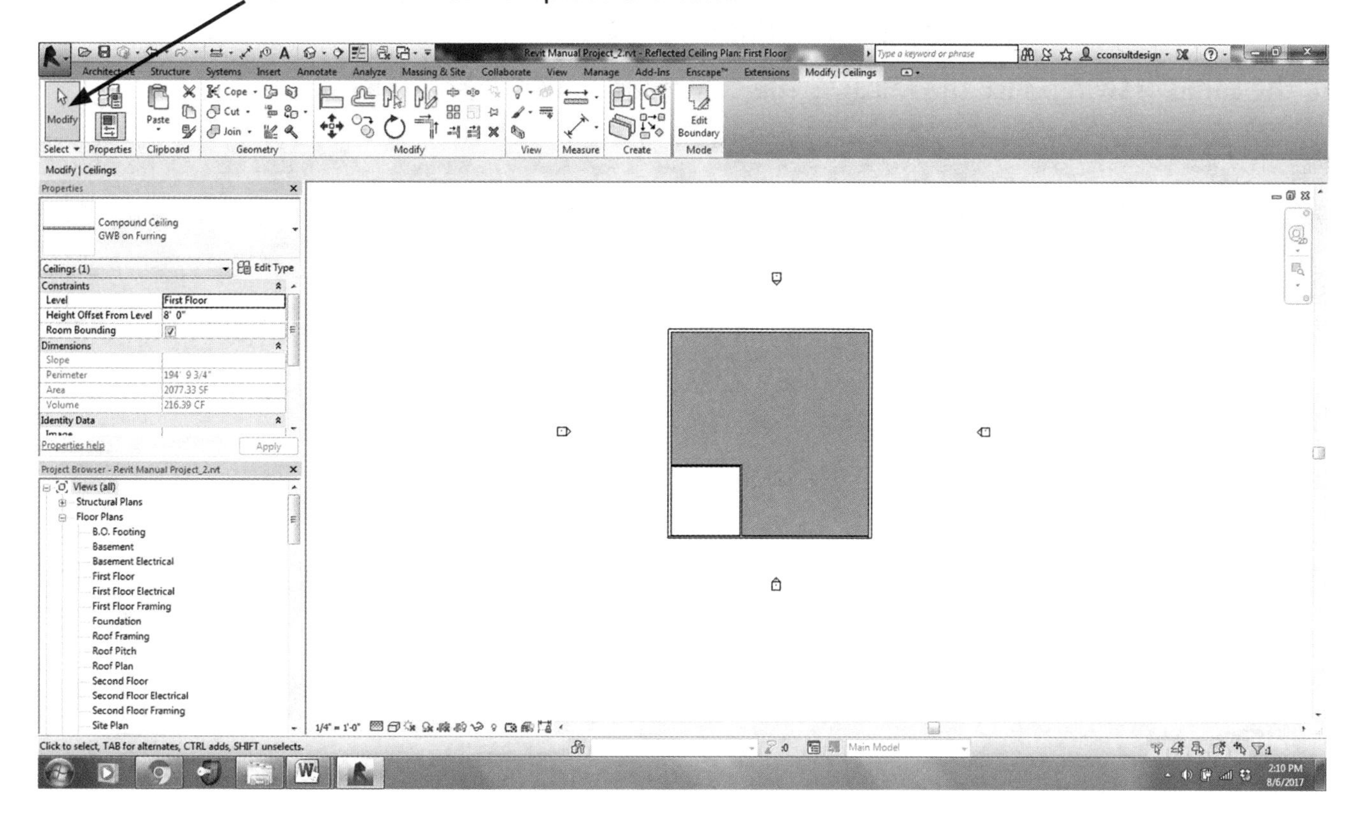

After placing the ceiling in the space, click on the **MODIFY** Arrow and the drawing will clear itself out.

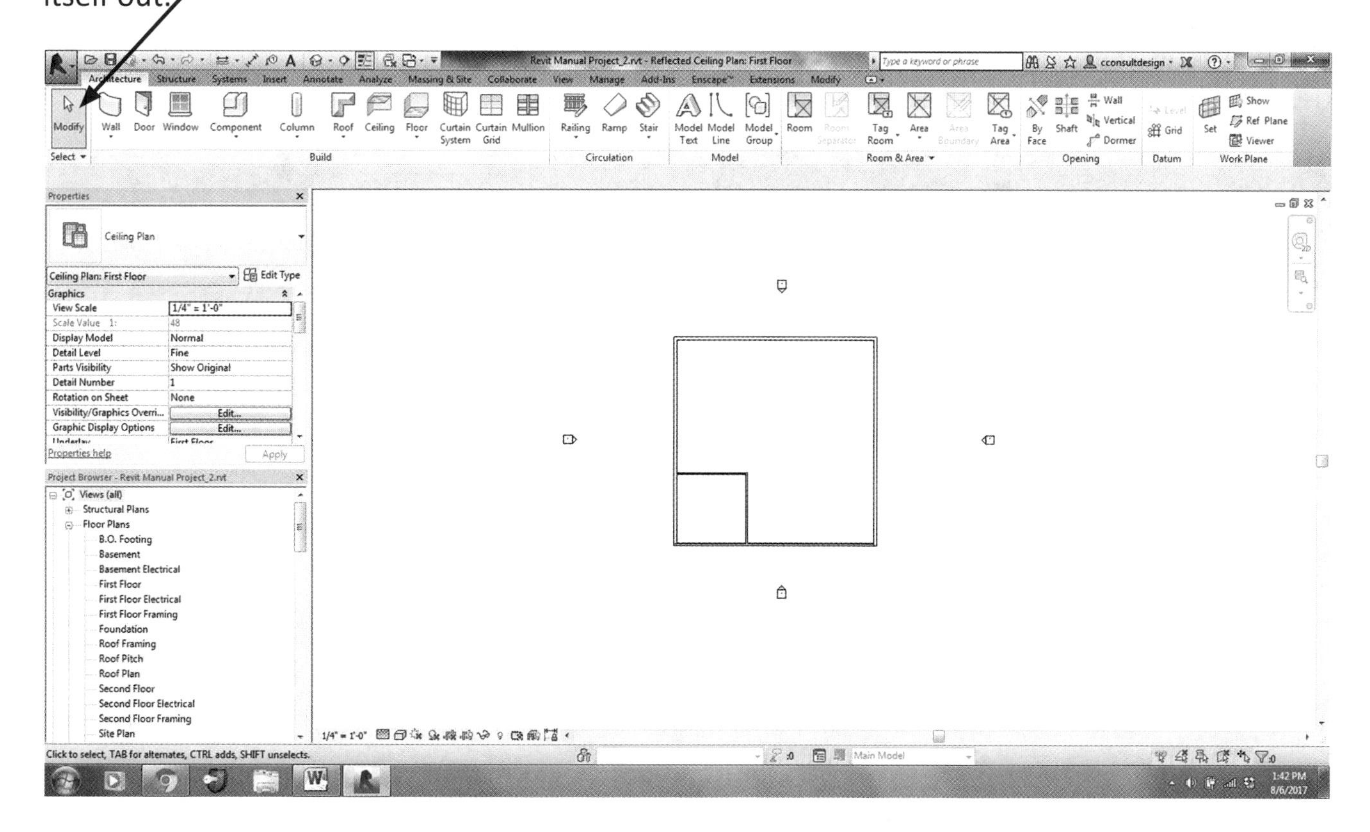

Since we have now placed one ceiling type in the main space of the house, let's use the same steps and place another ceiling type in another space.

In the same drawing, click the **CEILING** Tile command in the ribbon again.

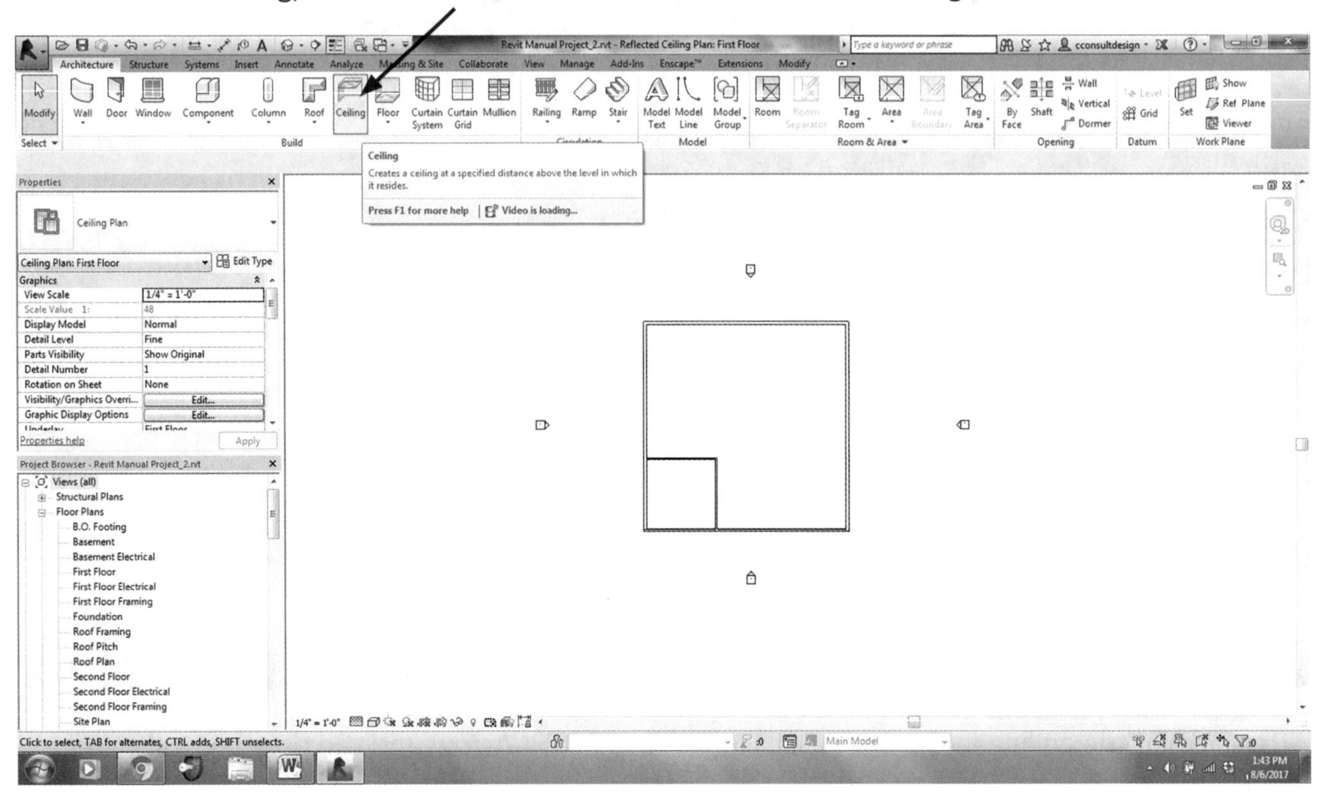

This will activate the Modify Ceilings command. In the Modify Ceilings command, the Properties Box changes so you can select a different Ceiling Style.

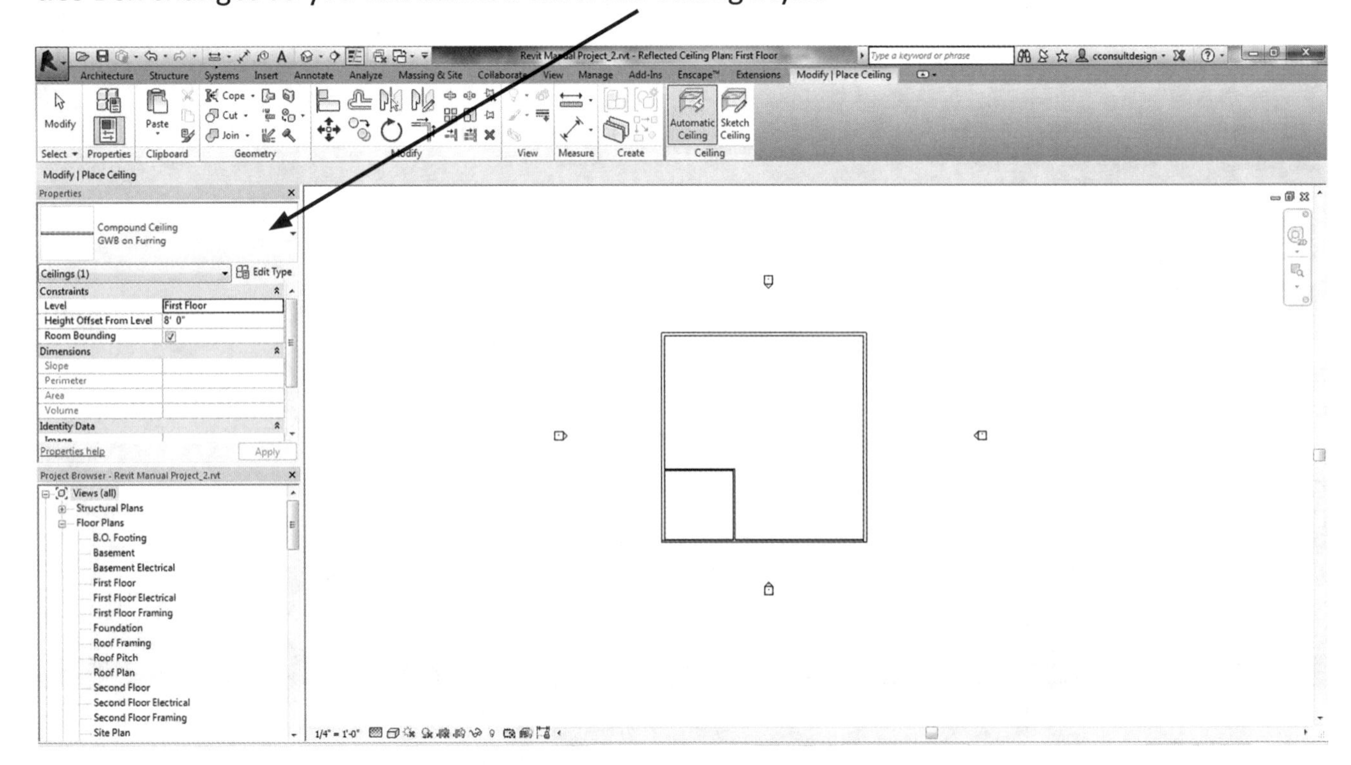

In the drop-down, select a different Ceiling Style. We will select the **2x4 ACT COMPOUND CEILING** Tile.

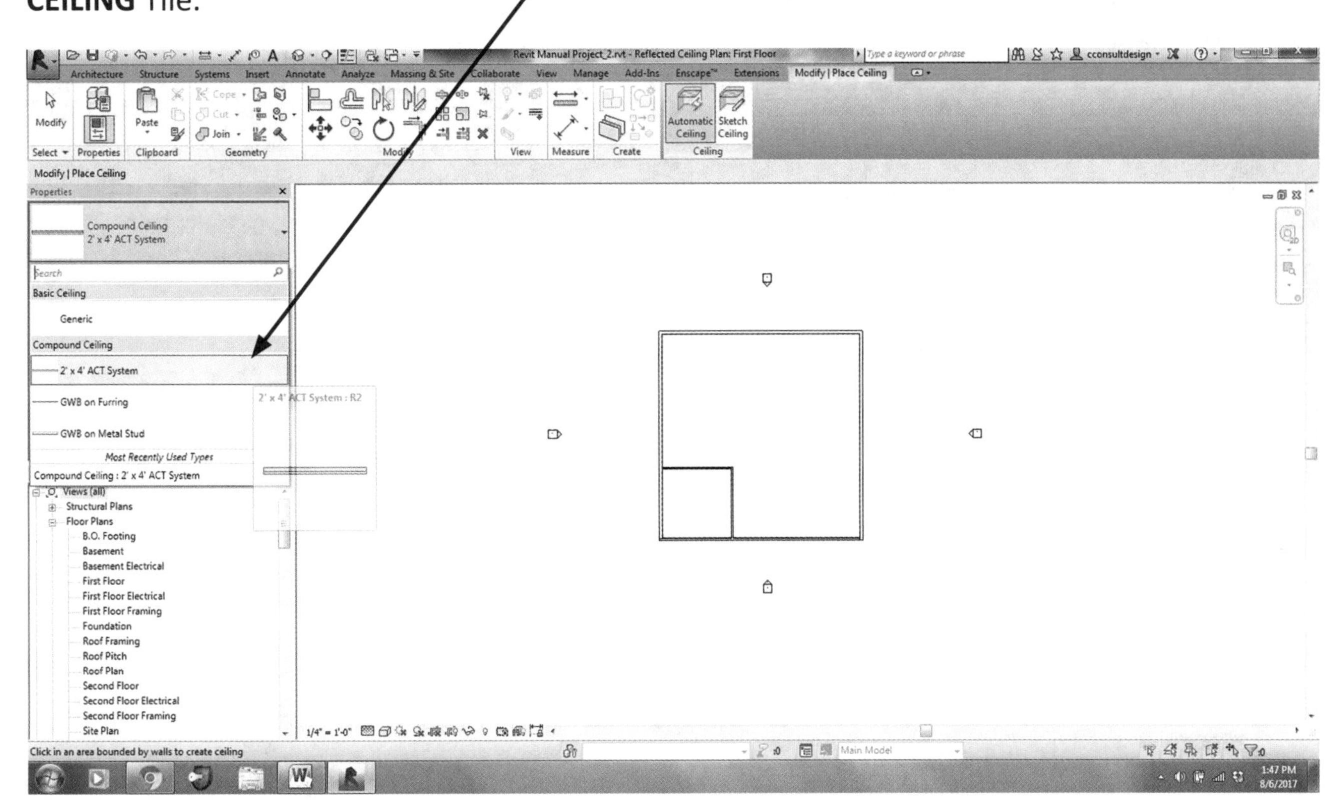

Once you have selected the different Ceiling Style then you will select the **SKETCH CEILING** mode to trace the location of the new ceiling.

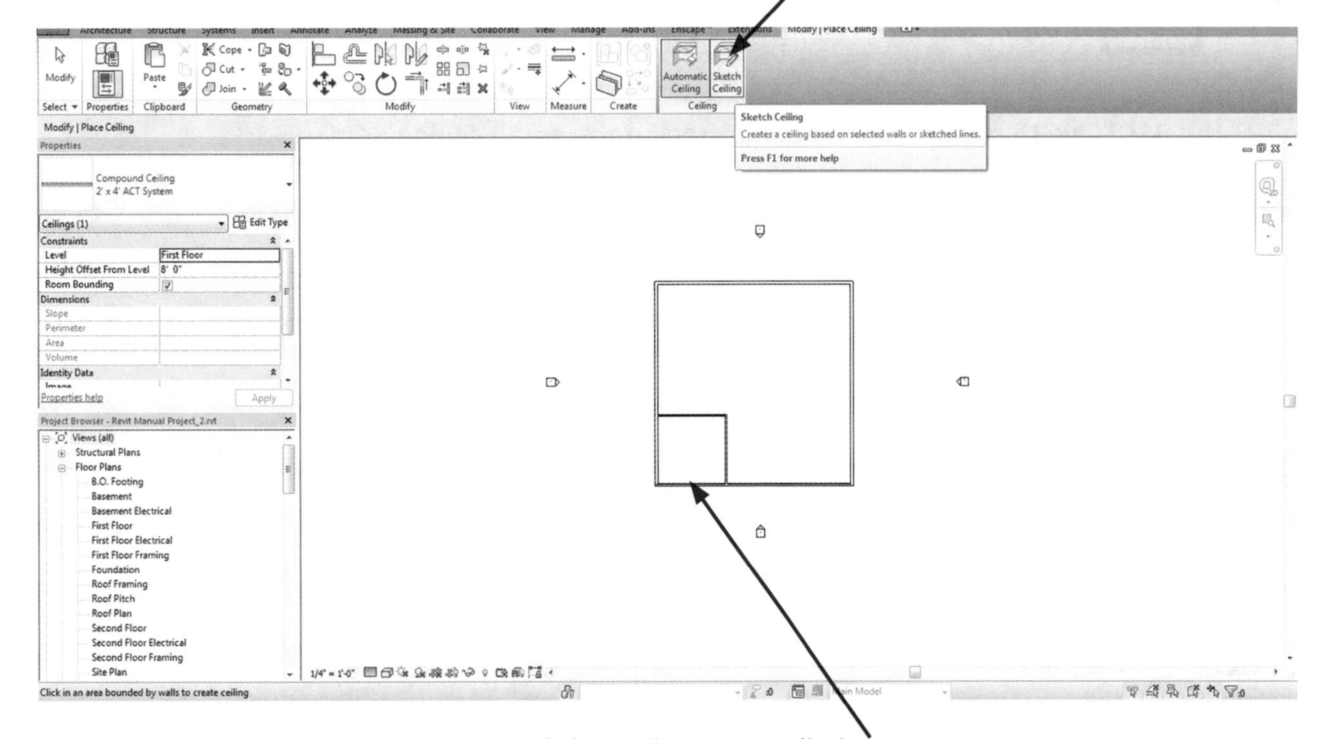

Here you will trace the new location of the ceiling you will place.

Once you select the Sketch Ceiling mode the ribbon will change to Modify Create Ceiling Boundary on the ribbon to allow you to trace your new location, with either the **RECTANGLE** or **LINES** commands.

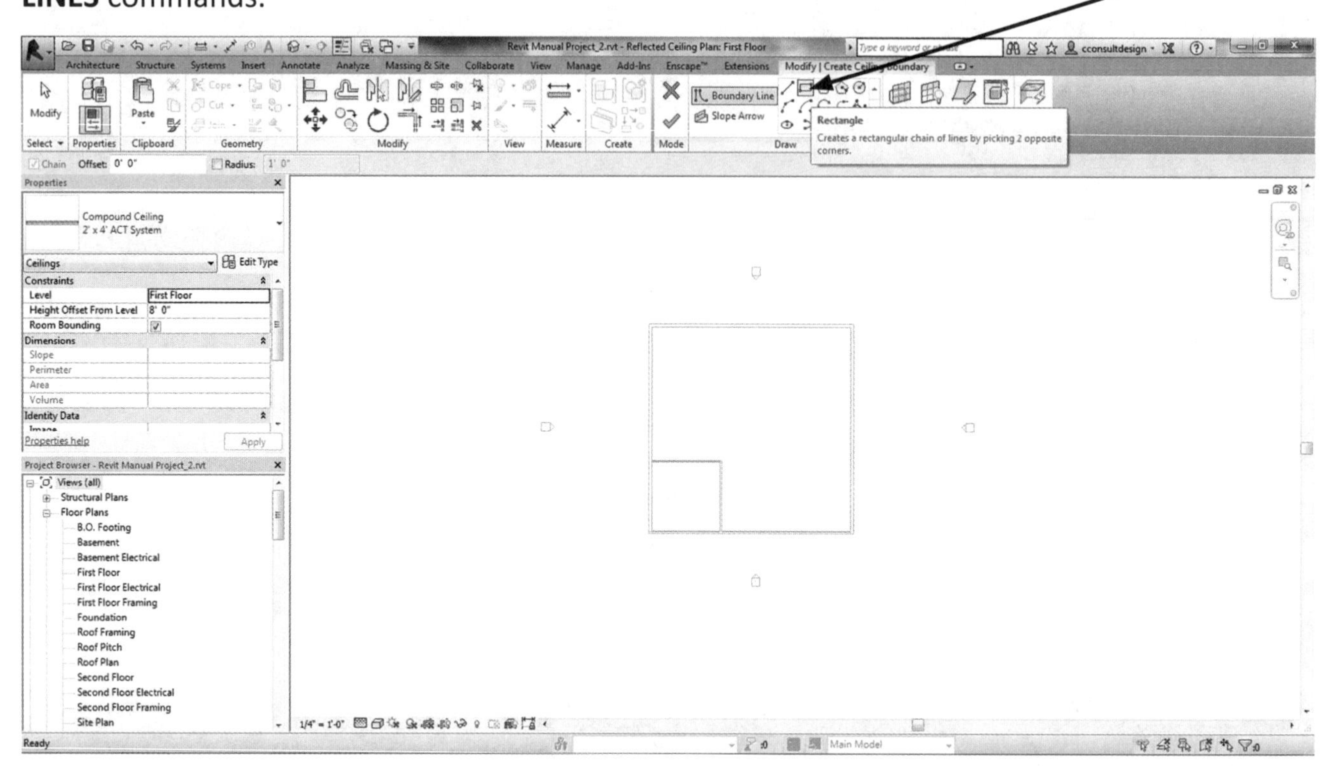

NOTES

If you need to zoom in to see where you are going to begin tracing the ceiling in a smaller space, you can use the wheel on the mouse and scroll in and out to get a closer view to become more precise.

Select the **RECTANGLE** command and begin tracing the small ceiling area.

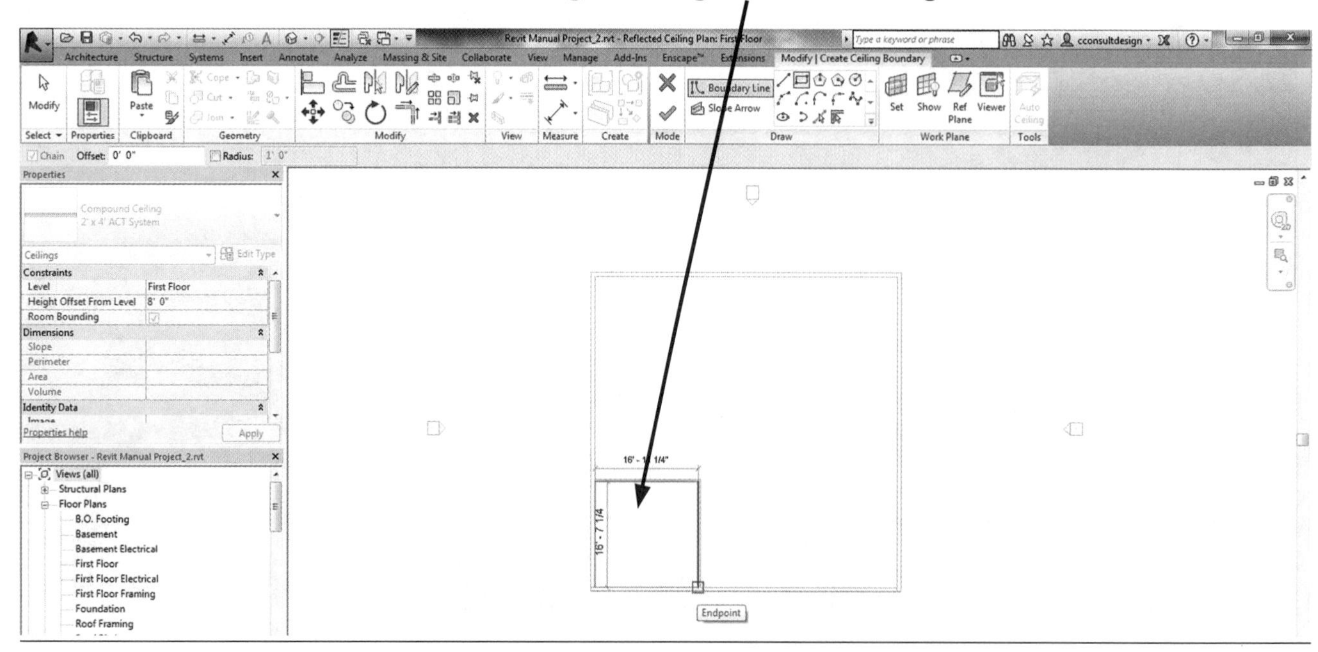

Once the new ceiling boundary is traced click on the green **CHECK MARK** to finish the command.

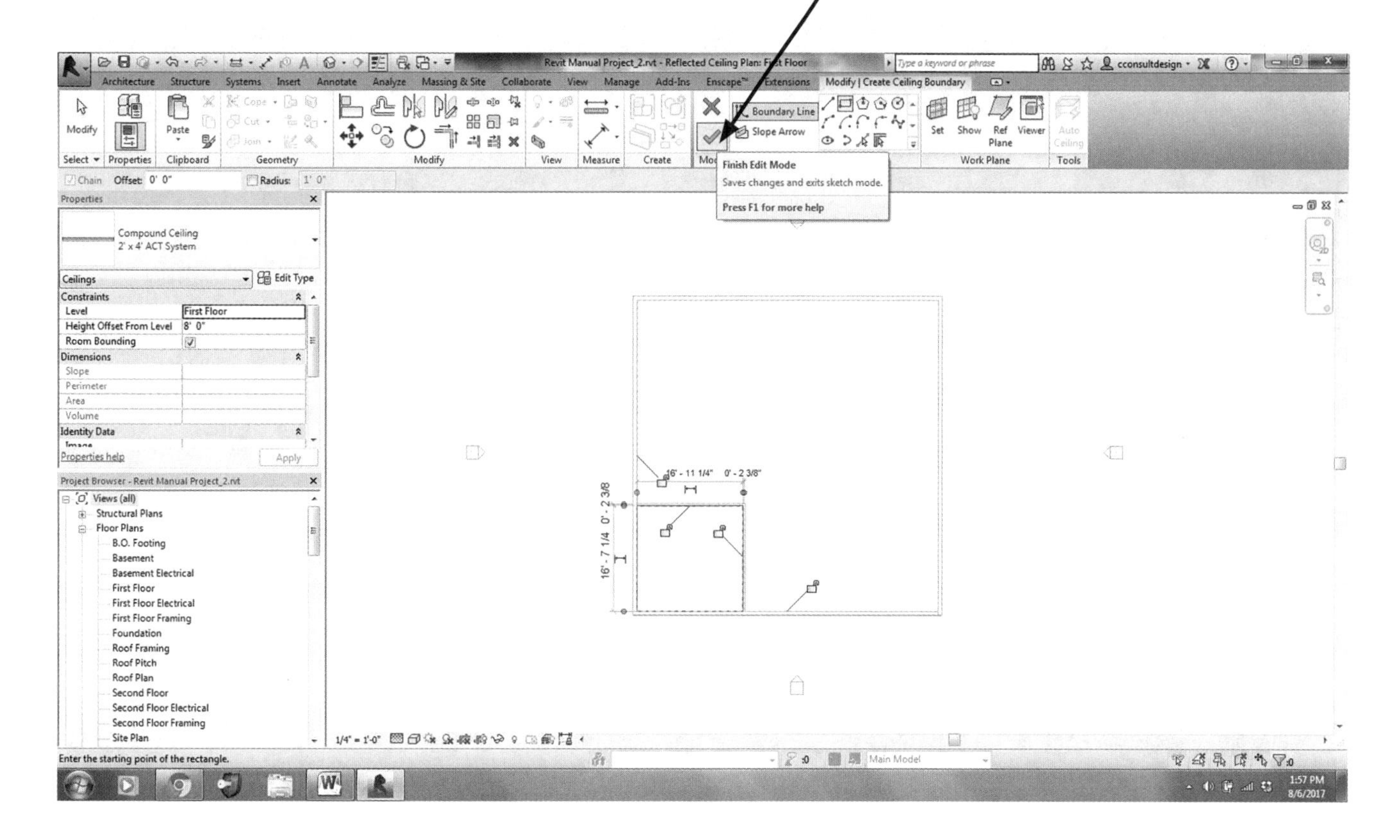

The completed drawing or ceiling will look like the following sketch with the new ceiling type installed.

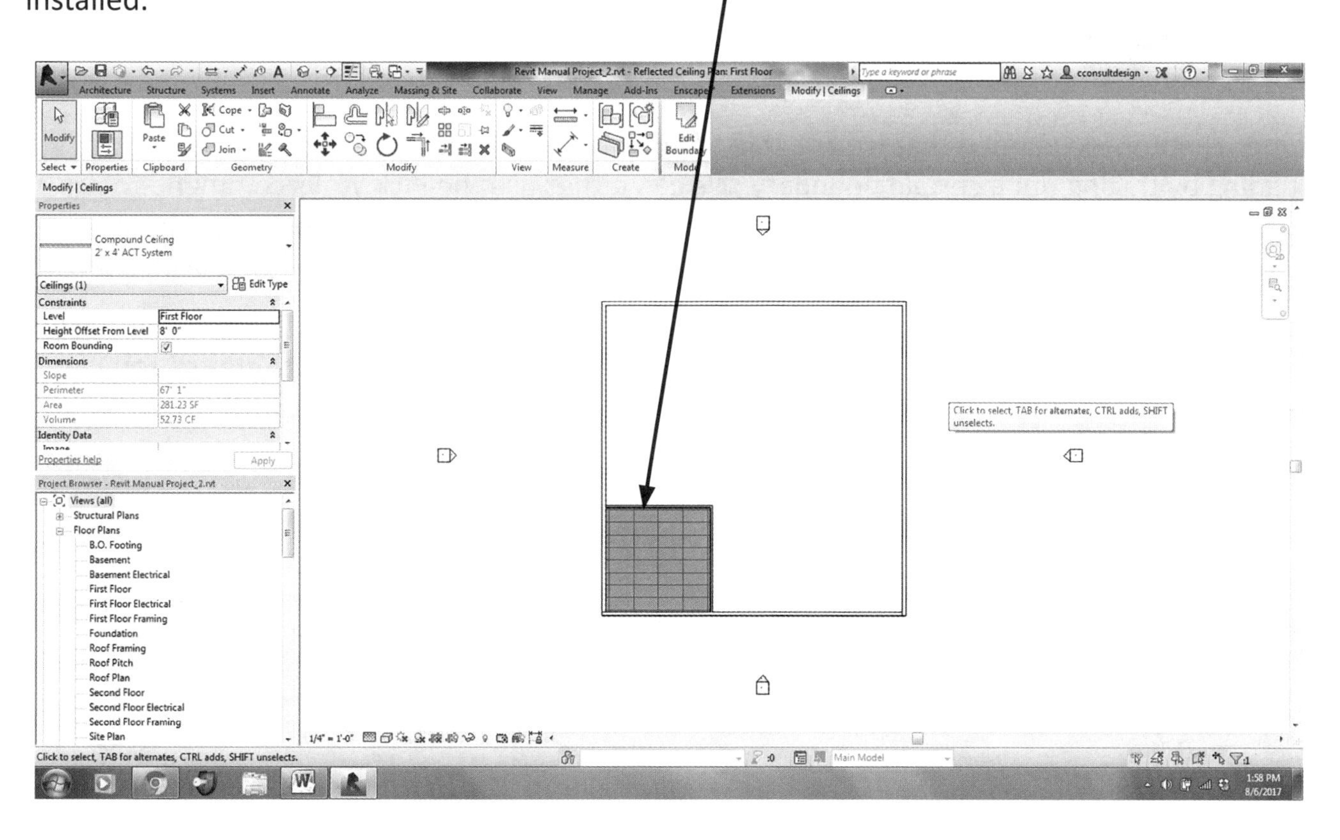

Click the **MODIFY** Arrow to clear the drawing and to begin another command.

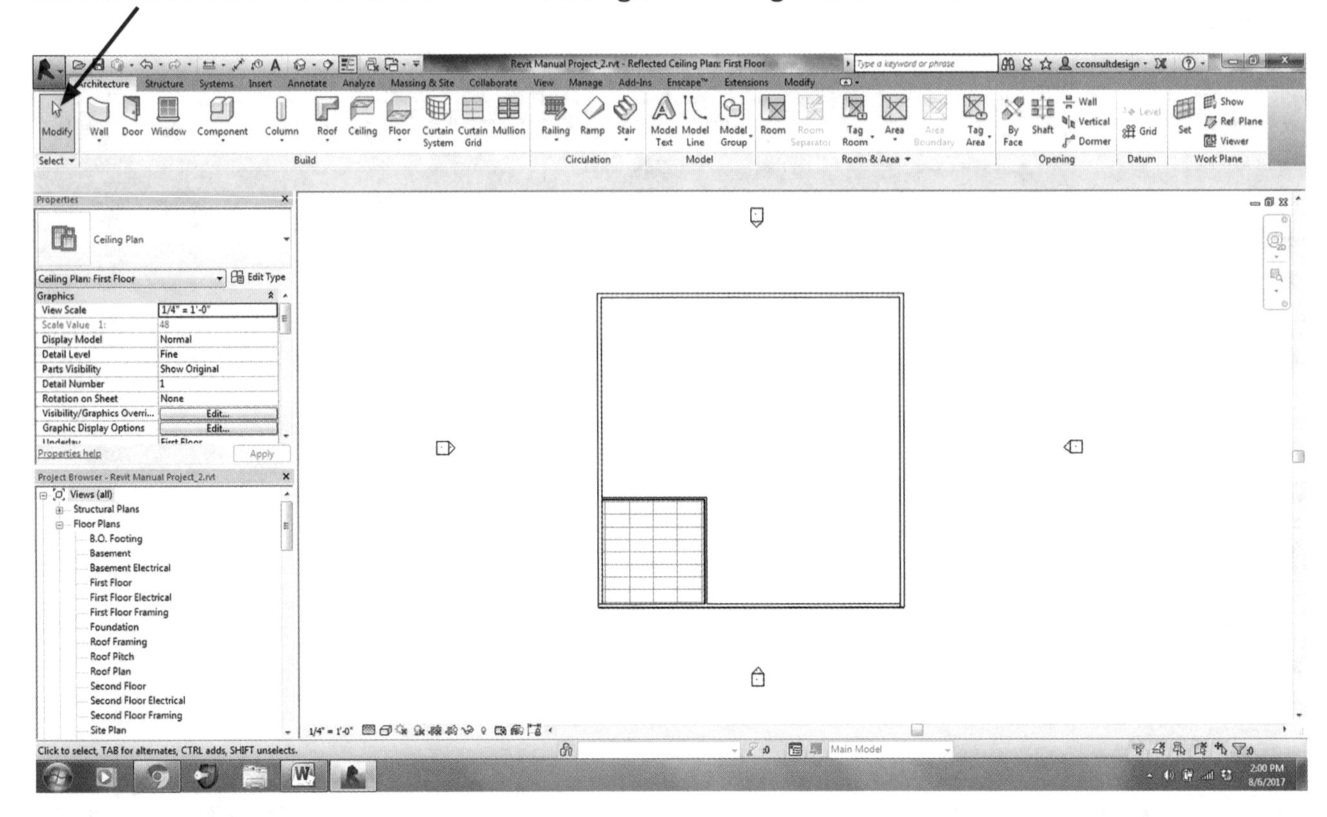

REMINDER

Do not hit the* ESC *Key, as this will undo everything you've just done, and you will likely need to start over again.

In this chapter, we learned that we are able to add different types of ceilings in a house or building by tracing out a certain boundary. We can either use the Pick Walls command to quickly select a particular area, or the Automatic Sketch to select the entire space. This makes ceiling design more efficient.

Let's take the next step in Chapter 7, *Creating Levels or Datum Lines in Revit*, where you will learn about adding a second level to your house or building, which helps with stairs. You'll learn much more about stairs in Chapter 8.

CHAPTER 7

Creating Levels or Datum Lines in Revit

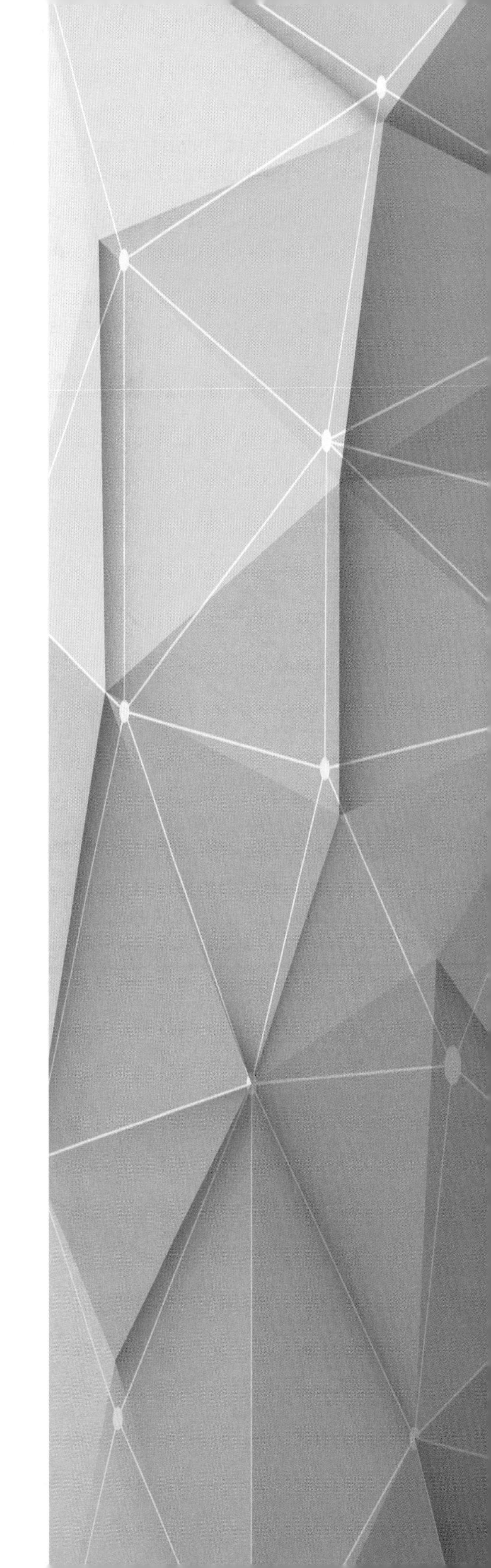

Revit defaults to having a Level created in elevation views. These Levels can be modified by changing the heights, or by adding additional levels.

Use the following steps to make sure that your levels are added to your project in the Project Browser and will correspond to the other views all at once.

The first step would be to open your project to the First Floor, or be sure you are in the First Floor plan view. In the Properties Box, scroll down to the Elevations (Building Elevations).

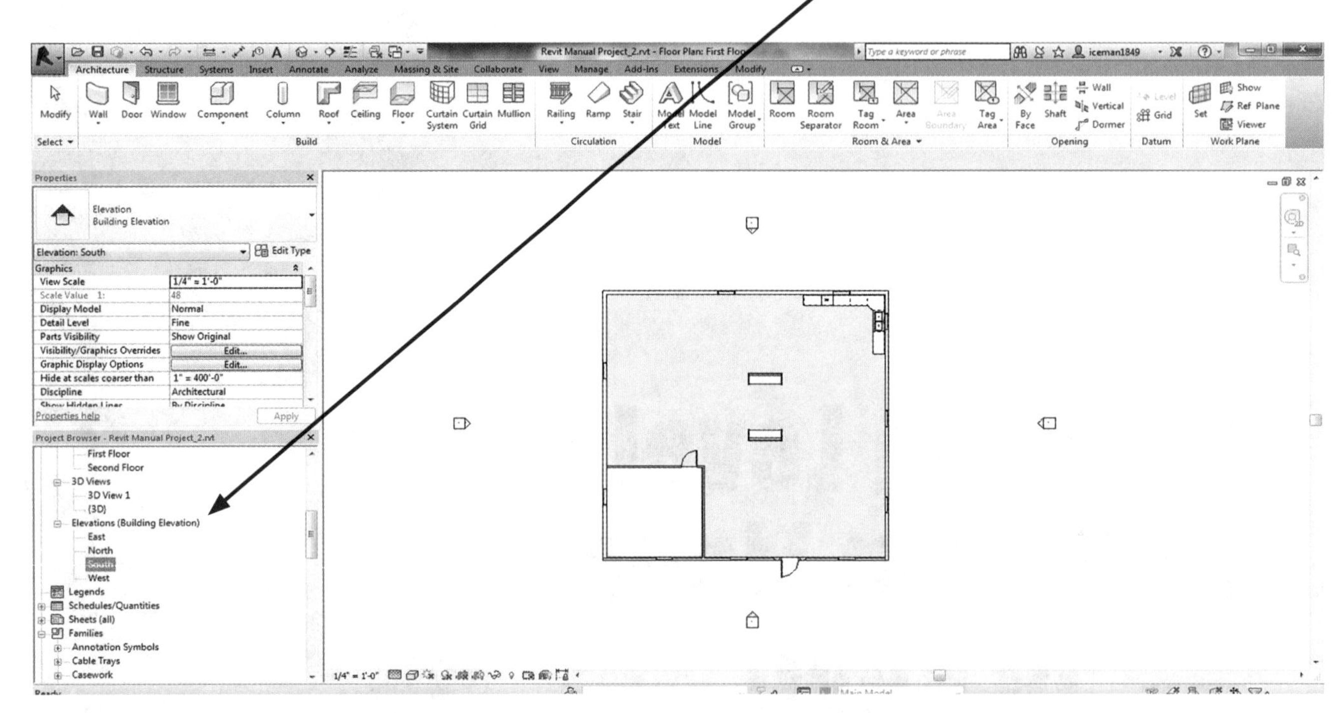

Click on one of the Elevation views. (I usually select the **SOUTH** elevation to get a better elevation view). Double-click on an Elevation view name, and the view will look like the following:

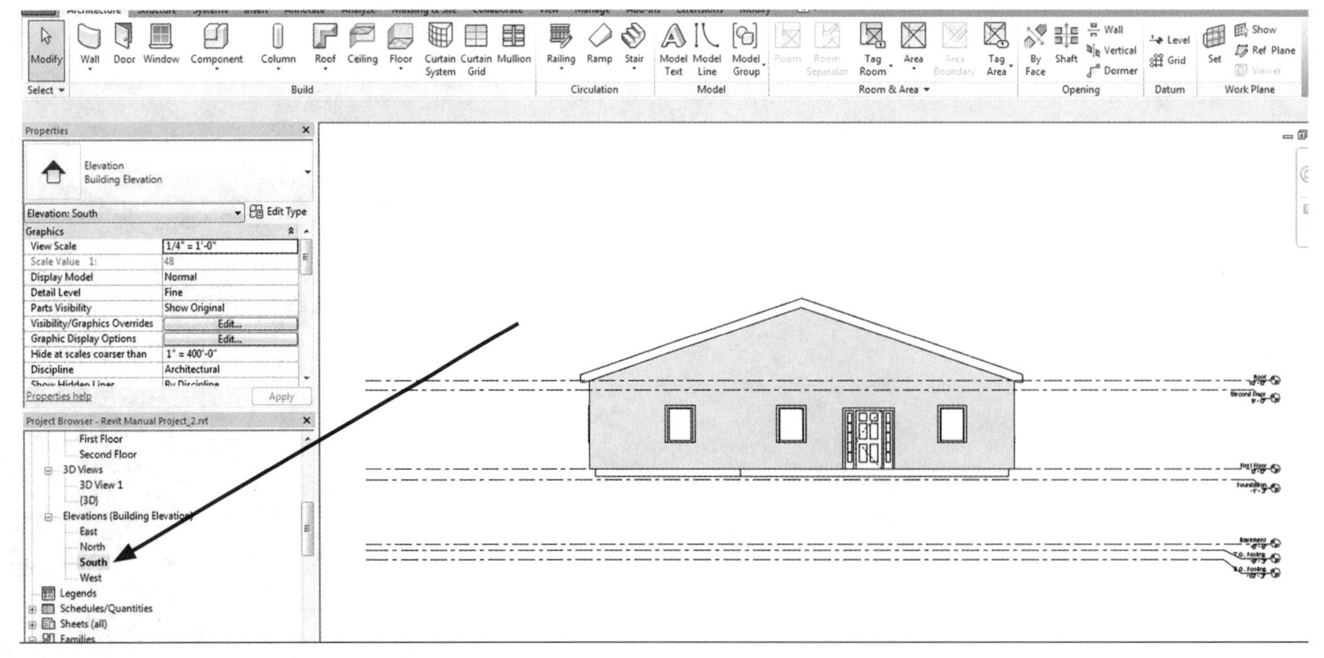

Once the view is opened, you can add a new Level to your elevation.

Make sure you are in the Architecture Tab and click on the **LEVEL** tile in the ribbon.

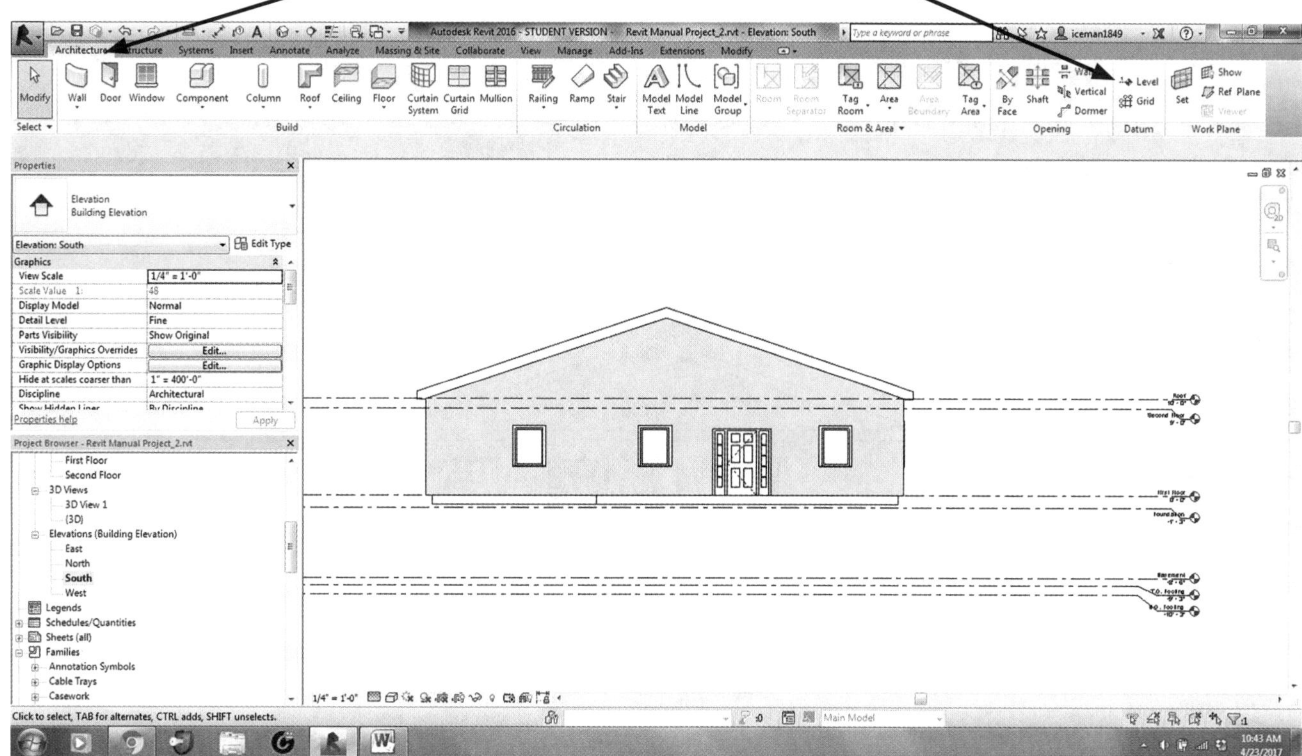

Once you have selected the **LEVEL** tile, the ribbon will change, and you will be able to add or draw a new Level in your view.

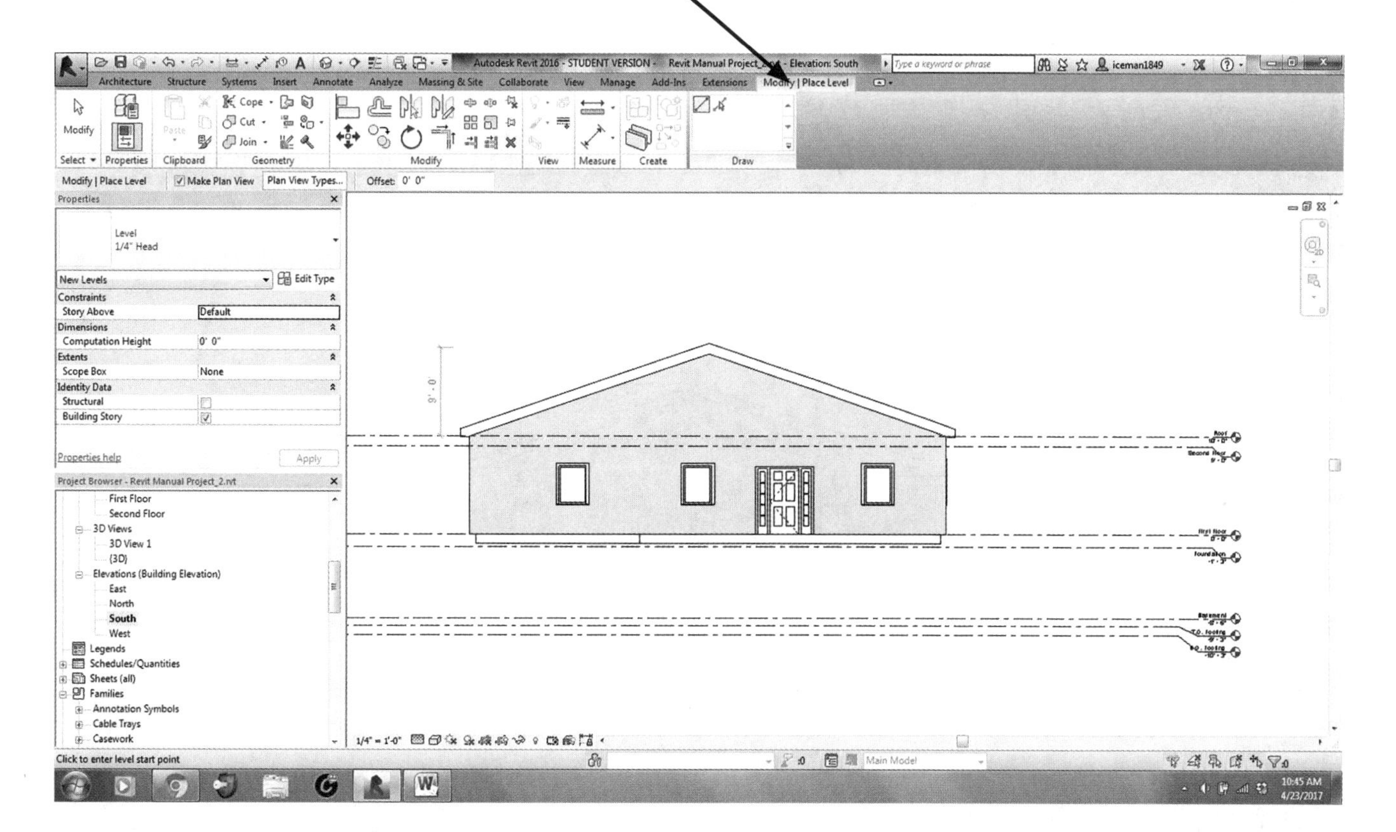

Once you bring your cursor down into the Drawing Window, notice that the Level defaults to a dimension (measurement) connected to a Level line.

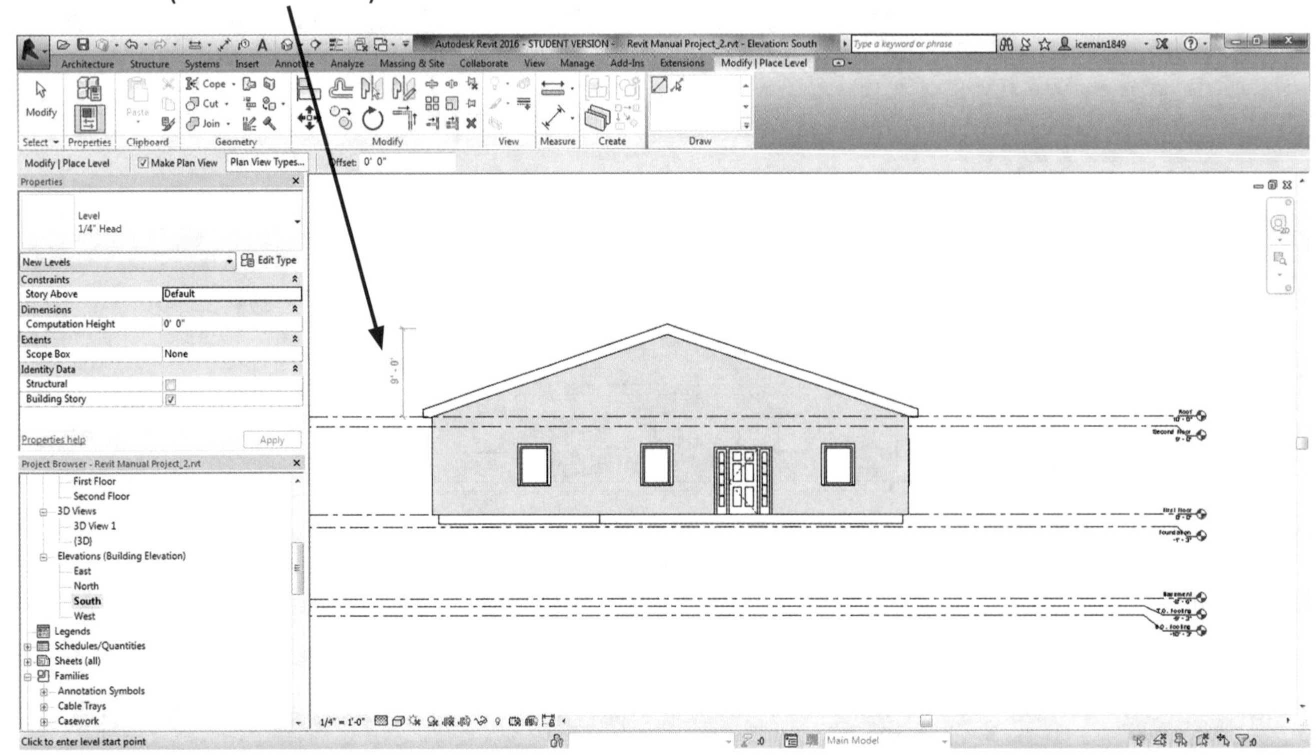

This allows for quick placement. Then you can modify the name of the Level and the height.

Begin to draw the Level line from the left side and drag the cursor to the right side. The Levels will automatically line up with the other side.

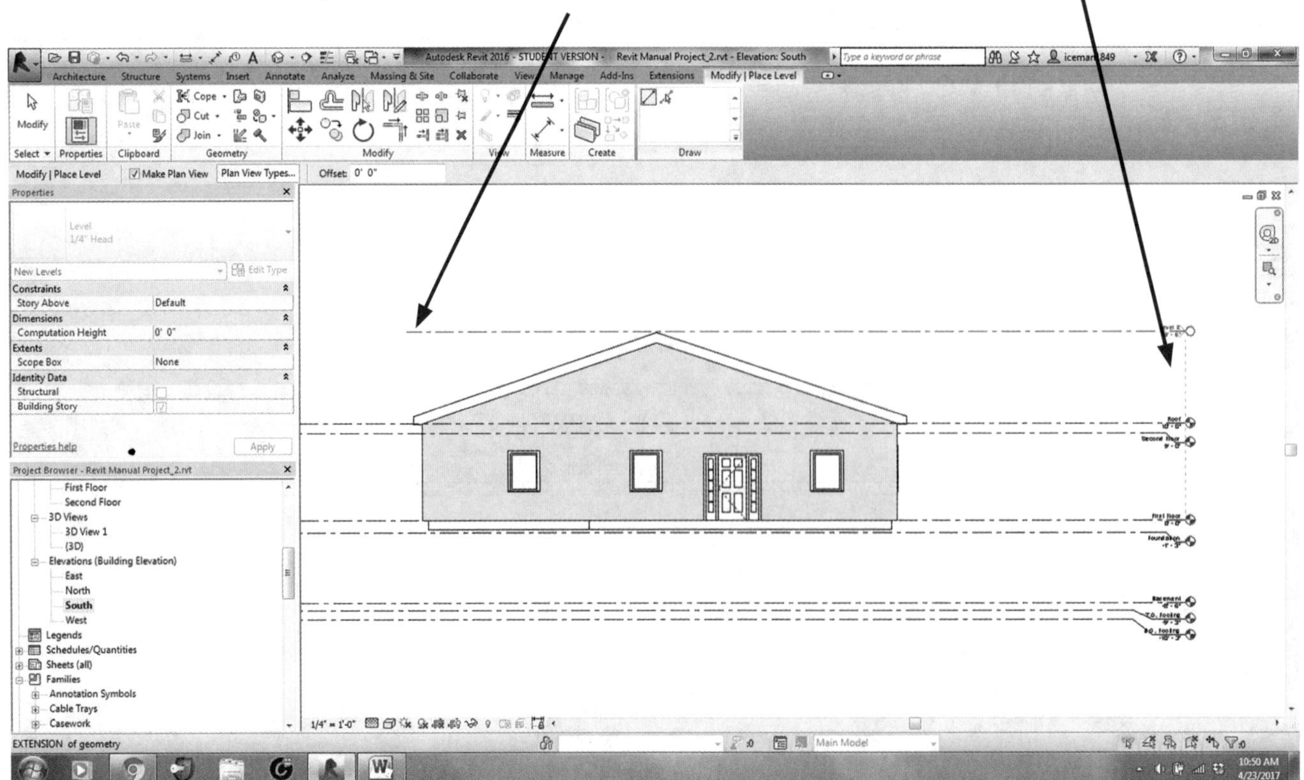

After clicking to end the movement, Revit defaults to begin drawing another Level line, or you have the option to end the command. To end the command, click the **MODIFY** Arrow.

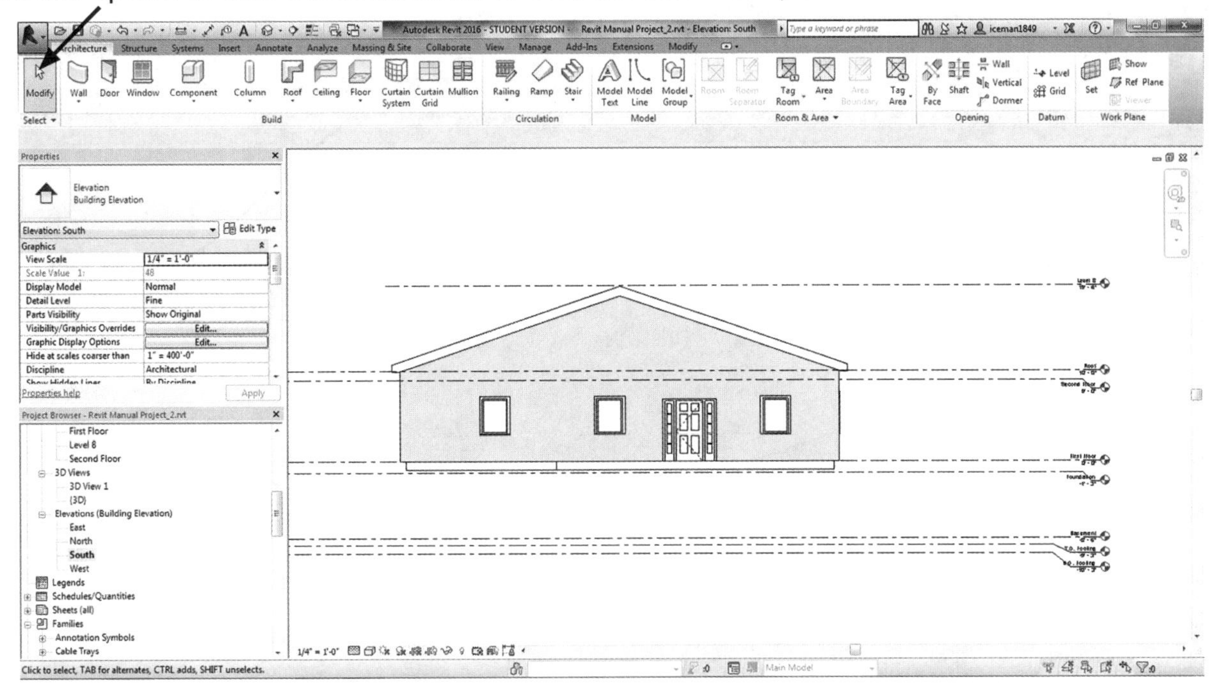

REMINDER

Do not click the ESC Key, or you will end more commands, and likely need to start over.

After ending the command, you can change the name and the height of the Level line created.

NOTES

You can zoom in to the level you want to change using the wheel on your mouse.

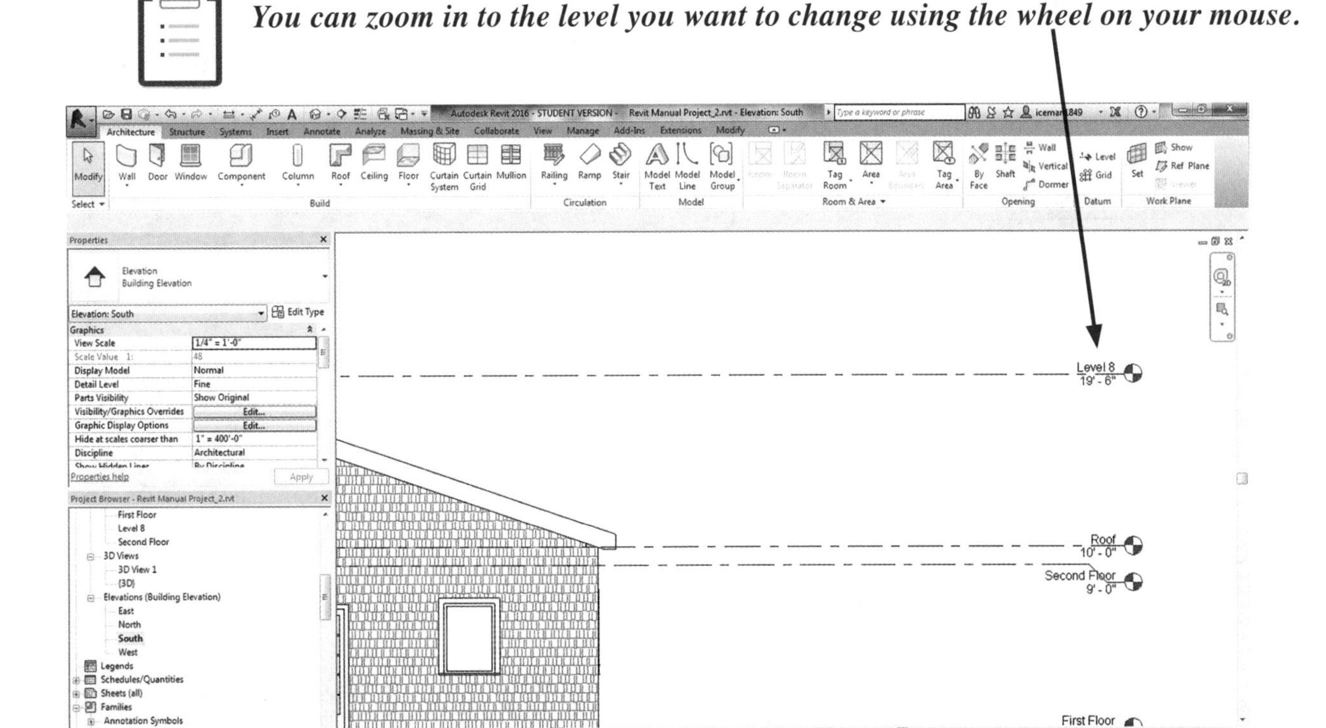

Double-click on the Level name, and change the name.

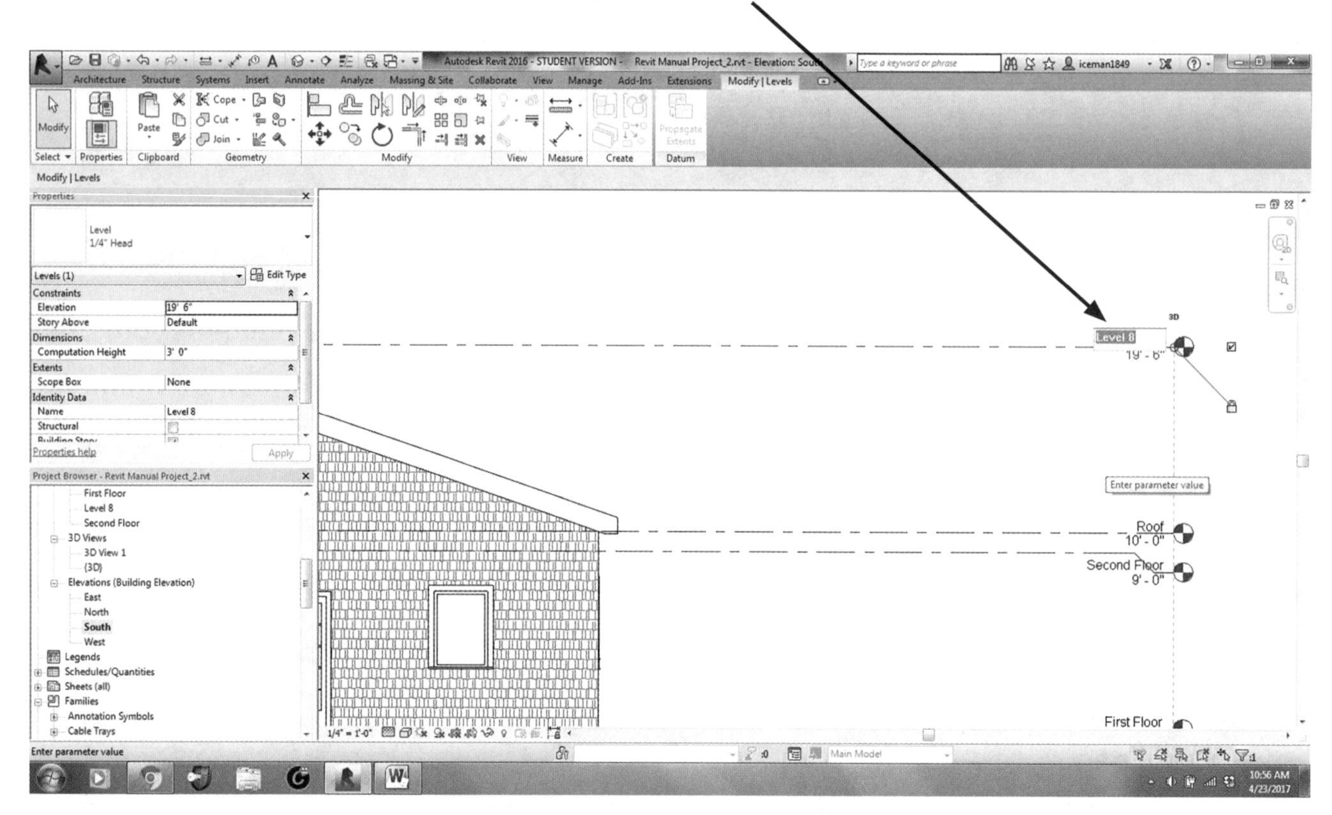

Once the name is changed, you can then change the height (if desired), and hit the **ENTER** key. A dialog box will appear with, "**WOULD YOU LIKE TO RENAME THE CORRESPONDING VIEWS?**" Click **YES**.

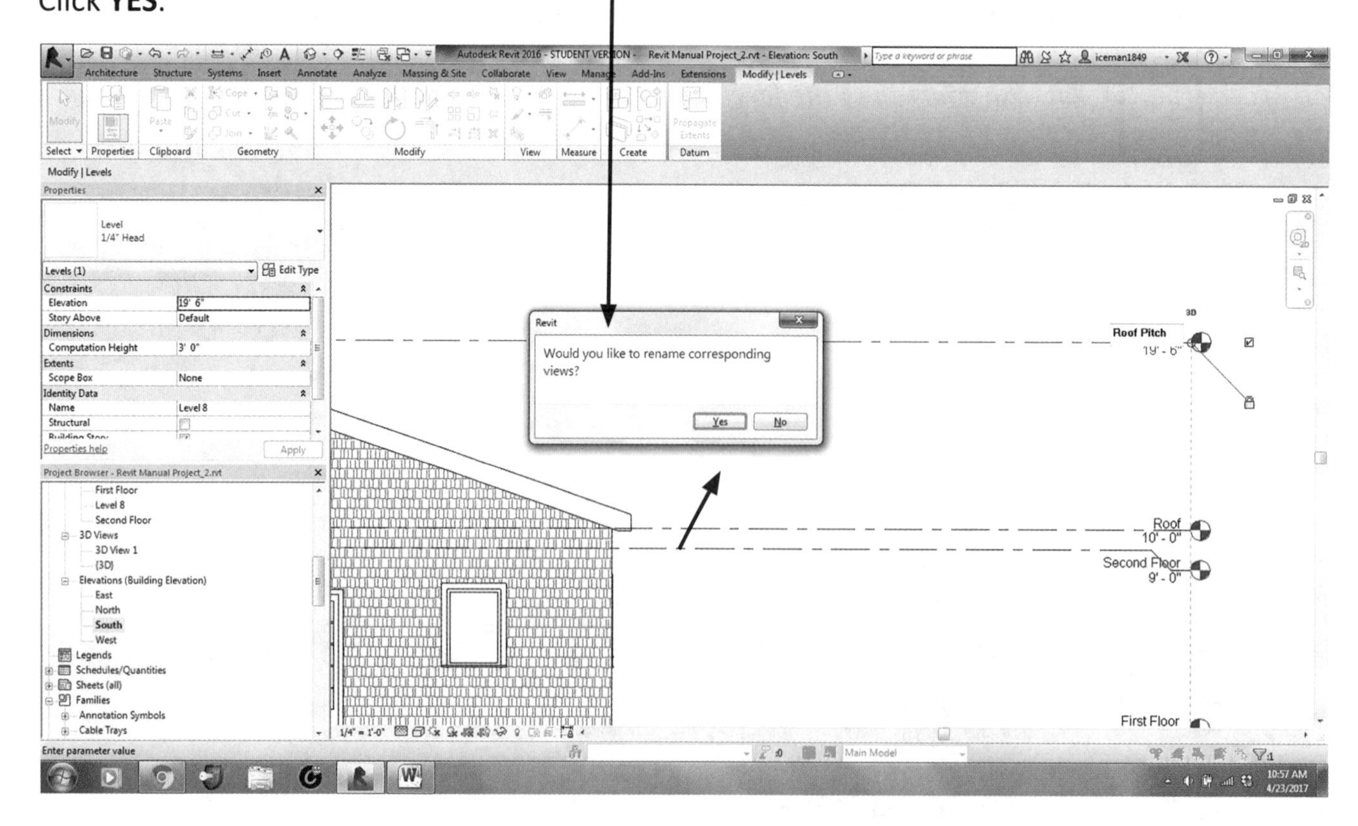

This will change the name of the new Level you added in your Floor Plans and also add to your Ceiling Plan views. Scroll in your Properties Box and see the new name view added.

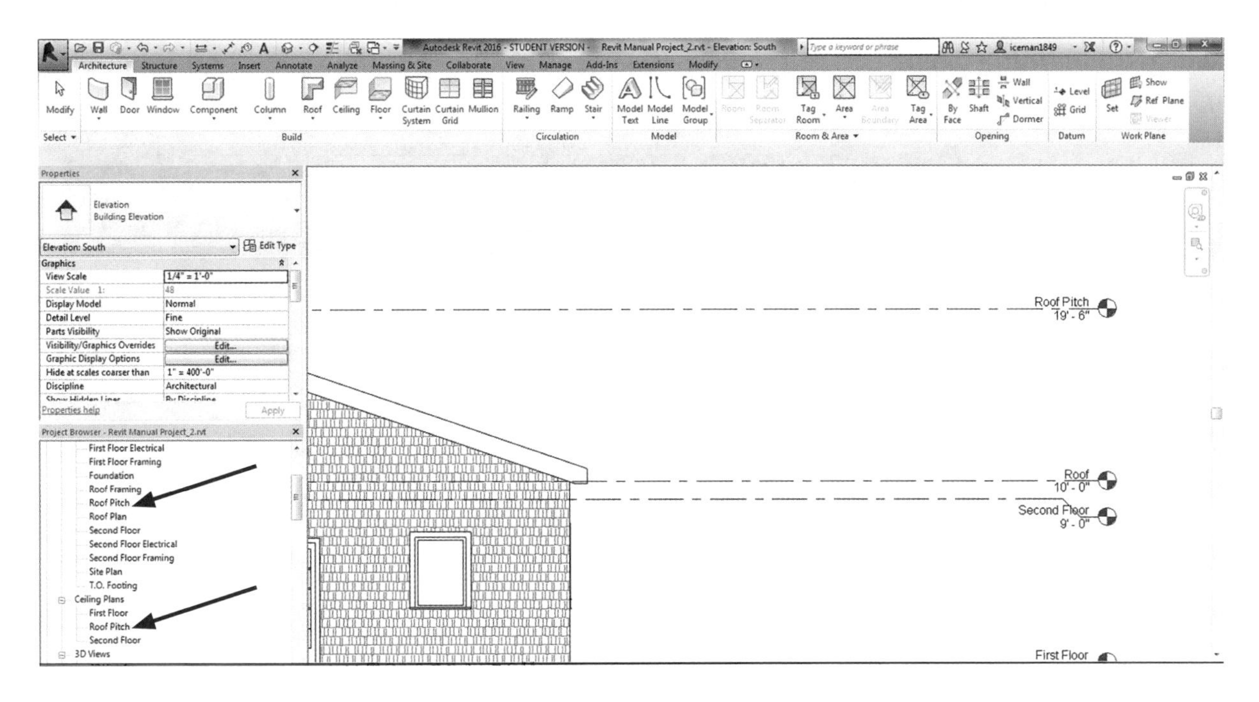

REMINDER

If you zoom out with the wheel on the mouse, you can see the complete view of the elevation.

In this view, you can change the elevation heights, change heights of footing placements or basements, or add other reference levels.

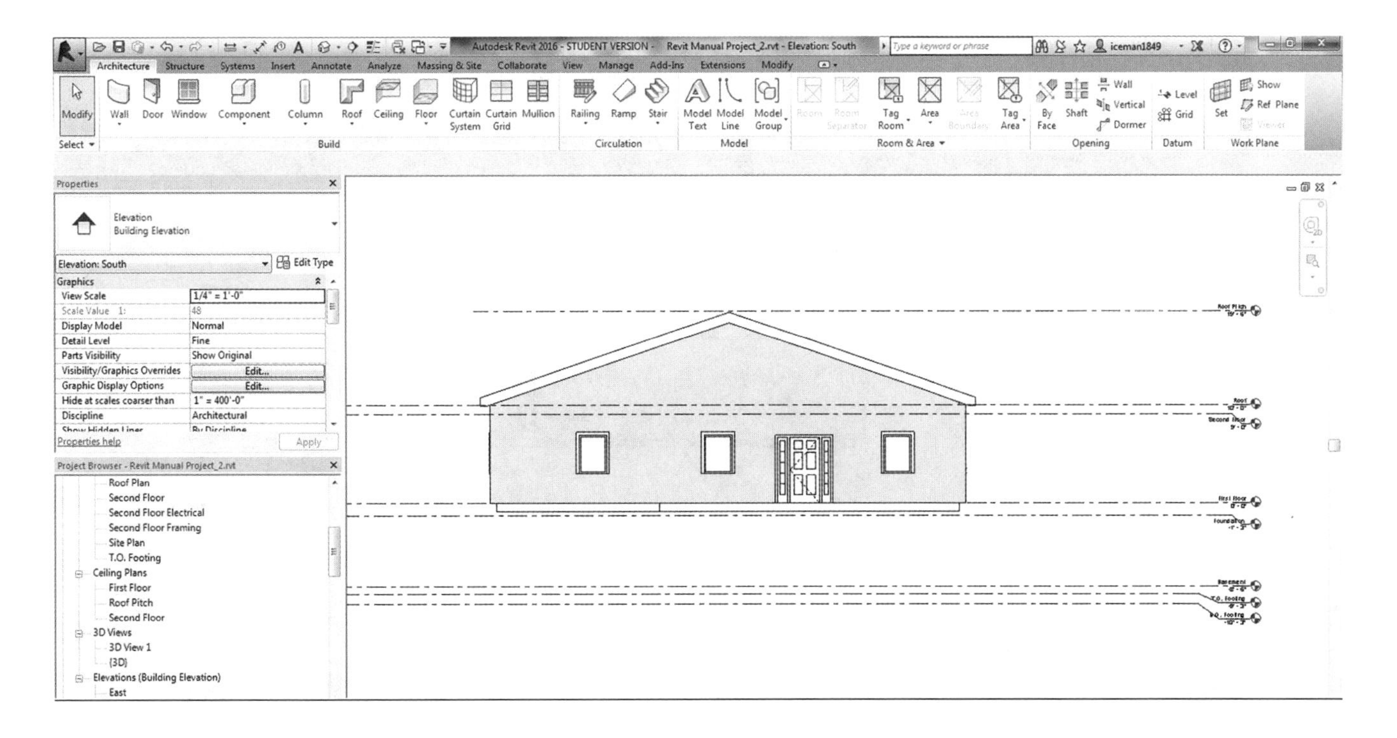

If you are not adding a basement to your house or building, you can click on the Level name and remove it by simply clicking on the level in the drawing selection window and clicking **DELETE** on your keyboard.

You will get a warning dialog box reminding you that if you delete this view then other views associated with that level will be deleted also.

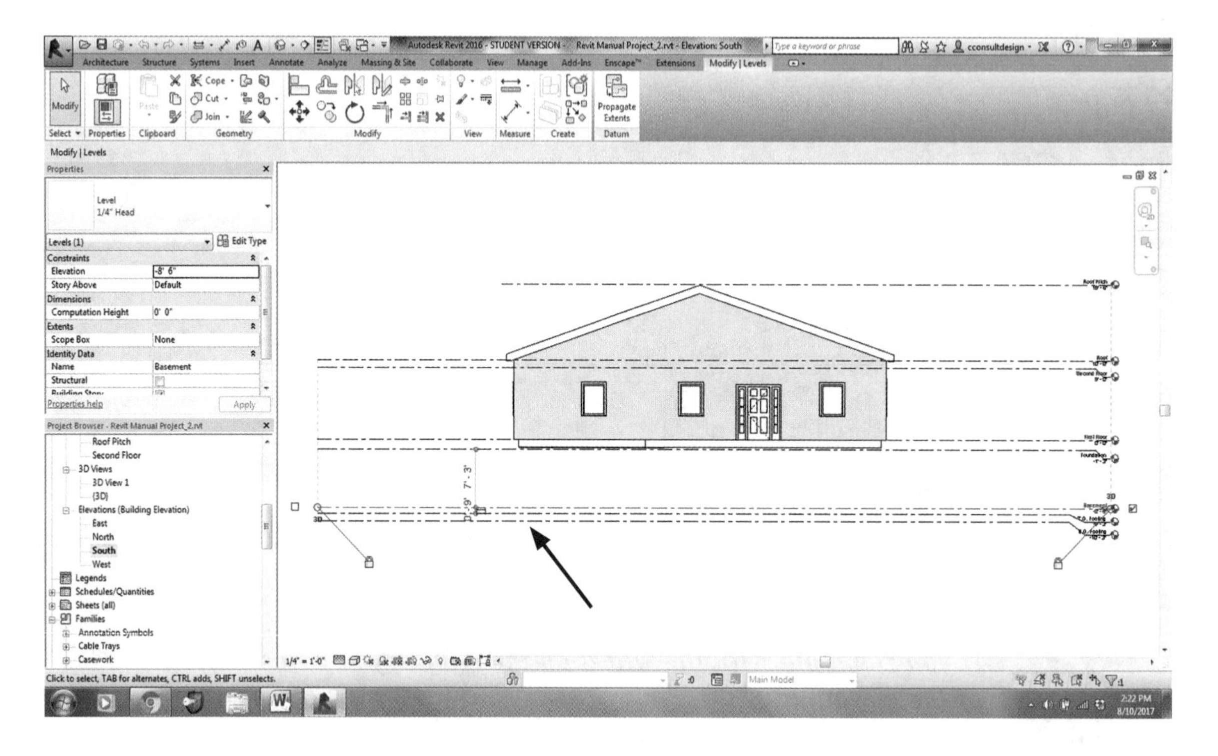

If you are sure you want to delete that level, click **OK.**

The level will be removed, and you can adjust the other levels to the specific heights you desire.

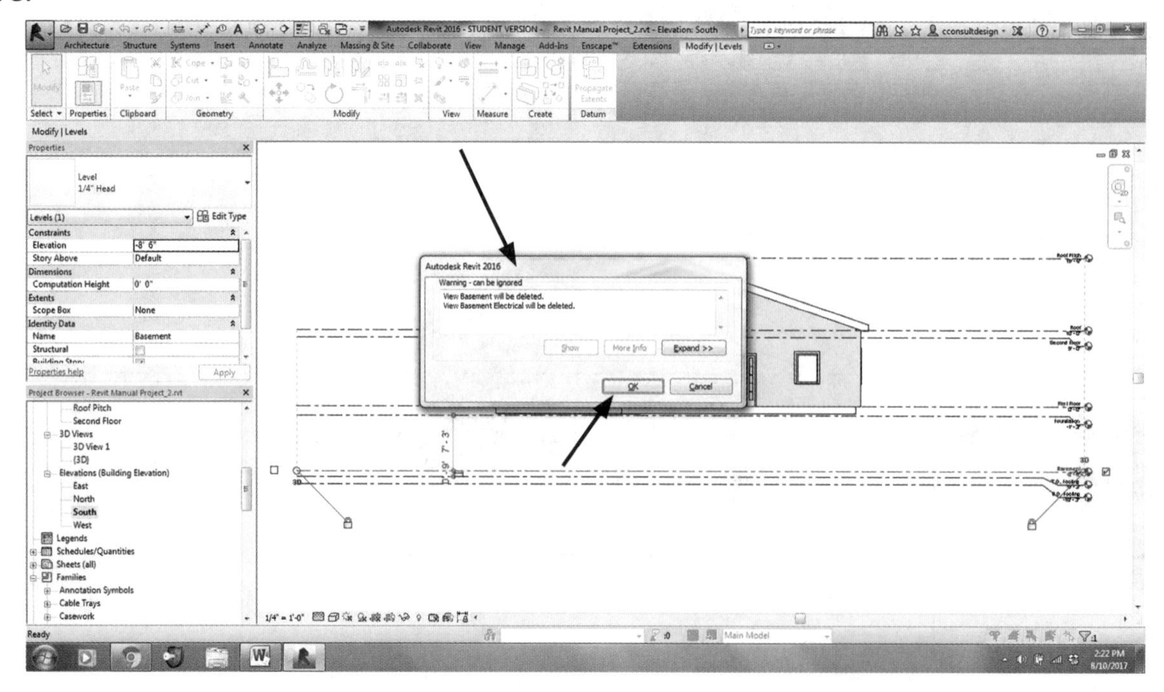

In this chapter, we learned that adding levels to just one side of the house or building elevation will affect the entire model. Adding new levels to just one side of the model will add the newly placed level to all of the other three elevations. This drastically cuts down on the time spent designing the different levels, and shows how much easier it is to add a second floor or multiple stories to your project.

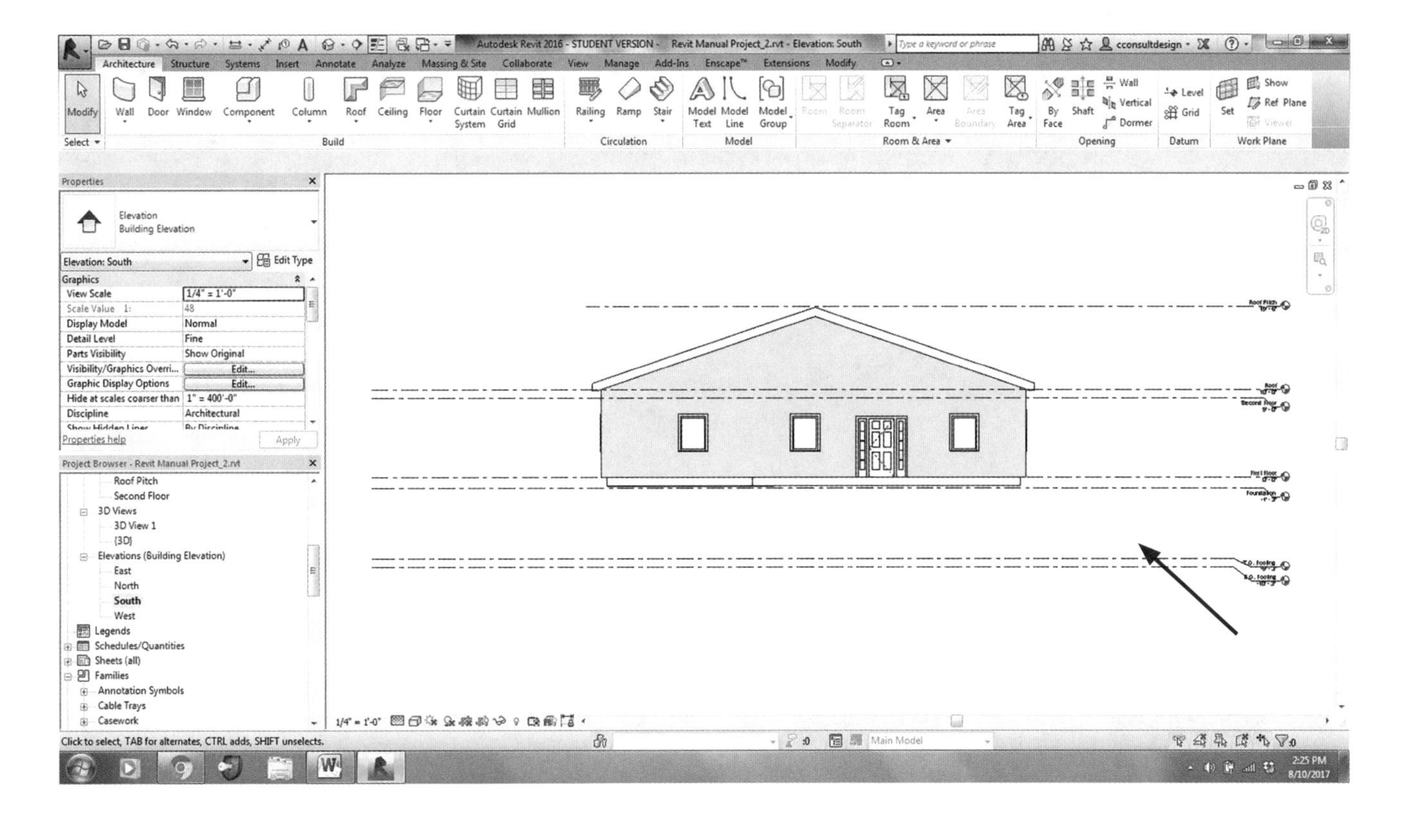

Now let's take a look at Chapter 8, *Placing Stairs in Revit*.

CHAPTER

8

Placing Stairs in Revit

PLACING STAIRS IN REVIT

Placing a set of stairs in Revit doesn't have to be a difficult operation. Stair placement is where a lot of users get aggravated. They generally first draw the area where the stairs are to be placed, like a stair chase or stairwell, then try to create the stairs in that little space.

The following steps will take out the worry of trying to place stairs in the small space and allows you to create them with ease, then move them into place. Since Revit automatically knows the heights from floor to floor (because of the levels you created), that part is already accomplished, and you don't have to worry about it.

Let's begin. Make sure you are on the First Floor Plan view in the Architecture Tab, and select the upper section of the **STAIR** tile, on the ribbon.

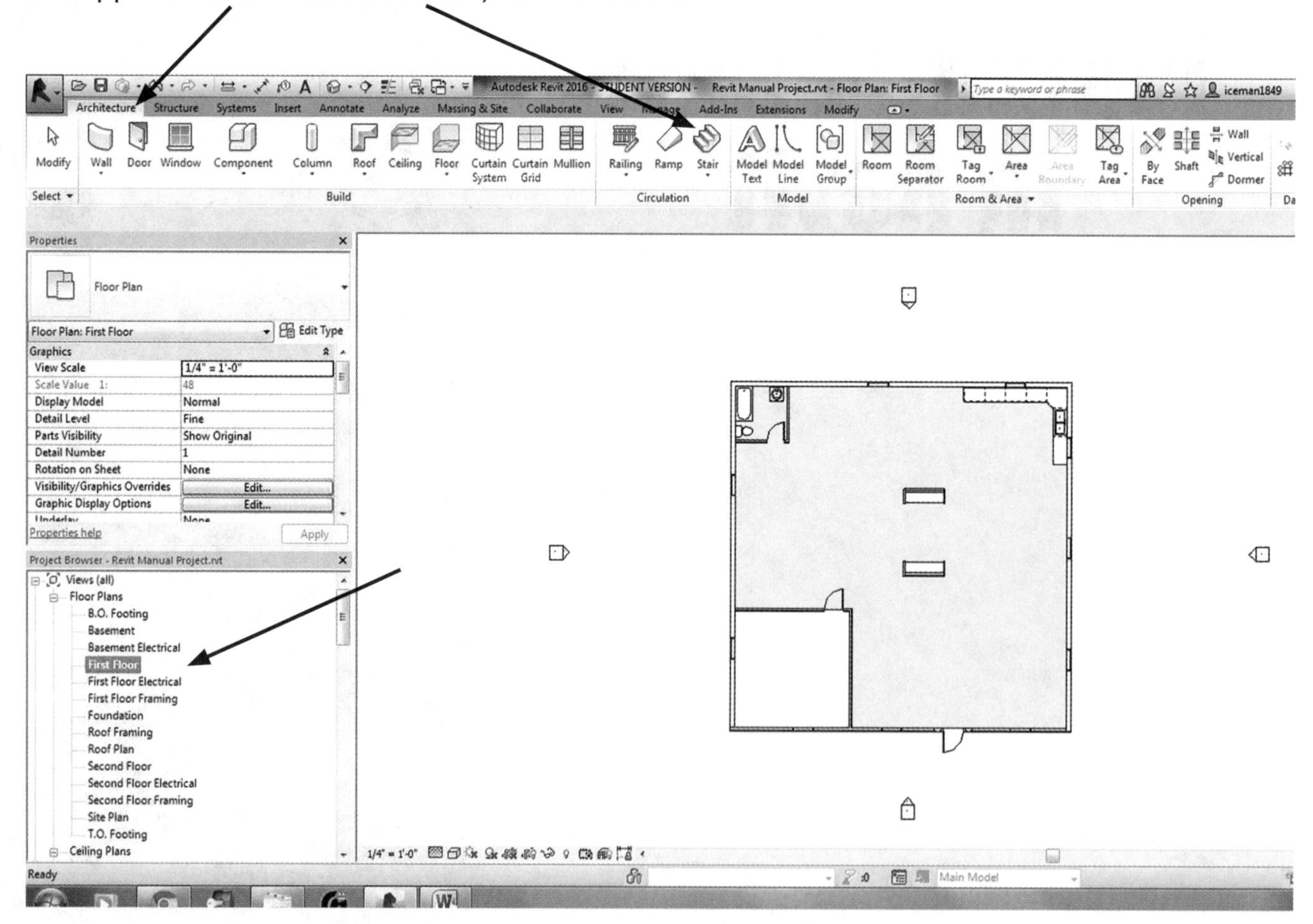

Notice the ribbon will change to the **MODIFY STAIRS** option.

The Options Bar defaults to Location Line: **RUN CENTER, RUN WIDTH**, with an **AUTOMATIC LANDING** checked.

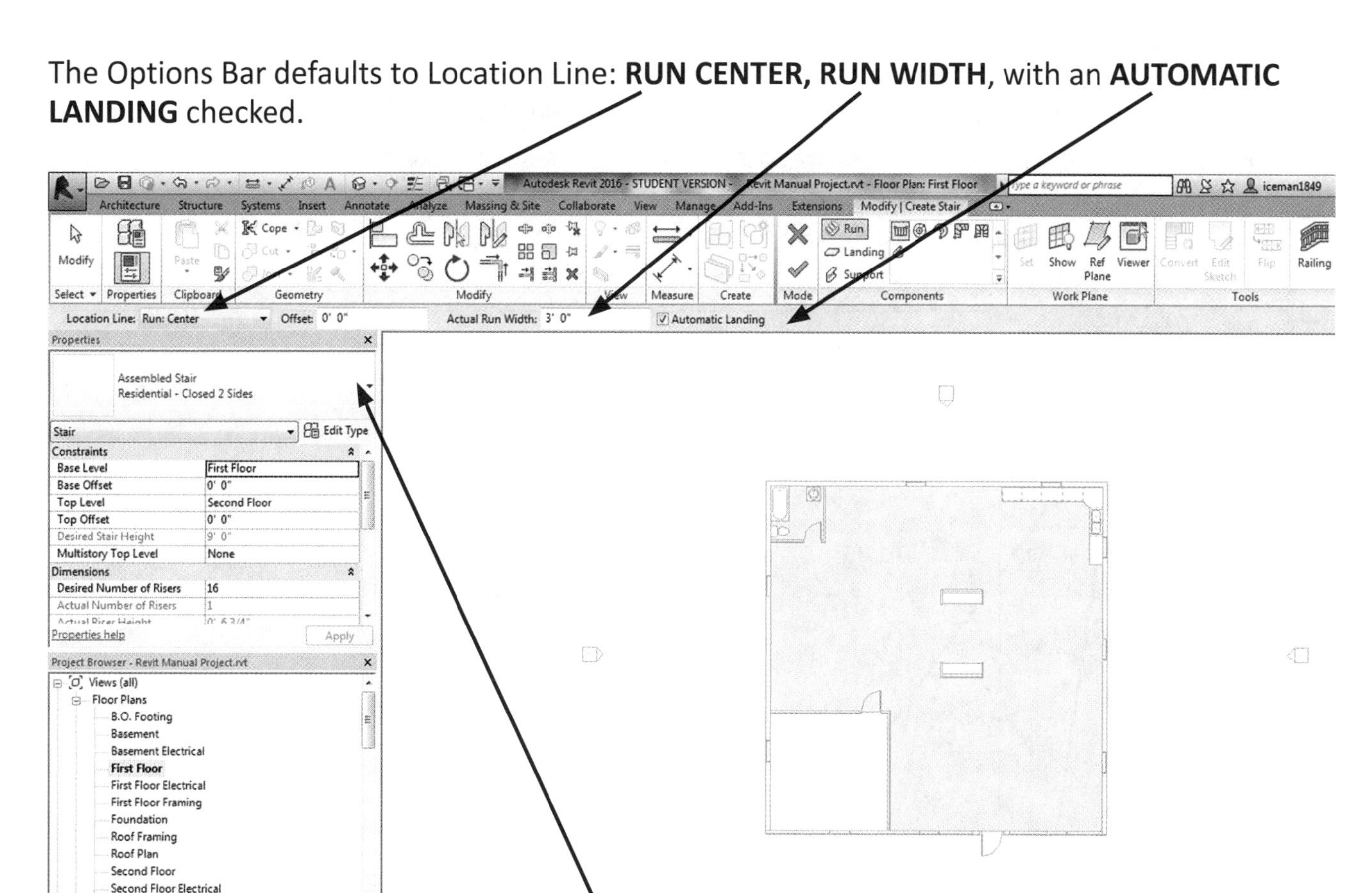

Notice that this stair in the Properties Box defaults to **CLOSED 2 SIDES**, meaning there are handrails on both sides. The drop-down box can change these Stair types.

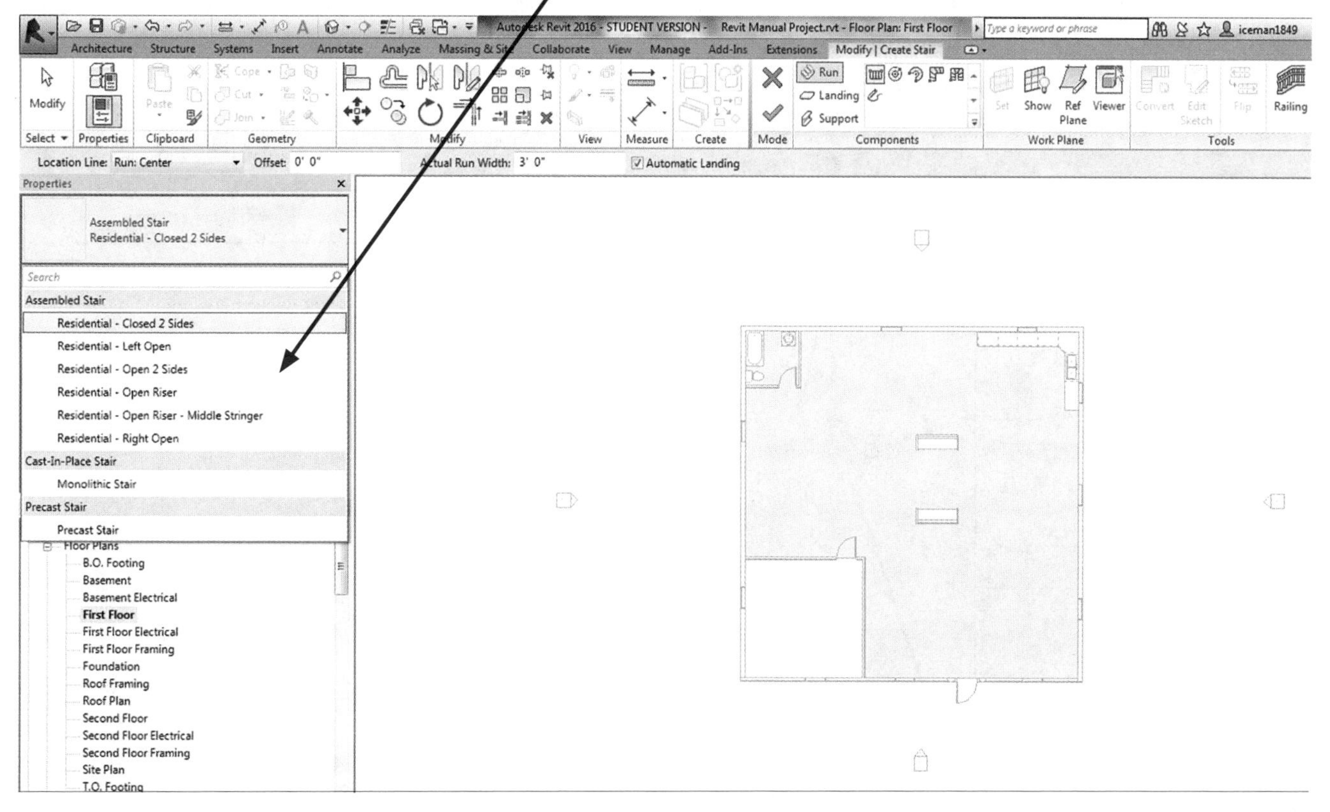

Once you select the stair type you desire, then you can begin to sketch your stairs. In this example, we will use the **ASSEMBLED RESIDENTIAL – CLOSED 2 SIDES**.

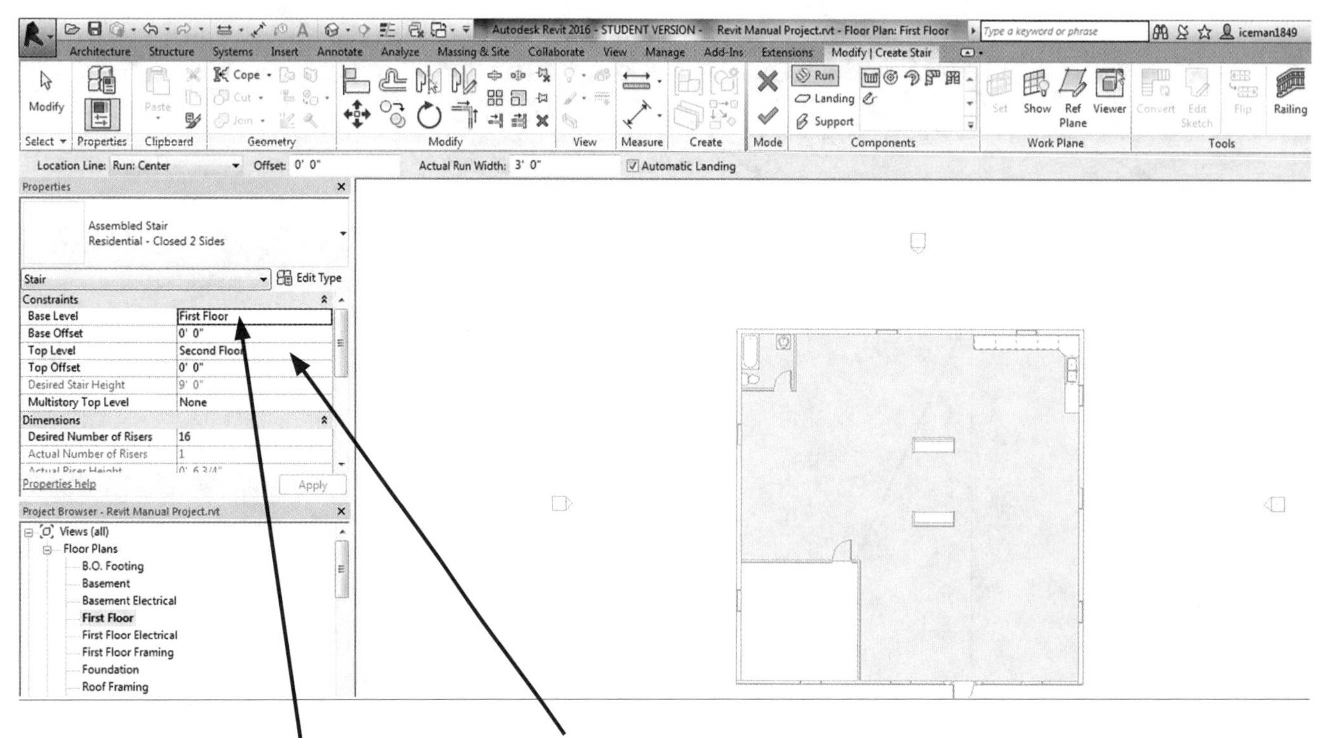

Notice that the Base Level and the Top Level are named by the level that was created earlier in the elevation views.

Now, instead of sketching the stairs in the house or building, the best and most efficient way to sketch the stairs to the begin *outside* of the project.

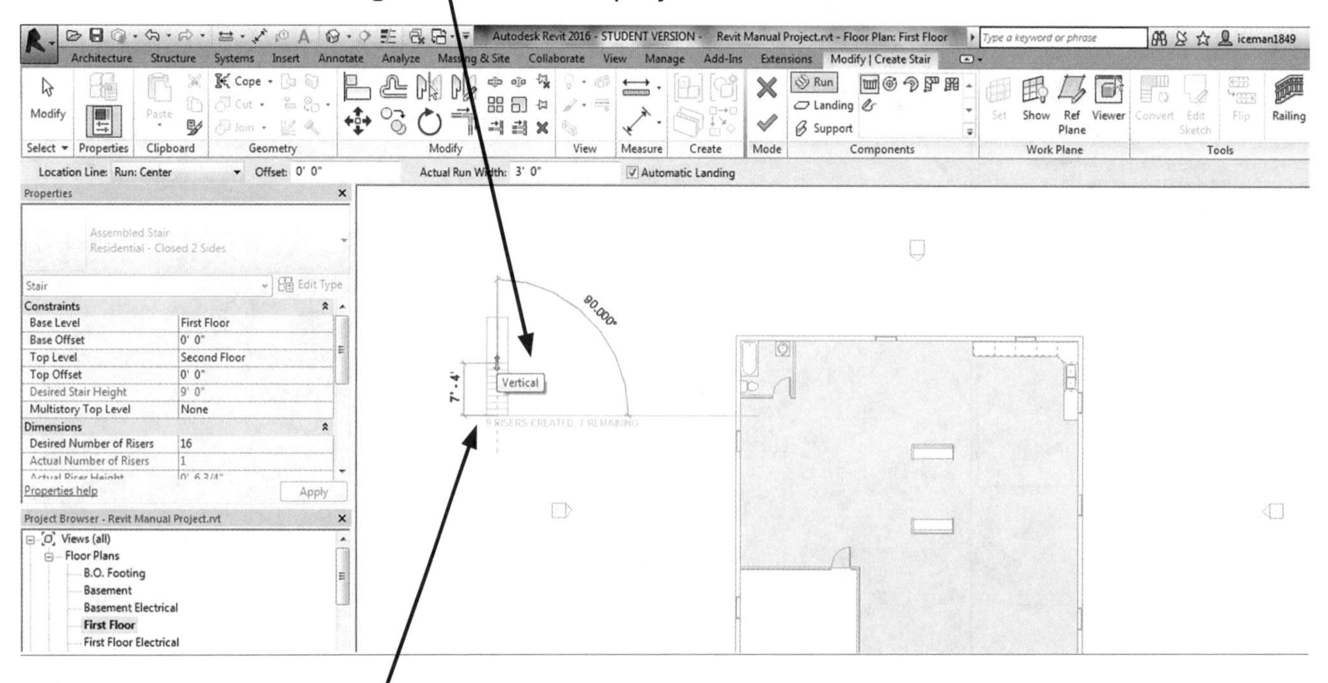

The Location Line is the run center of the stairs, as the levels have already been preset in Revit by the designer. Revit automatically calculates the number of risers from floor to floor.

Risers are the height of the step that you raise your foot up to in order to go to the next step. So Revit knows through the Building Code that there is a maximum height that the step height can be (no more than 8 ¼" high). And Revit will calculate the correct number of risers from the first floor to the top of the next floor (*not* to the ceiling). Revit does the calculation/formula for you so you won't have to worry about the aggravation.

At the bottom of the stair sketch, you will see the the number of risers already done, and the number of risers remaining. Then you can sketch the correct number of risers going up, and then sketch the risers for the other side. First click with the mouse to begin the location, then drag the steps in the direction you want to go.

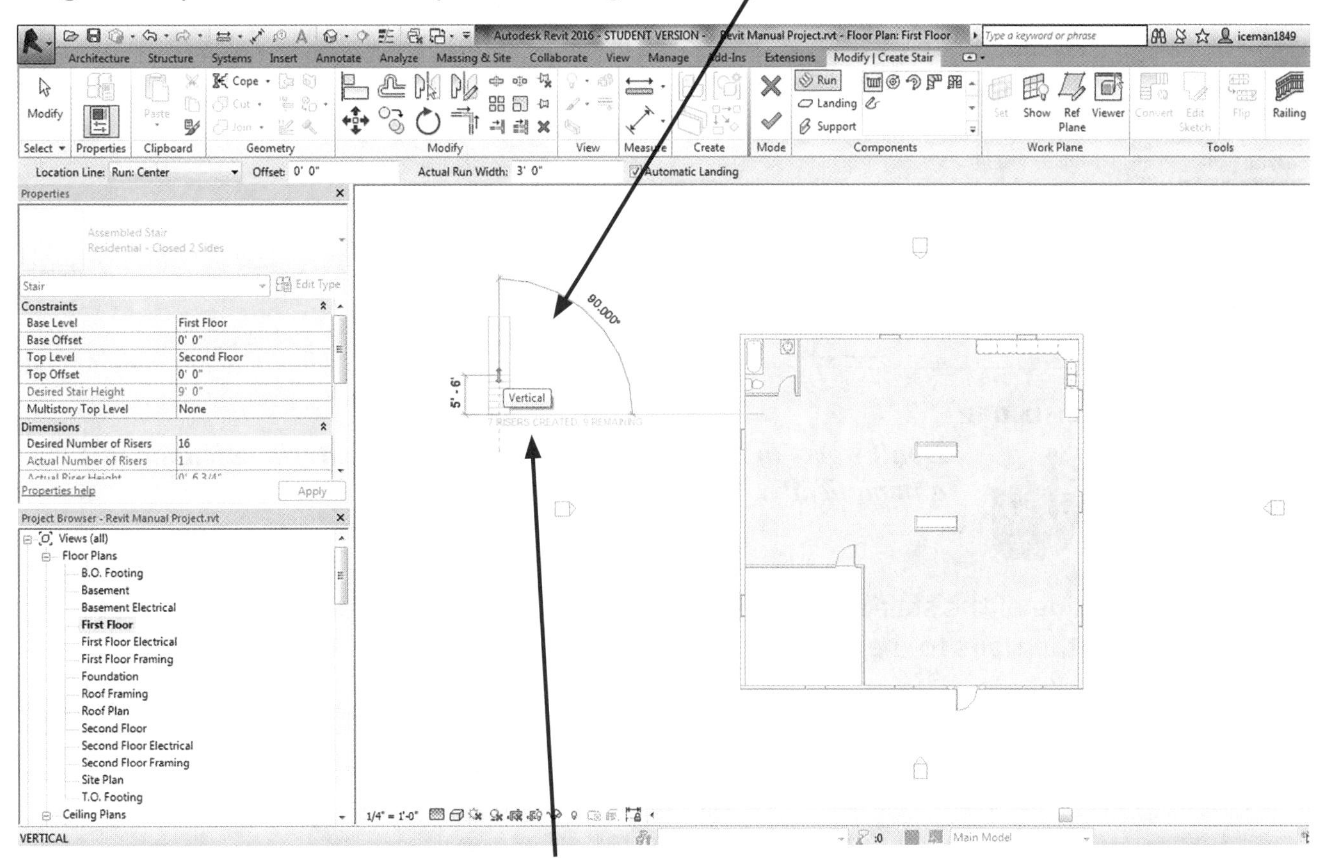

As you begin to place your stairs you will see that a statement appears explaining that you have a number of risers completed, meaning that you have created these steps and you have these number of steps remaining to continue to draw to get to the next level. This can allow you to continue to go straight up with a set of stairs or make a "U" shape set of stairs. Revit guides you to let you know how many more steps you have to complete to reach the next level or the next floor.

TIPS

Zoom in to view the sketch of the stairs to read the risers information.

REMINDER

You can use the wheel on your mouse to zoom in and get a more detailed view at any point in your design.

Once you have sketched the first side of the stairs, if you are going to create a "U" shape set of stairs, then pick a good distance away from the first set of steps and click with the mouse to begin the next set of risers going in the other direction.

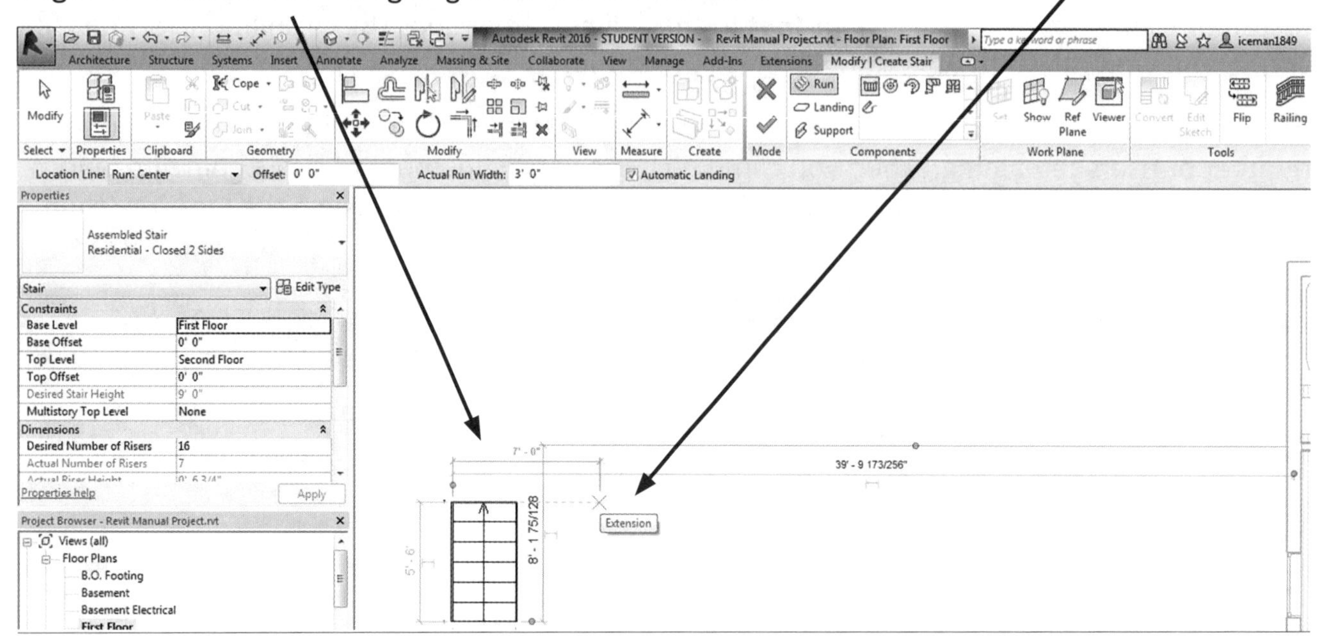

REMINDER

Don't worry too much about the distance. Revit will calculate this automatically.

When the other side of the stair sketch is selected, the landing will automatically be created once you extend the stairs to the last riser.

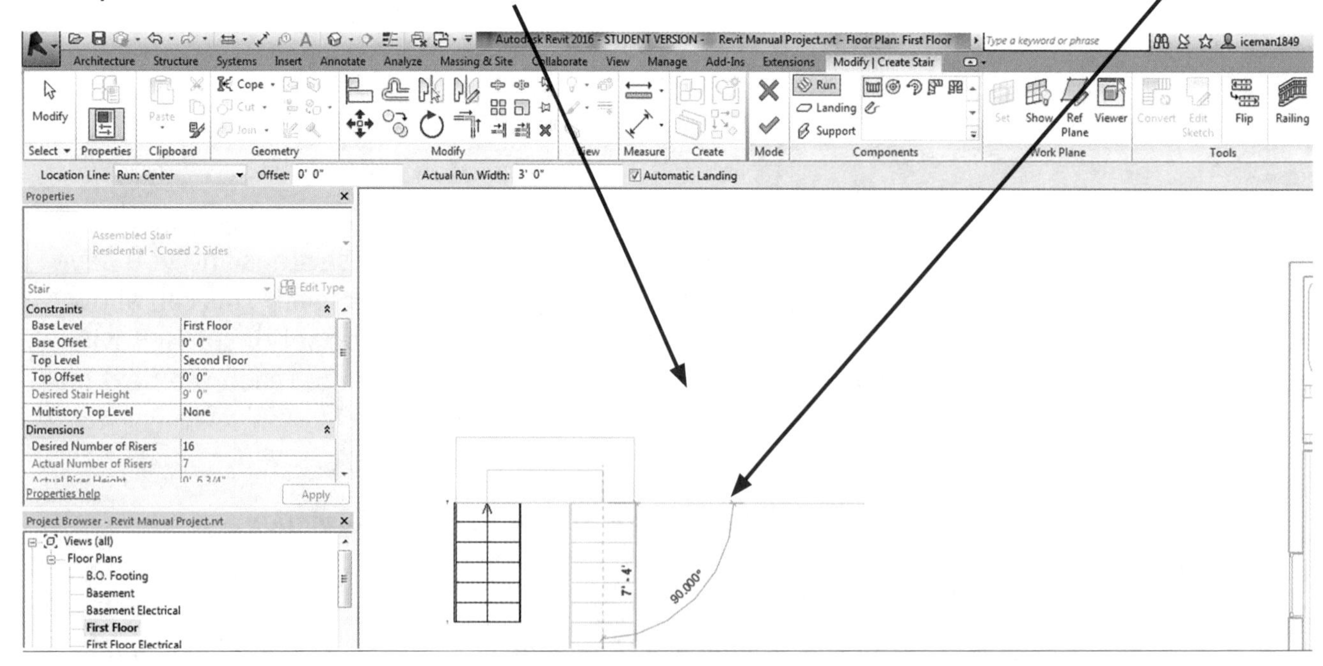

When the stair sketch is complete, the stairs will default back to the **MODIFY STAIRS** Tab in the ribbon.

WARNING

Do not click the green CHECK MARK at this point. This will complete the stair sketch, and will allow you to make modifications to the stairs, like the spacing between the risers, etc.

There are different ways to change the total width of the stairs. The simplest way is to click on a riser or flight and highlight just one side of the stairs, use the arrow keys on the keyboard to slide the riser or flight of stairs closer to the other flight of steps.

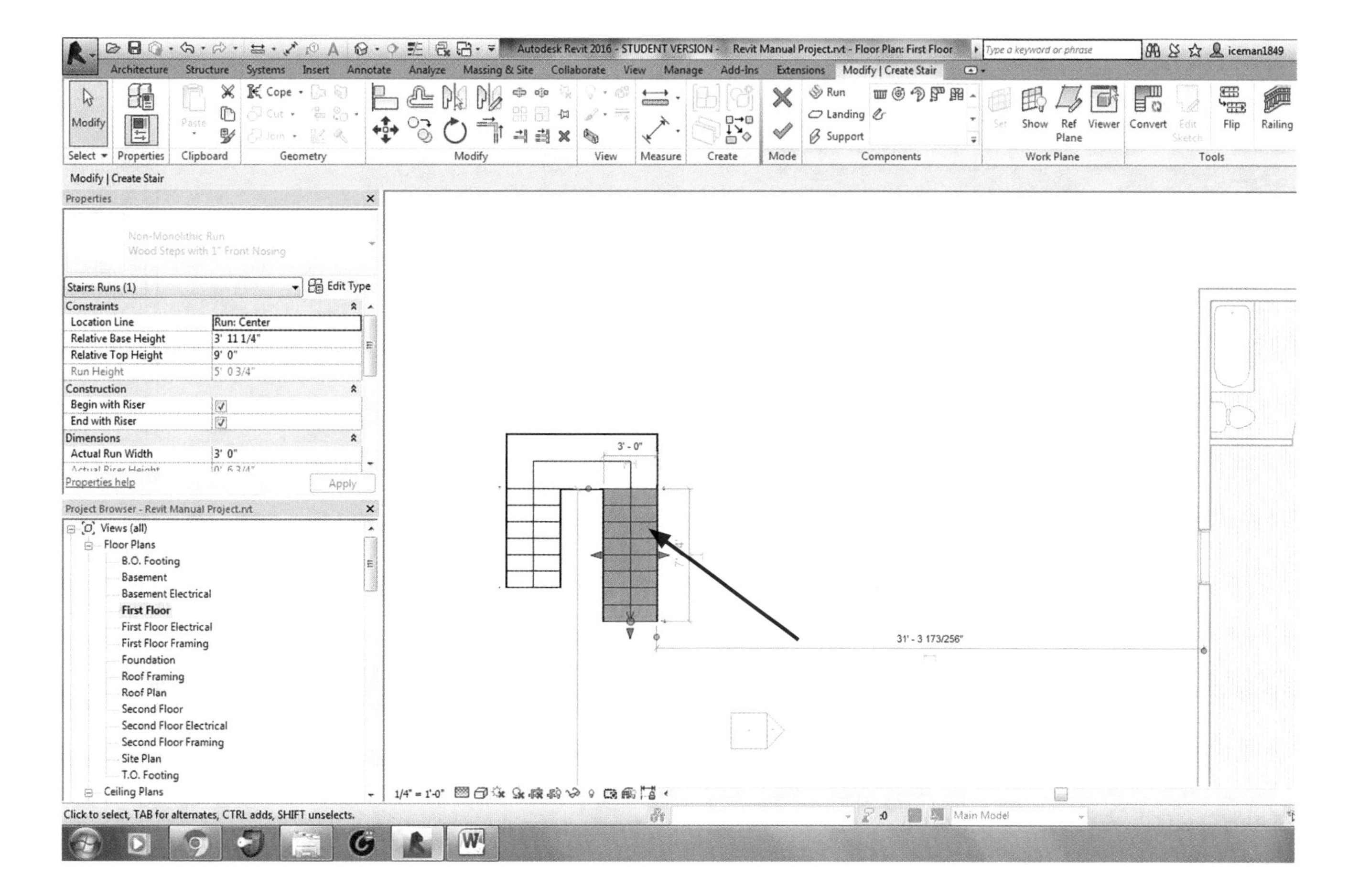

ESSENTIAL REVIT FACT

If you happen to click the green* CHECK MARK*, the stairs will appear complete.

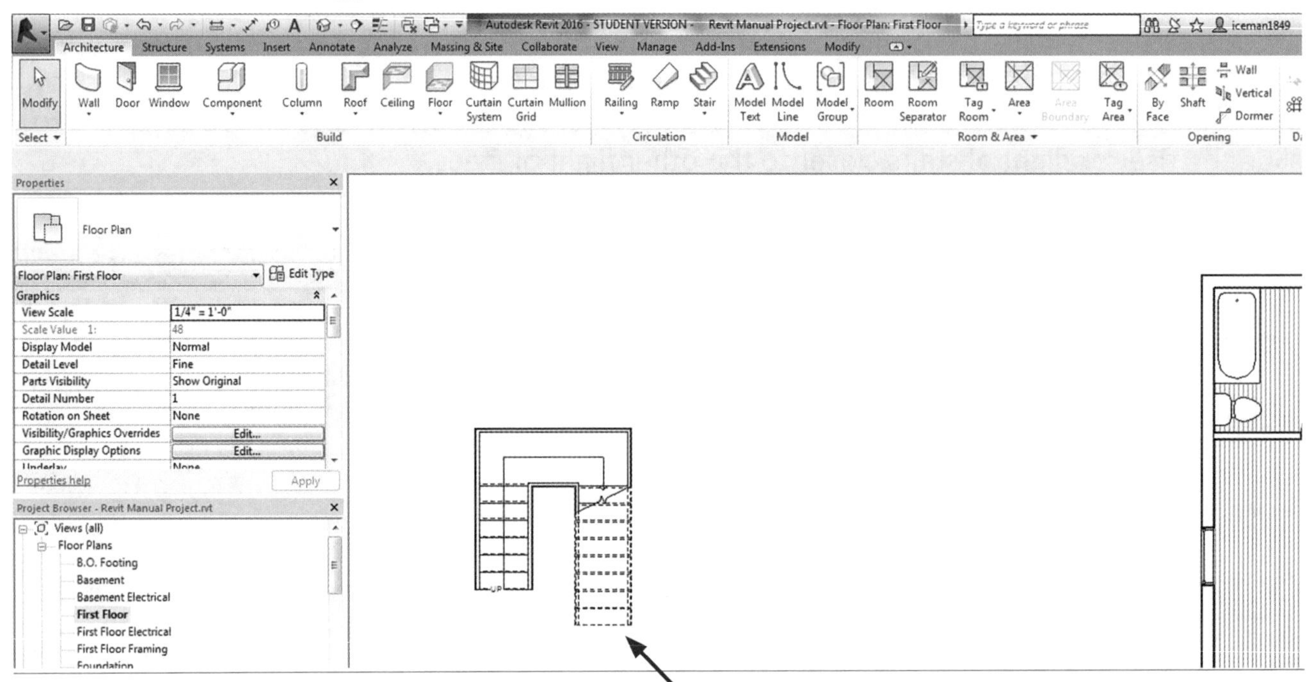

TIPS

Don't panic. If this happens, just click on a riser,* not *the handrail, to highlight the stairs, and the ribbon will change. Then you can proceed to select the* EDIT STAIRS *tile in the ribbon.

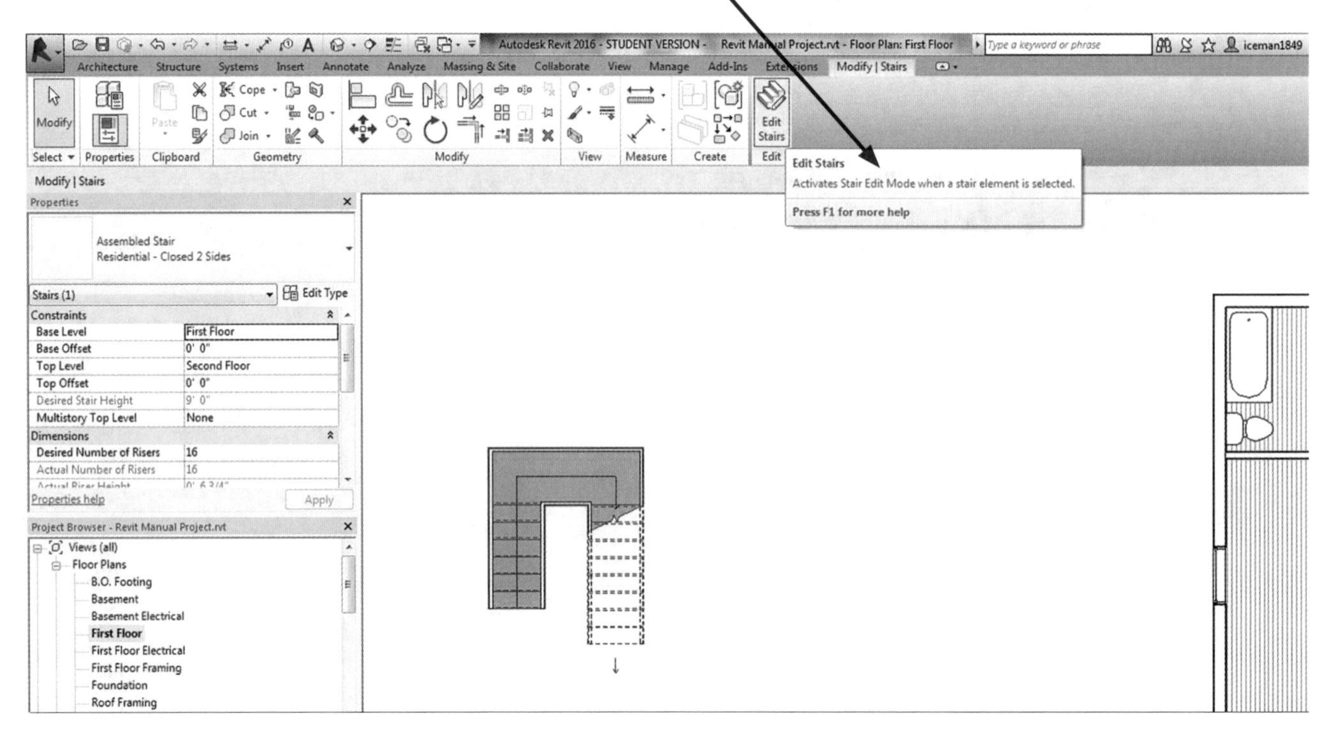

Once you are back in the **EDIT STAIRS** mode, you can modify the stair width. Click on a riser (or a flight), and activate the stairs. Here you can use the **NUDGE** command to change the width.

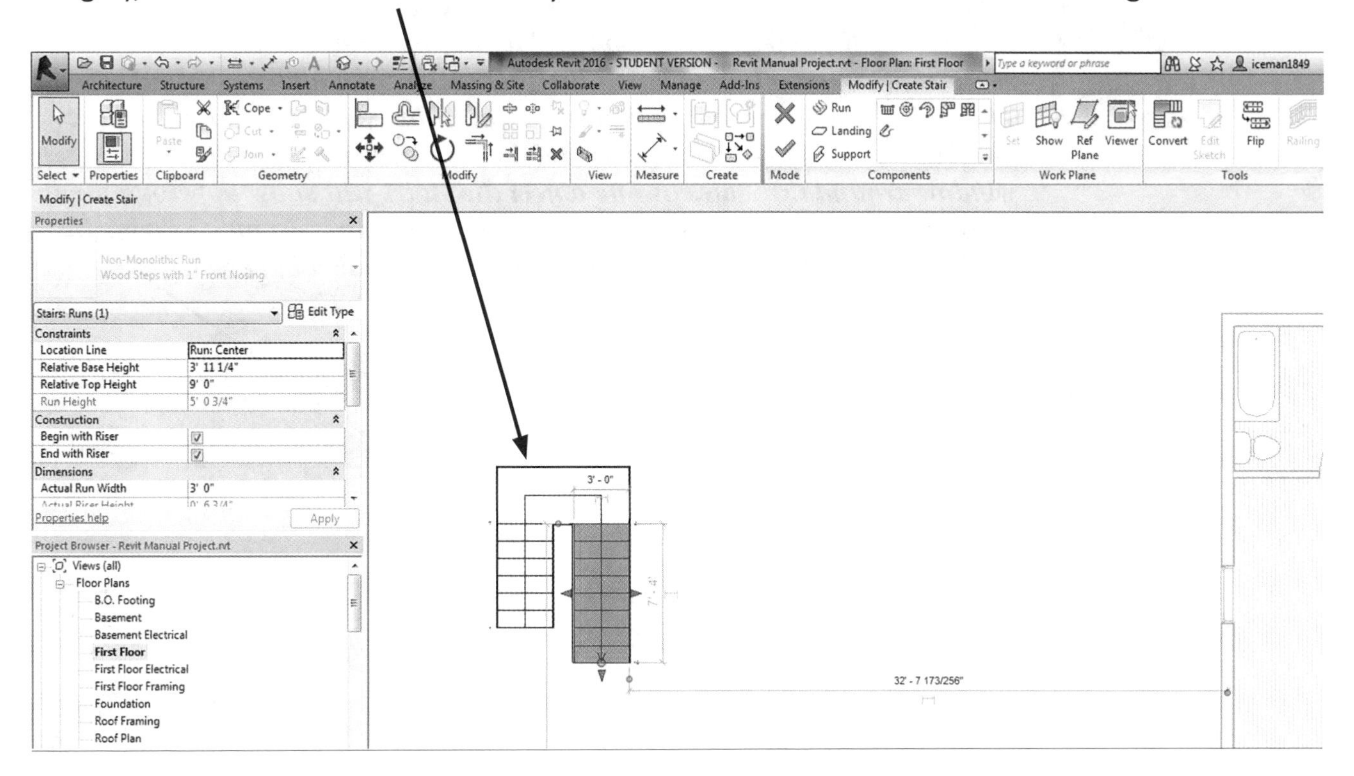

When the width is adjusted to the correct, desired size, then you can click the green **CHECK MARK** to complete the stair design.

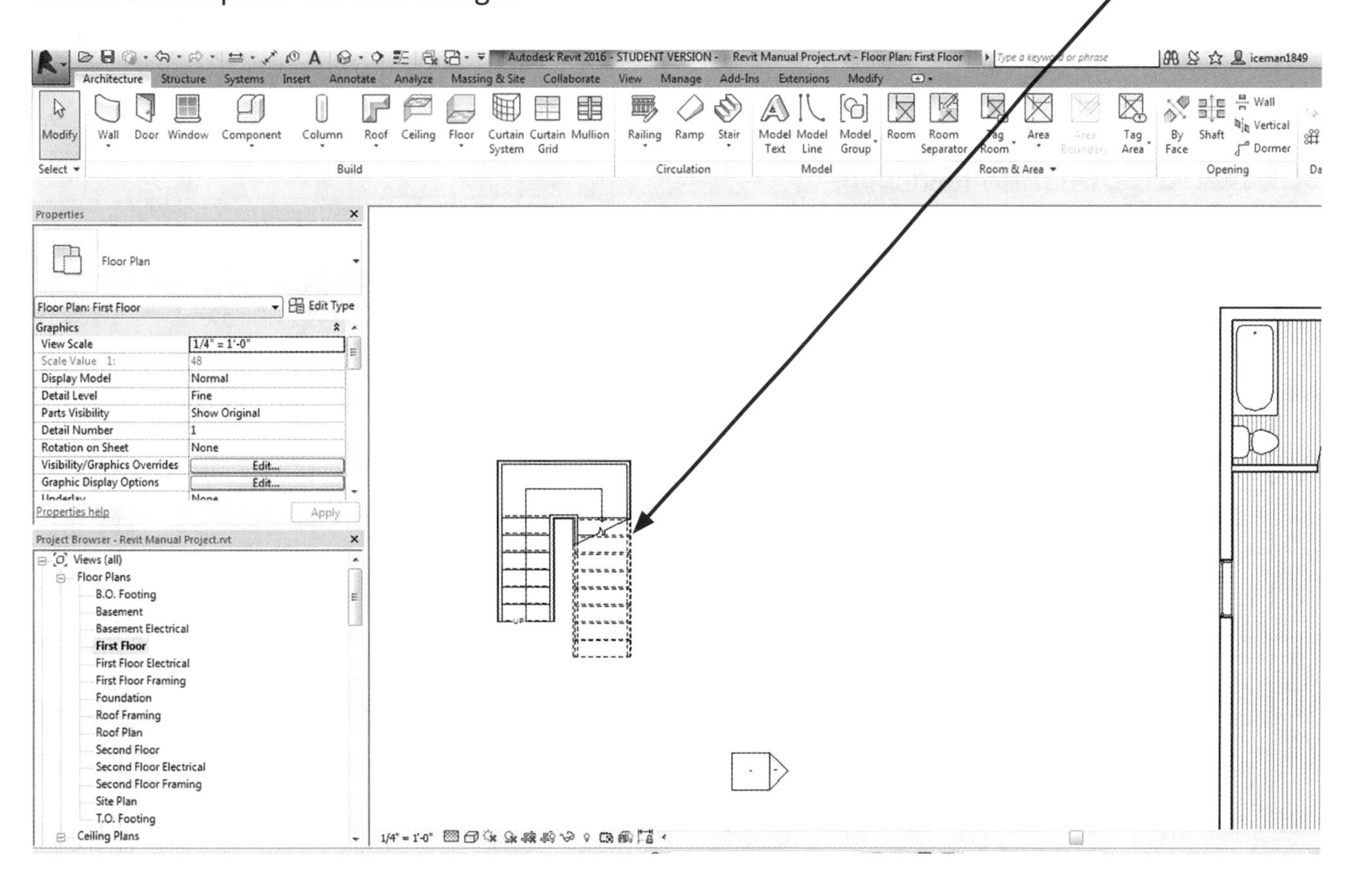

Once the stairs are complete, they may need to be rotated.

WARNING

Do not put a window around the complete set of stairs. A pin will appear. This means that the stairs are pinned to that location. When you draw a set of stairs in a certain location they act almost like a door or window. The stairs become hosted in that location. In the stairs format it is called pinned; the stairs will become pinned in that location. If you draw a window around the stairs as one object the stairs will show up with a pin in the middle and you will not be able to move the stairs or modify them. Just click one set of risers as described in the instructions above.

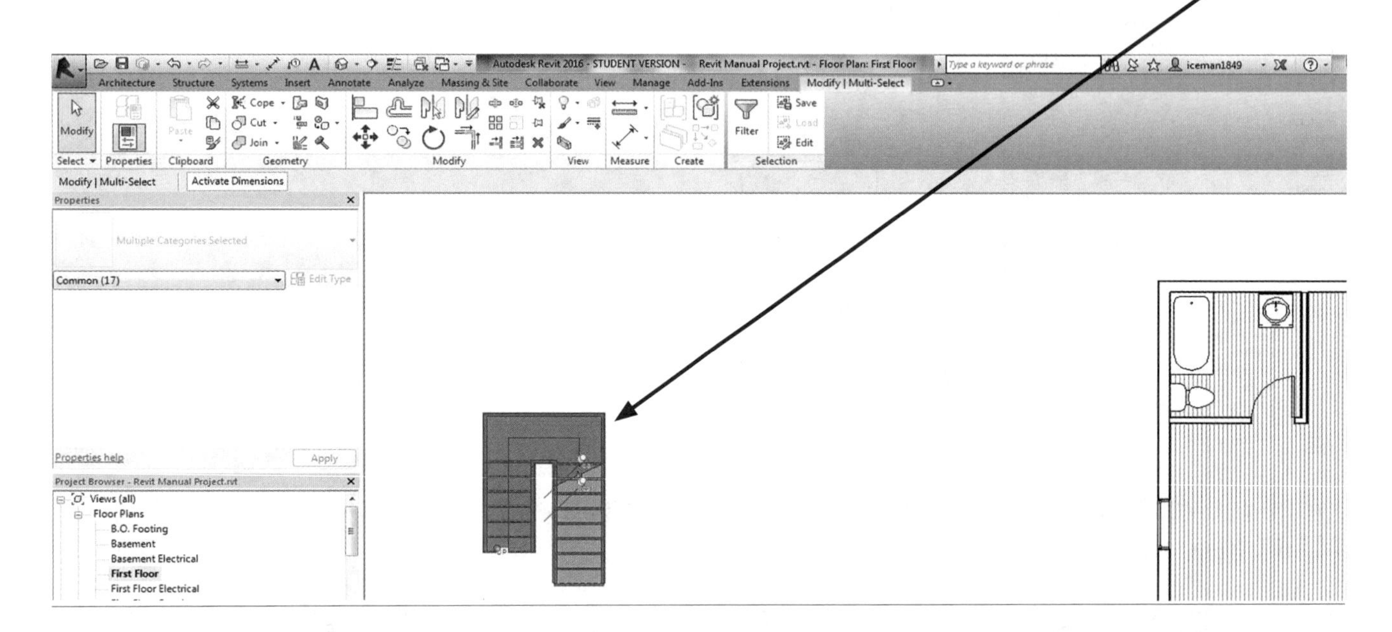

To rotate the stairs, click on a riser (or flight) to highlight the risers and the railings. This will activate the **MODIFY STAIRS** ribbon.

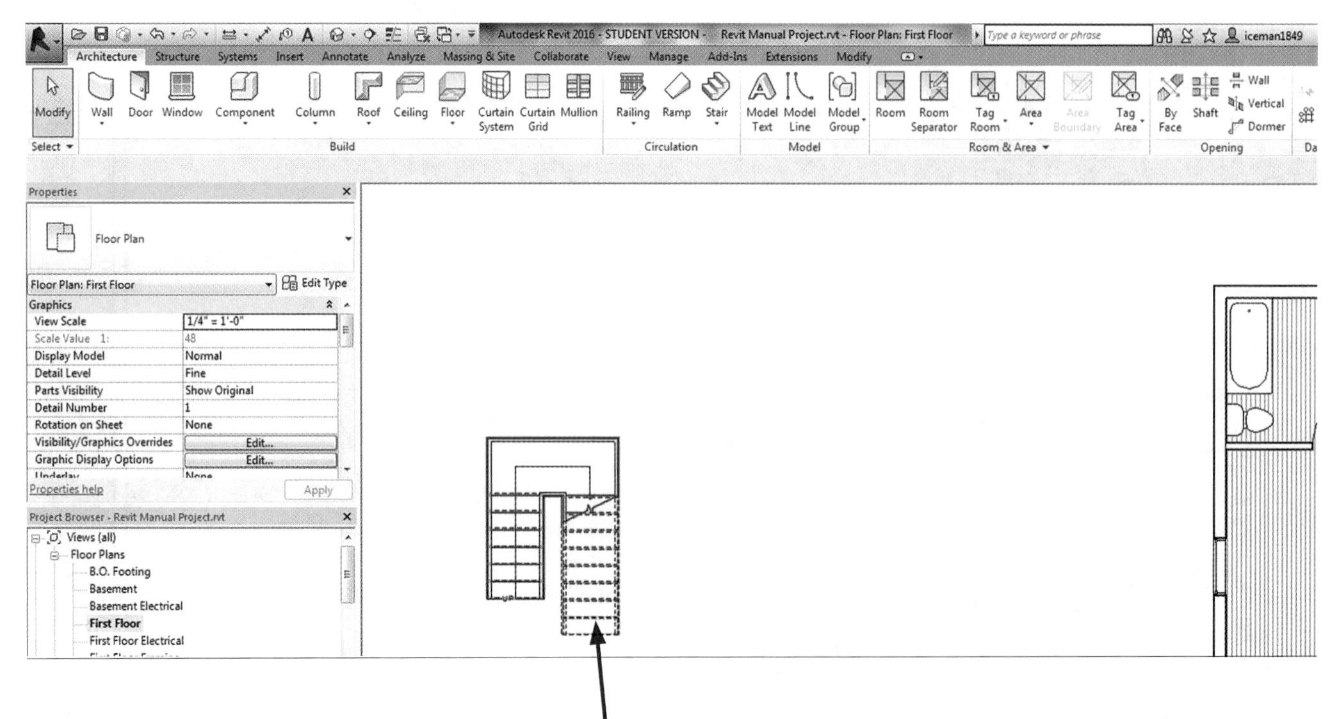

Once on the **MODIFY STAIRS** ribbon, click on the **ROTATE** command.

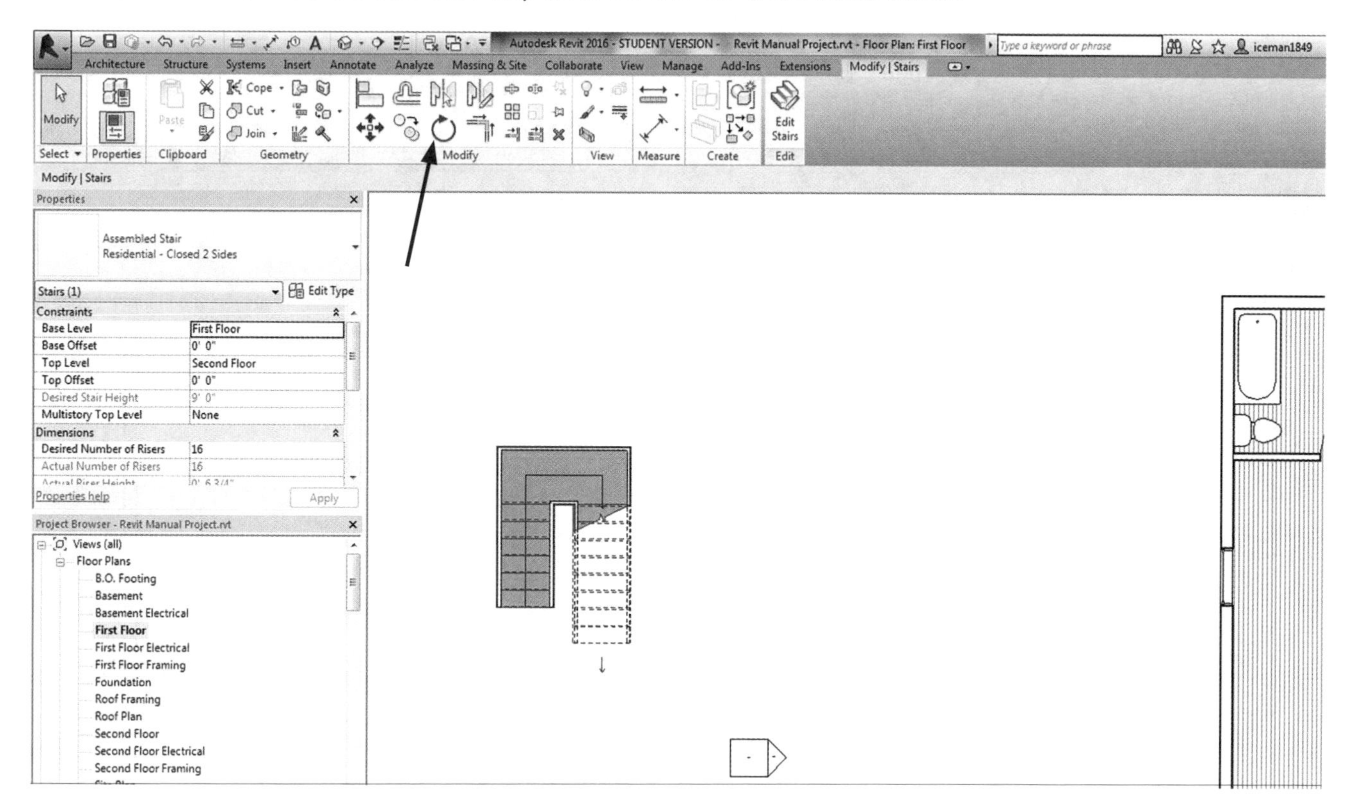

The Rotate command will appear in the middle of the stairs. To extend the rotate reference line in a straight line, click your mouse button once.

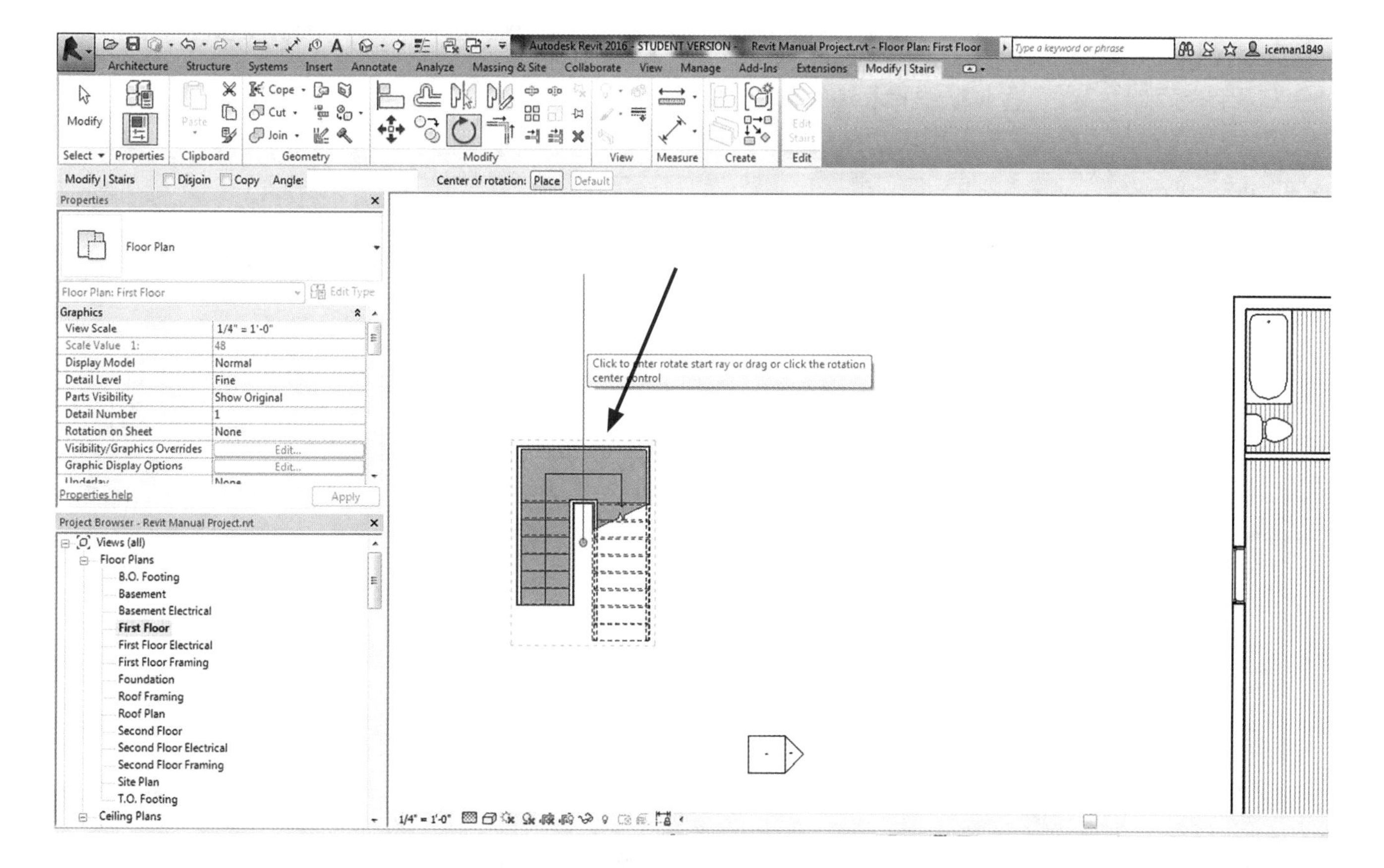

Then rotate in the direction you desire the stairs to be placed.

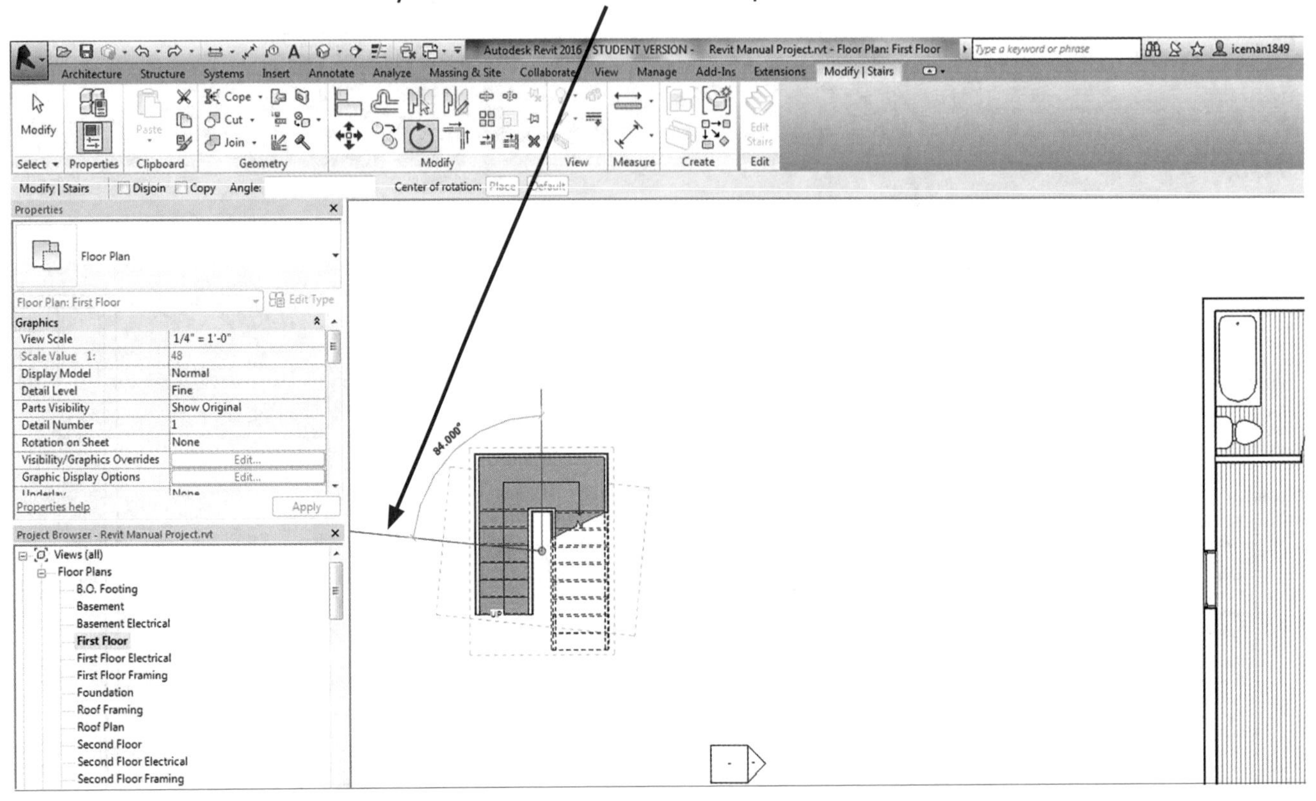

Type in the desired angle, and hit the **ENTER** key.

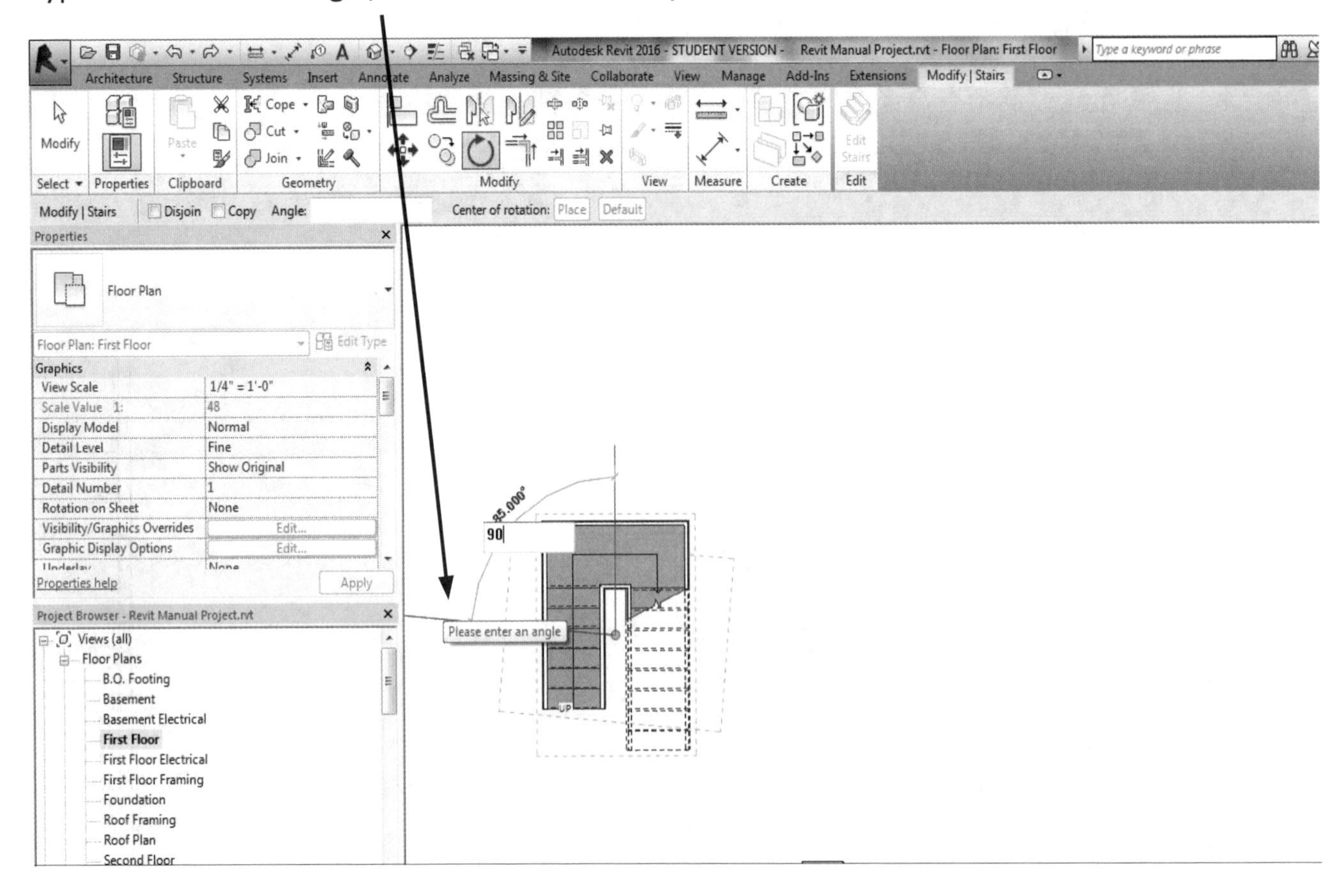

The following screen shows the completed stairs and angle desired to be placed in the project location.

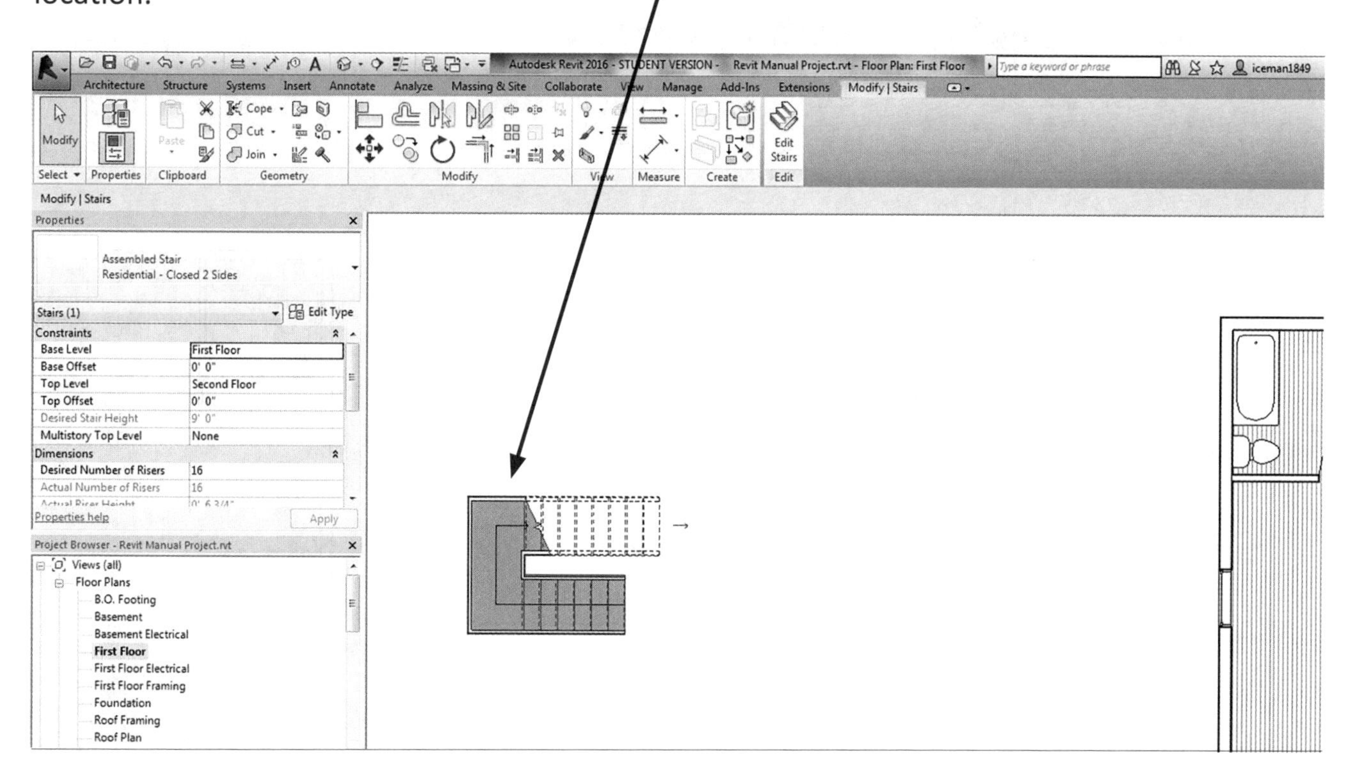

While still in the EDIT STAIRS command in the ribbon, click on the **MOVE** command, and select a corner of the stairs to make it easier to relocate the stairs.

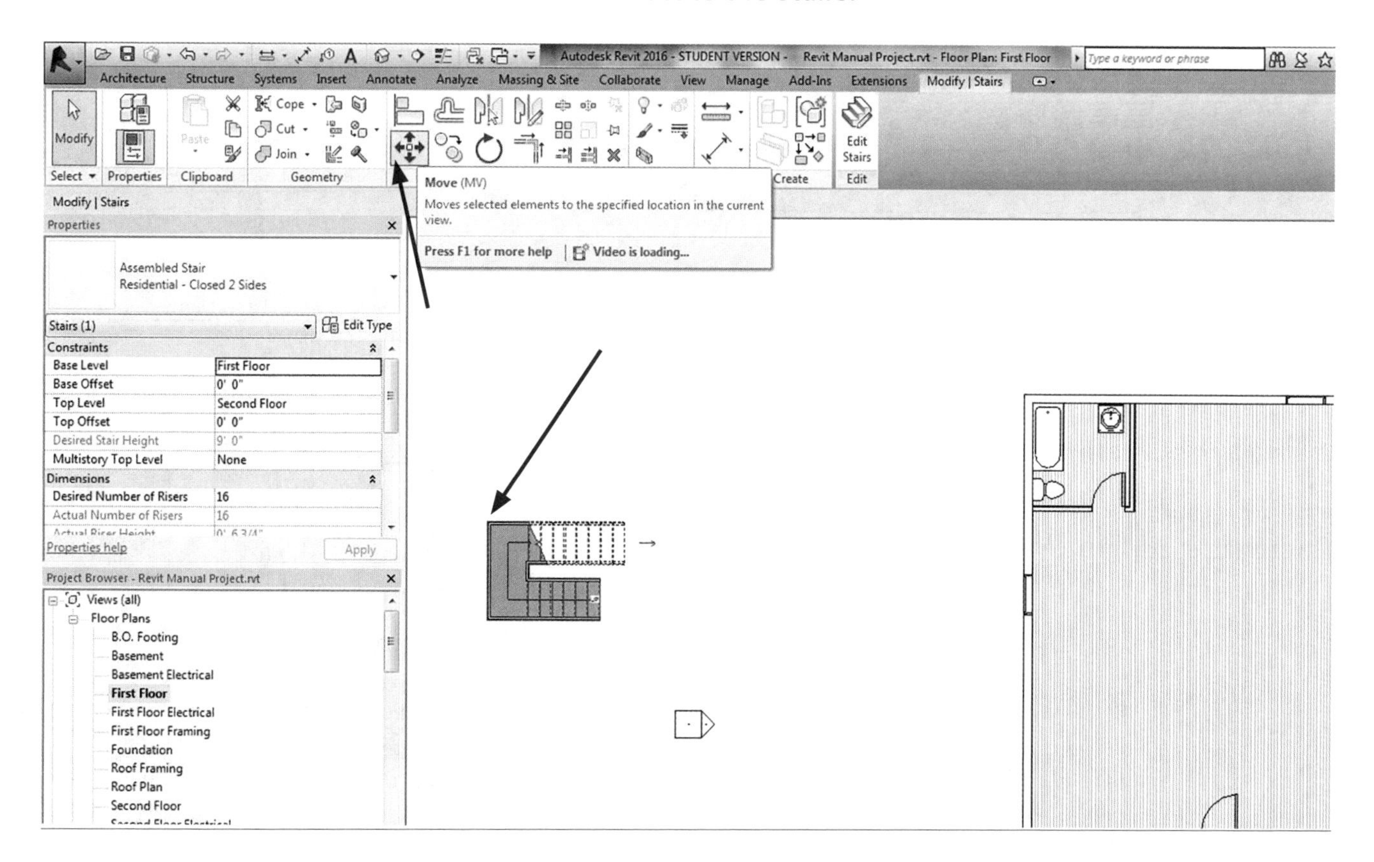

Click on the corner of the stairs, and drag them to the new location.

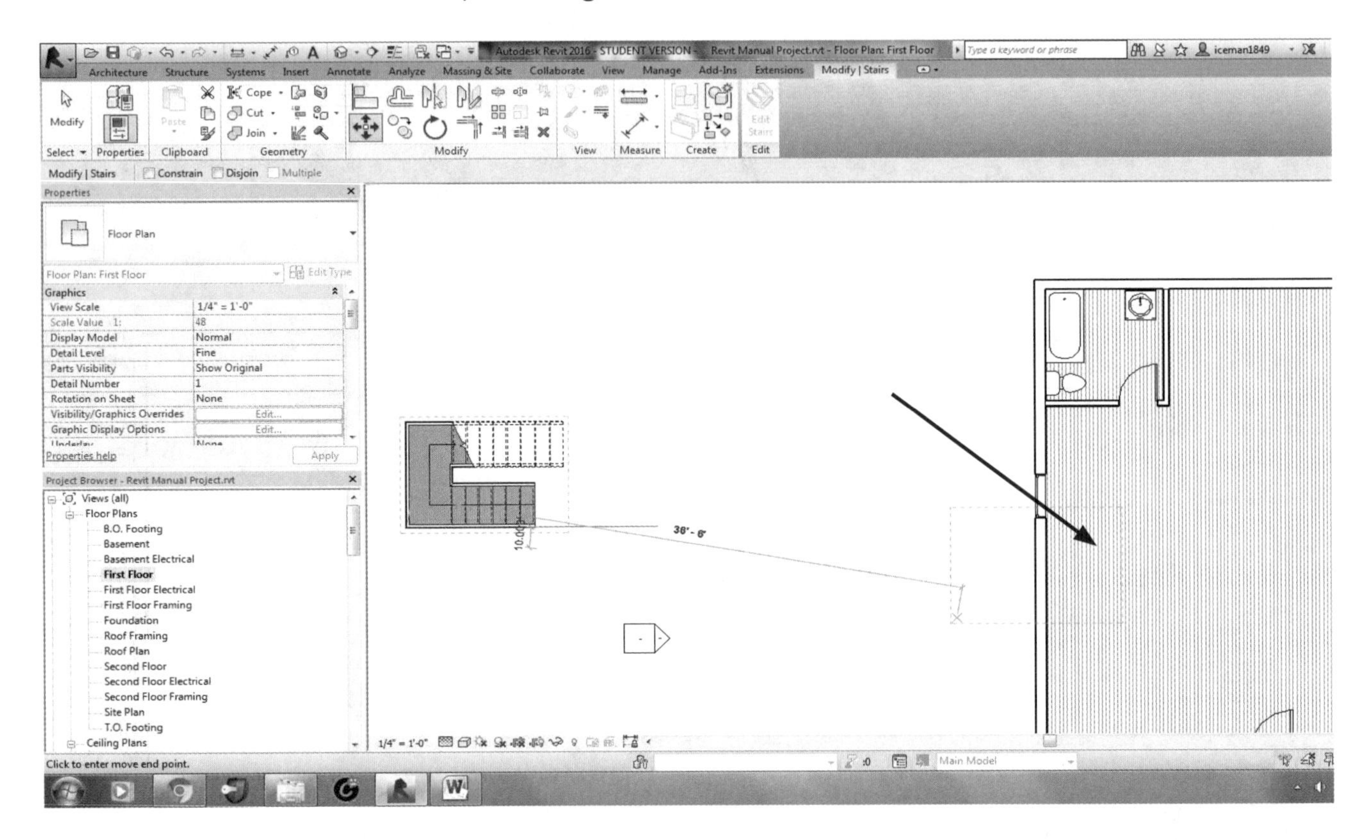

This way of creating stairs is much easier than sketching the stairs in the house and building.

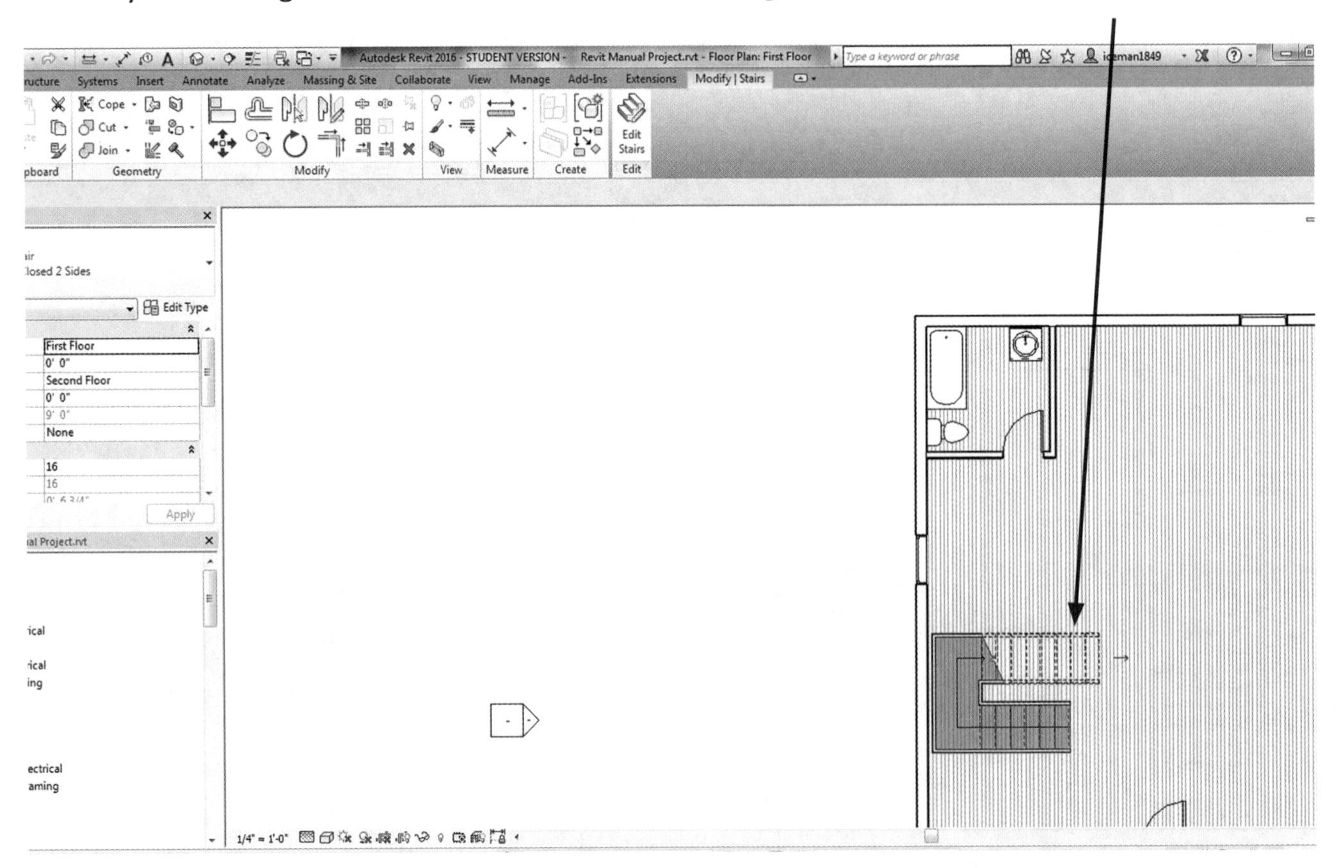

After the relocation, then you can draw the walls around the stair placement, using all the steps you learned in Chapter 3.

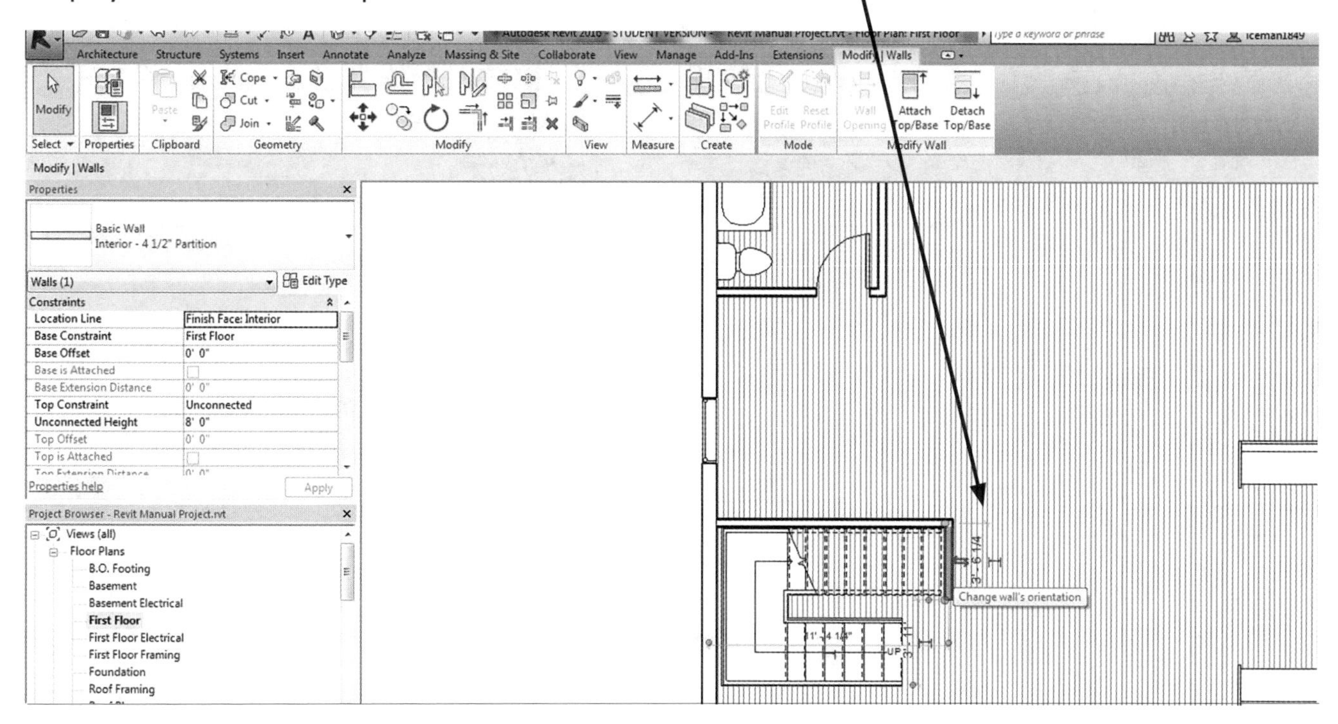

ANOTHER WAY TO CHANGE THE STAIR WIDTH

Here is another way to change the stair width after you have already drawn and completed the stairs. Click on the stairs and once they are activated, click on the **EDIT STAIRS** Tile.

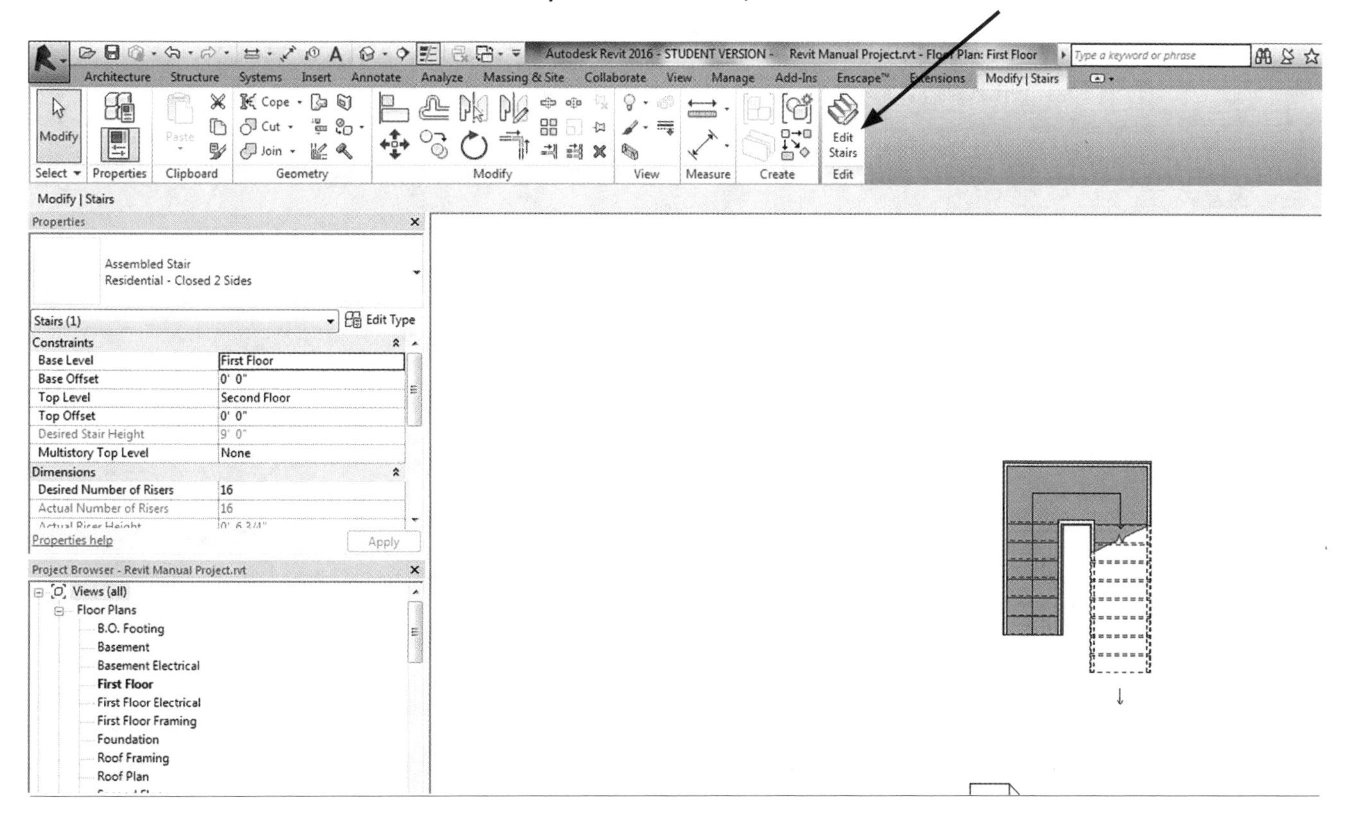

This will allow you to edit the stair width. Click on one of the riser sides. This will activate the temporary dimensions.

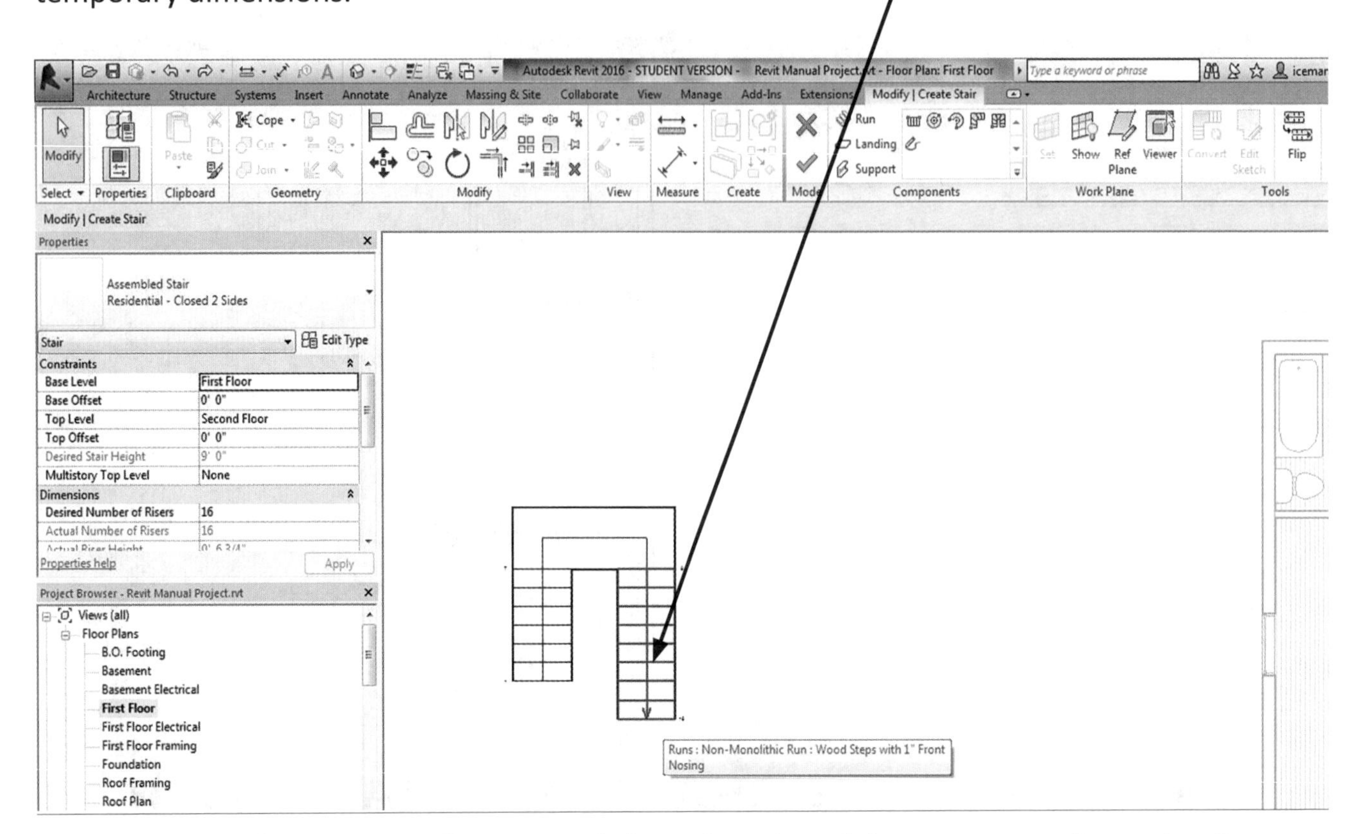

The temporary dimensions will appear. Click on the grips of the temporary dimension lines, and drag the extension line to the outside width of the stairs.

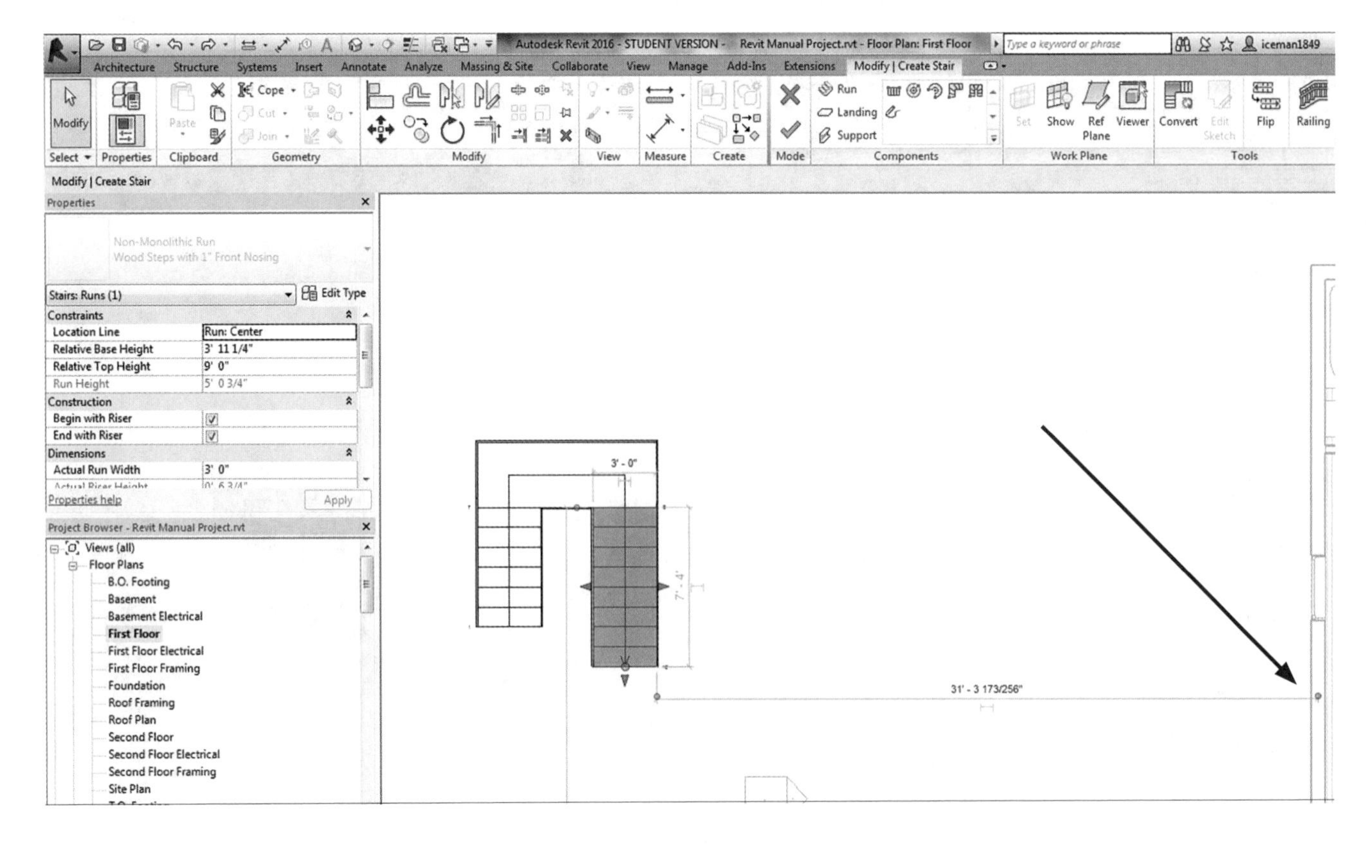

Then drag the extension line to the other side of the stair width.

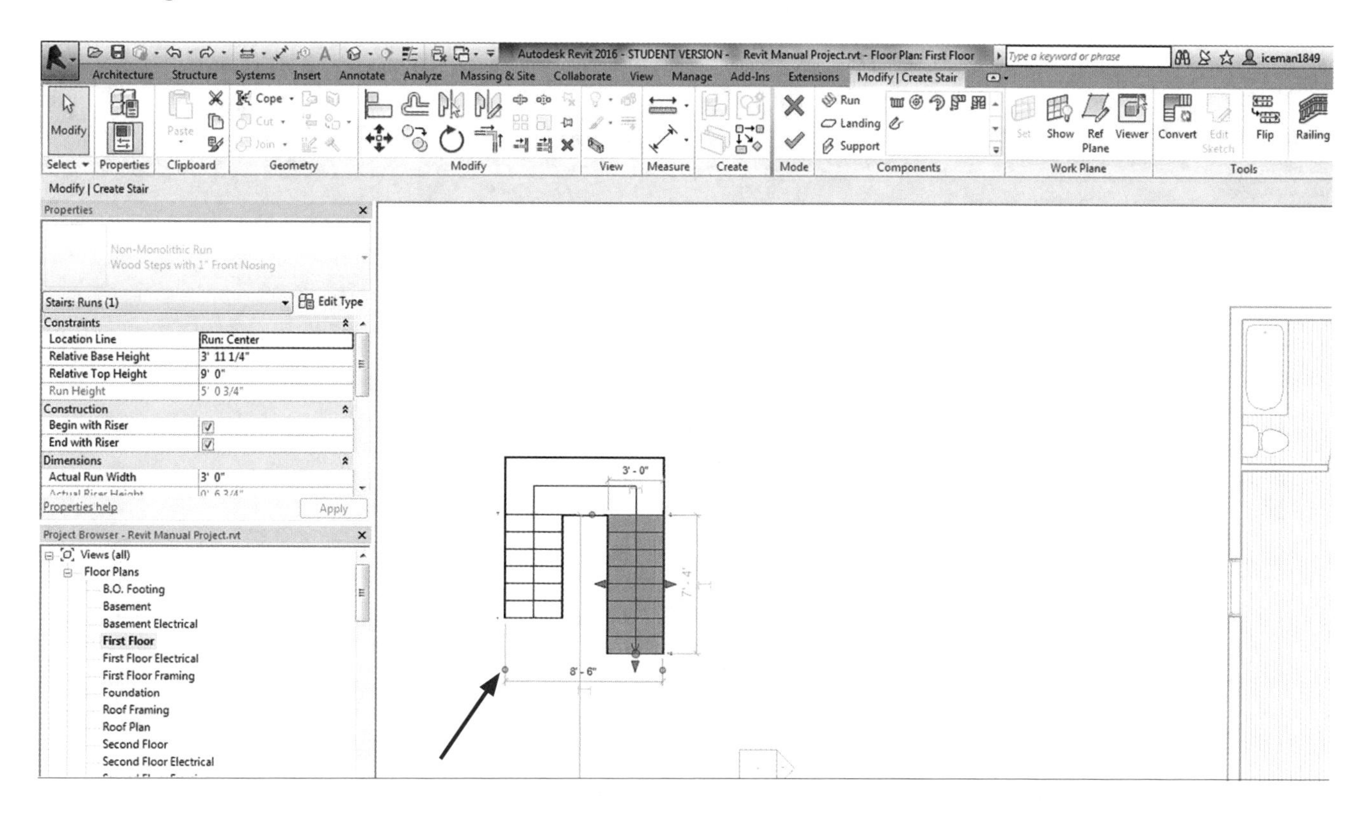

Then you can double-click on the **TEMPORARY DIMENSIONS** inside of the dimension line to enter a new length.

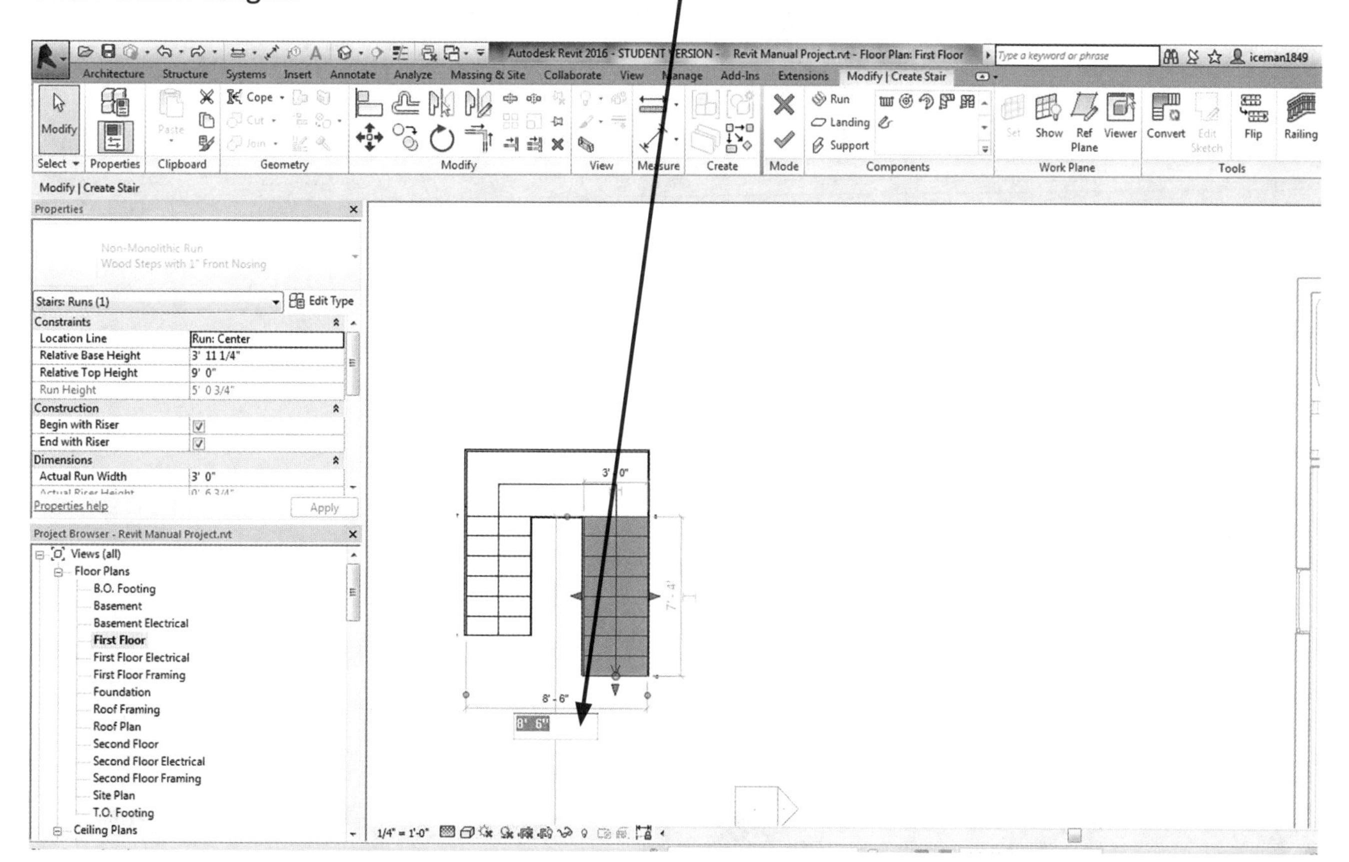

Then hit the **ENTER** key.

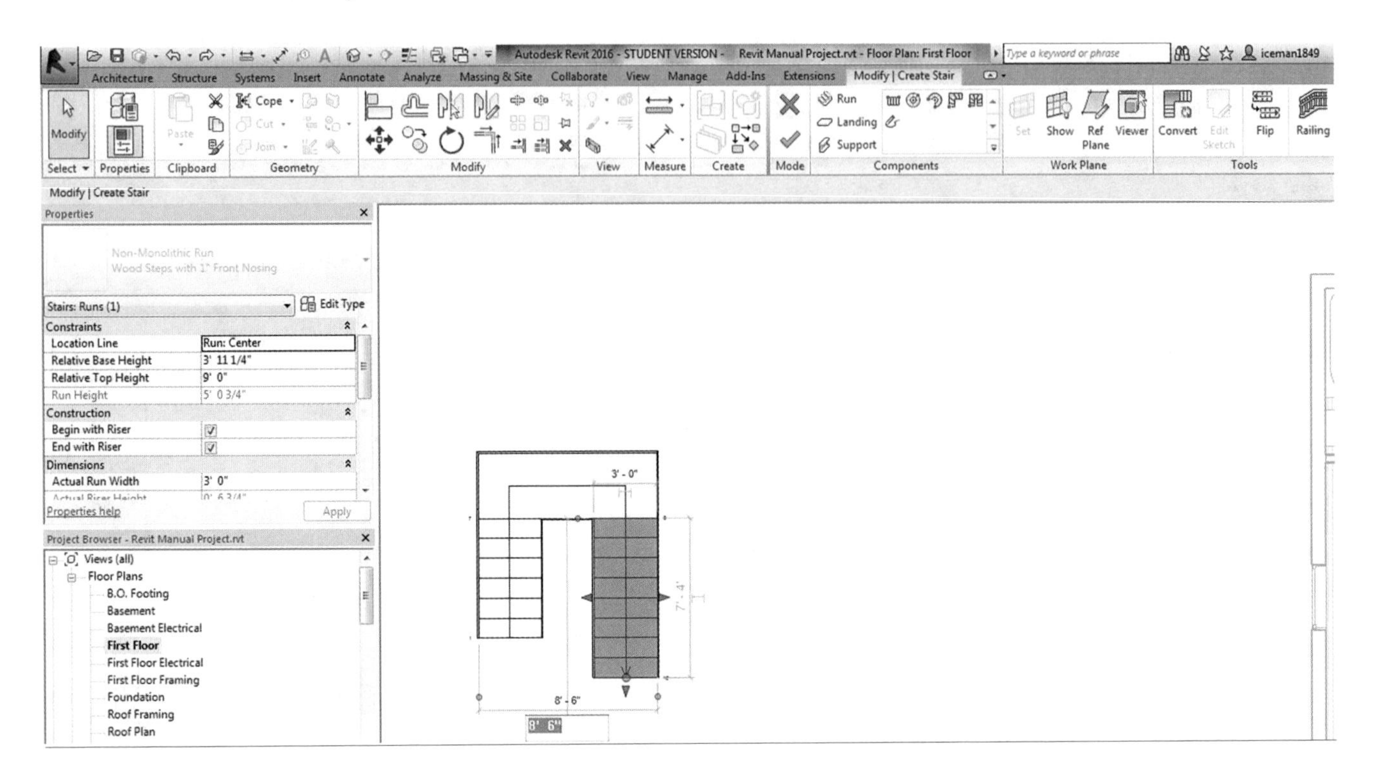

Next, click the **green CHECK MARK** to complete your stair design.

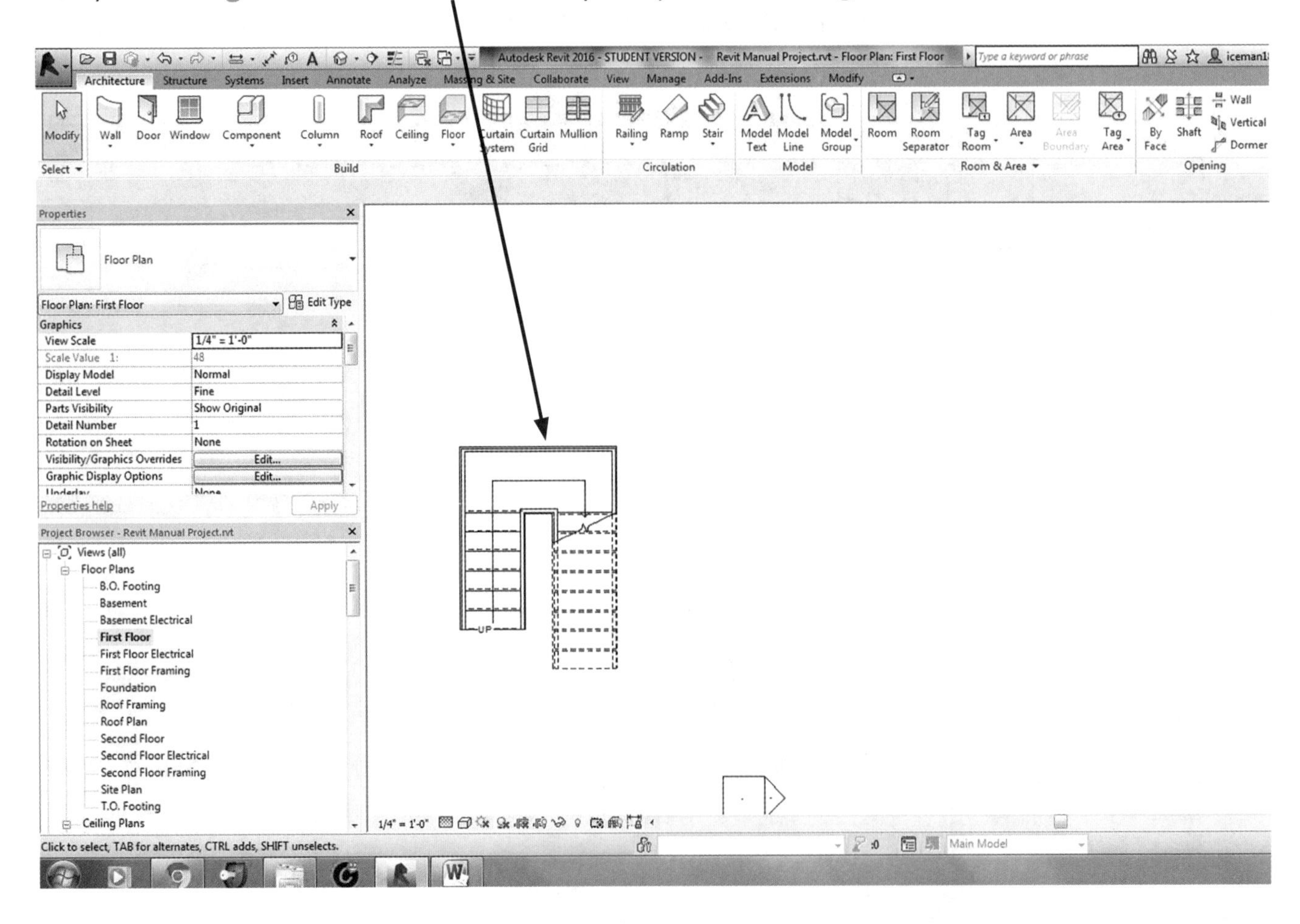

Then rotate the stairs into the desired position. Click on the riser, then the **ROTATE** command.

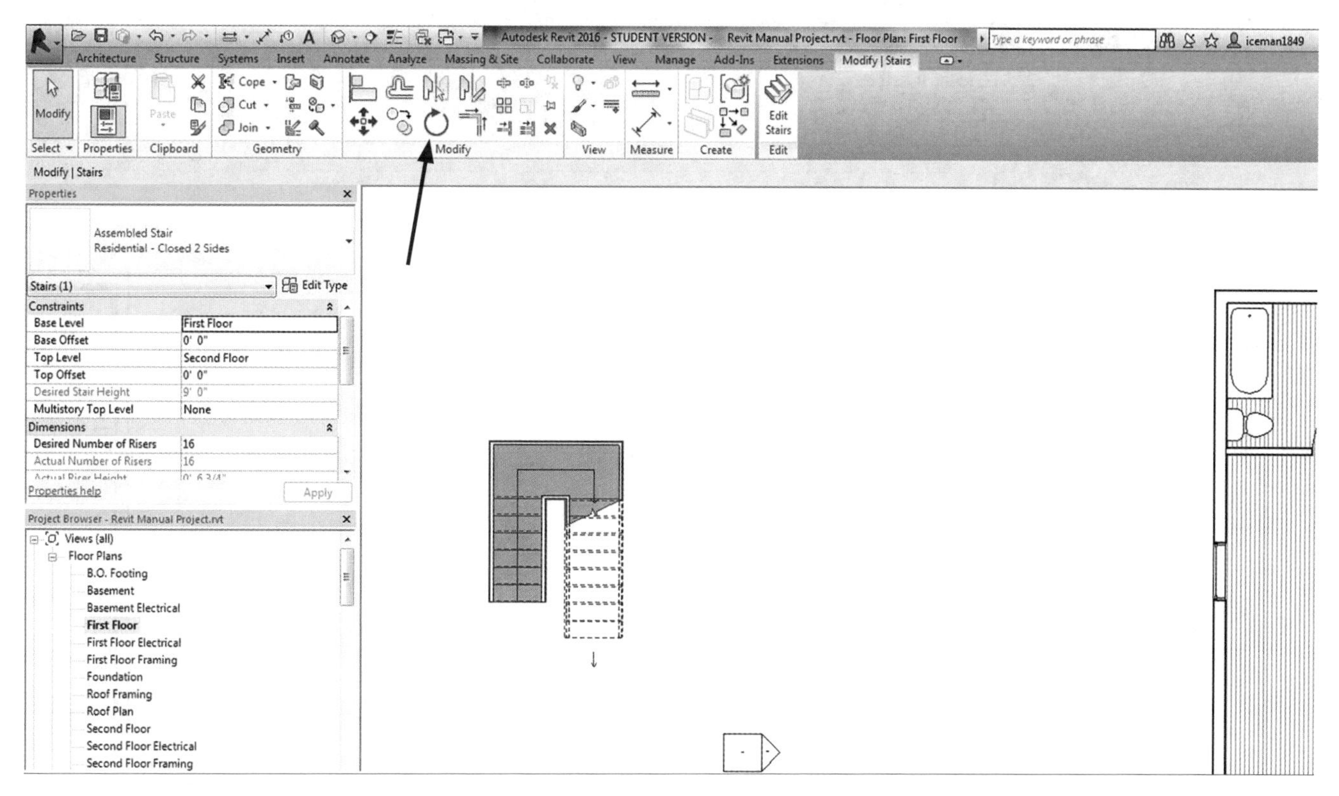

Rotate the reference line, then type in the desired angle for rotation.

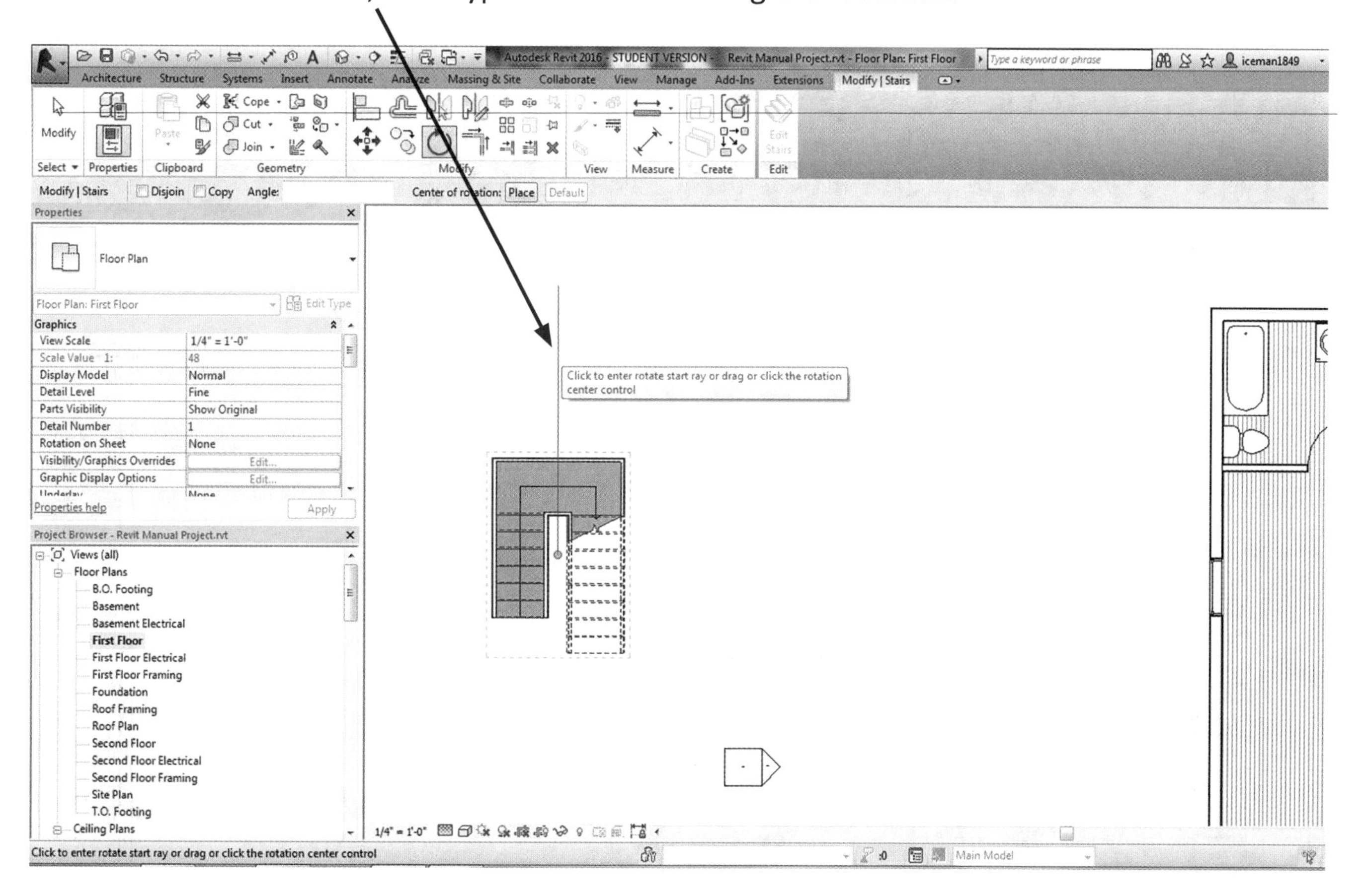

Type in your desired rotation angle.

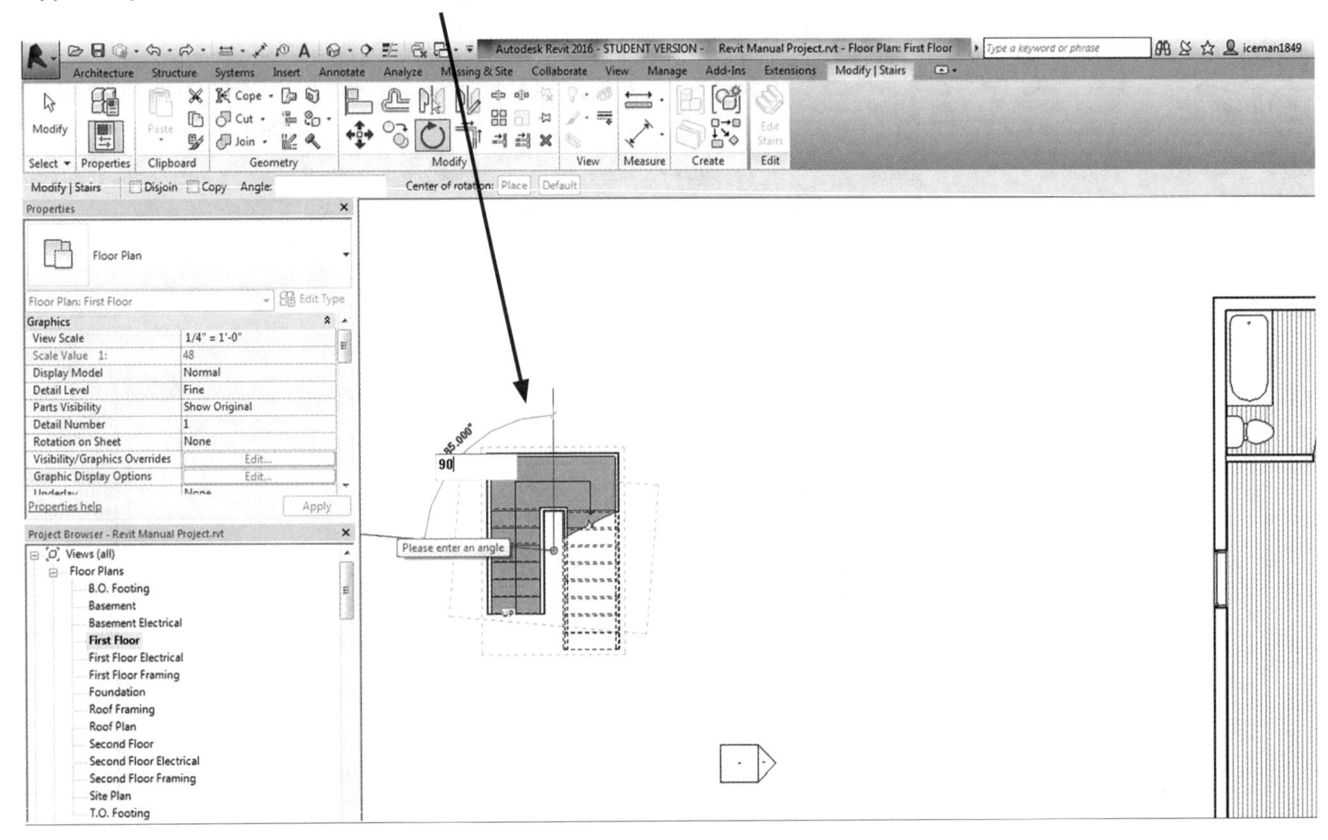

You will see the completed stairs and angle desired to be placed in the project location.

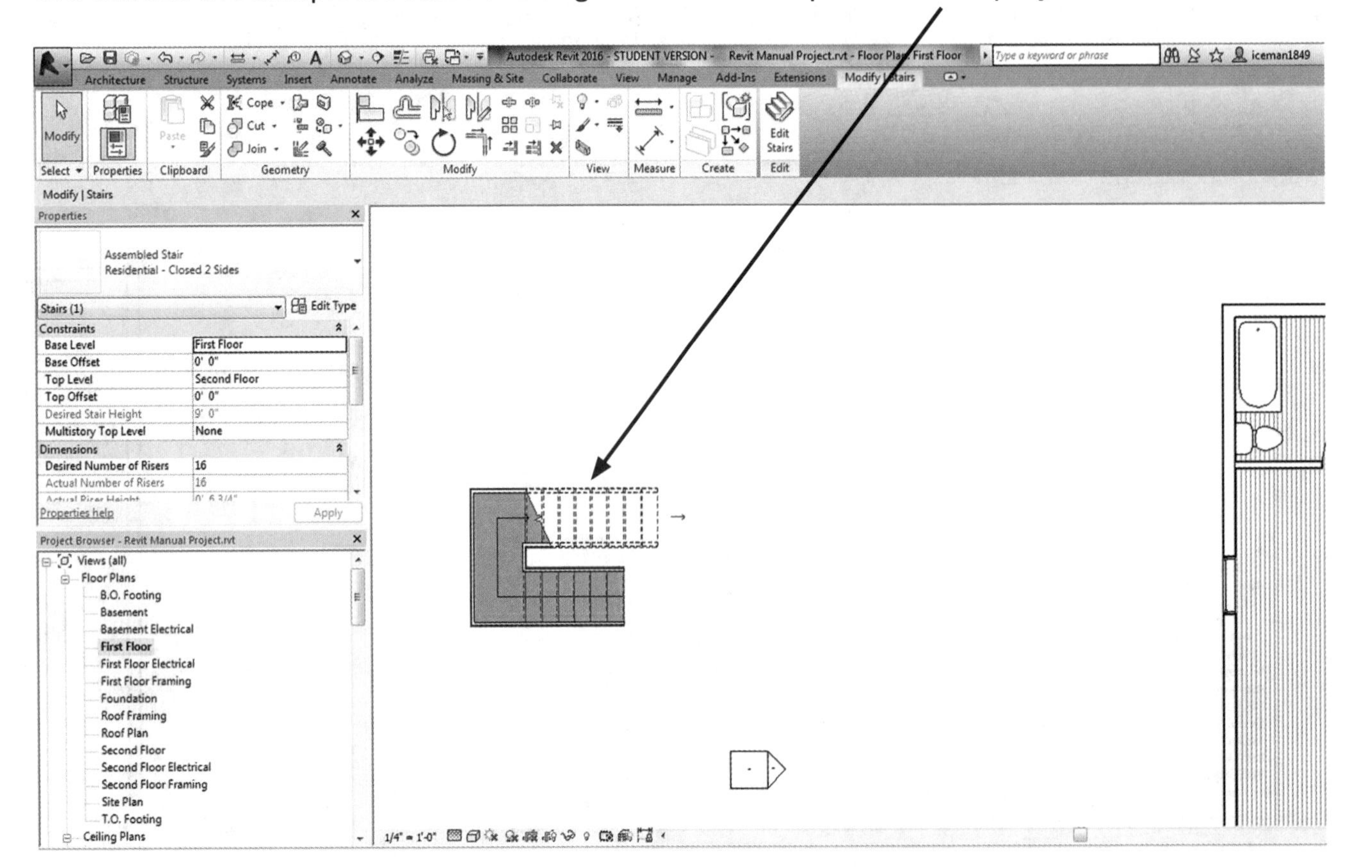

While still in the Edit Stairs command in the ribbon, click on the **MOVE** command, and select a corner of the stairs to make it easier to relocate the stairs.

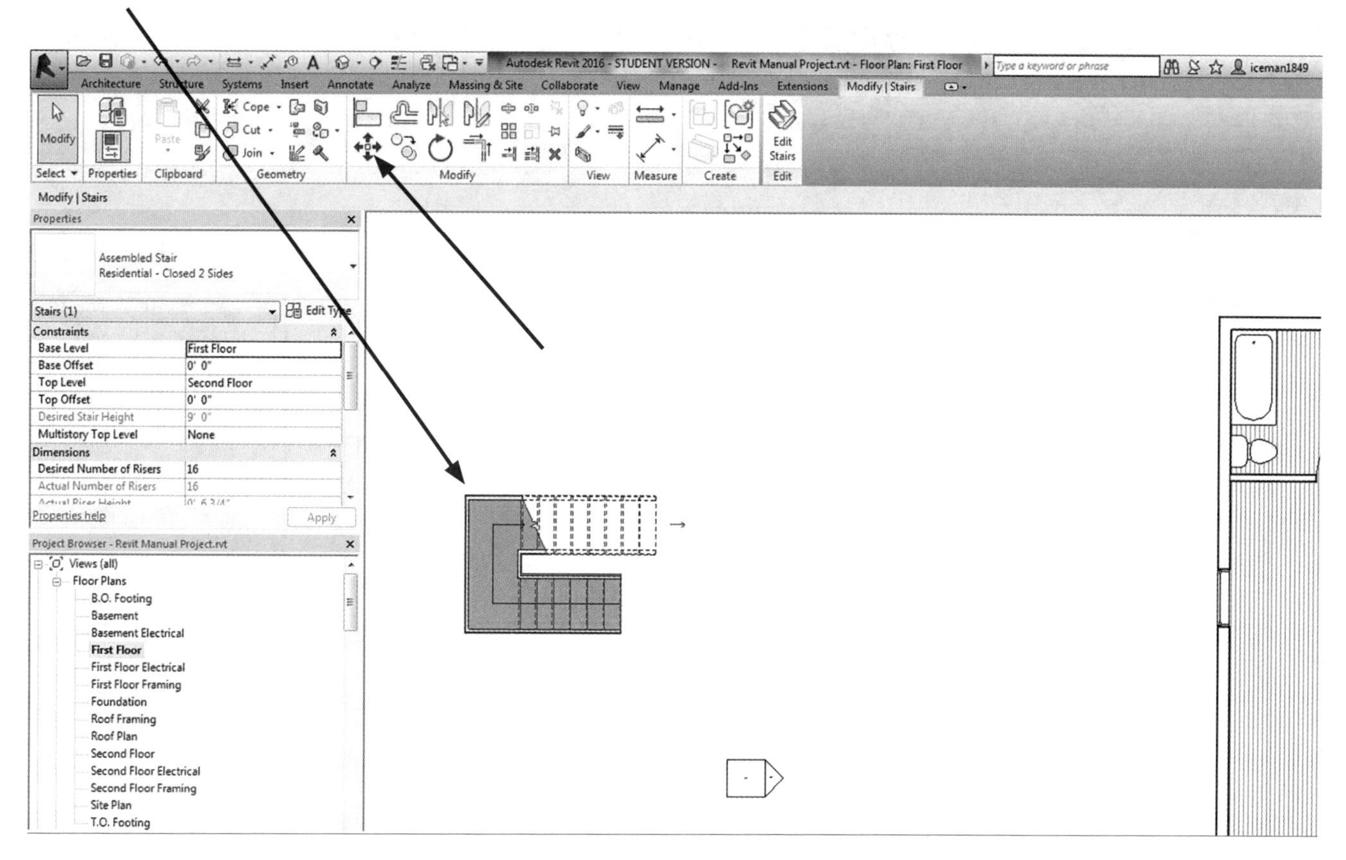

Then click on the corner of the stairs, and drag them to the new location.

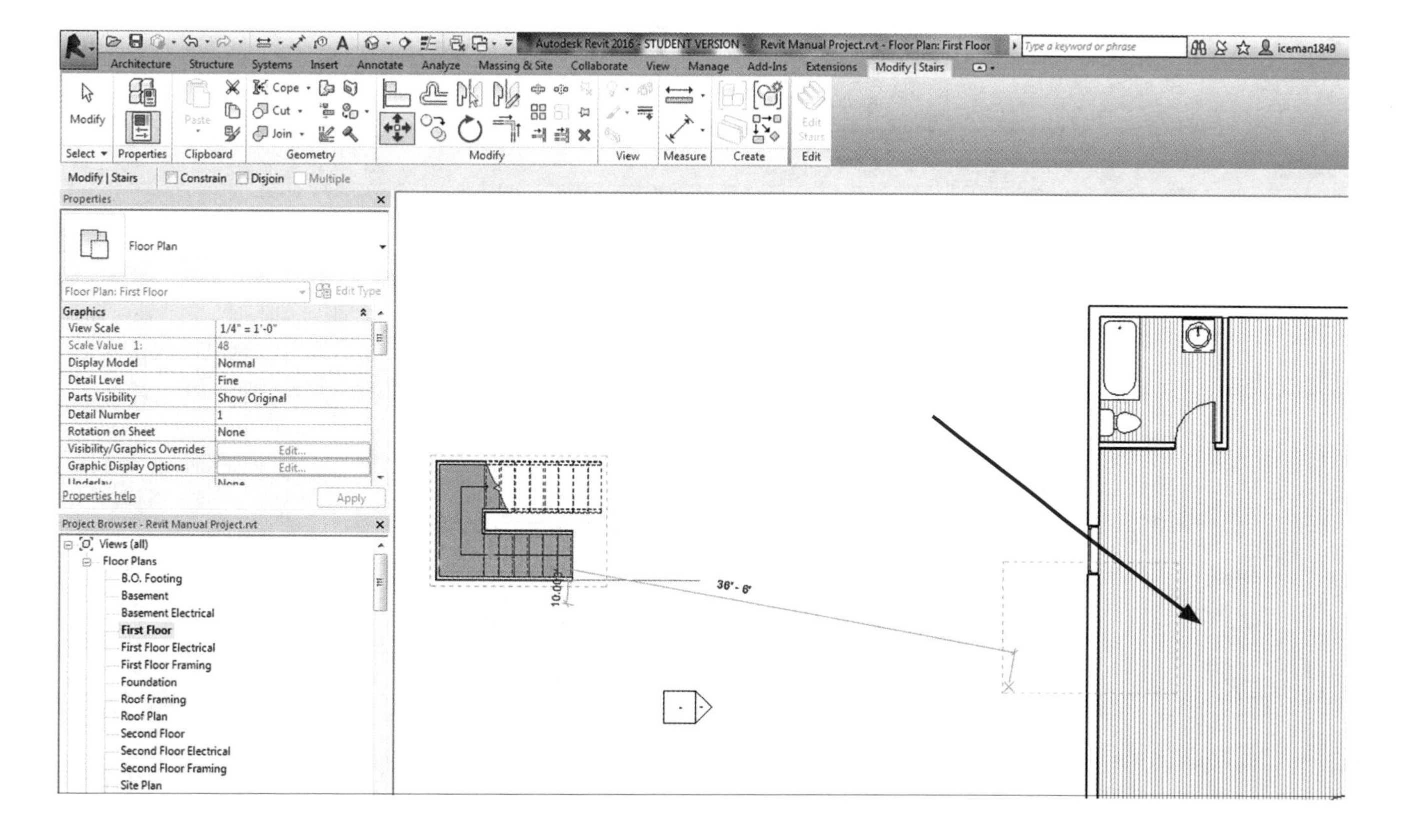

As stated earlier, this is much easier than sketching the stairs in the house and building.

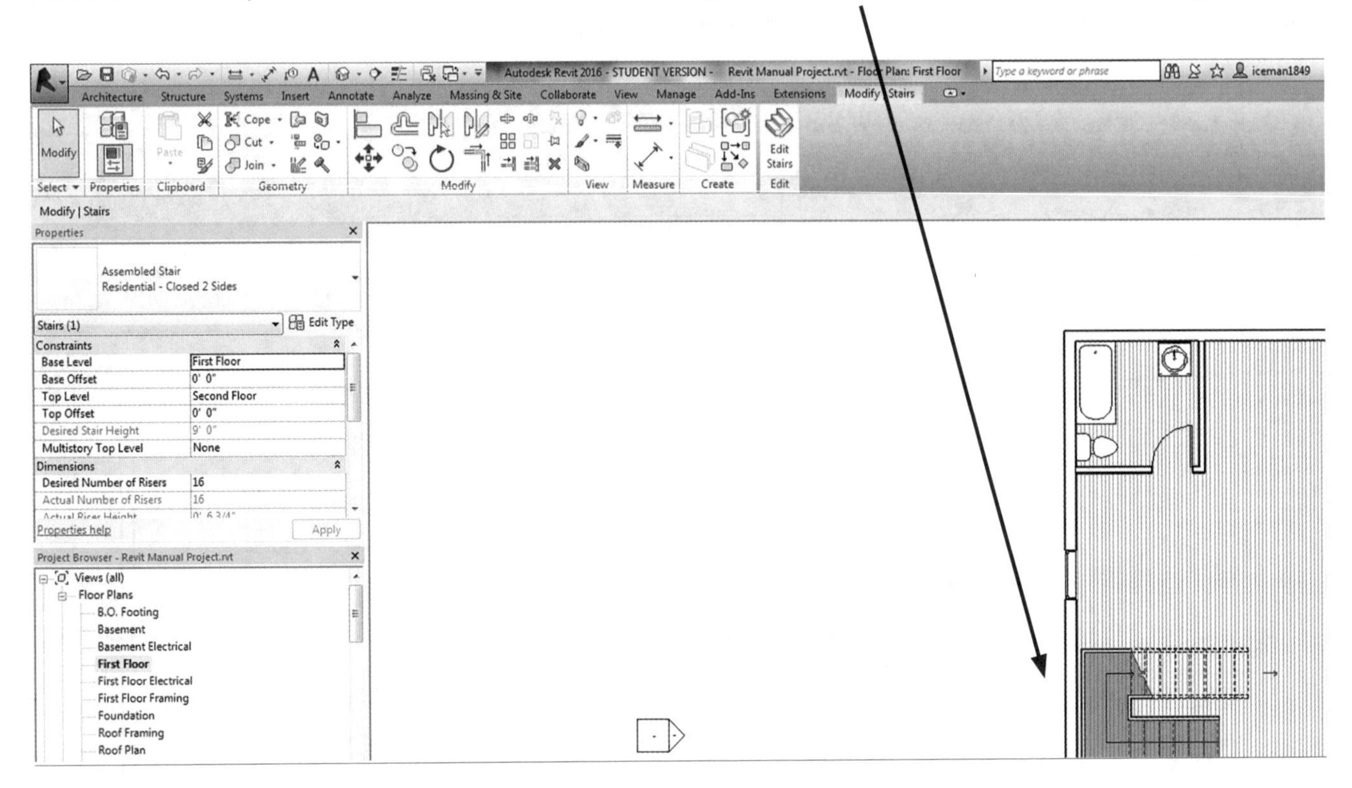

After the relocation, you can draw the walls around the stair placement using the steps you learned from Chapter 3. Drawing a wall around the stairs will close off the side of the stairs and then close off the underside of the steps so you can't see under the stairs.

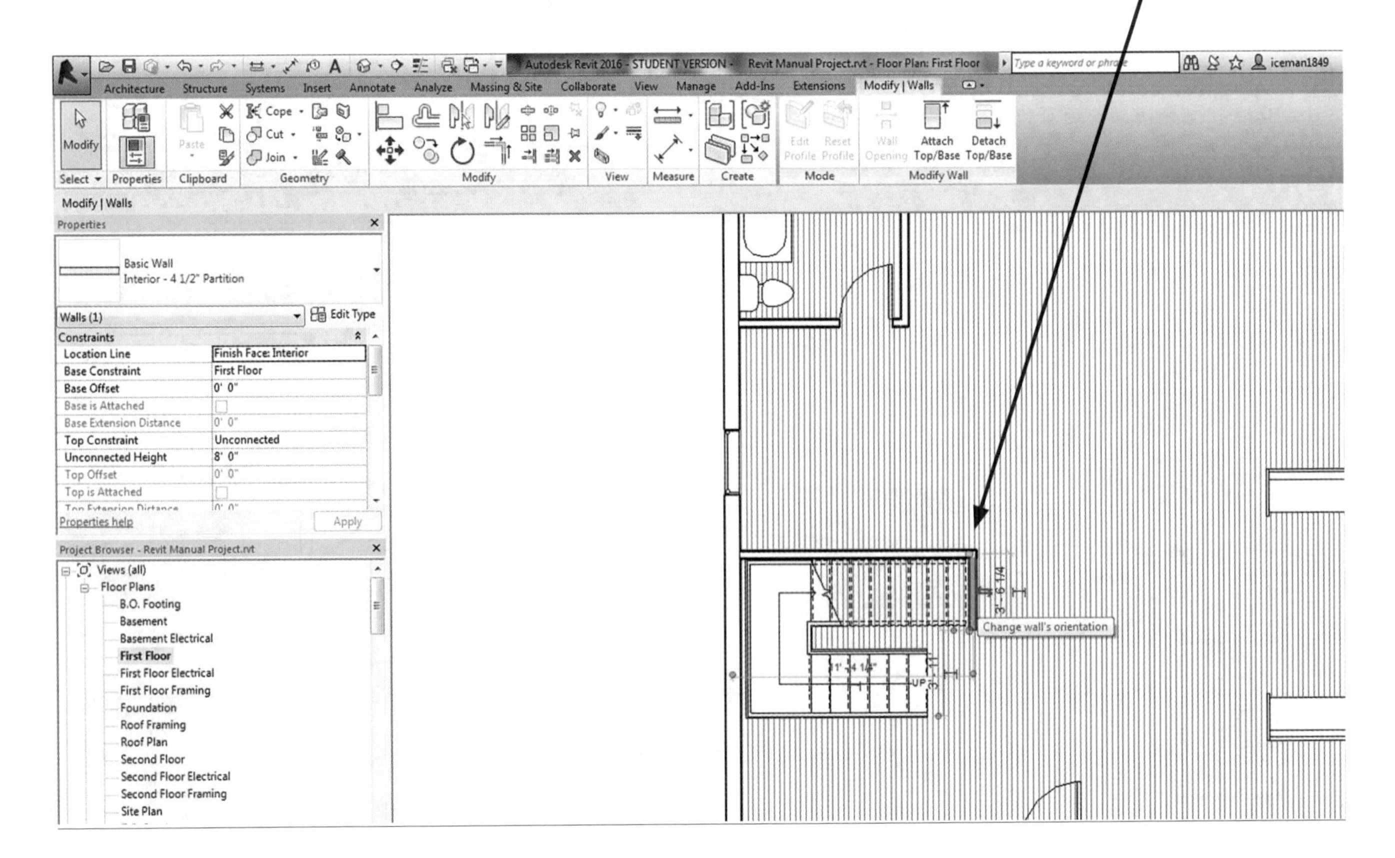

Once the walls are drawn, click the **MODIFY** Arrow.

REMINDER

Do** not **hit the** ESC **Key, as you will cancel other commands, and likely undo a lot of your previous work.

The following screen shows the completed stair layout.

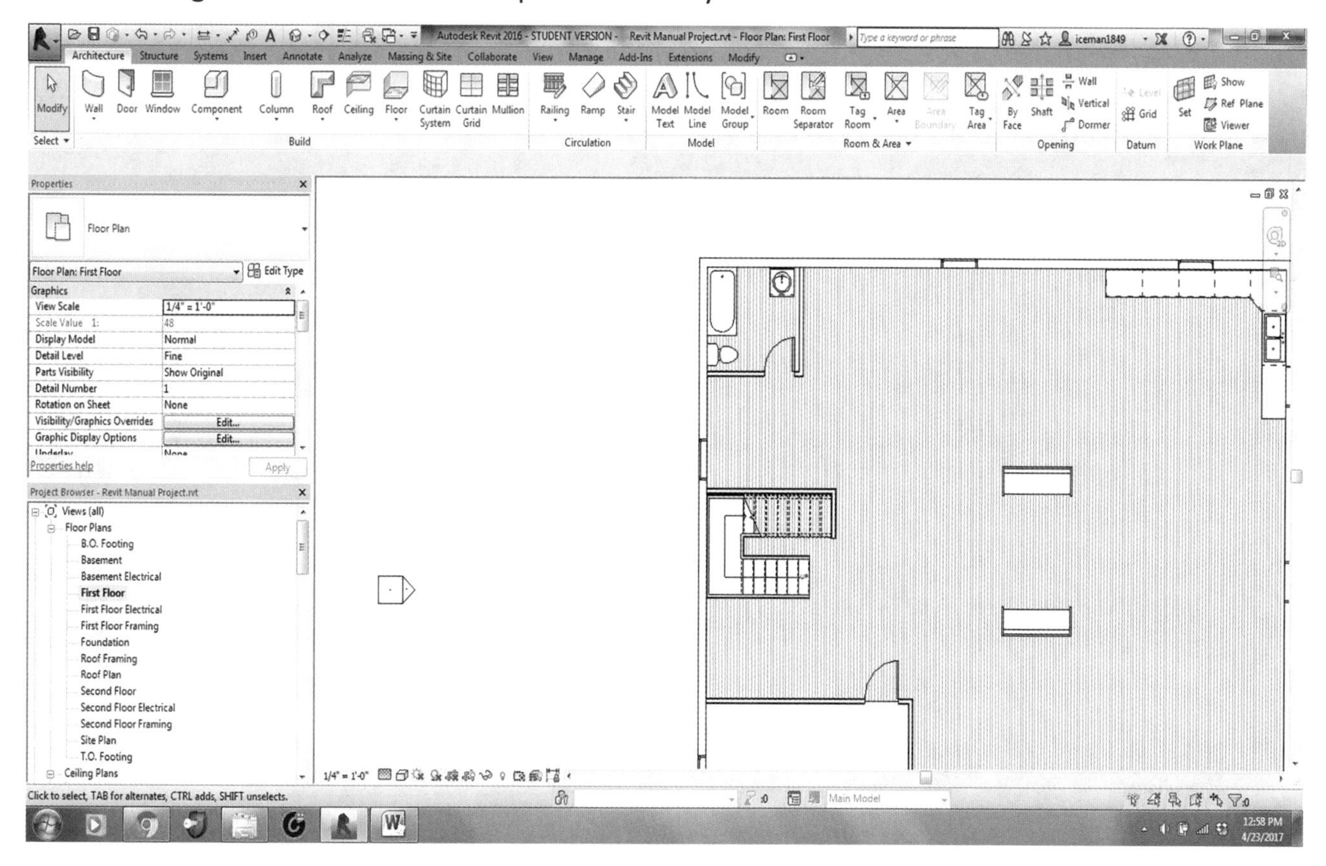

In this Stairs Chapter, Revit assists in creating stairs for you by using the International Residential Code so you as the designer don't have to do the floor-to-floor calculation. This allows you to enjoy the layout of your house or building. The important thing is to just read the screen. Revit will tell you what you are drawing and you will know where you are.

Now let's take the next step to Chapter 9, *Placing Kitchen Cabinets, Counter Top Caseworks and Sinks.*

CHAPTER 9

Placing Kitchen Cabinets, Counter Top Caseworks, and Sinks

PLACING CABINETS

When placing cabinets in your project, first be sure you are on the correct floor plan.

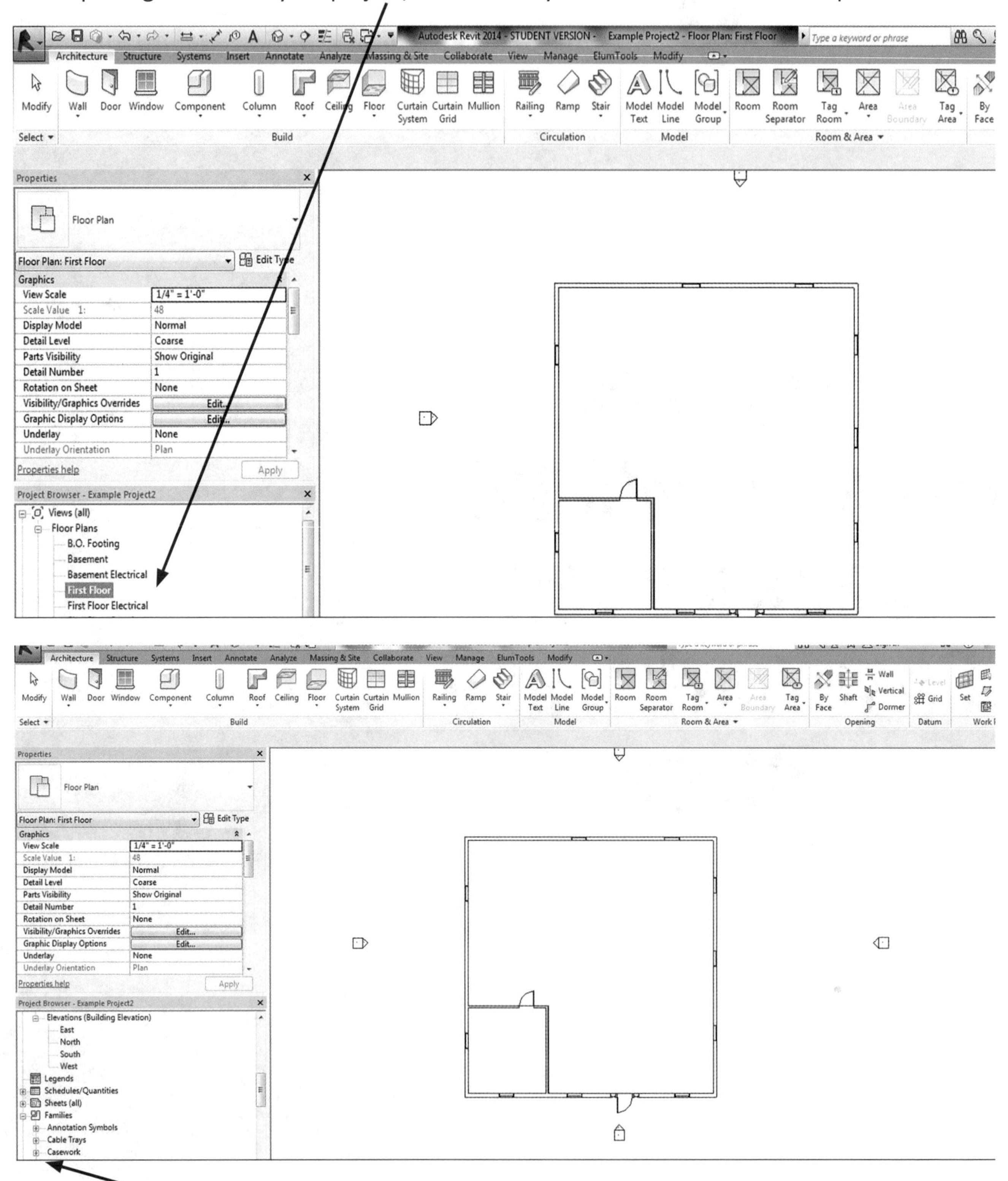

In the Project Browser Box, scroll down to the families category, and then scroll down to the **Casework** category.

Expand the Casework tree to look at the cabinets available and as you expand the cabinet you want, you will see the sizes that are available.

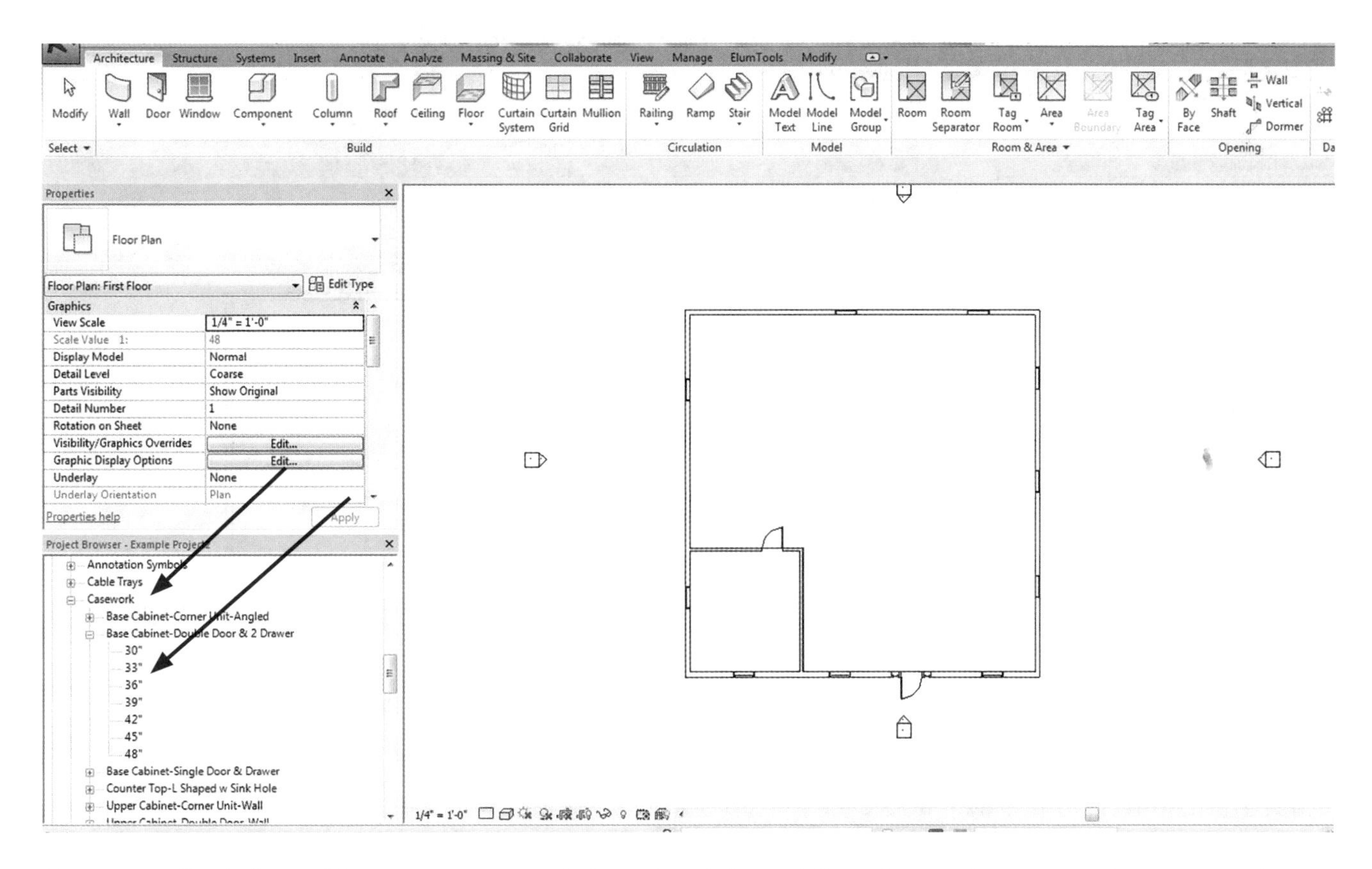

Once you find the cabinet you want, click on it while holding down the mouse button, and drag the cabinet into your design. Note that the curser is located in the back of your cabinet.

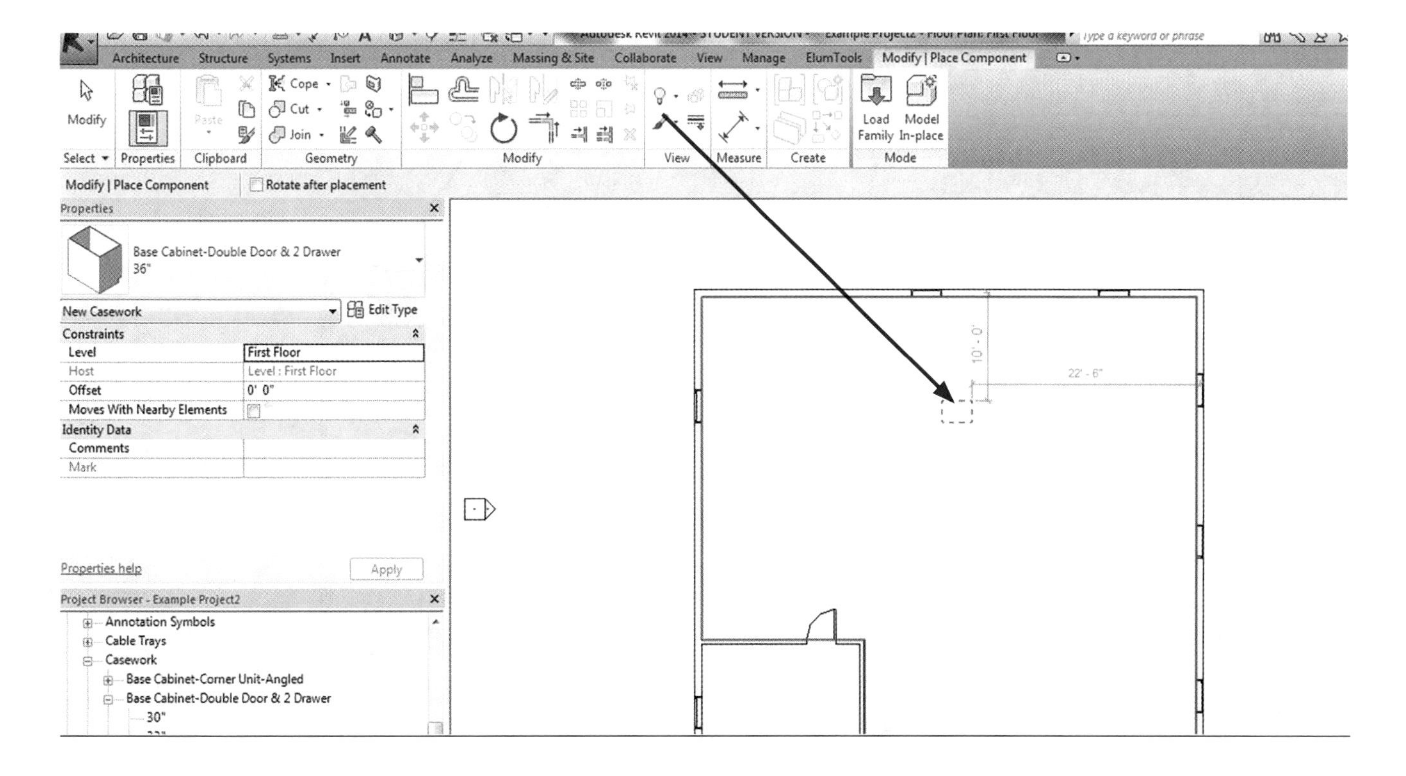

If the cabinet is not in the right direction, click the **SPACE BAR** and it will rotate into the correct direction.

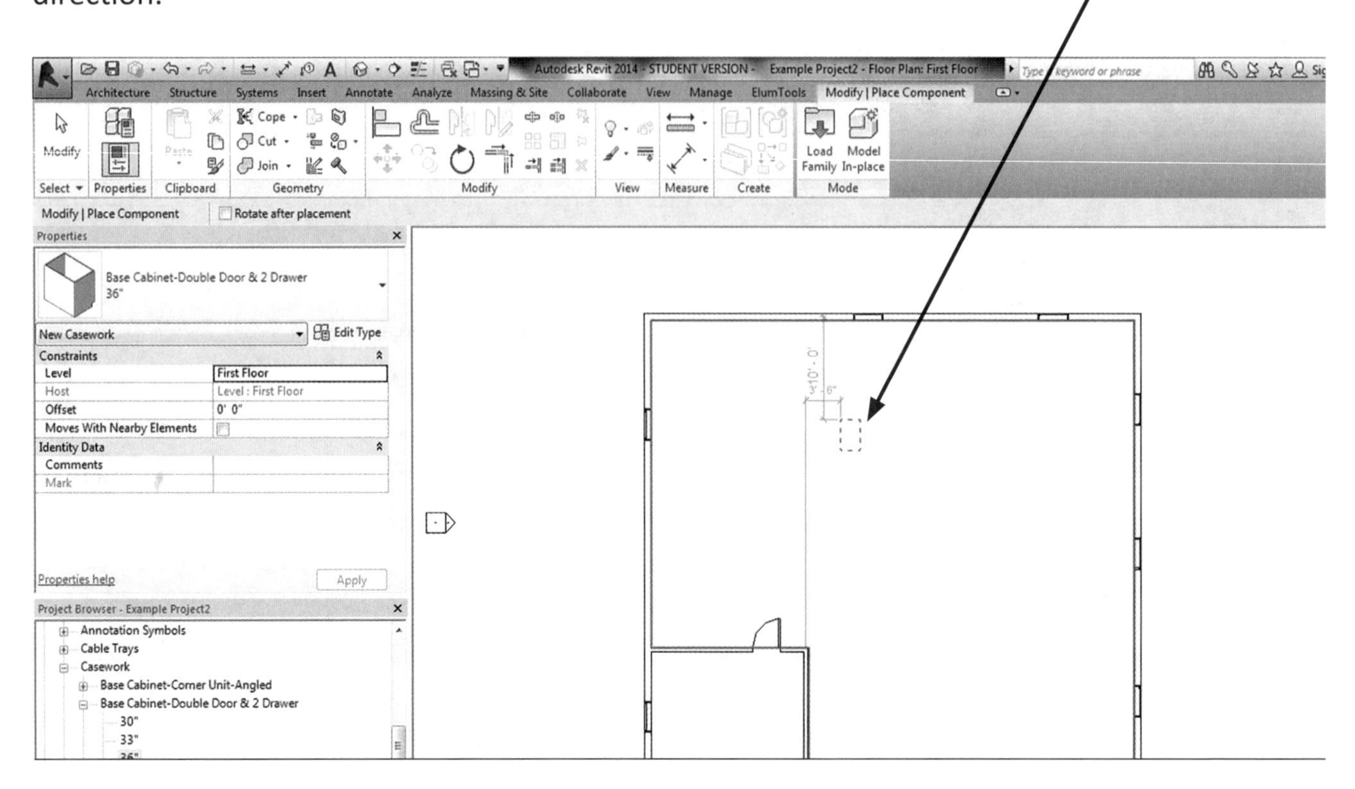

Place the cabinet on the wall and click the cursor, then select another cabinet and continue until you are complete with the kitchen layout.

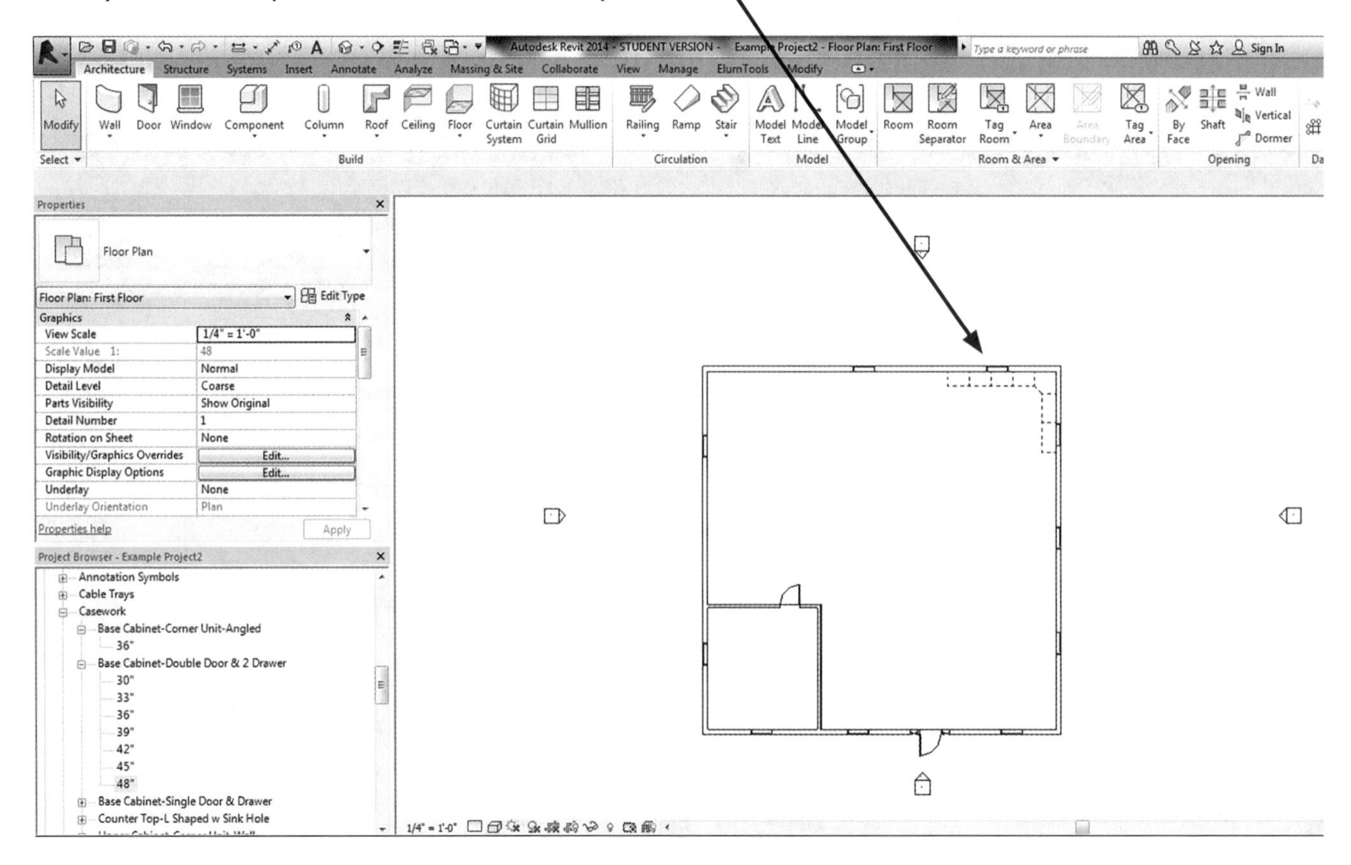

PLACING COUNTER TOP

After placing the cabinets, you will need to place the counter top. In the Project Browser, select the **COUNTER TOP**.

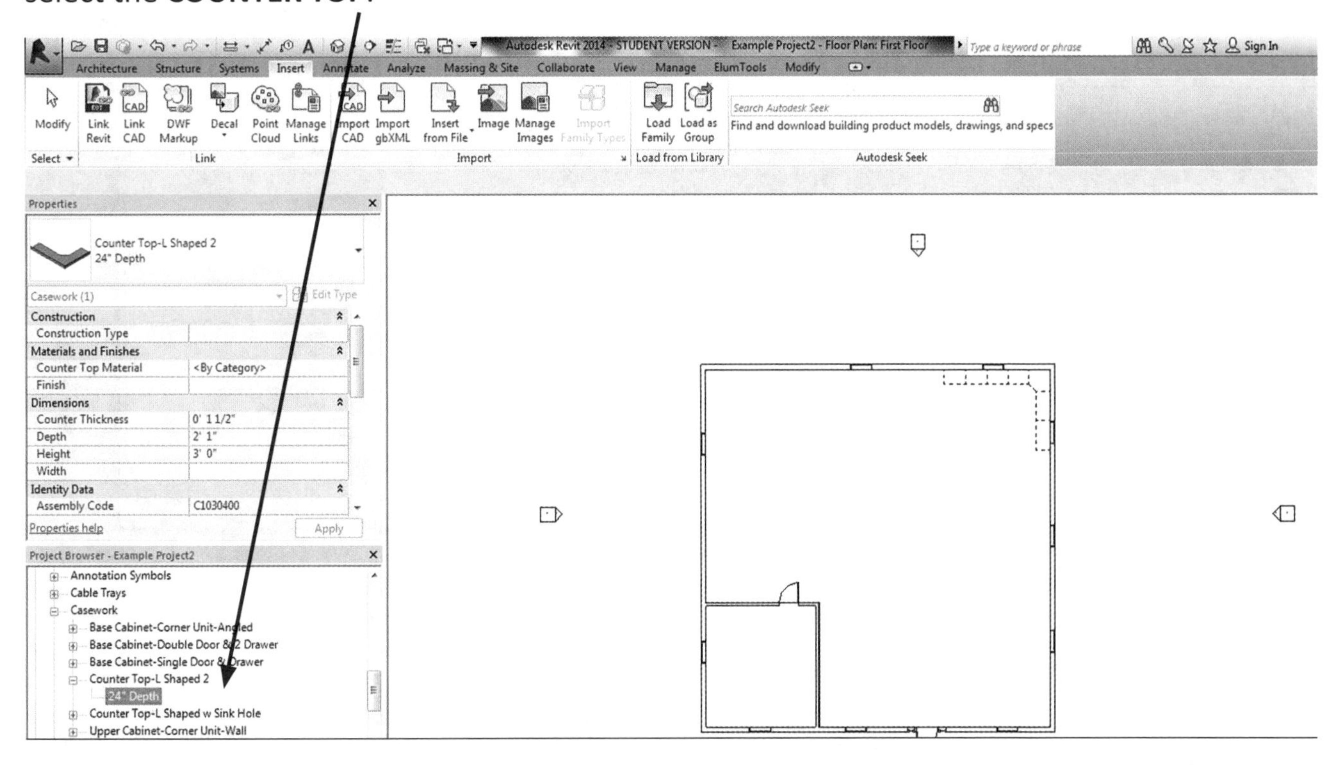

If the counter top you need isn't shown in the Project Browser then you will need to load the correct one for your project.

Click on the **INSERT** Tab, and click on **LOAD FAMILY**.

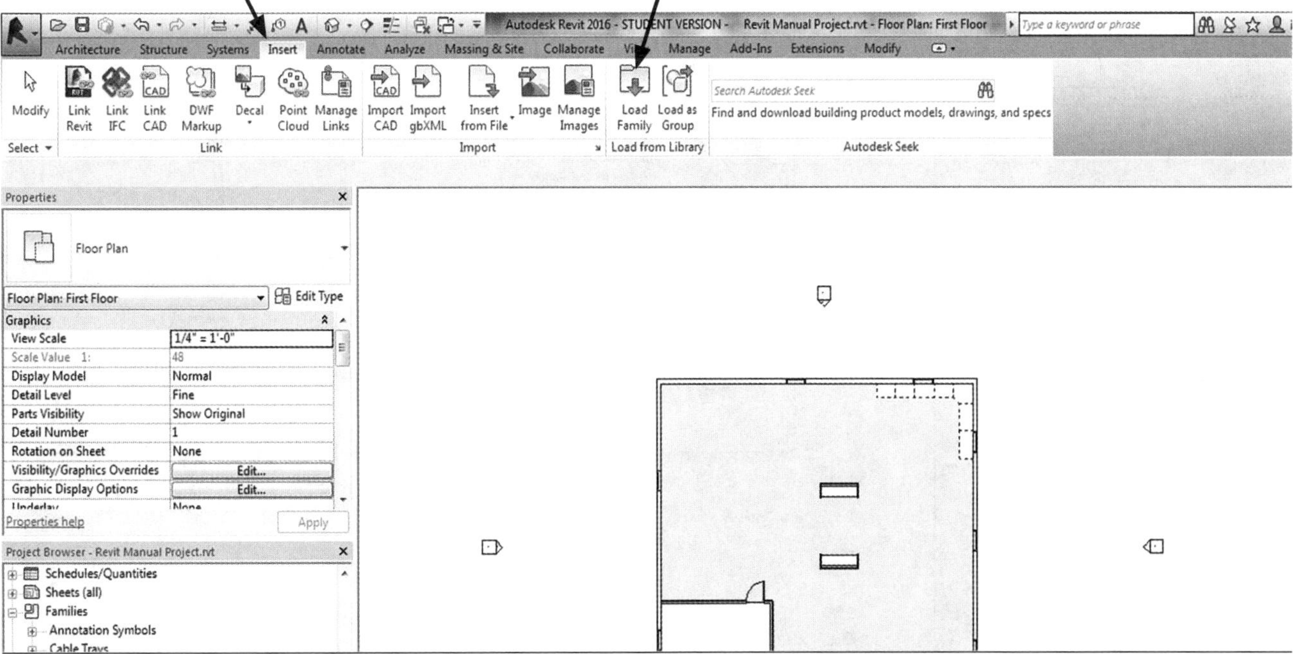

Here you will get a Dialog Box with load family template folders that are stored in the Revit Software.

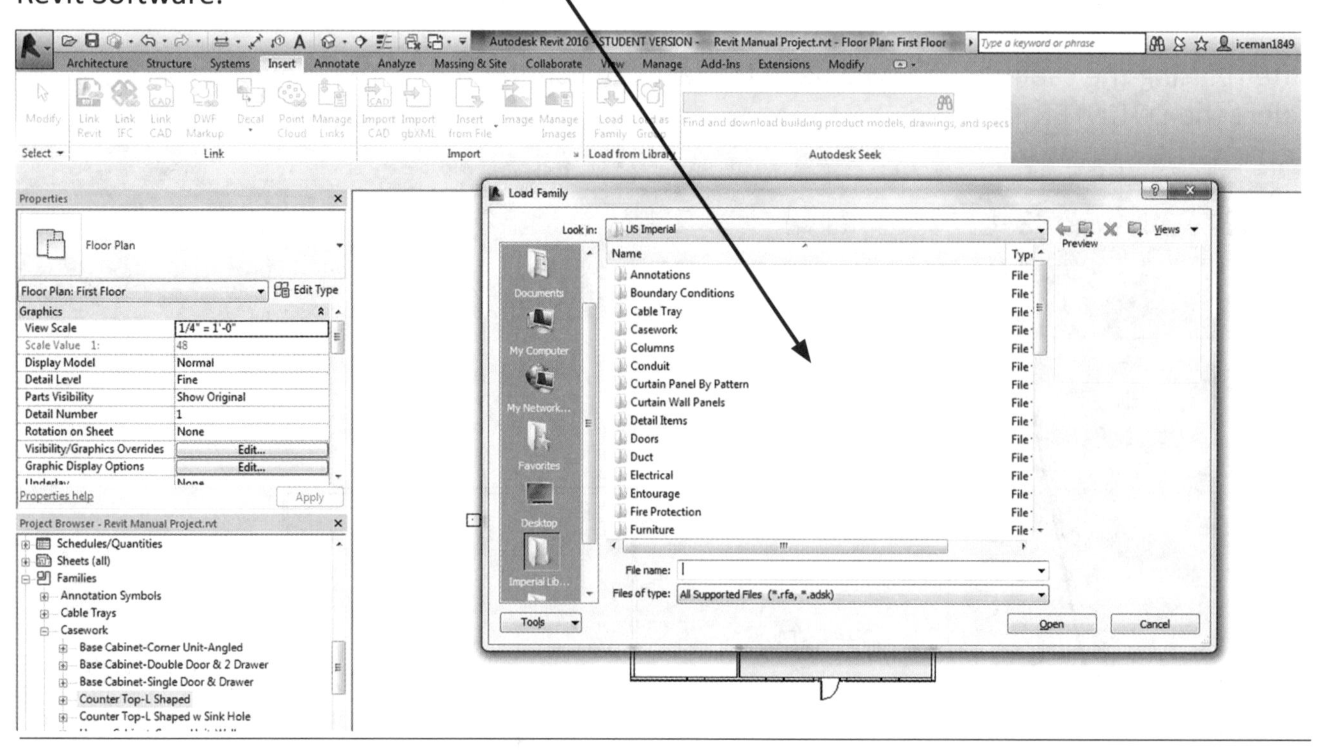

Next click on **CASEWORK,** and open that folder.

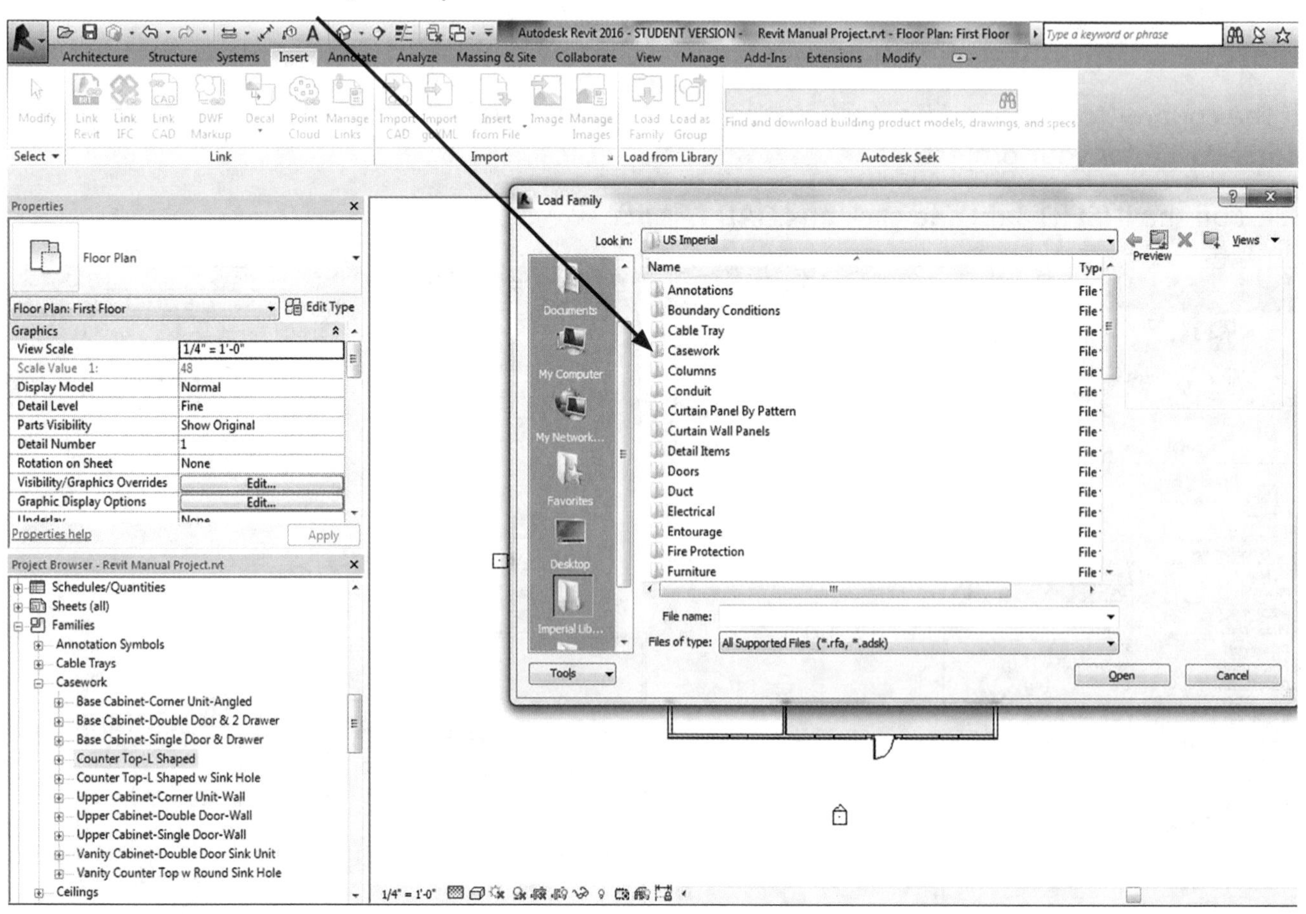

Then click on the **COUNTER TOPS** folder.

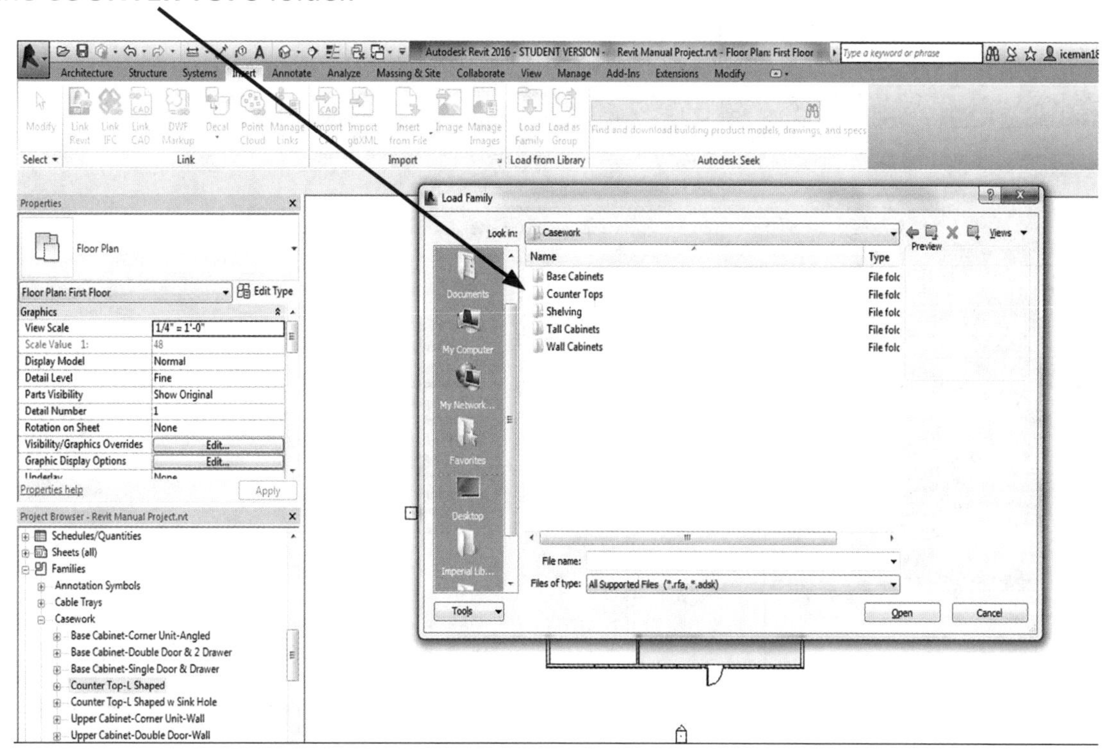

Then you will select the counter top that best suits your design layout, or application.

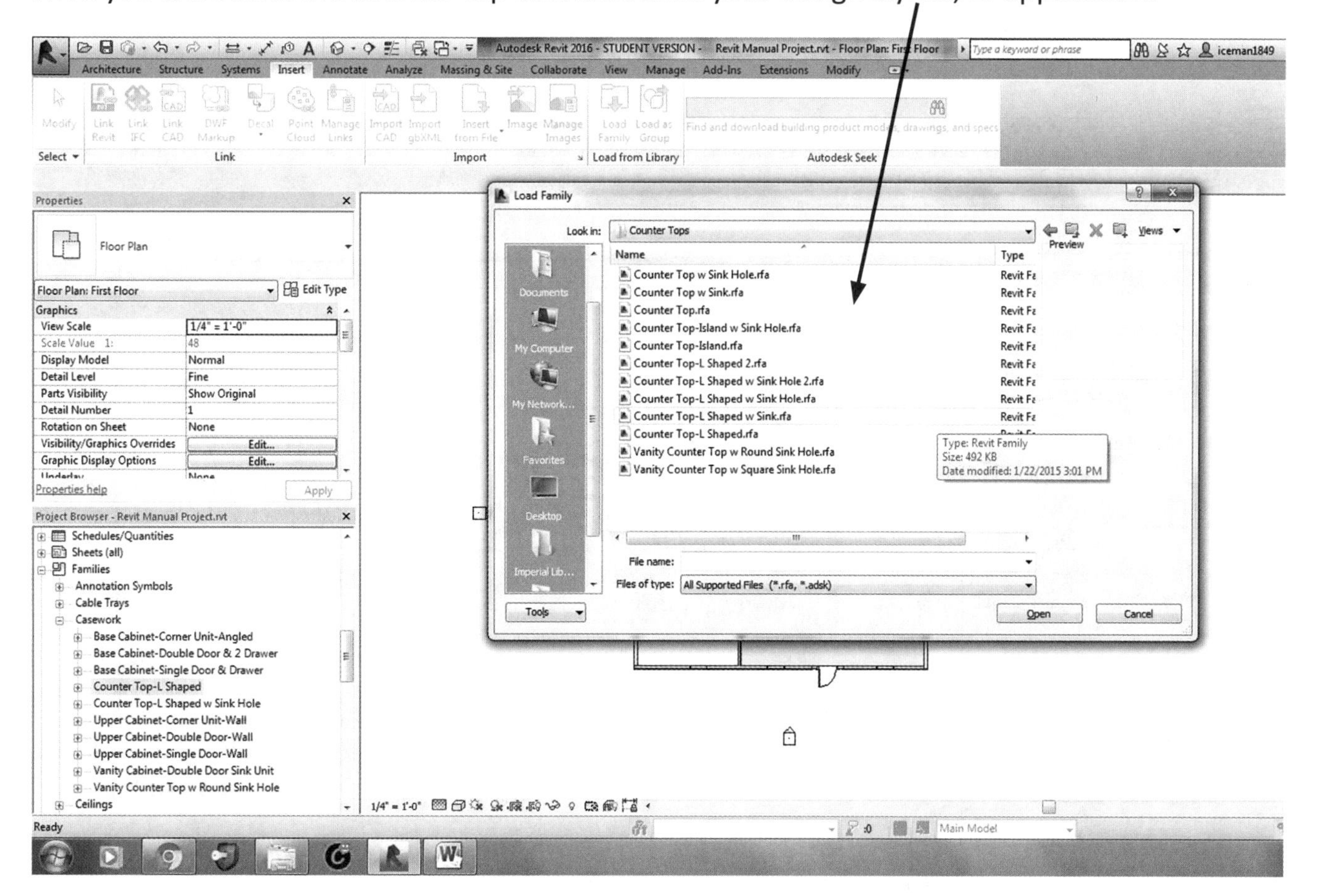

You can select the counter top with the angle cabinets and with a sink hole. You will get an image of what the family will look like in the window on the right.

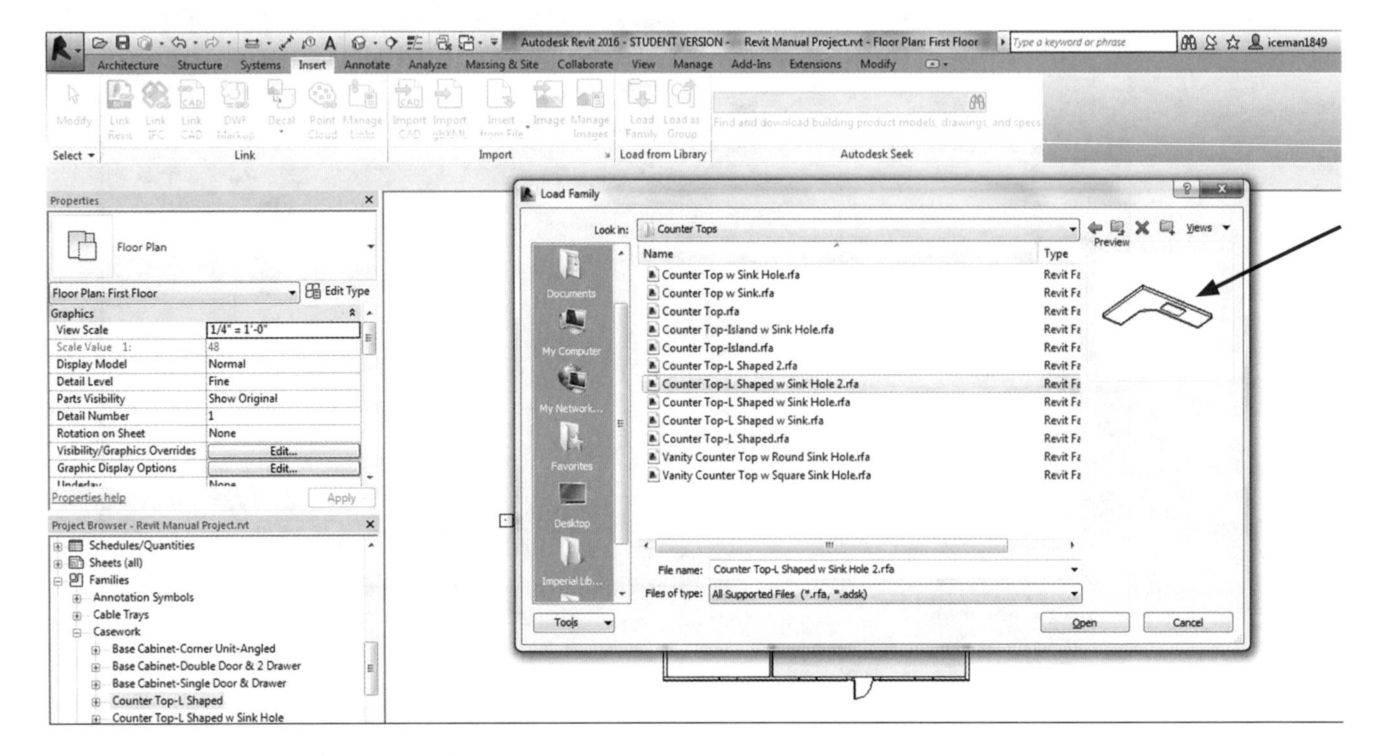

Once you have selected the counter top you want for your design, click **OPEN** and the counter top will be loaded into your project in the Project Browser.

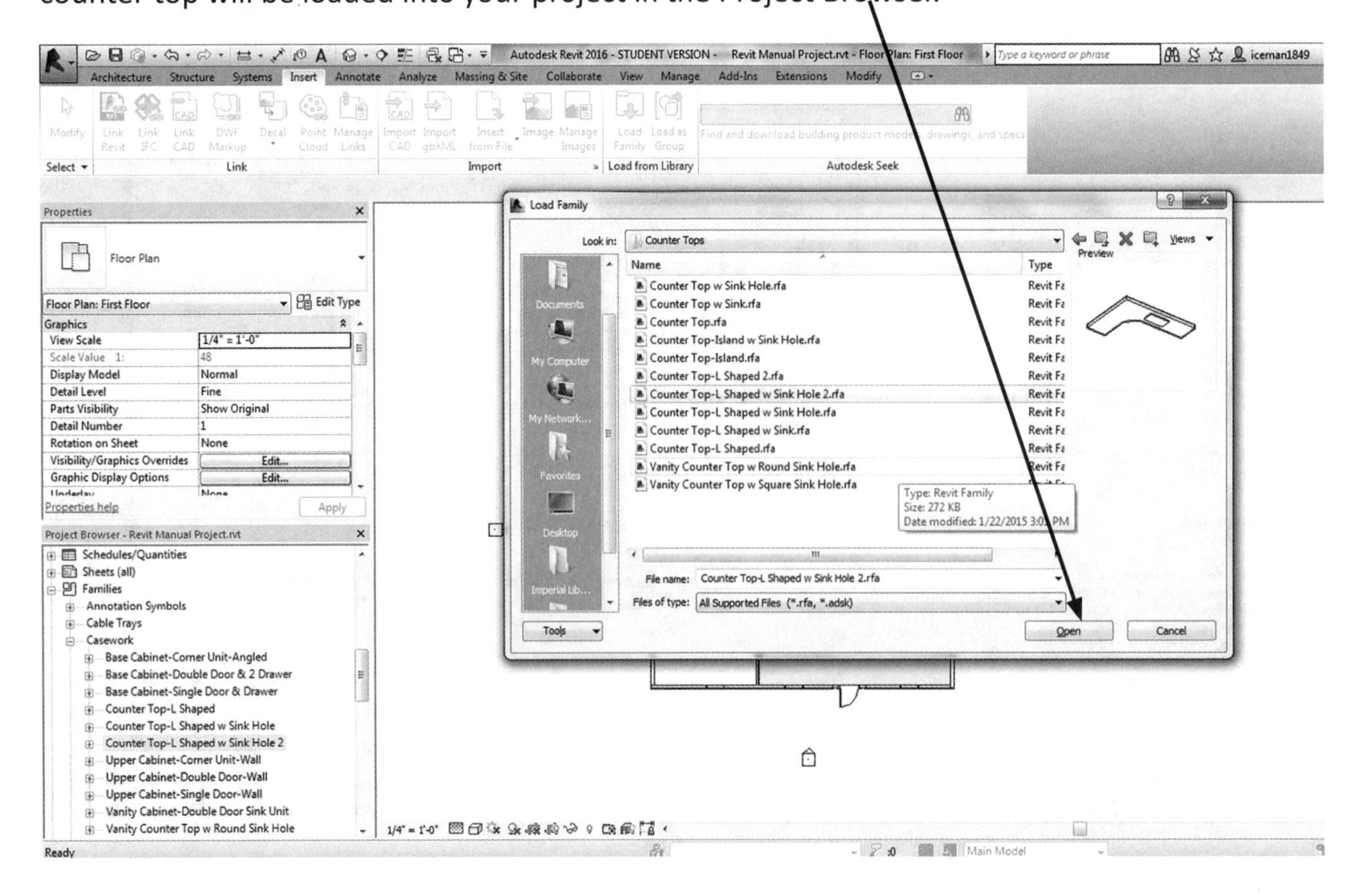

The Counter Top is now loaded in your Project Browser.

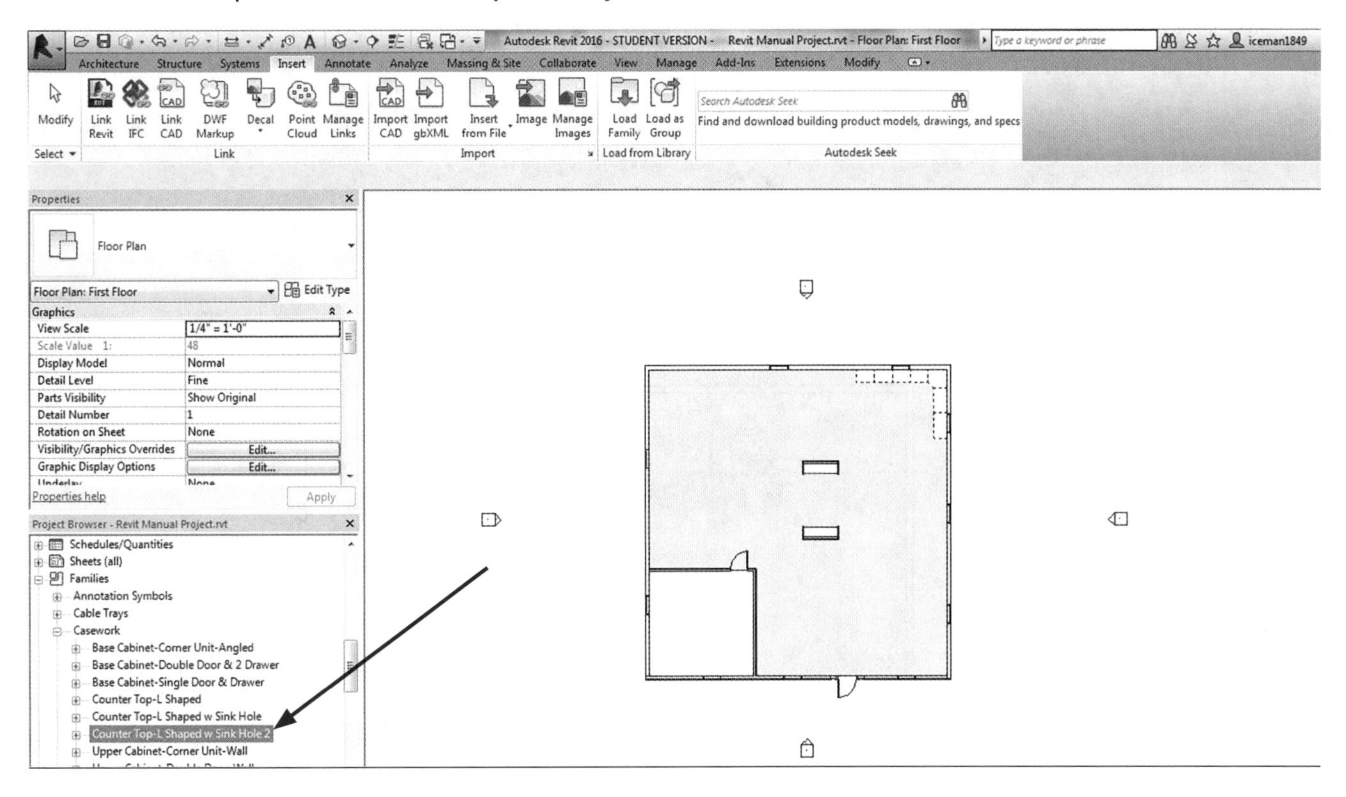

Once the counter top is loaded into your project, then you will click on the counter top name and expand the tree (using the "+"— this is the *only* way you will be able to select the family in order to place the object in your drawing).

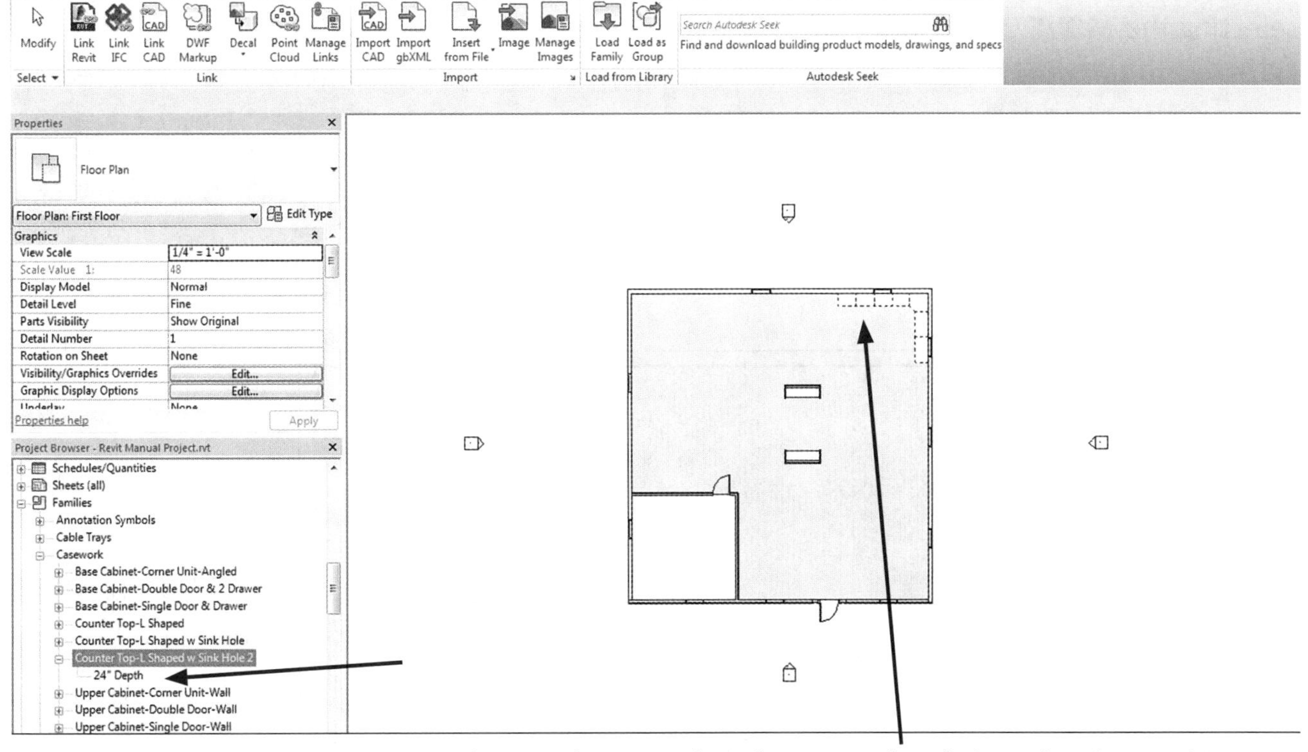

At a later time, you can remove a base cabinet and replace it with a dishwasher located in specialty equipment family category.

Click once on your counter top and drag the counter top into place. As you bring it into place, if it isn't in the correct direction you can click the **SPACEBAR** and the counter top will rotate to place it in the correct position.

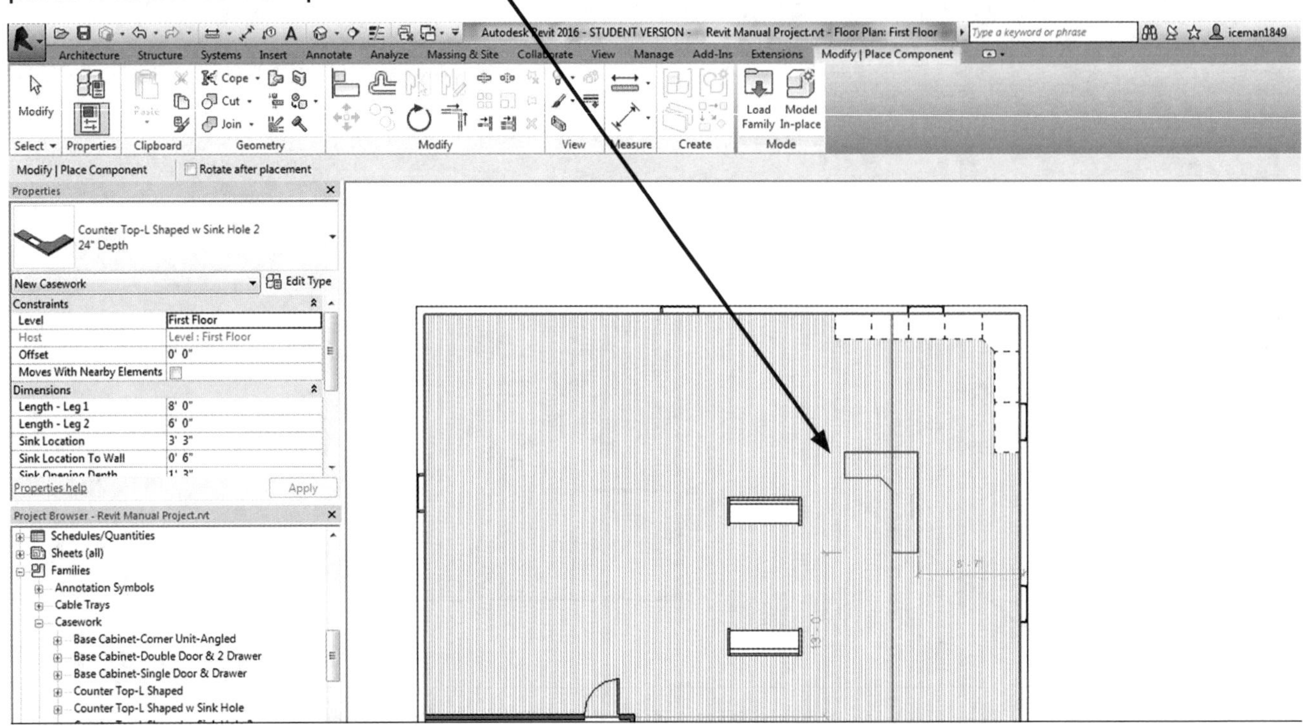

You can even mirror the counter top if you need to depending on the direction (refer to Chapter 20, *Revit Modify Commands*).

Place the counter top in the corner of the cabinets by selecting the **COUNTER TOP** and then selecting the **MOVE** Command. Pick the corner of the counter top and drag it onto the corner of the cabinets.

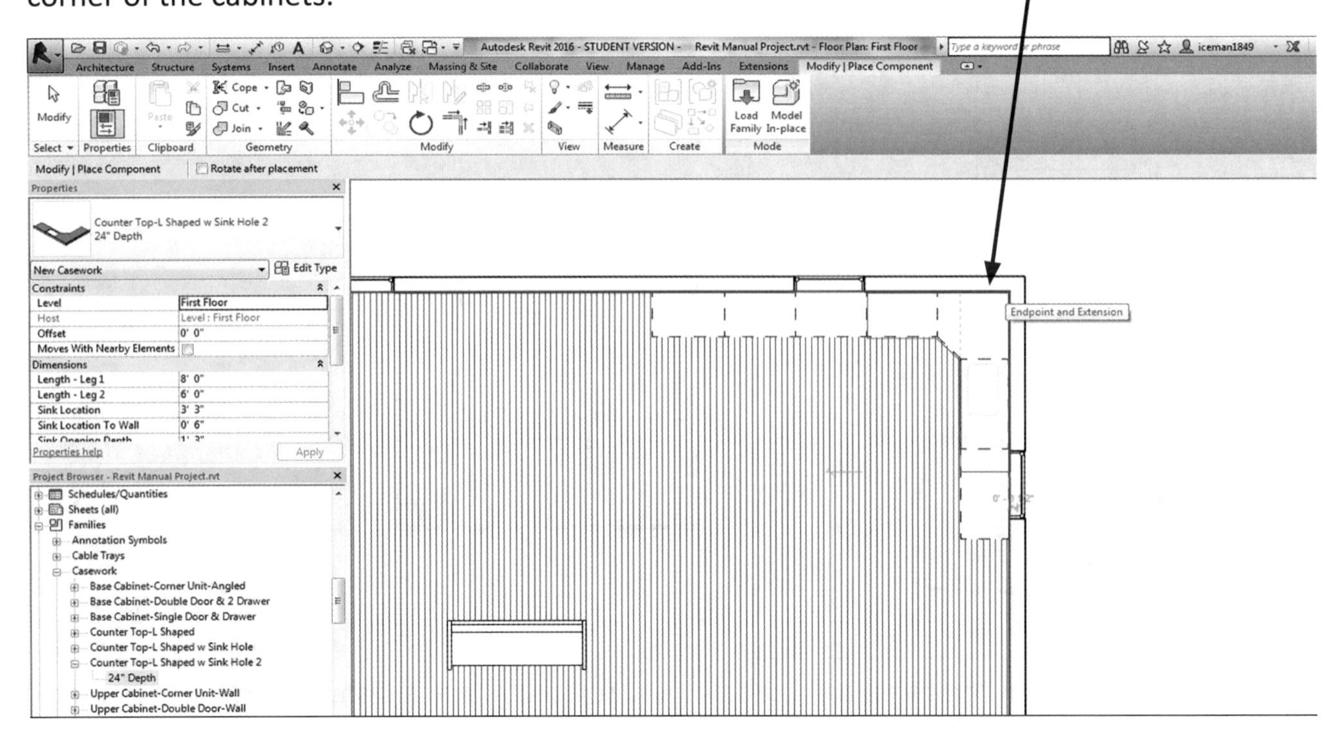

Once you have placed the counter top, Revit automatically has another counter top attached waiting to place another one, so click the **MODIFY** Arrow to end the command and *not* the **ESC** Command.

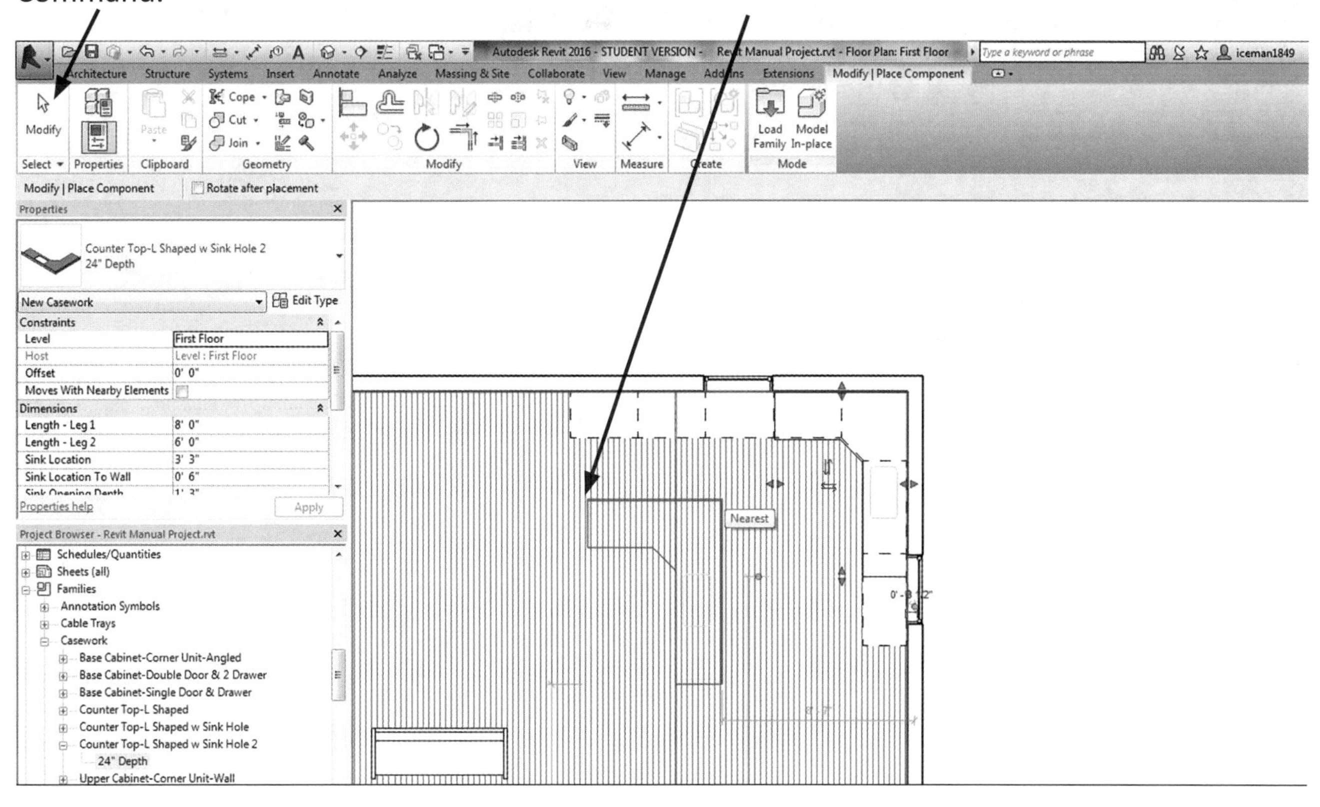

After hitting the **MODIFY** Arrow it clears the command.

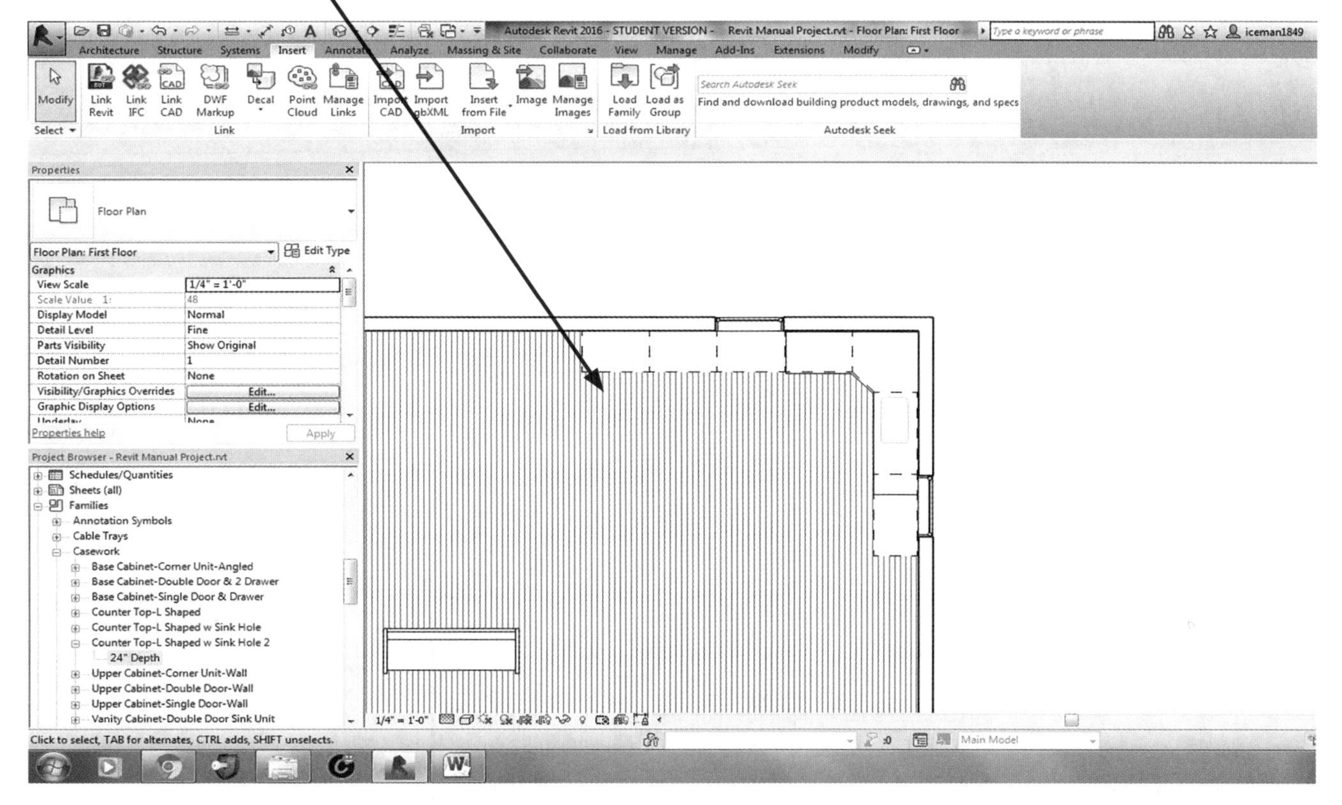

After you have clicked the **MODIFY** Arrow, you will have to reselect the counter top again to activate the adjustment arrow tabs so you can stretch the counter top to the right size.

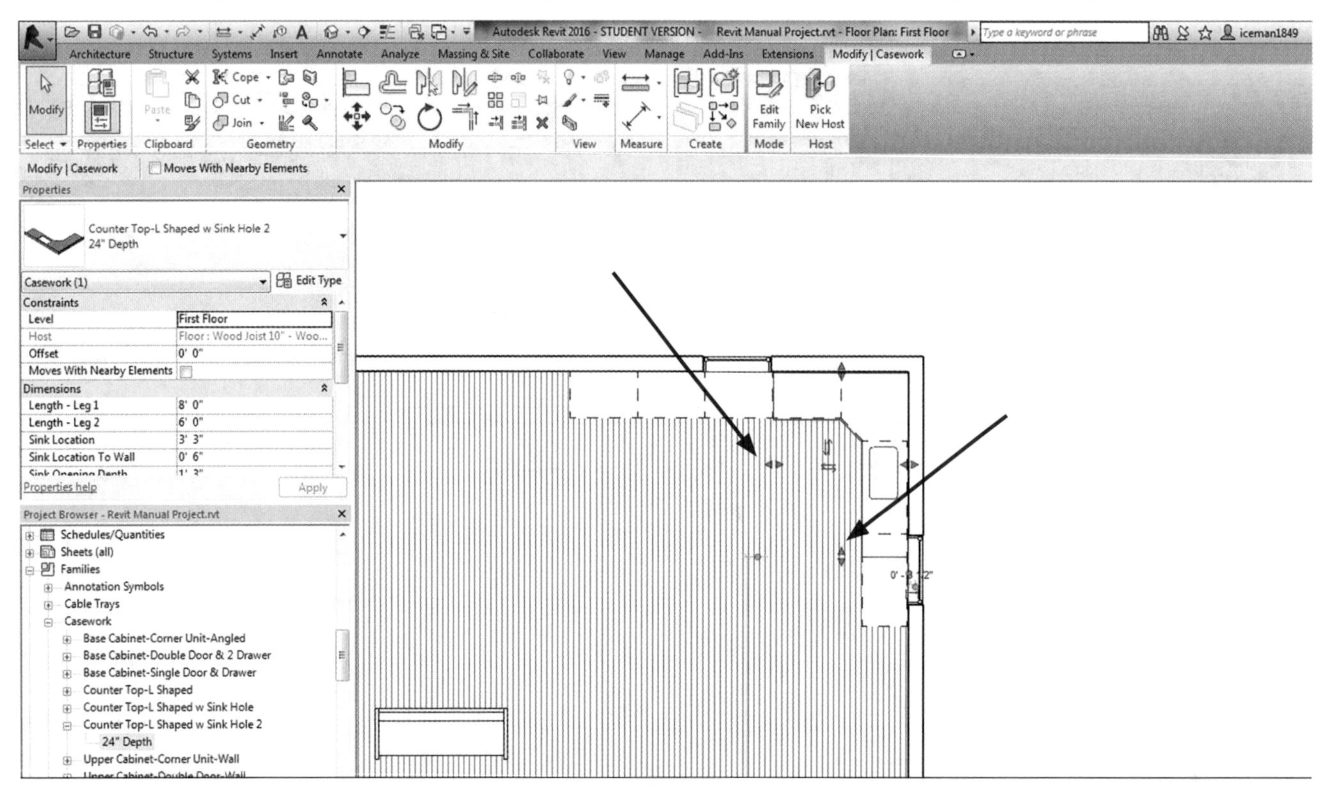

Click or grab onto the **GRIPS** and pull the arrows in one direction for the end of the counter top to move. You will notice dashed lines so you can see the edge of your counter as you stretch it.

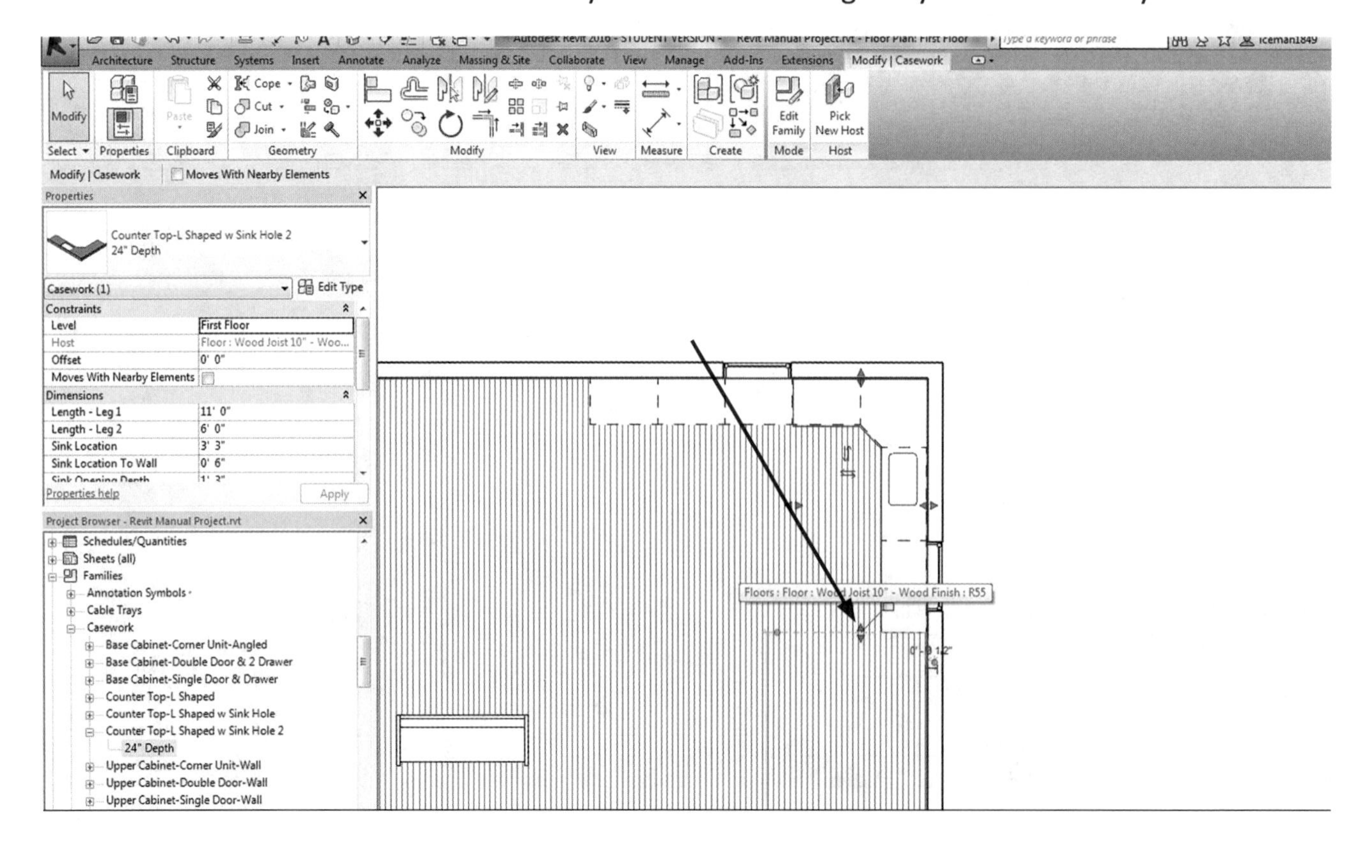

This allows you to size the counter top correctly. Both sides of the counter top will be adjusted.

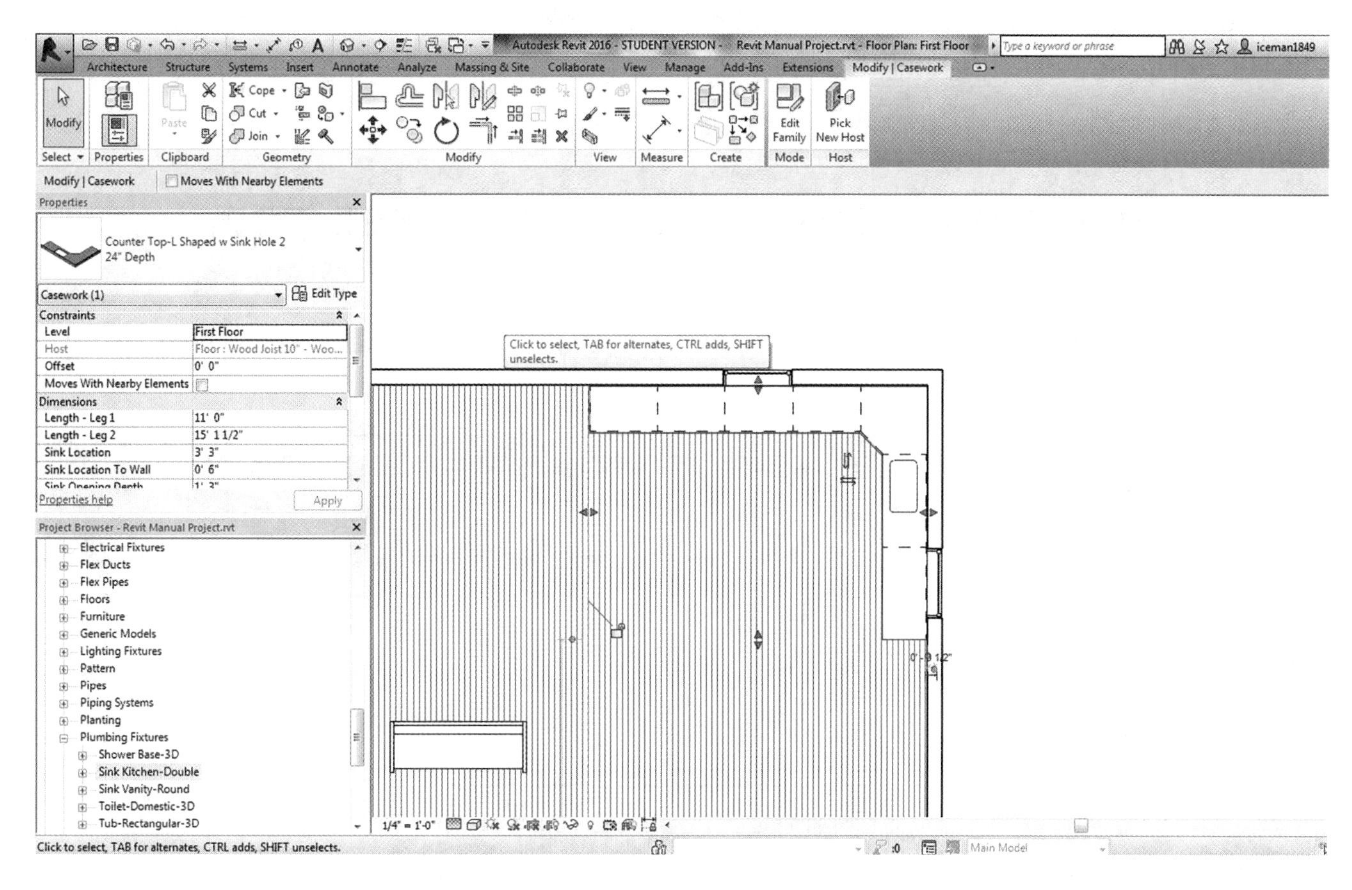

Once the counter top is correctly sized you will click the **MODIFY** Arrow to end the command and you are finished with placing the counter top.

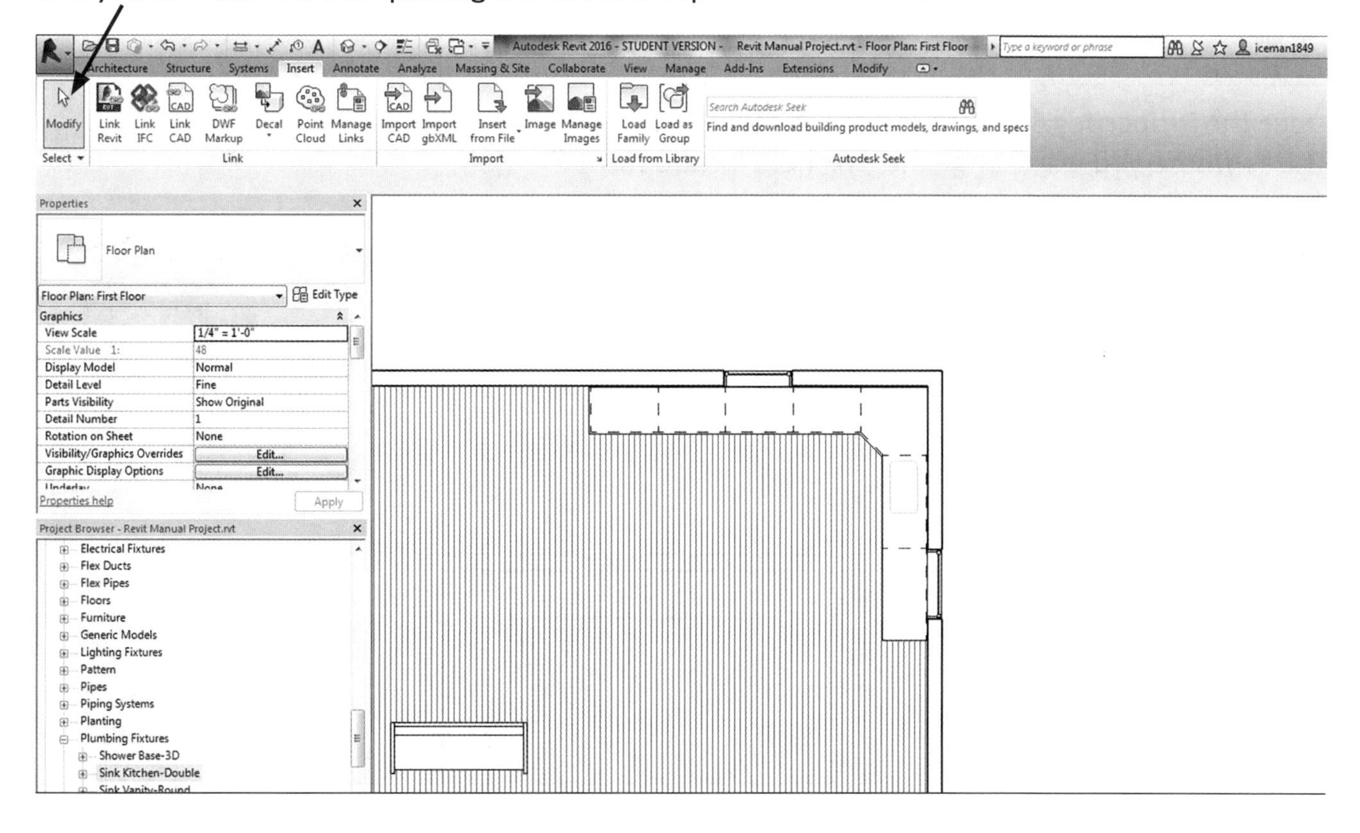

Next step you will place a sink in the sink hole in the counter top. After placing the counter top, you may notice that the sink hole is not centered, this means that you will have to adjust the sink hole location in the counter top.

PLACING A SINK IN THE COUNTER TOP

You will first click on the **COUNTER TOP** to activate it, then the Properties Box will change on the left of the screen.

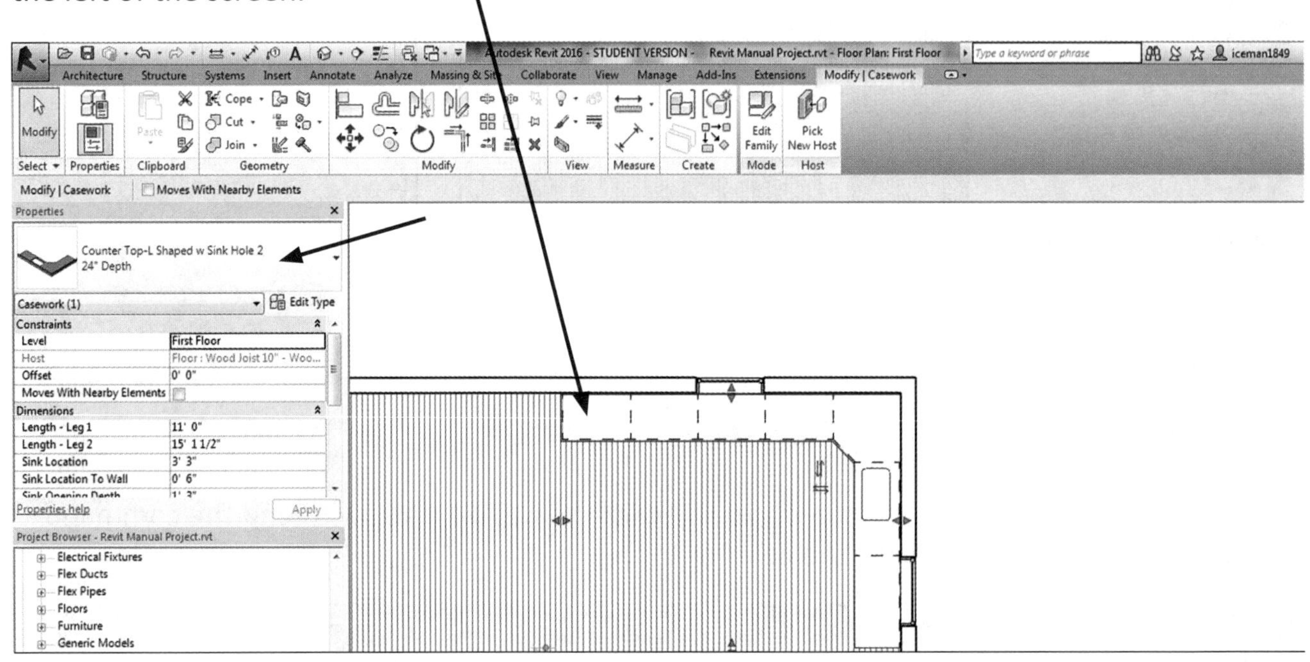

Once the Properties Box changes, in the middle you will notice a section called Dimensions. This is where you will change the sink hole location.

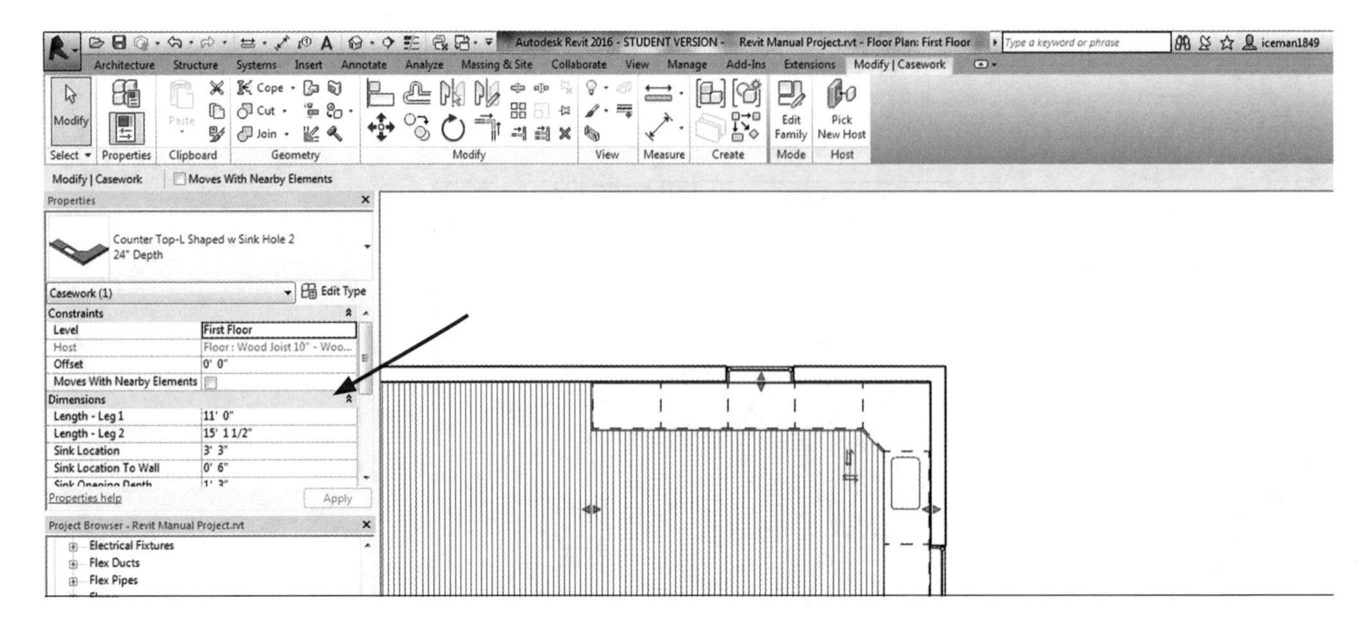

Locate the line that reads: Sink Location, this will move the sink to the left or right on the counter top.

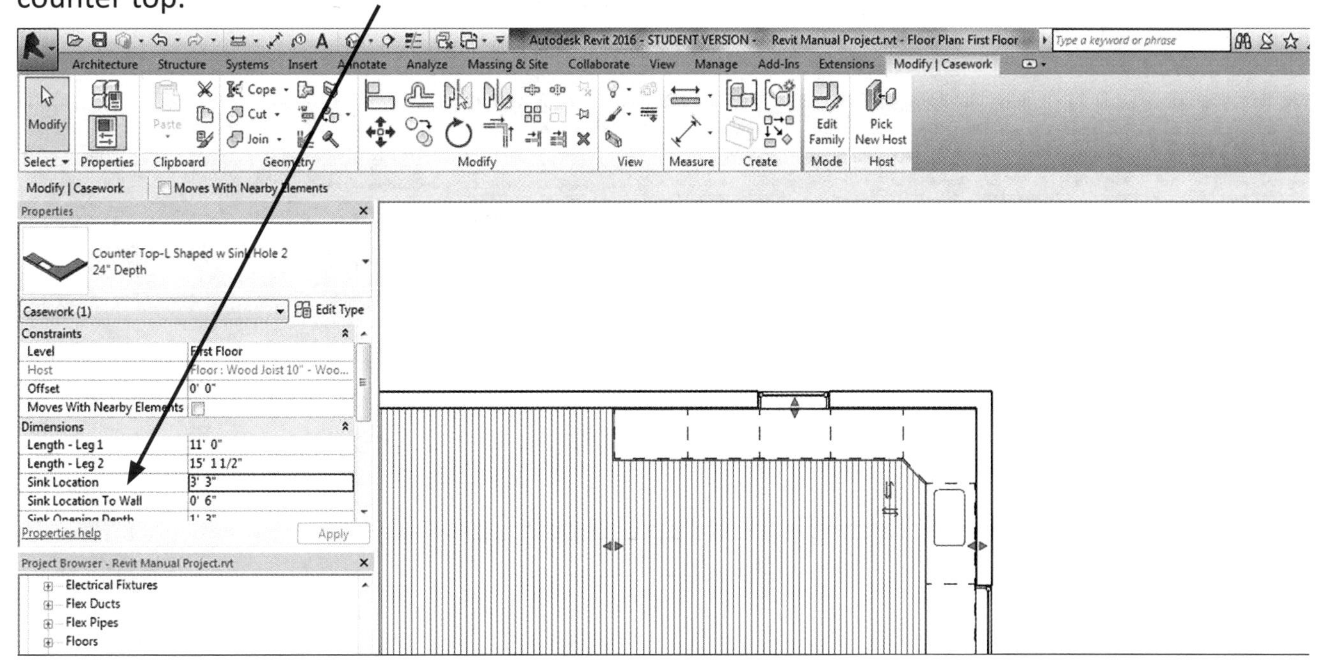

Click in the measurement of the Sink Location and change the measurement. Then click **APPLY** and the sink hole will slide either to the left or the right. (Apply will be visible once you change information in the Properties Box.)

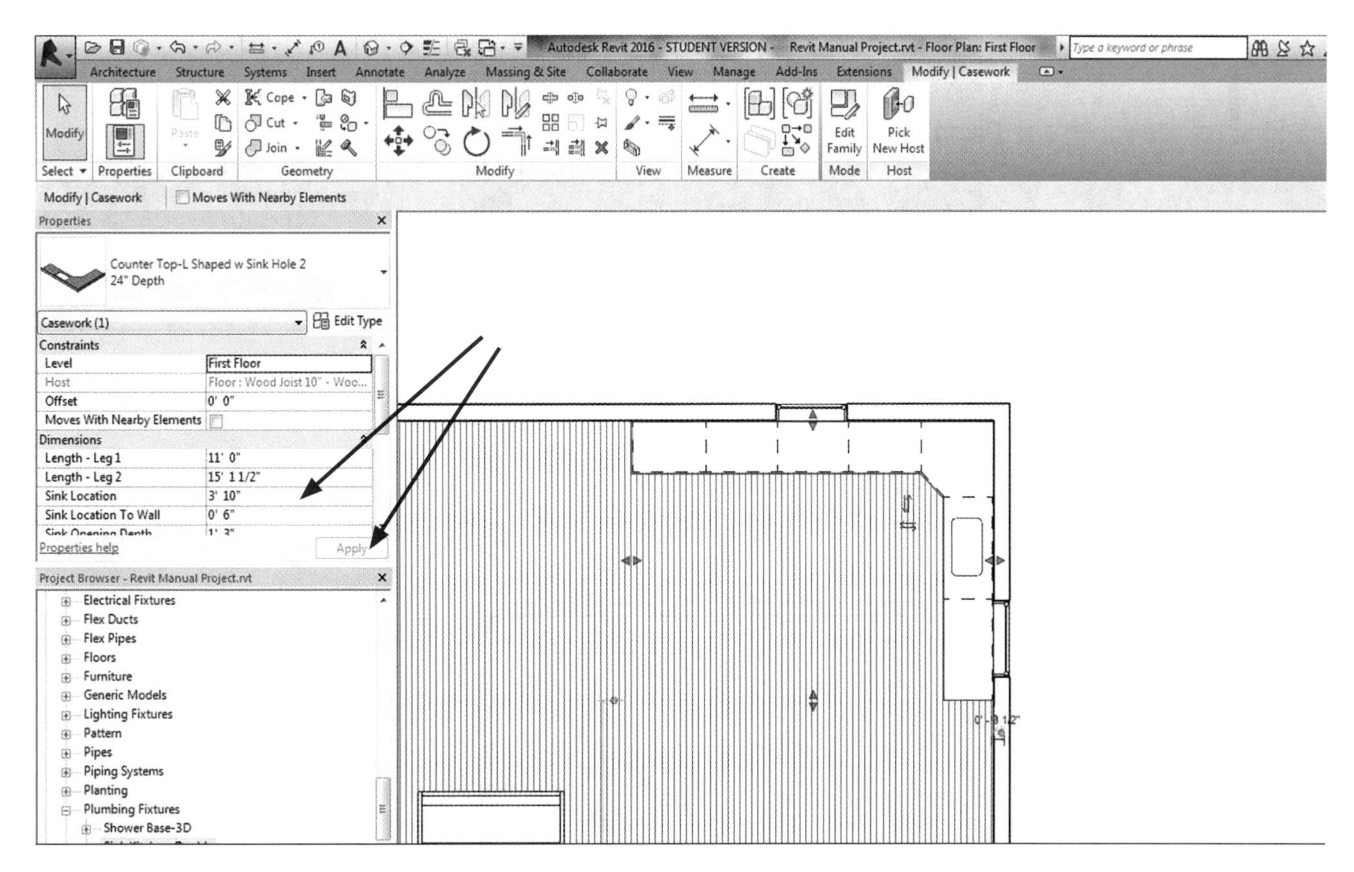

Keep changing the measurement until the sink is in the desired location.

Once the sink hole is placed where you desire it to be, you can place a sink. Just like the previous steps in placing casework and doors and windows, in the Project Browser, scroll down to the Families category.

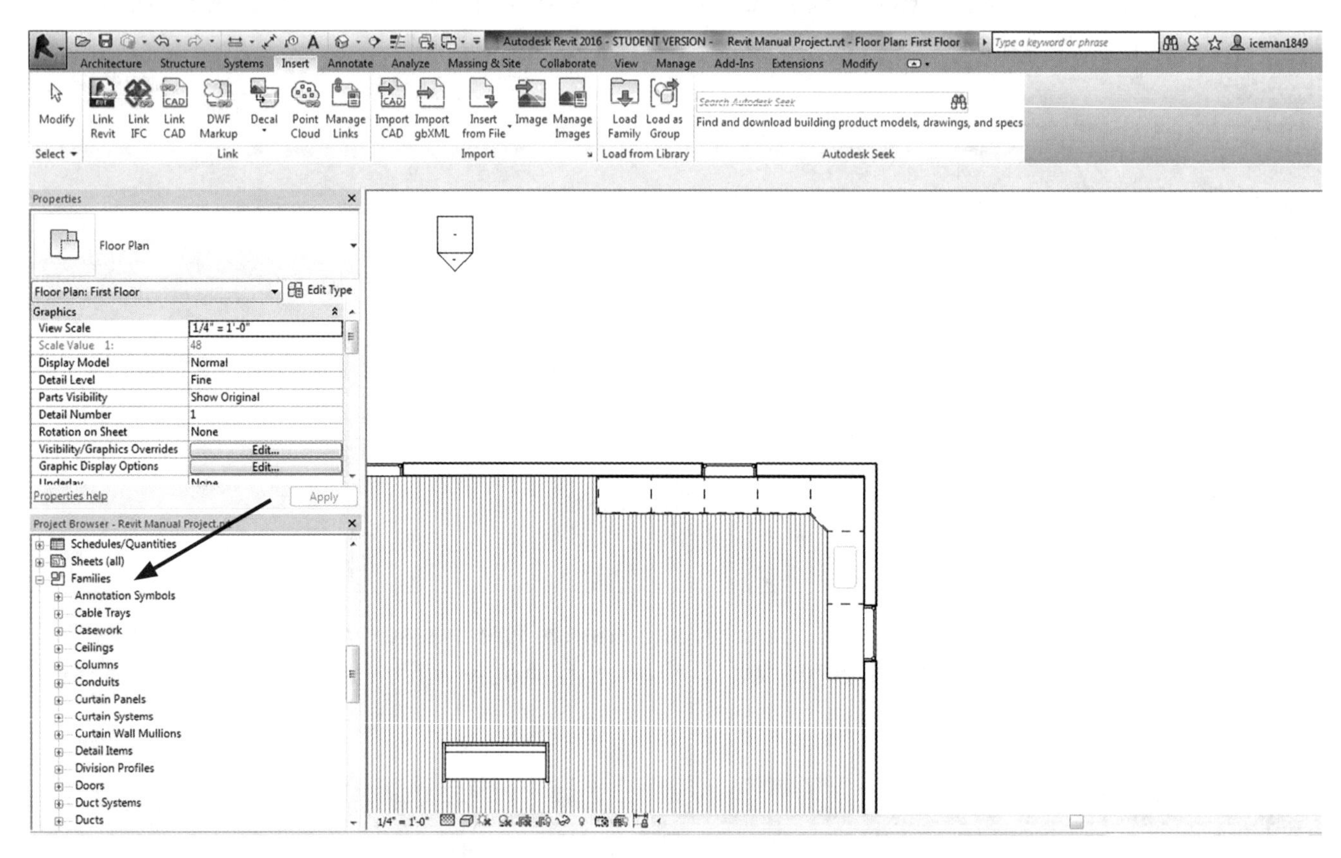

Next scroll down to the Plumbing Fixtures.

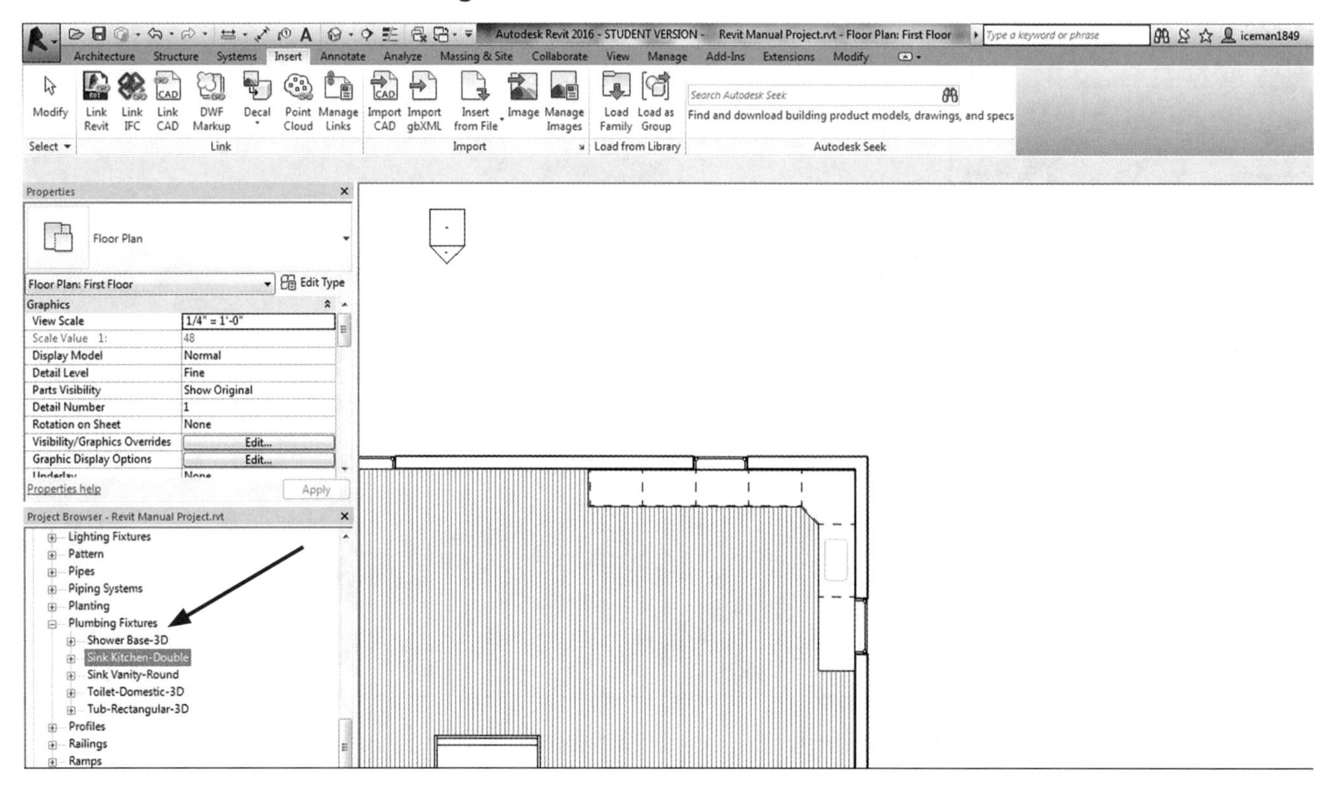

Expand the plumbing fixtures tree ("+", if not already expanded), locate the Sink Kitchen-Double and expand that tree ("+"). This will be the *only* way you can access or drag the sink into the drawing or on the counter top.

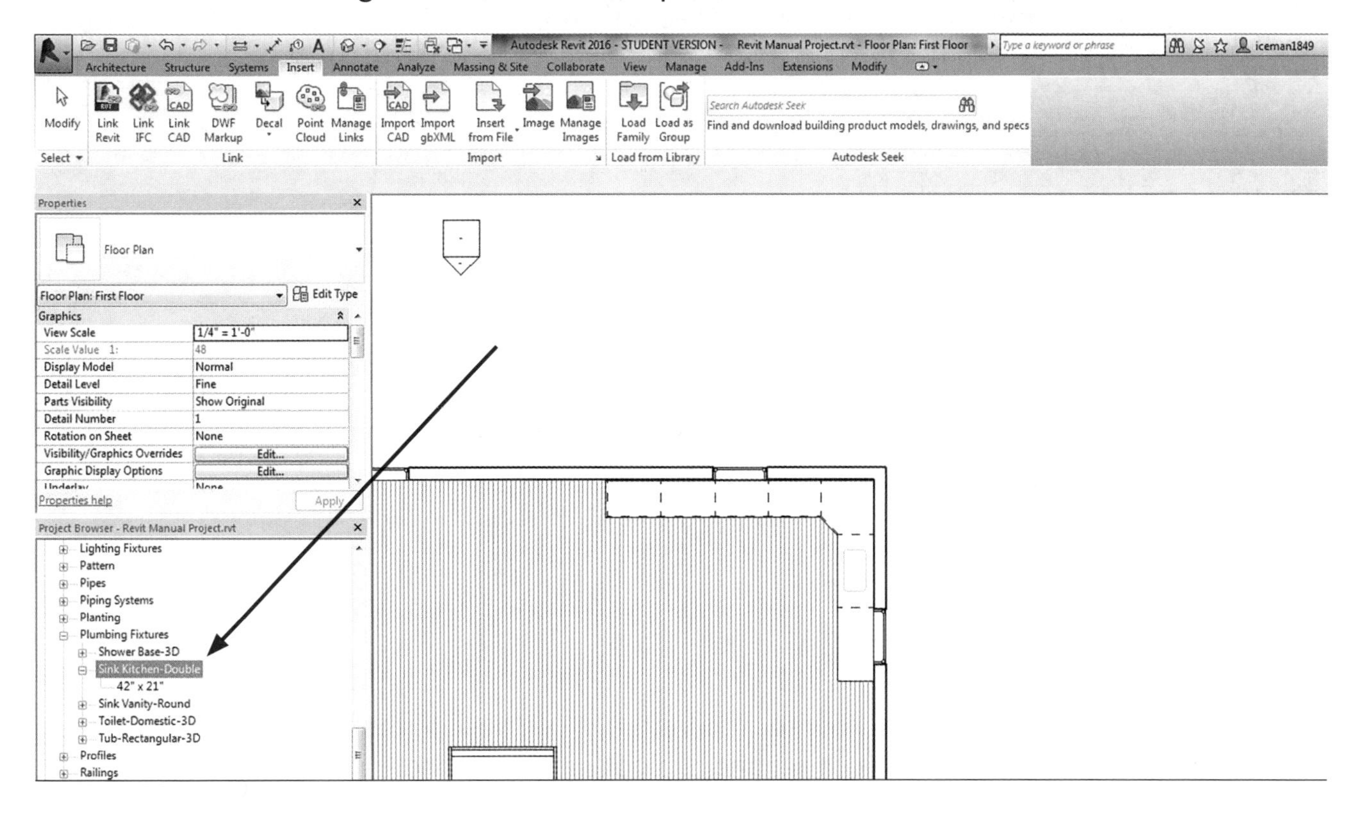

Once you have expanded the Sink Kitchen-Double, click and hold the mouse button down and drag the sink over to the top of the counter top sink hole.

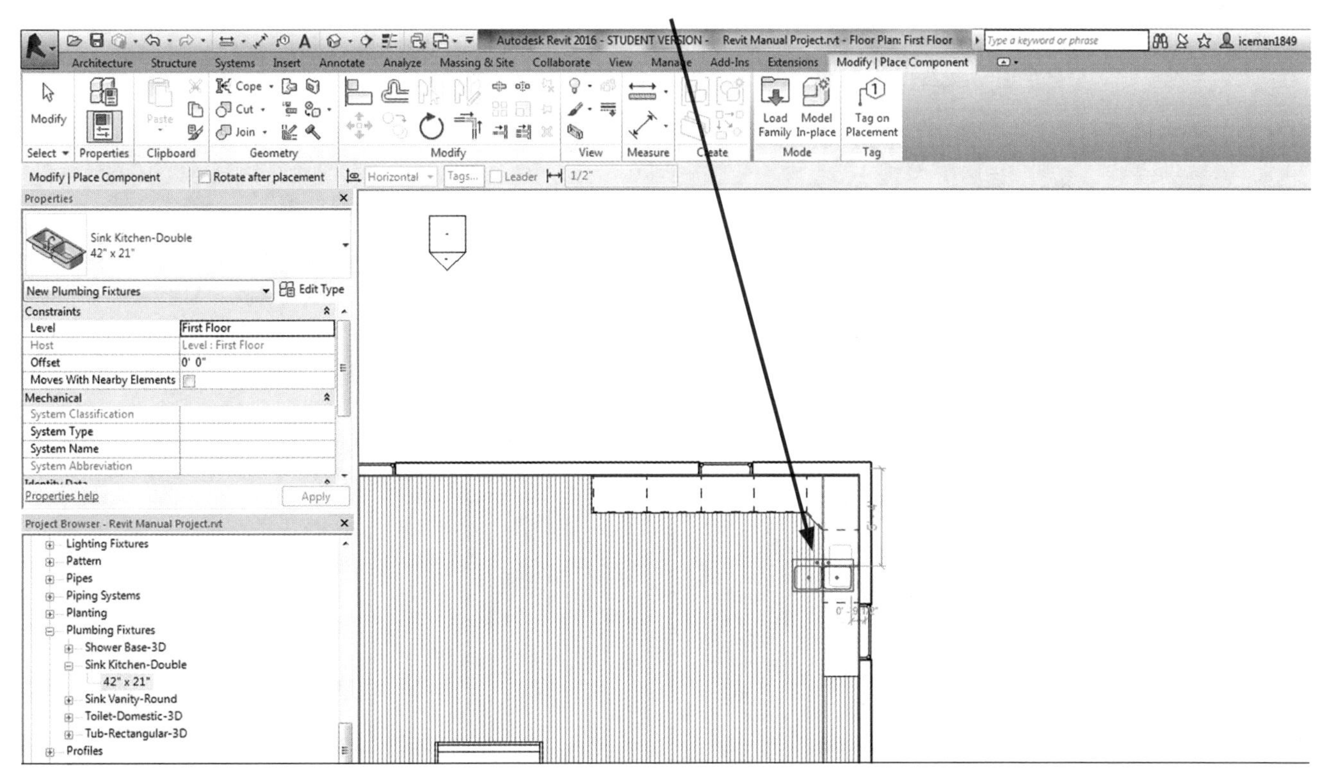

Once over the sink hole you can click the **SPACEBAR** and the sink will rotate until it is in the correct position over the sink hole, then click the mouse to place it.

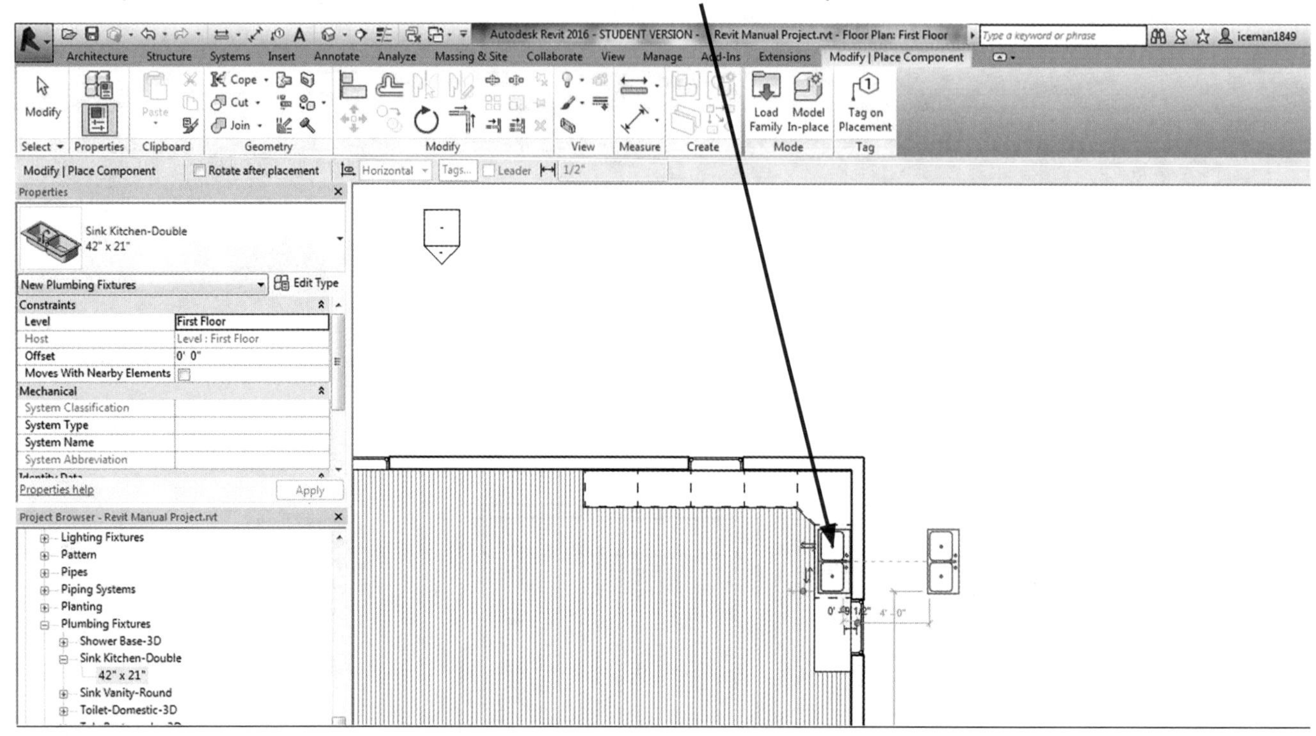

After placing the sink, just like the counter top before, Revit has another sink ready to place, so click the **MODIFY** Arrow to end the command. Do not hit the **ESC** Key.

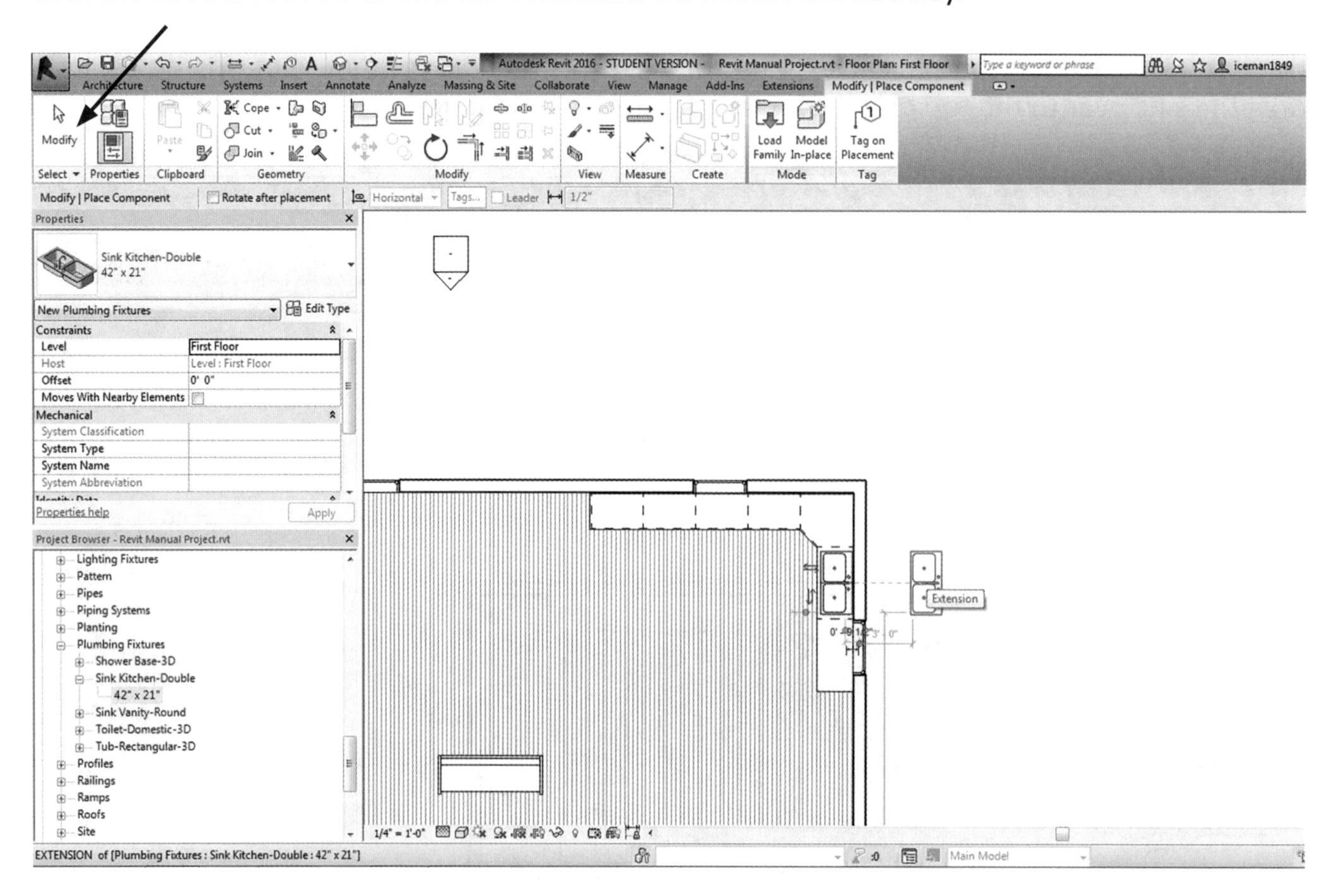

After hitting the **MODIFY** Arrow, you will notice the second sink will automatically disappear and your design will look like the screen shot below.

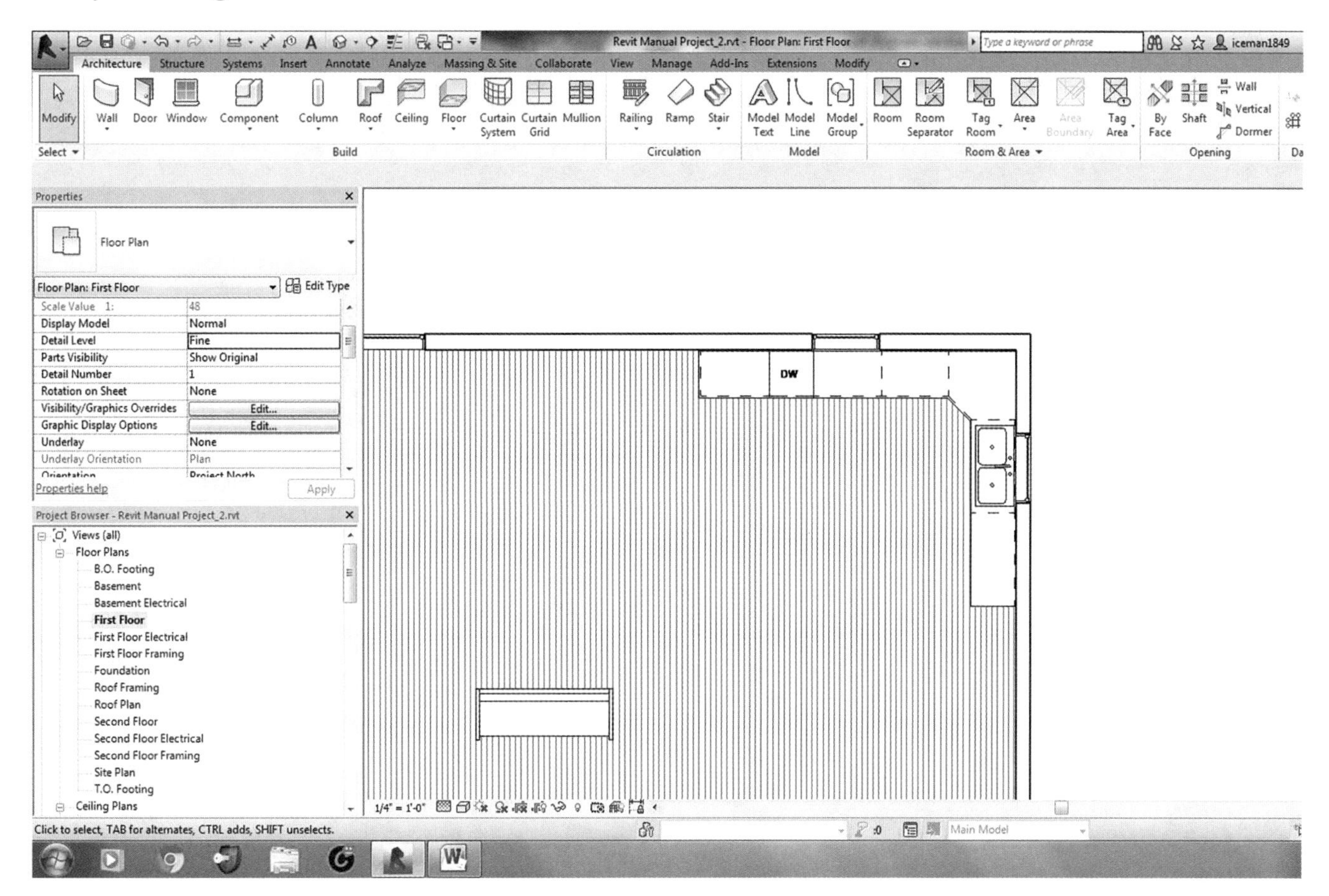

Now you have placed casework, a counter top, and a sink in your kitchen design (with dish-washer).

Now that you have placed a kitchen in your house or building, you can see how easy it is to either lay out your new kitchen and then in the future remodel it or even redesign someone else's kitchen layout. Revit is here to assist in making design and layouts easier for the user.

Let's continue the momentum in Chapter 10, *Inserting Bathroom Fixtures, Caseworks, and Sinks*.

CHAPTER 10

Inserting Bathroom Fixtures, Caseworks, and Sinks

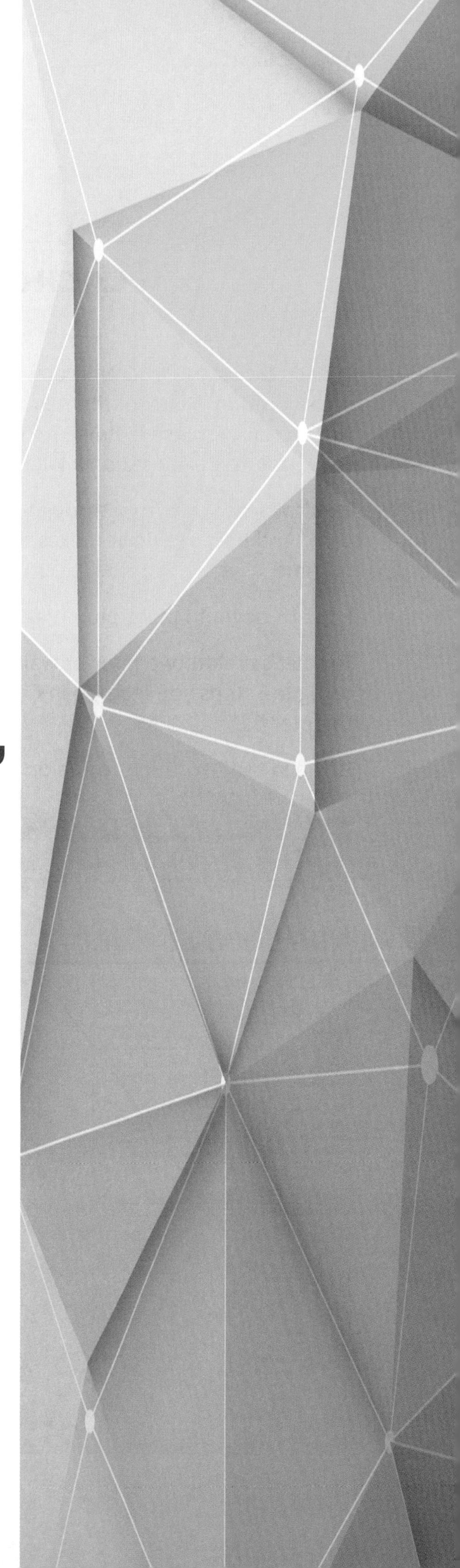

PLACING BATHROOM FIXTURES

When placing bathroom fixtures in your house or building, there are many factors to take into consideration. Many of them involve where you will be placing the fixtures, because some of the fixtures you use in Revit are wall-hosted, just like the doors and windows, which means they have to be placed on a wall. A good example of this is a tub.

Don't worry. We will go through this information step-by-step, like we have in the other chapters. And because Revit makes these tasks so easy, you will become an advanced designer in no time.

So, let's begin to place plumbing fixtures in the bathroom.

First let's create two interior walls at the corner of your house or building, roughly 8'-0" x 8'-0", using the steps you learned in Chapter 3. Place a door in the room using what you learned in Chapter 4.

Begin in the Architecture Tab on the First Floor in the Project Browser Box.

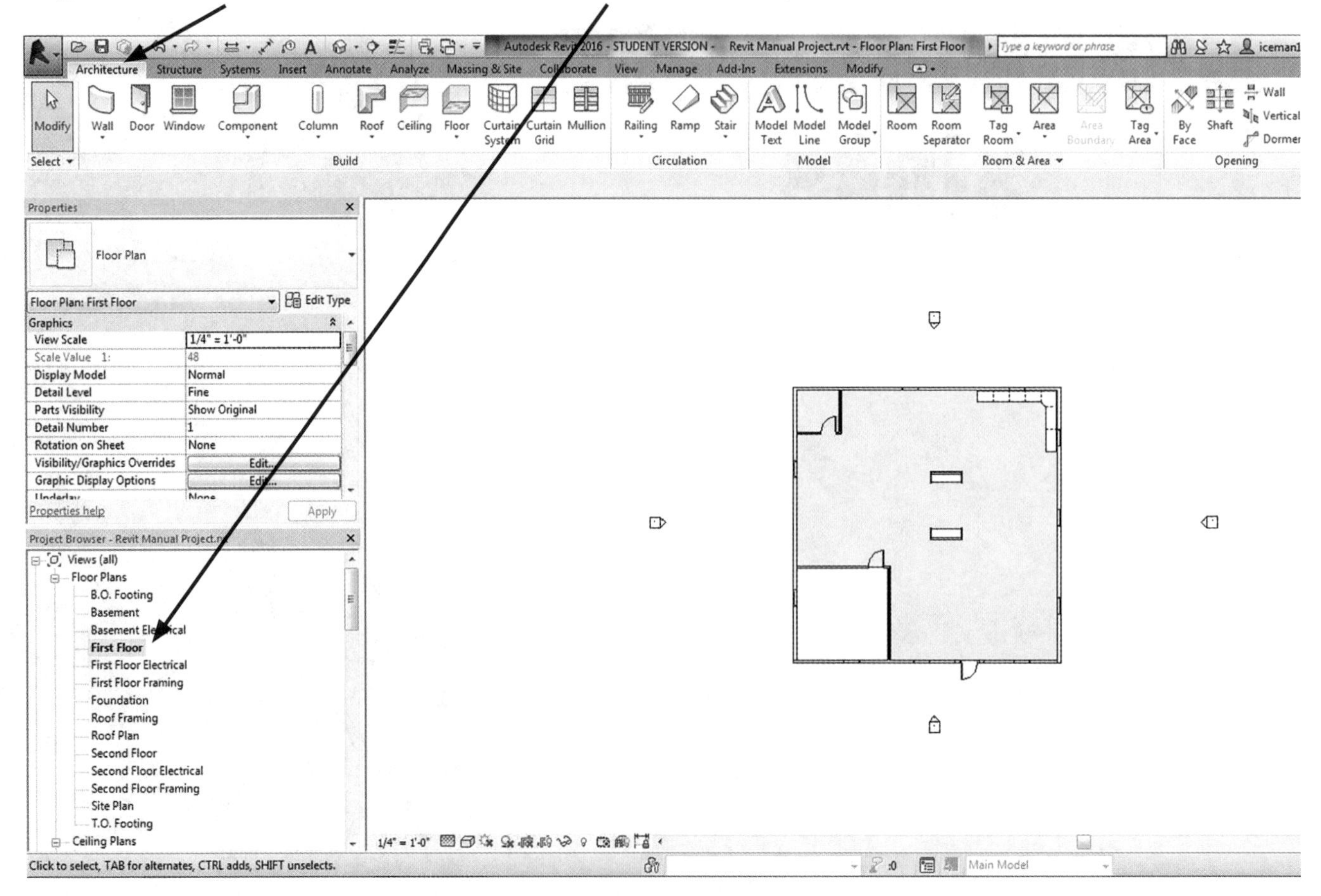

Next, scroll down in the Project Browser Box to the Families category.

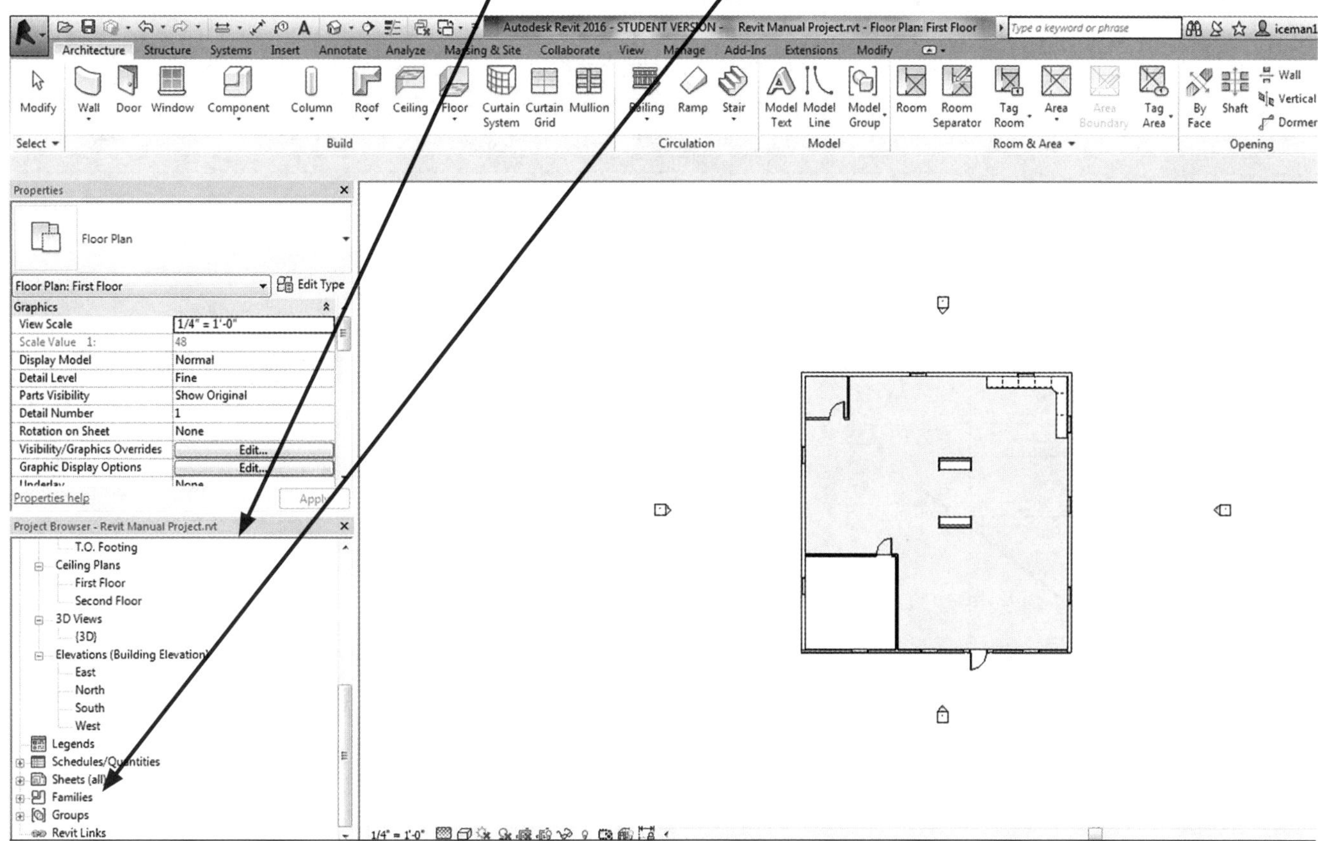

REMINDER

If the + symbol isn't expanded, click on it to expand the tree.

The following screen shows the expanded tree for the Families category:

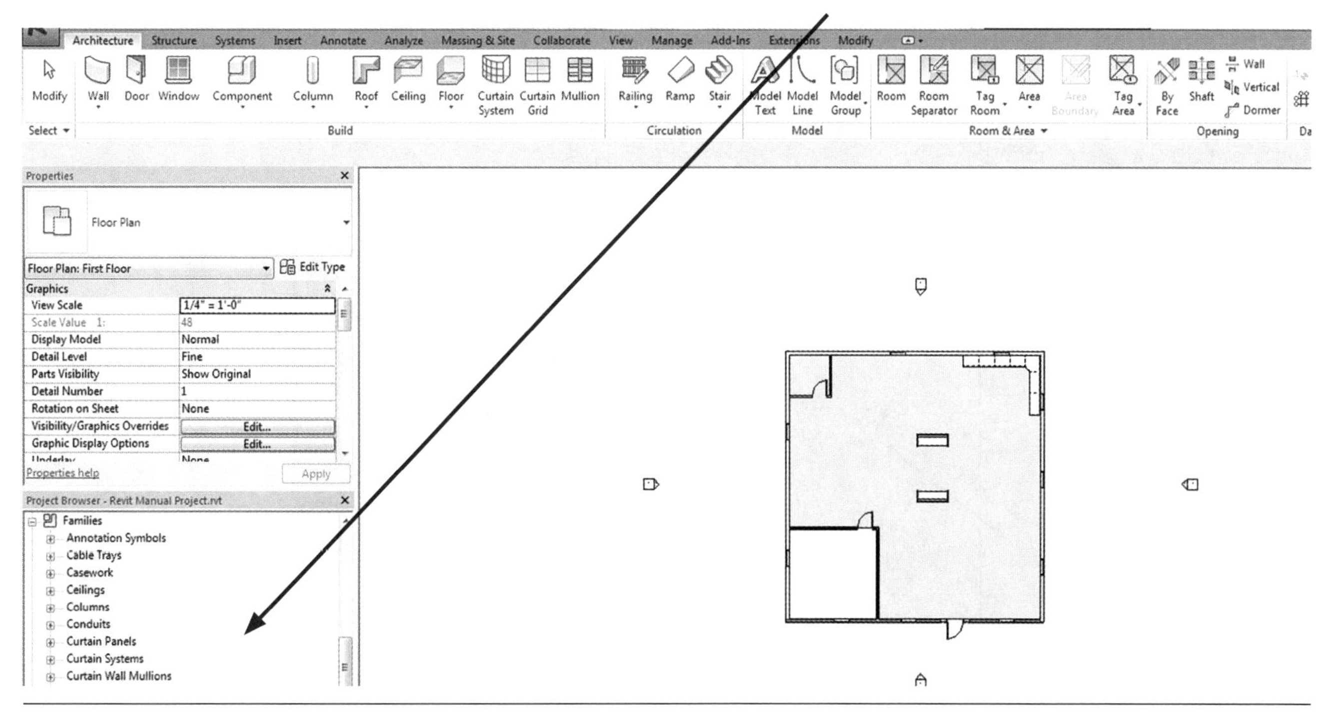

Once the tree is expanded, scroll down to find **PLUMBING FIXTURES** (listed in alphabetical order), and expand the **PLUMBING FIXTURES** tree.

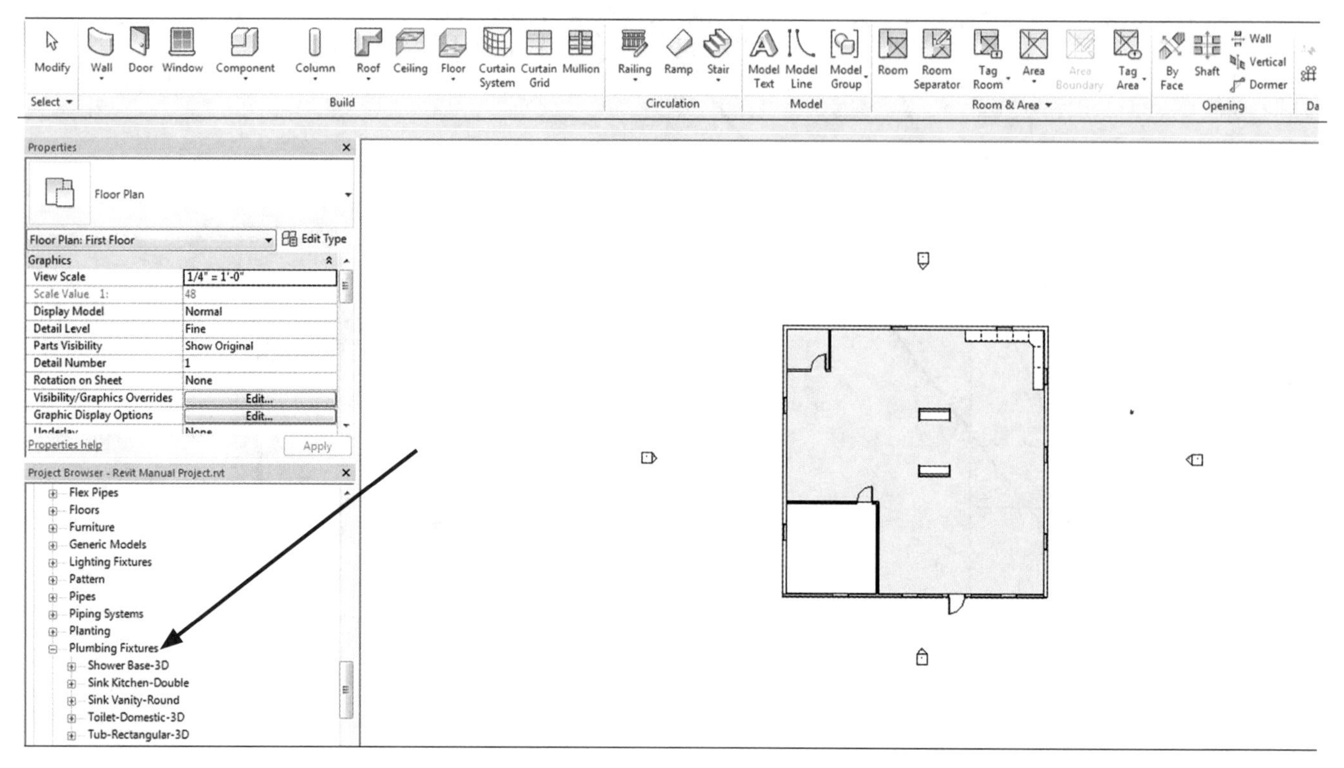

Locate **Tub-Rectangular-3D**, and expand that tree (you will *have* to expand the tree because that is the *only* way to select that family).

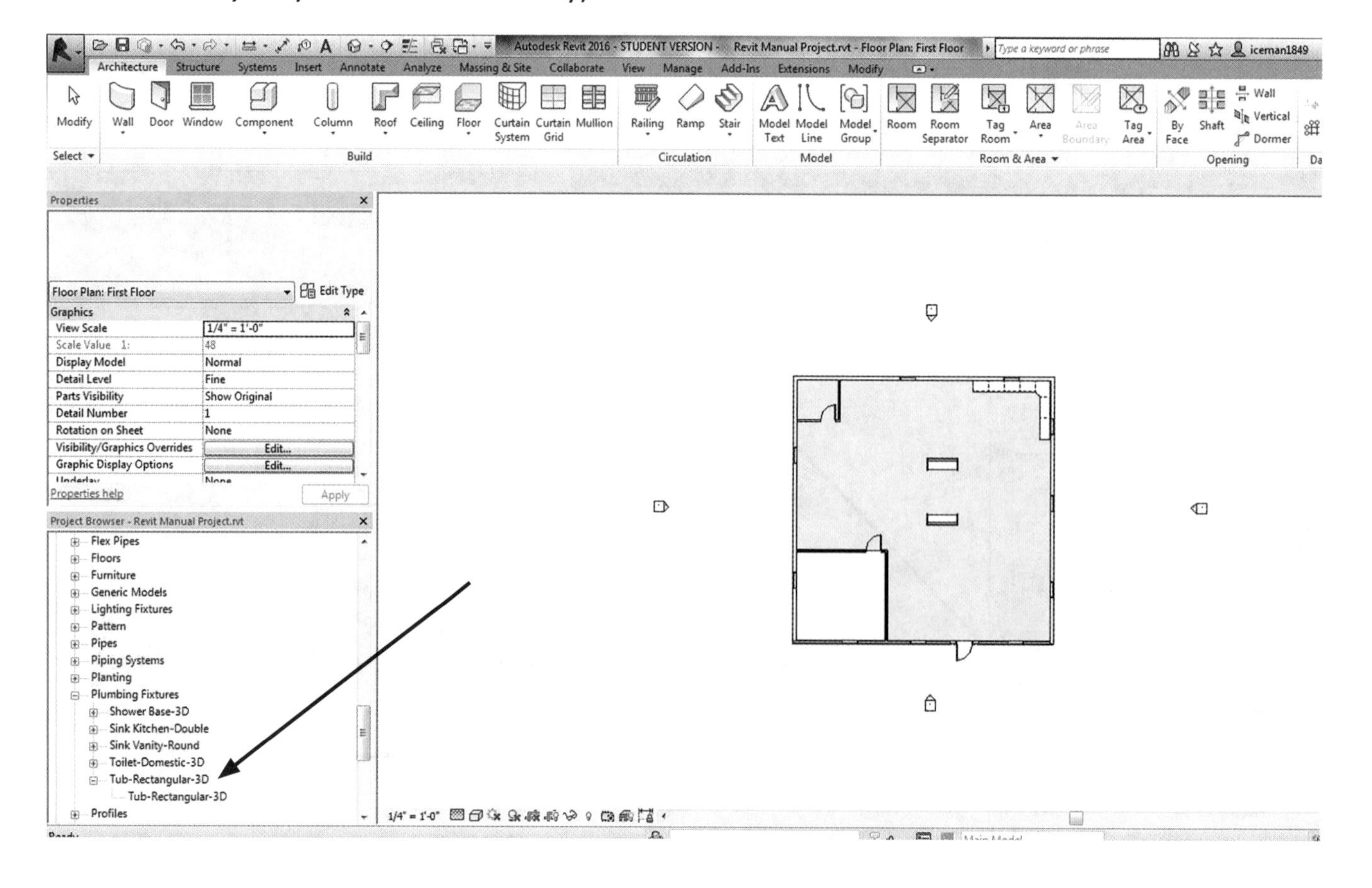

With the Tub-Rectangluar-3D (the 3D part means you will be able to see the object in 3D views, camera, and perspective views), you will need to place it on a wall and your cursor *must* be on a wall before the tub will show in your drawing.

Use the left button on the mouse and click on the **TUB-RECTANGULAR-3D** family. Hold down the mouse button, drag the tub into your drawing, and place it in your room.

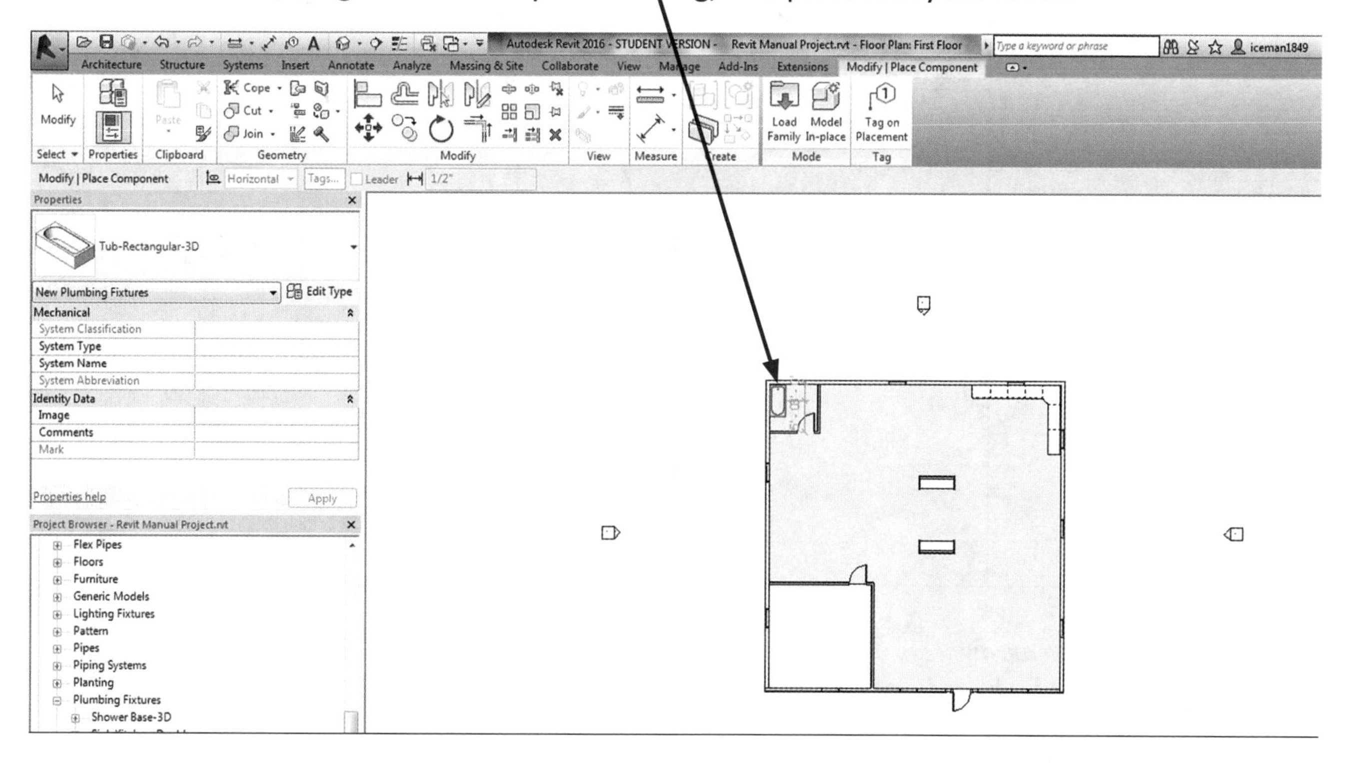

Use the wheel on your mouse to scroll or zoom into the corner and get closer so you can see *exactly* where you want to place the tub.

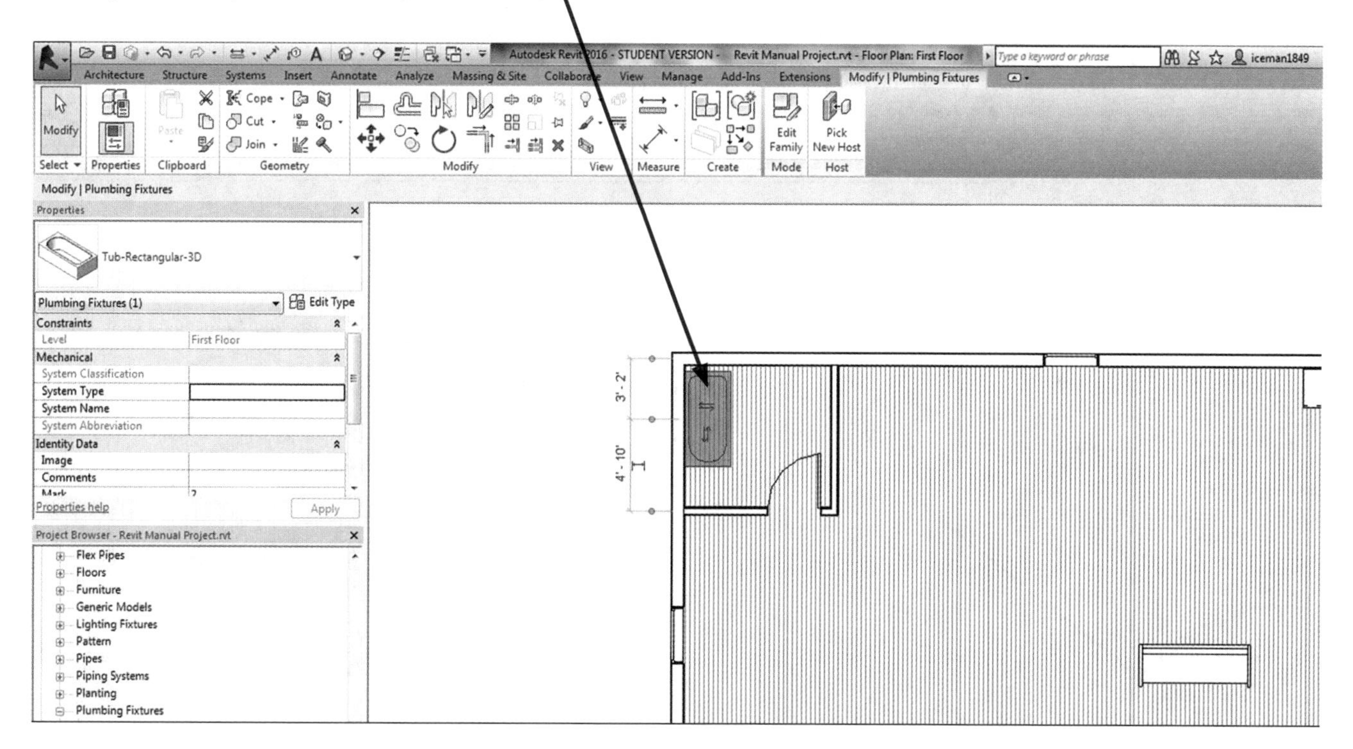

When you select the wall where you want to place the tub, click one more time to place it. Don't worry about it not being in the corner. You can click on the tub again and use the **ARROW** keys to nudge the tub into the corner of the wall.

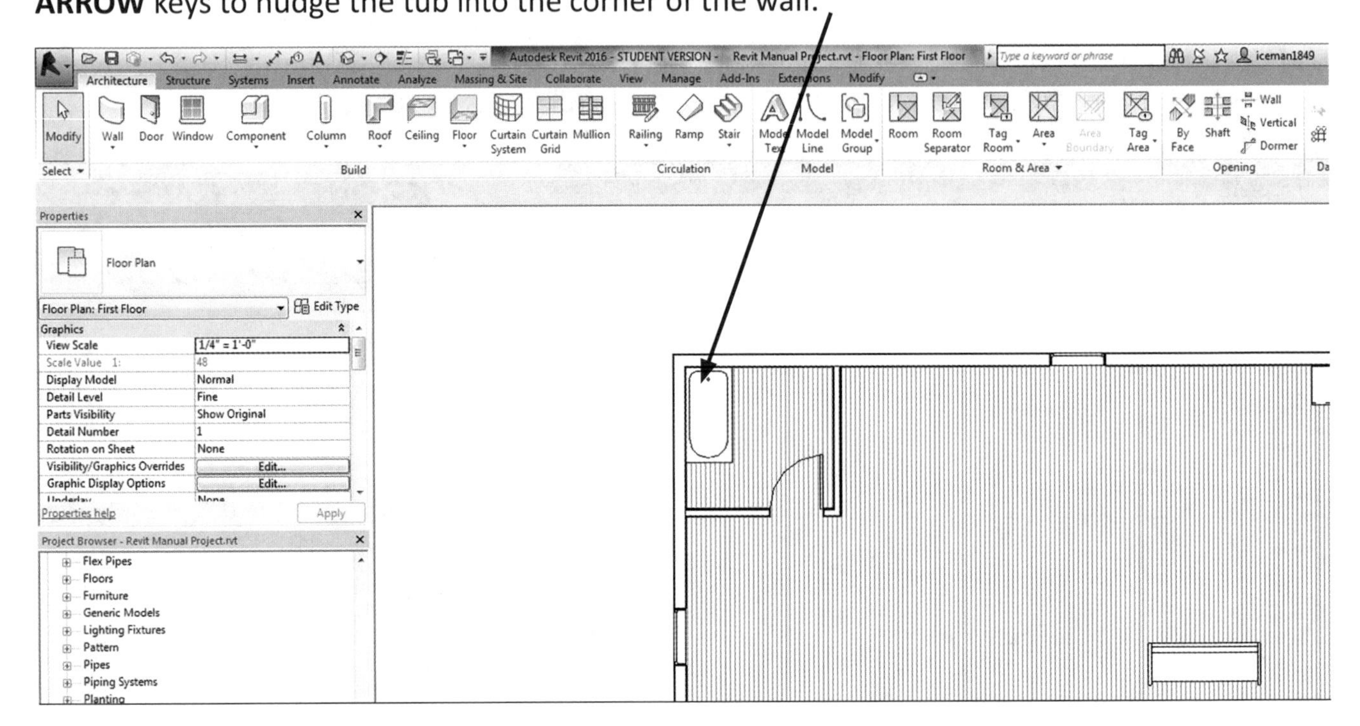

NOTES ***The arrow keys you use to nudge objects are located below the Shift Key on the keyboard.***

Once the tub is in the correct position, you can then place the toilet in the bathroom. In the Project Browser Box, you should still be in the Plumbing Fixtures category.

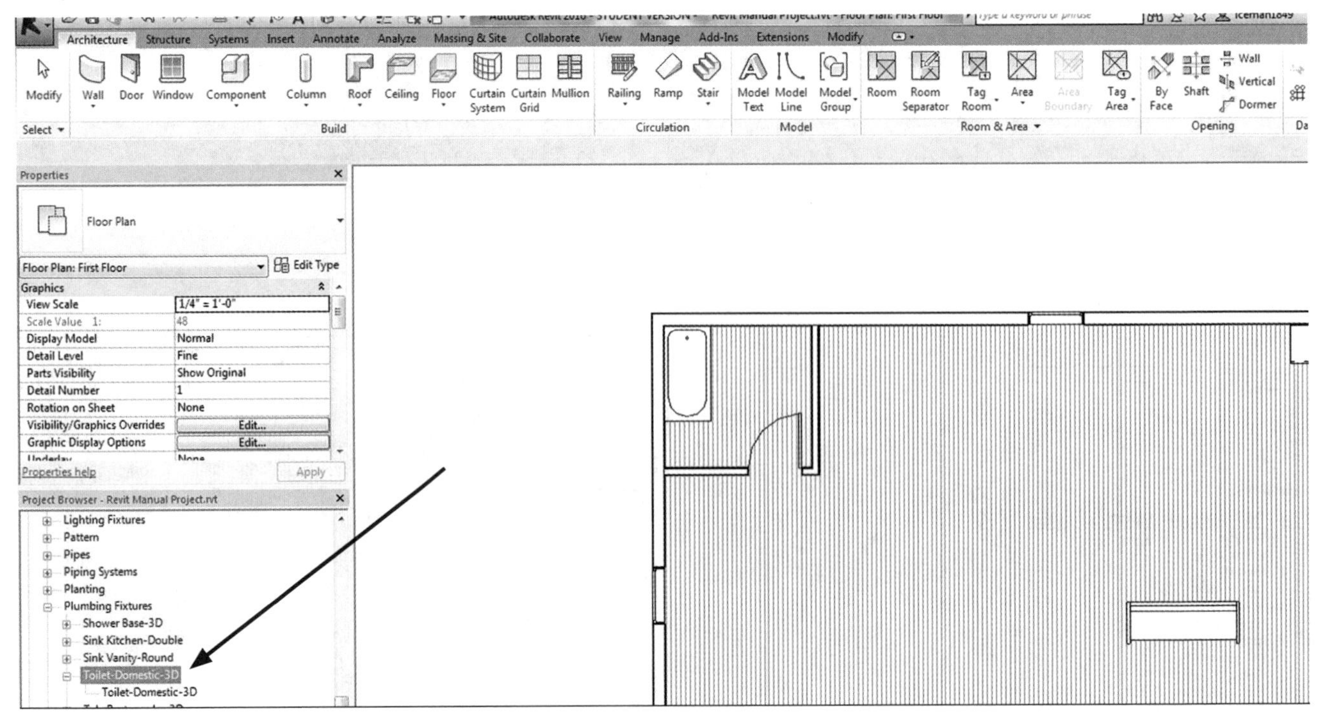

REMINDER

You should still be zoomed into the area where you placed the tub, so you can get a closer view of your location for the toilet.

Locate Toilet-Domestic-3D and expand that tree (that's the *only* way to select that family type).

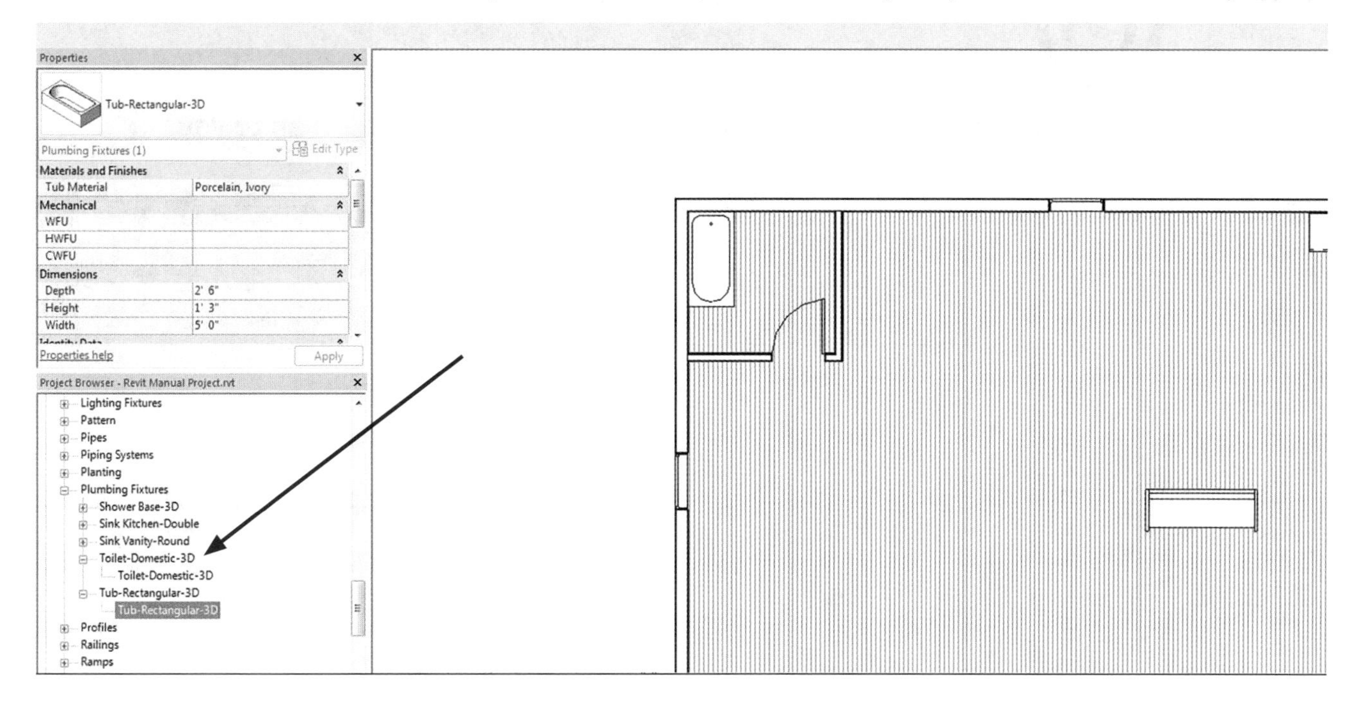

Using the left button on the mouse, click on the **TOILET-DOMESTIC-3D** family, hold down the mouse button, drag the toilet into your drawing and place it in your room.

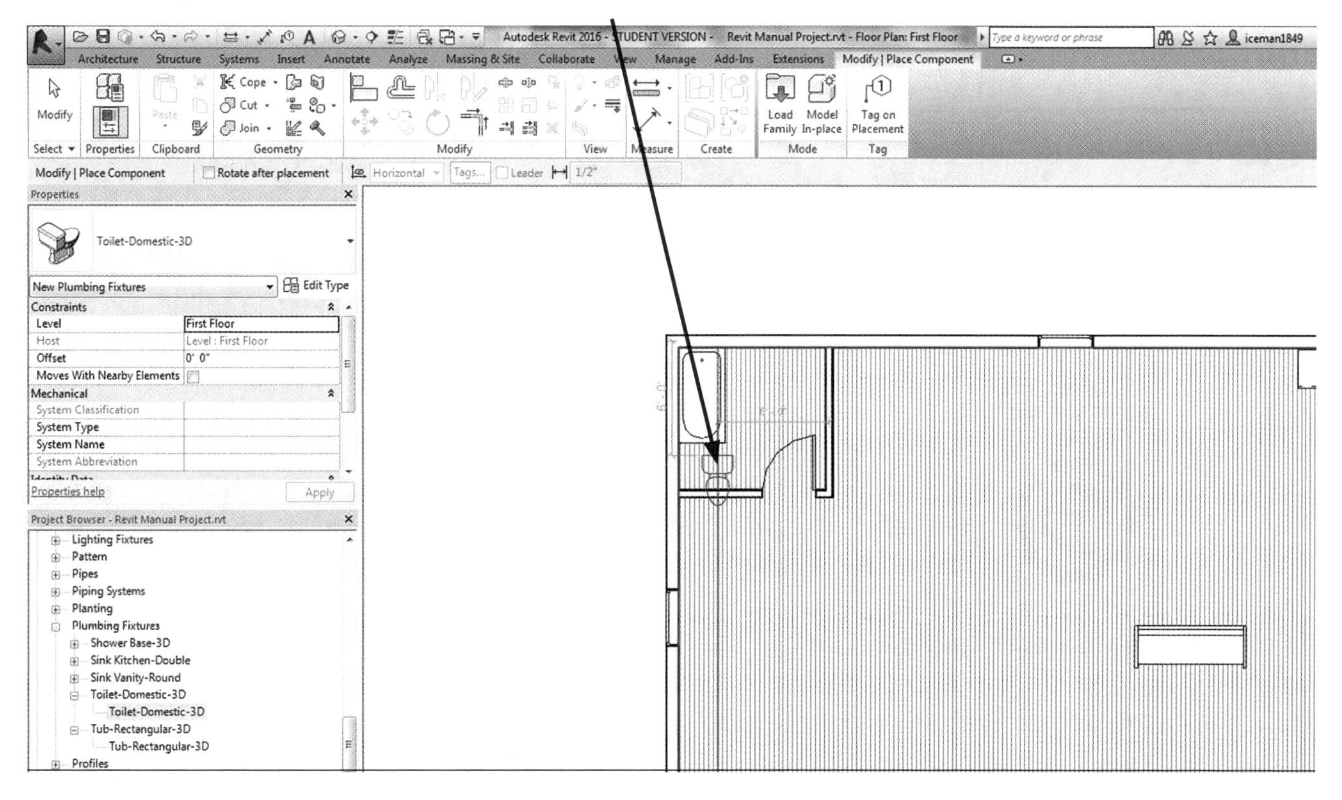

When you are placing the toilet in a space and you want to rotate the object, you can hit the **SPACEBAR**, and the object will rotate.

ESSENTIAL REVIT FACT

Toilets are not wall hosted, like tubs. So you can see them immediately when you bring them into your house or building.

Once you have placed the toilet in your room, click the mouse and it will snap to your location.

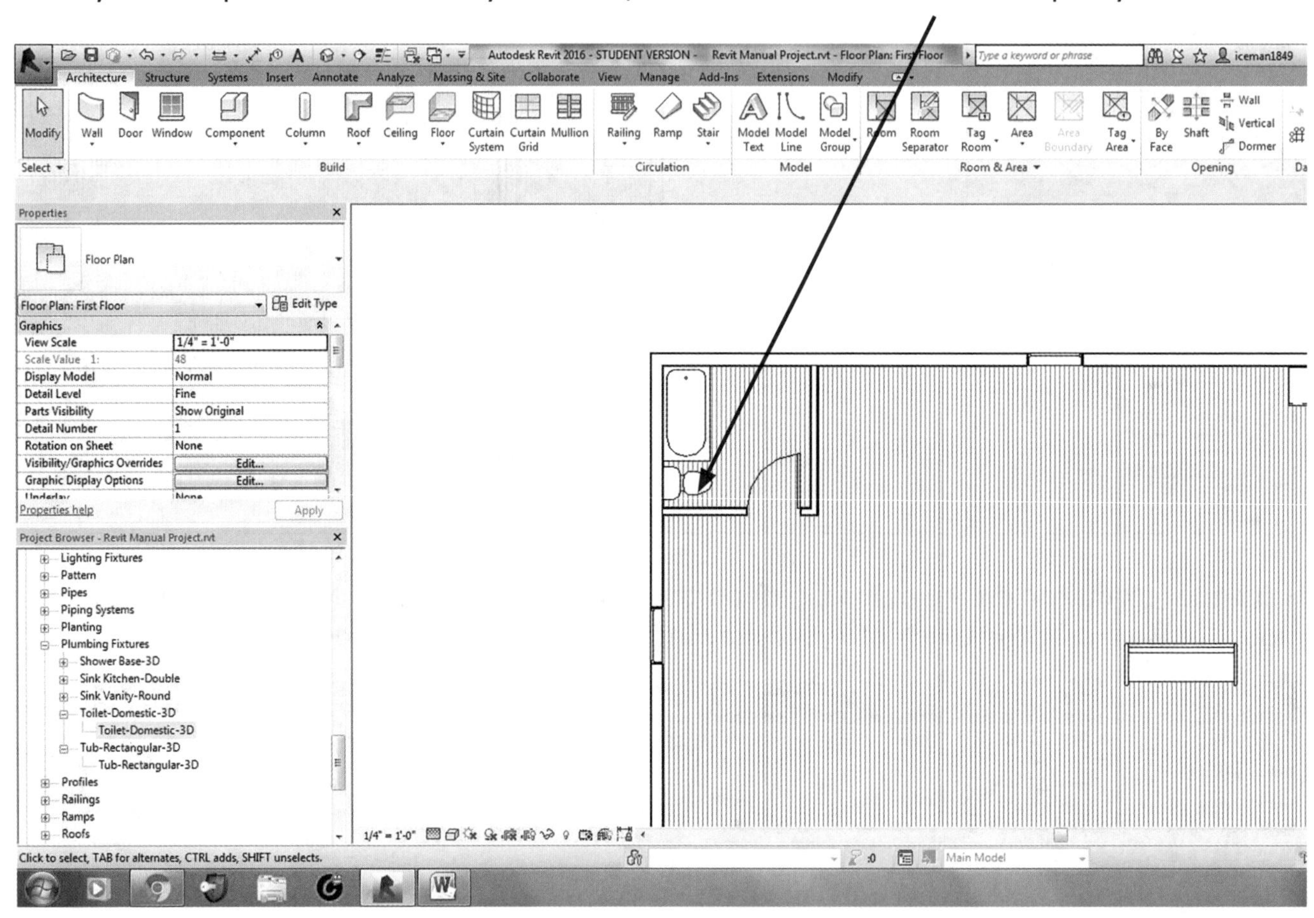

You can then click on the toilet and nudge it with the **ARROW** keys to move the toilet into a specific location.

ESSENTIAL REVIT FACT

You will get periodic warnings from Revit asking if you want to save your project or drawing. Be sure to heed these warnings, or better yet, get into the habit of regularly saving your work on your own. Otherwise, you can lose all the work that you have already done for your design.

To make the bathroom complete, you can add a vanity cabinet, counter and a sink. Following what you learned in Chapter 9 on placing cabinets and caseworks, you will select a vanity cabinet from the casework families and a counter top with a sink hole. Once you have expanded the tree, it's better to get in a habit of closing the tree to minimize space in the Project Browser.

Once you have placed and sized both the vanity and the counter top, you should scroll down in the Project Browser to the Plumbing Fixtures category again and select **SINK VANITY-ROUND**.

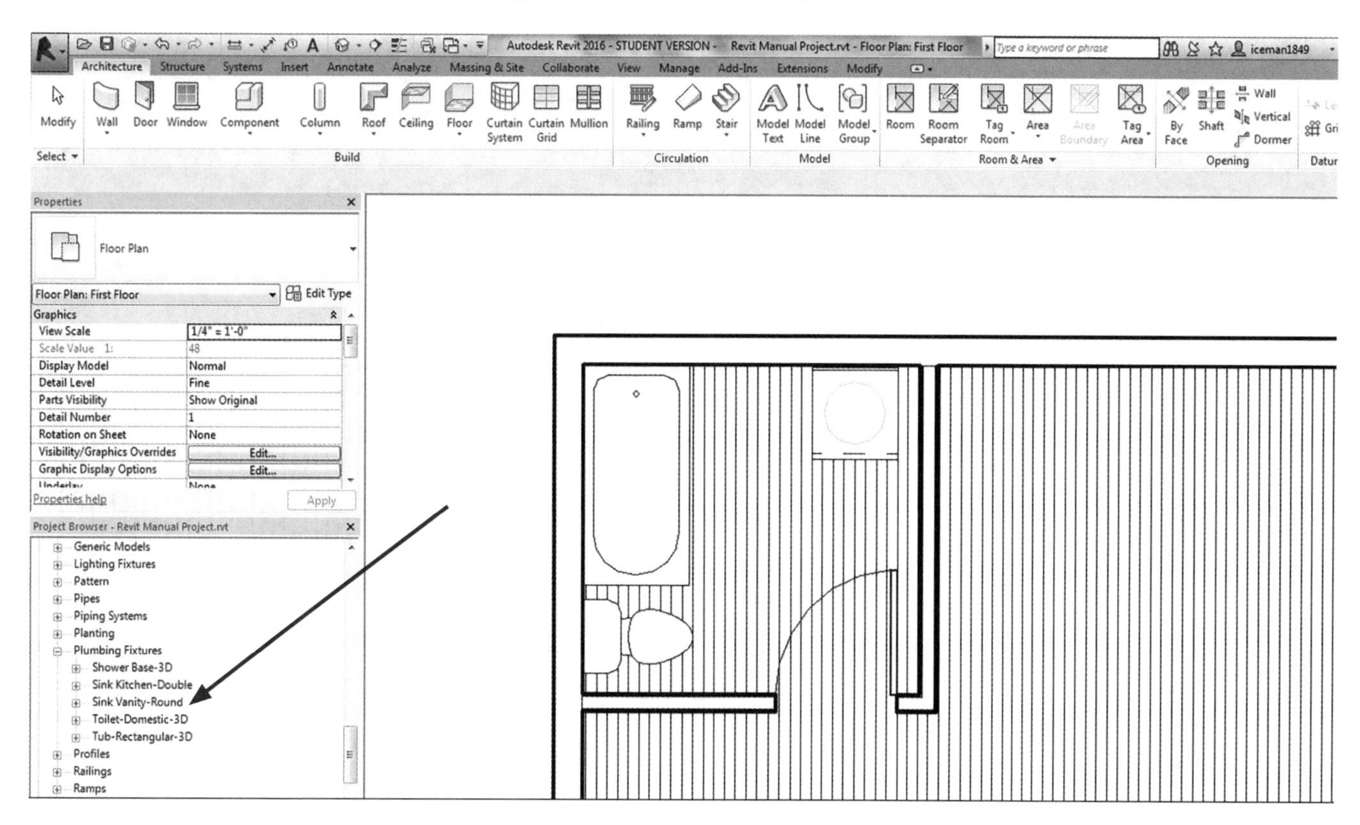

Expand that tree to open the family, and there is a **19" x 19"** round sink. Select that sink.

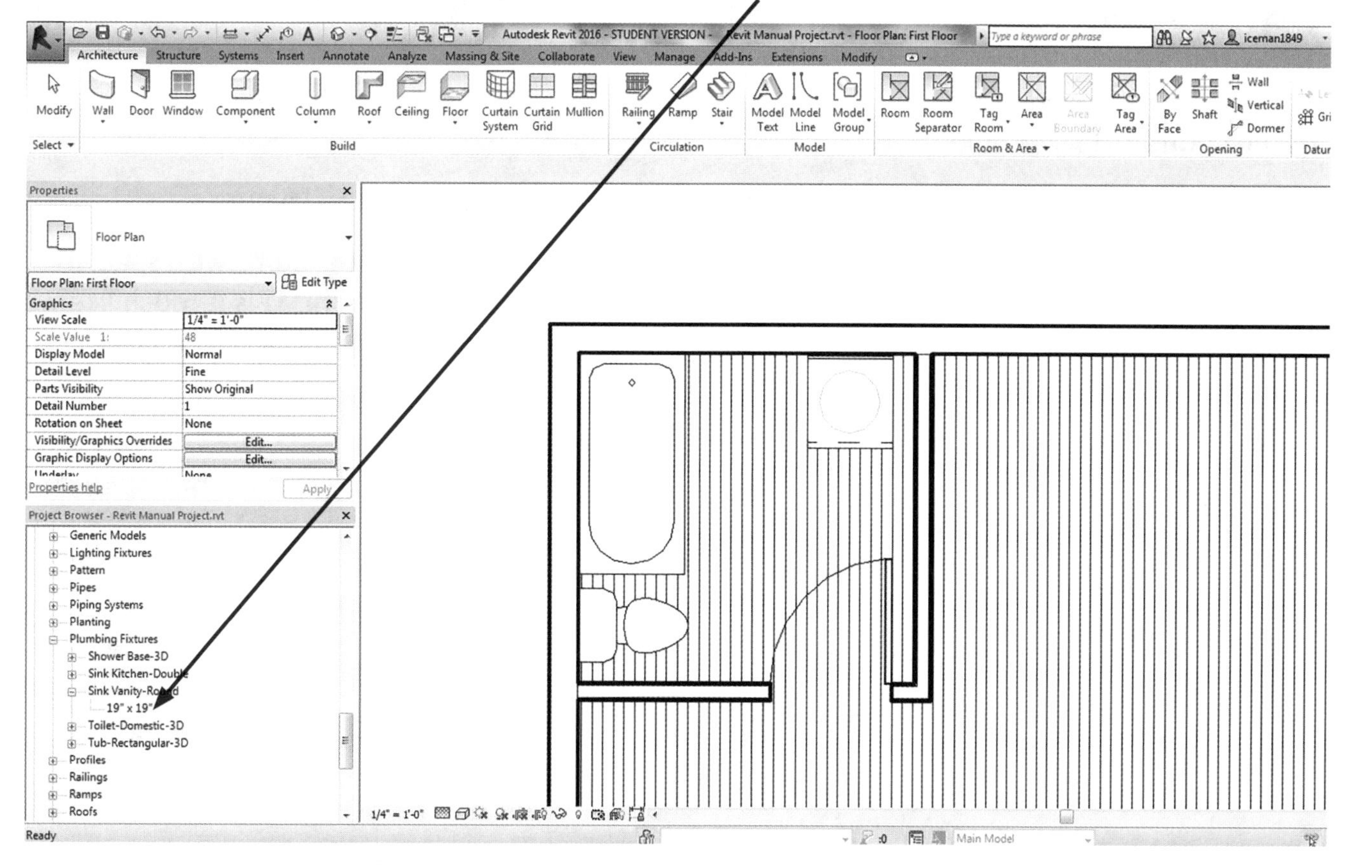

Click once on the **19" x 19"** sink, drag it over to the sink hole in the counter top, and place it in the middle of the hole.

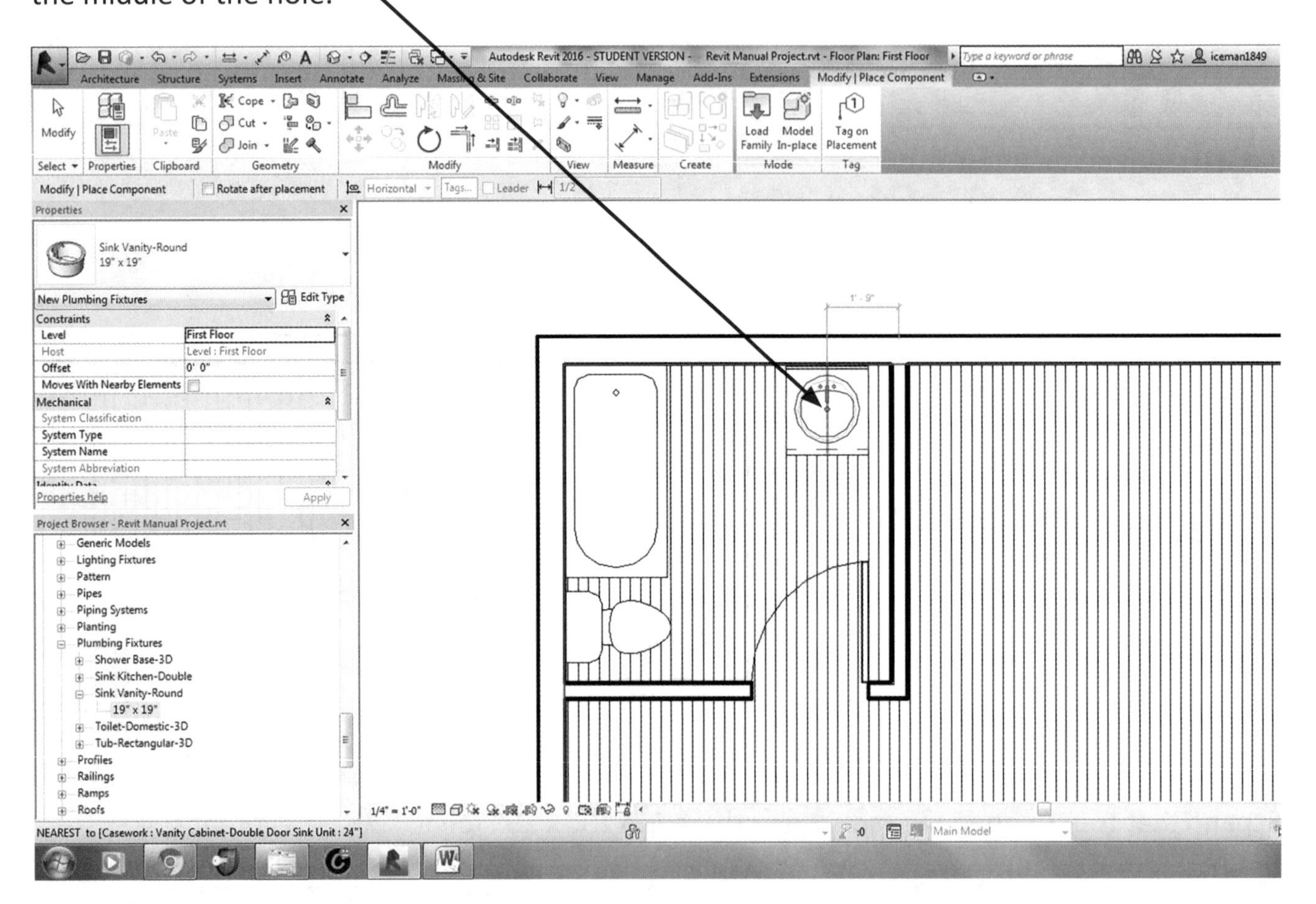

You will see a shaded reference line that will let you know that you are placing it in the center of the countertop sink hole.

Congratulations! You have just created a bathroom complete with fixture layouts.

As we have shown in the previous chapters, Revit makes it easy to create rooms with special features, items and in this case, plumbing fixtures to complete a bathroom. This is much like we did in Chapter 9 when we placed a sink in the kitchen. This gives you the sense of how a solid house/building will begin to come together.

Let's move on to Chapter 11, *Placing Air Conditioners and Furnace Units.*

CHAPTER 11

Placing Air Conditioners and Furnace Units

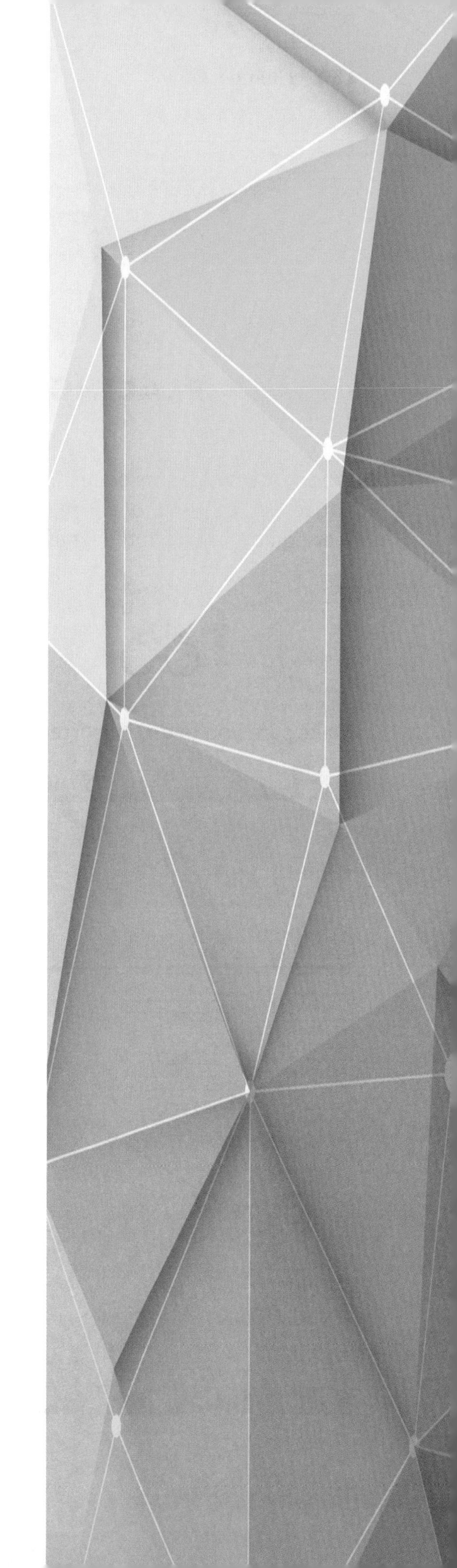

PLACING AN AIR CONDITIONER UNIT

When placing an outside Air Conditioner Unit, it is listed as Mechanical Equipment. You will notice that there is *not* a Family category listed in the Project Browser, so you will need to load that equipment from the Revit Family.

You can follow the steps in Chapter 17, Inserting Objects from the Revit Library and the Internet.

Make sure you're on the correct floor plan, the Architecture Tab, First Floor Plan.

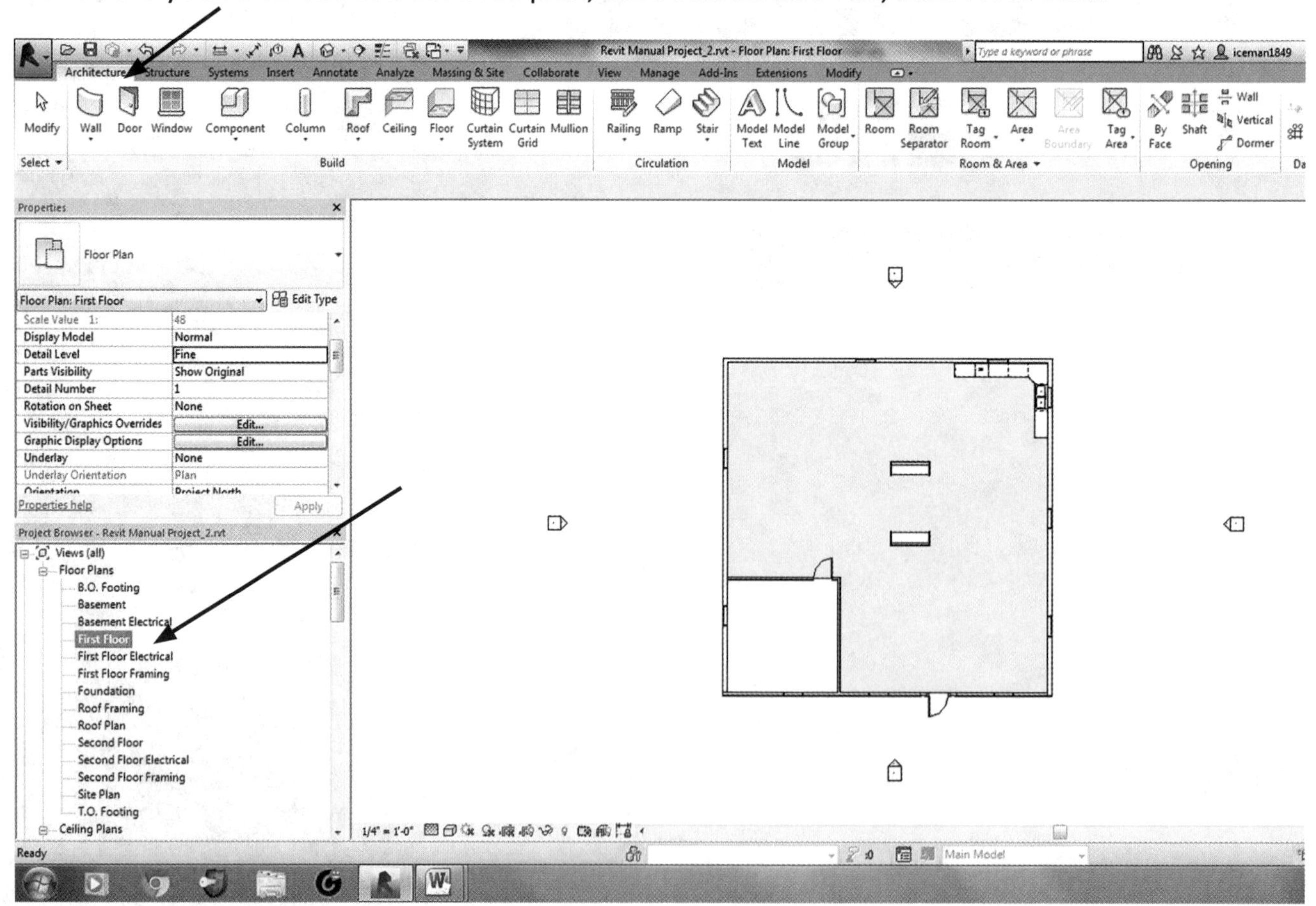

Click on the **INSERT** tab, and click on the **LOAD FAMILY** tile.

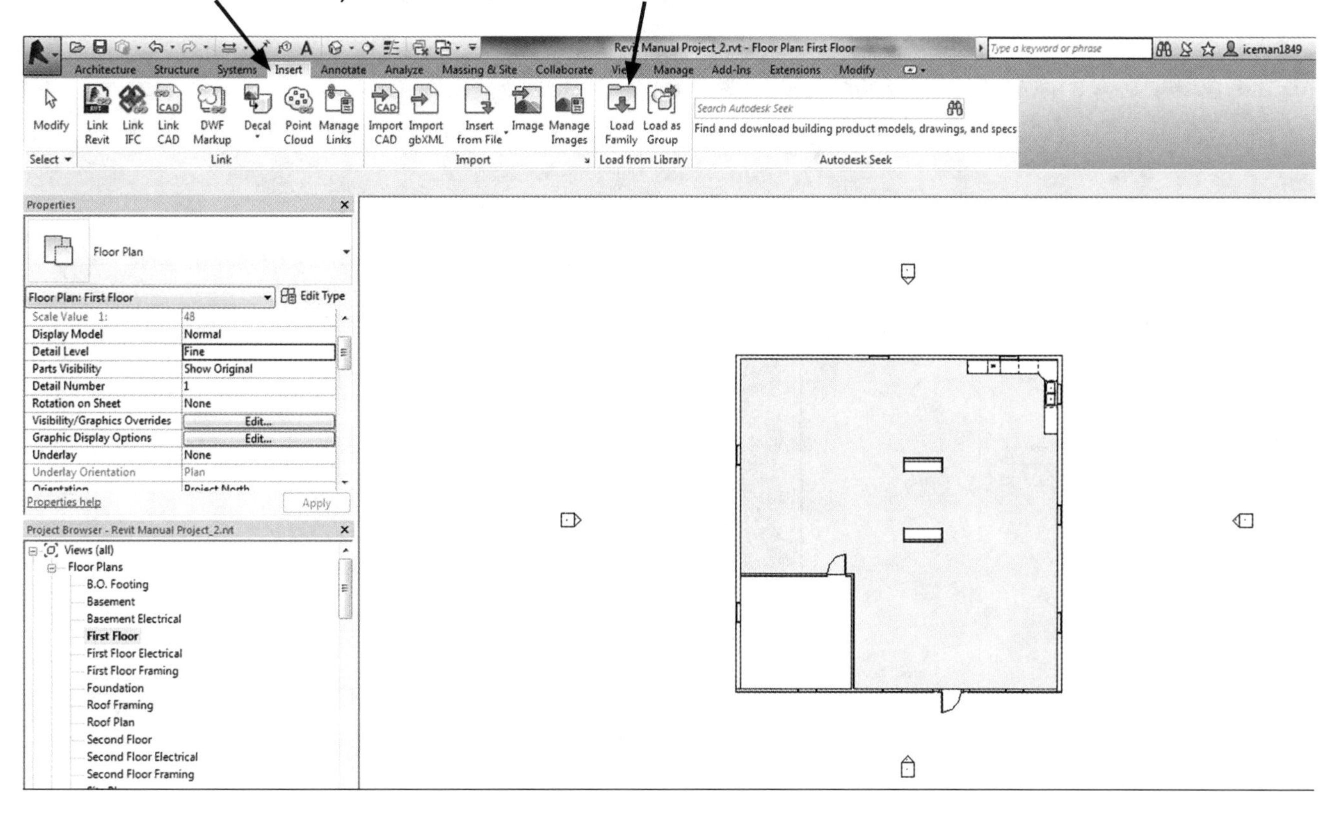

The Load Family dialog box will open. Then scroll down to the Mechanical folder.

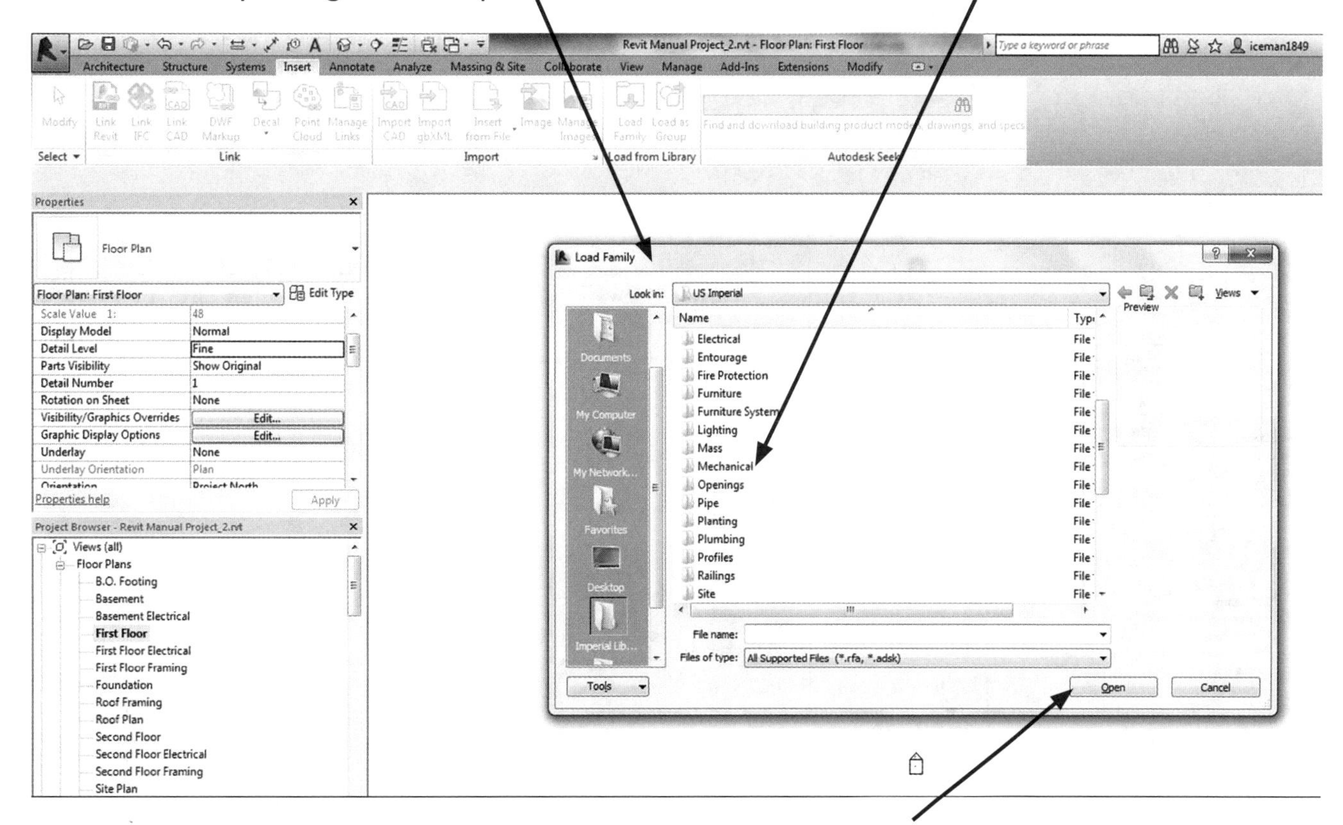

Click **OPEN**, and the Mechanical folder will open, or you can double-click on that folder.

Then click on the **ARCHITECTURAL** folder, and open it.

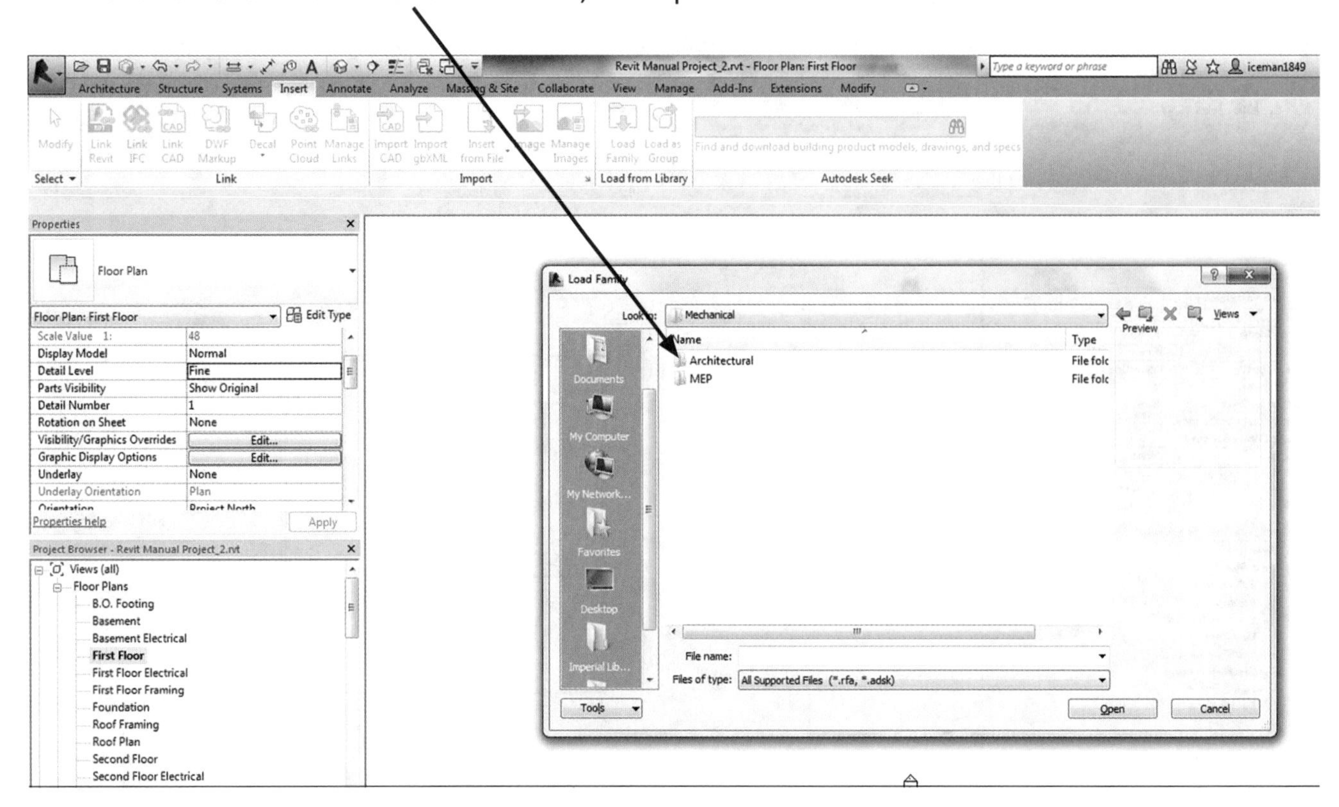

Next, click on the **AIR-SIDE COMPONENTS** folder, and open that folder.

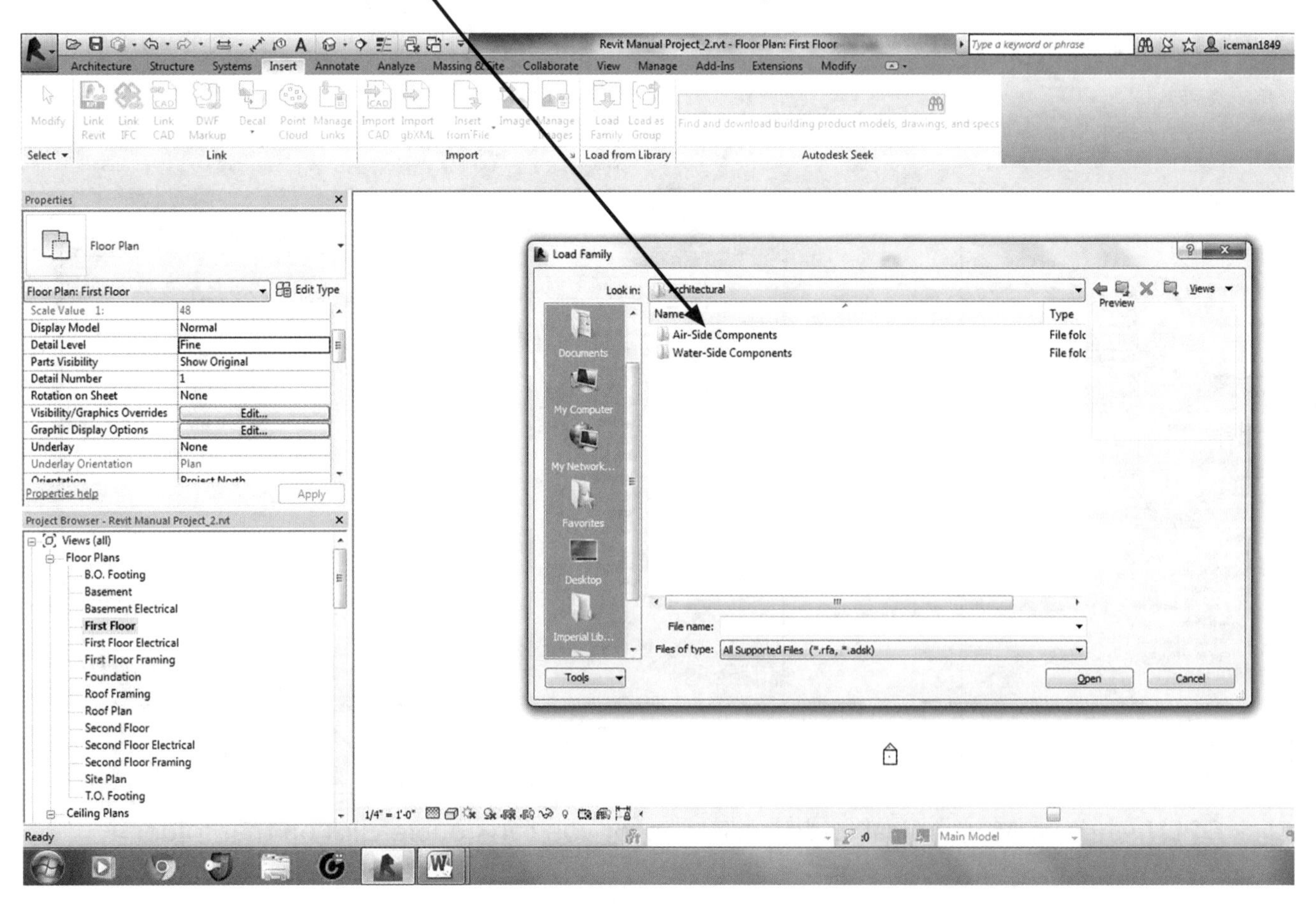

Then click on the **AIR CONDITIONERS** folder, and open it.

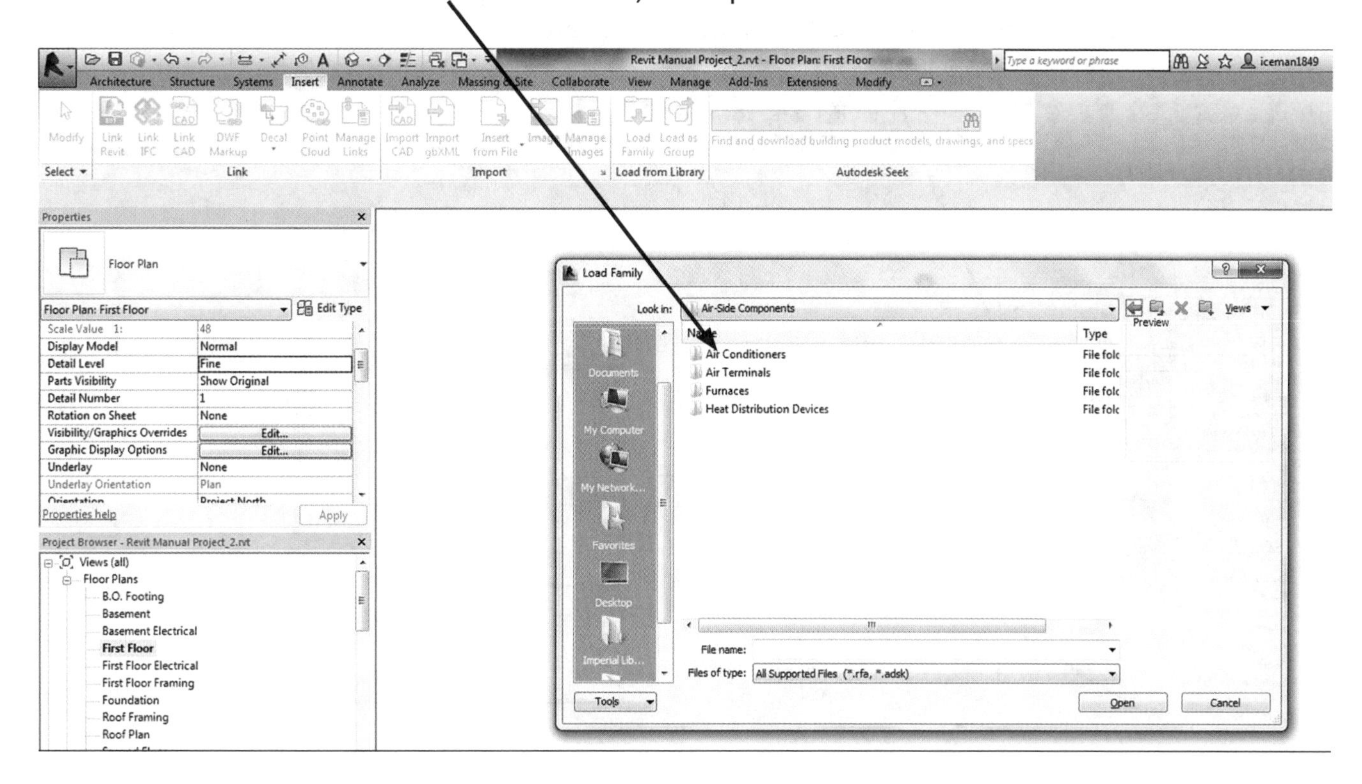

Finally, click on the **AIR CONDITIONER-OUTSIDE UNIT** family, and click **OPEN**, or double-click on the Family.

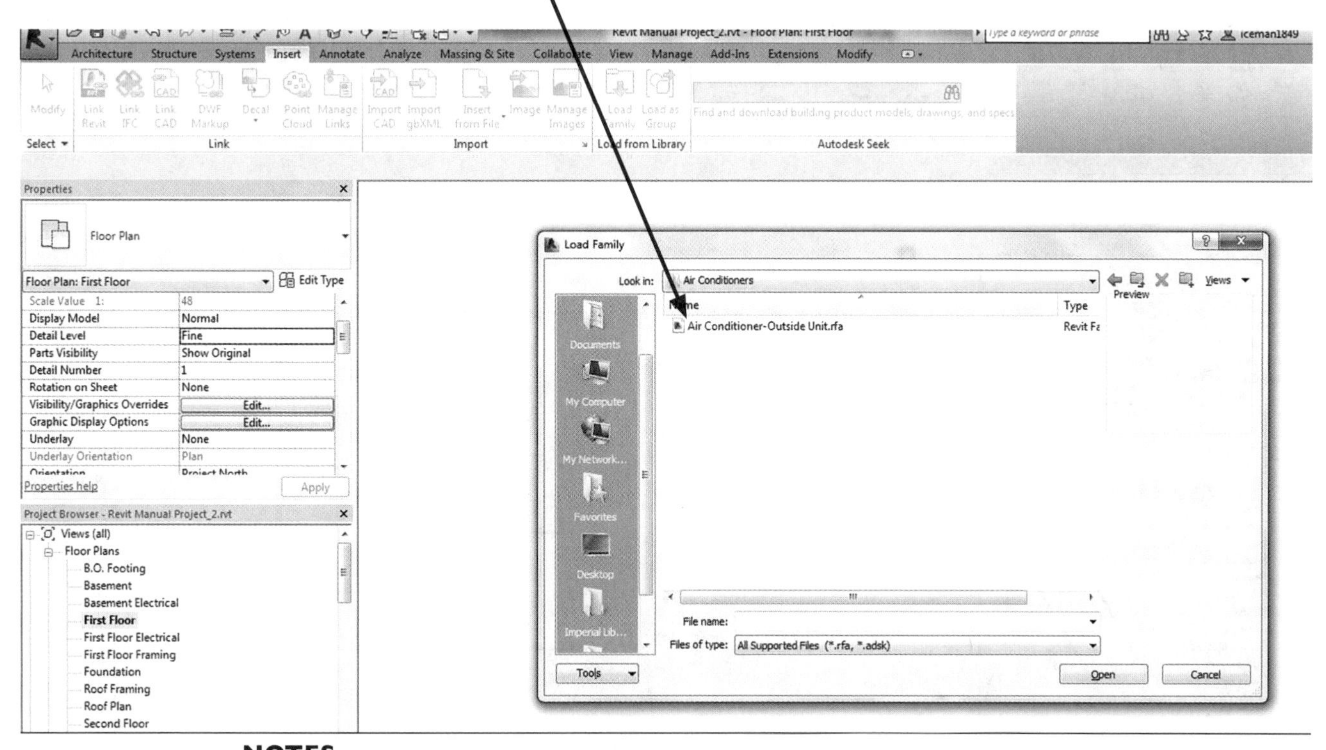

NOTES

This unit goes *outside* ***the house or building.***

The Air Conditioner-Outside Unit will then be loaded into the Families category in your project in the Project Browser under the Mechanical Equipment category.

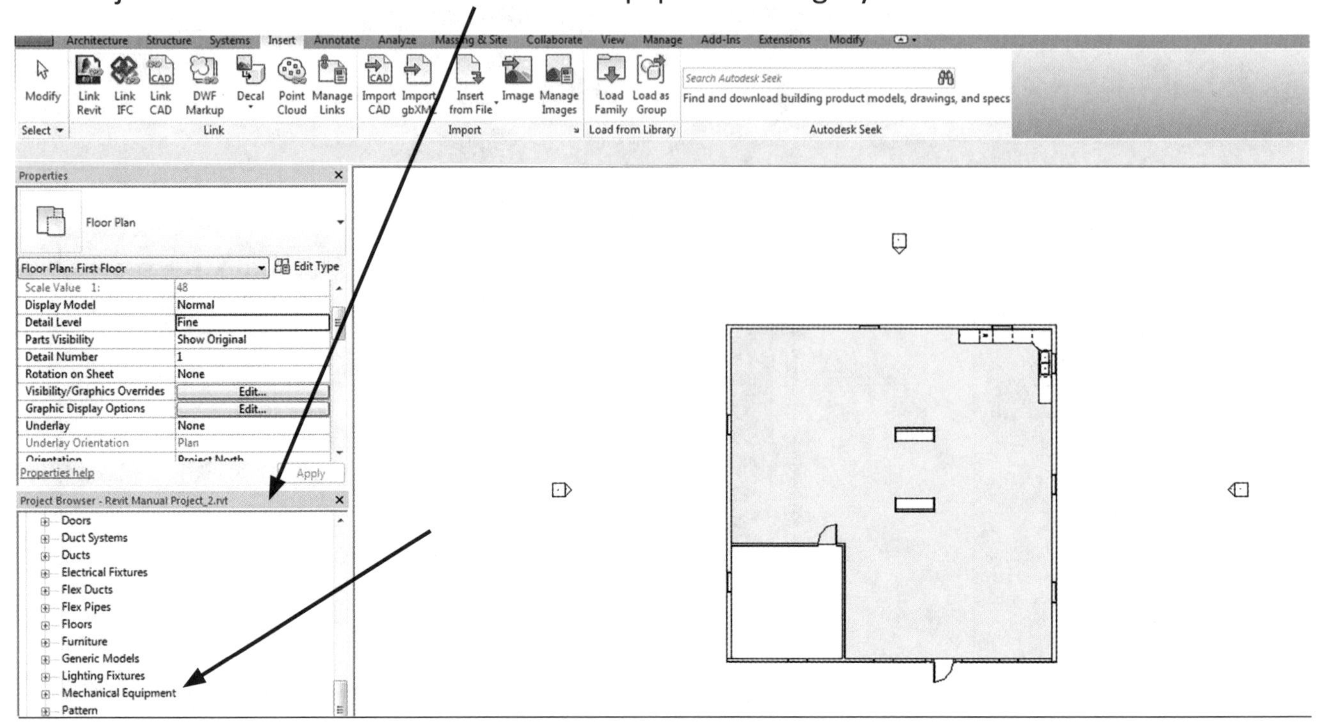

Expand the **MECHANICAL EQUIPMENT** category tree, then expand the **AIR CONDITIONER-OUTSIDE UNIT** tree.

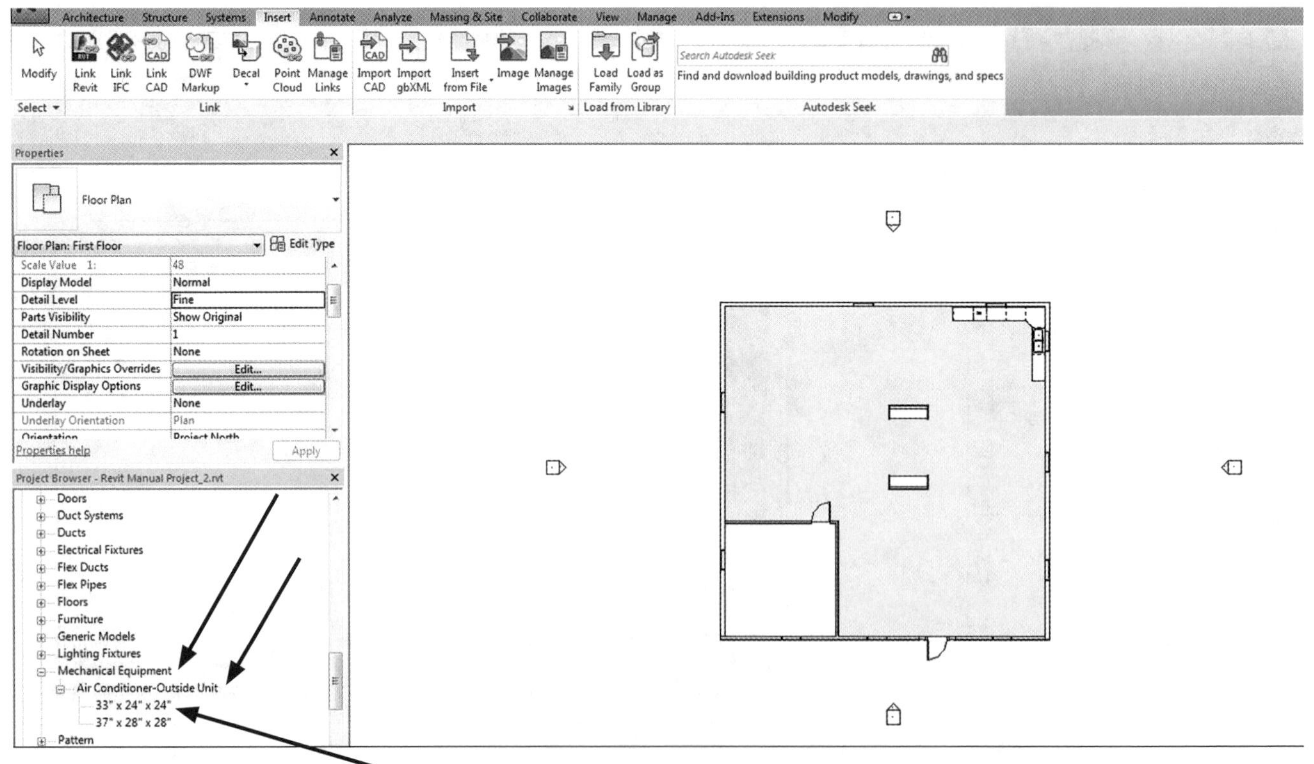

Here you will see different sizes of air conditioners from which you can select to place outside your house or building.

To place the air conditioner, using your mouse you can click on the size you desire and drag it to the side of your house or building that you prefer. Once you select the size, you will see the Air Conditioner in the Properties box.

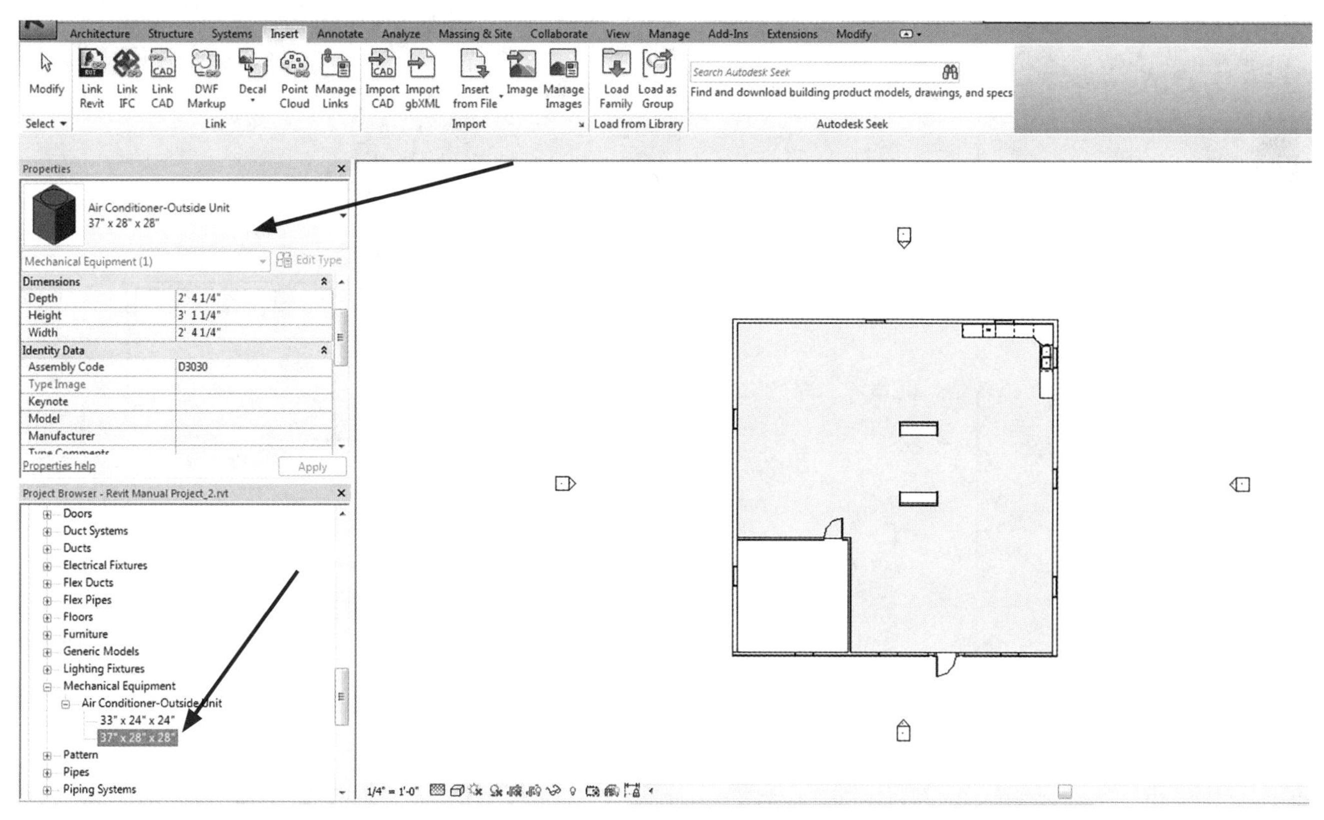

Then you will drag the air conditioner into your project location.

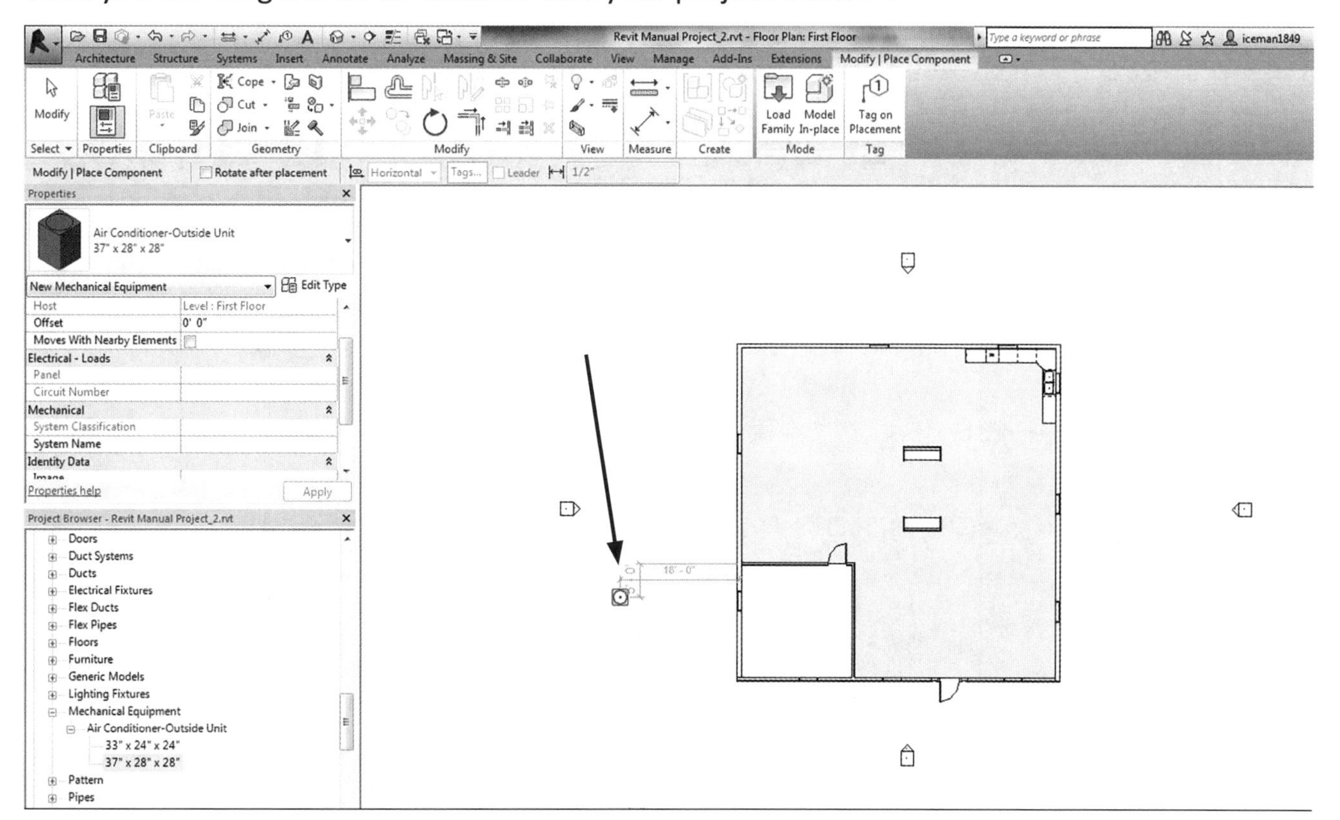

You can then place the air conditioner wherever you want it to be.

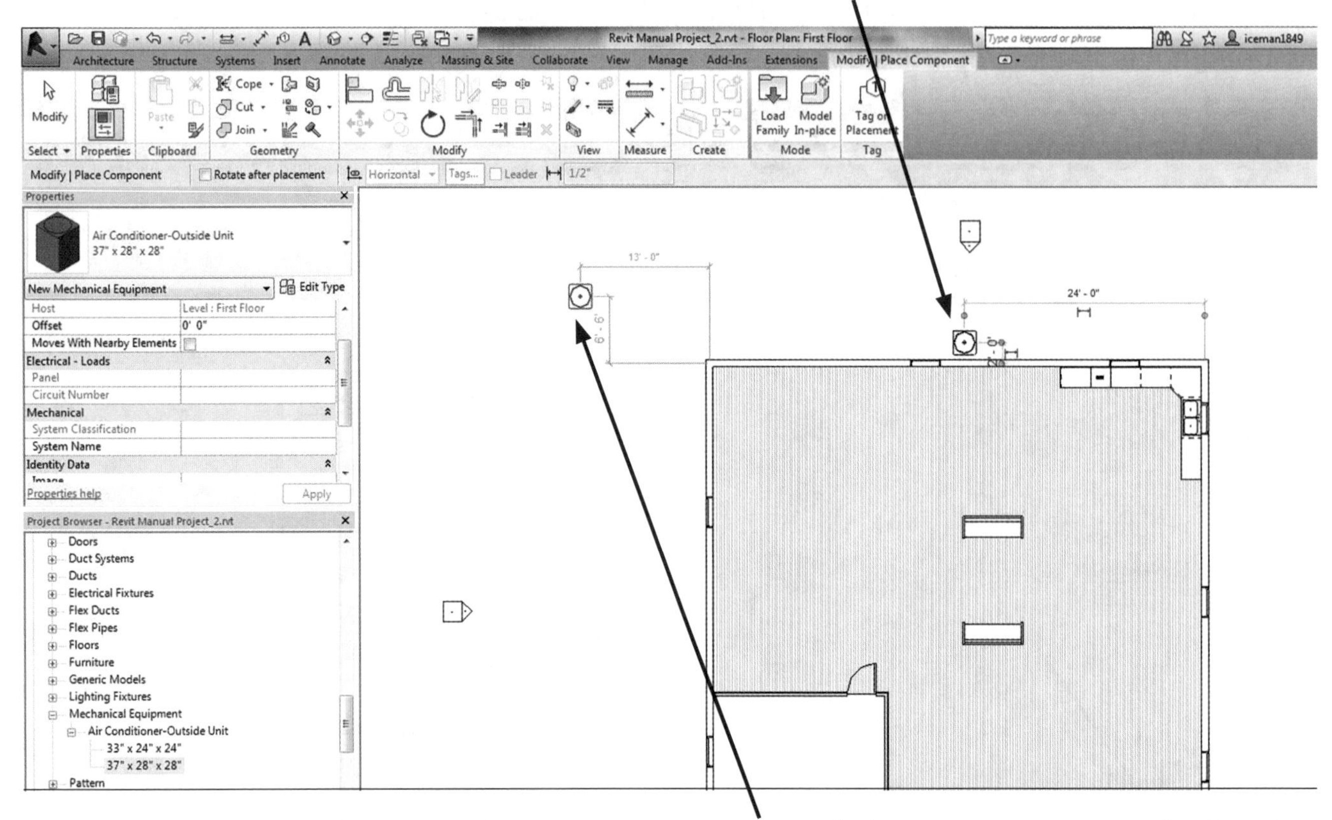

Once you have placed the air conditioner, Revit automatically has another one attached to the cursor, ready to place if you so desire. You will need to click the **MODIFY** Arrow to end the command.

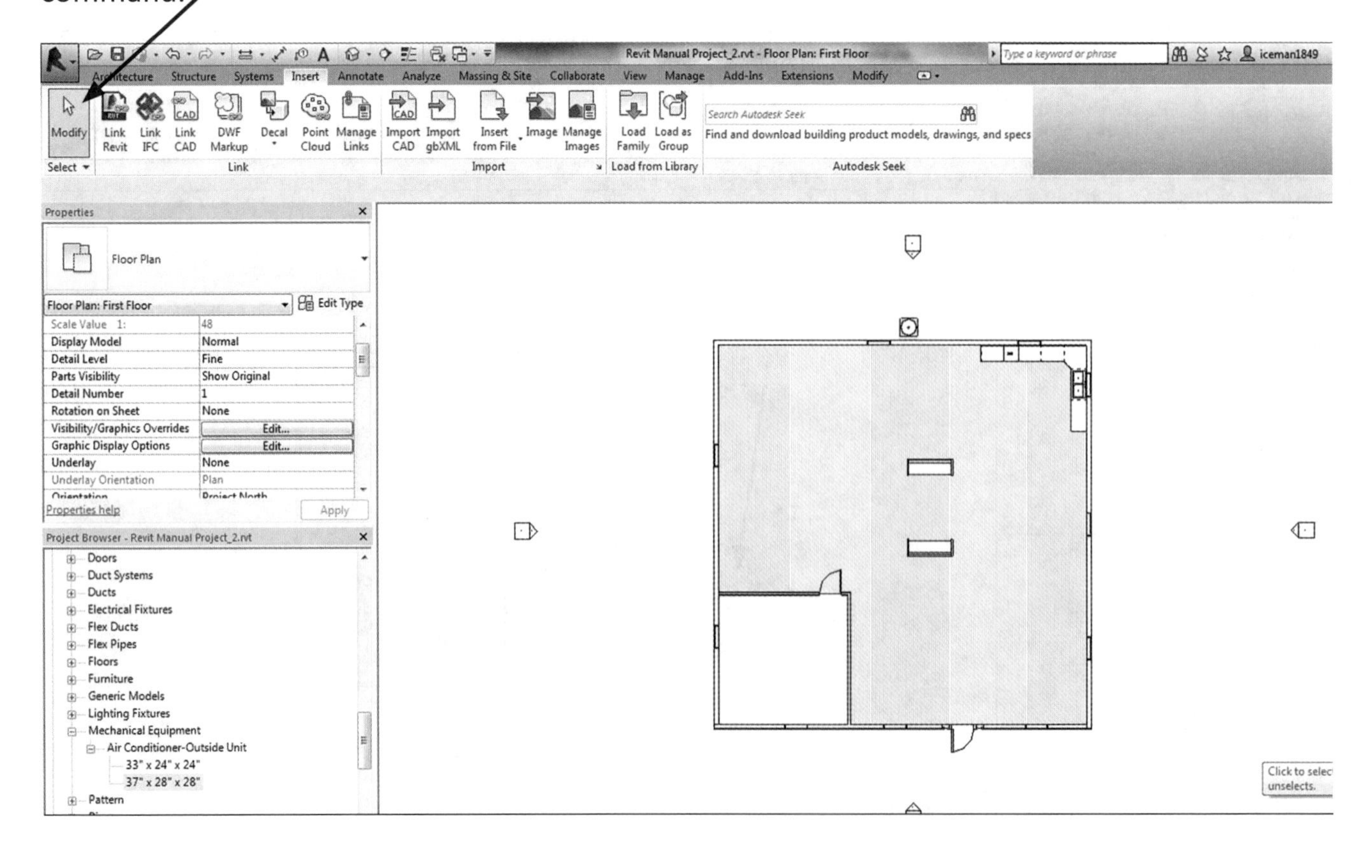

REMINDER

Do** not **hit the** ESC **Key, or you will end other commands and potentially undo a lot of your previous work.

PLACING A FURNACE

When placing a furnace in a house or building, just like the air conditioner, that equipment is listed as Mechanical Equipment. You will notice that there is *not* a Family category listed in the Project Browser. Therefore, you will need to load that equipment from the Revit Family.

LOOK AHEAD

To load the appropriate equipment, follow the steps in Chapter 17, Inserting Objects from the Revit Library and the Internet.

Make sure you're on the correct floor plan, Architecture Tab, and the First Floor Plan.

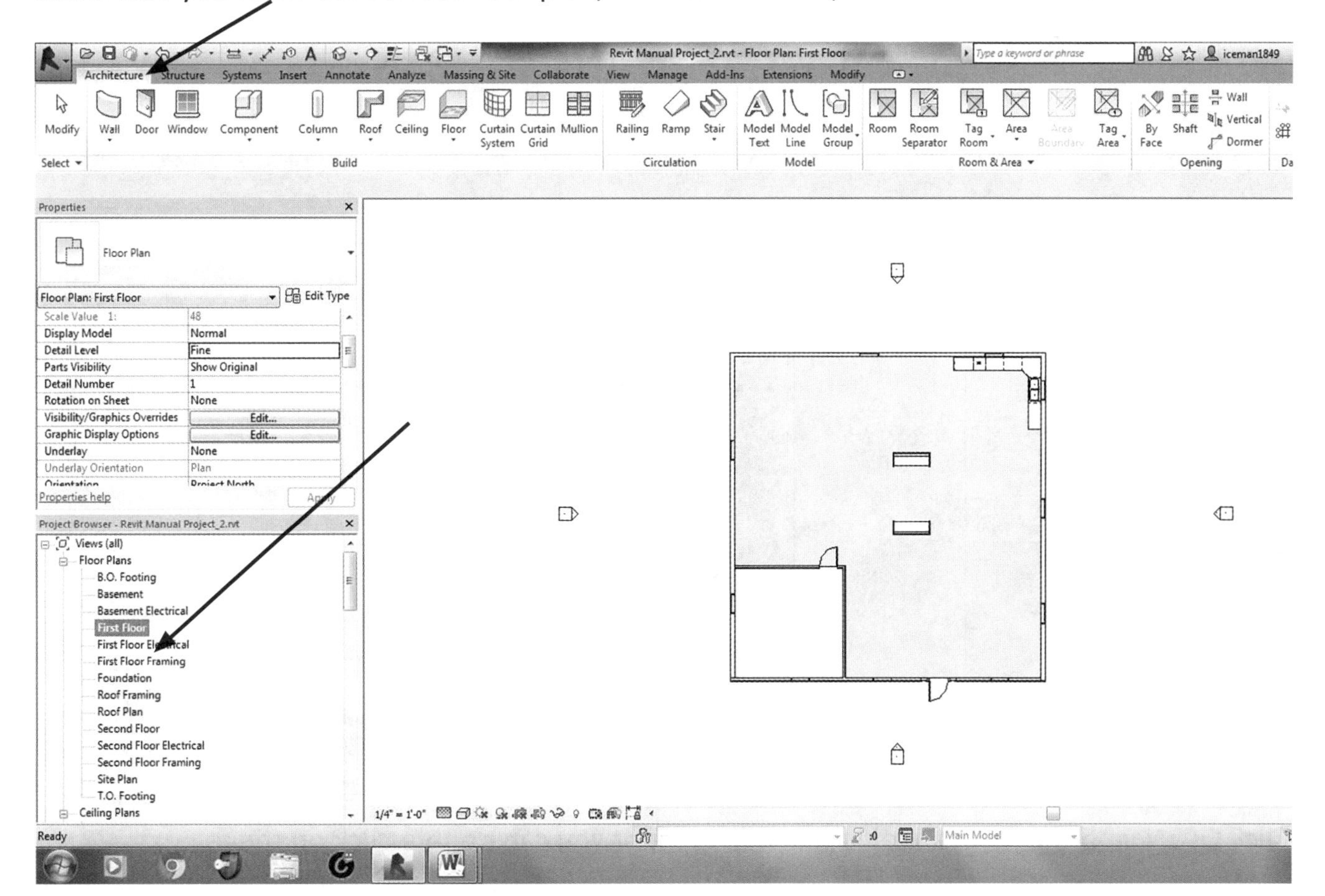

Click on the **INSERT** Tab, then click on the **LOAD FAMILY** tile.

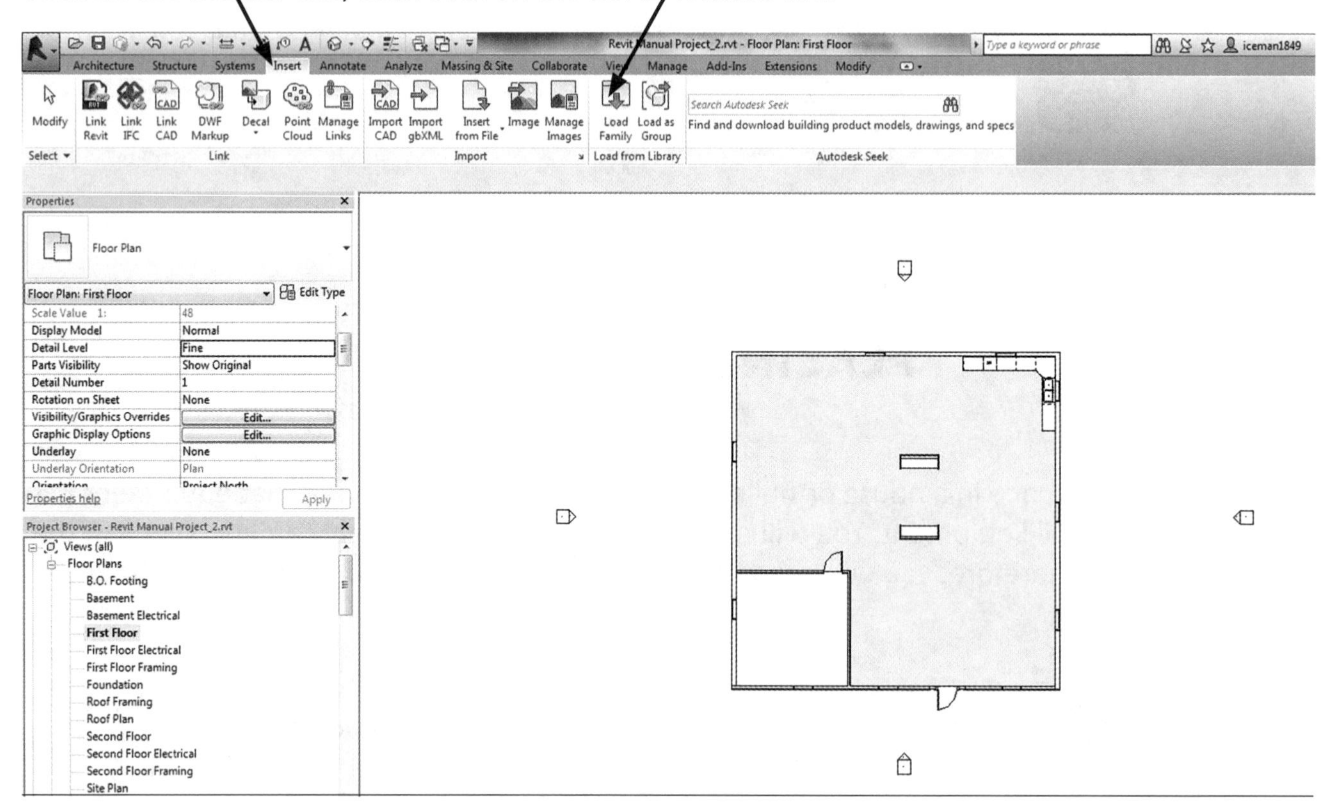

The Load Family dialog box will open. Scroll down to the Mechanical folder.

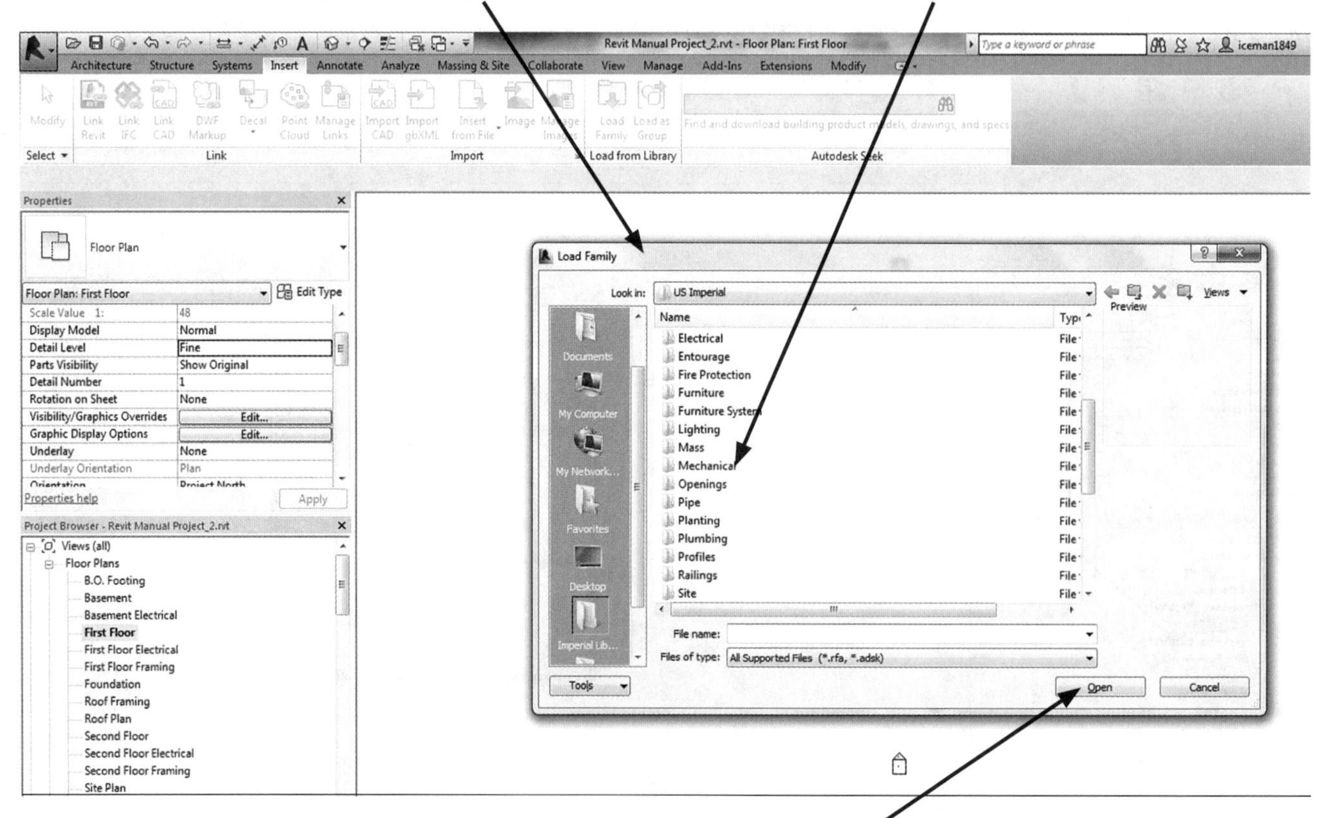

Click **OPEN**, and the Mechanical folder will open, or you can double-click on the folder.

Next, click on the **ARCHITECTURAL** folder, and open that folder.

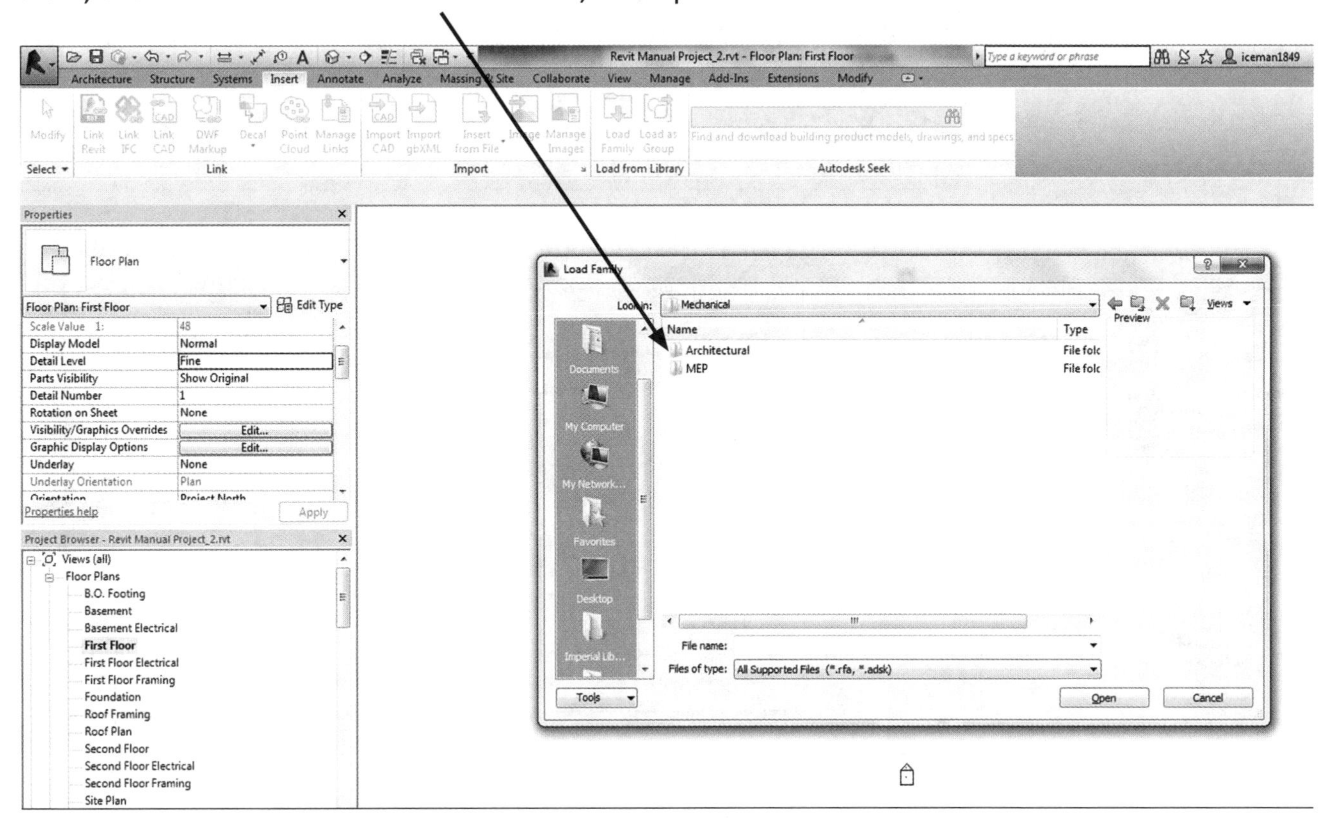

Then click on the **AIR-SIDE COMPONENTS** folder, and open that folder.

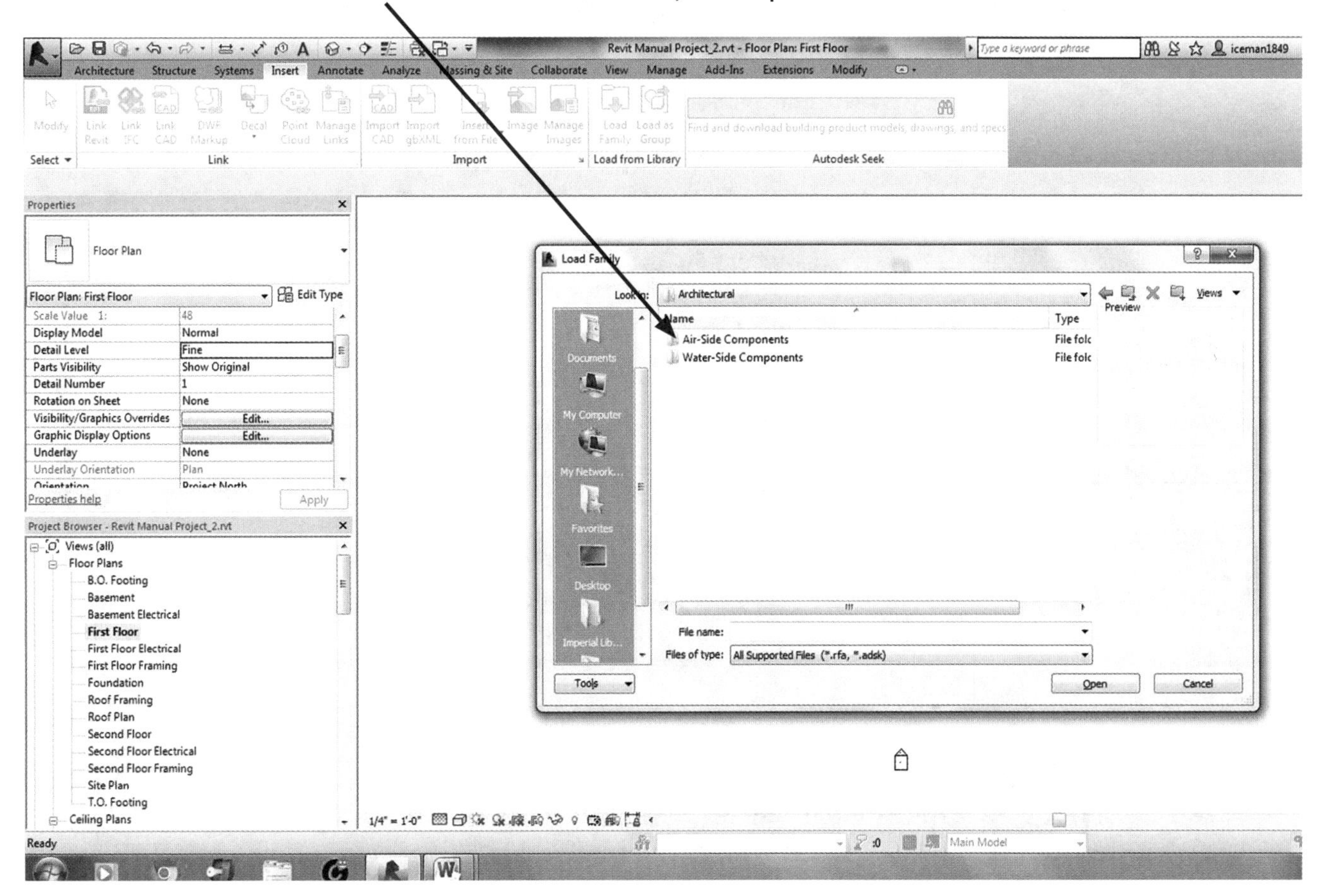

Next, click on the **FURNACES** folder, and open it up.

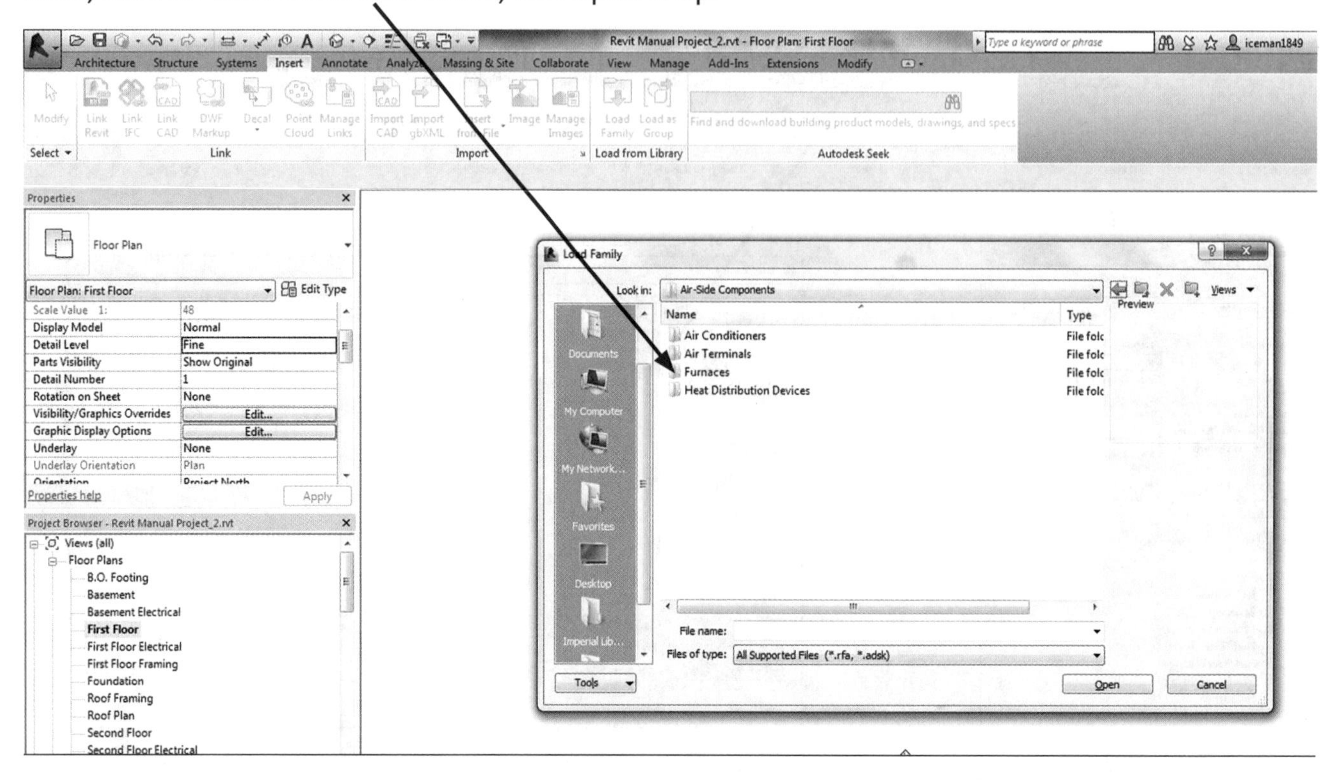

Finally, click on the **FURNACE** family, and click **OPEN**, or double-click on the **FAMILY.**

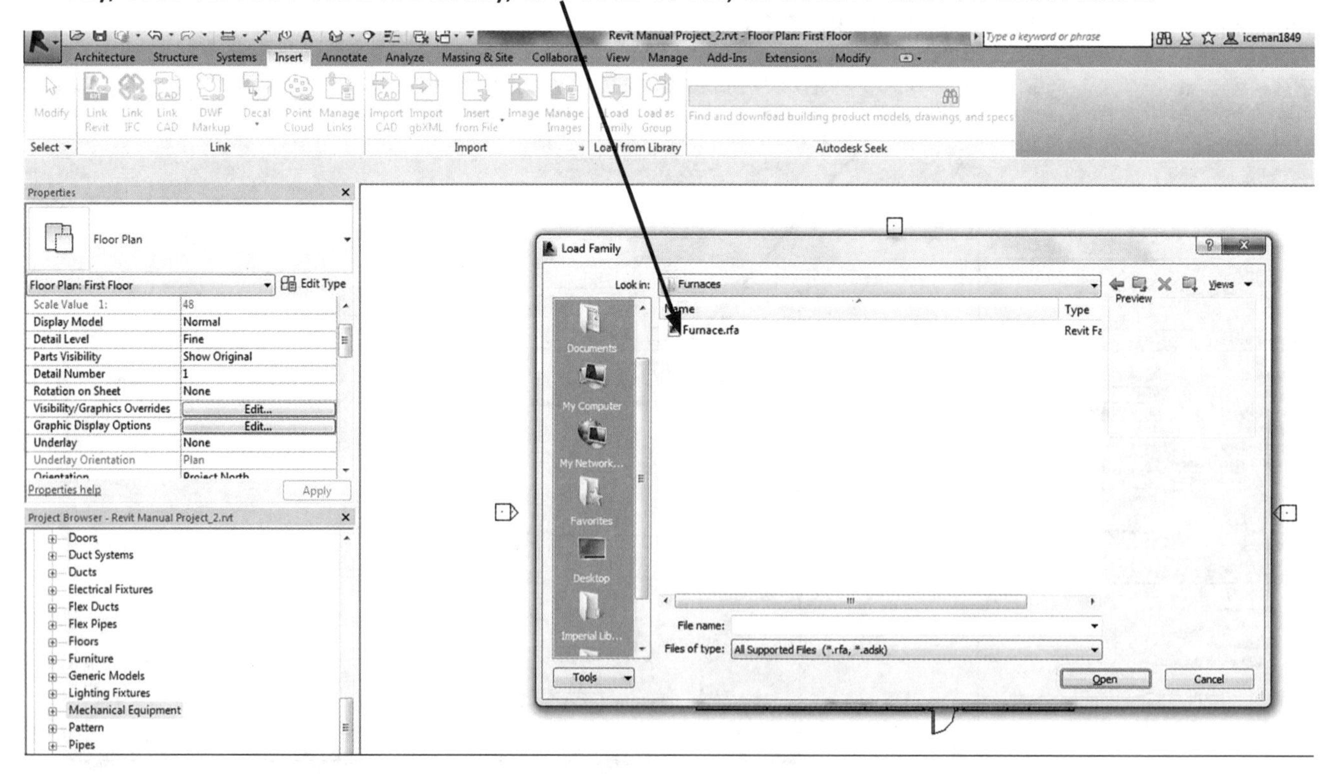

NOTES

This unit goes** inside **the house or building.

The furnace will then be loaded into the Families category in your project in the Project Browser under the Mechanical Equipment category.

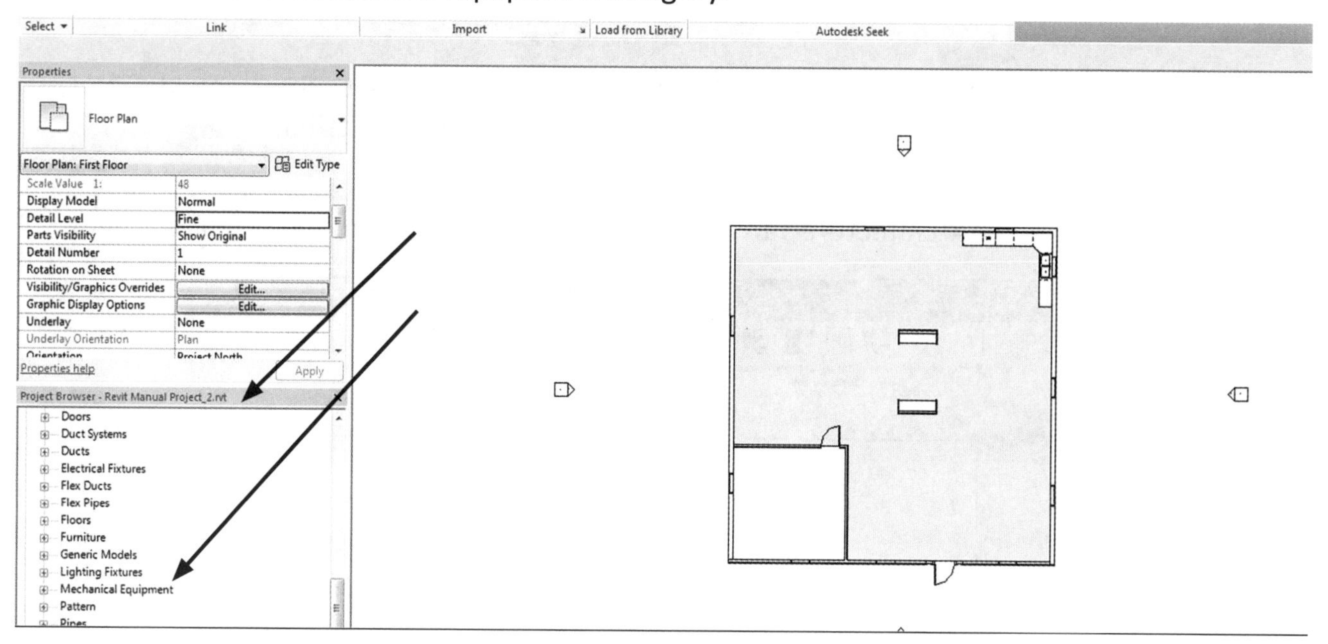

REMINDER

To expand any tree, you can use the + symbol.

Expand the **MECHANICAL EQUIPMENT** category tree, then expand the furnace.

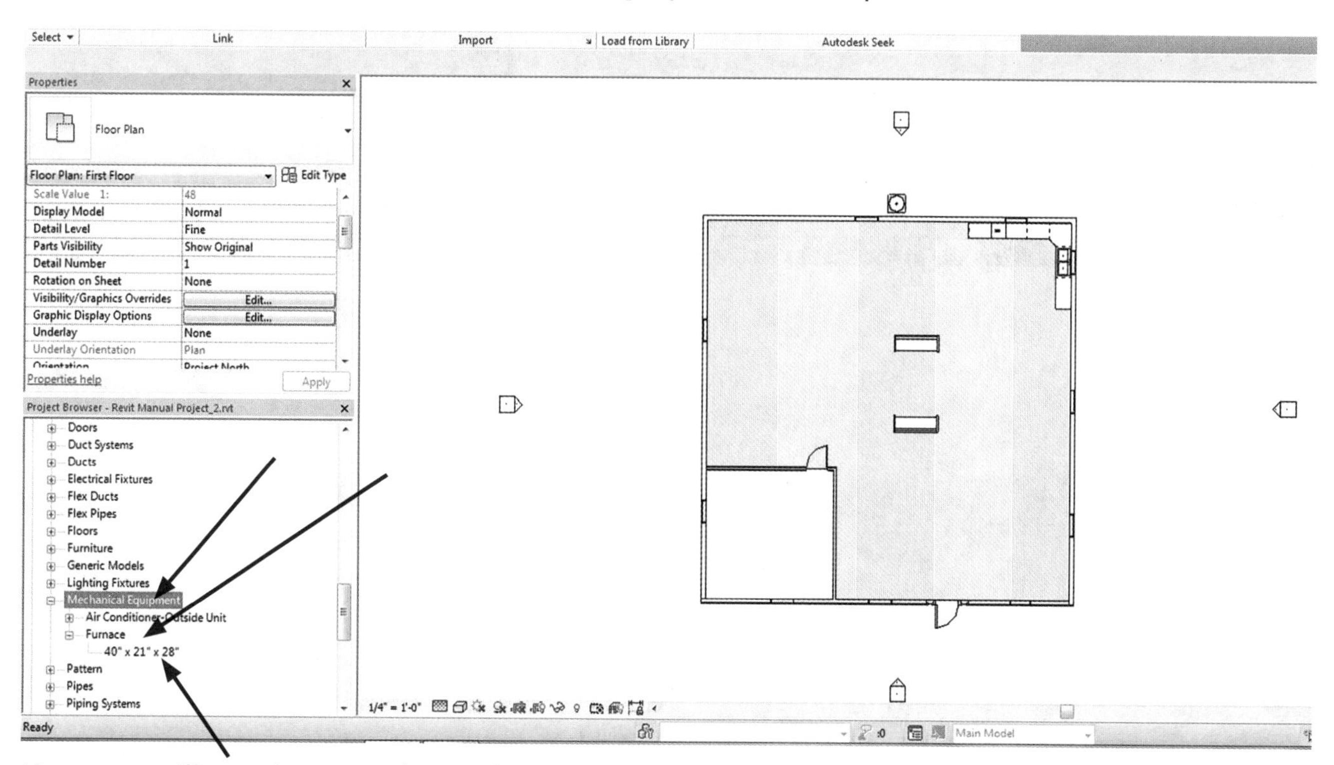

Here you will see that Revit has only one default size. If you need a different size you can visit Chapter 14, and follow those steps to download another type of Furnace from the Revit Distribution Library or Internet.

After you have loaded the furnace into your project, you should then design a closet or mechanical room to enclose the furnace.

Follow the steps you learned in Chapter 3, *Beginning a Drawing*, to place interior walls, and create the appropriate sized space for the furnace. Let's make one together.

Make sure you are in the Architecture Tab and on the correct floor plan, the First Floor plan.

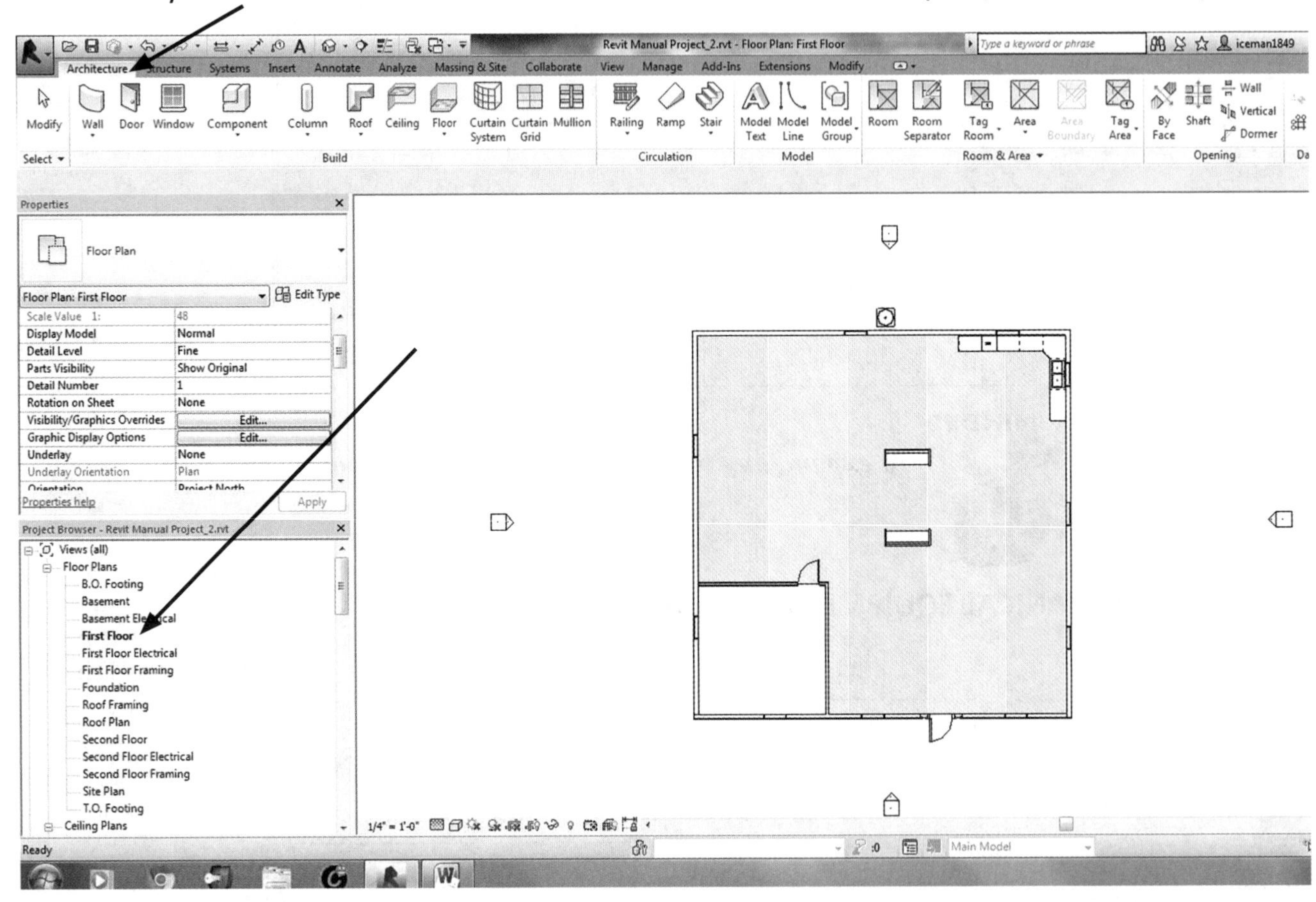

Click on the **WALL** tile and select the **INTERIOR** wall that you would like to design.

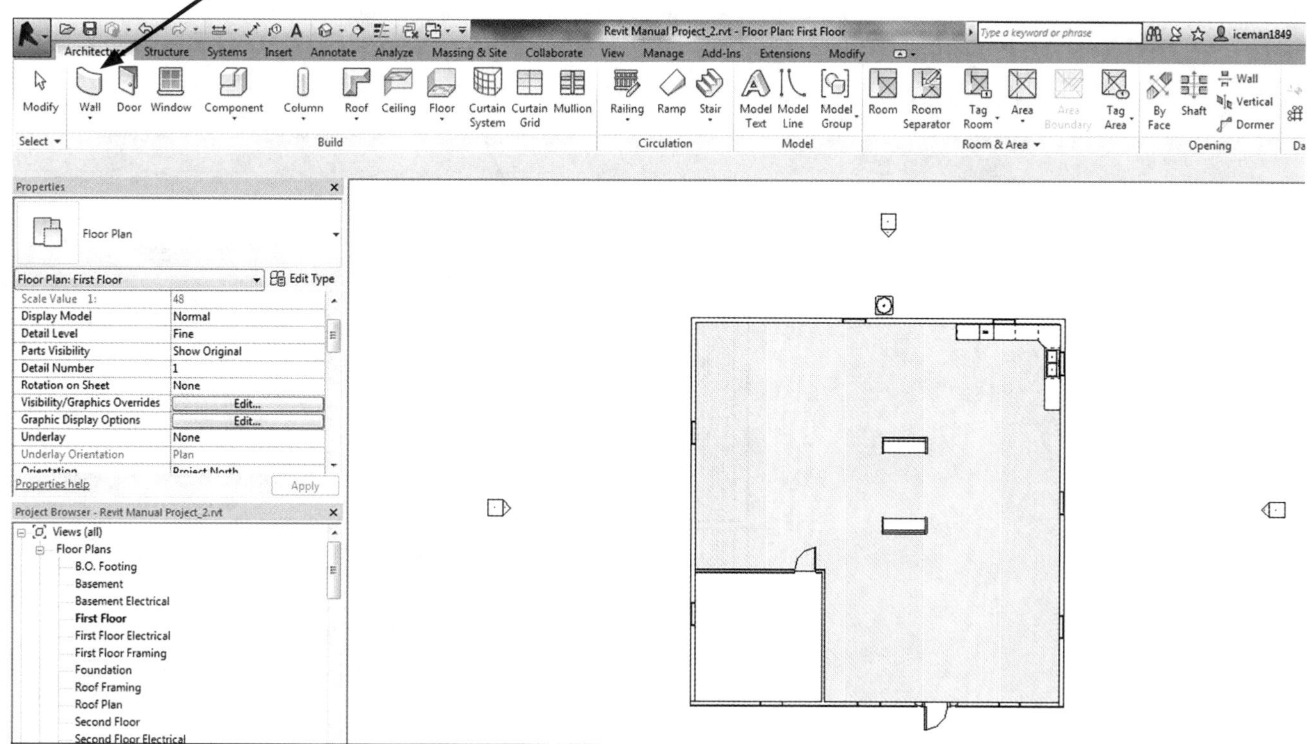

I usually create a wall with a 4 ½" interior partition for furnace rooms, because the furnace is usually in a closet or small room off by itself for maintenance. The typical walls around the equipment are usually 4 ½" thick, just to hide the sight of the mechanical equipment and the sound.

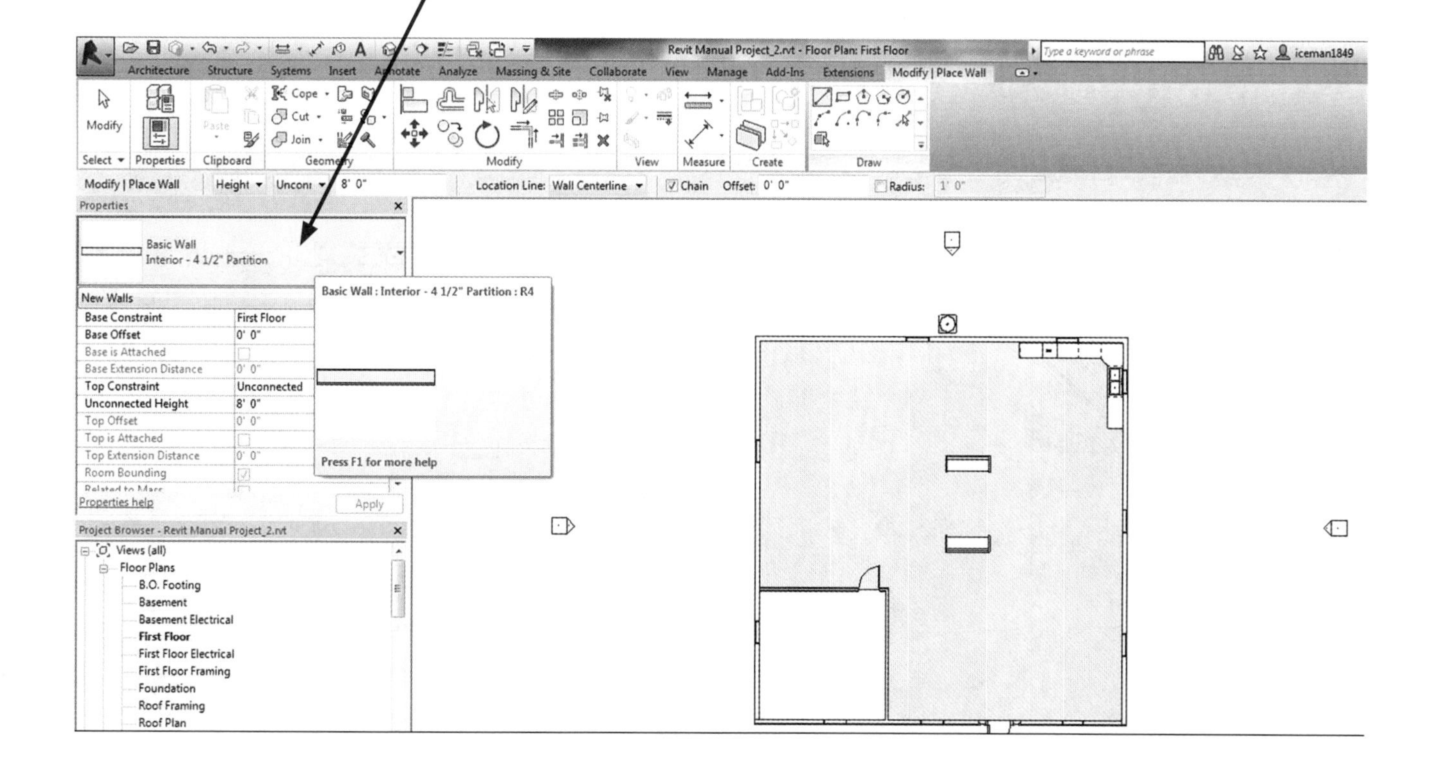

Then you should decide where you want to place or design the furnace or mechanical room.

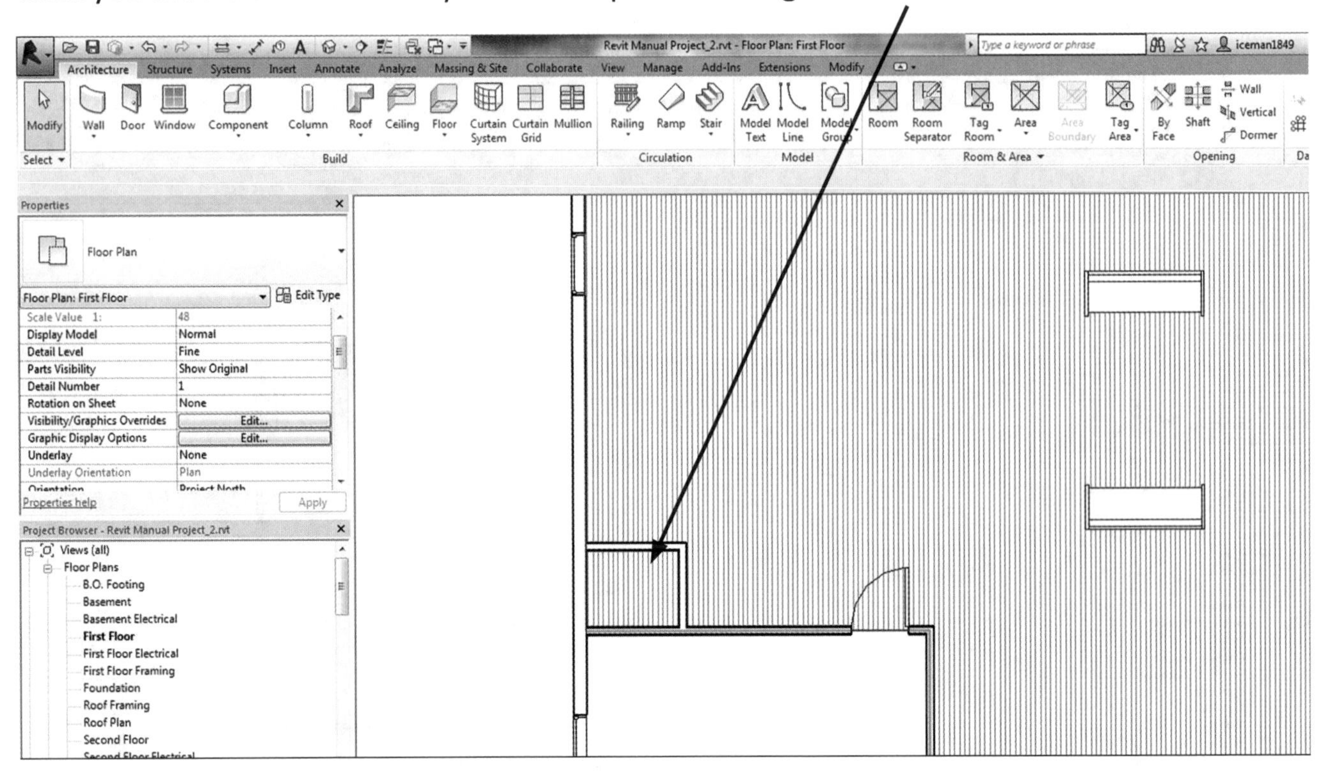

Next you need to create a door for access to the furnace room. To do that, you can follow the steps you learned in Chapter 4, *Placing Doors and Windows.*

Make sure you are in the Architecture Tab and on the correct floor plan, the First Floor plan.

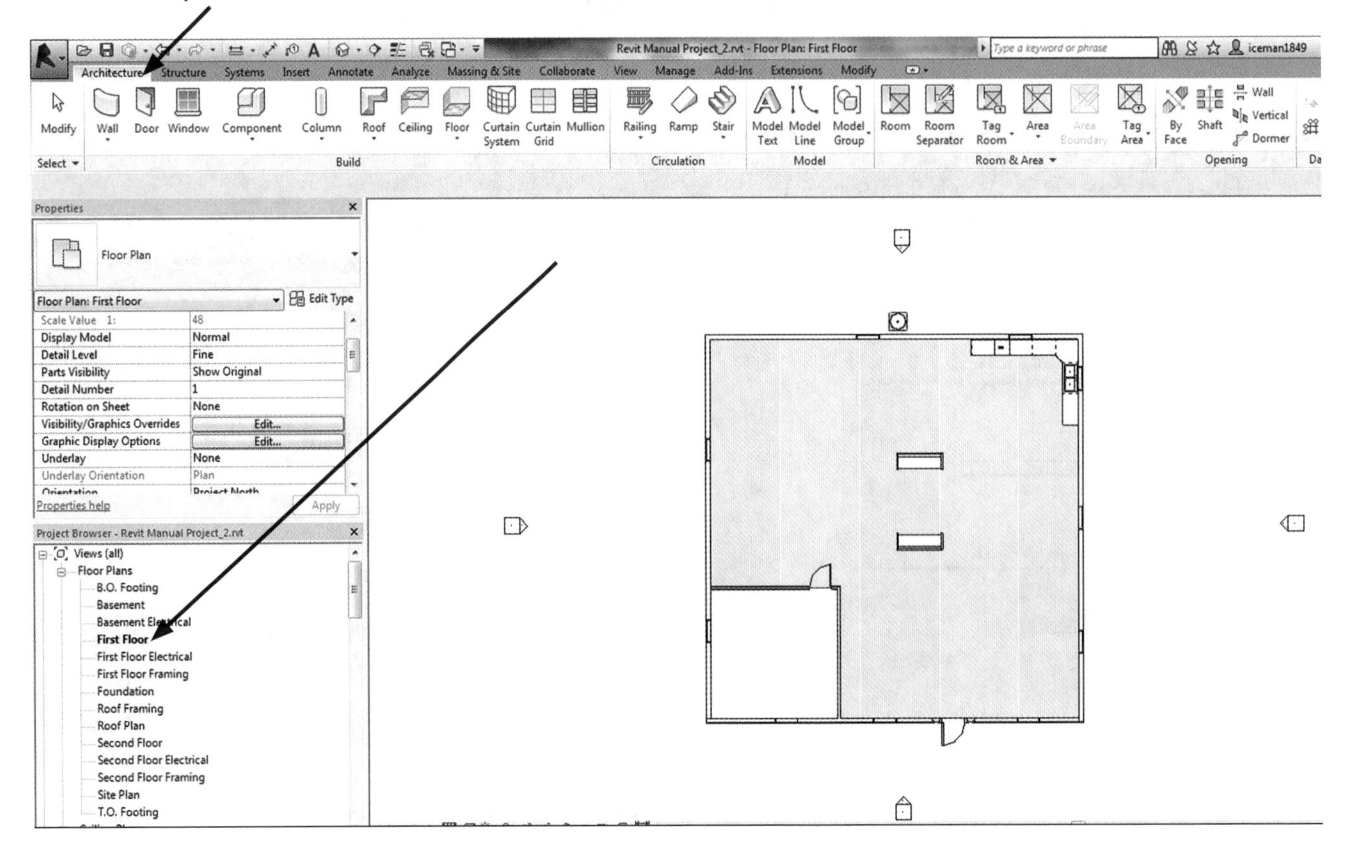

Click on the **DOOR** tile and select a door size that you would like to place in the room.

TIPS

Usually the doors for furnace rooms are no larger than 30" wide.

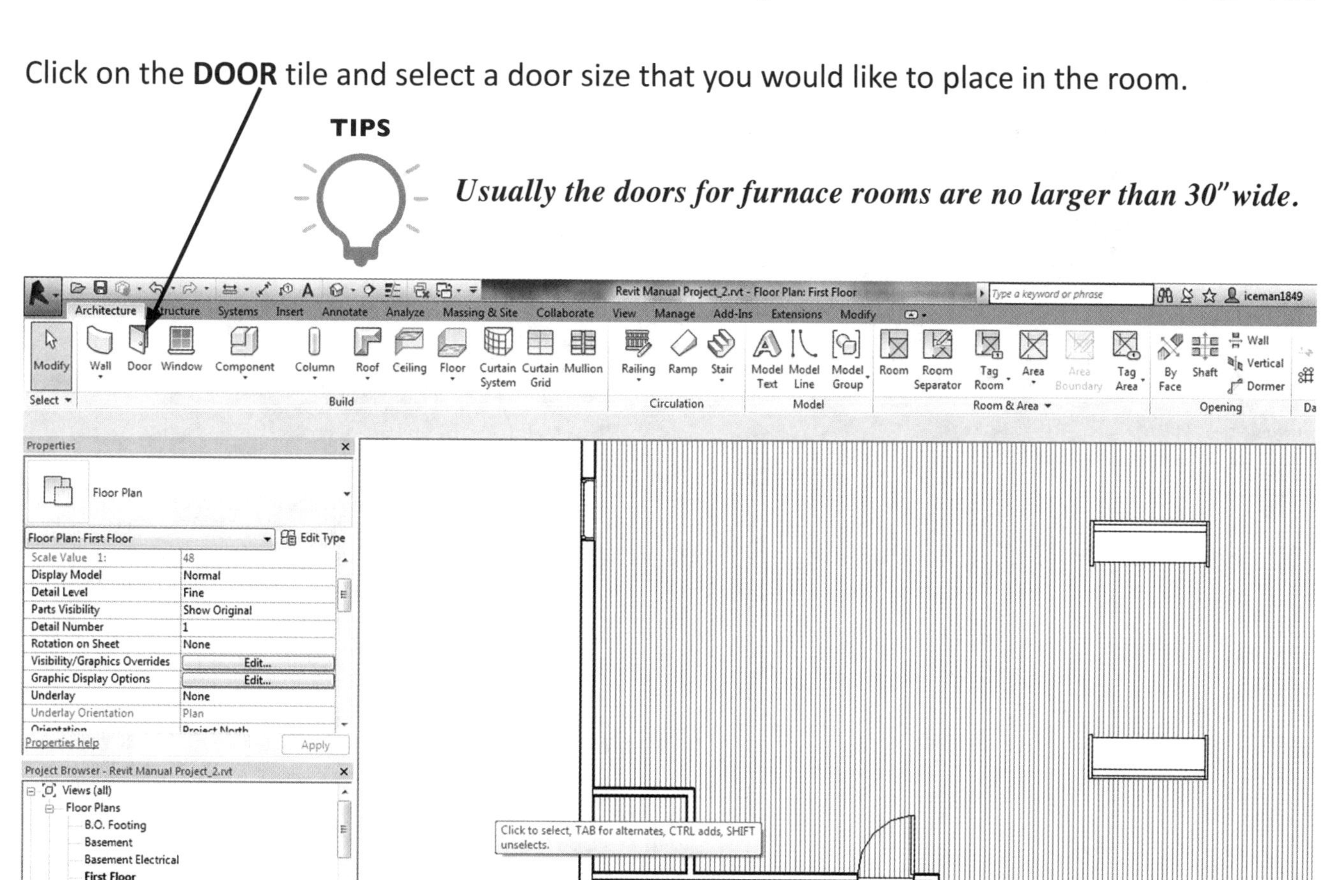

The door size will appear in the Properties Box.

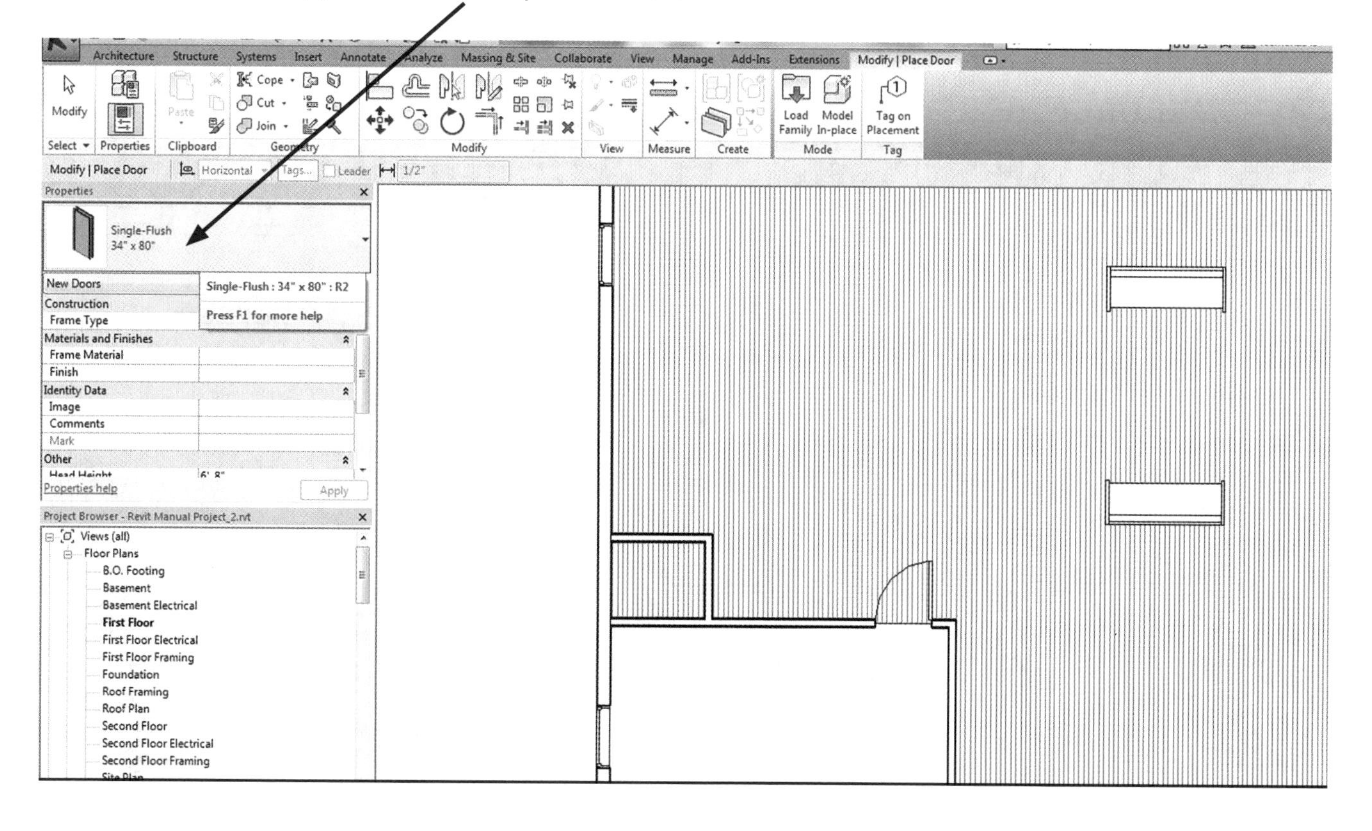

Drag the door onto the wall.

REMINDER

Remember that doors *must* ***be hosted to a wall.***

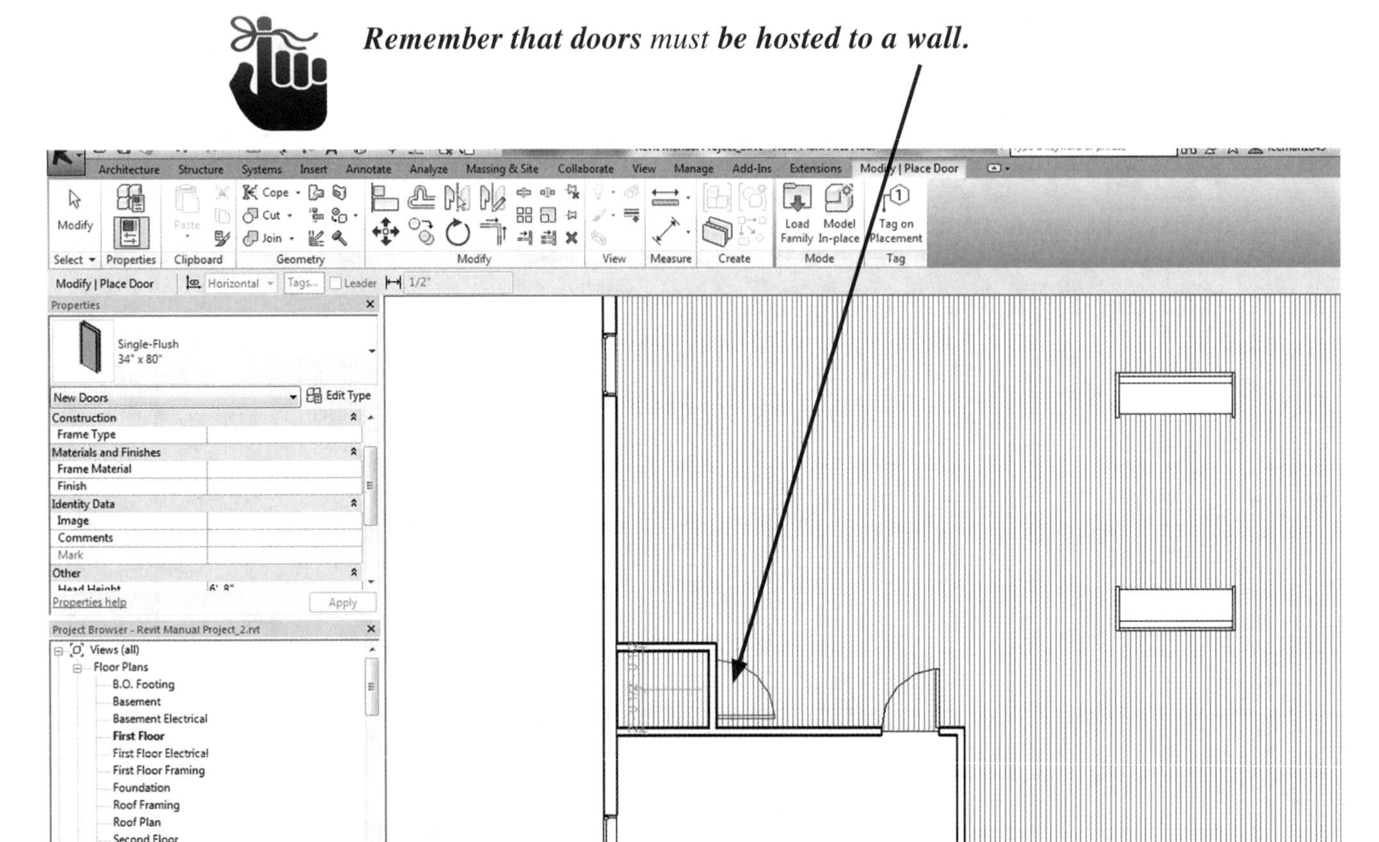

Click the door into the wall and the door will automatically cut the door opening and place the door, as seen in this screen shot:

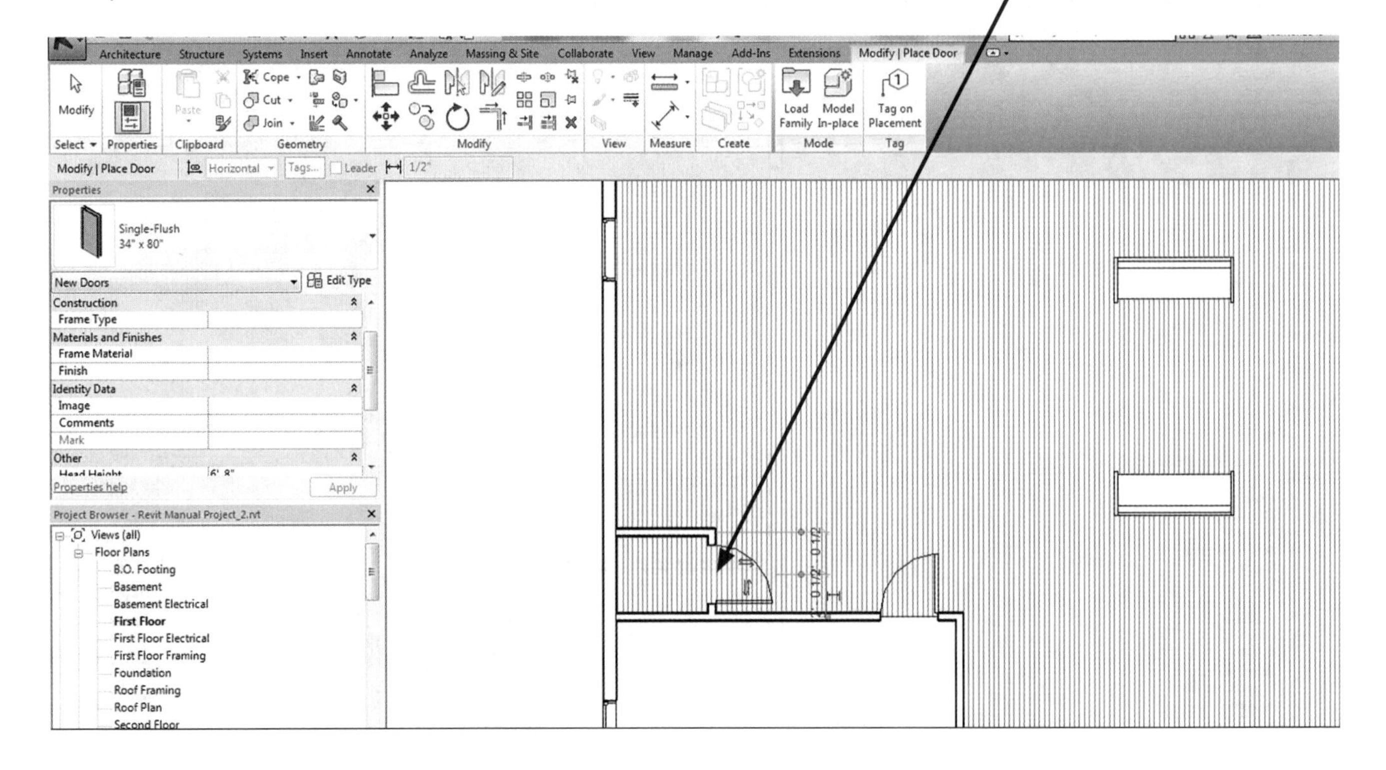

Once this is completed, you're still in the Place Door command, so you'll need to click on the **MODIFY** Arrow to end the command.

REMINDER

Do** not **hit the** ESC **Key, or you will end other commands that you don't want to lose.

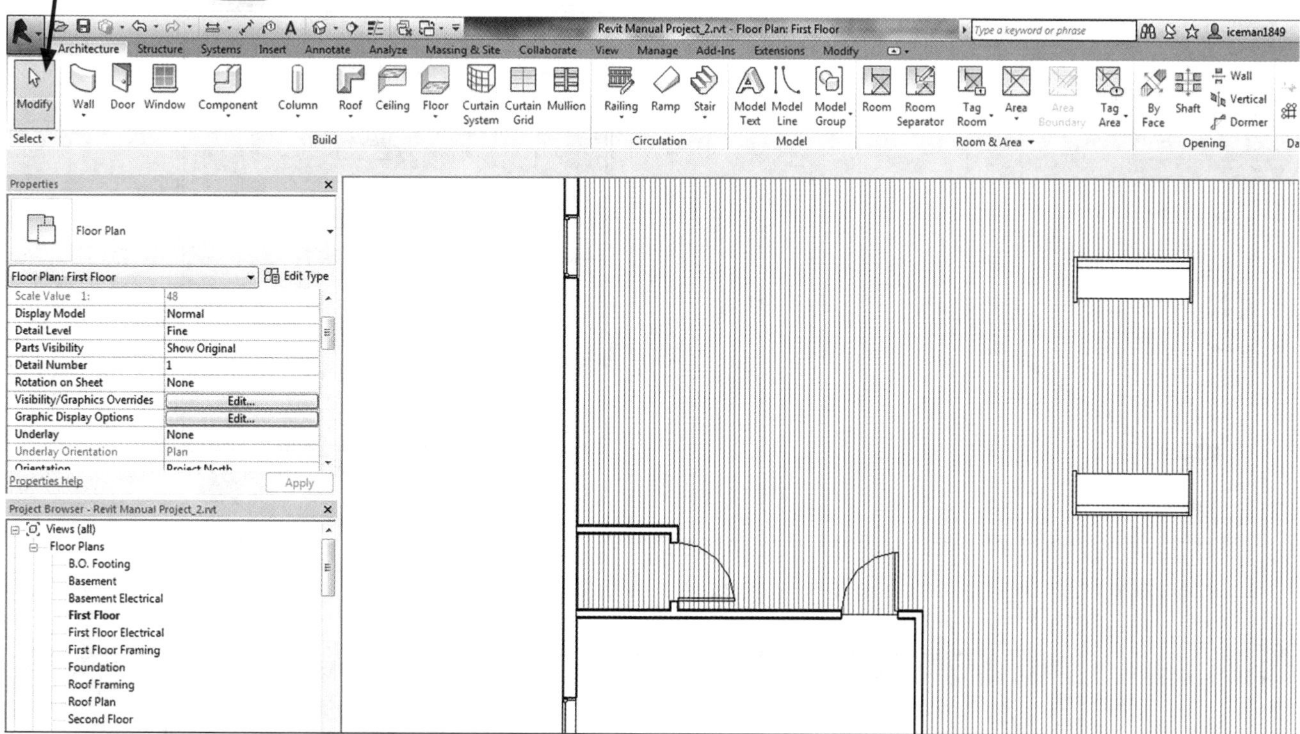

Now you can place the furnace you just loaded into the Families category.

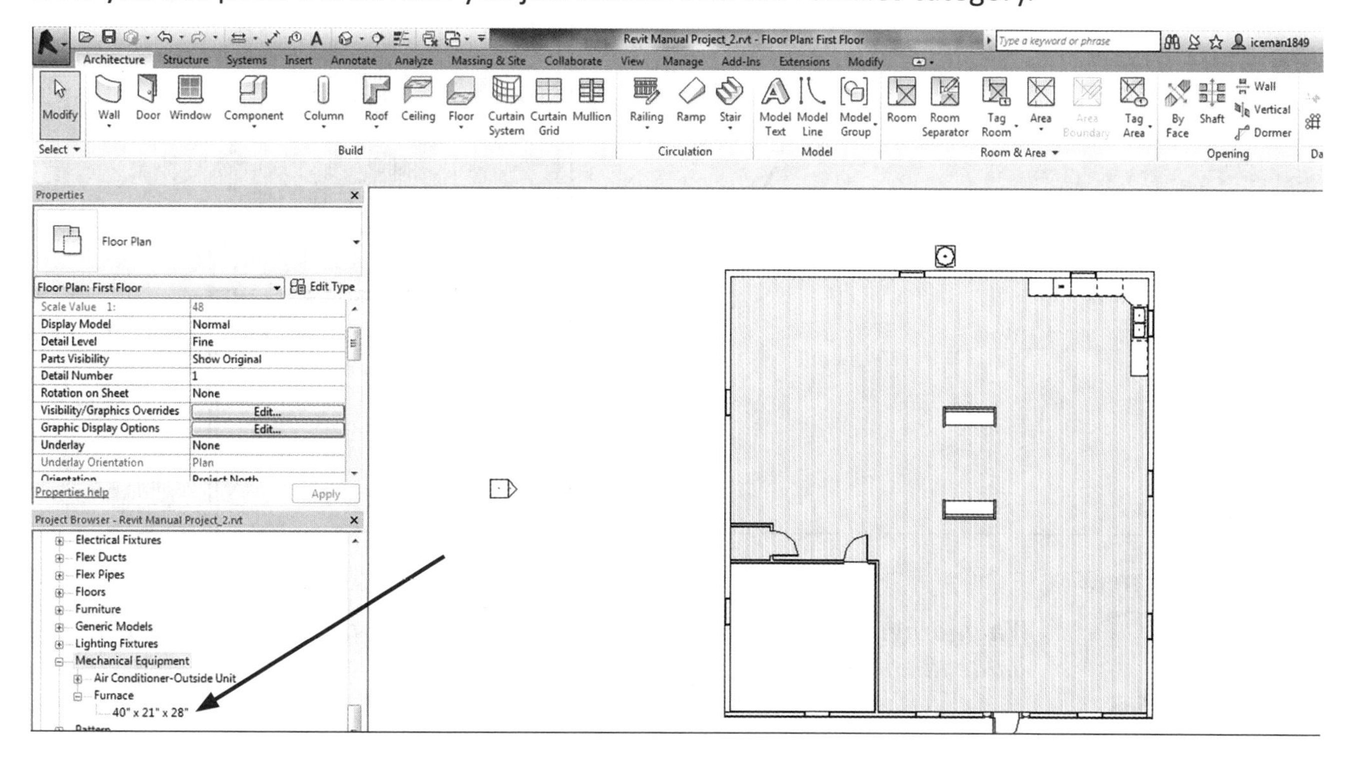

To place the furnace, use your mouse and click on the size you desire, then drag the furnace in the closet or room you created. Once you select the size, you will see the furnace in the Properties Box.

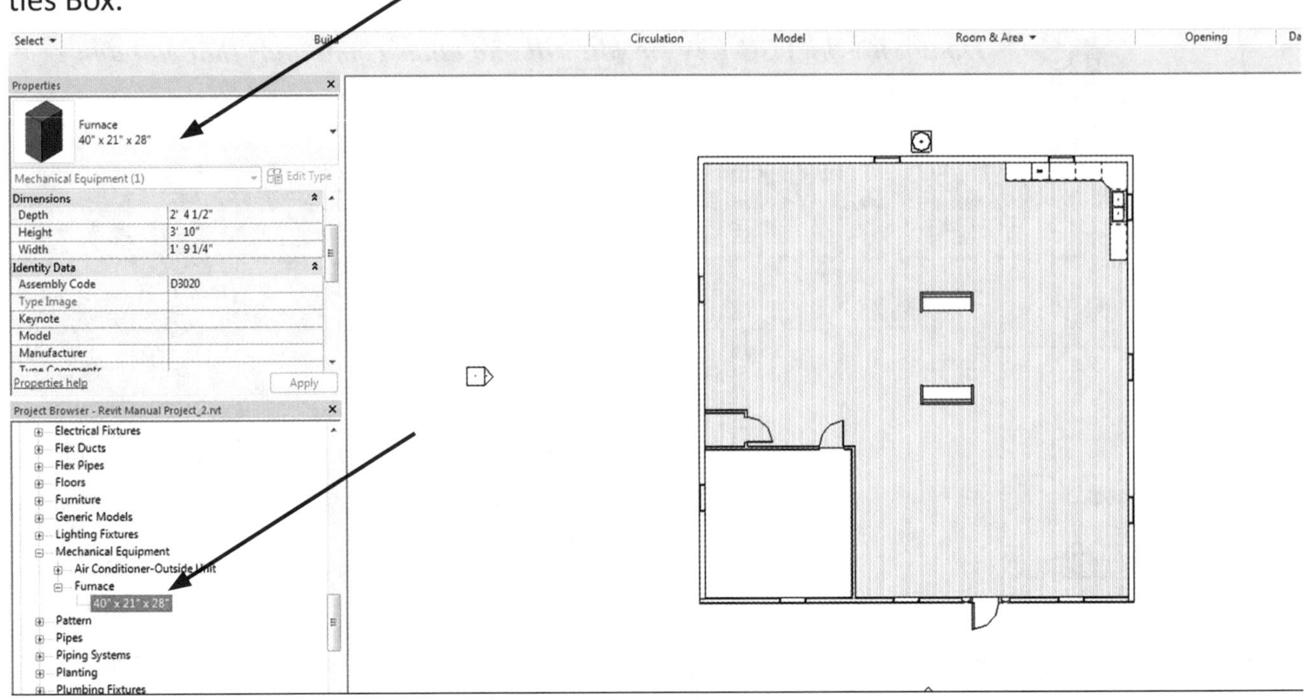

Then you can drag the furnace into your project location.

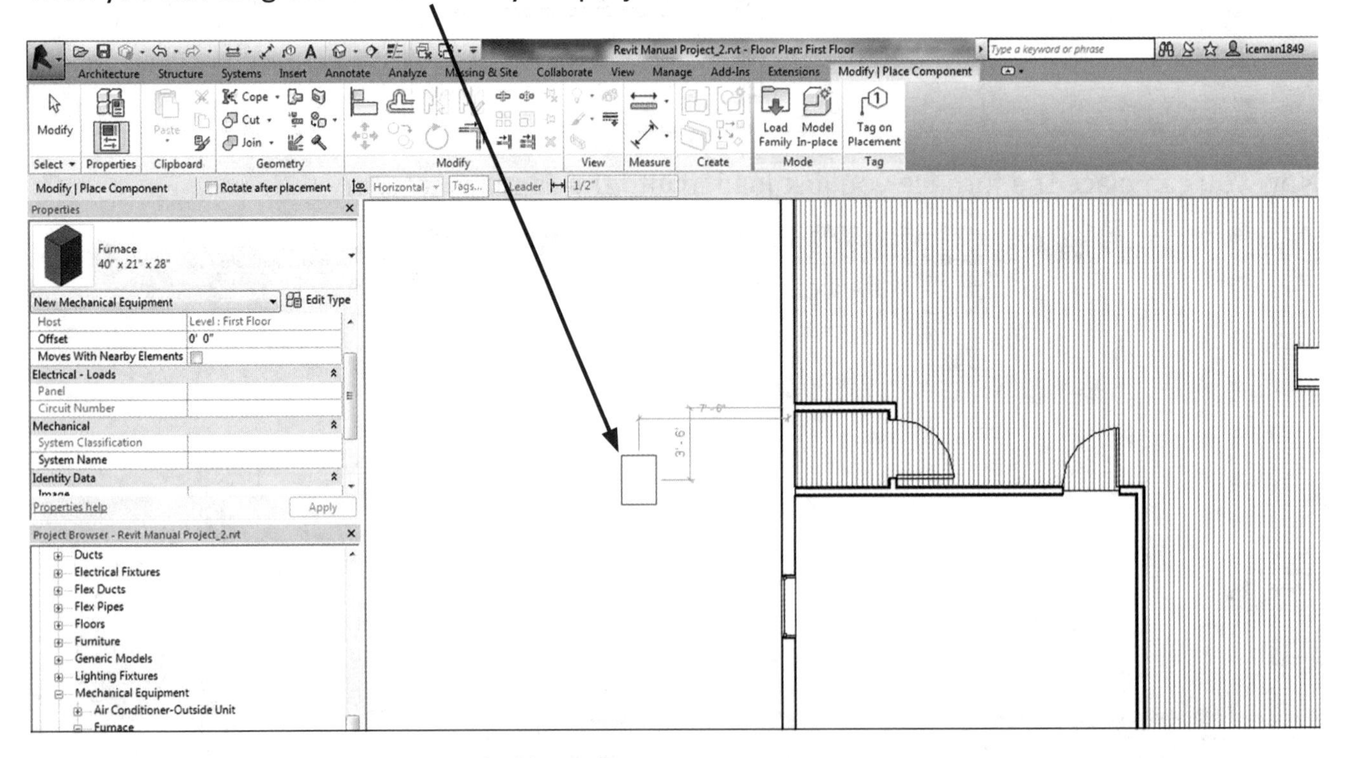

TIPS

When dragging the furnace into your project, click the* SPACEBAR *to rotate the object.

As you can see in this screen, you can rotate the object by clicking the **SPACEBAR.**

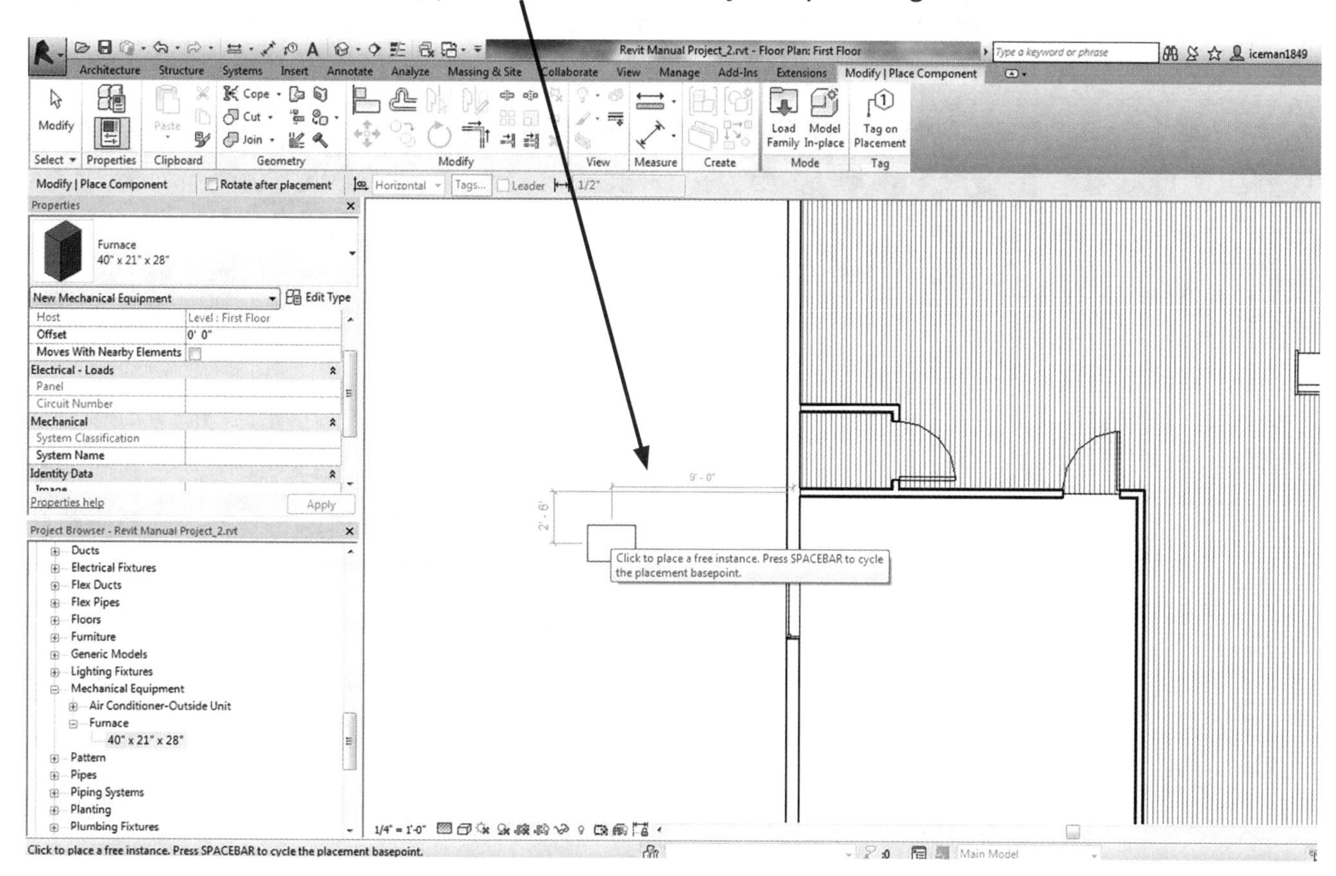

Then you can place the furnace in the designated area in your house or building.

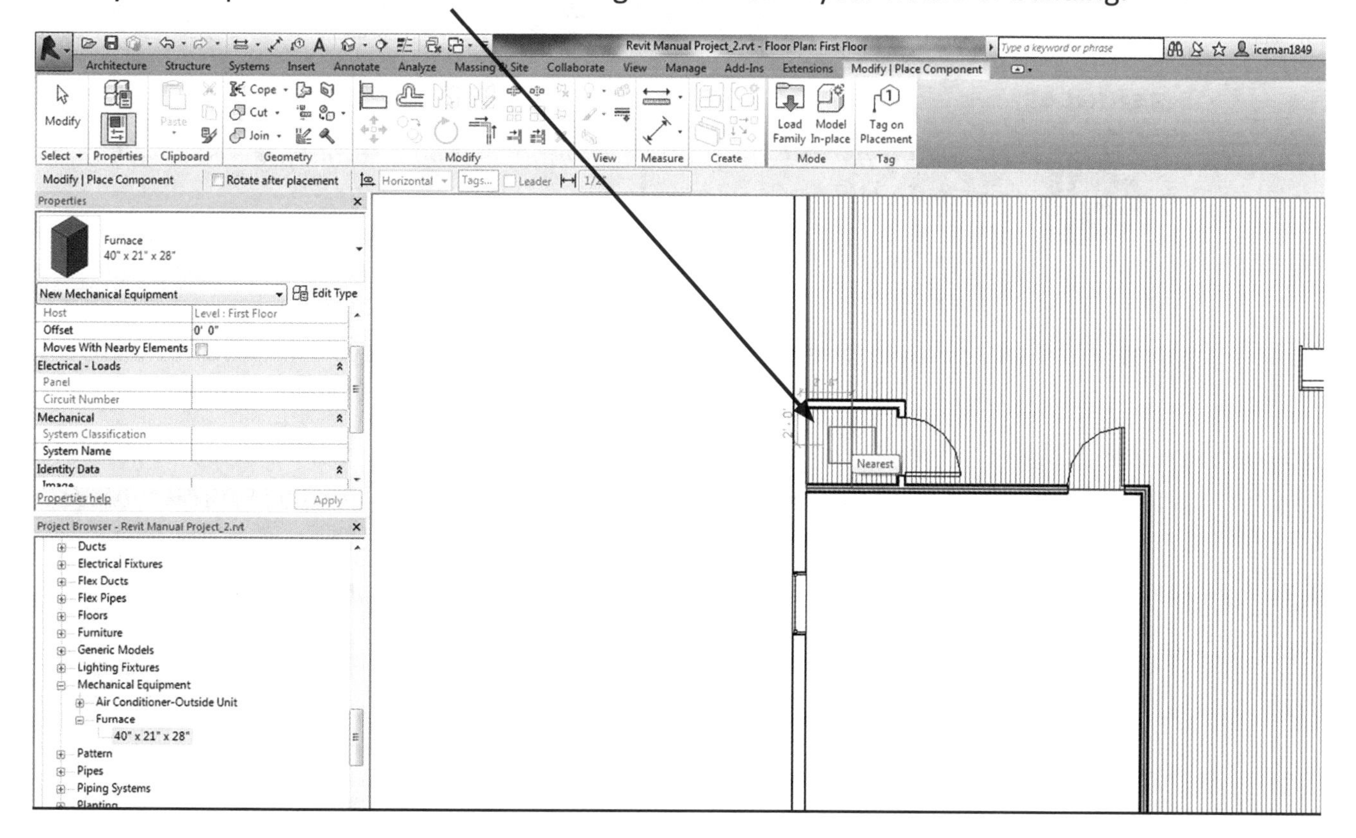

After placing the furnace, Revit defaults to placing a second furnace, because you are still in the Modify Component command. You will need to click on the **MODIFY** Arrow to end the command.

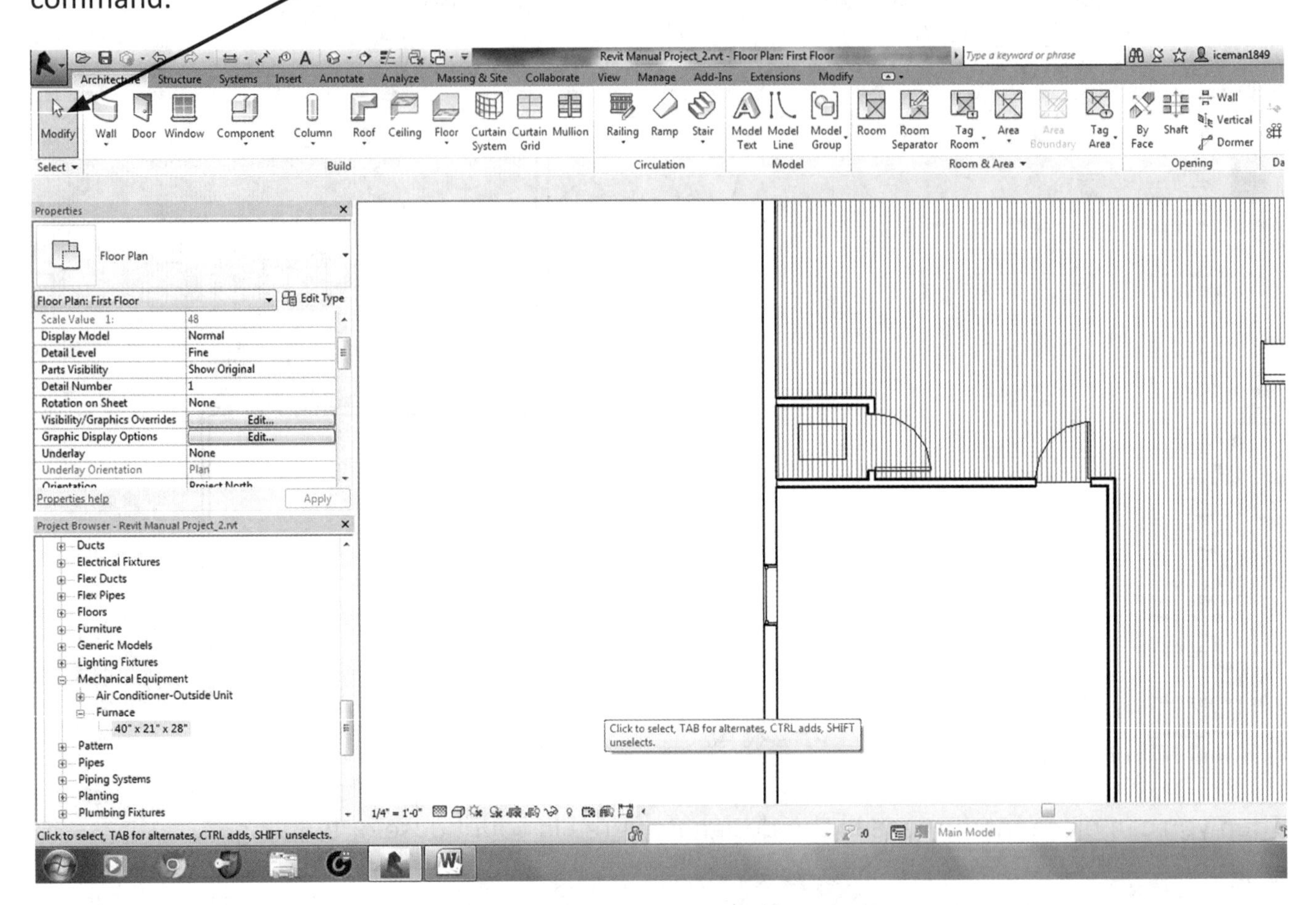

REMINDER

Do not hit the* ESC *Key. You will undo commands and previous work that you definitely don't want to lose.

Congratulations! You have now added an air conditioner and furnace system to your house or building.

Now we are able to see that the Revit software not only can design the architectural layout of your house or building but can go even further in adding some mechancial components to your design to give it that real life look.

After the heating and cooling has been added, let's look at Chapter 12, *Placing Furniture*.

CHAPTER 12

Placing Furniture

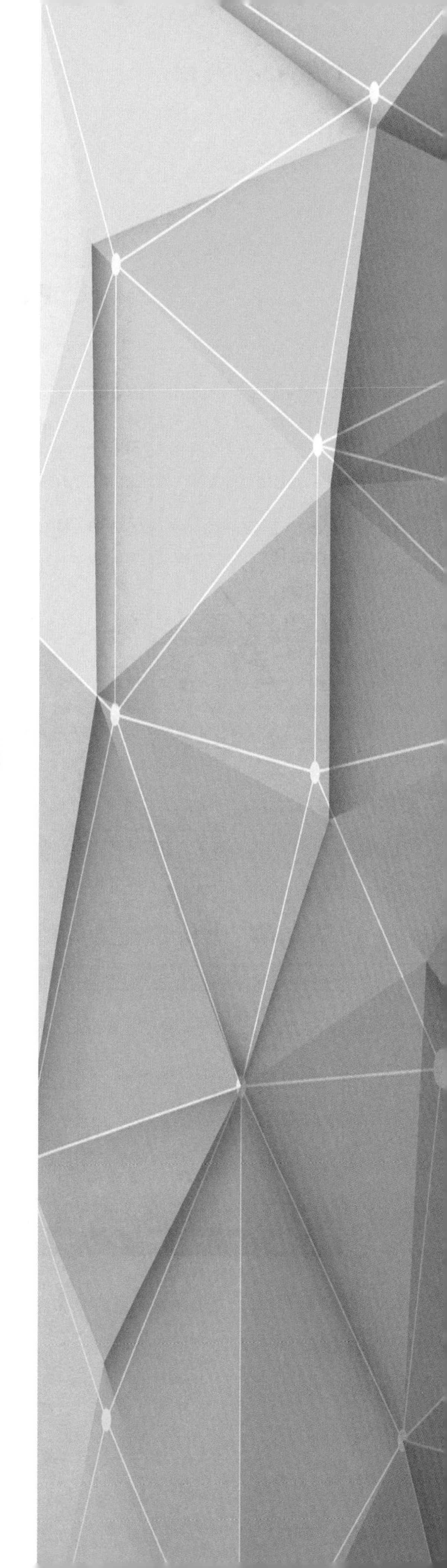

PLACING FURNITURE

In Revit, you have the ability to place furniture such as chairs, sofas, end tables, coffee tables, beds (of different sizes), dining room tables and even accessories such as lamps and rugs (just to name a few things) in a residential home. For a commercial building, you can add cubicles, file cabinets, desks, task chairs, copiers, computers, etc.

When placing furniture in your design, make sure you are in the Architecture Tab and on the Floor Plan on the one you are working on.

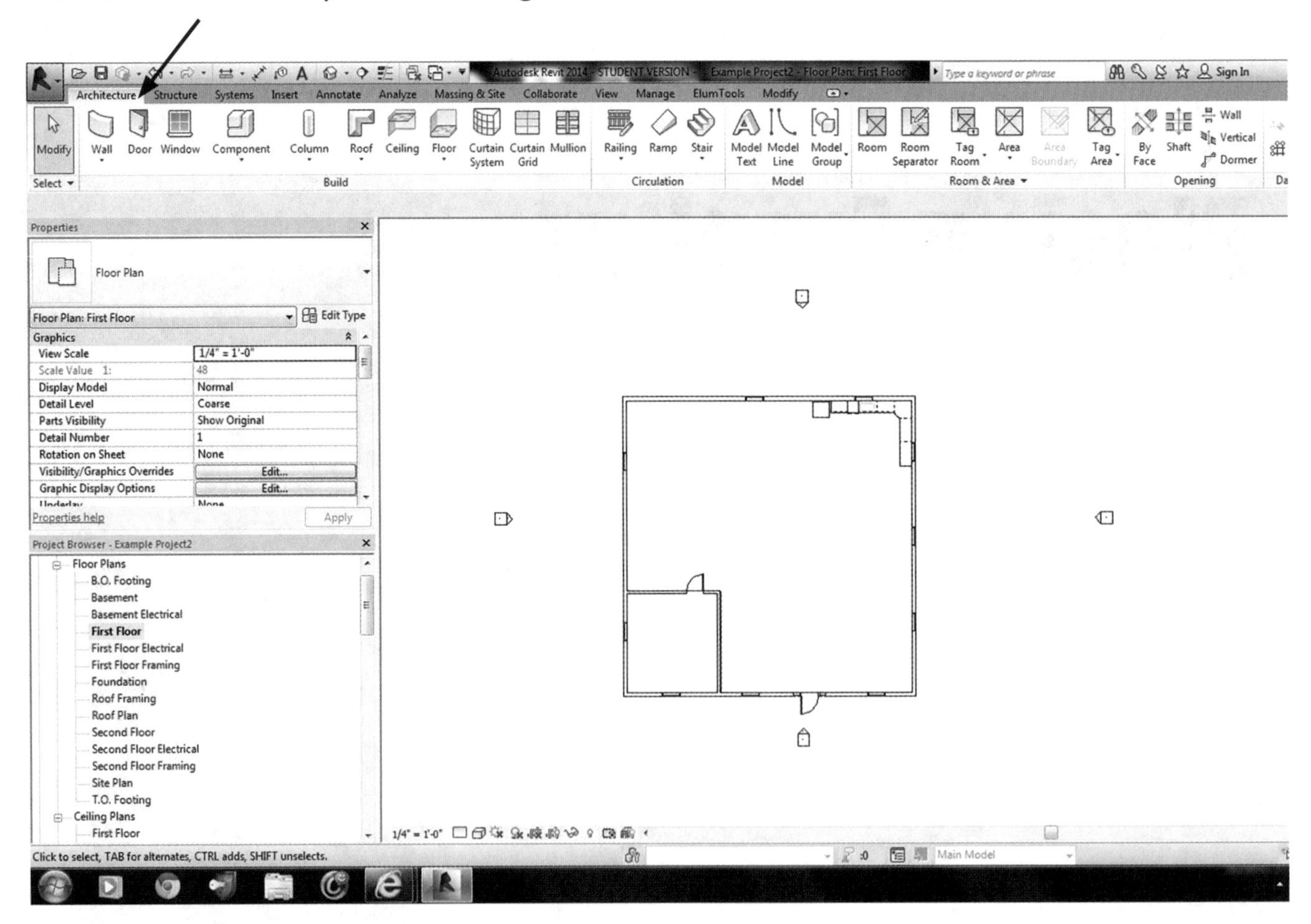

Then go to the Project Browser and select the **FURNITURE** category.

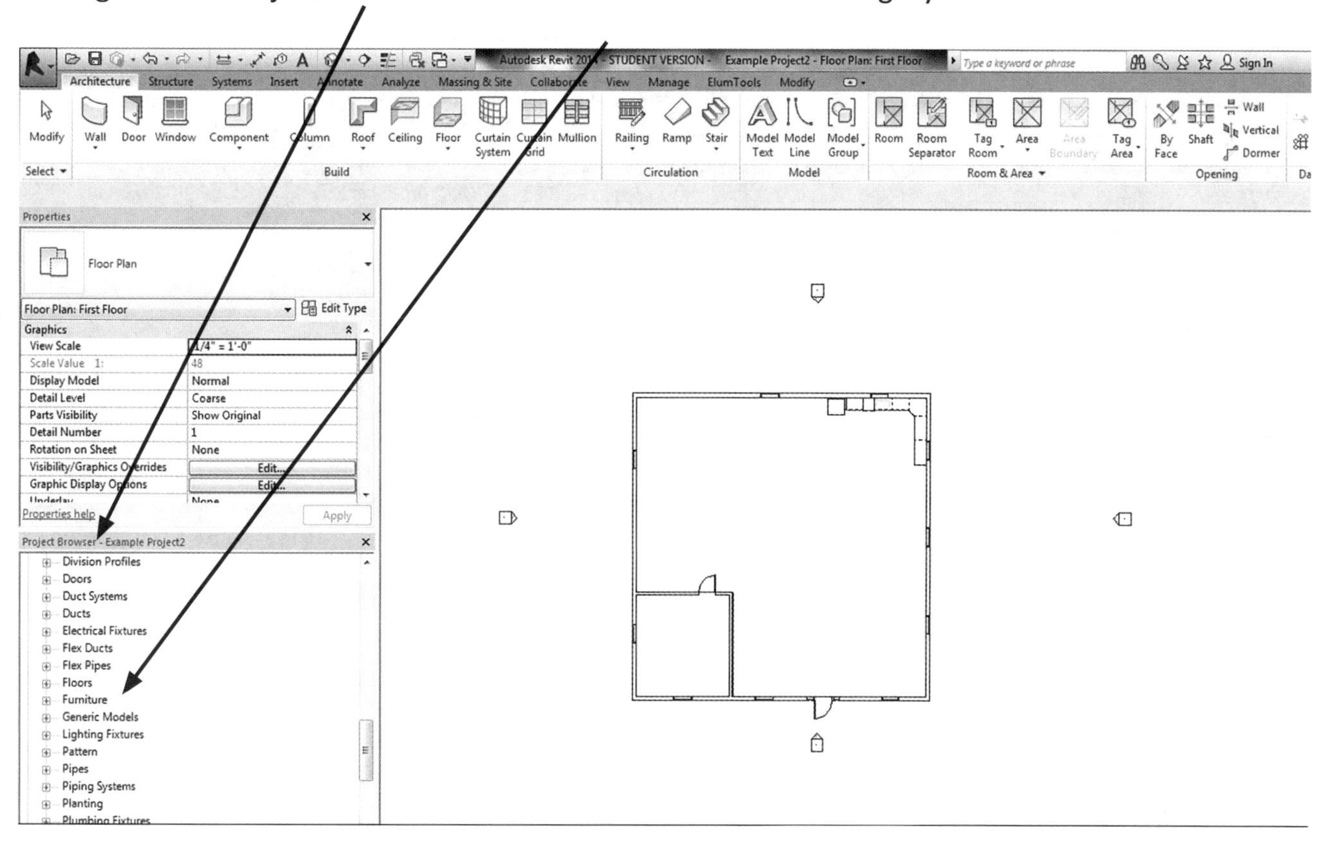

Open the **FURNITURE** category, and select the furniture you want to bring into your design.

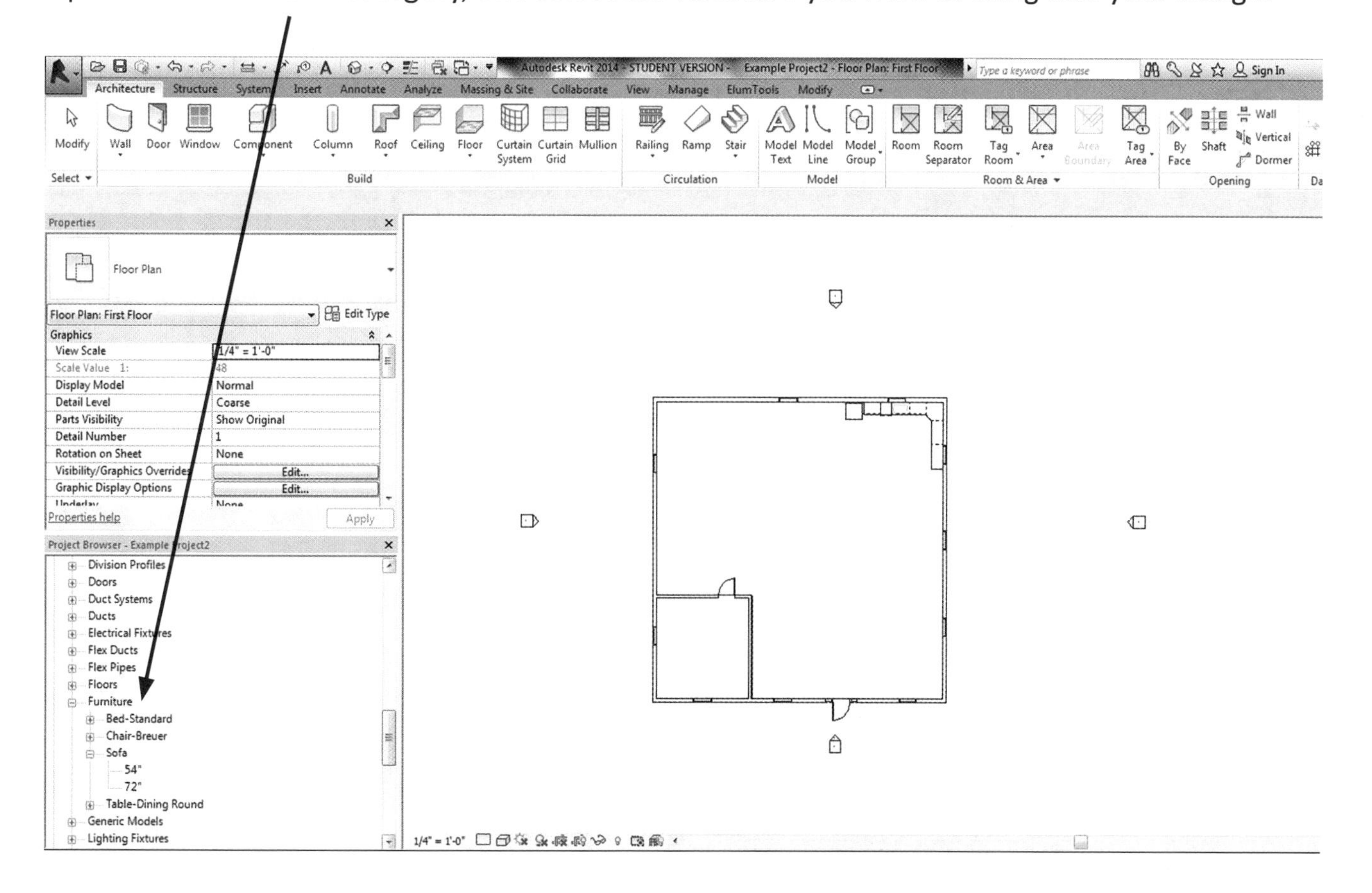

Hold down the mouse button, drag the furniture into your drawing, and place it where you want.

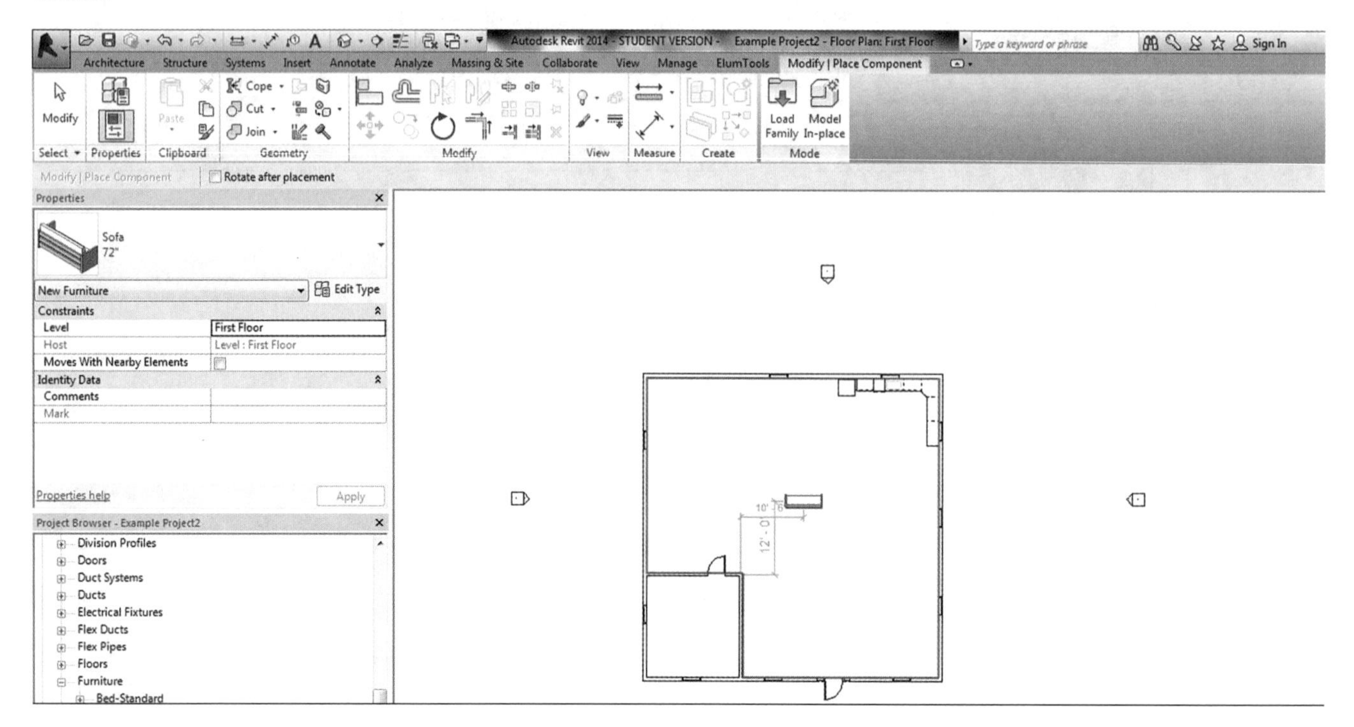

NOTES

Clicking your* SPACEBAR *before you place the furniture will rotate it in a different direction.

REMINDER

Do* not *hit the* ESC *Key to finish the placement, or some of your work may be undone.

Click the **MODIFY** Button to complete the task.

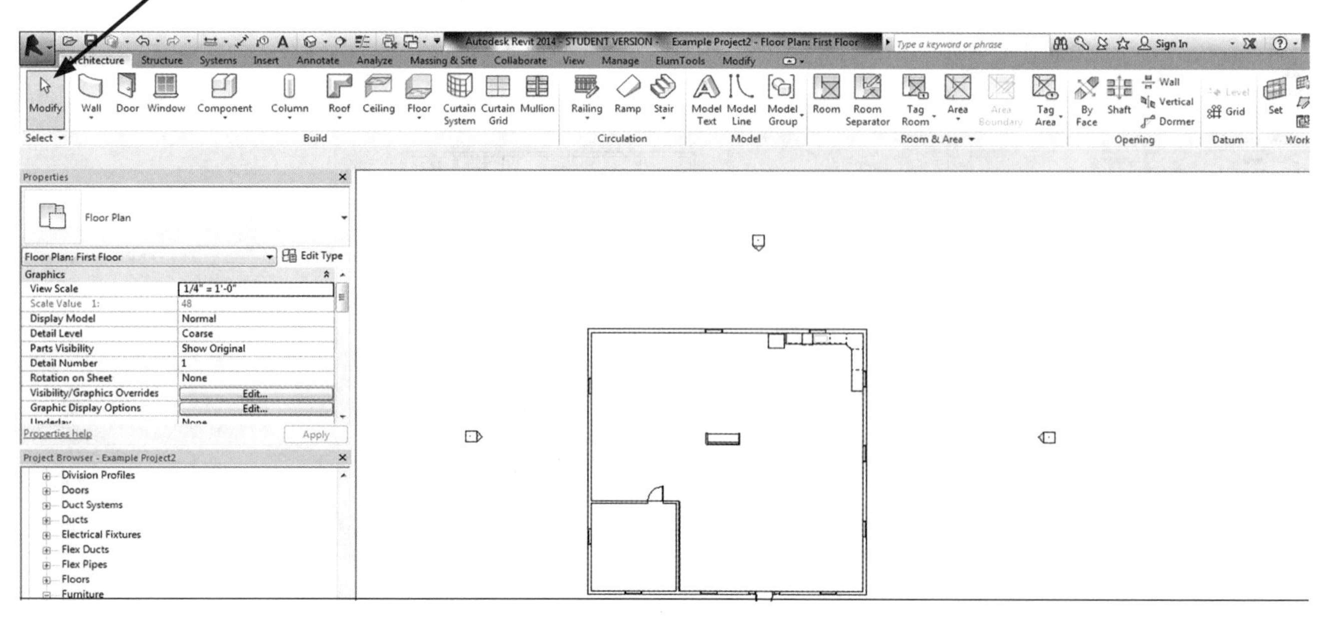

If you click on the piece of furniture you just placed, you can use your mouse and the **MOVE** command to drag it into a new position as seen in the next view.

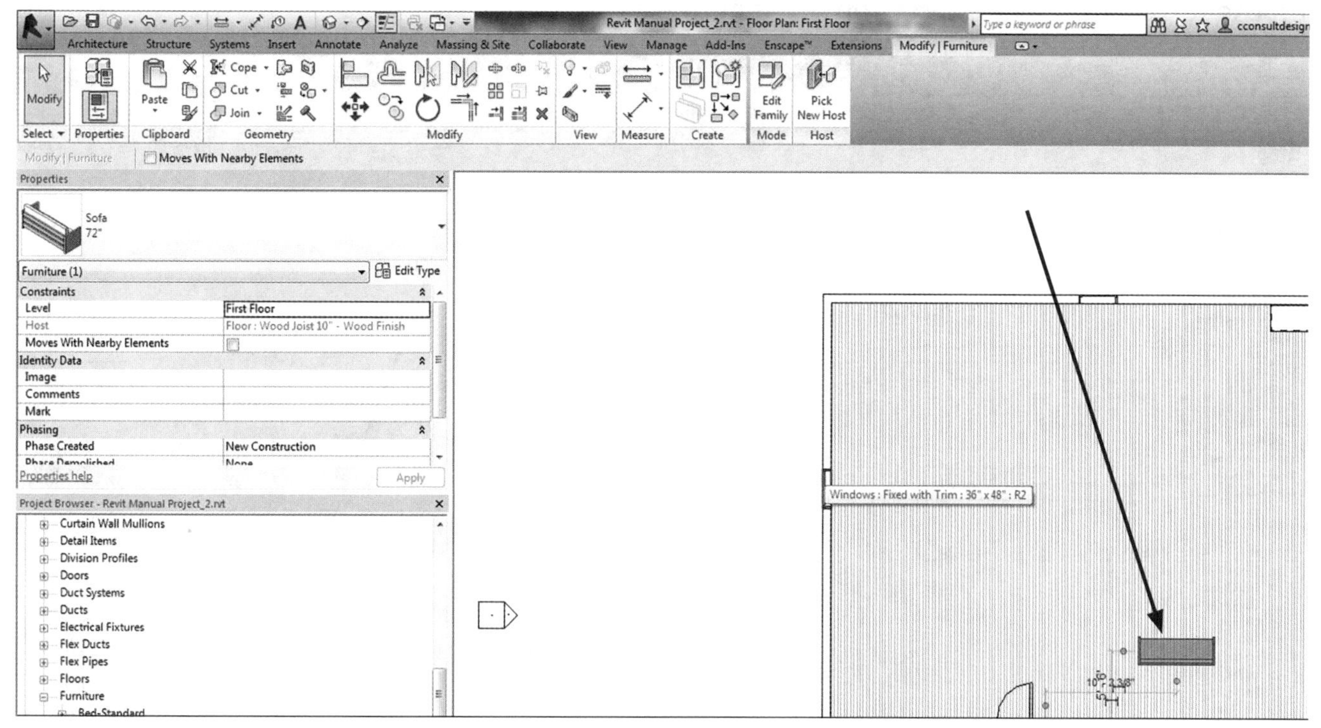

You can follow the same steps to add additional pieces of furniture into your drawing.

Notice you can add an additional sofa across from the original one by simply going to the Project Browser and dragging the same size sofa over *or* you can even pick a smaller size and place it in your space.

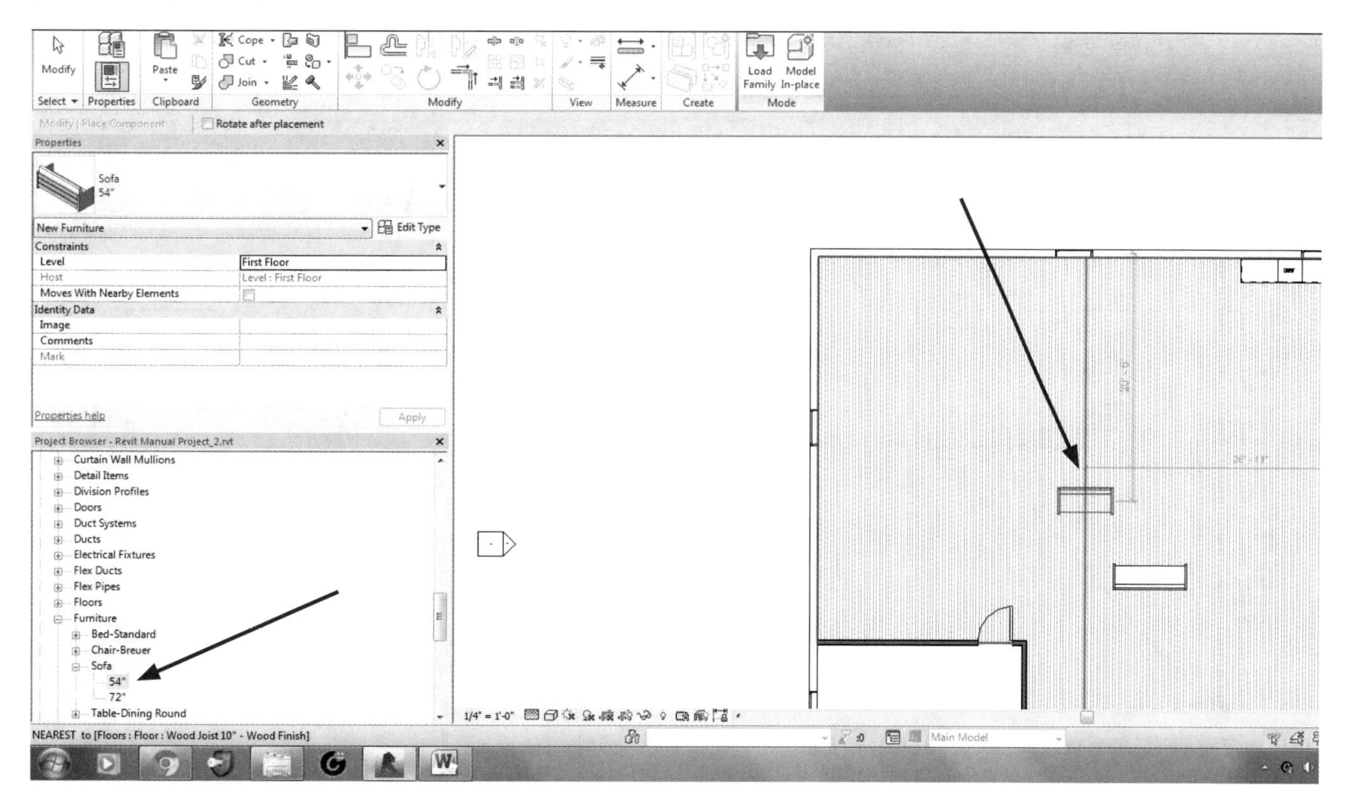

Once you bring the new sofa in the space, if you click the **SPACEBAR**, the sofa will rotate into a new position so you can see the desired furniture layout or direction. Then click with the mouse to accept the new direction of the furniture.

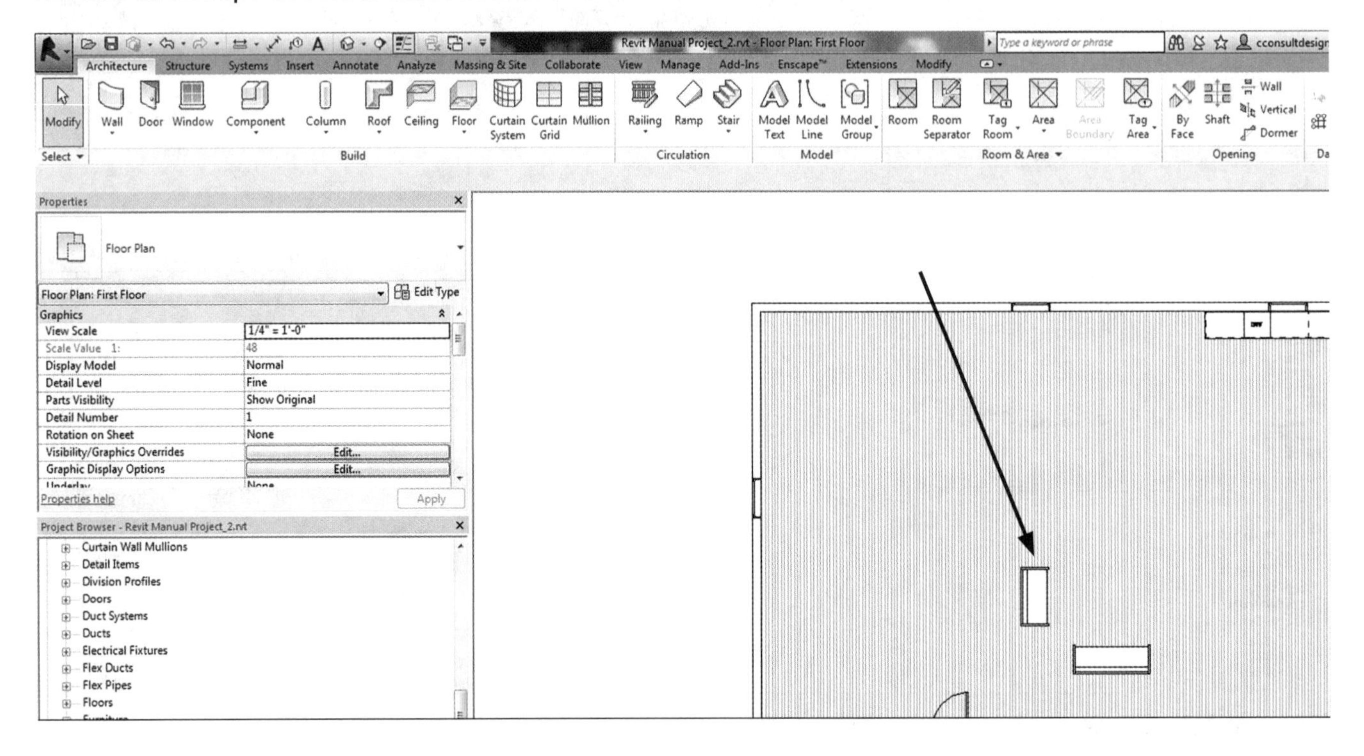

You can also add a chair to the room area from the Project Browser by following the same steps.

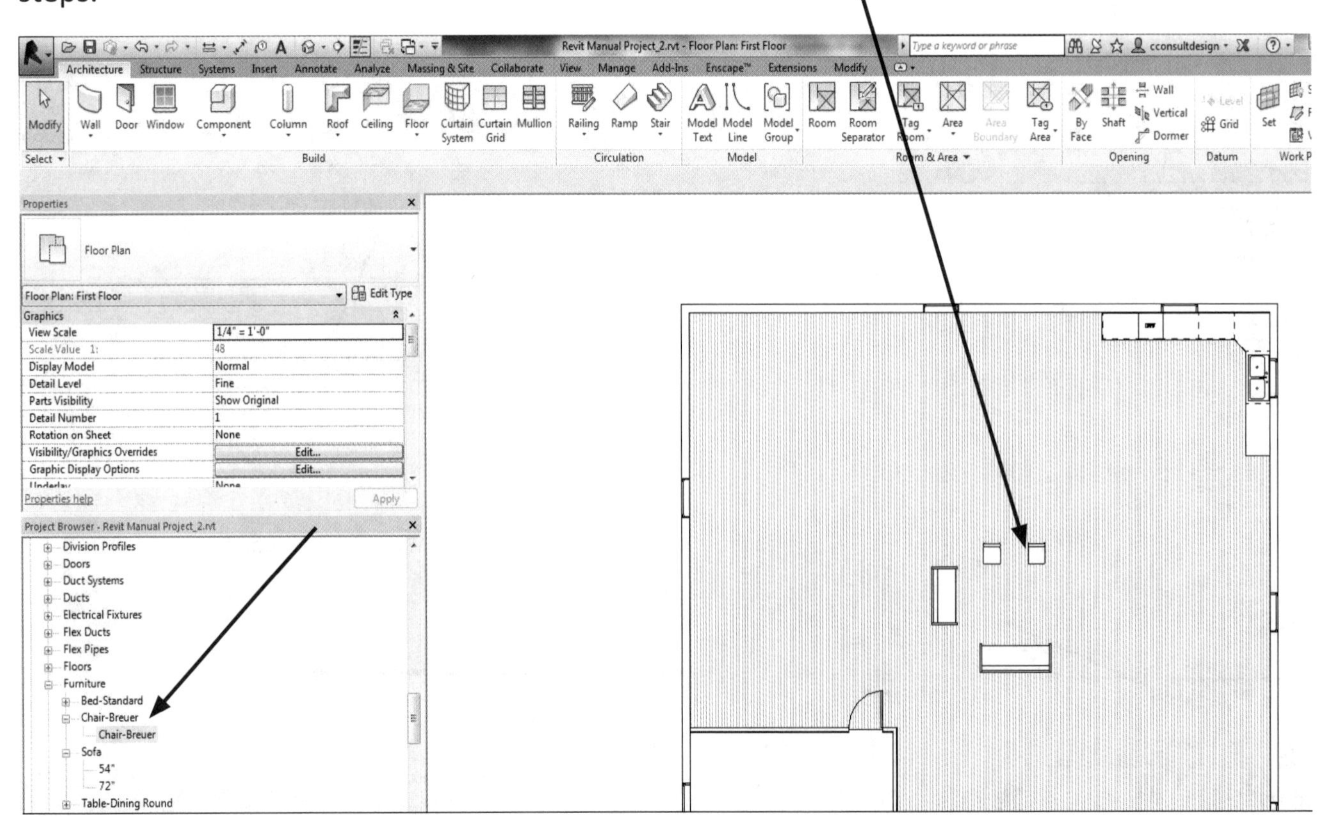

If you want to add a coffee table and end tables, etc., you will need to load additional furniture from the Furniture Library in Revit. Those steps are found in Chapter 17.

Now let's look at the small room on the bottom left. We can add a bed and a small sofa in that room. In the Project Browser, the Furniture category, select the **BED+** and expand the tree. You will see more than one size bed to select.

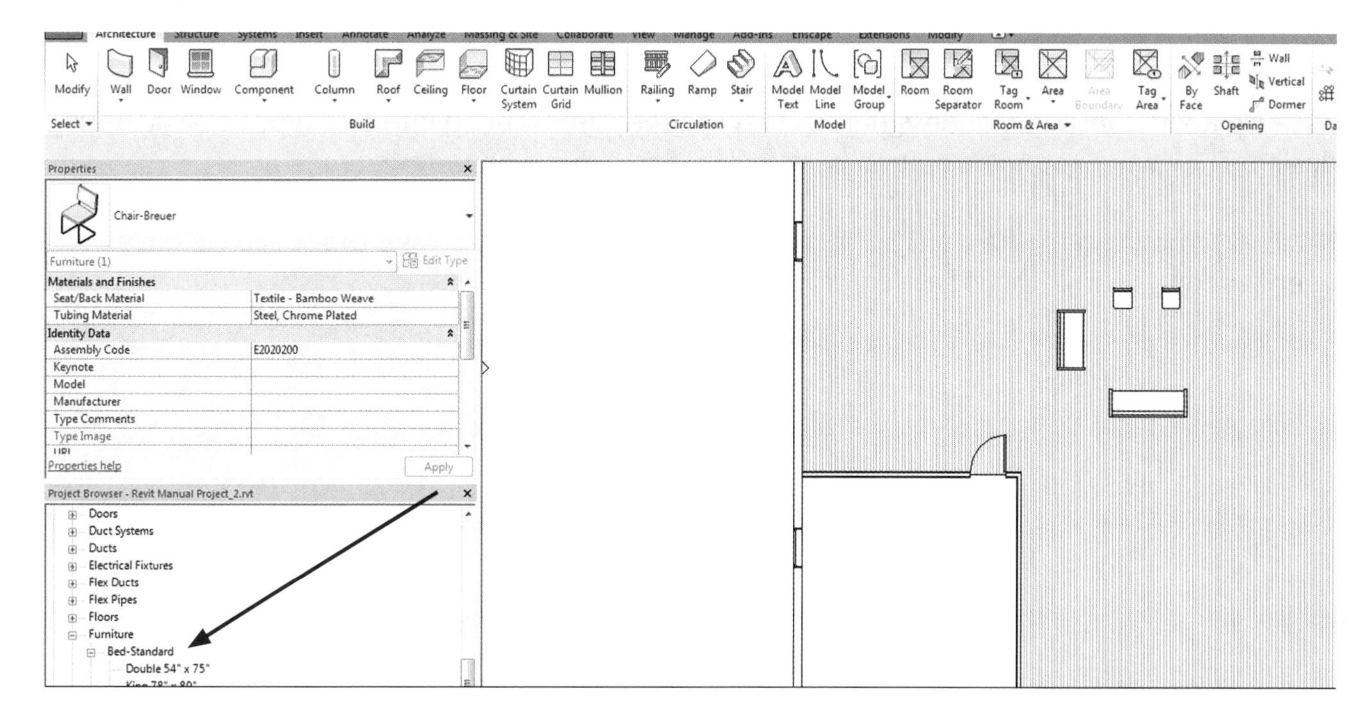

After you find the size bed you are looking for to place in the room, click the bed name and drag it into your drawing. Click the **SPACEBAR** to rotate the bed into the desired direction.

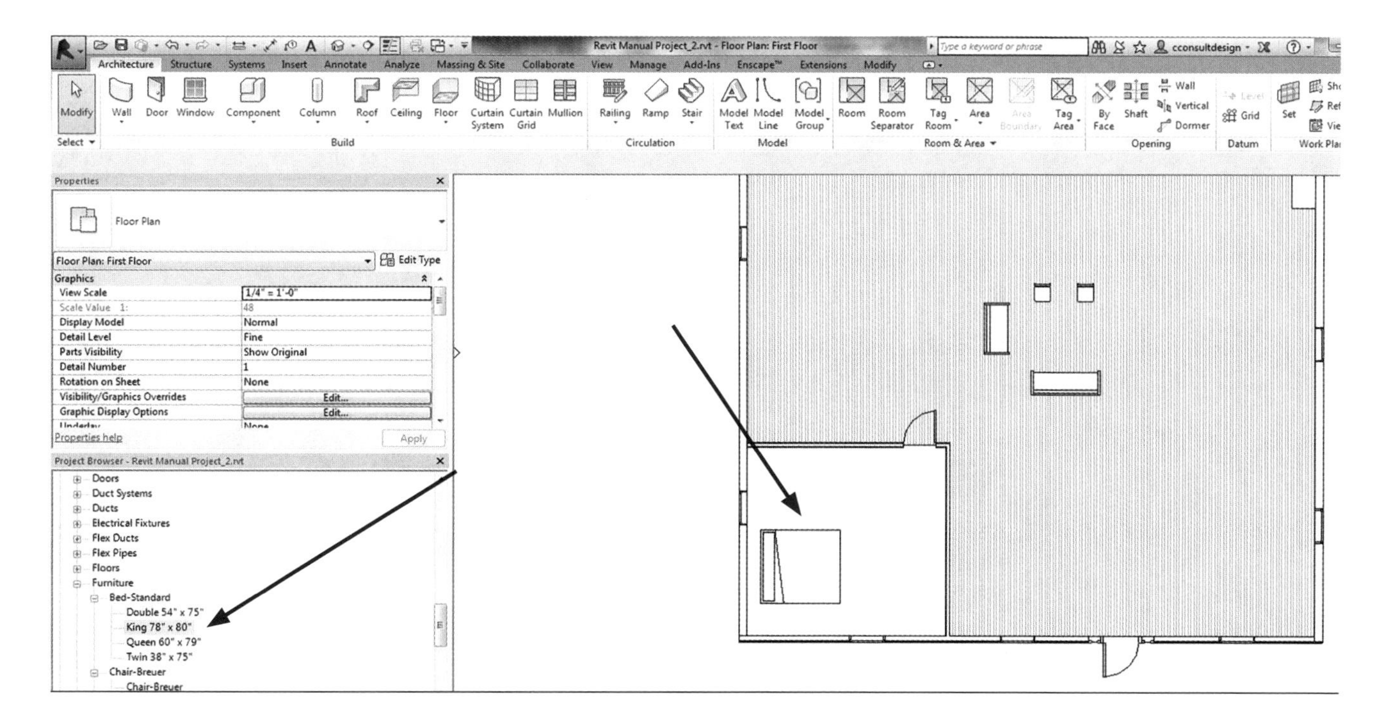

You can also follow Chapter 17, Inserting Objects from the Revit Library and the Internet, ***for more information on adding other furniture than what is already included in the Revit software.***

In this chapter, we were able to show the different types of furniture used in both residential and commercial settings. Revit has a very in-depth library already. If you find a piece of furniture or object such as a bed, it's often not just one type of bed. If you expand the folder, you will notice that Revit gives the designer a larger selection of sizes or styles to pick from. That's because Revit doesn't want designers to feel like they are trapped into one type of basic boring design. They want you to be able to be creative, and expand your designs.

Now we'll move on to Chapter 13, *Placing Ceiling Lights*, so we can see the space we have been working on in realistic shades.

CHAPTER 13

Placing Ceiling Lights

PLACING LIGHTS IN THE CEILING

Placing lights in Revit can consist of placing recessed lights in the ceiling, placing fluorescent lights in the ceiling, or putting outdoor lights on the side of the house or building. Here, as an example, we will be placing recessed lights in the ceiling in the kitchen area.

Once you have placed the ceiling in your design and you have placed cabinets in your kitchen, you will want to place lights. Make sure you are in the Floor Plan, First Floor View.

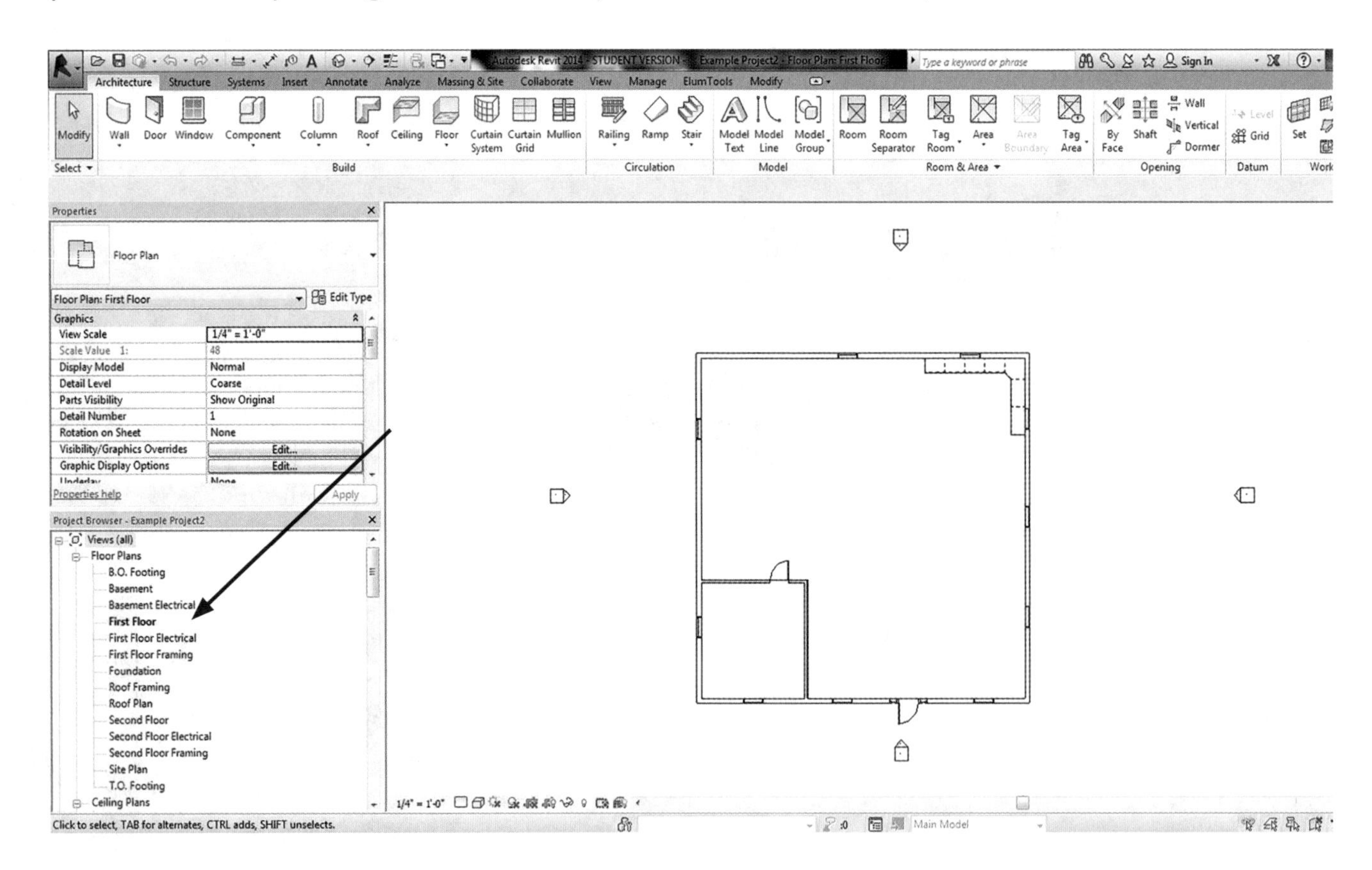

Scroll down to switch to the Ceiling Plan-First Floor.

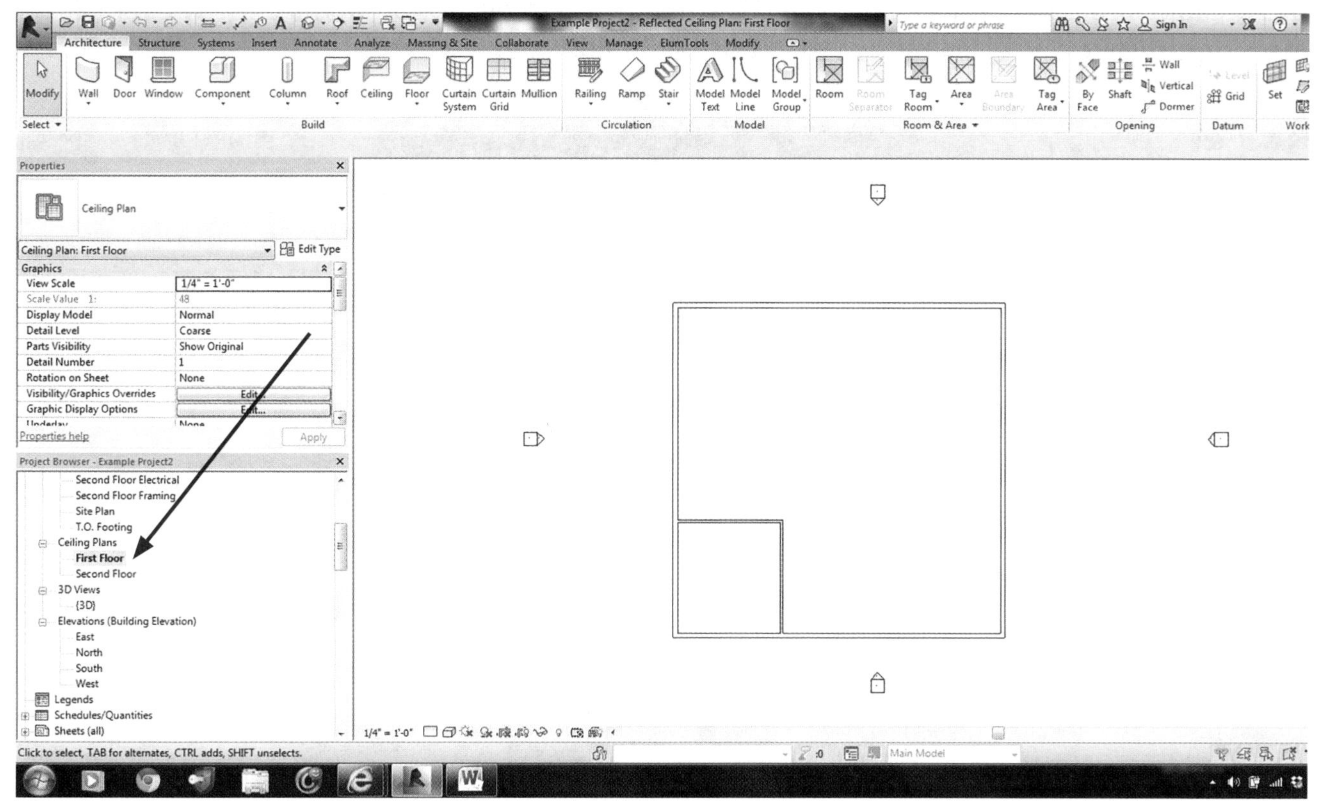

You will notice the cabinets or any furniture will *not* display, since the ceiling is in the way. So you need to make the underlay of the First Floor visible.

To do this, go into the Properties Box. Then click on the **UNDERLAY** button. Change the underlay from **NONE** to **FIRST FLOOR**.

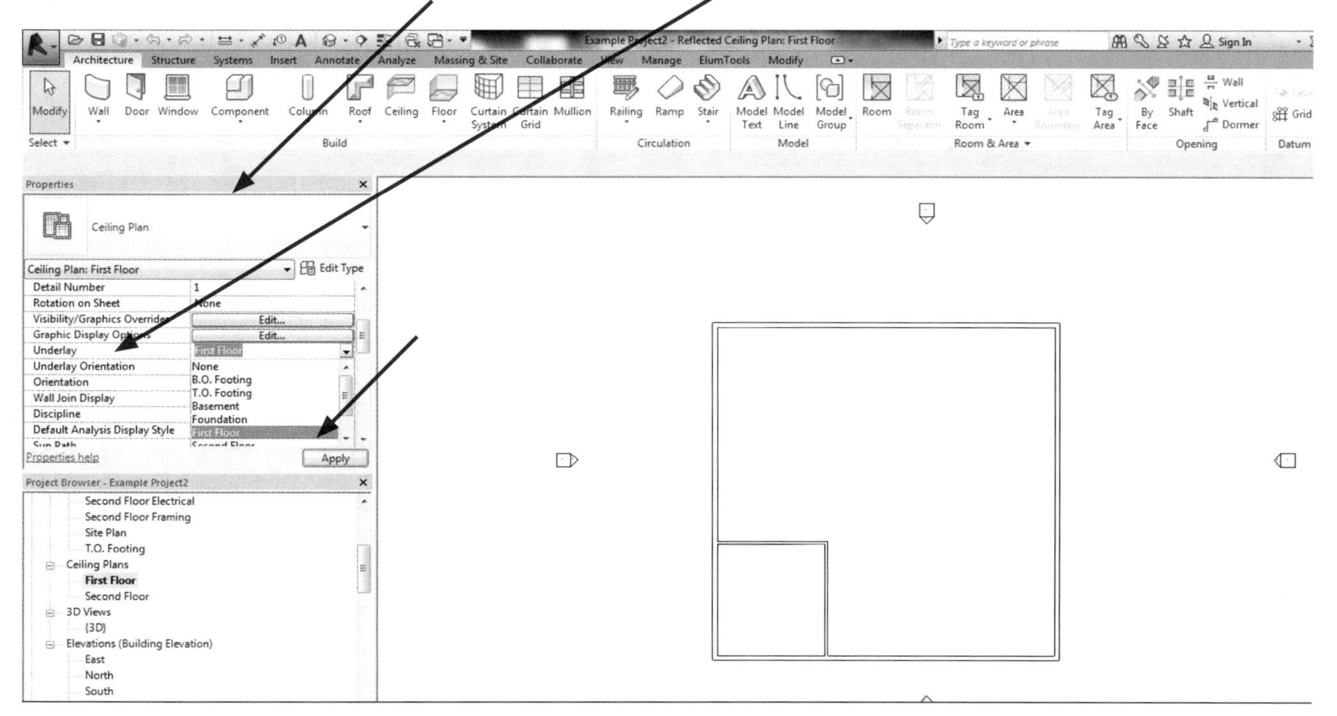

You should now see the First Floor Plan of the cabinets and everything else you placed in your design appear beneath the ceiling plan. This allows you to place your lights directly over cabinets or furniture and elsewhere in the house or building, exactly where you want them to appear.

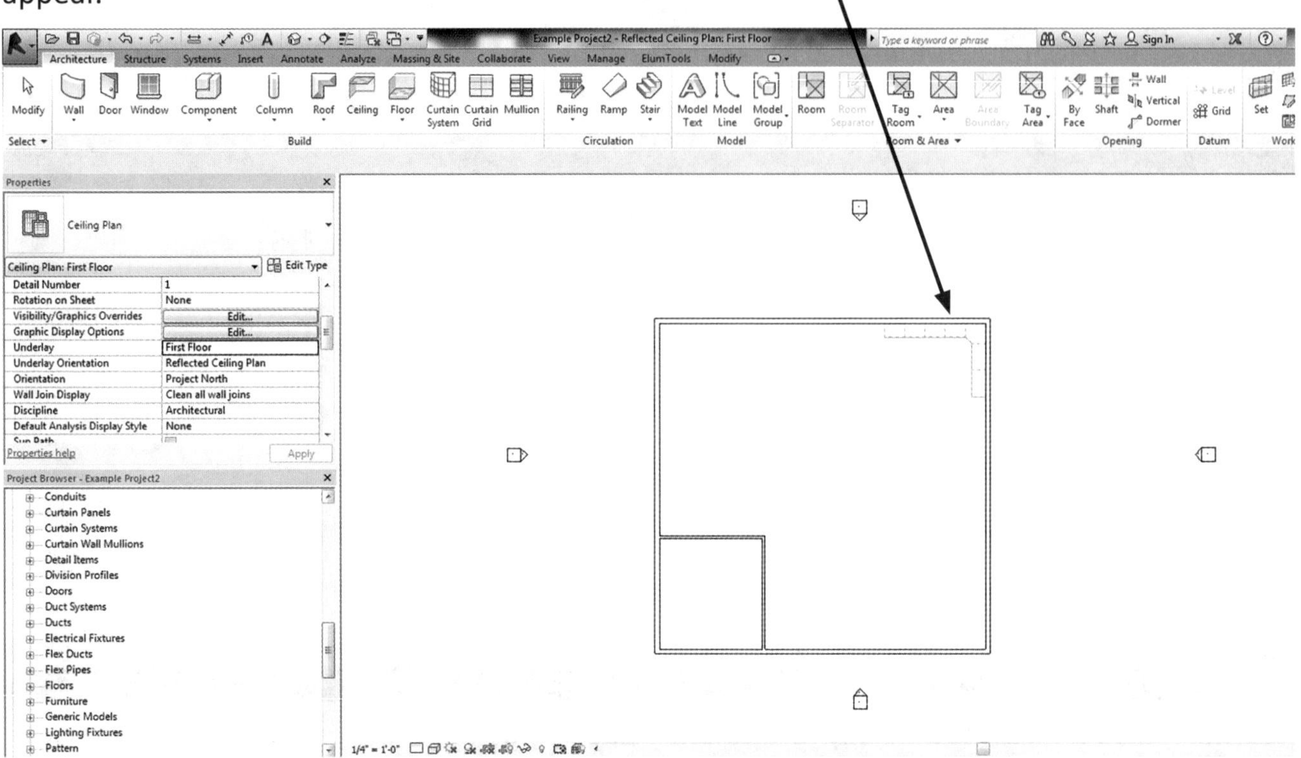

Once you see the cabinets, scroll down in the Project Browser to the Families Tab to the Light Fixtures category, and select the lights you want to place.

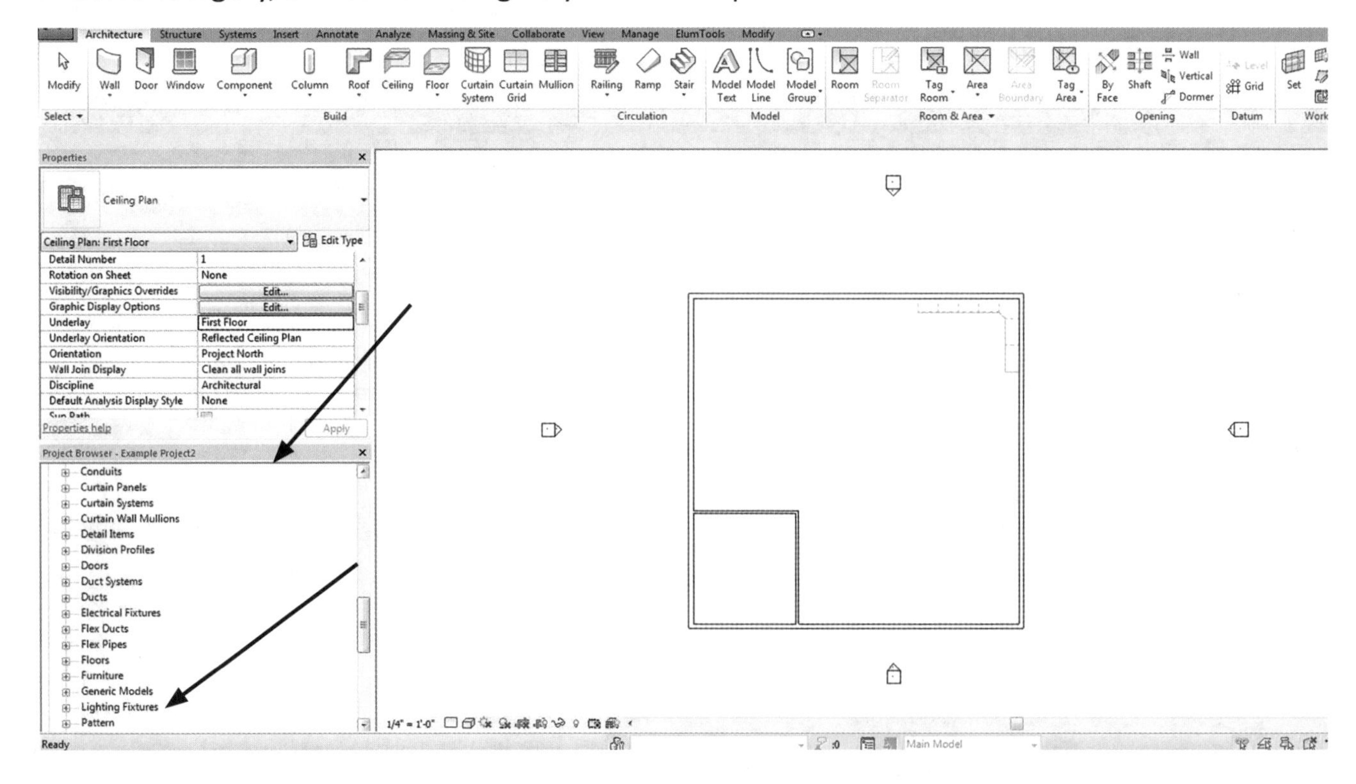

Open the **LIGHT FIXTURES** category, and select the types of lights you want to place in your design.

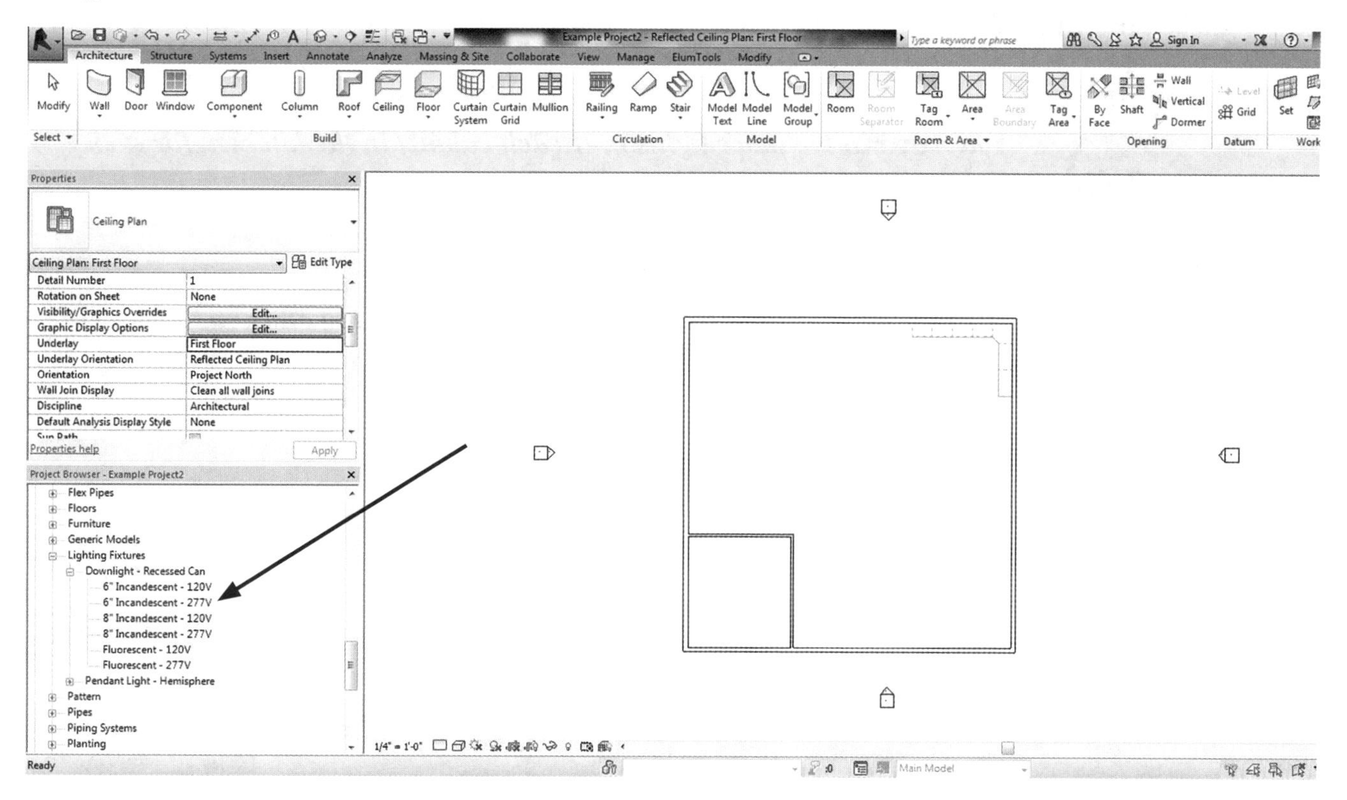

Once you have selected the type (or types) of light you want, click on it, hold down the button on the mouse, and drag the light into your drawing. Place it wherever you want it.

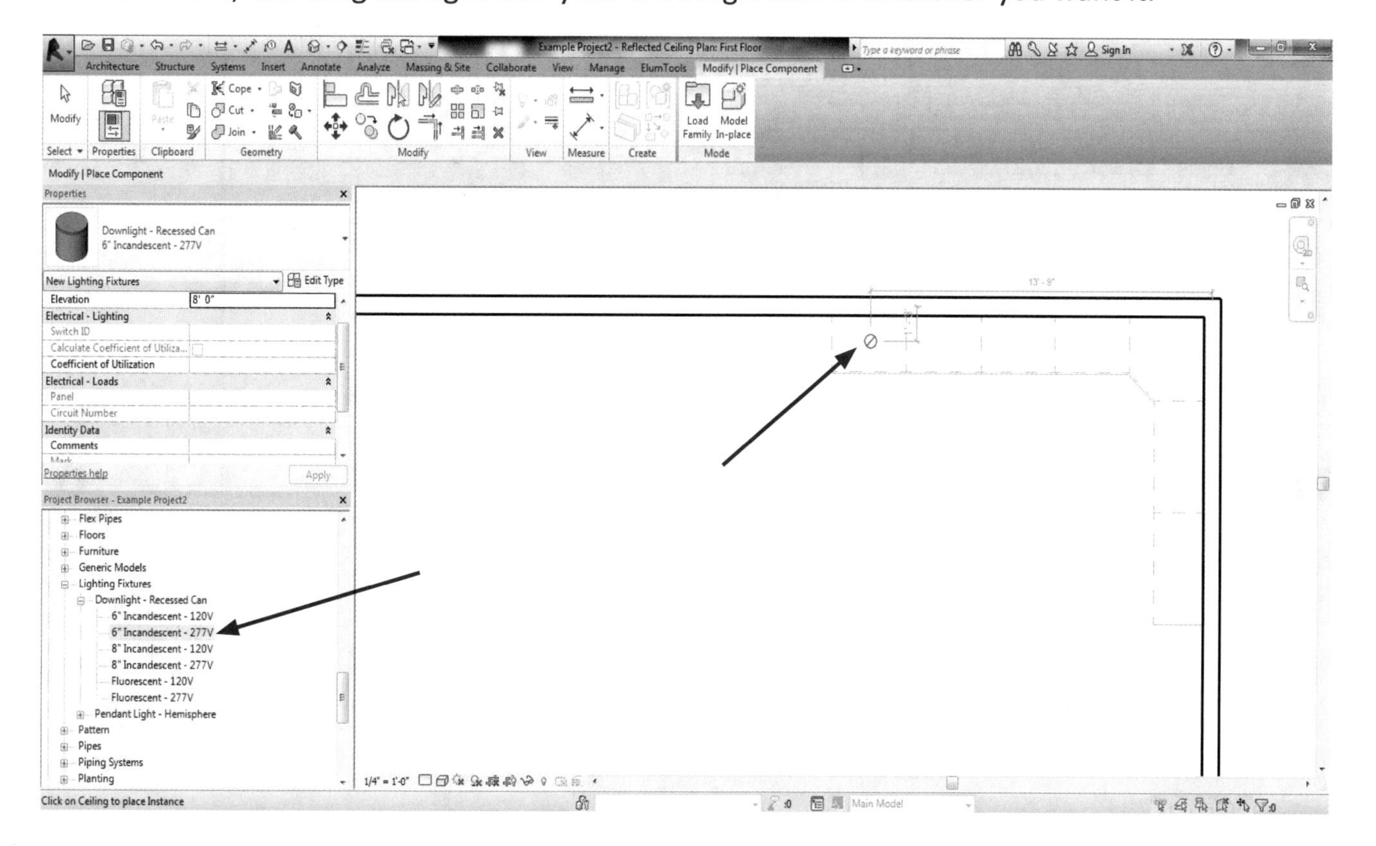

Once you pick the location, you then can copy and place more throughout the kitchen, or pick different types of lights to place in your design. Once you're satisfied with the overhead lighting, click the **MODIFY** button and you are finished.

EXAMPLE

You may want to place lights over each cabinet in the kitchen, for even lighting, and easier cooking.

TIPS

If you drag your light fixture over into your drawing and you do not ***see your fixture follow your cursor, chances are that your categories name is turned off.***

The first thing you must do is click the **MODIFY** Arrow, then type **VV** or **VG** for Visibility Graphics Override dialog box.

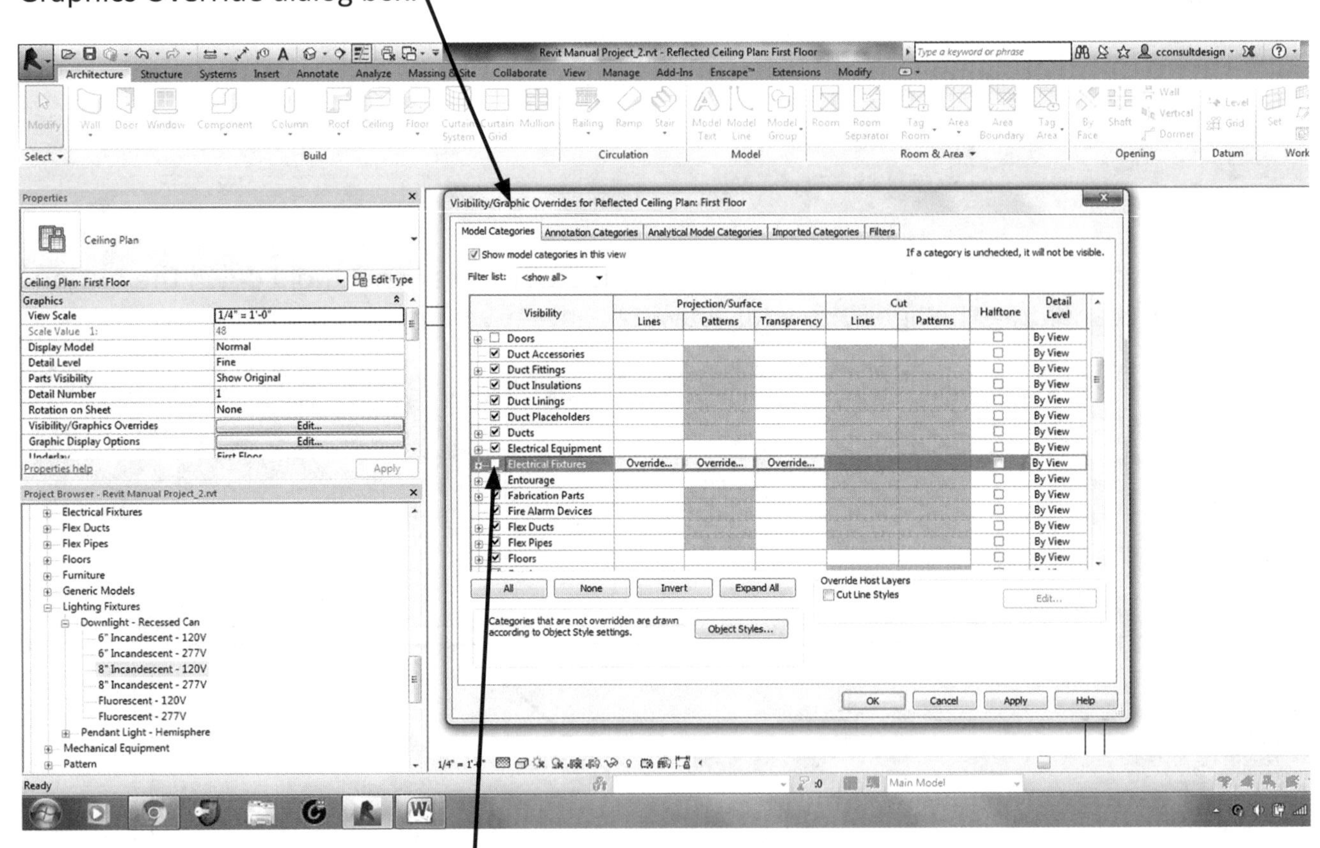

Scroll down until you find the Light Fixture name and place a **CHECK MARK** in the box. The light fixture will appear in your drawing.

The completed layout of lights placed over each cabinet will look like the following layout:

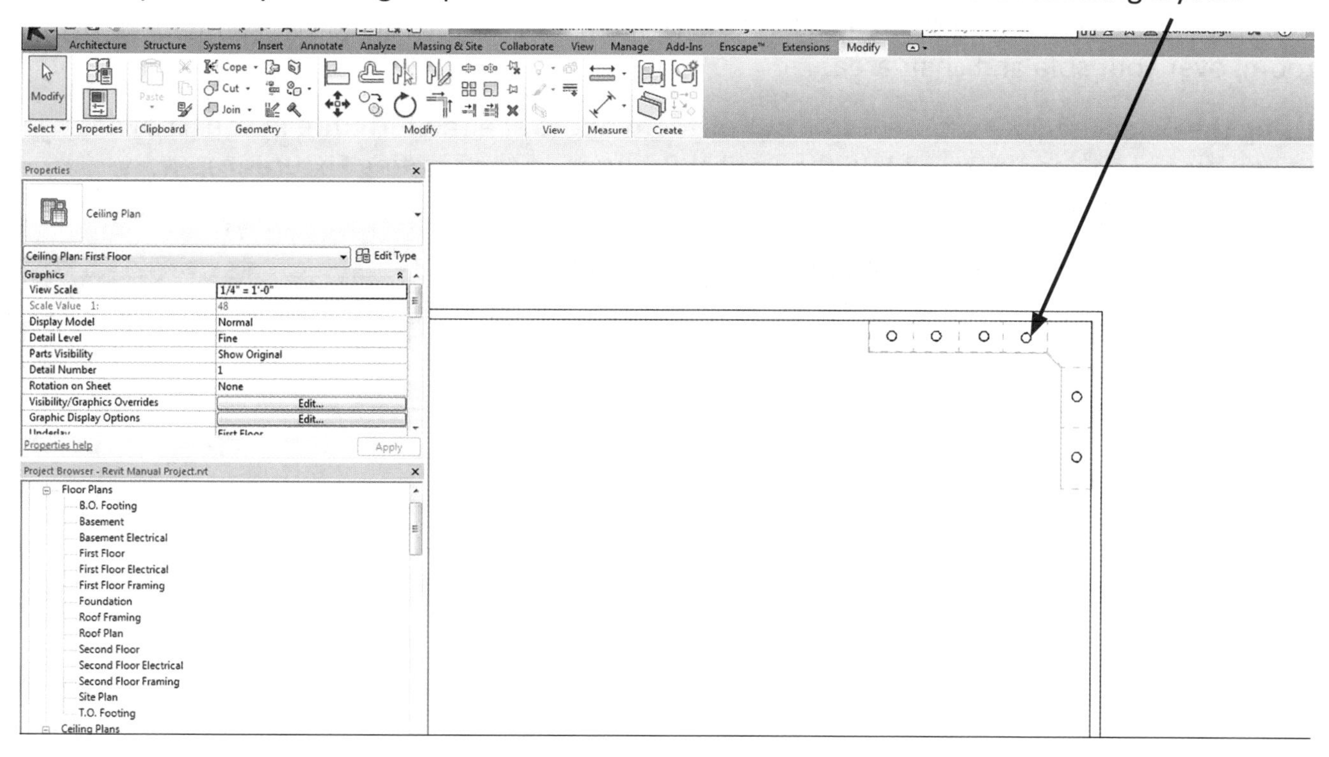

EXAMPLE

If you want, after you have completed the light locations over the cabinets, you can place additional lighting layouts elsewhere in the ceiling, such as the area around the kitchen.

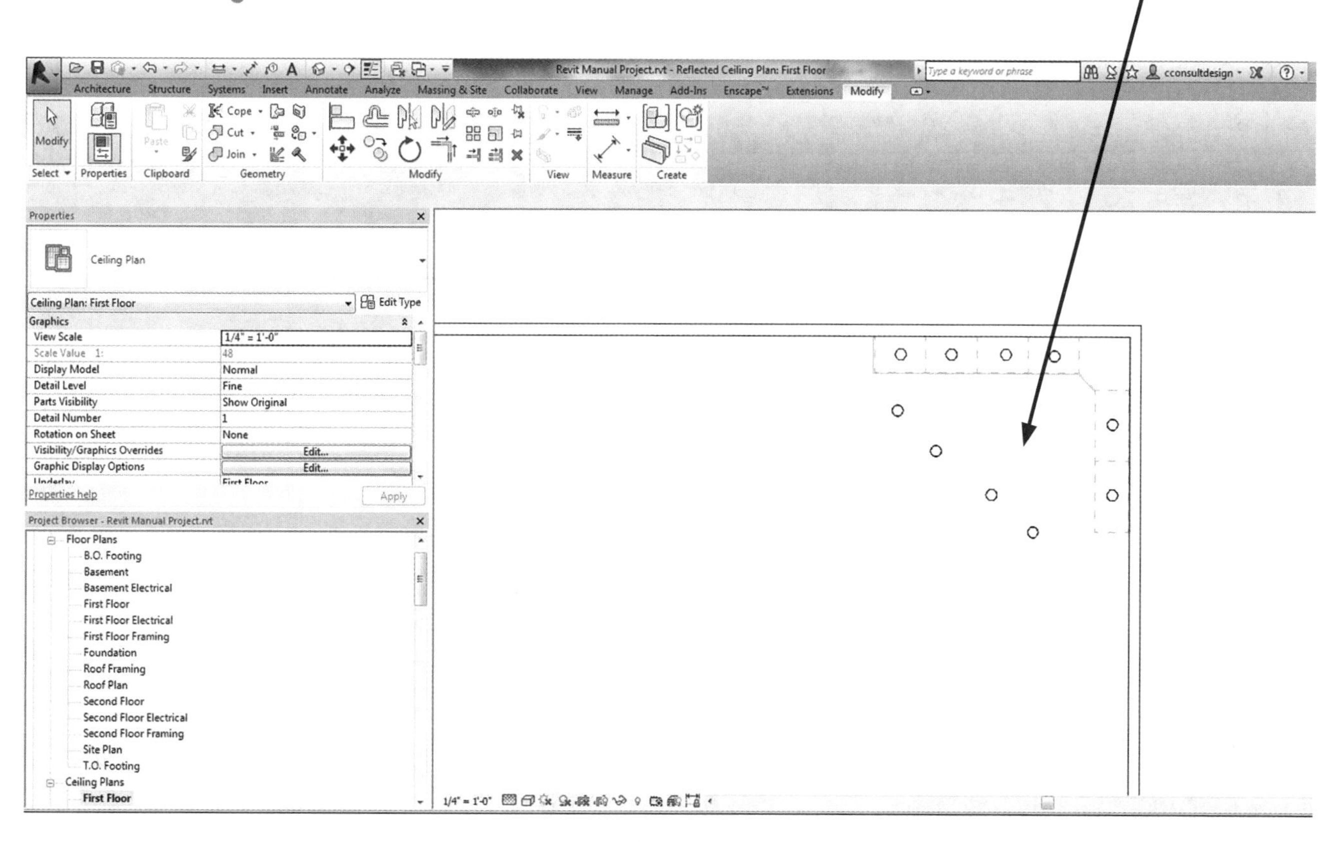

Let's see what more we can add to this design. After you have placed lights in your kitchen/ house, perhaps you decide that you'd like to place a hanging pendant light in your den or living room, over your sofa and chair area.

Using the wheel on your mouse, scroll out or zoom out to see the furniture in your drawing (if you placed any on the space). In the Project Browser, in the Lighting Fixtures category, expand the tree **(+)** next to Pendant Light – Hemisphere.

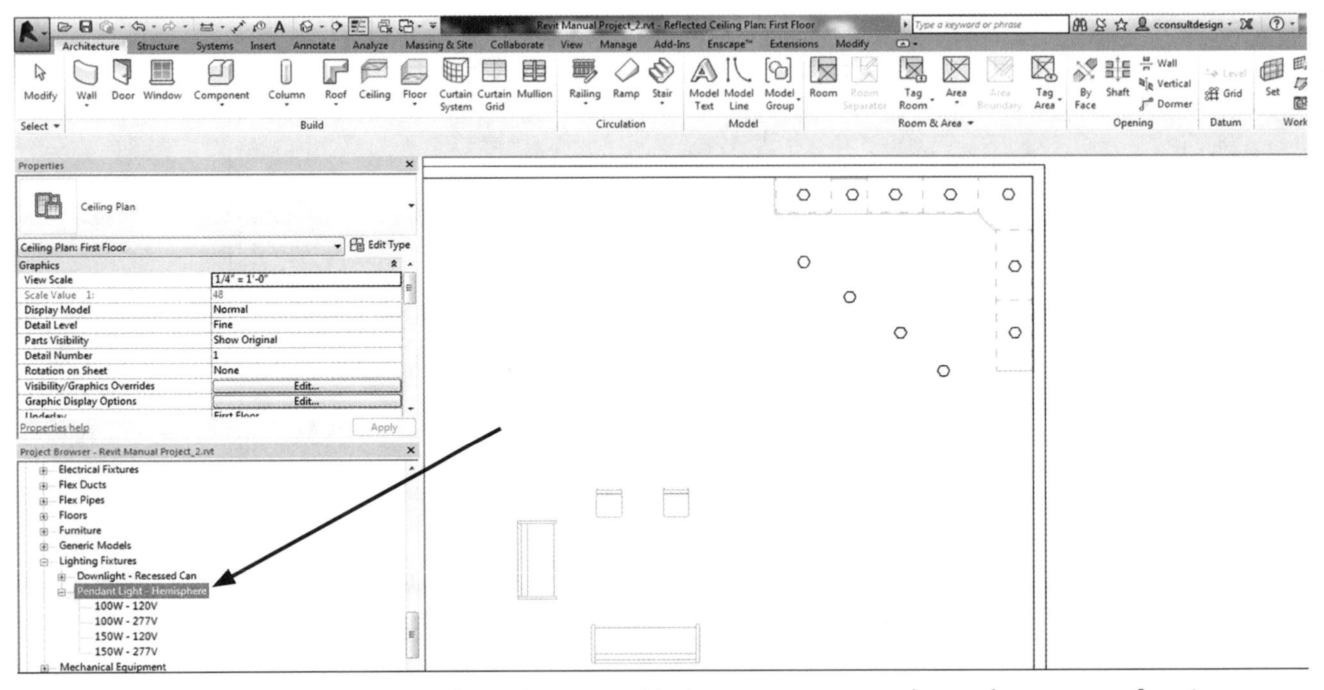

Once the category is opened, select the size of light you want to place above your furniture area. Using your mouse, drag the light fixture into position.

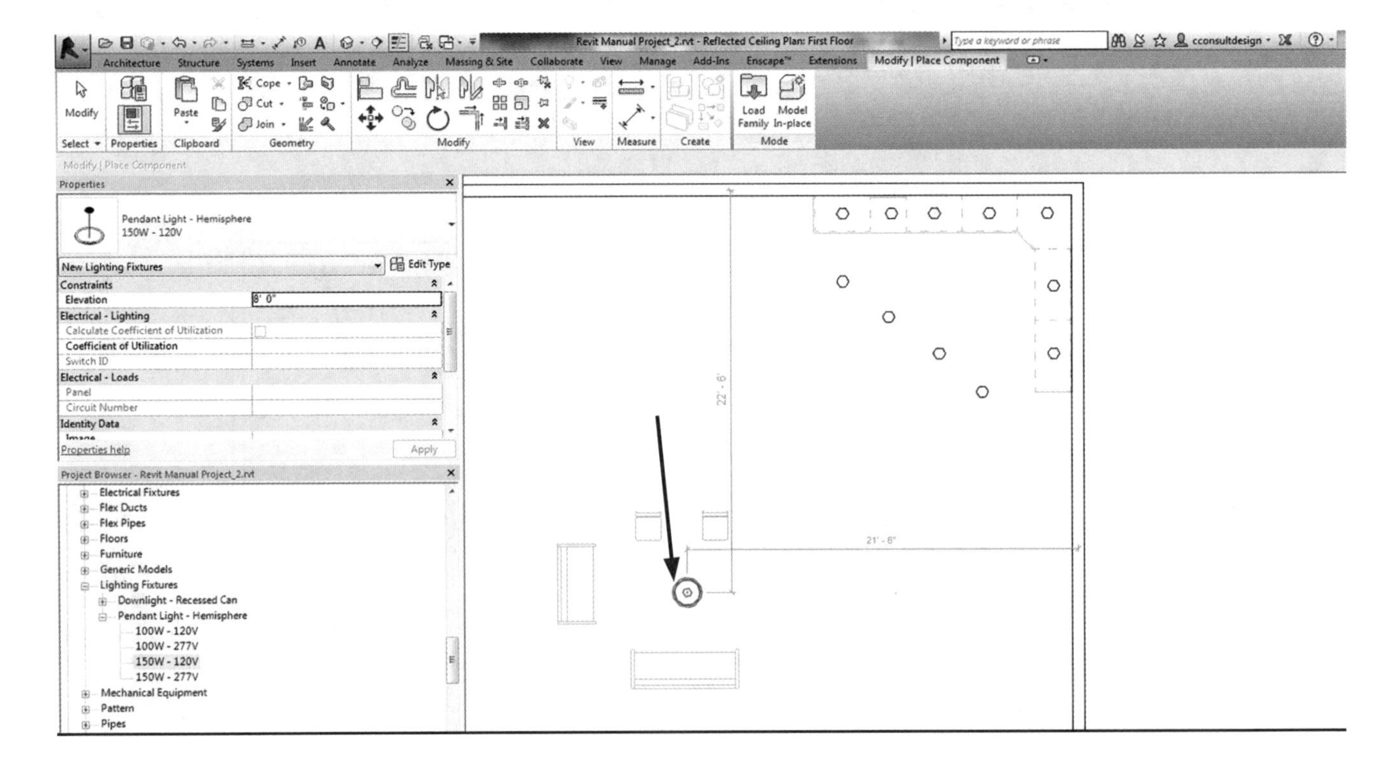

Here is the finished location for your lighting fixtures in your house layout.

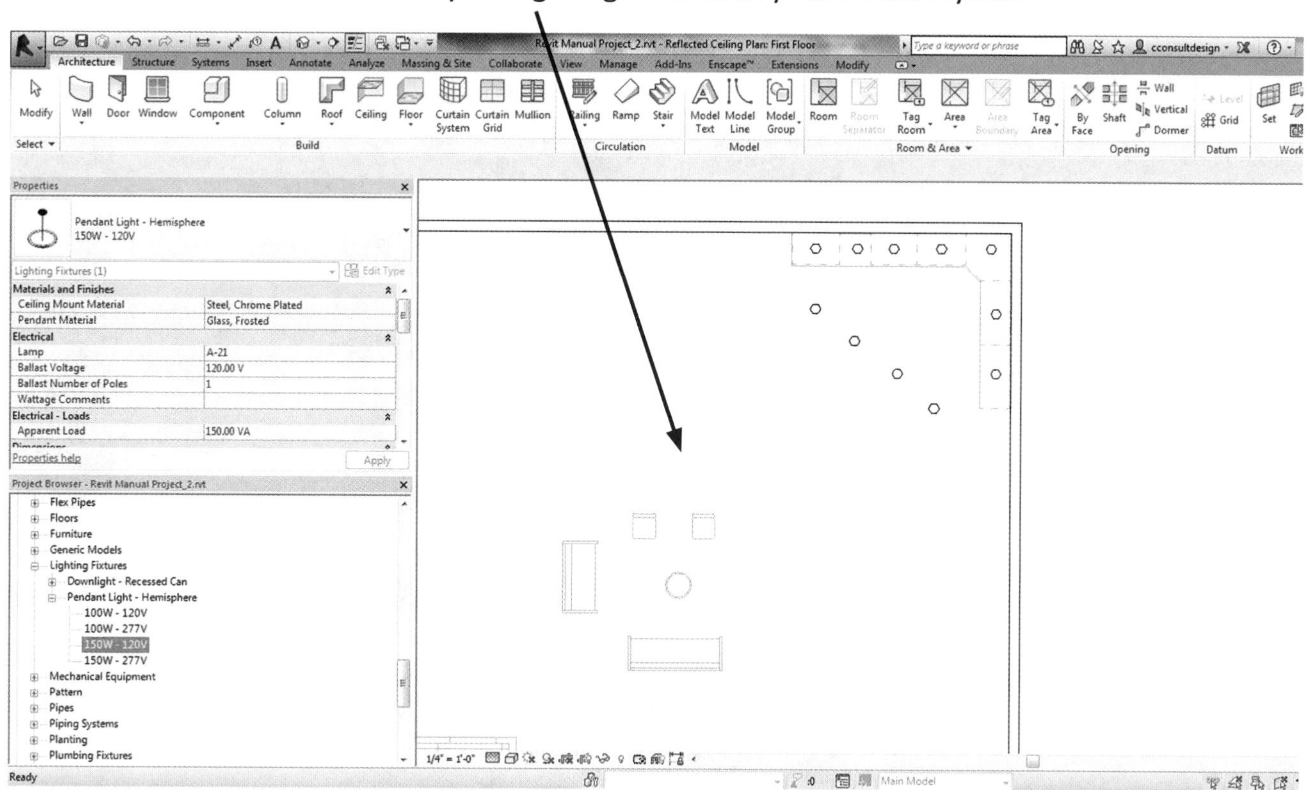

Placing lights in Revit will add that additional “stylish flair” to your décor, and let you have the-feeling that you are designing your house or building the way you would want it to look. Please remember the Lighting Fixtures you see in the Project Browser are *not* all that Revit has to offer. You can go to the Revit Library in the Insert Tab on the ribbon, or load other Revit objects and lights from outside manufacturers. Just follow the instructions in Chapter 17, *Inserting Objects from the Revit Library and the Internet,* for more information.

Now your design is really beginning to come together. So, before it rains, let’s go to Chapter 14, *Creating Roof Designs*, and cover your beautiful new layout.

CHAPTER 14

Creating Roof Designs

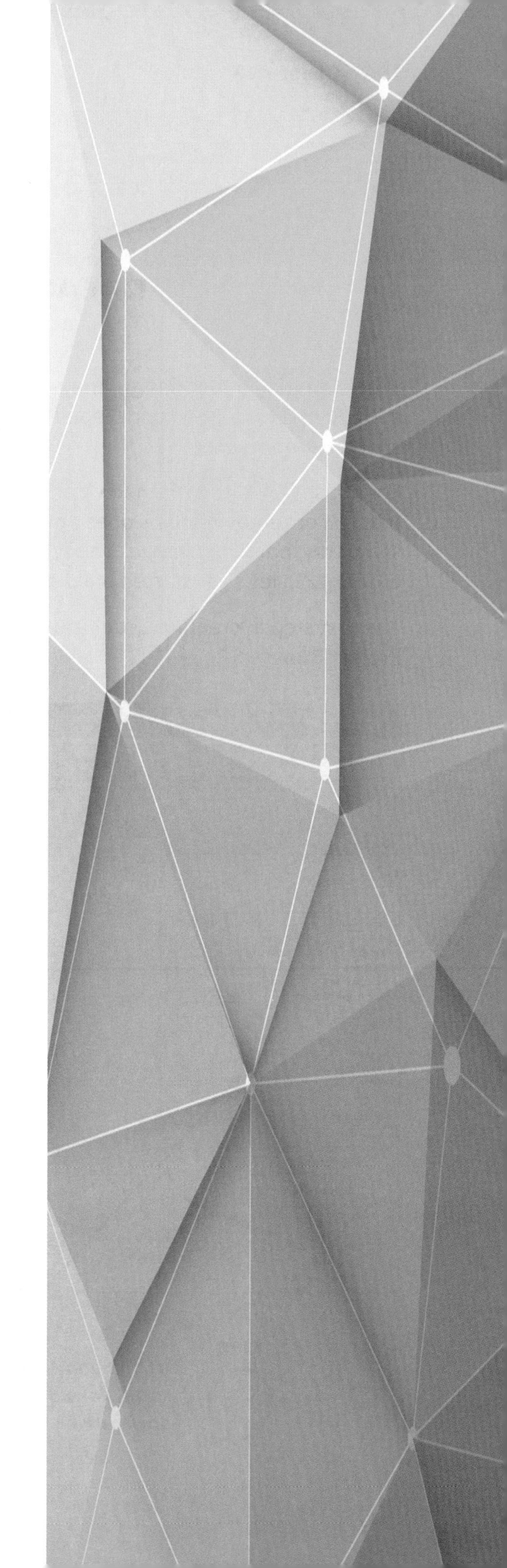

CREATING ROOF DESIGNS

When creating a roof design, there are many different roof types that can be created, such as a Hip Roof, Gable Roof, and Sloped Roof. We will go over creating each type of roof in this chapter, step-by-step.

For houses, all the roofs in Revit are designed the same way, but with minor changes to the pitches, or removing the slopes or pitches from certain sides the roof. Each roof will start the same way, but with changes you will be able to create a different effect, and style and be more efficient. So let's get started building a roof for your design project.

The first step in creating a roof is that you *must* make sure you are in the Roof Plan in the Project Browser.

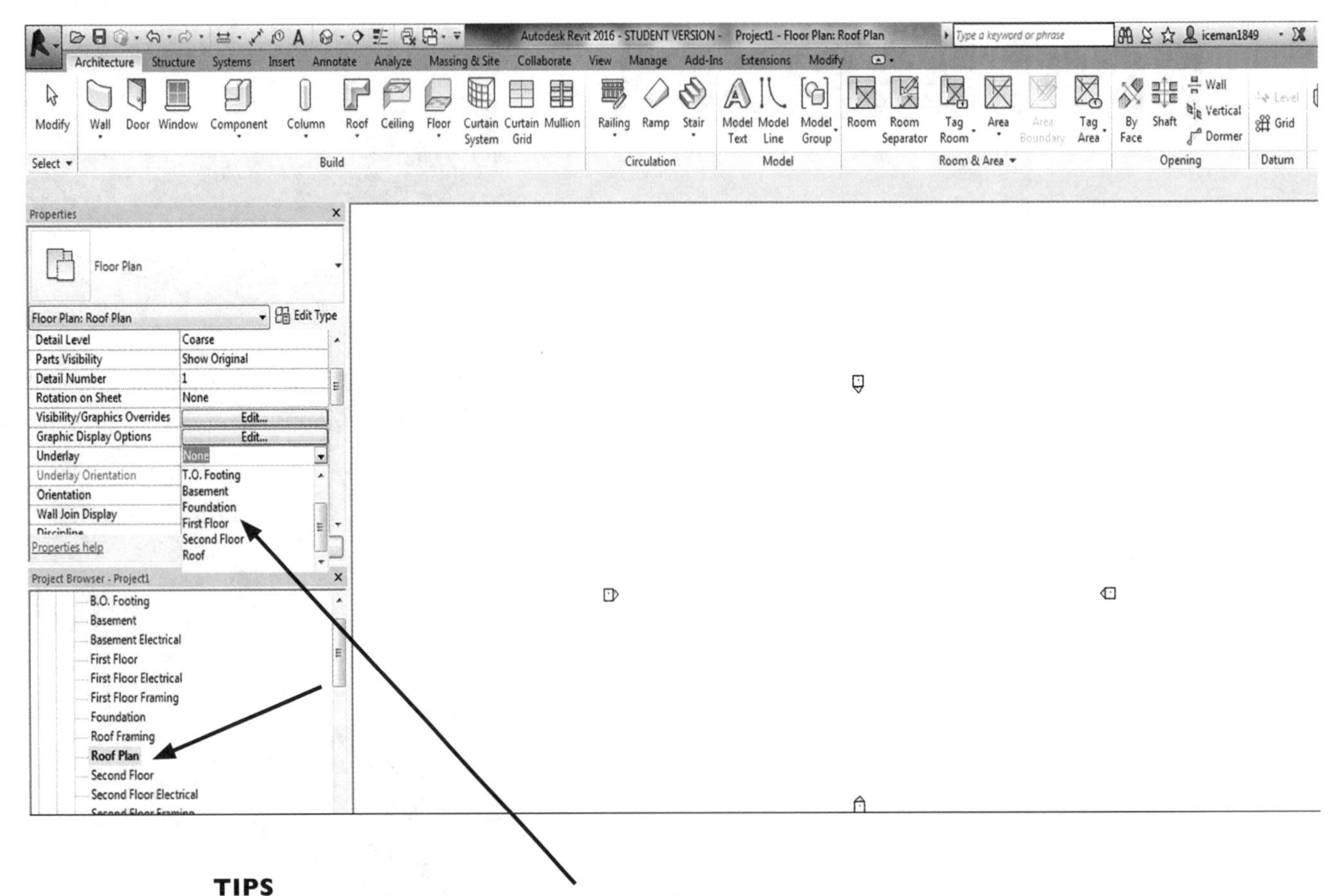

TIPS

If you don't see your house or building in the Properties Box on the left, scroll down to the Underlay Box and click the box to the right. The drop-down arrow will appear, and you can click on FIRST FLOOR.

Your house or building plan will appear greyed out so you can see to place your roof plan.

NOTES

You may see other objects, such as lights or ceilings, and fans. That's fine.

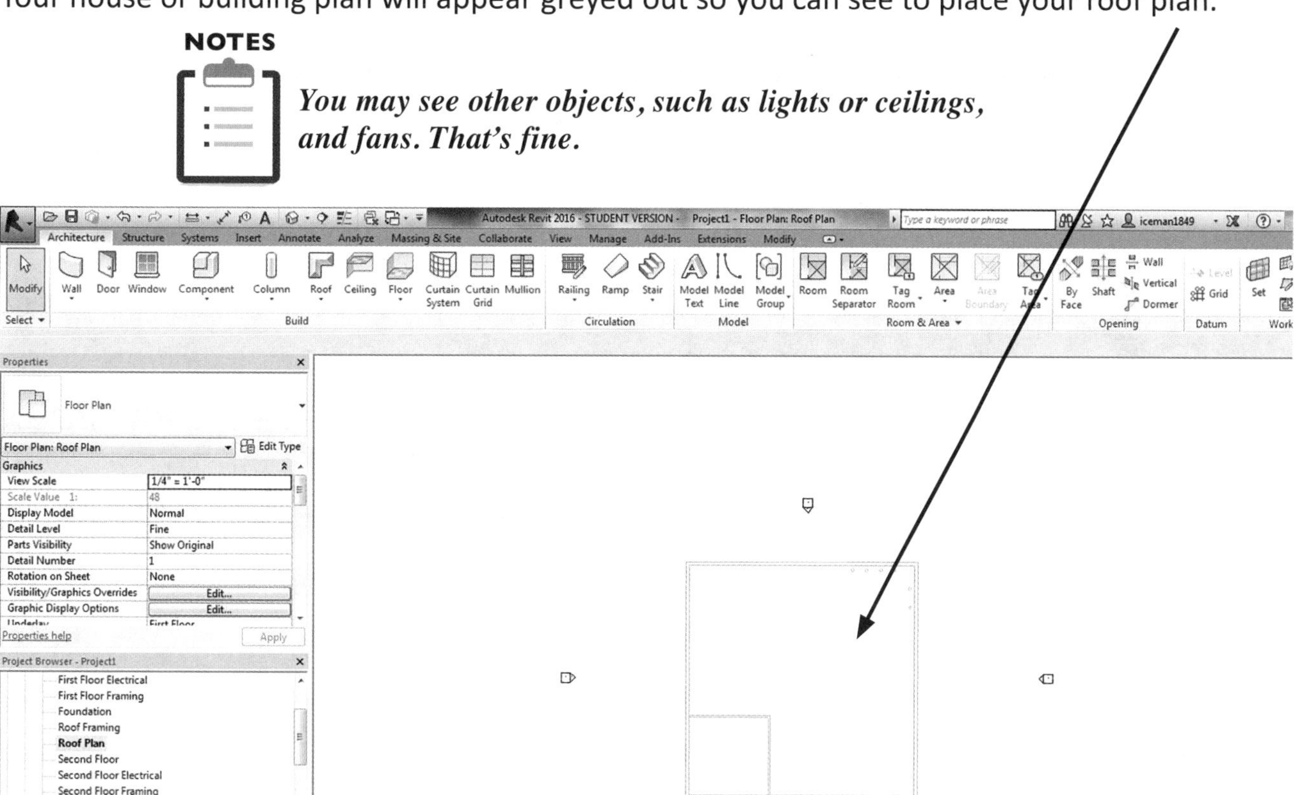

Depending on whether you have a one- or multiple-story home, you will need to make sure your Roof Level is set correctly.

Go to the Project Browser, scroll down to the Elevations section, and click on the **SOUTH** elevation to change the view.

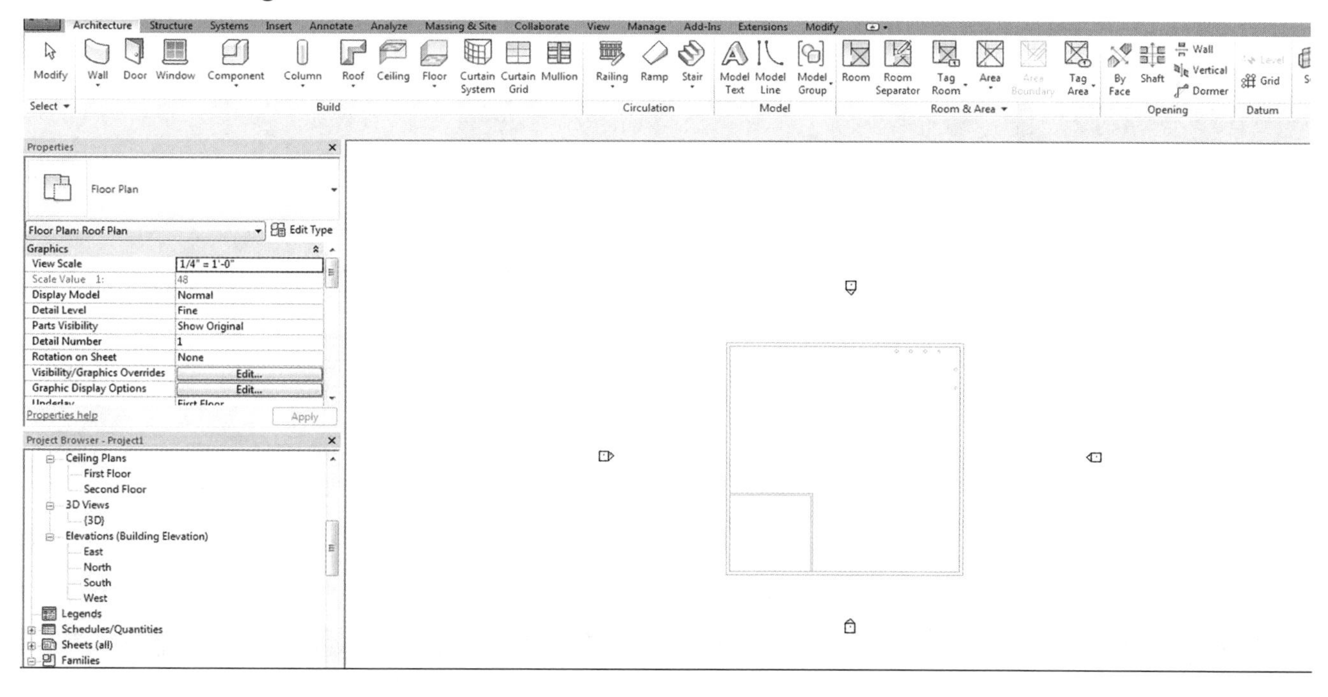

Reset Levels to 10 ft.

The view should look like the following screen shot:

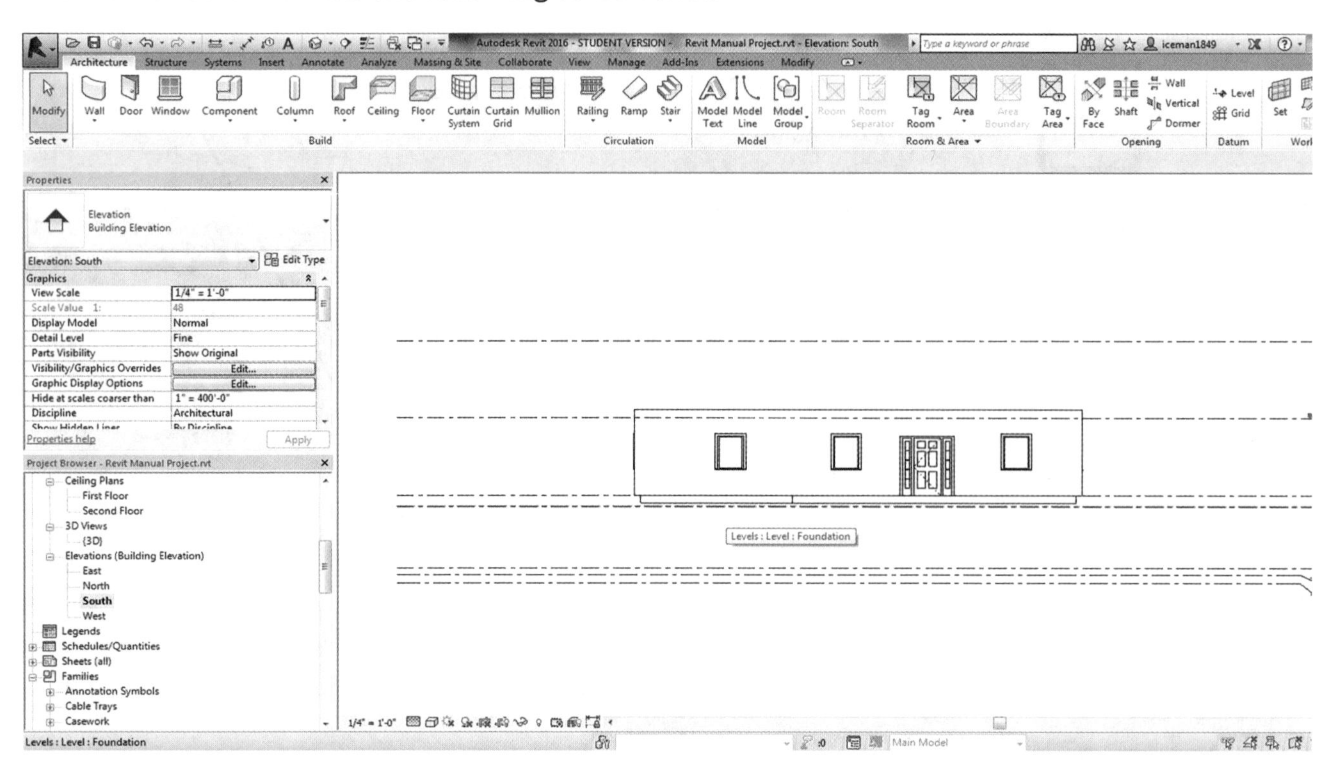

The next step is to make your roof height correct. The roof height means the base or bottom of your roof *not* the top or ridge of your roof. Therefore, depending on how high you made your exterior walls, you will need to change the height of your roof on your Level Line or Datum Line (see Chapter 7, *Creating Levels or Datum Lines in Revit*).

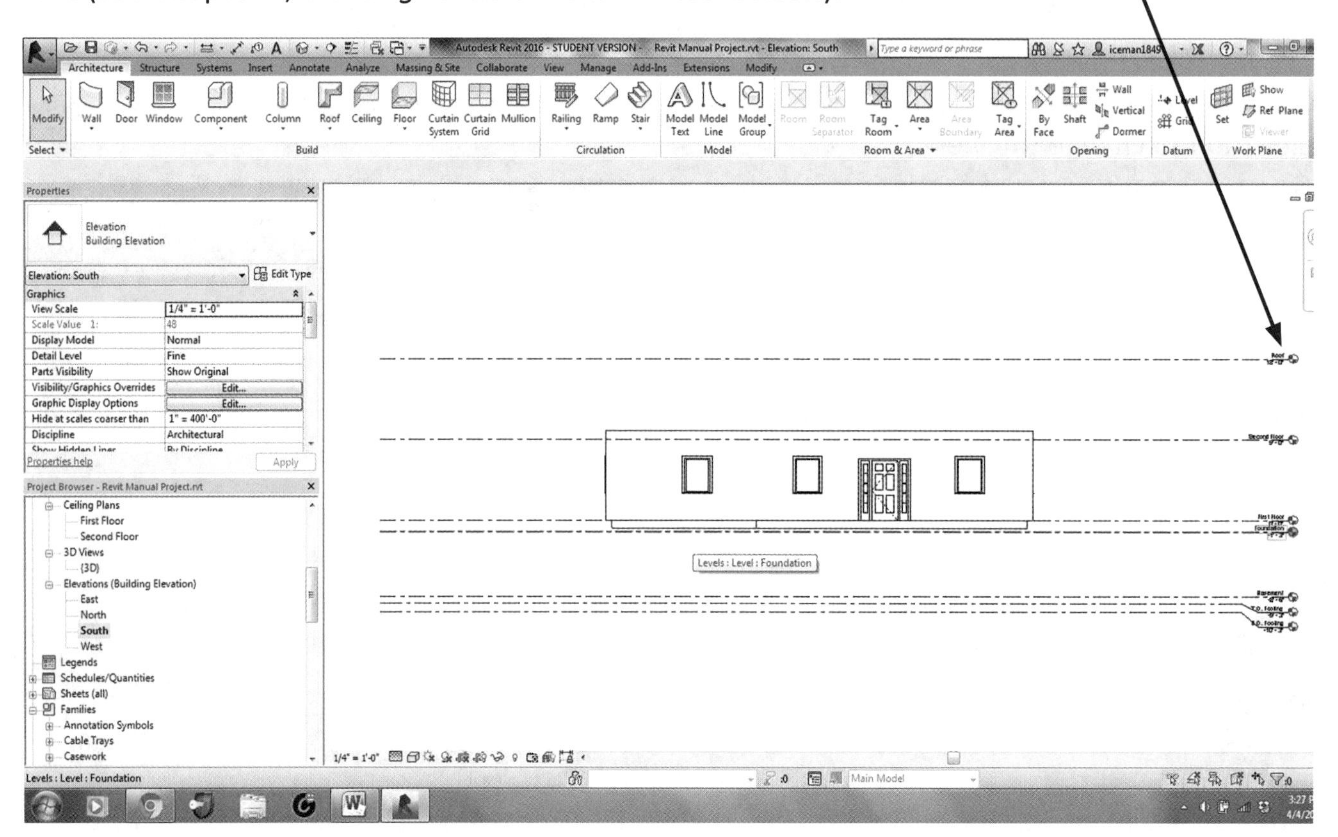

Click on the number under the roof name and change it to the height of your walls. That will become the base of your roof.

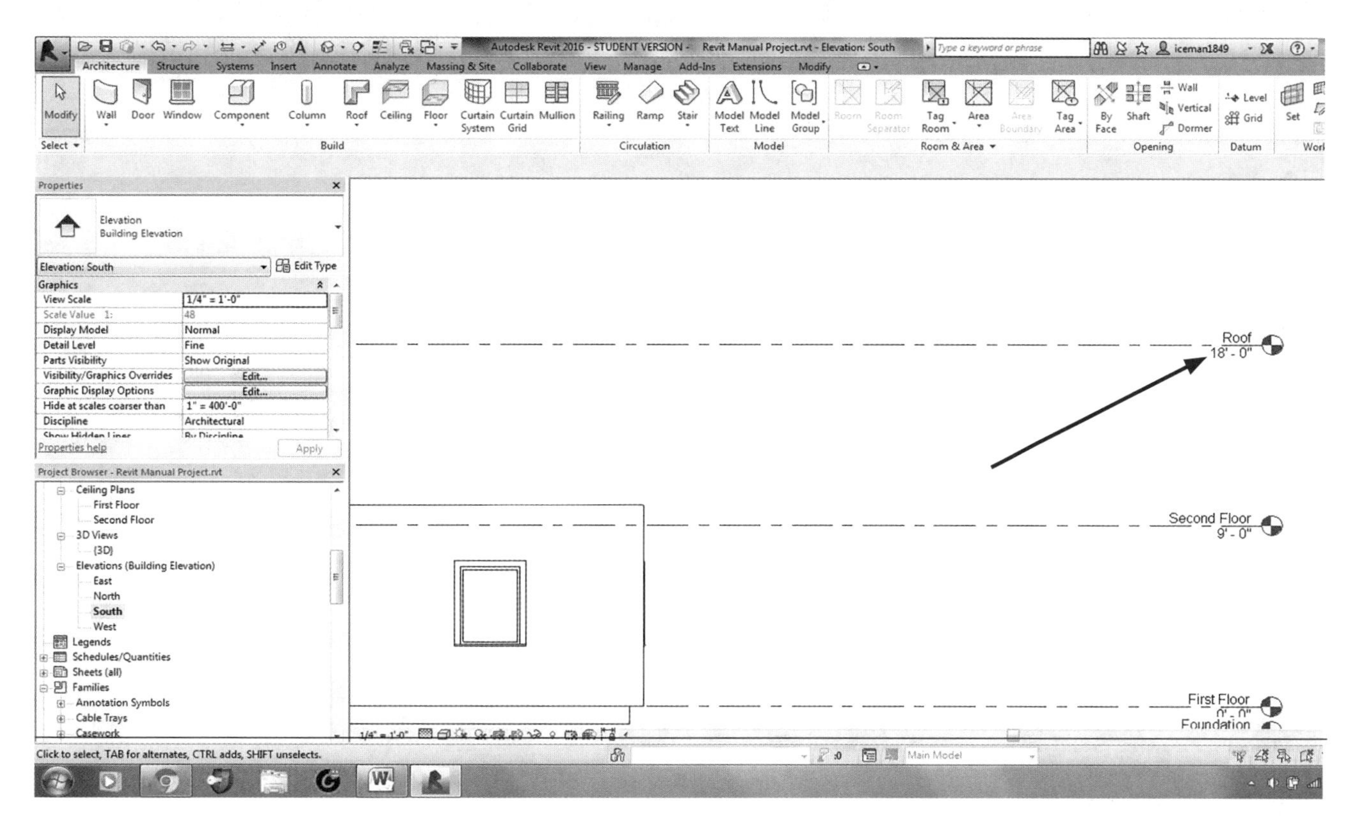

Change the height under the roof name by double-clicking the **DIMENSION**, and typing in the box.

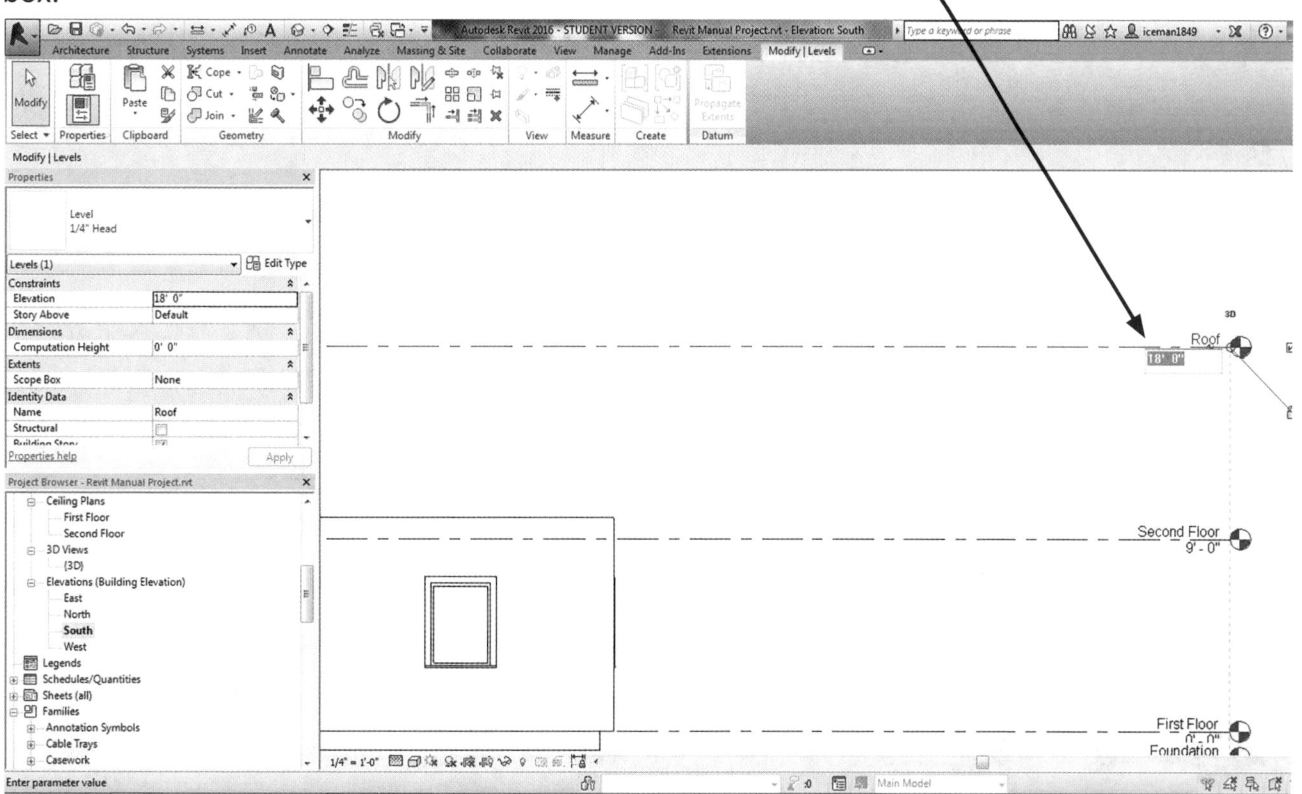

As a result, the Level Line or Datum Line will move either up or down on its own.

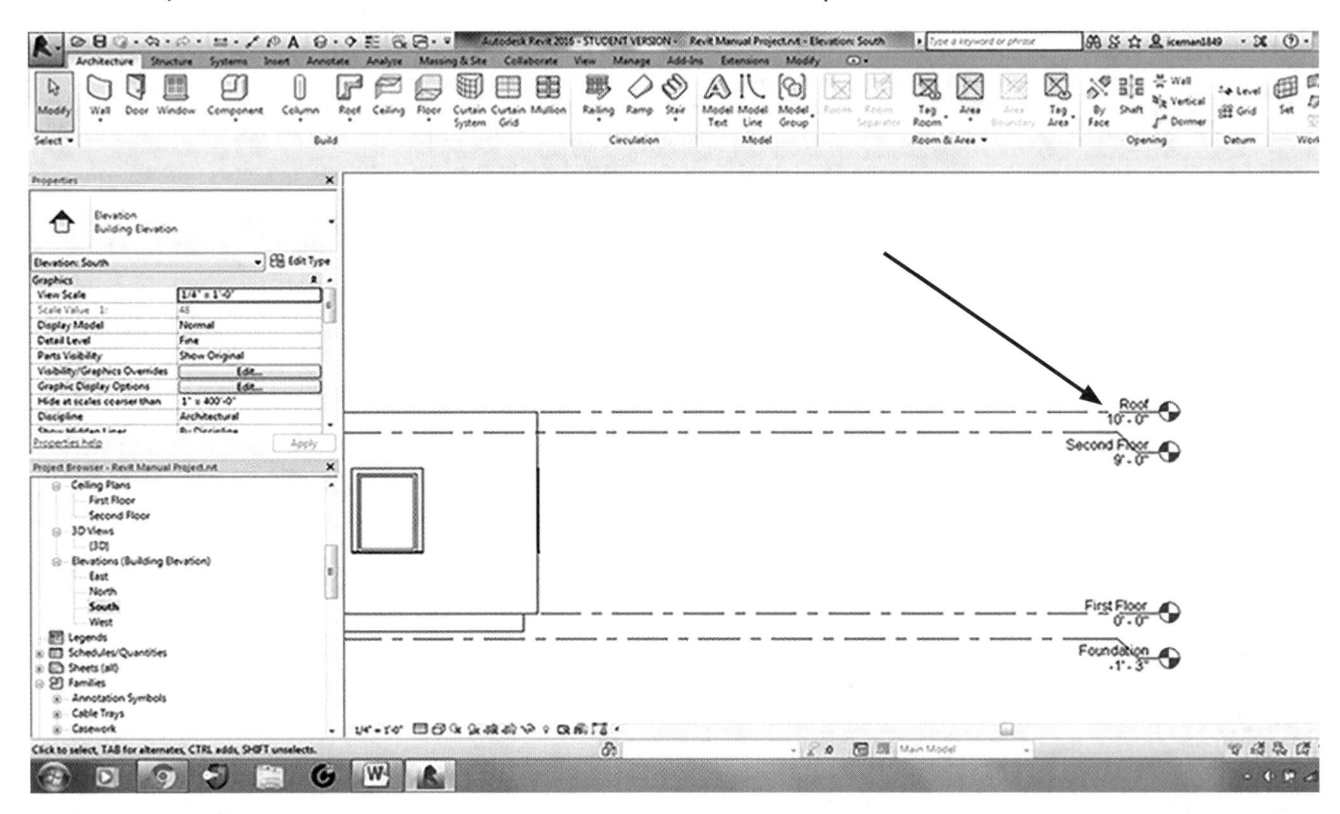

Once this is completed, you can zoom back out to center the house or building.

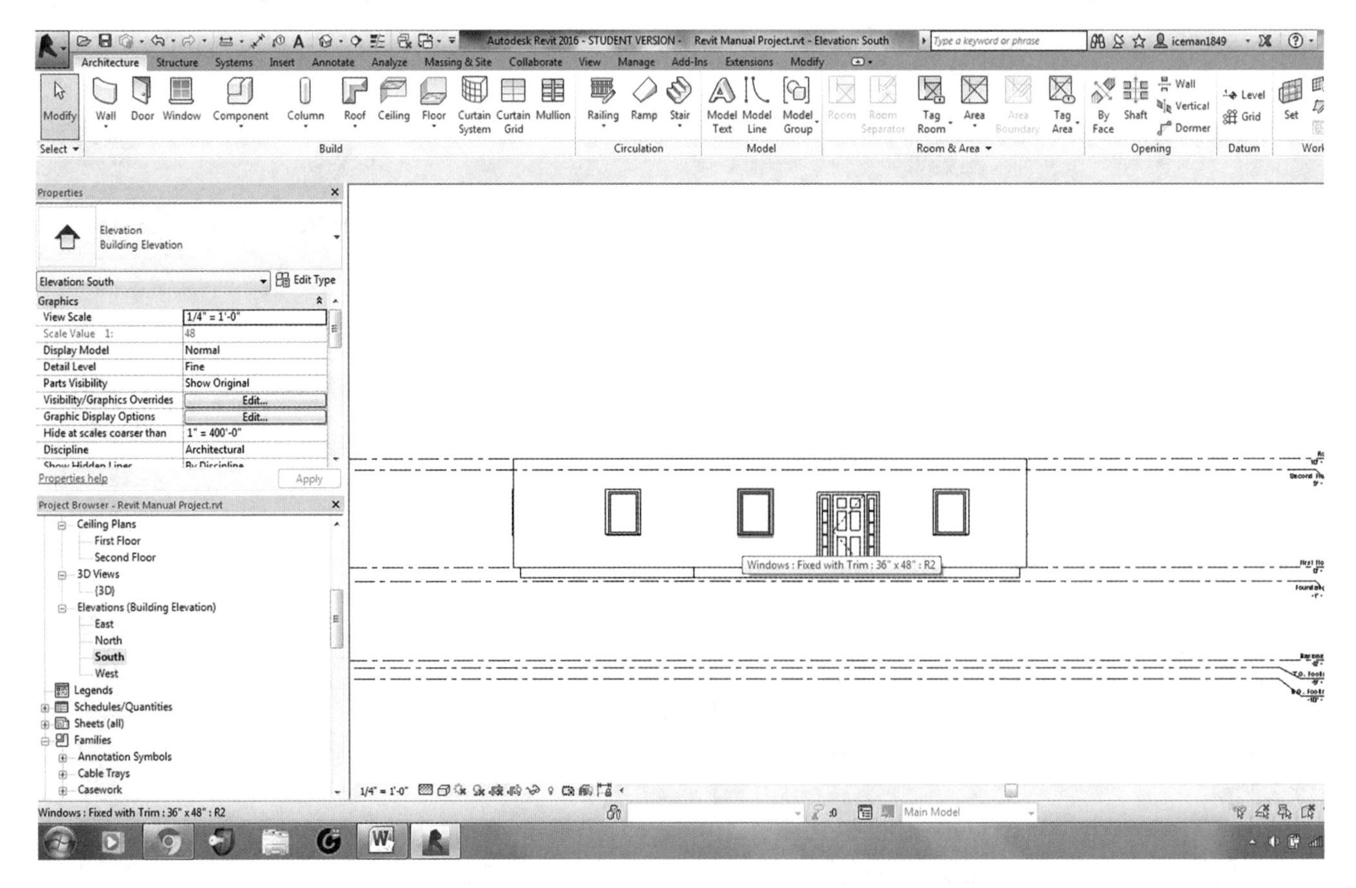

Once you have adjusted the wall height and the roof levels have been corrected, you can switch to the Roof Plan in the Project Browser Box by double-clicking the **ROOF PLAN** name.

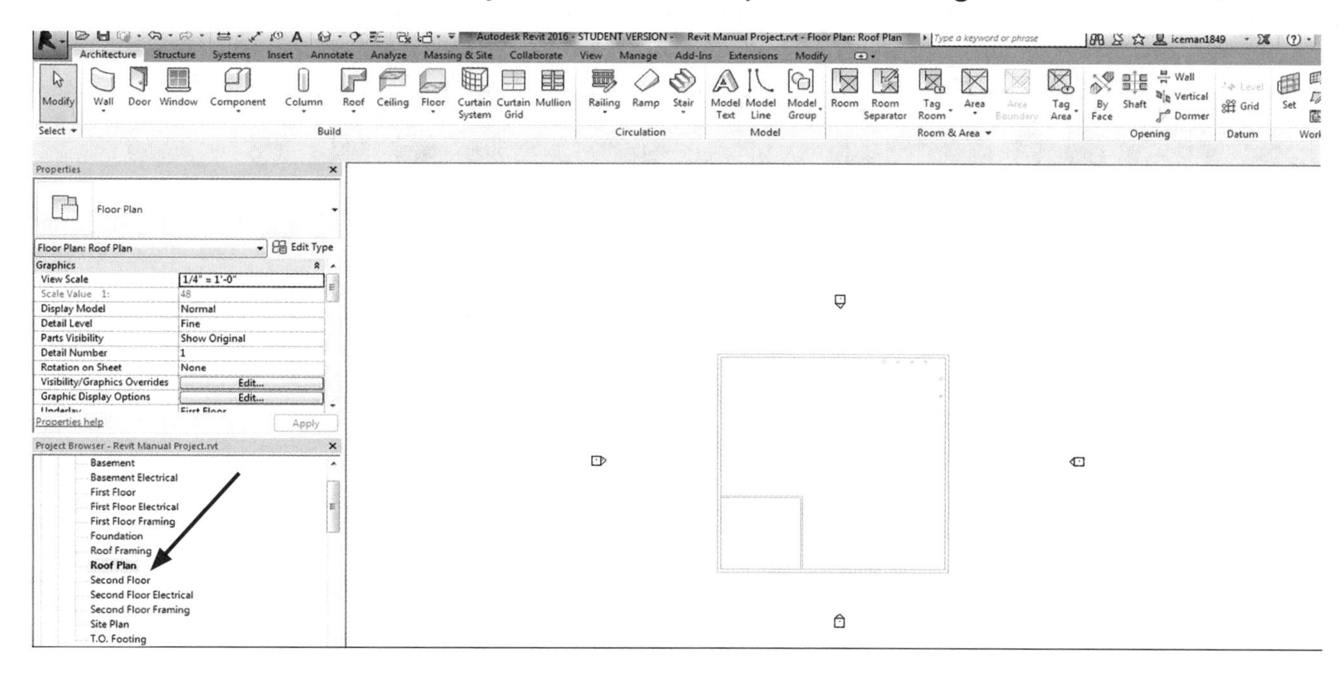

Here we will begin going over steps to create the different types of Roof Plan Layouts.

There will be a right triangle symbol on the sides of the roof. The right triangle symbol indicates the roof has a slope or a roof pitch.

HIP ROOF LAYOUT

In the Ribbon, be sure you are in the Architecture Tab, then click on the **ROOF** Tile Tab.

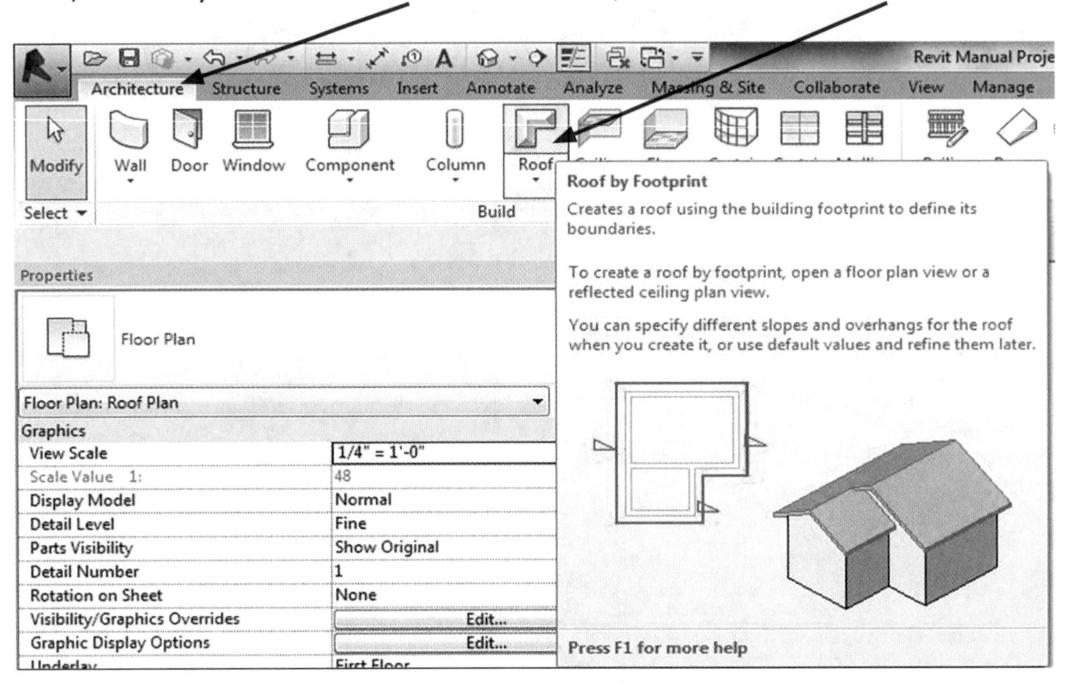

NOTES

The Hip Roof Layout has a pitch on all *4 sides.*

TIPS

If you select the top of the tab, it will default to the ROOF BY FOOTPRINT *selection. That's where we'll begin.*

The other way you can get to the Roof By Footprint selection is by selecting the drop-down arrow and clicking on **ROOF BY FOOTPRINT**.

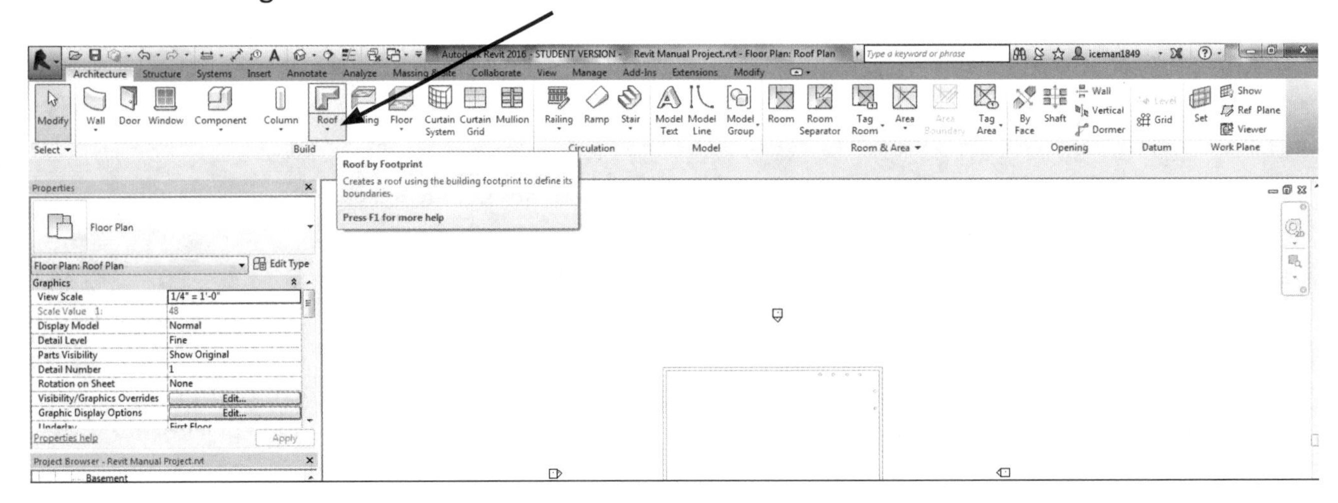

The drop-down box appears, and you can select the **ROOF BY FOOTPRINT** tool.

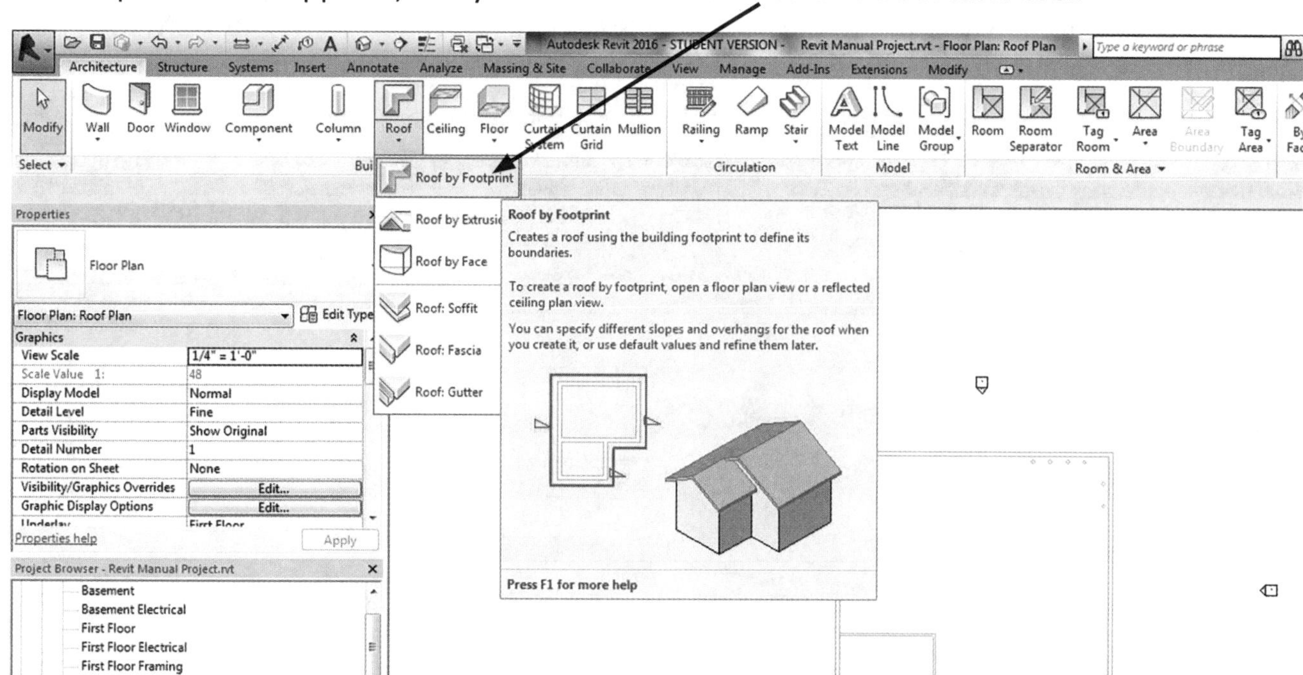

Once the Roof by Footprint tool is selected, the Ribbon will change and you will be in the Modify/Create Roof Footprint Mode.

NOTES

The Options Bar** automatically **defaults to a 1′-0″ overhang, and the DEFINES SLOPE ***Box is checked (which means all the sides have a slope which is ideal for a Hip Roof).***

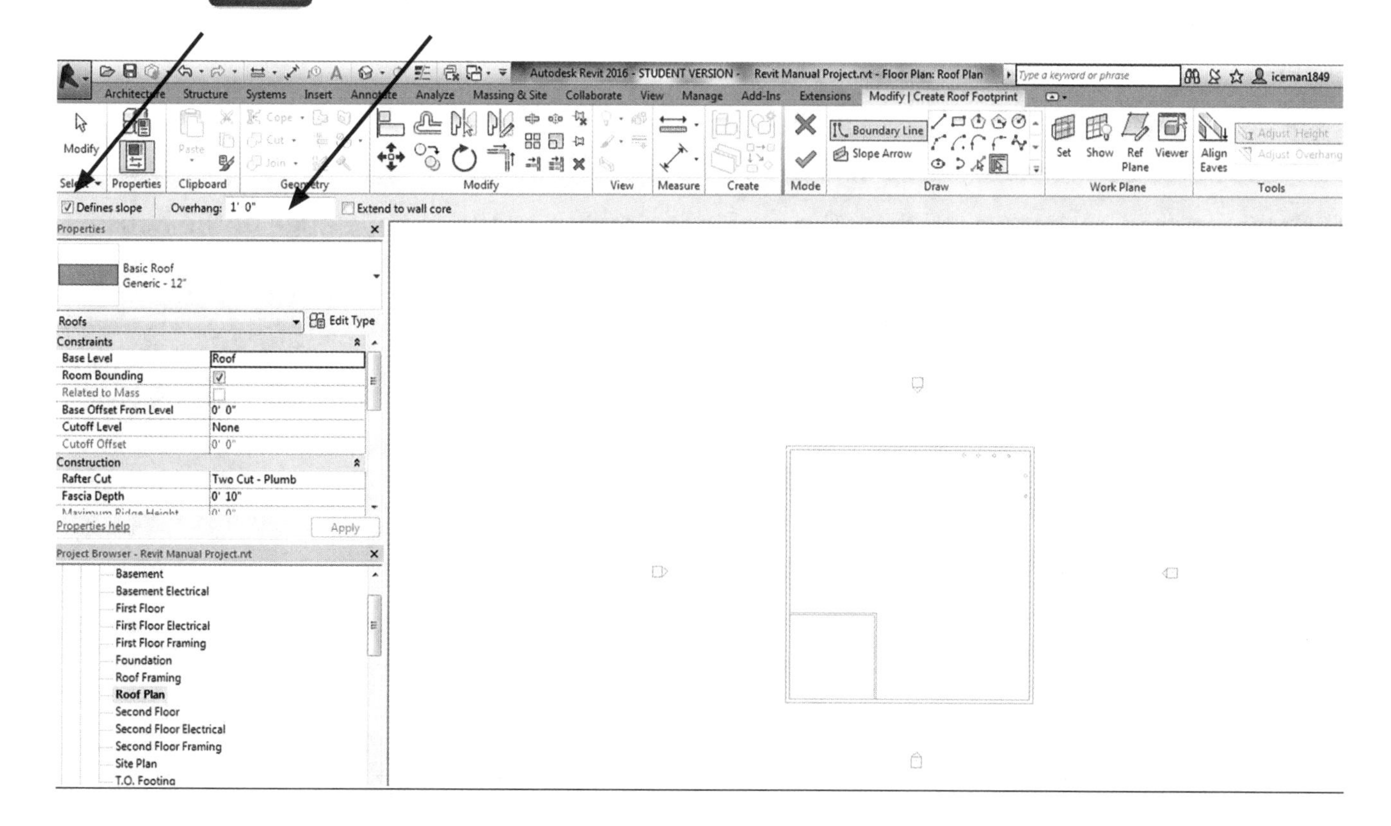

Next, you need to pick a roof type in the Properties Box, such as **ASPHALT SHINGLES ON RAFTERS** (this will allow you to see the roof shingles in your elevation views), or **GENERIC** (made of basic roof material where you won't see any shingle design in your elevation views), if you don't know exactly what type of roof you're looking for.

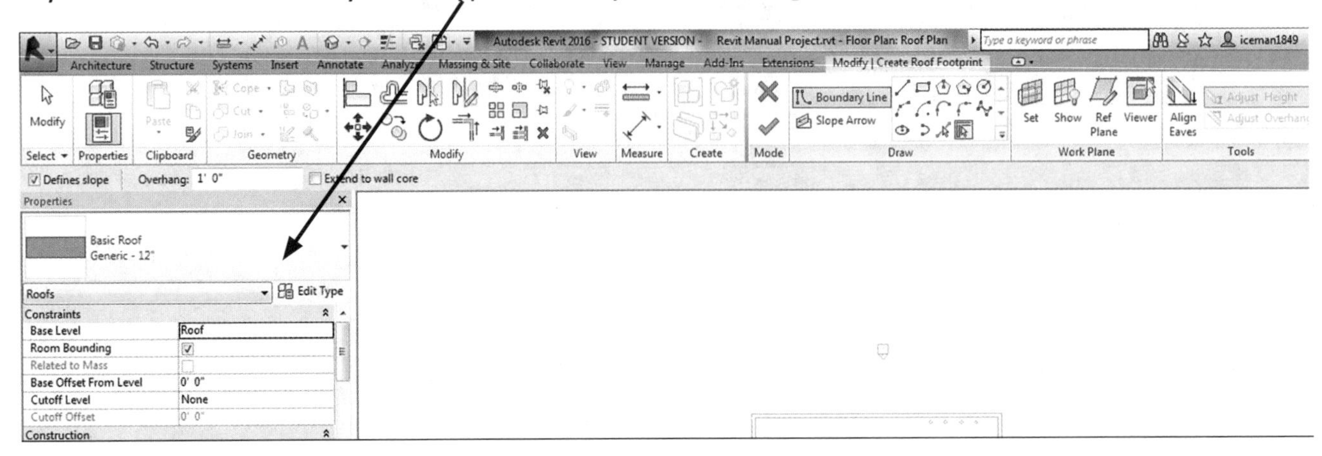

Revit defaults next to drawing the roof by allowing you to pick walls as an efficient way of moving around the house or building.

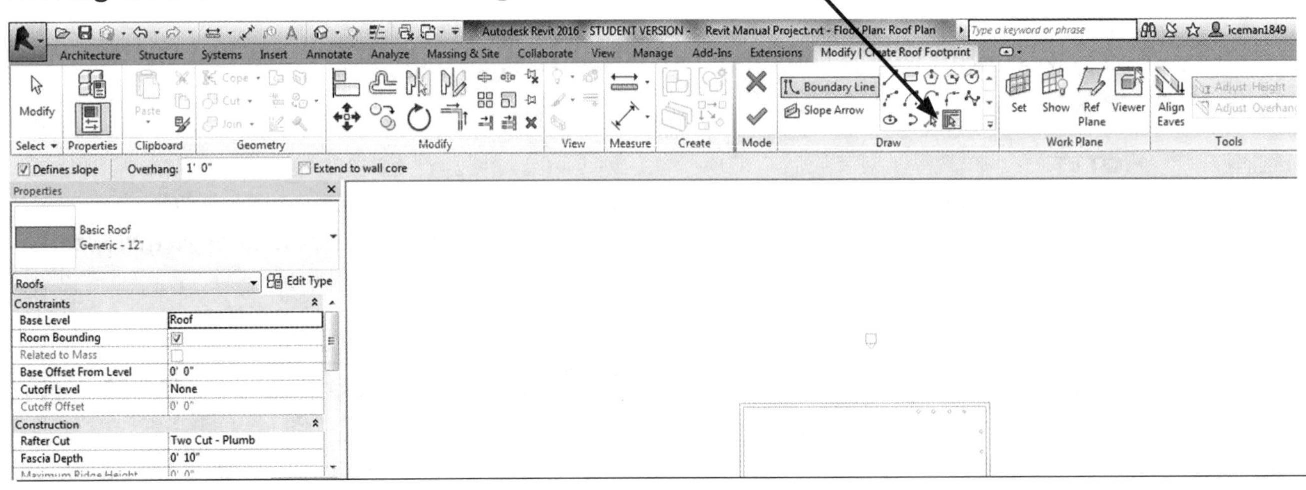

The quickest and easiest way to draw the roof is to simply pick the outside edge of the house or building. You will notice the 1'-0" overhang is automatically placed.

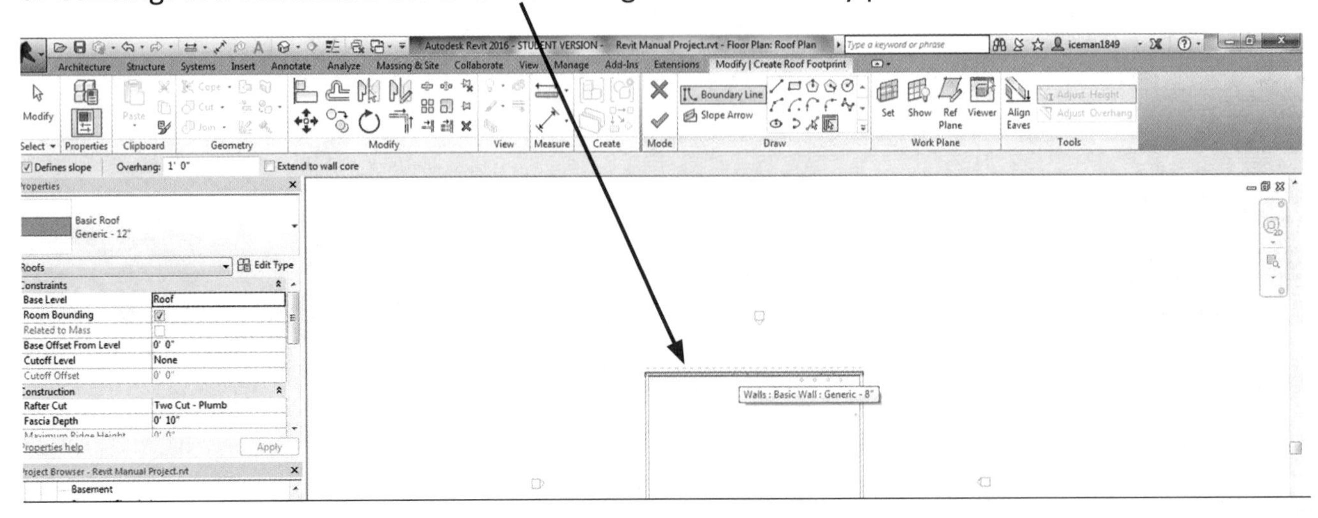

The finished Roof Footprint will look like the following view:

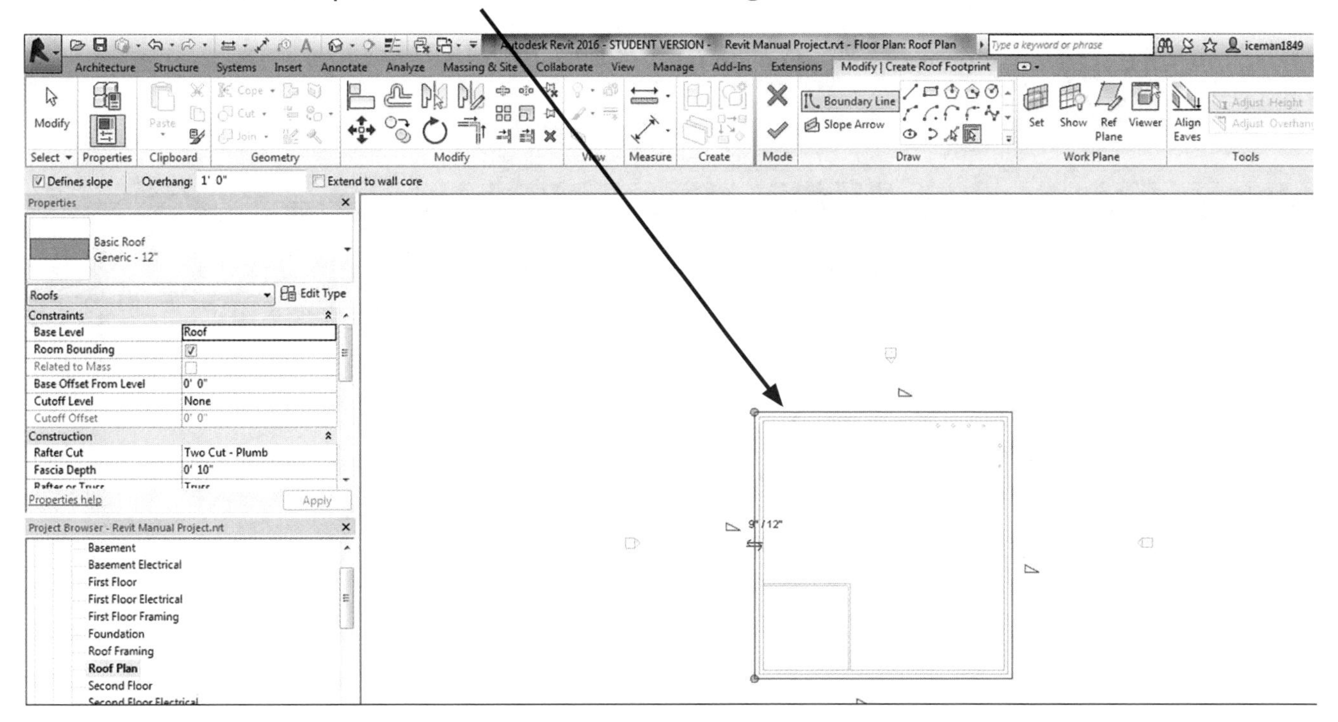

NOTES

Notice that the Defined Slope numbers are automatically placed on all 4 sides with the Pitch Slopes noted.

With the Roof Boundary placed, and the defined slope noted, Revit defaults to a 9/12 slope. If you'd like, you can click on the **DEFINES SLOPE** number and change the slopes to adjust the roof pitch.

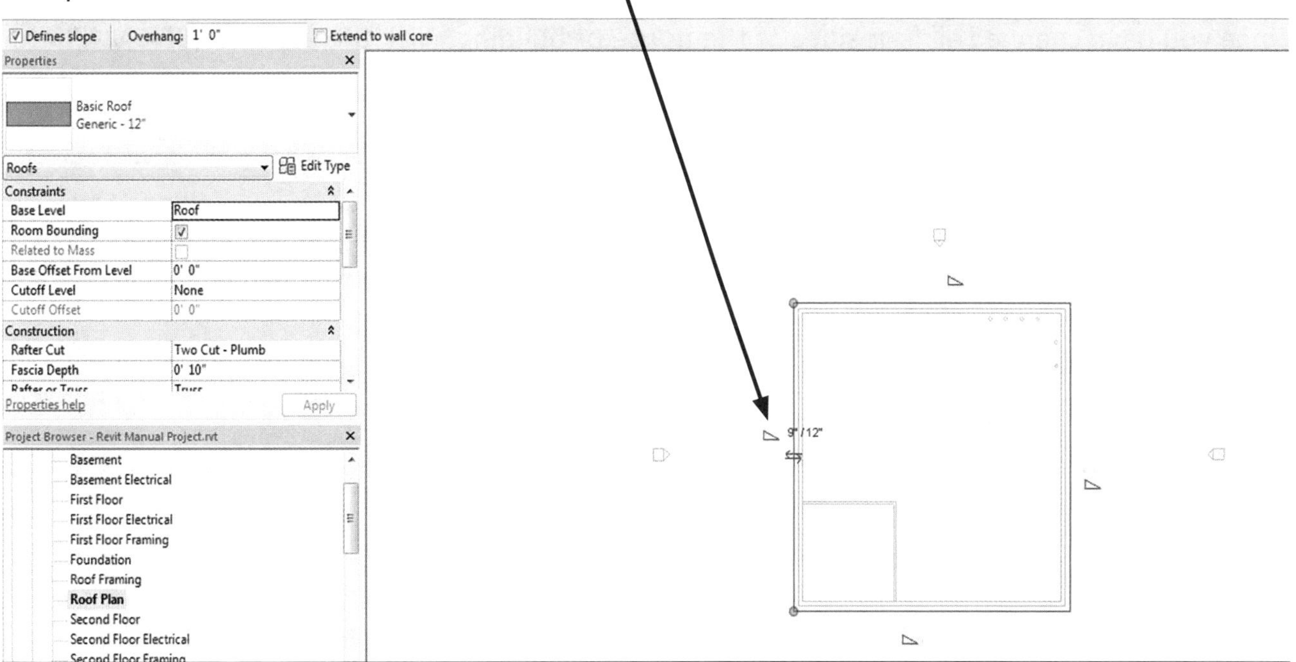

EXAMPLE

Zoom in and click on the pitch number once to change the roof pitch.

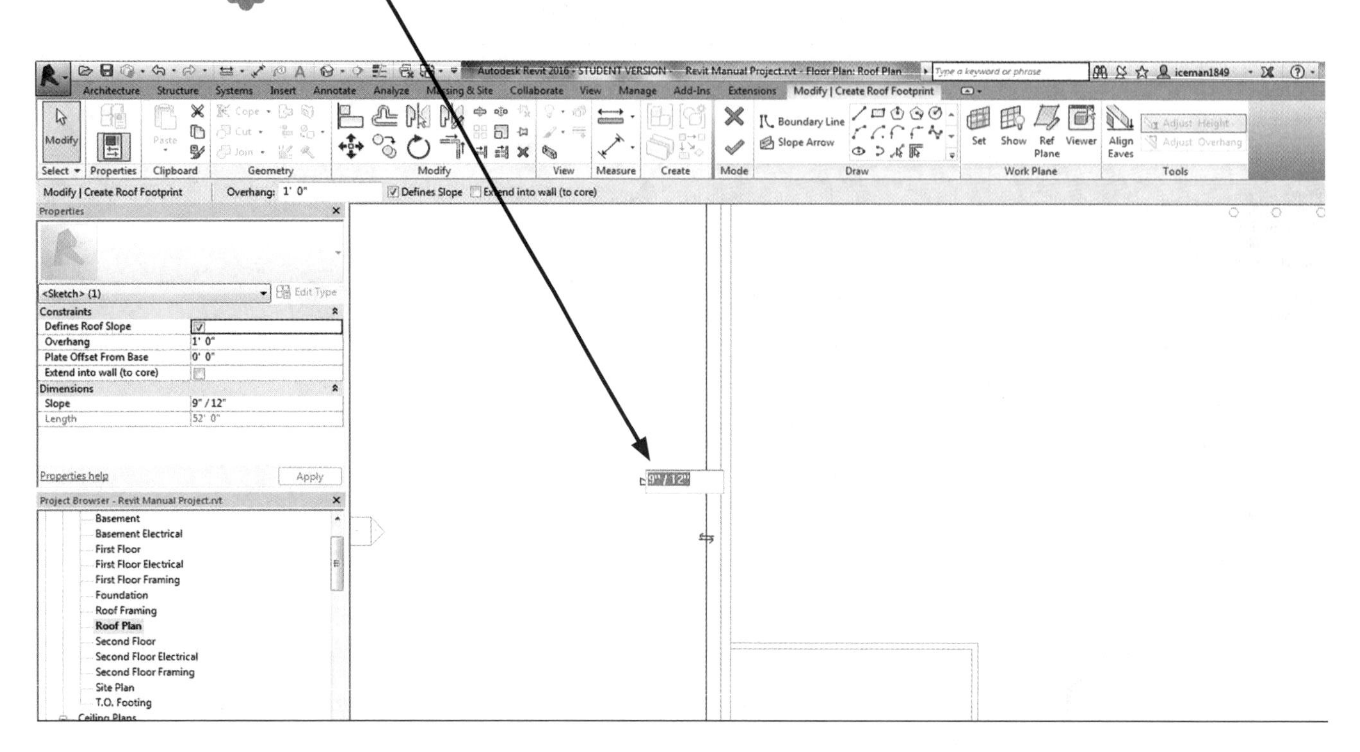

TIPS

Be sure to do this all four (4) sides to match the design.

Once you have changed all four sides of the house or building, click the green **CHECK MARK** to accept or complete the process.

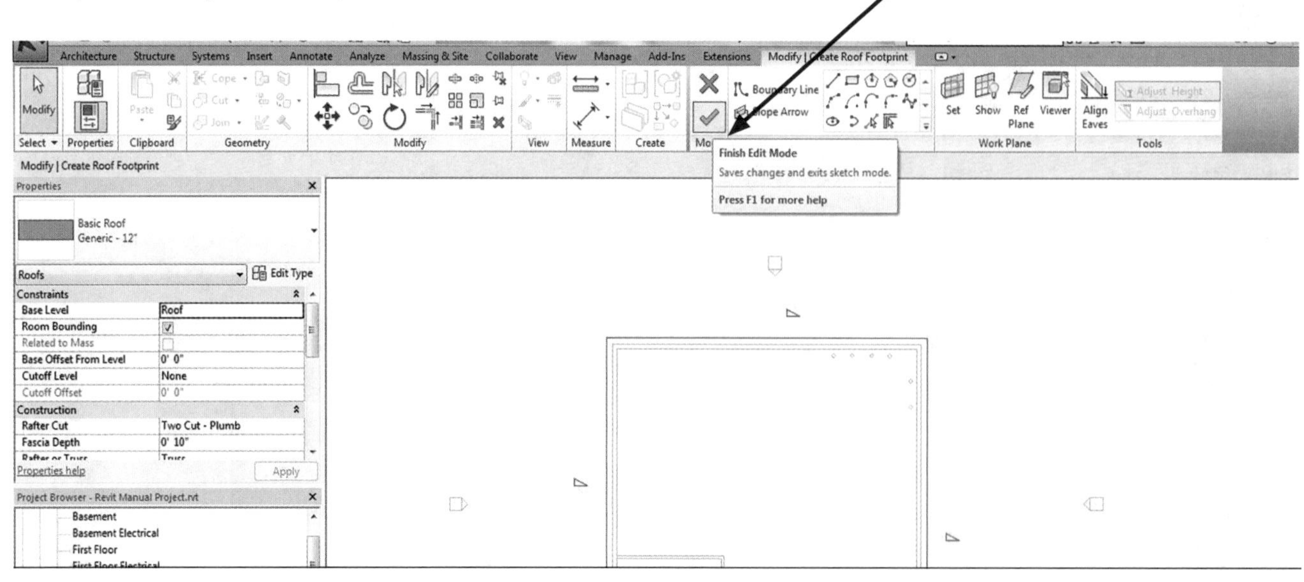

WARNING

A warning will appear about having the walls attach to the roof. Click* NO. *This will allow you to make changes to the wall heights later, if necessary.

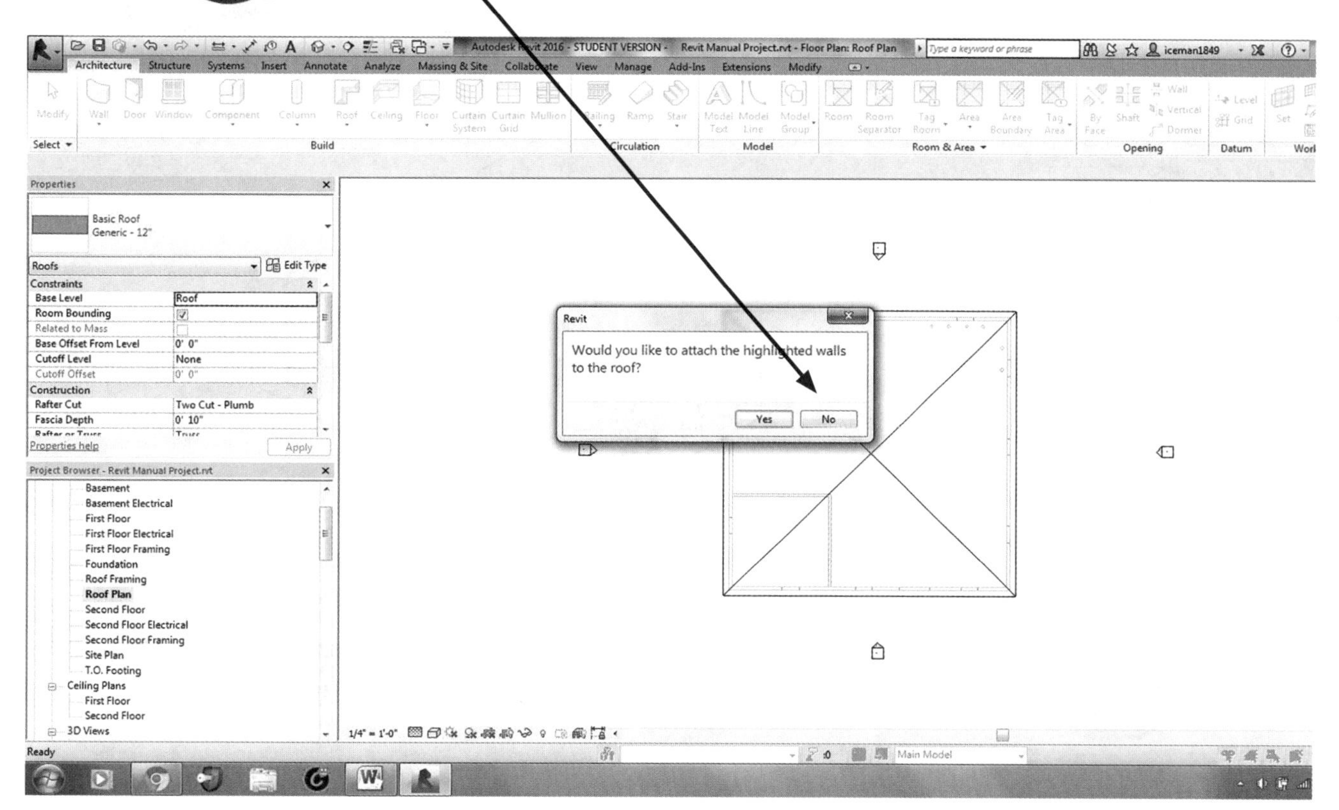

The screen that follows shows the Finished Roof Footprint:

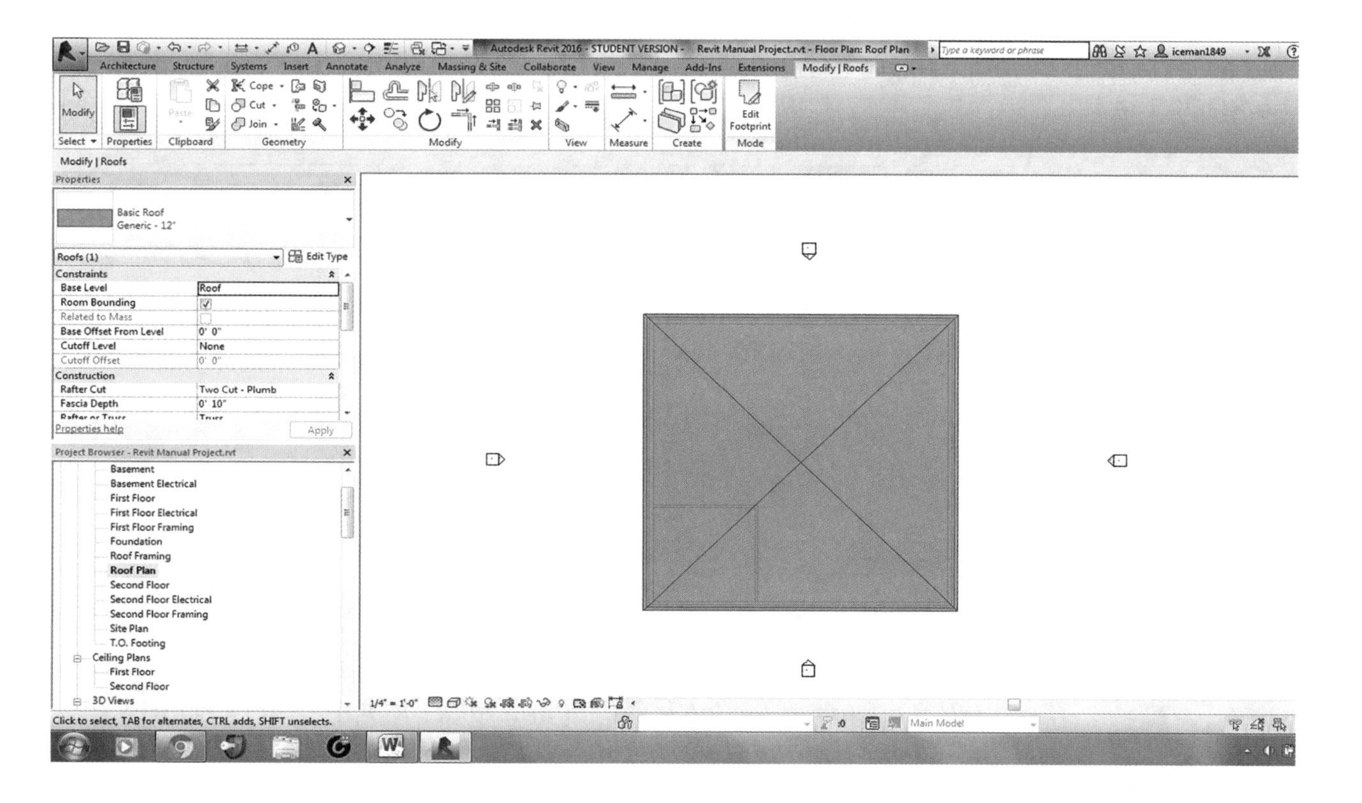

You can then click on **3D** to see your finished roof design.

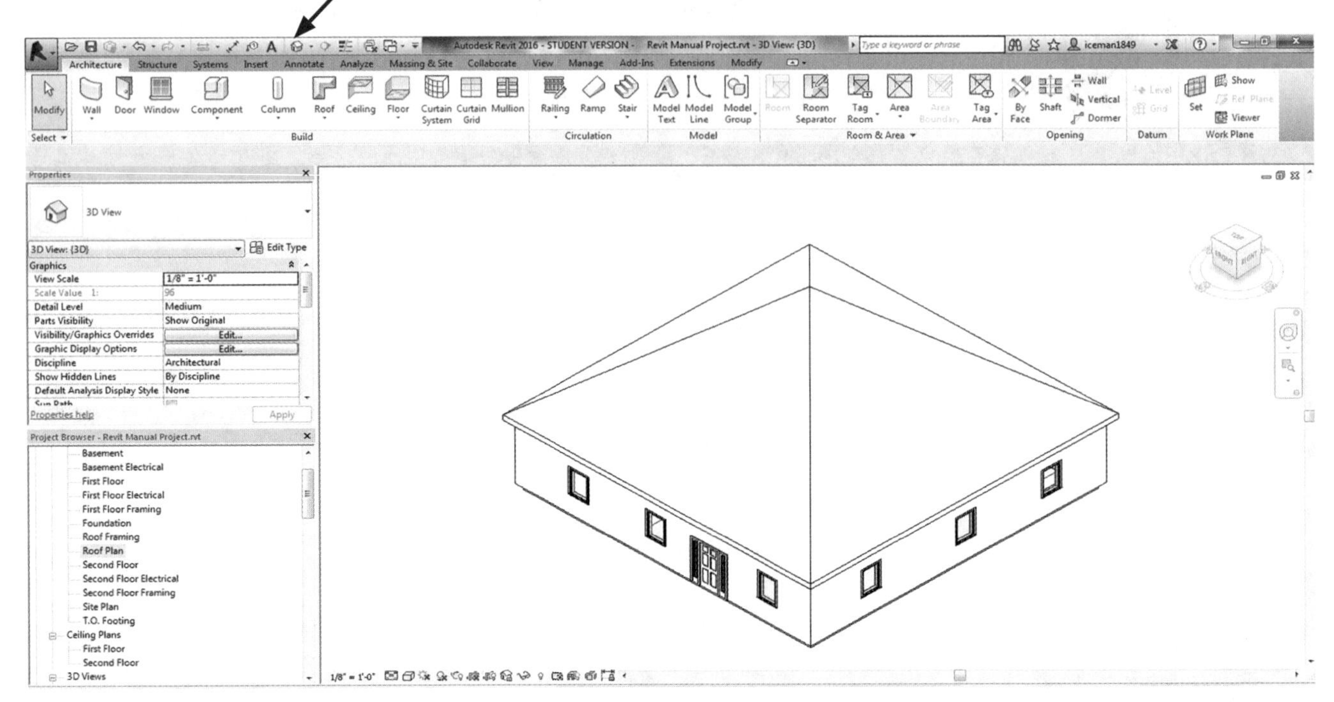

REMINDER

Just as we discussed in the beginning of the chapter, make sure you have all the correct steps started before you start creating the roof layouts.

GABLE ROOF LAYOUT

The Gable Roof has roof pitch on *only* two (2) sides, instead of all four like the Hip Roof. To include a Gable Roof in your design, first make sure you are in the Architecture Tab, then click on the **ROOF** Tile Tab.

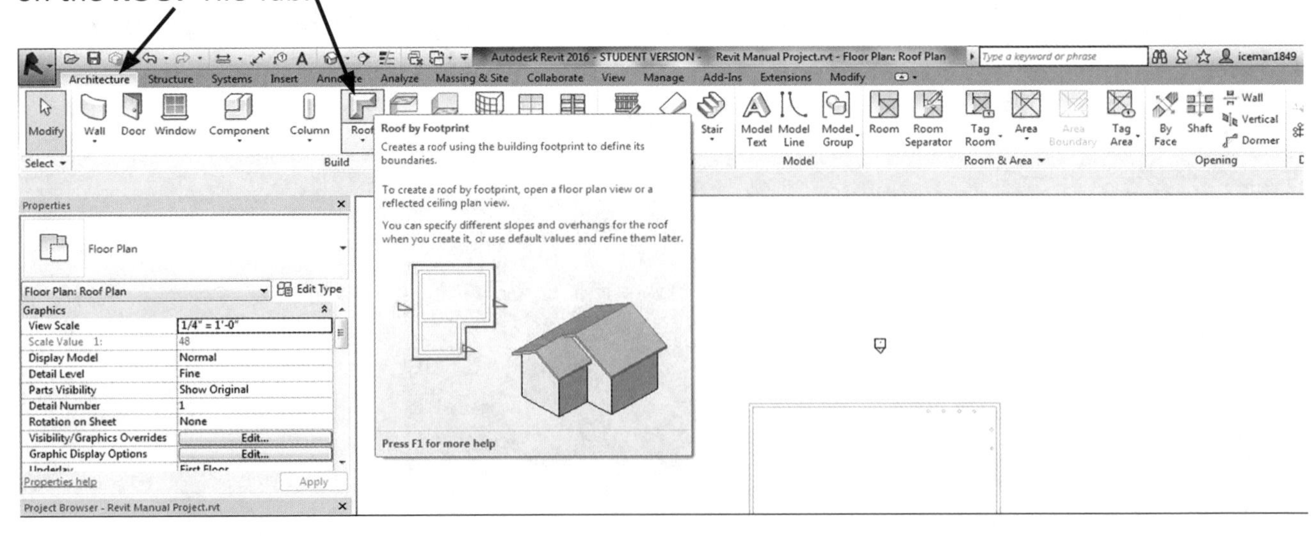

REMINDER

If you select the top of the tab it will default to the ROOF BY FOOTPRINT ***selection. That's the one you want to begin with.***

Just like in the steps for the Hip Roof, another way to get to the roof selections is to click on the **DROP-DOWN ARROW** and select **ROOF BY FOOTPRINT**.

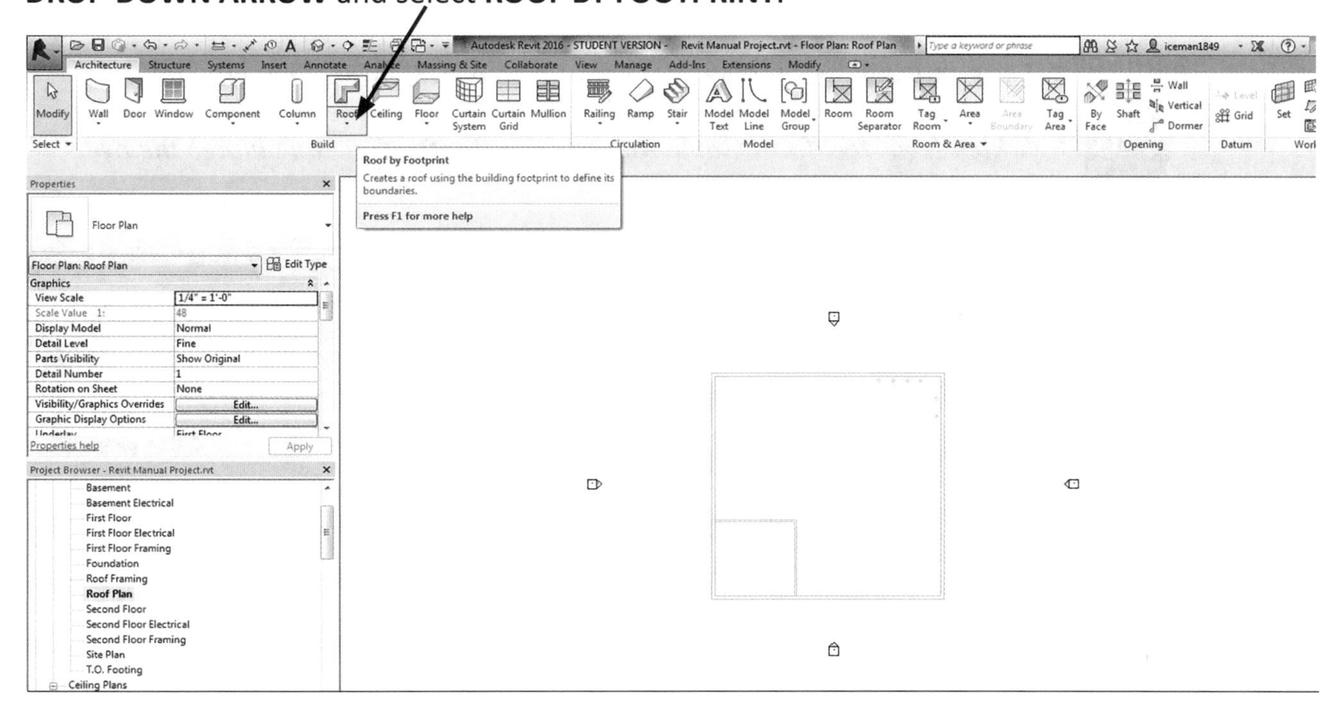

When the drop-down box appears, select the **ROOF BY FOOTPRINT** tool.

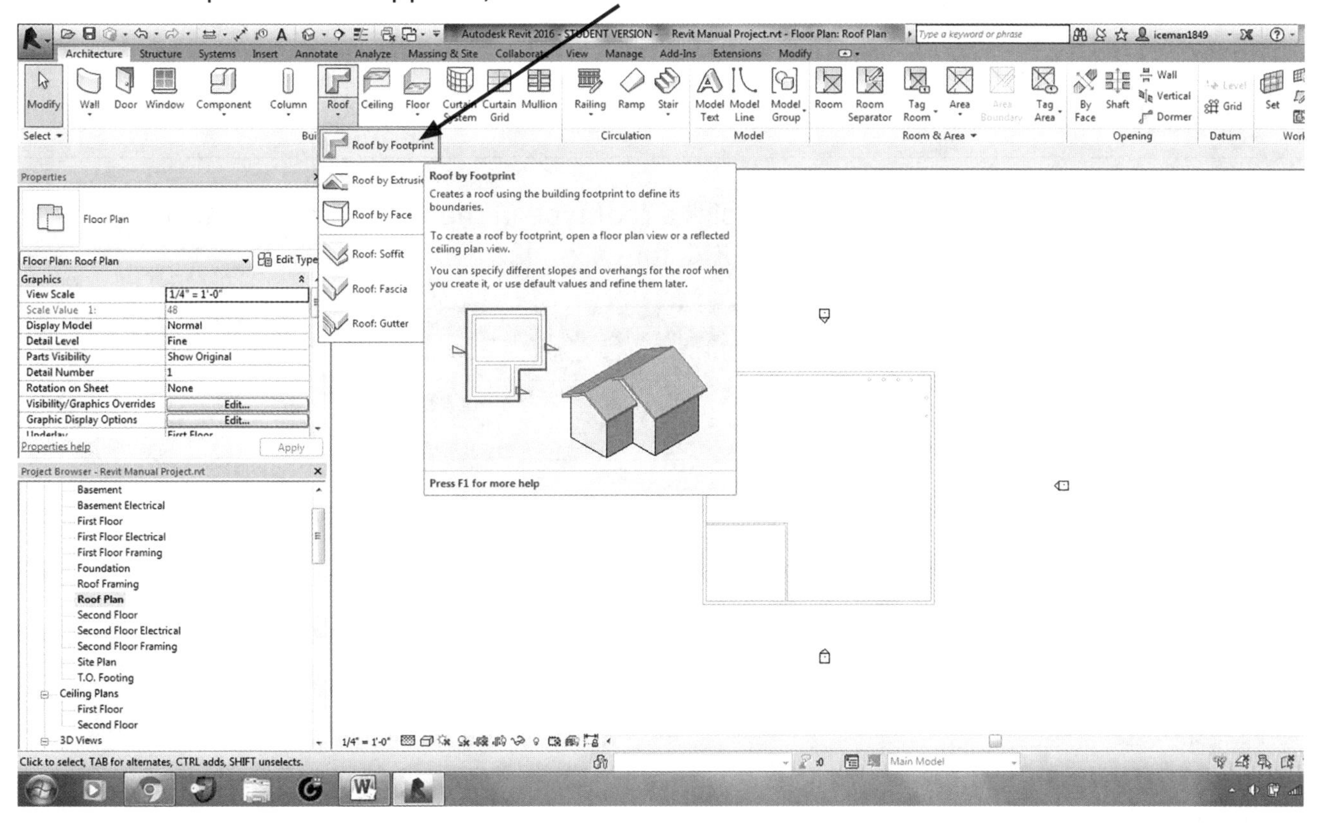

Once the **ROOF BY FOOTPRINT** tool is selected, the ribbon will change and you will be in the **MODIFY/CREATE ROOF FOOTPRINT** mode.

REMINDER ***The Options Bar*** automatically ***defaults to a 1'-0" overhang, and the*** DEFINES SLOPE ***box is checked (which means*** all ***the sides have a slope (which is perfect for a Gable Roof or Hip Roof). You can change the overhang, if you need to.***

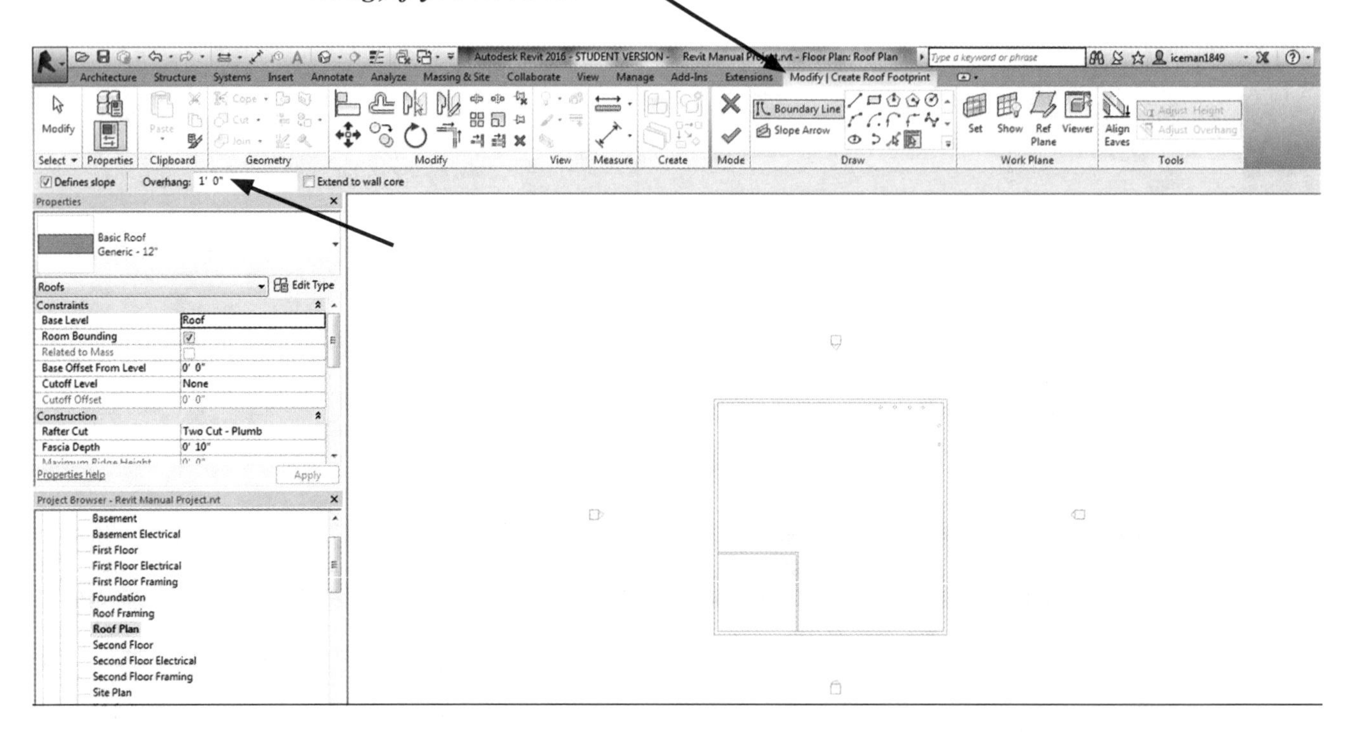

To change the overhang, first you need to pick a roof type in the Properties Box, such as **ASPHALT SHINGLES ON RAFTERS,** or **GENERIC,** for example.

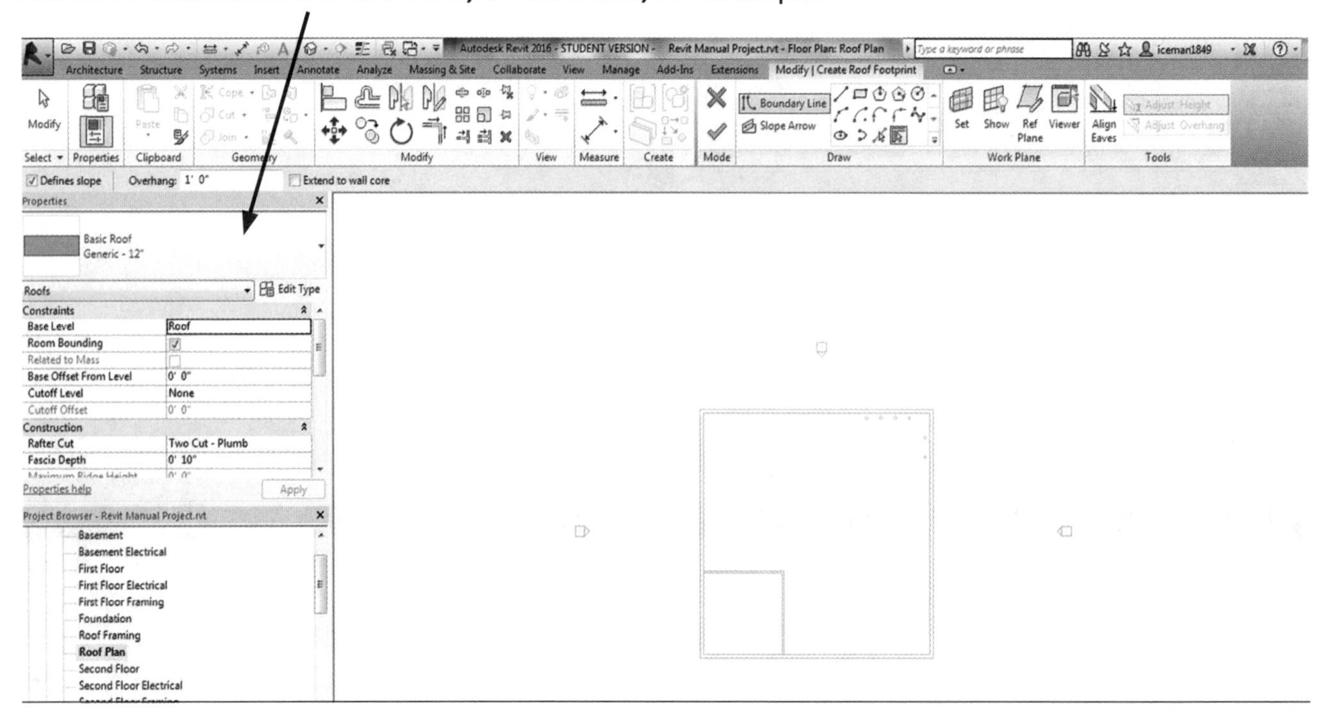

Revit defaults next to drawing the roof by allowing you to pick walls as an efficient way of moving around the house or building.

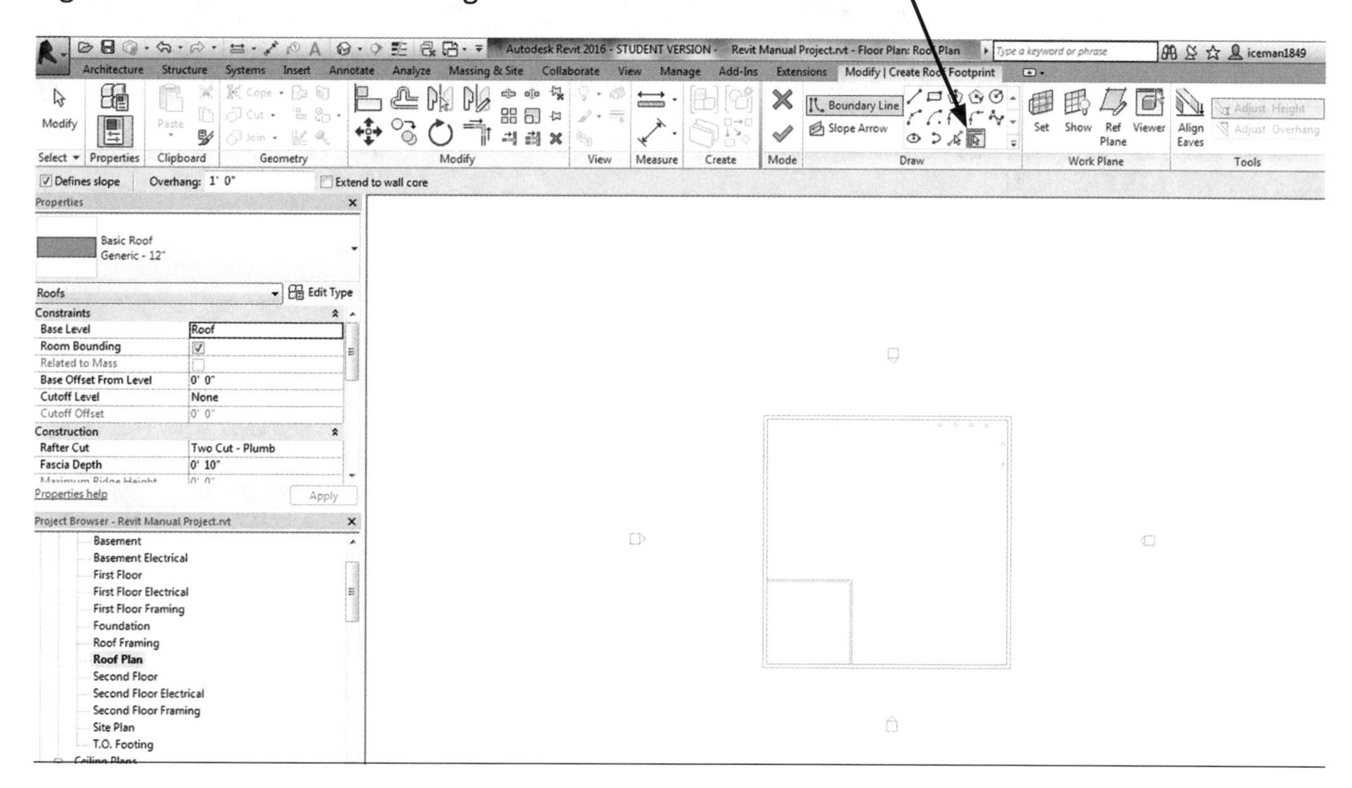

Pick the outside edge of the house or building, and you will notice the 1'-0" overhang is automatically placed.

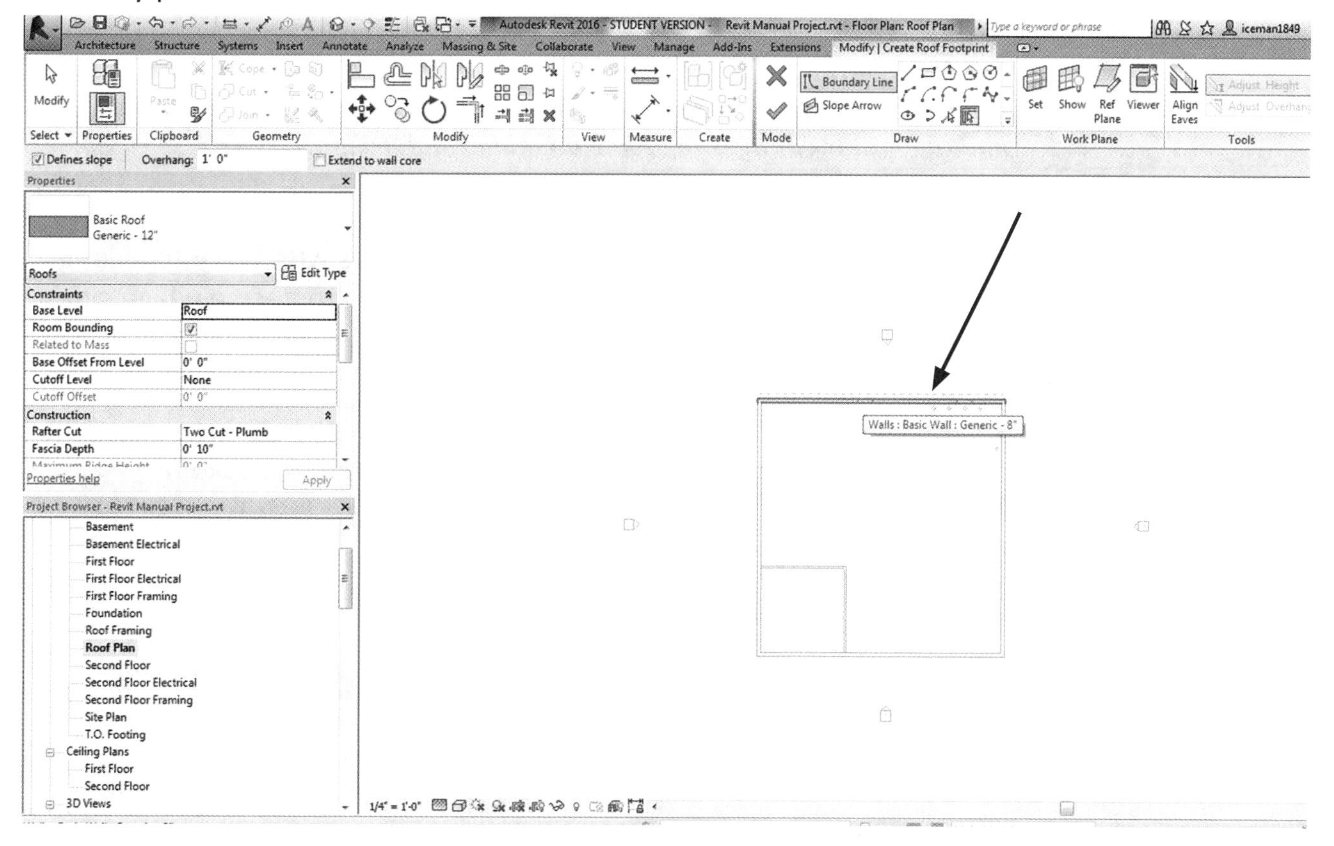

The finished Roof Footprint will look like the following view:

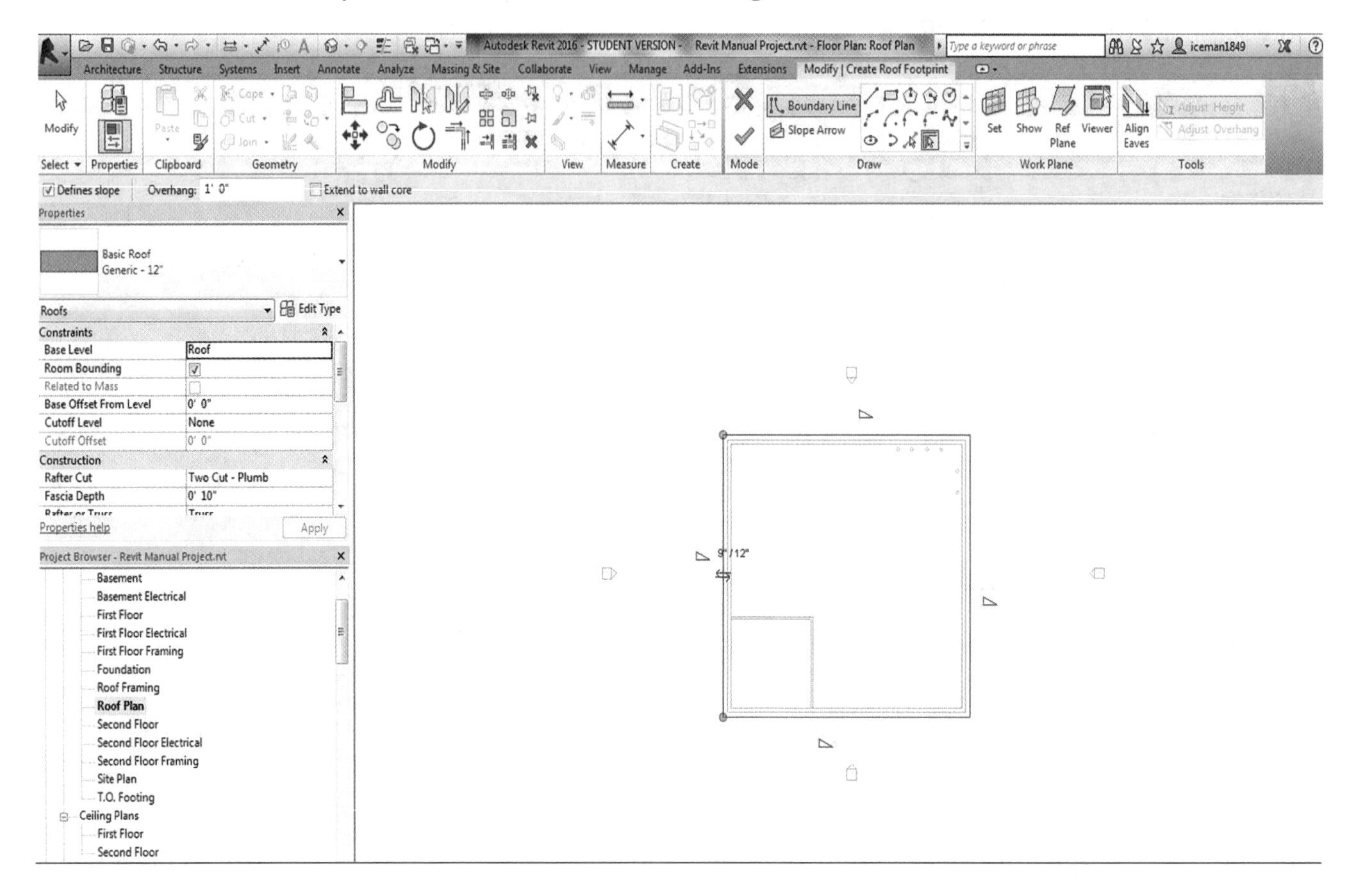

WARNING

Don't be alarmed if this layout and roof design looks the same as the Hip Roof. We are going to change the roof design in a couple of more steps, making minor changes to create a Gable Roof design.

Also notice the Defined Slopes are automatically placed on all four sides with the Pitch Slopes noted. Since a Gable Roof has only two sides with a slope, you will have to decide which sides should maintain the slopes, and which to turn off.

TIPS

After the roof has been placed, you do not ***want to click the*** green CHECK MARK ***to accept the finished roof. Instead, you want to click the*** MODIFY ***button and then select the two sides from which you want to remove the slope.***

REMINDER

Do not ***hit*** ESC ***Key. That will cause other commands to become undone.***

Then click the magenta line on the side of the house or building where you want to remove the slope direction. The line will turn blue or a different shade.

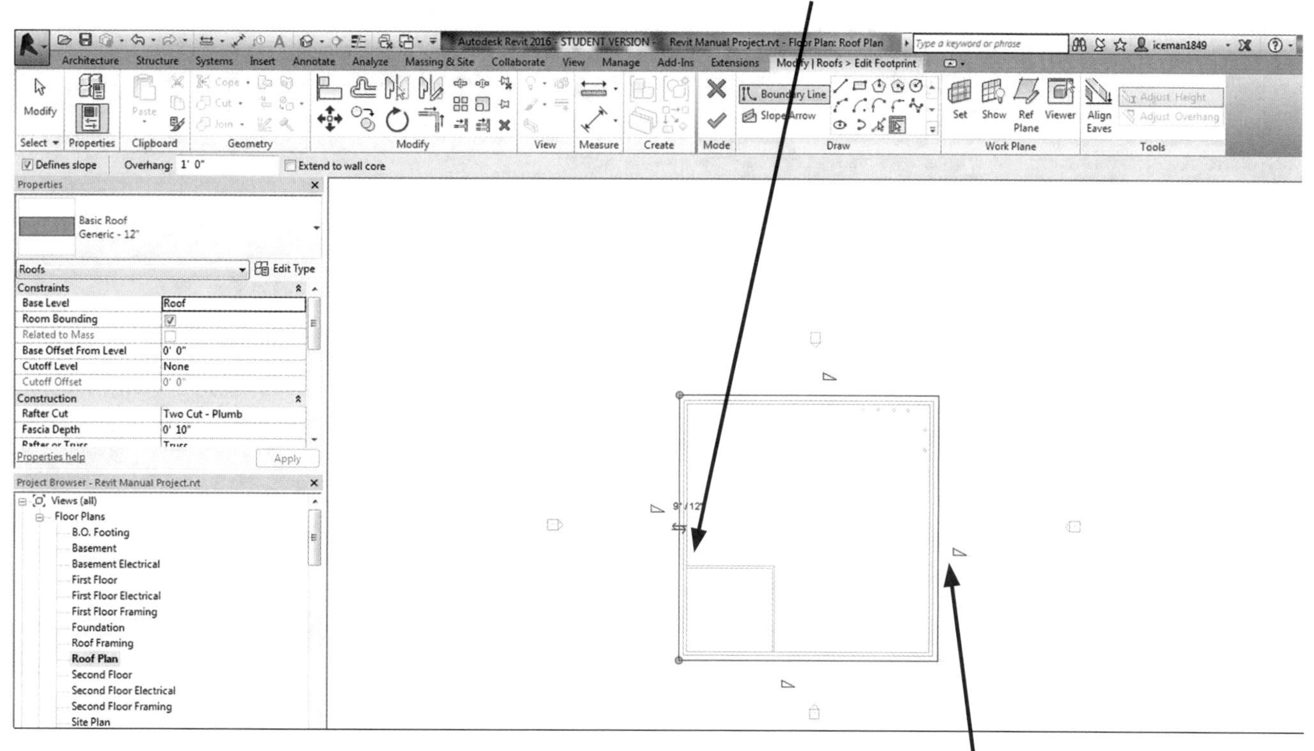

Next, select the opposite side of the house roof or building and do the same thing.

Once you have selected both lines on the outside of the Roof Footprint, in the Options Bar, uncheck the **DEFINES SLOPE** Box. This will turn off the slope of the roof.

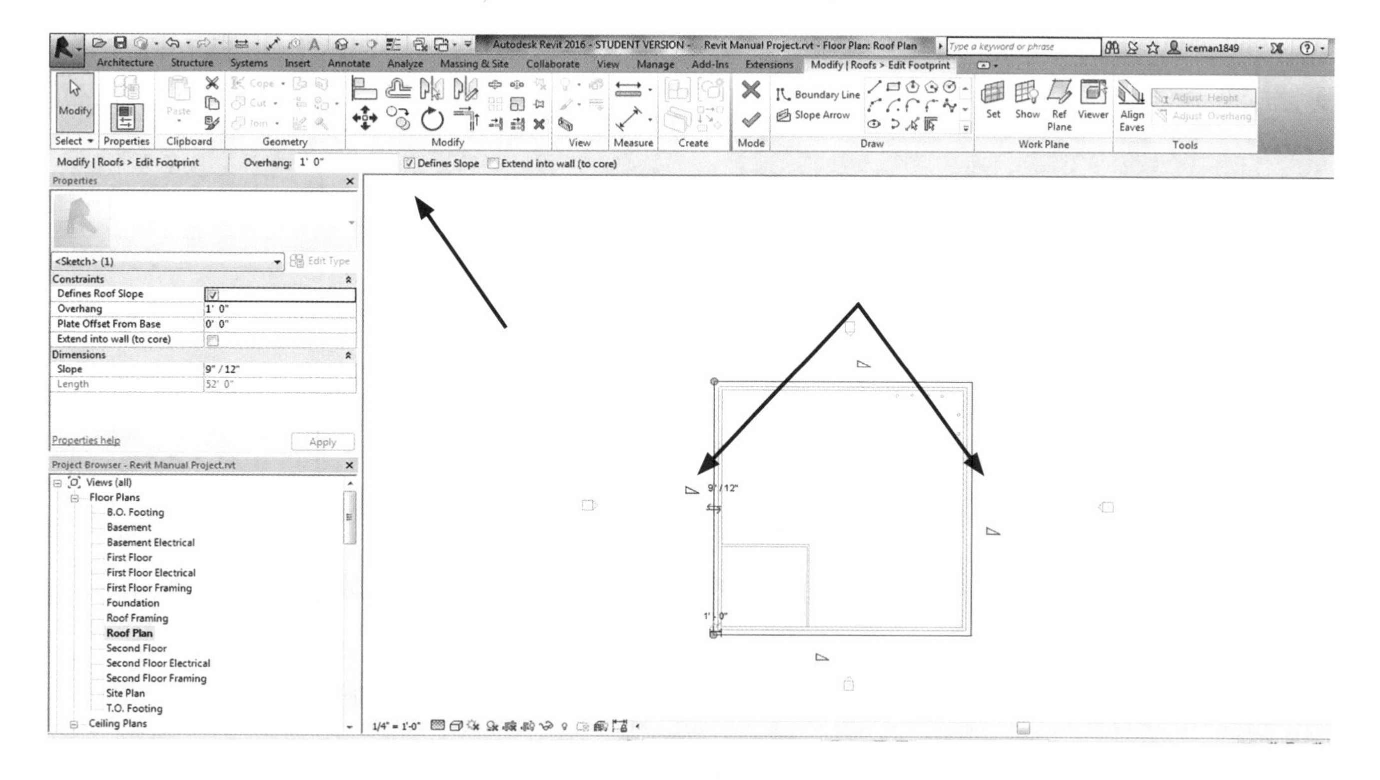

Once the **DEFINES SLOPE** Box is unchecked, click the green **CHECK MARK** to accept or complete the roof design layout.

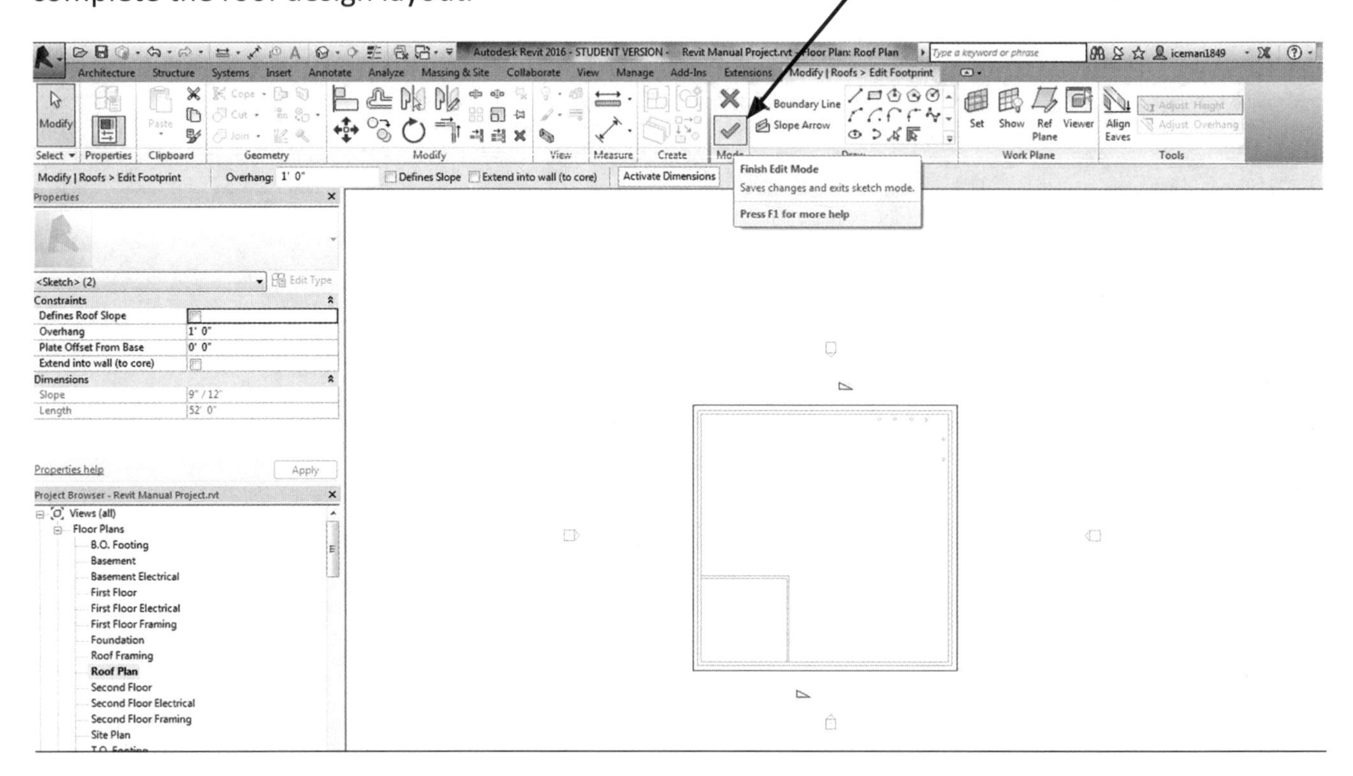

REMINDER

Once again, you will get a warning window asking if you want to attach the highlighted walls to the roof.

Click **NO,** so you can modify the walls or roof at a later date without having to detach the roof.

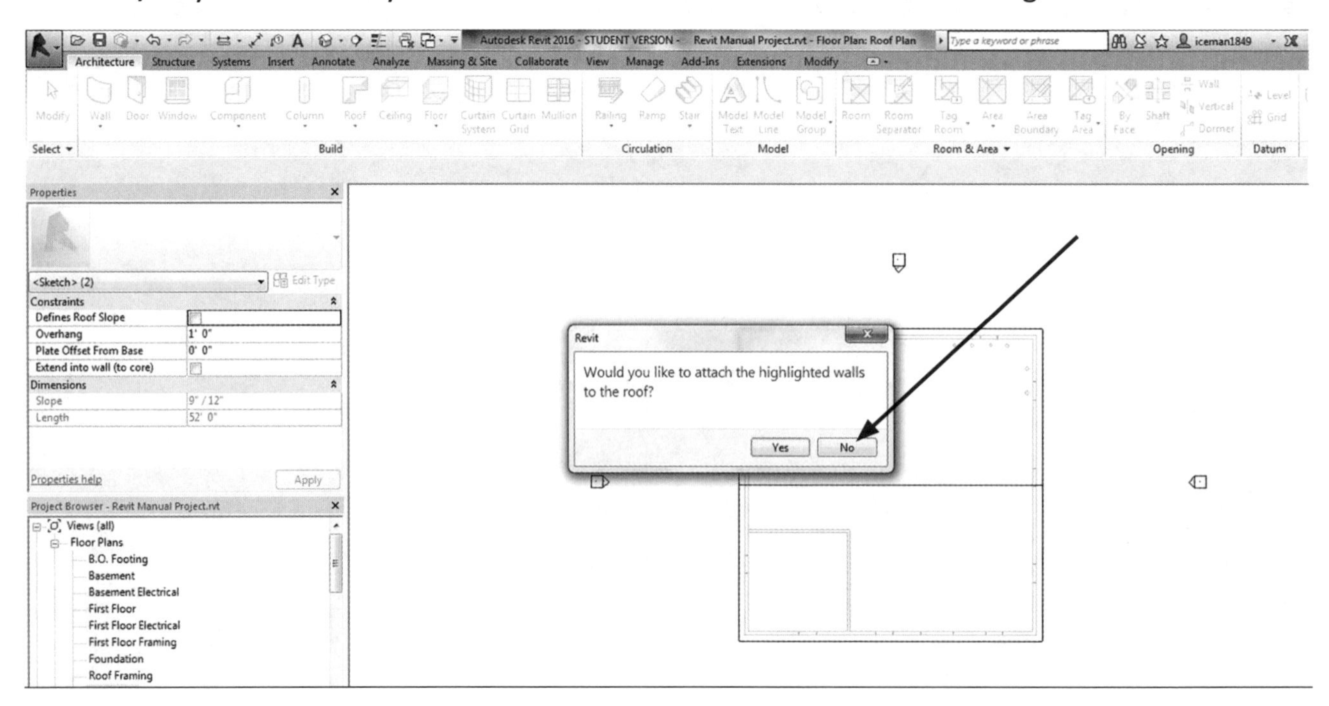

NOTES

Notice the line going through the middle of the roof. That is the ridge line where your roof folds down towards the ground for water to run off the roof.

Then click on **3D** on the *QAT* (*Quick Access Toolbar*), and look at the finished design layout.

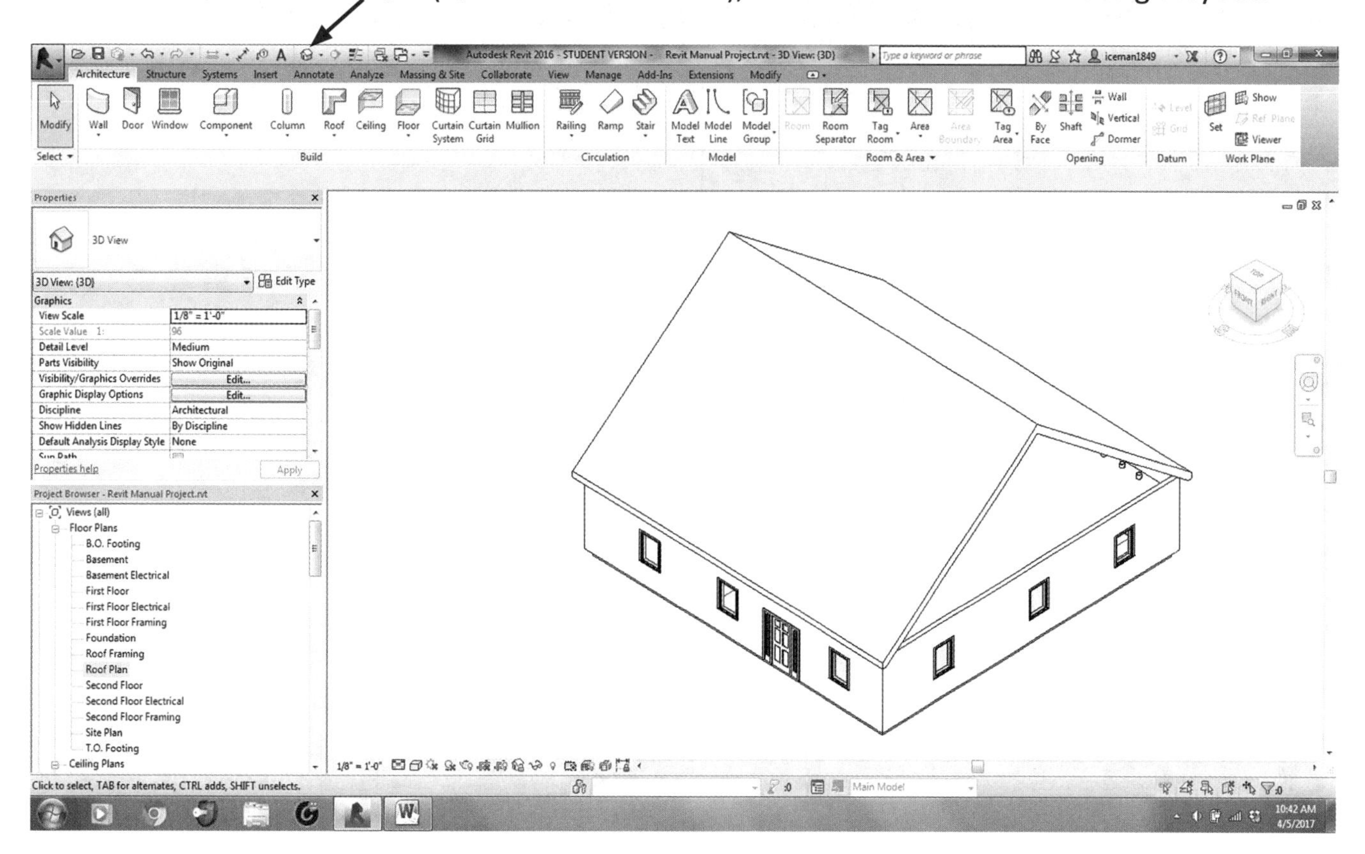

TIPS

The Quick Access Toolbar is located all the way at the top of your screen.

TIPS

Just as in the beginning of the chapter, make sure you have all the correct steps started before you start creating the roof layouts.

SLOPE ROOF LAYOUT

The Slope Roof layout has only one (1) side with a slope (instead of the two-sided slope in a Gable Roof, or all four sides, like a Hip Roof) .

In the ribbon, be sure you are in the Architecture Tab, then click on the **ROOF** Tile Tab.

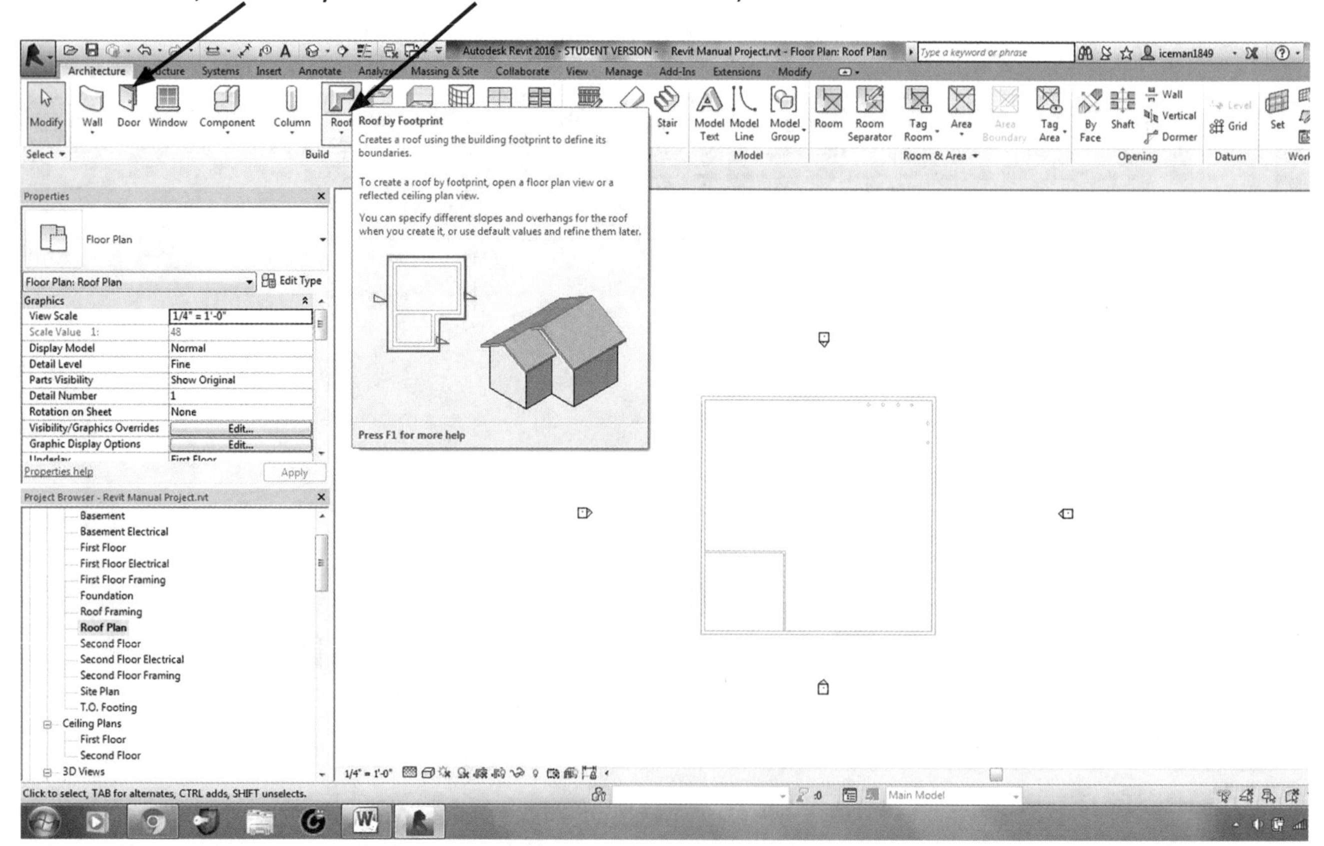

REMINDER

If you select the top of the tab it will default to the ROOF BY FOOTPRINT *selection. That's the one you want.*

Alternatively, you can click the drop-down and select **ROOF BY FOOTPRINT.**

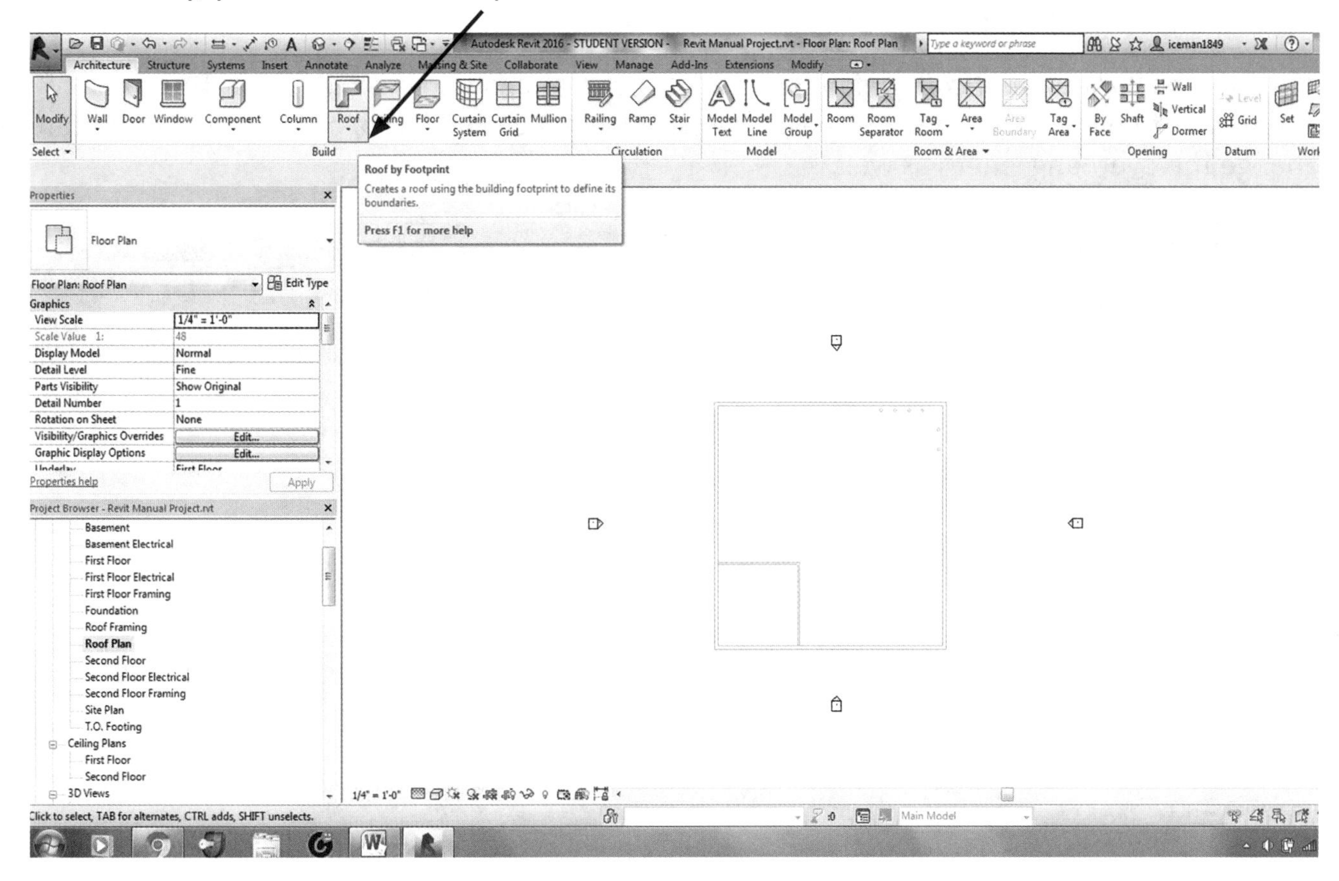

The drop-down box appears, and you should select the **ROOF BY FOOTPRINT** Tool.

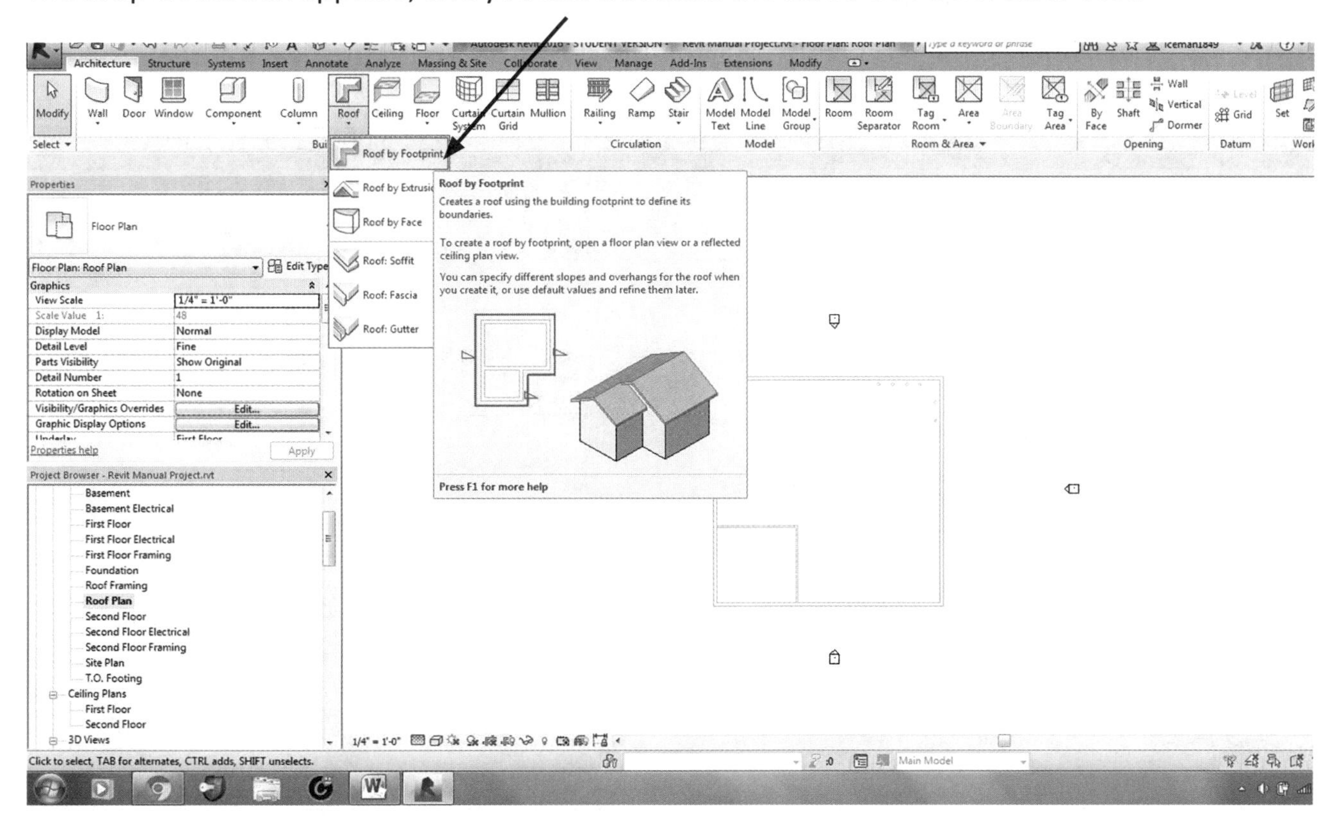

Once the **ROOF BY FOOTPRINT** tool is selected, the Ribbon will change, and you will be in the Modify/Create Roof Footprint Mode.

The Options Bar *automatically* defaults to a 1'-0" overhang, and the **DEFINES SLOPE** box is checked (which means all the sides have a slope (ideal for a Gable, Hip, or Shed Roof). You can change the overhang length as well.

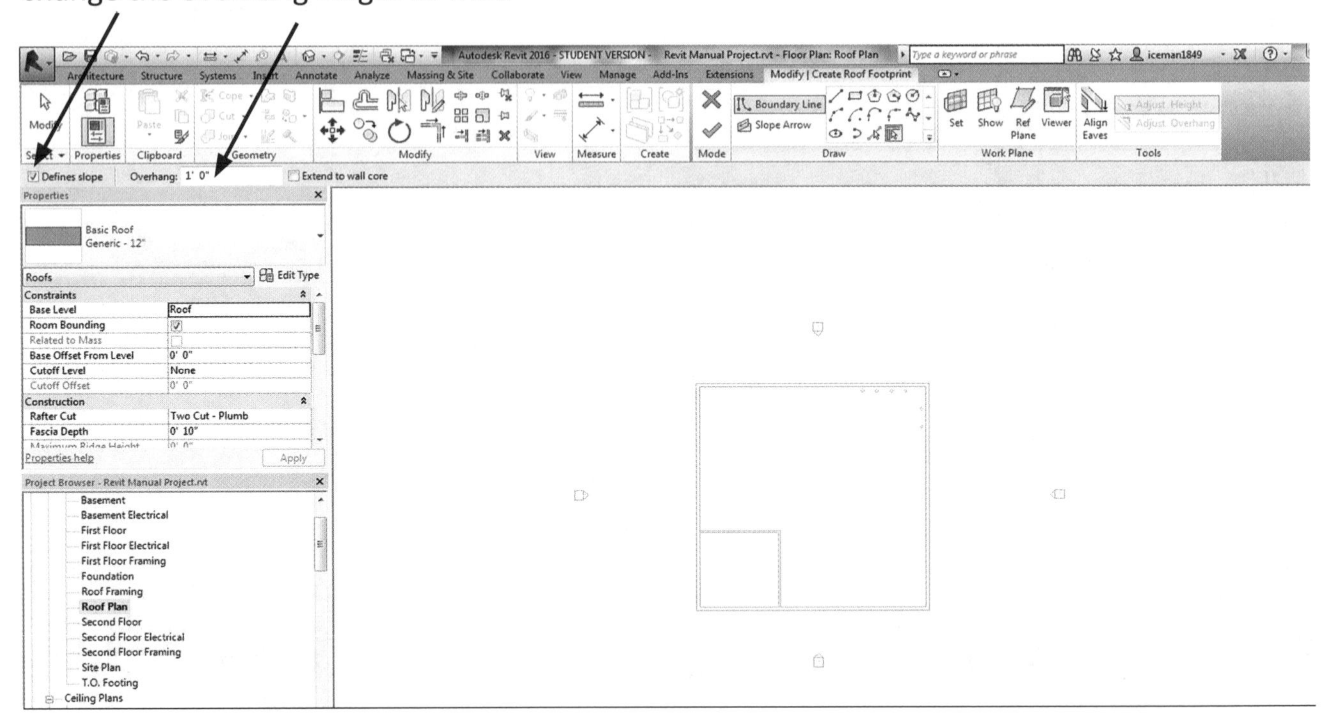

First, you need to pick a roof type in the Properties Box, such as **ASPHALT SHINGLES ON RAFTERS,** or **GENERIC** if you don't know which roof type you'd like to incorporate into your design.

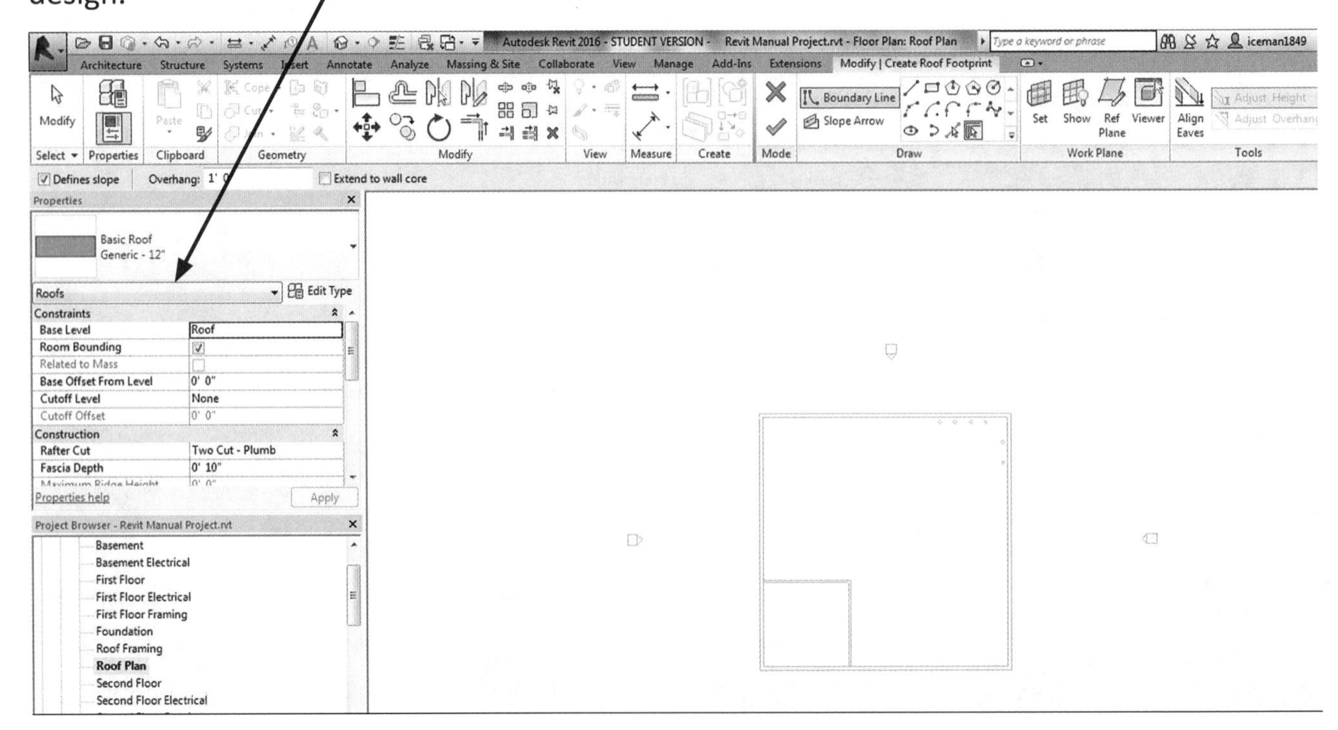

Revit defaults next to drawing the roof by allowing you to pick walls as an efficient way of moving around the house or building.

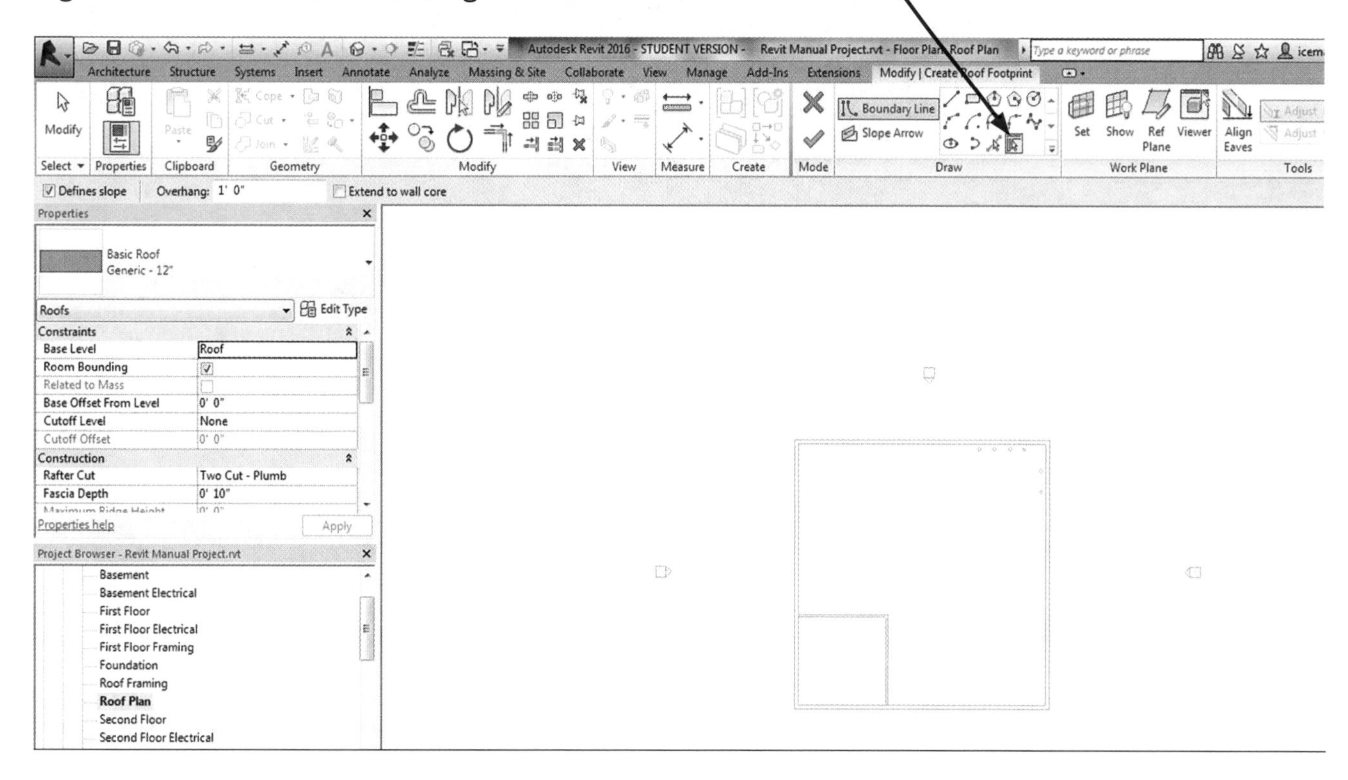

As you did with the other roof types, you can simply pick the outside edge of the house or building, and you will notice the 1'-0" overhang is automatically placed. This saves a great deal of time.

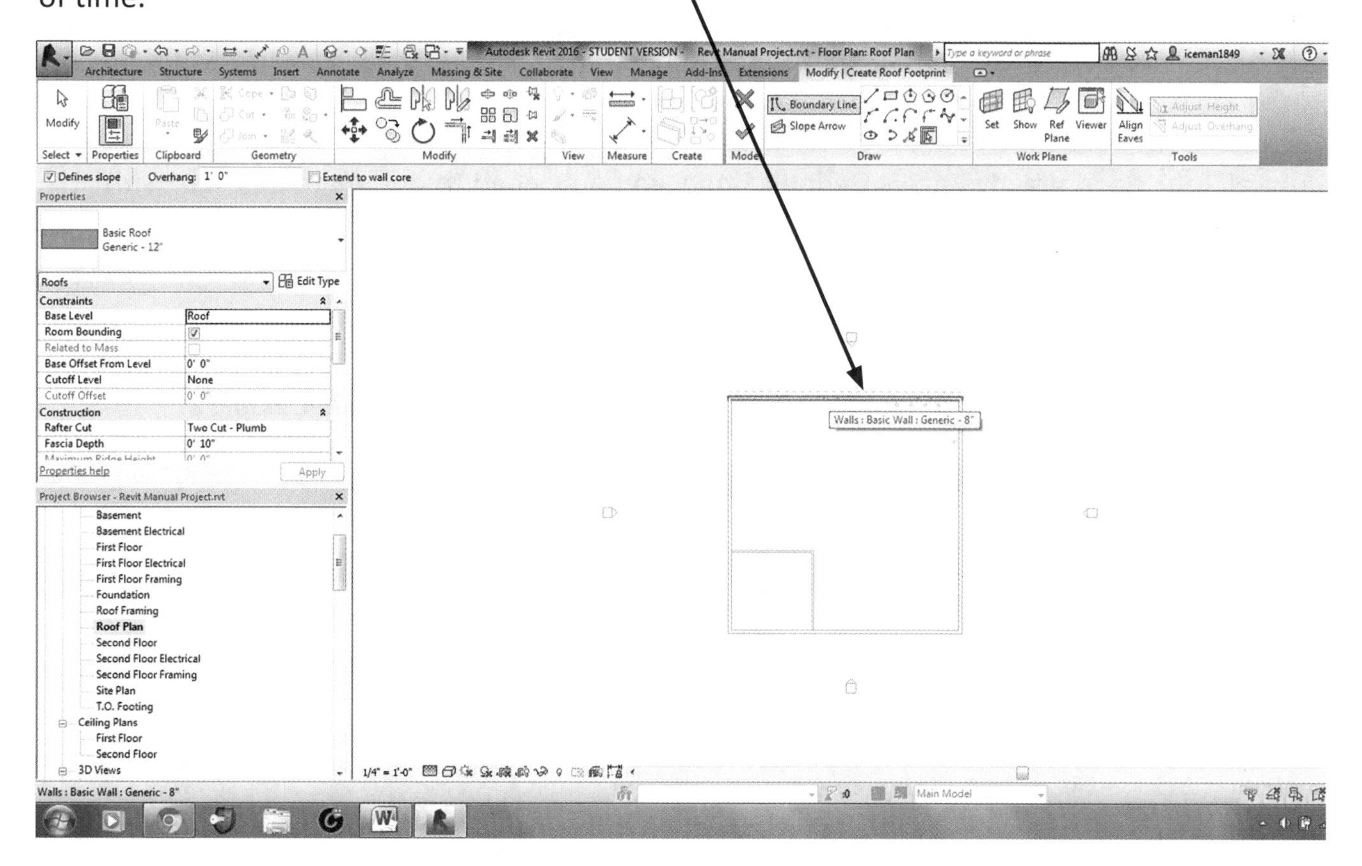

The finished Roof Footprint will look like the following:

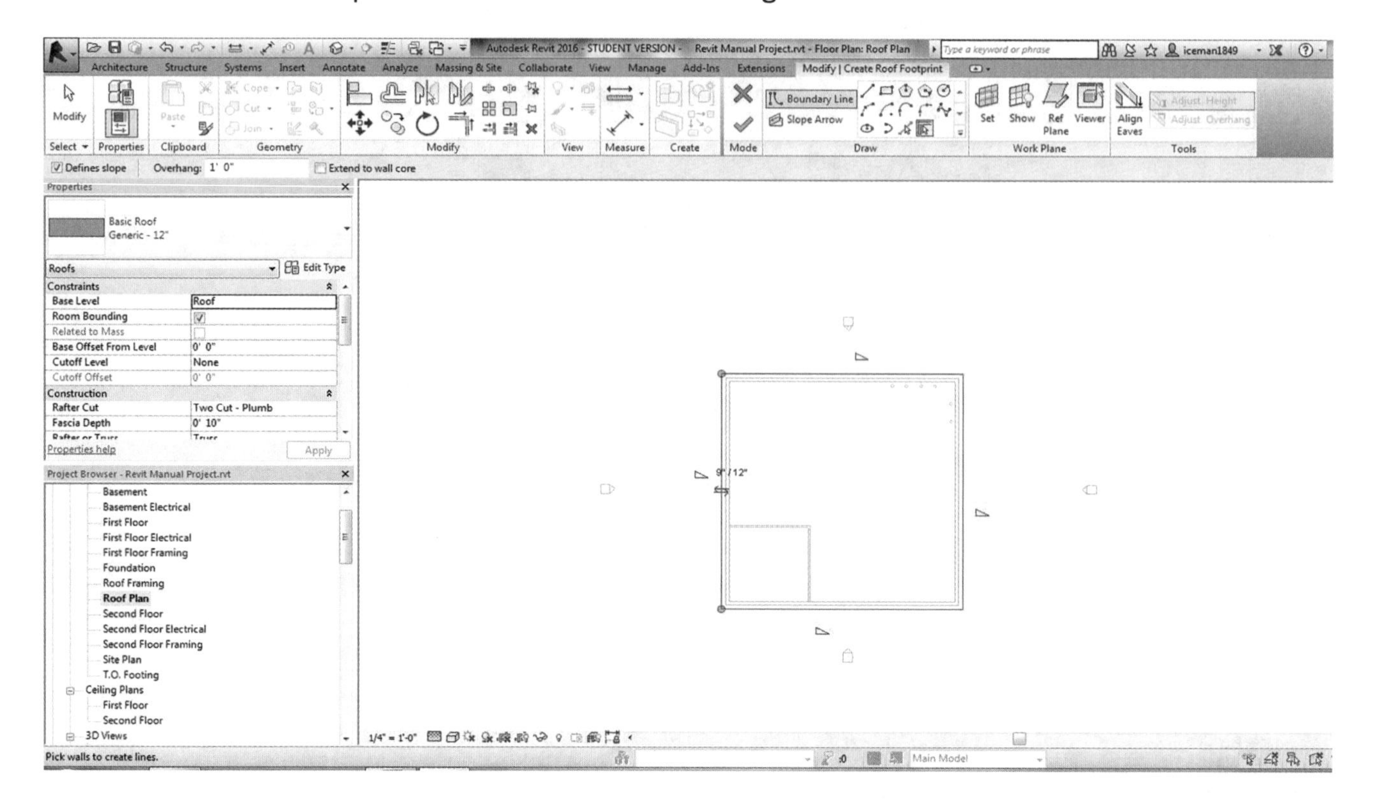

NOTES

Notice that the Defined Slopes are automatically placed on all four sides with the Pitch Slopes noted. As noted in the previous roof design, the Gable Roof, Revit defaults to having all 4 sides with* DEFINED SLOPES *activated, so the start of your roof designs will all look the same from the beginning. Don't be alarmed. This is very common.

Since all four sides already have a Defined Slope, you only need to designate one side with a pitch or slope for a Shed Roof. You can decide which side to keep the slope on, and which three to turn off.

REMINDER

Do* not *hit the* ESC *Key. That will cause other unpleasant issues.

Begin by selecting the first slope line you want to remove by clicking on the magenta line on the side of the house or building from which you want to remove the slope direction. Once you click on the magenta line, it should turn blue, or a different shade.

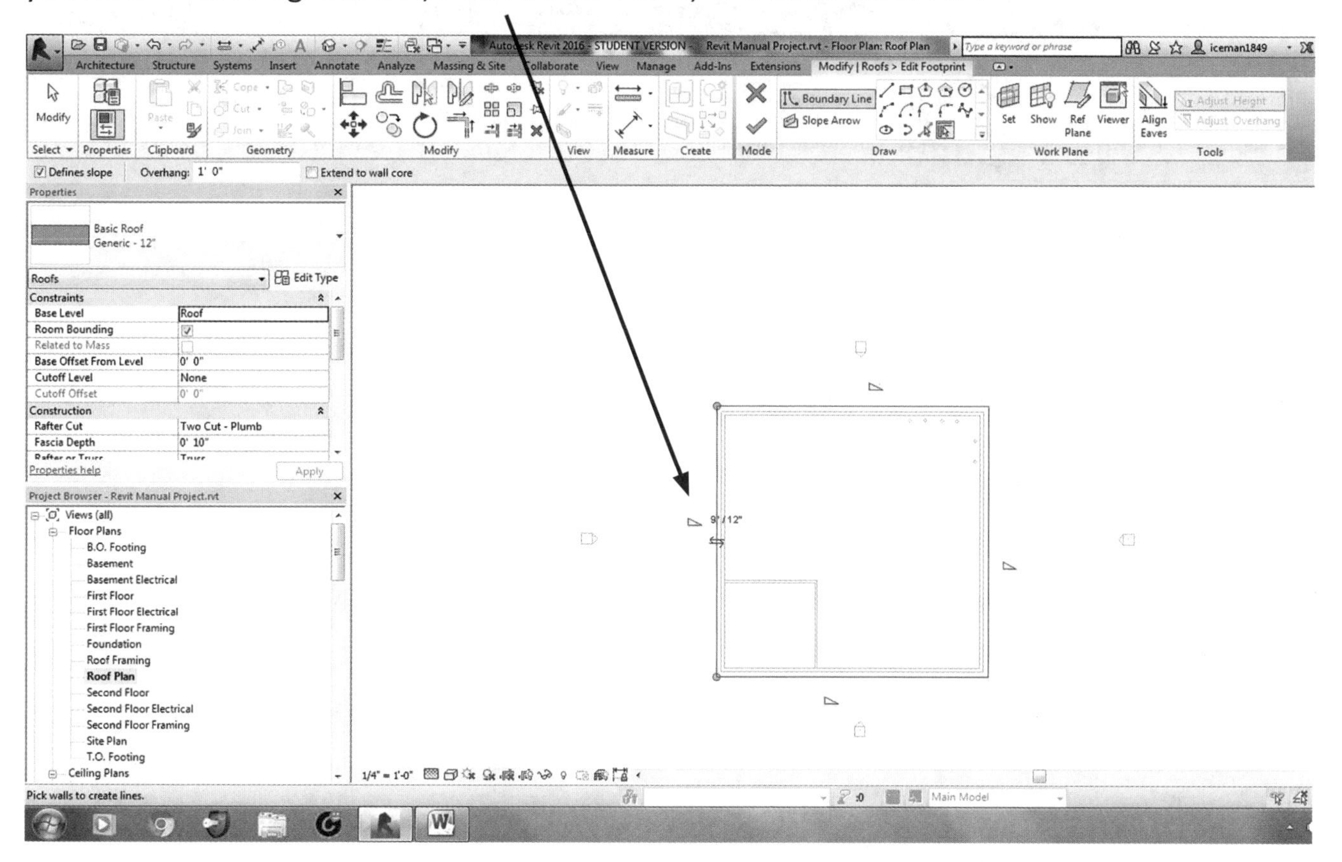

Next, select the *other* two sides of the house roof or building to remove the slope directions from these.

NOTES

The magenta lines should once again turn a different color or shade.

Once you have selected the three sides that you do *not* want to have a Defined Slope, go to the Options Bar, and uncheck the **DEFINES SLOPE** Box. This will turn off the slope of the roof.

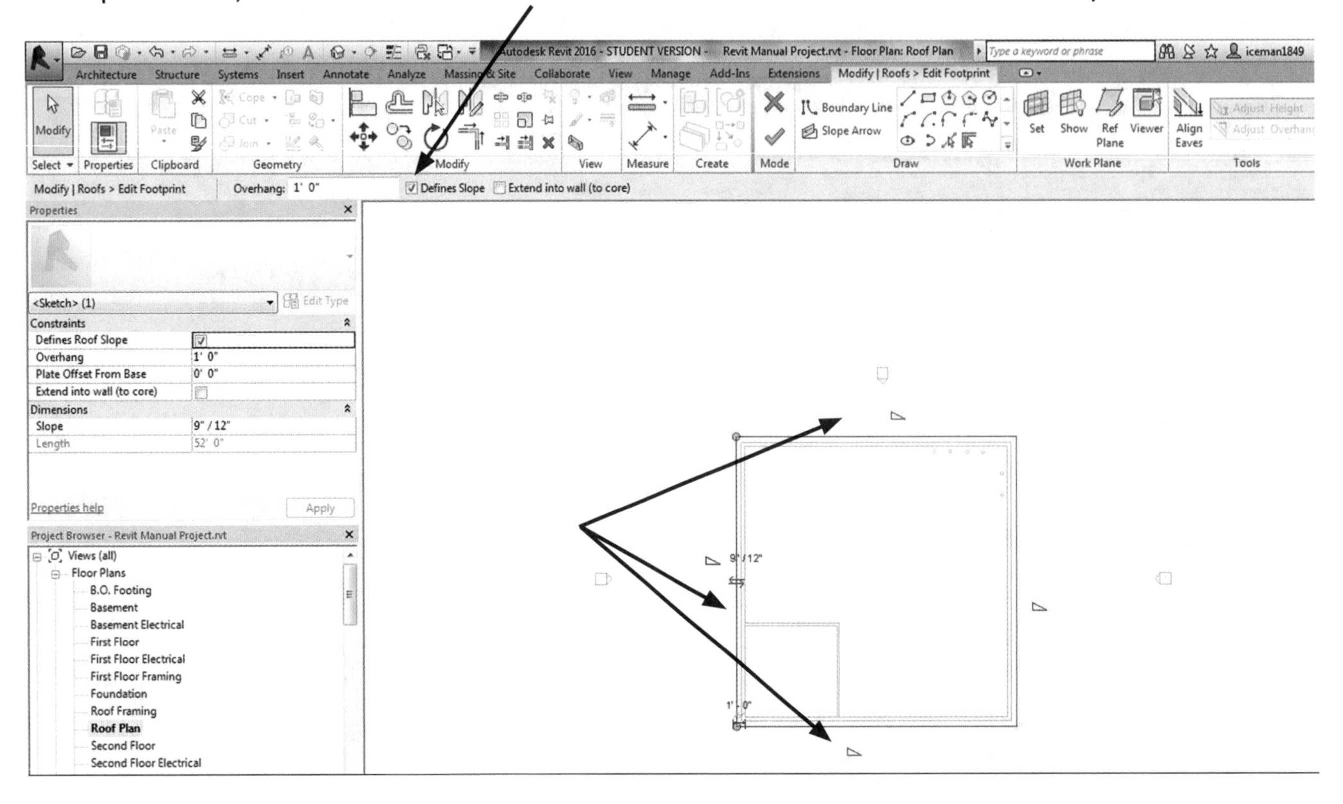

Once the **DEFINES SLOPE** Box is unchecked, click the **green CHECK MARK** to accept or complete the roof design layout.

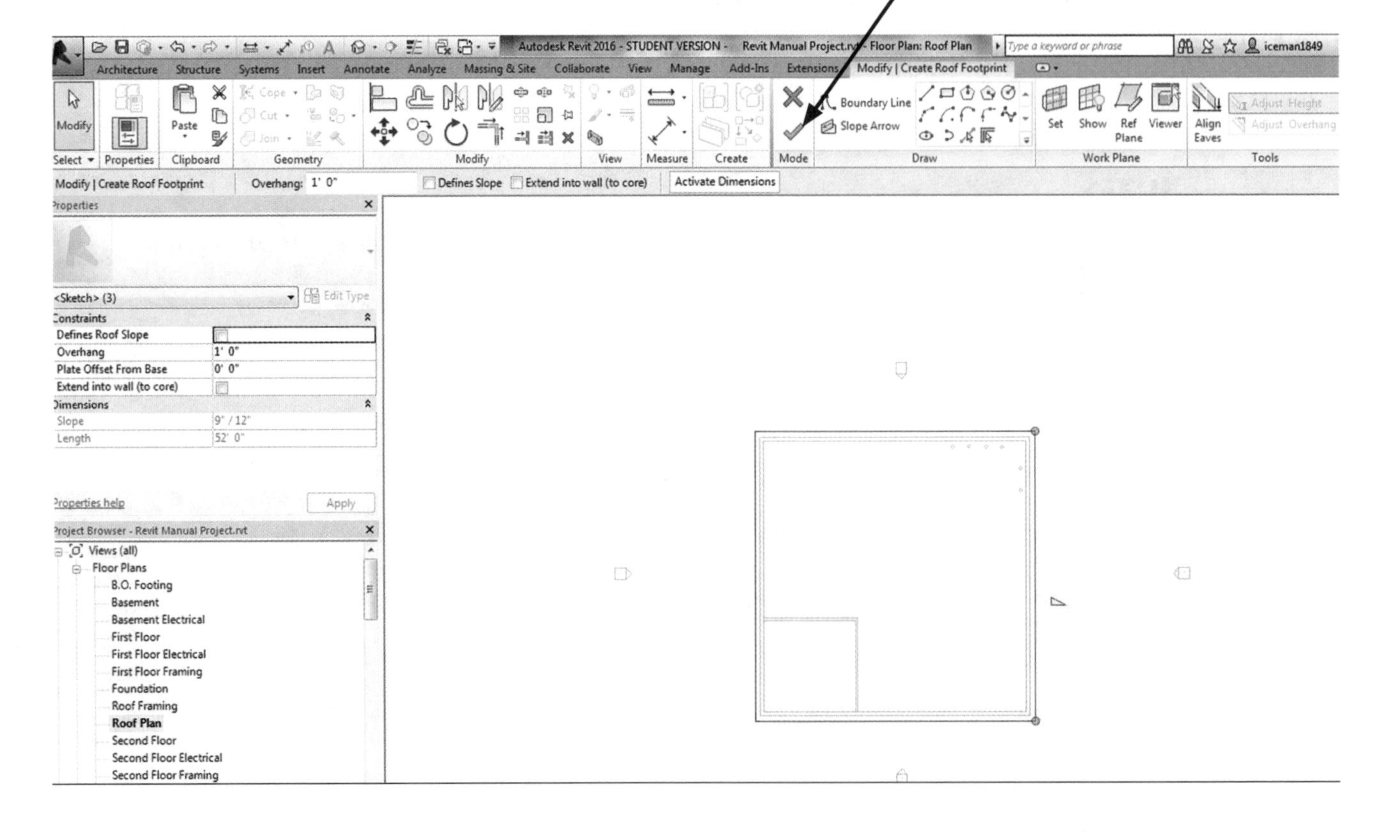

Once again, you will get a warning window asking if you want to attach the highlighted walls to the roof. Click **NO**. This will allow you to modify the walls or roof at a later date, without having to detach the roof.

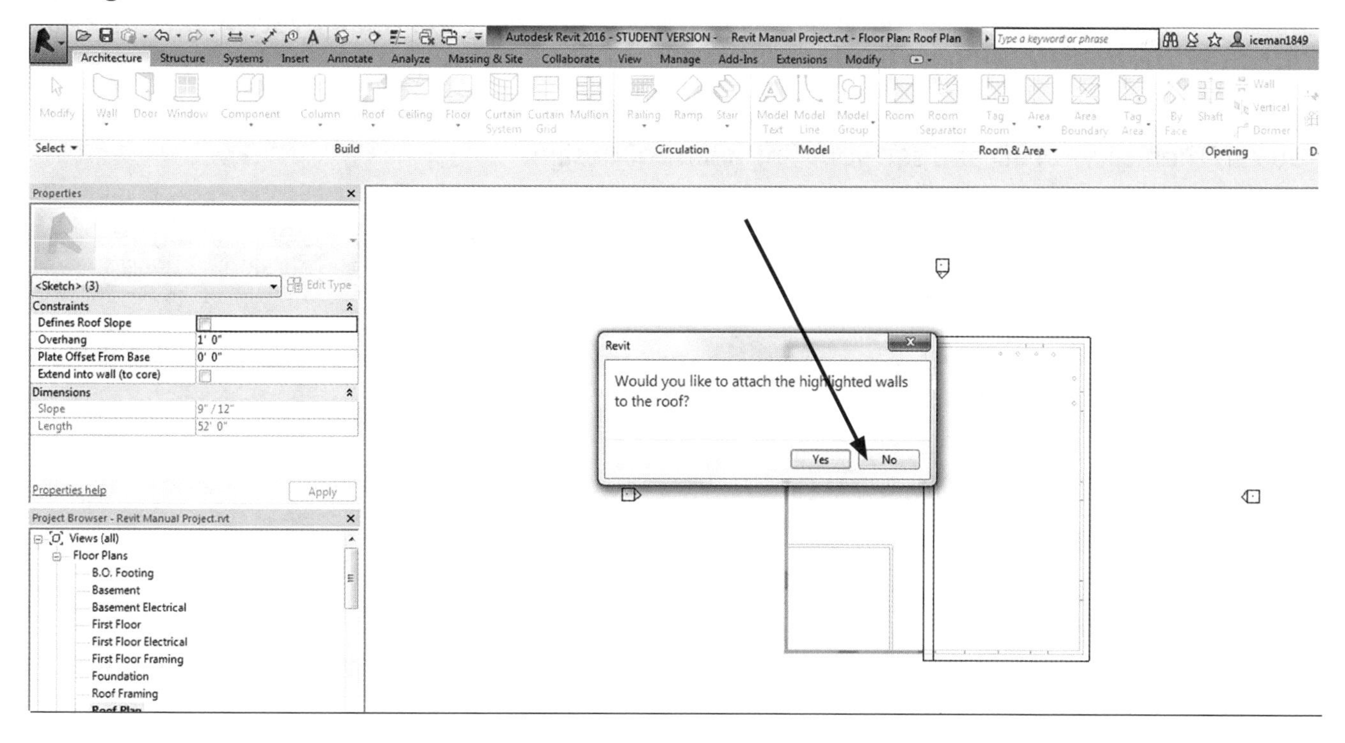

WARNING

With a Shed Roof, you will only see a partial of the roof because the roof is now at an angle of a slope.

The finished Roof Footprint should look like the following:

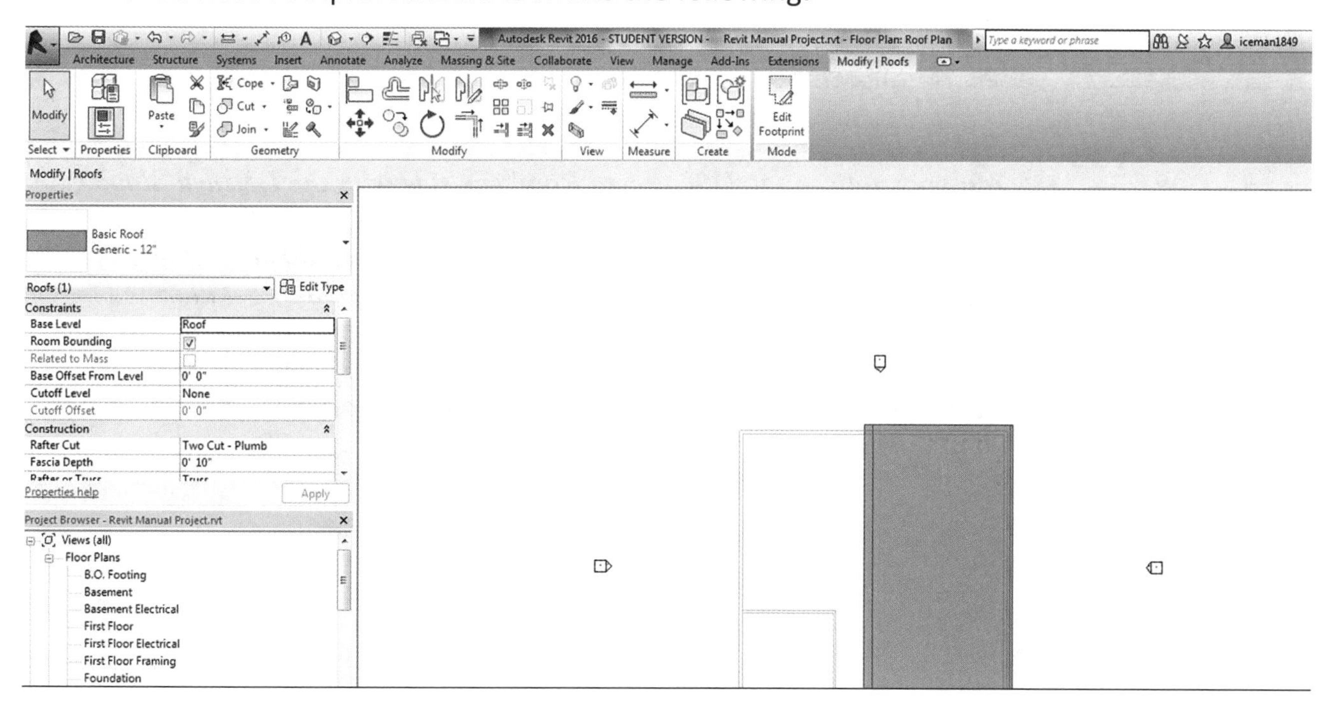

NOTES

The reason you are only seeing *half* ***the roof compared to the other roof designs is that the Slope Roof is only connected to*** *one* ***side of the house or building and is at a steep angle. Therefore, you can only see half of the roof.***

You can then click on **3D** and see your finished roof design.

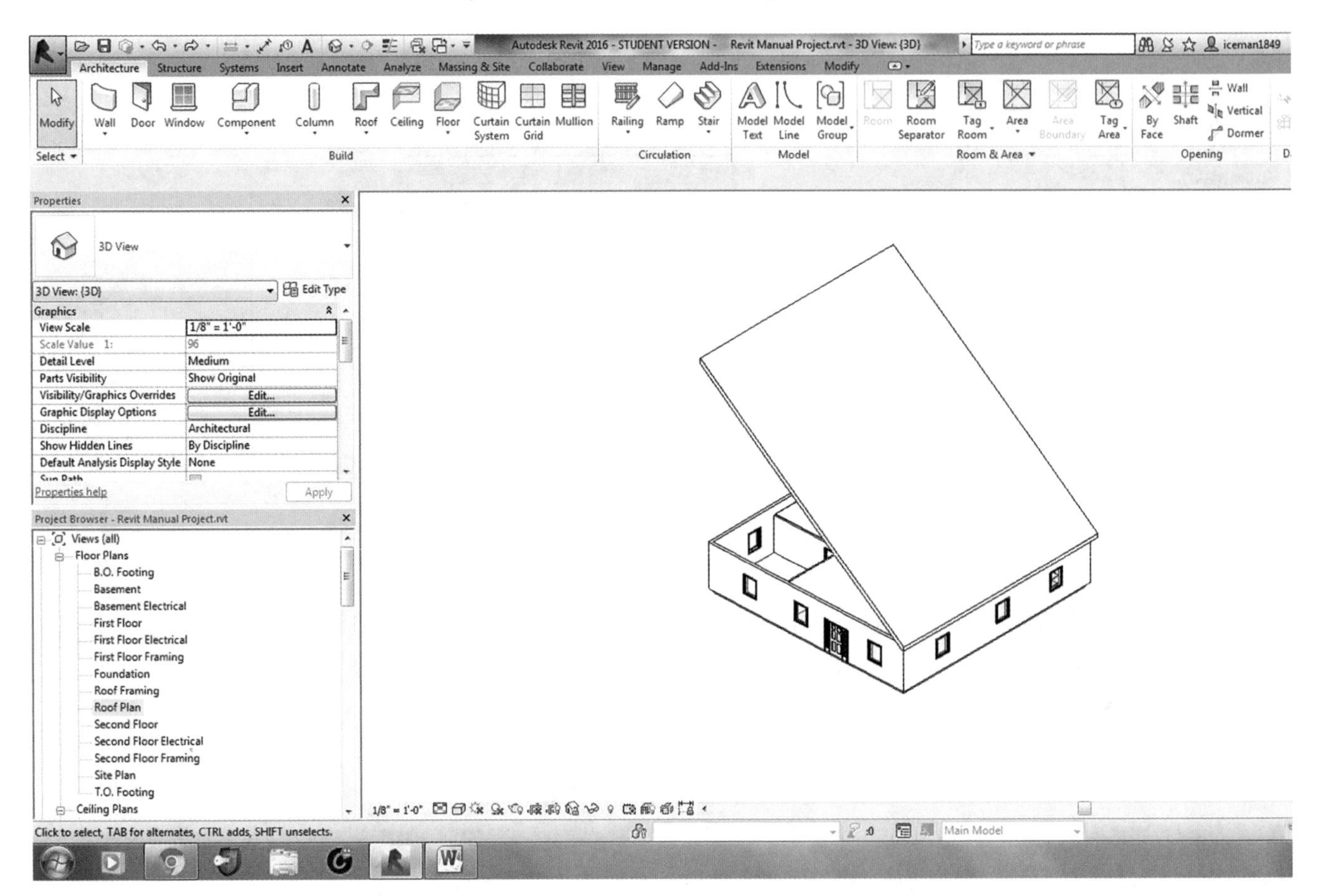

TIPS

You may notice that with a Shed Roof, the side with the Defined Slope is connected to the house or building, and the sides that don't have any pitch or slopes might be *really* ***high in the air. This is an indication that you will need to adjust the pitch or slope of the roof. Please don't get alarmed or shocked when this happens, it is very common. We are going to walk you through the steps to lower the pitch. You can do this using some simple steps, which makes you realize that Revit is very powerful and time-saving all at once.***

While you are still in the 3D view, simply click on the **ROOF**.

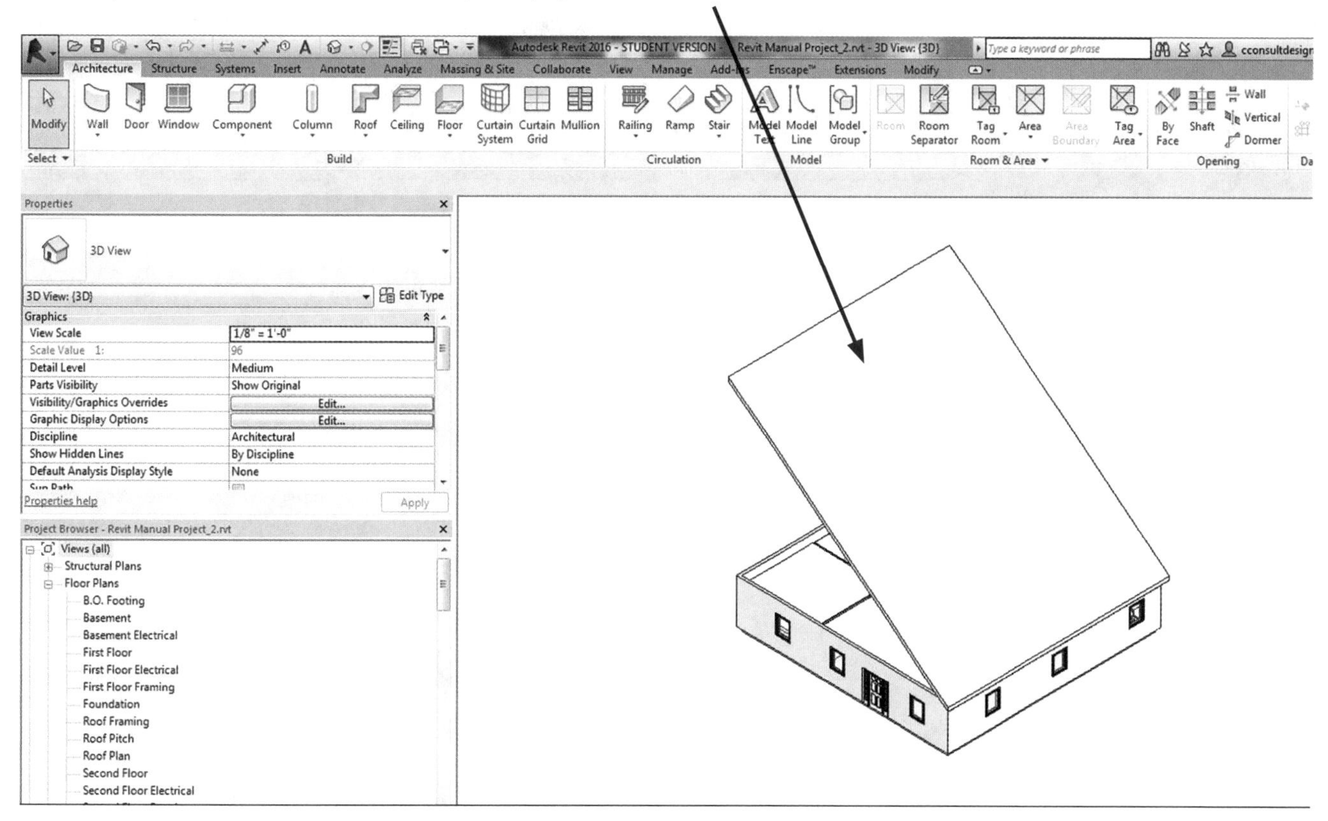

Then click on the **EDIT FOOTPRINT** tile in the ribbon.

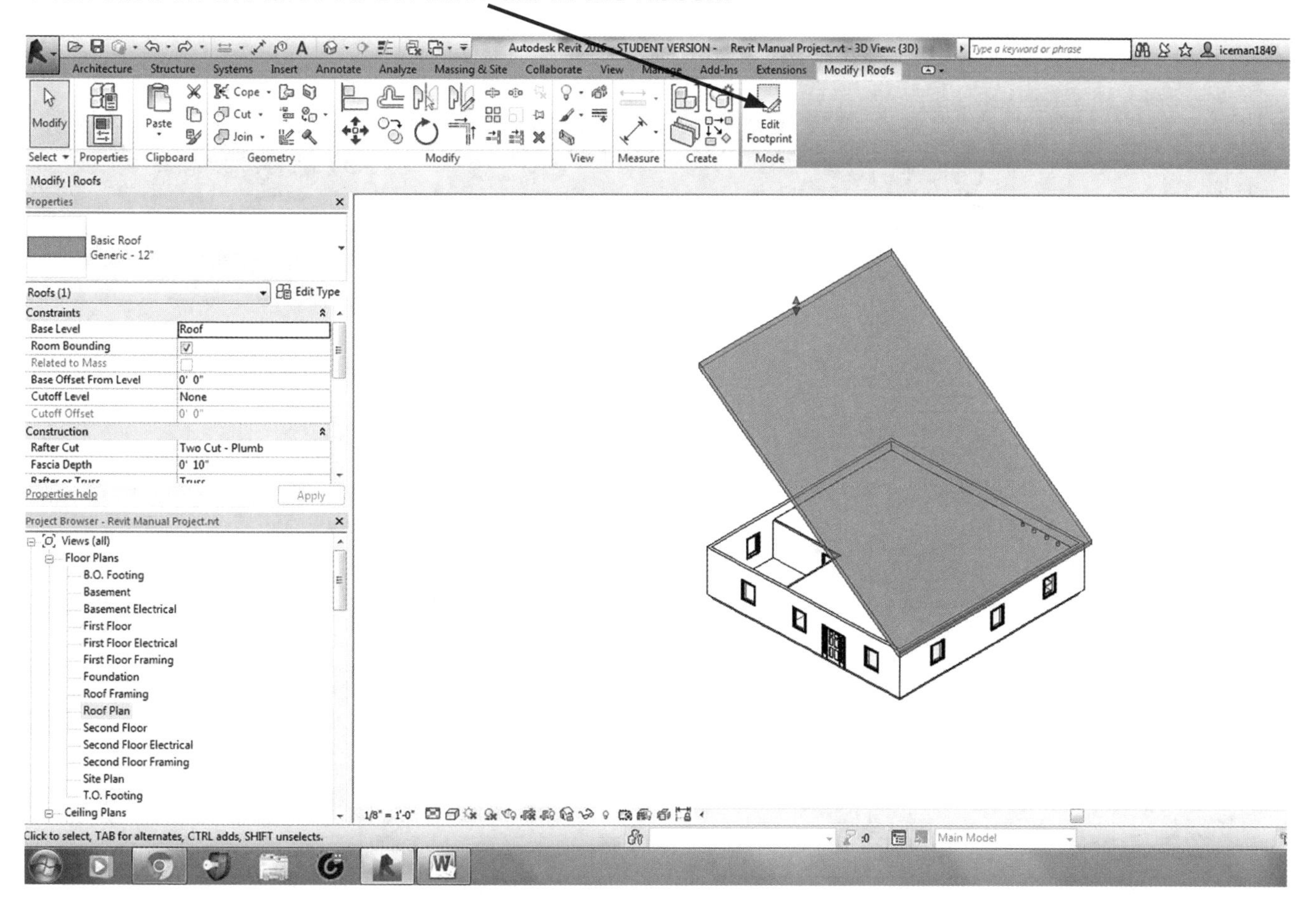

It will become a line drawing of your roof sketch.

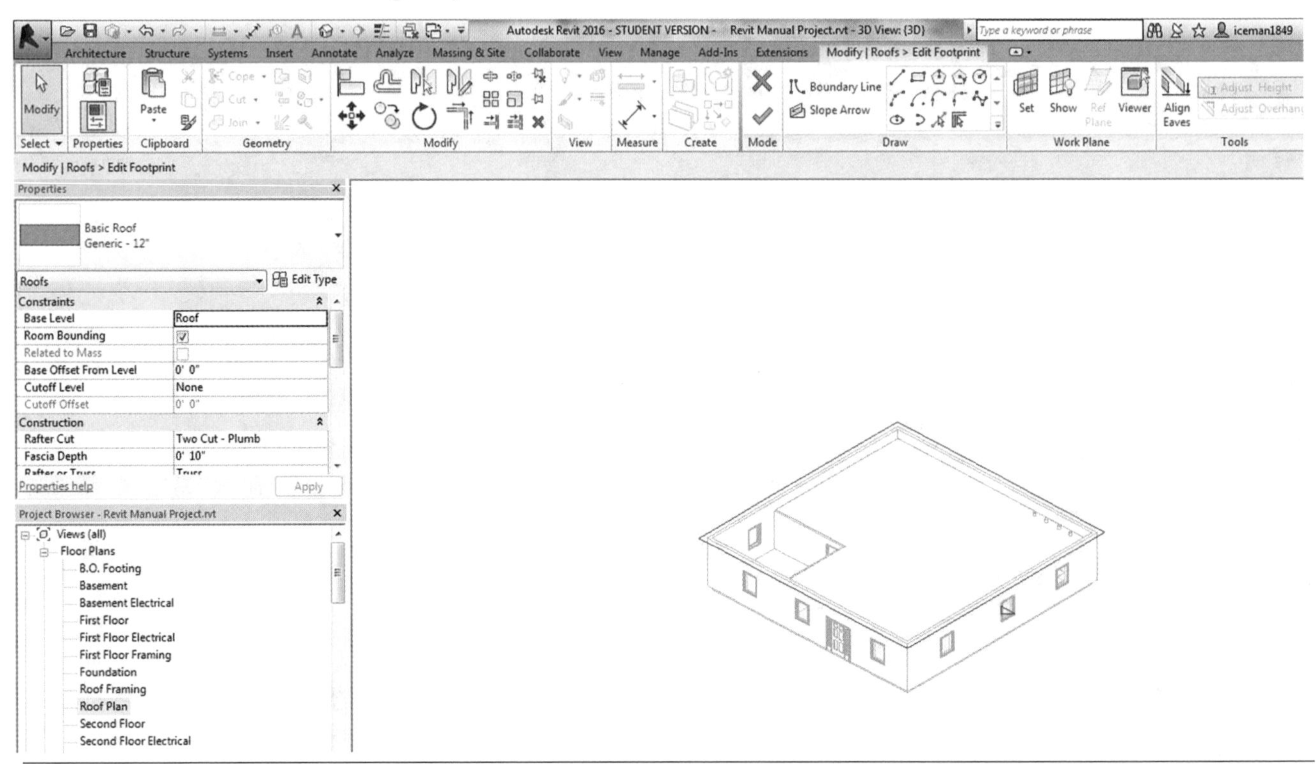

Once the Roof Footprint appears, you will see a Right Triangle symbol which is the slope symbol. Click on the **SLOPE** symbol to activate it.

Once activated, the slope angle will appear—**9/12**.

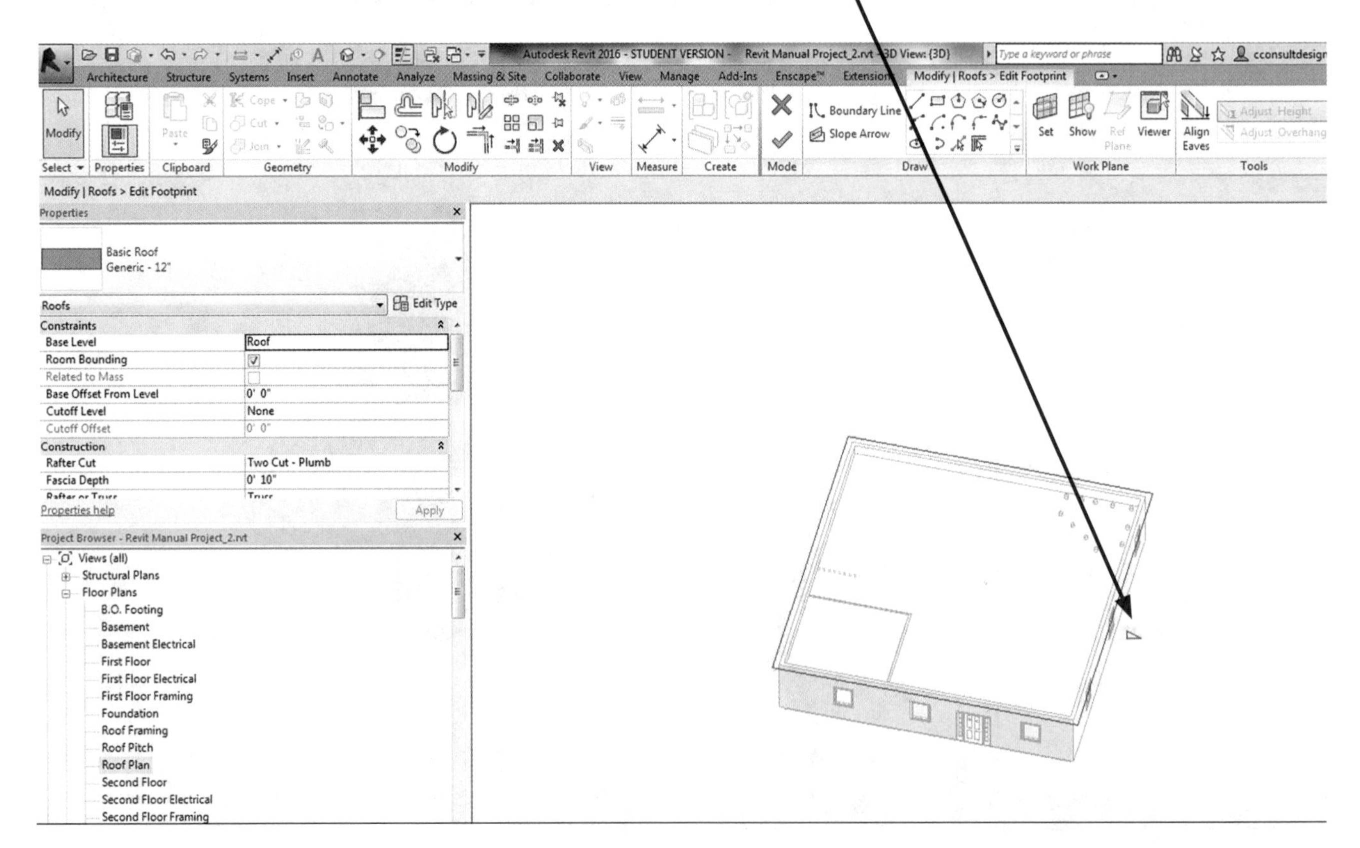

You can now change the slope angle by clicking on the number, and typing in whatever angle works for you.

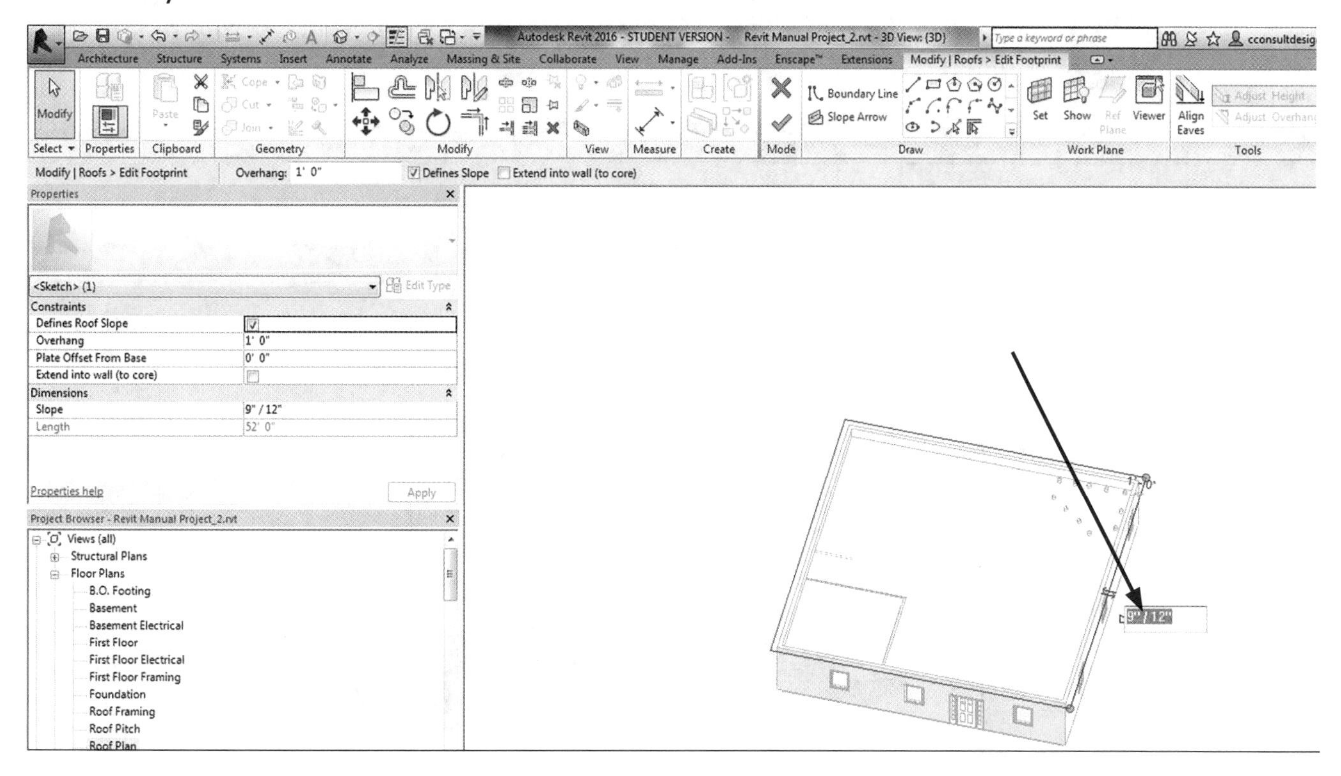

Click **ENTER** and then click the green **CHECK MARK** to accept or complete the new Roof Design Layout.

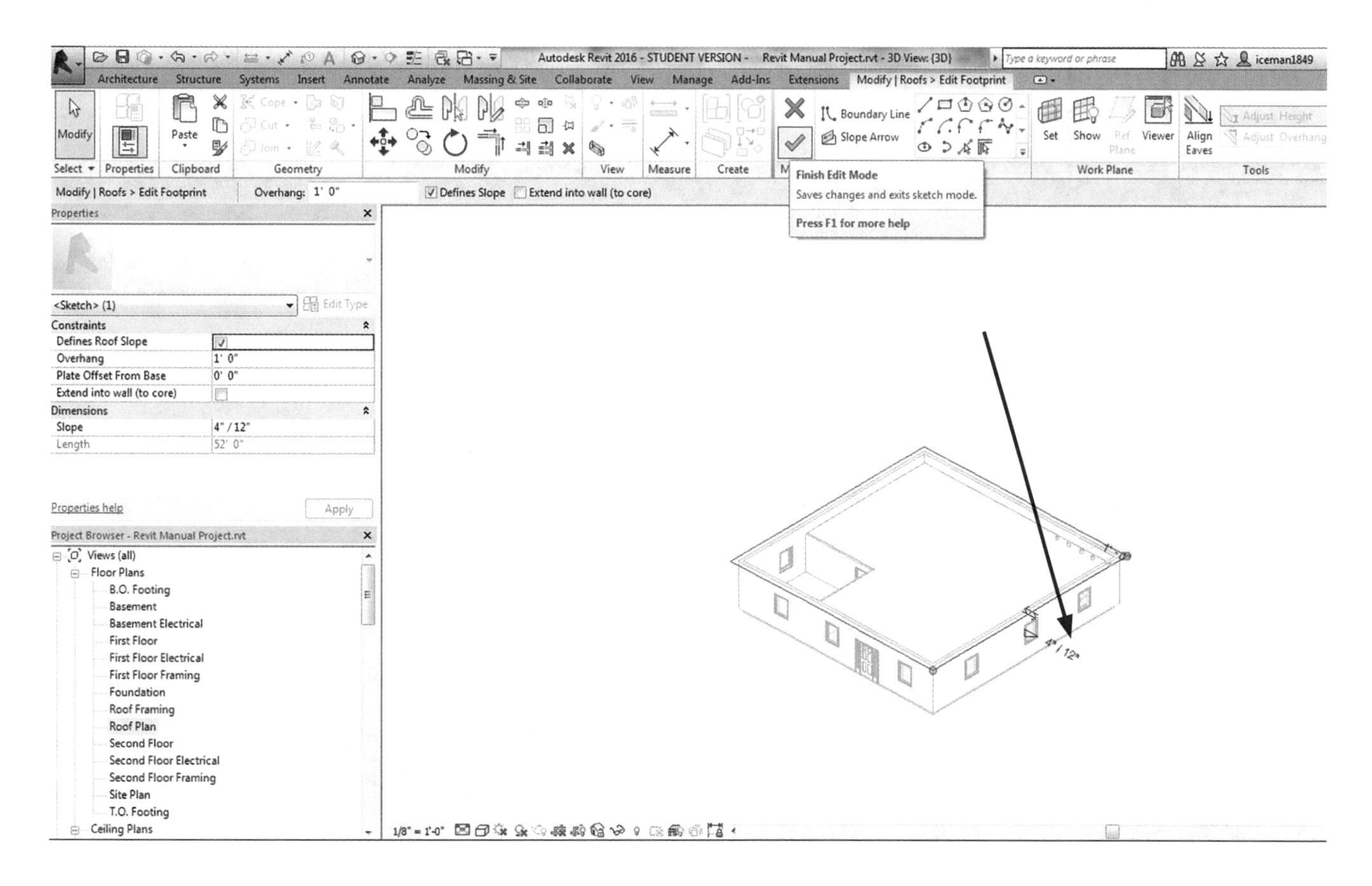

REMINDER

You will get a warning window asking if you want to attach the highlighted walls to the roof. Click* NO*, so you can modify the walls or roof at a later date.

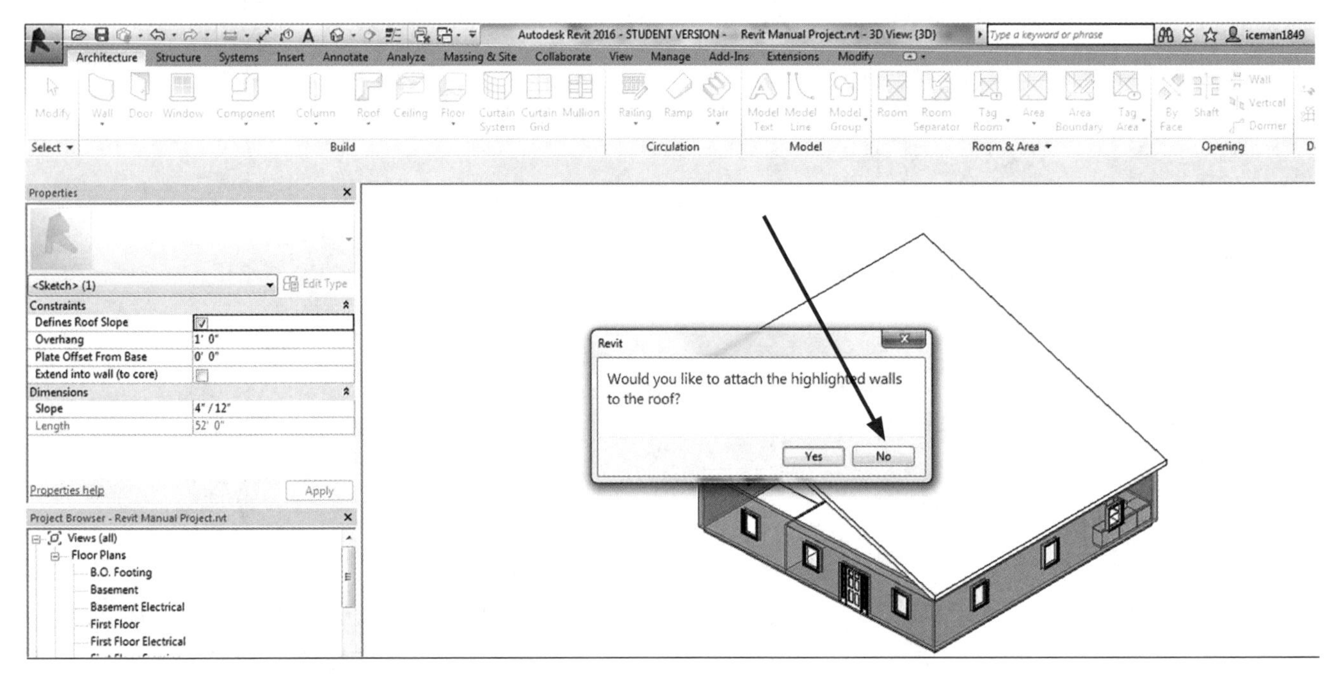

You will see the new slope or pitch angle in your 3D view.

NOTES

You don't have to click on the* 3D *button, because you by-passed that step.

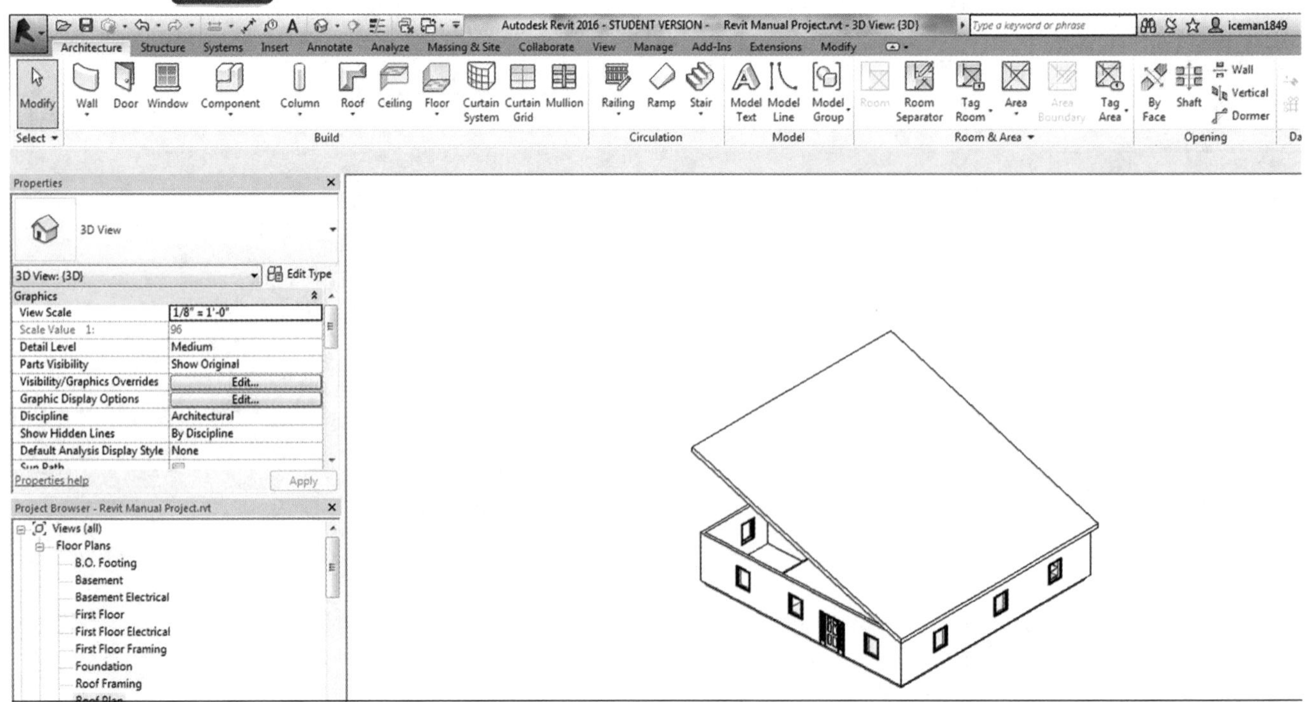

Another way to change the Roof Footprint is to click on the **ROOF PLAN** in the Project Browser.

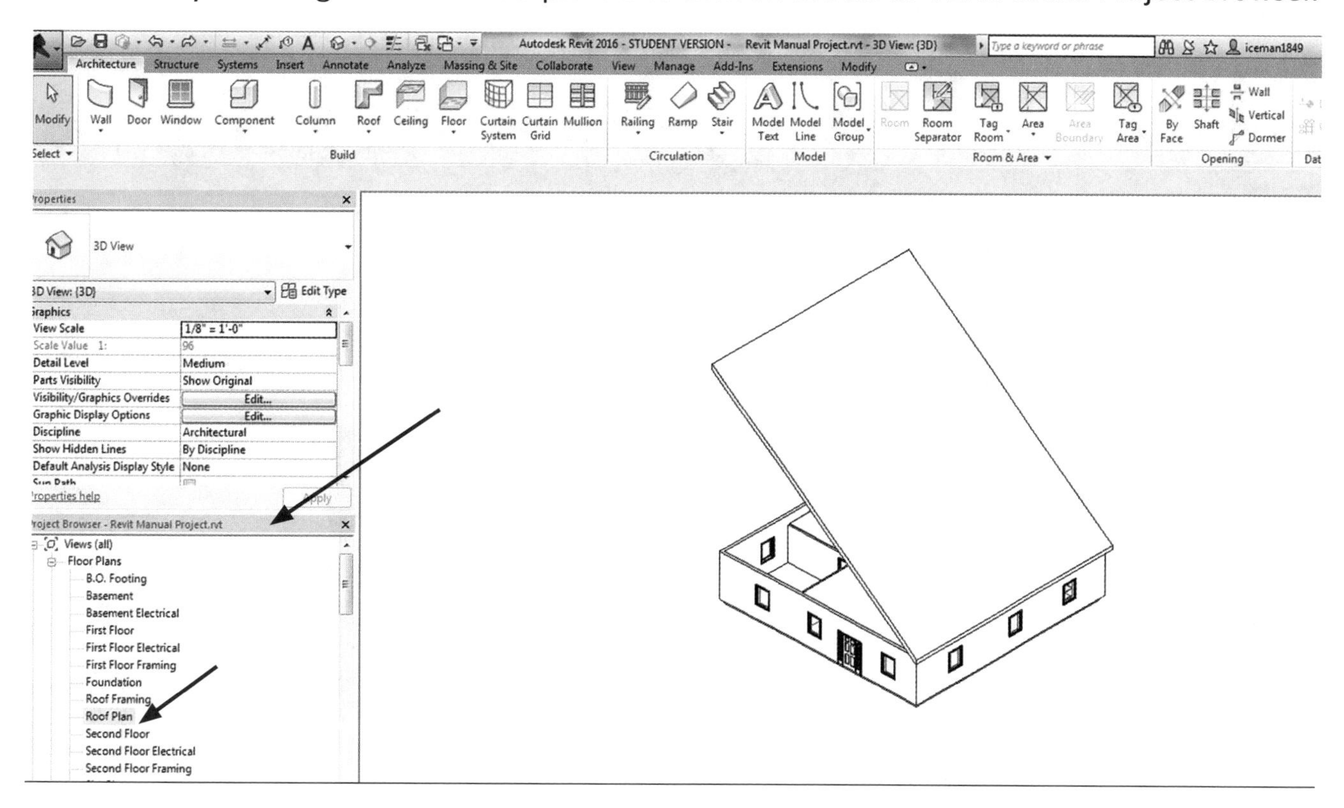

This will switch the view to a Roof Plan View, so you can select the roof itself.

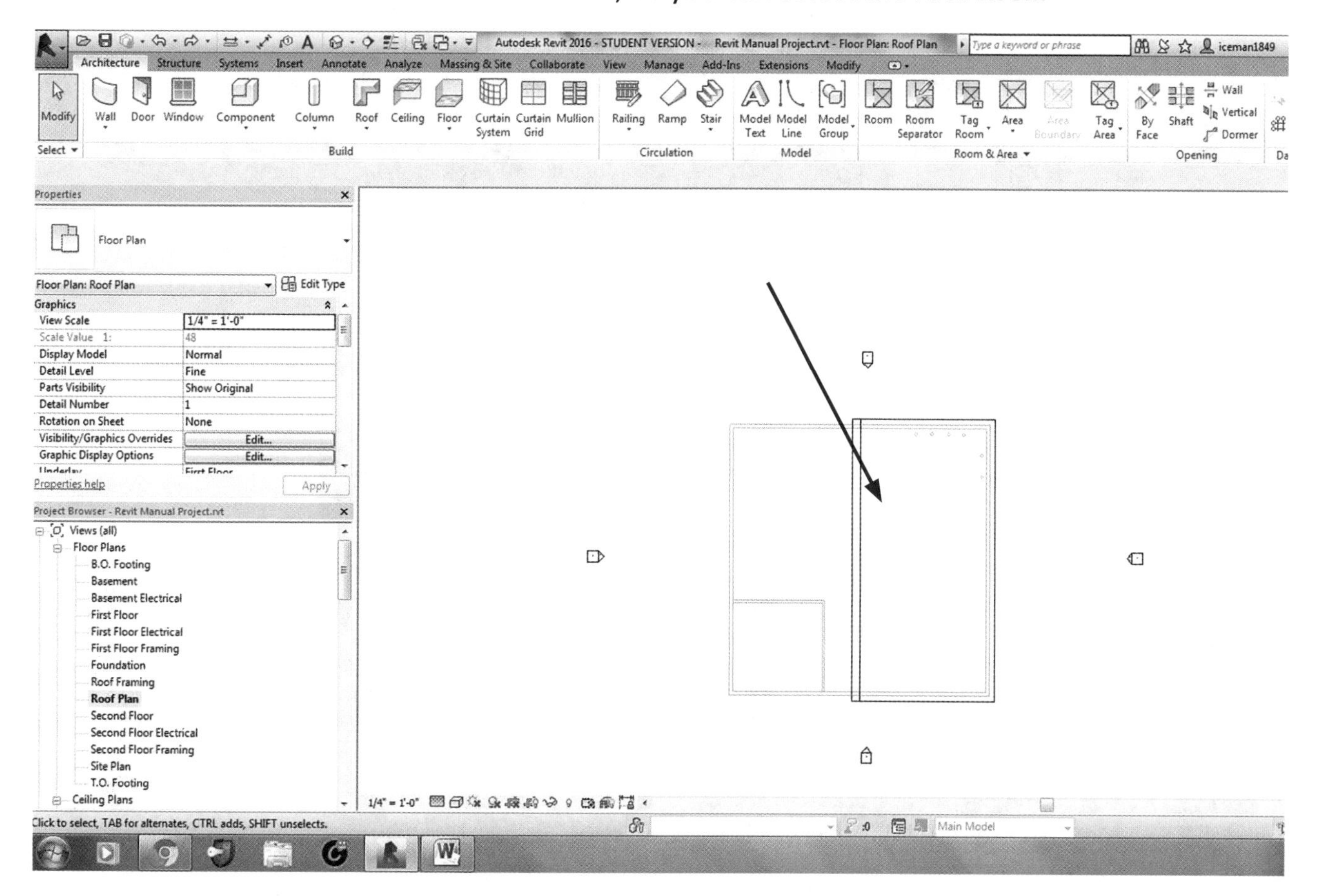

Click on the roof, and it will turn blue or a different shade.

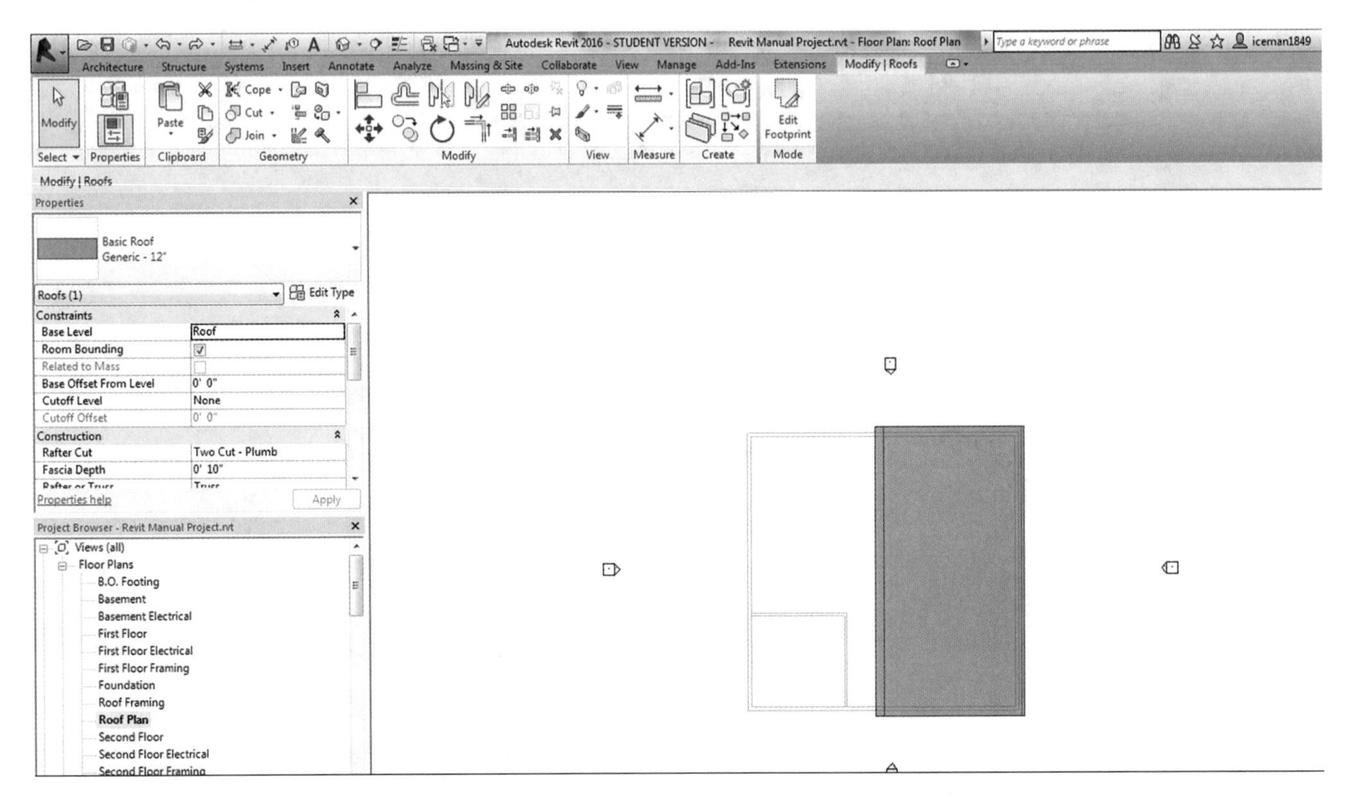

When you click on the roof, it will become a line drawing of your Roof Footprint. Then click on the **EDIT FOOTPRINT** tile in the ribbon.

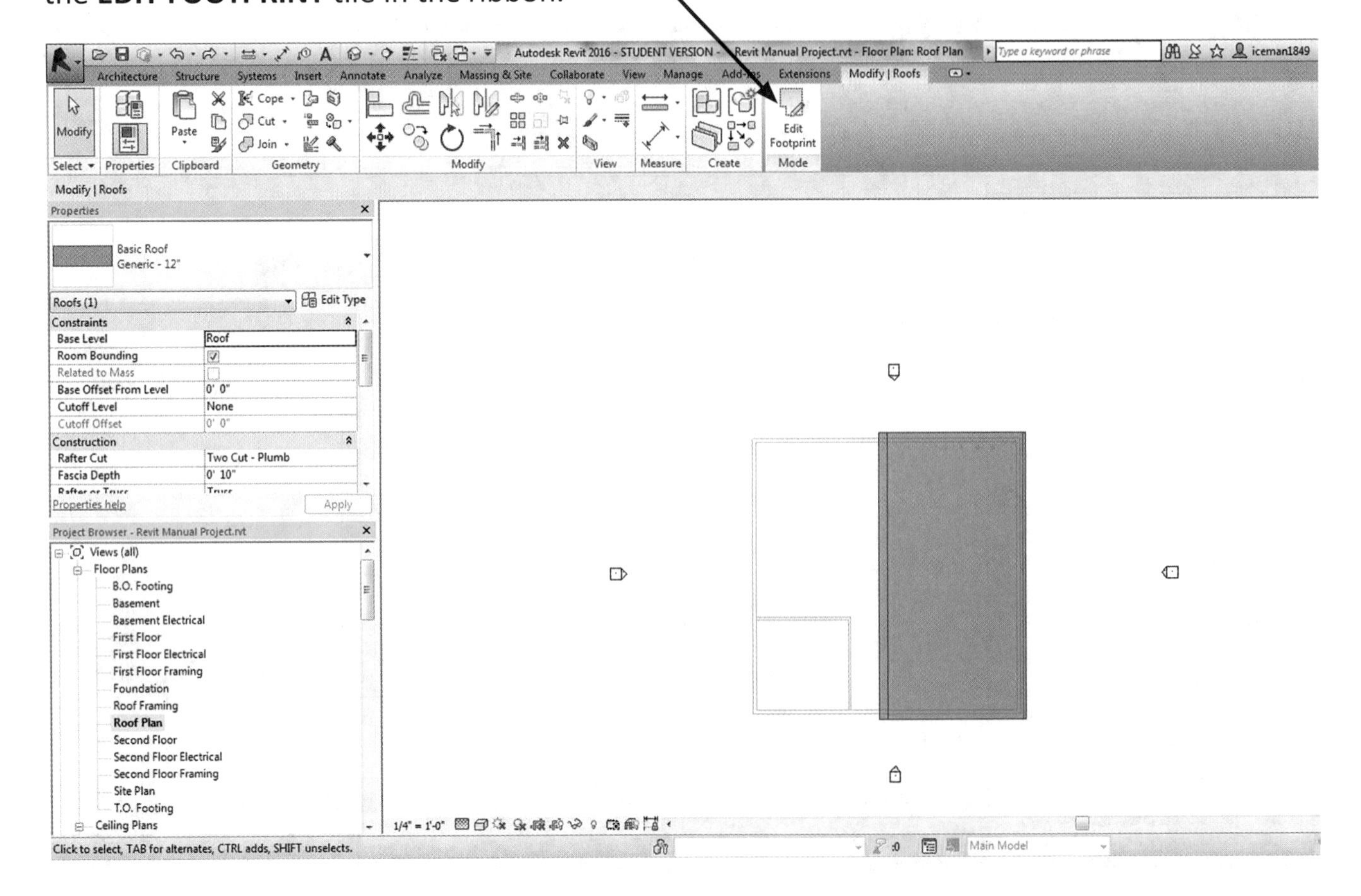

The line drawing will appear:

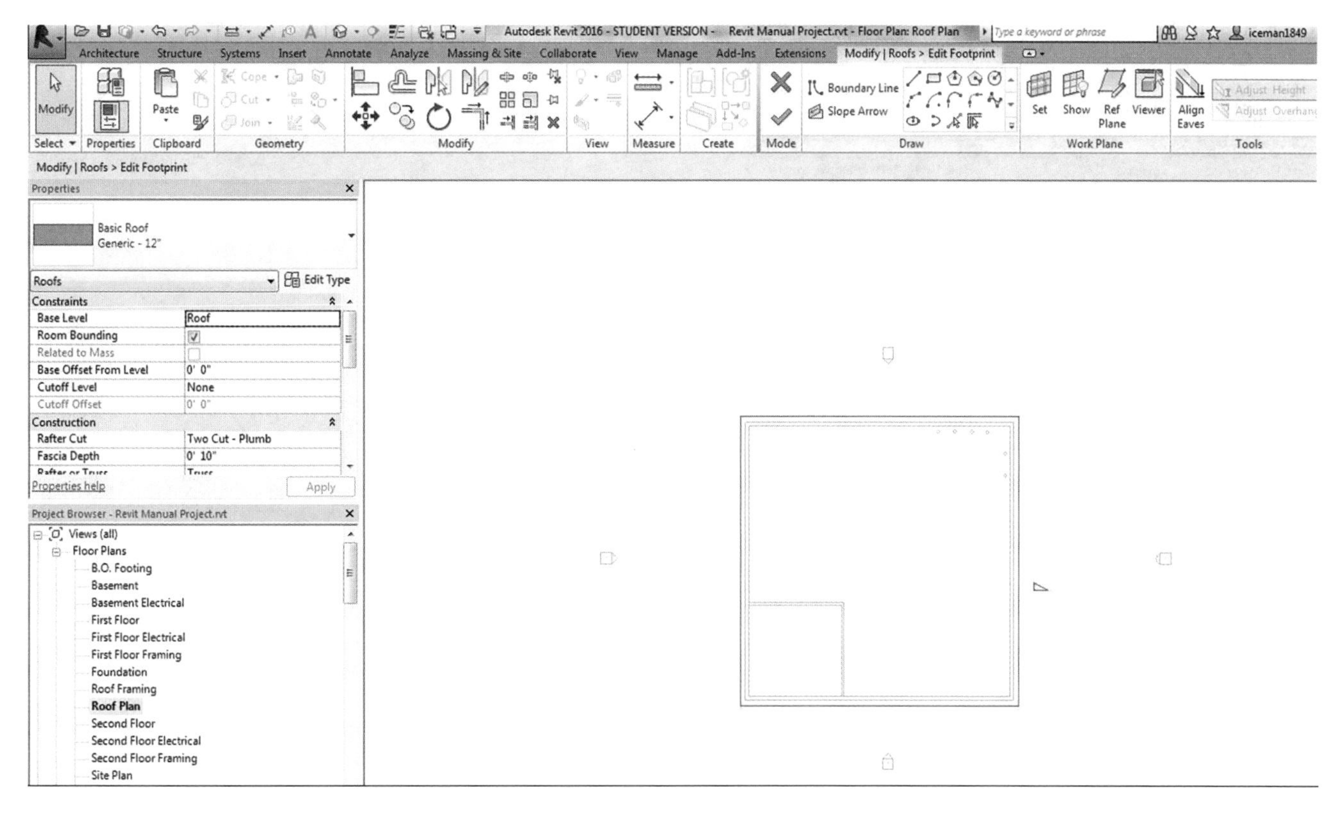

You will then see the slope or pitch symbol (a right triangle). Click on the line with the slope or pitch symbol.

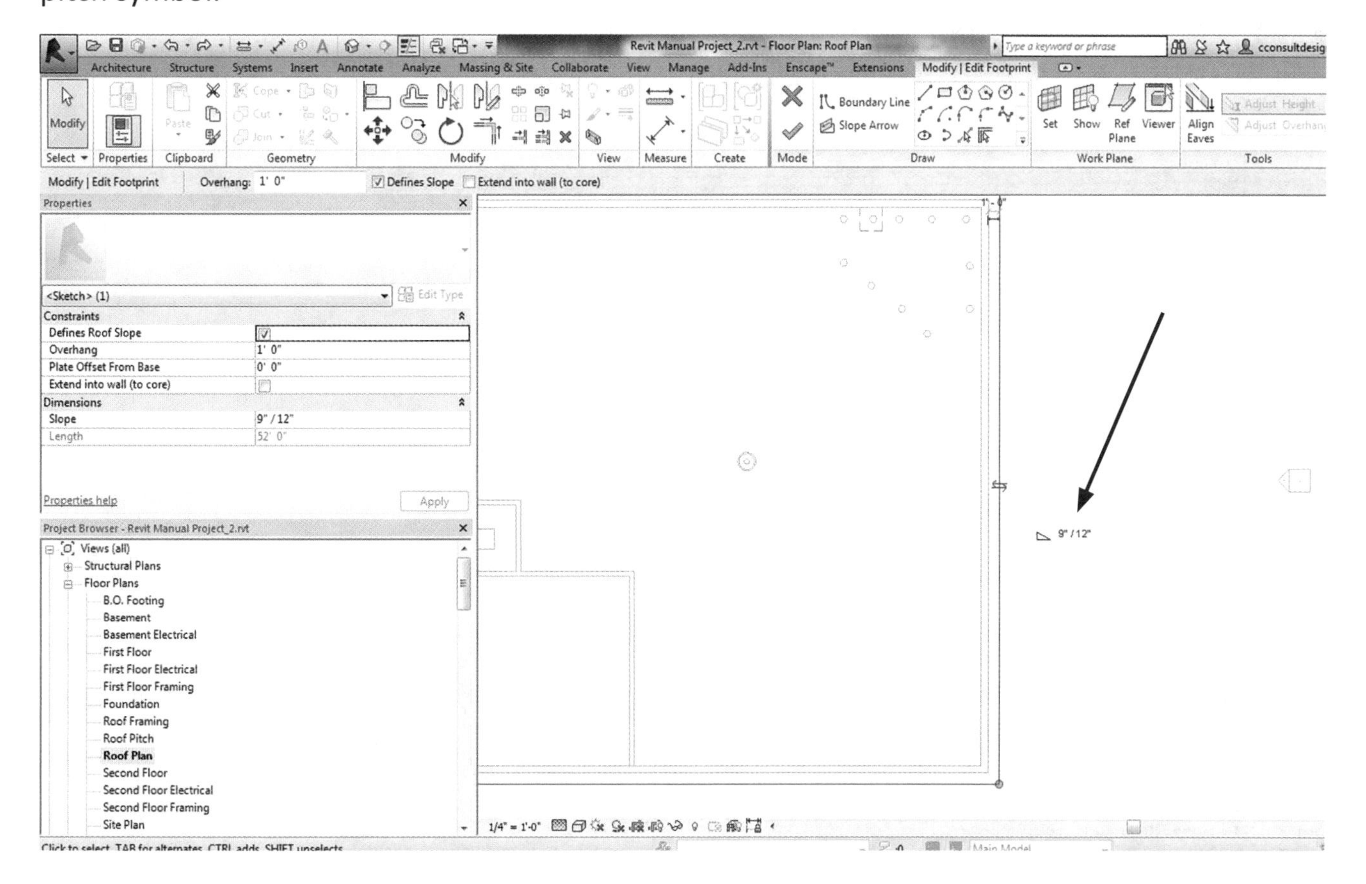

Now you can change the slope angle by clicking on the number and typing a new slope angle.

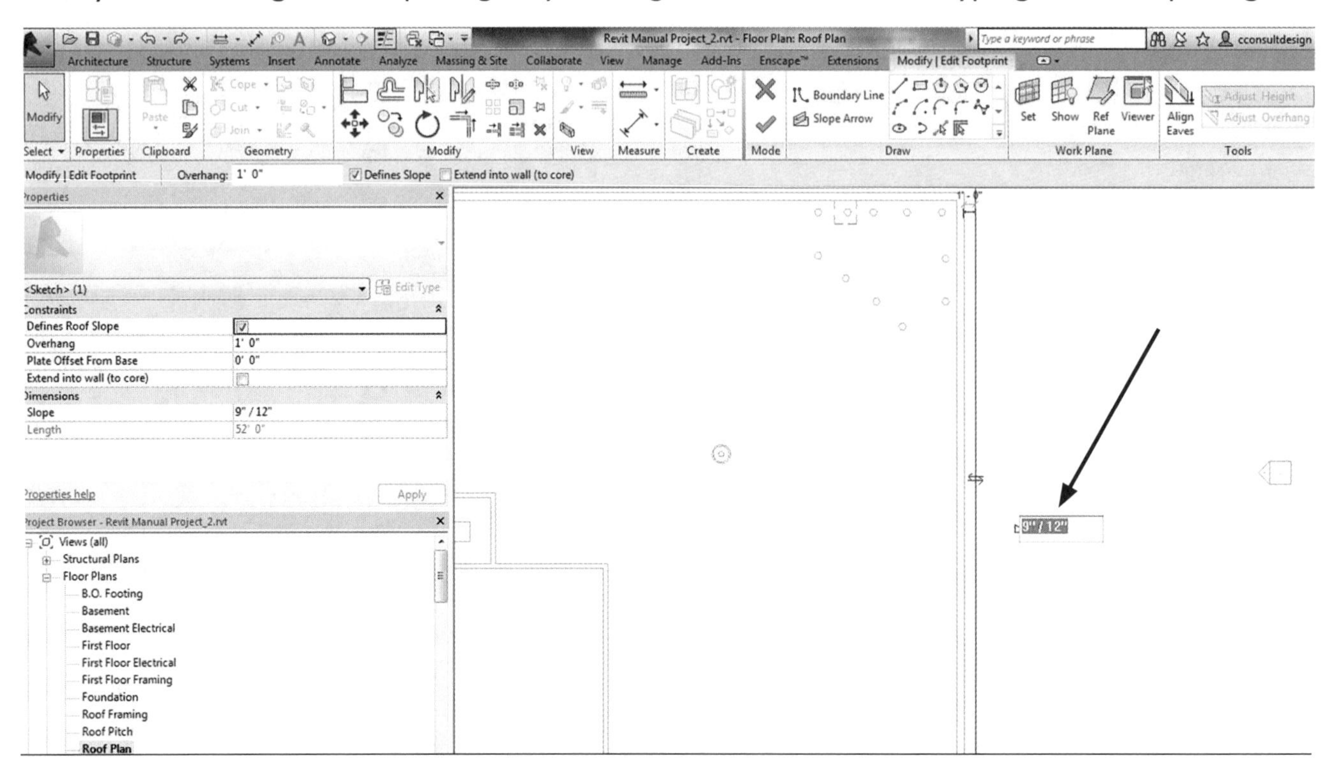

Click **ENTER** to accept the new slope or roof pitch and then click the green **CHECK MARK** to accept or complete the new Roof Design Layout.

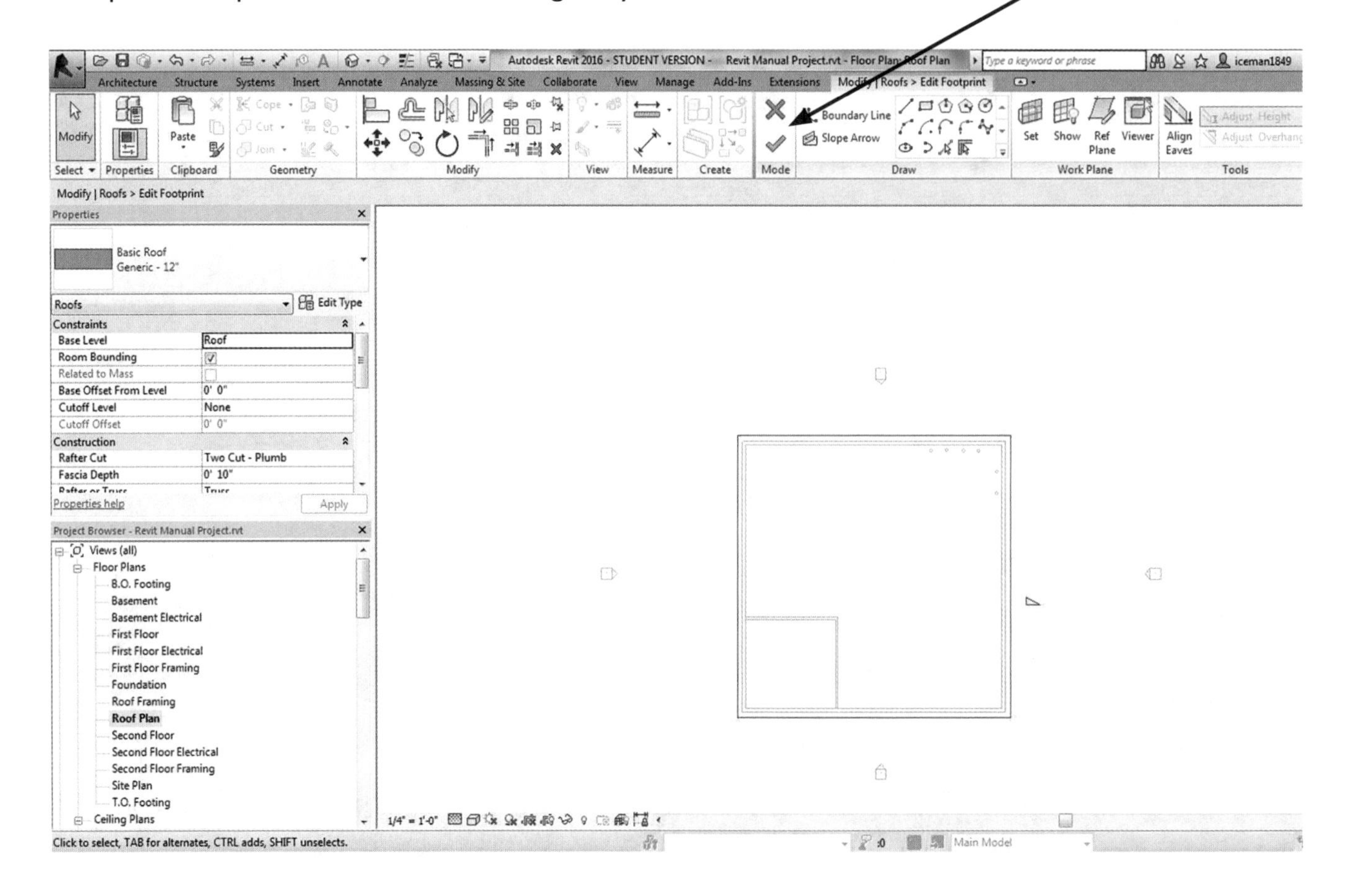

REMINDER

You will get a warning window again asking if you want to attach the highlighted walls to the roof.

Click **NO,** so you can modify the walls or roof at a later date.

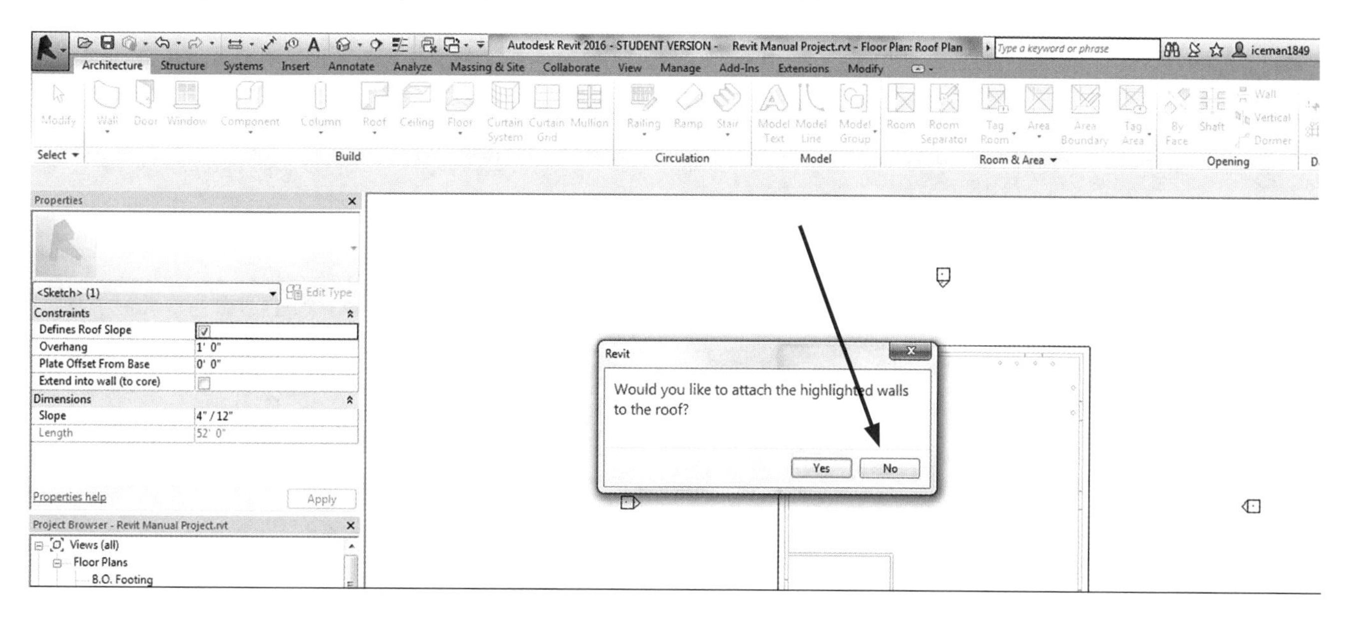

The finished roof will be highlighted or shaded, and since the slope or pitch isn't so steep, the roof shows up over more of the house or building.

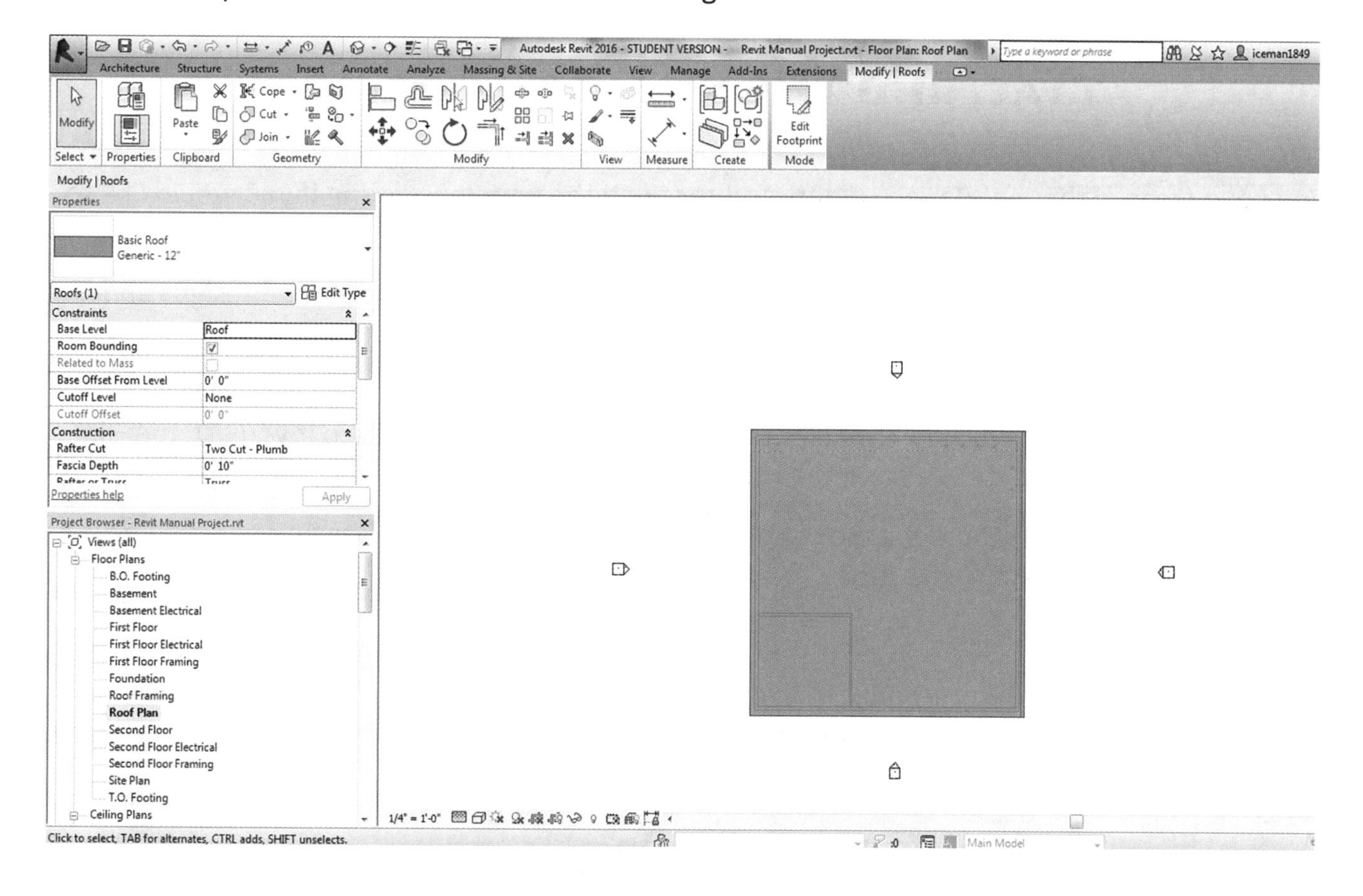

Click the **3D** view, and you will see the new slope angle or pitch angle that was created.

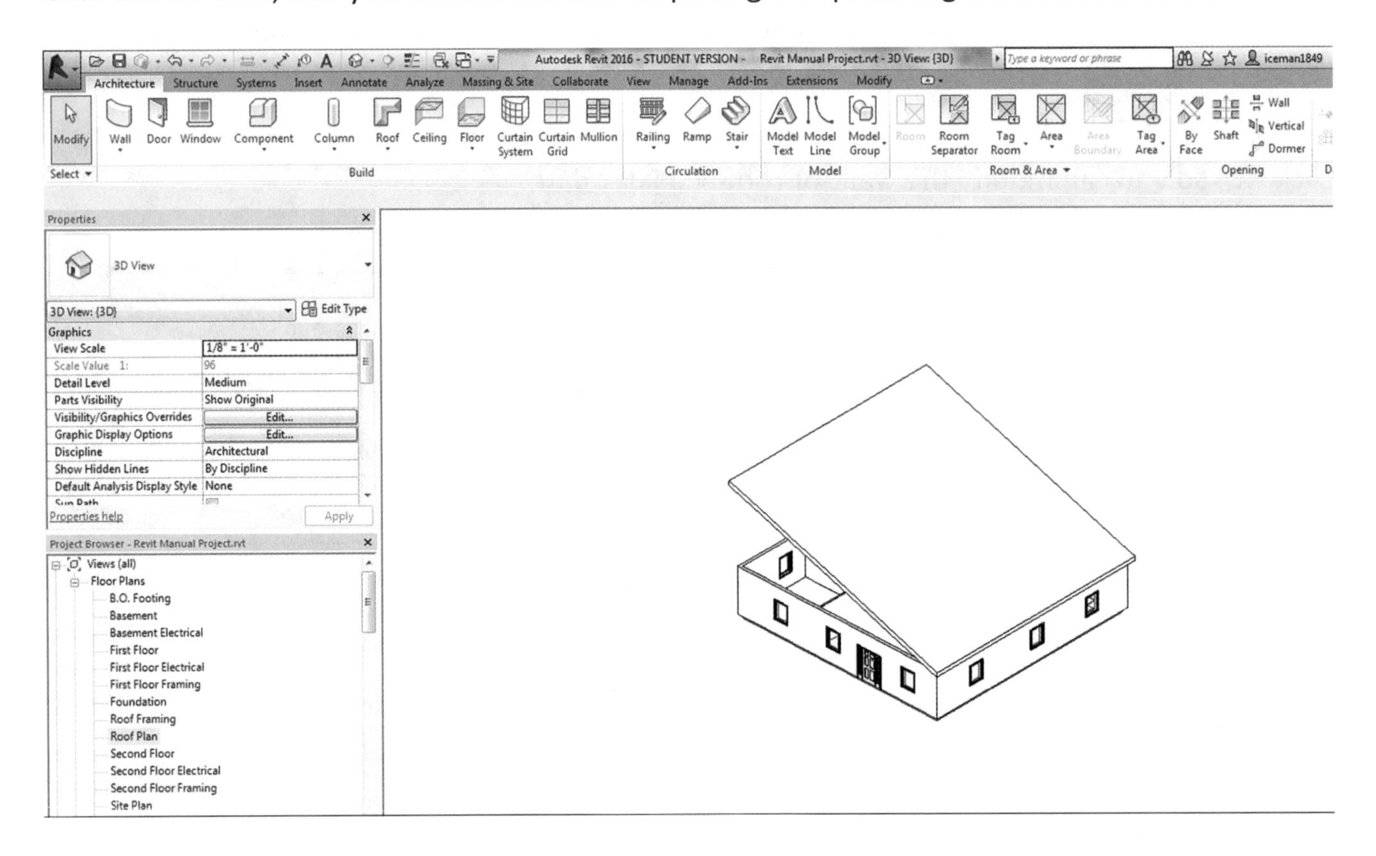

ATTACHING THE WALLS TO THE ROOF

You have created the roof design layouts and received the warnings asking if you would like to attach the highlighted walls to the roof.

It is time to attach the walls to the roof. If you are happy with your chosen roof designs and want to continue with your project, you can attach the walls to your roof so there are no openings in your house or buildings.

No matter which roof style you have created, the steps are the same (even if you use the Hip Roof, and you can't see the wall and the roof connection).

First, we will begin with the Gable Roof Attachment. Find the side that has an opening where the wall doesn't connect to the roof.

TIPS

You may have to select the ELEVATION VIEWS ***in the Project Browser to see which side that will be.***

Click on each elevation view to see how each of the side views will look to determine which view has an opening between the wall and the roof.

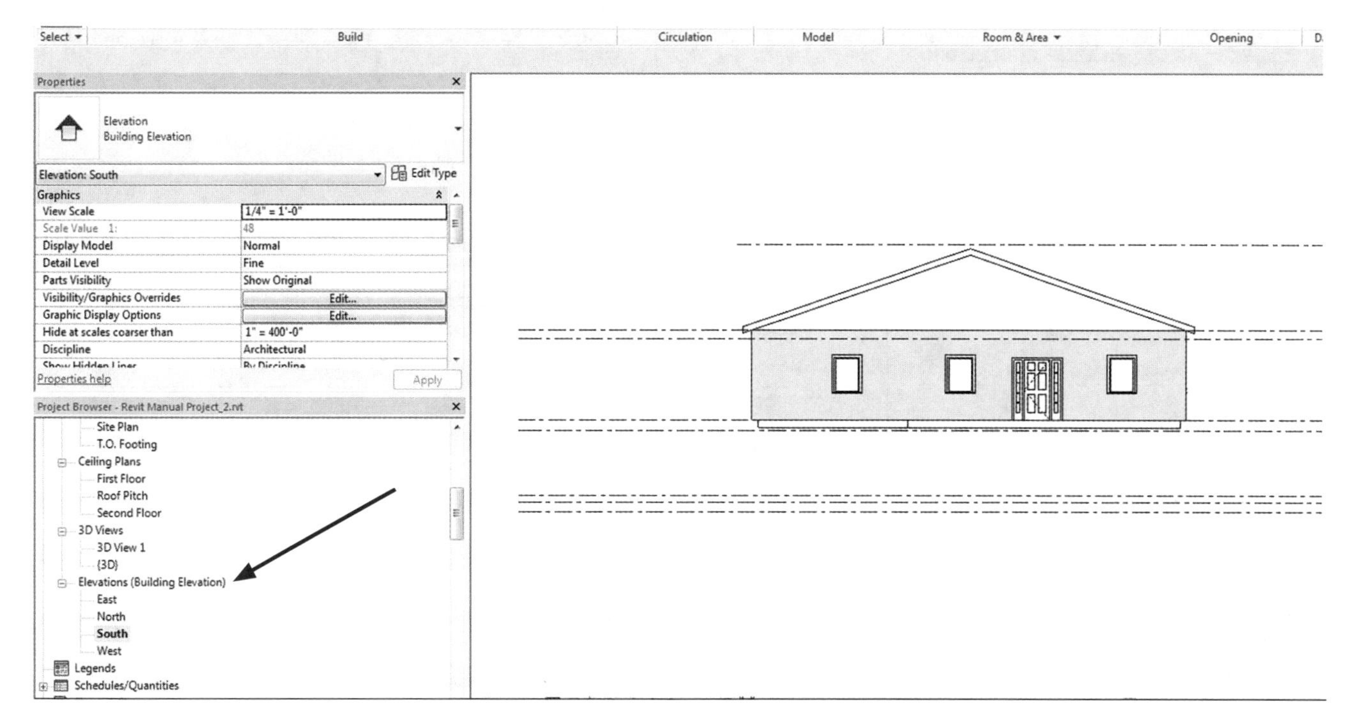

The opening will be a lighter shade than the house or the house itself will be darker or a different color.

TIPS

The easiest way to find the side where the walls don't attach the roof or where you have an opening between your wall and your roof is to go to the bottom of the screen and click on the* VISUAL STYLE *Button for the elevation views.

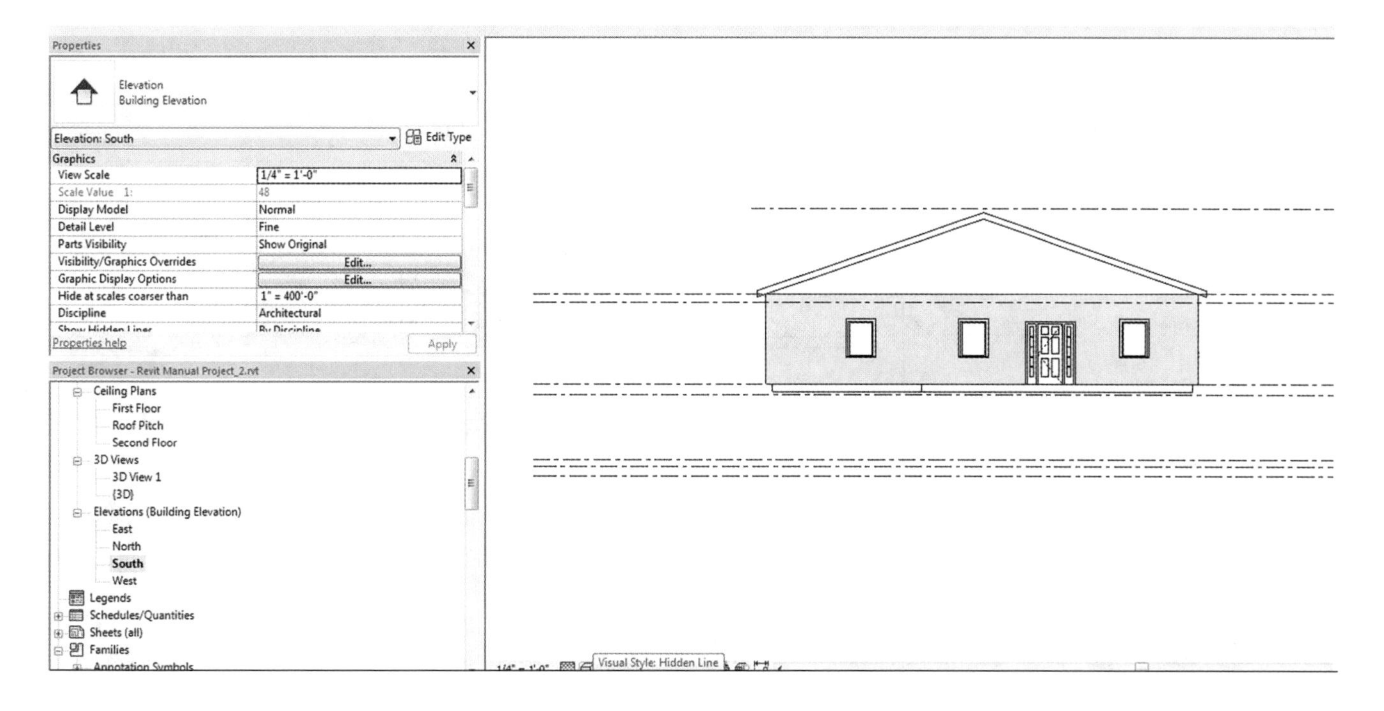

Then click on **CONSISTENT COLORS** and the colors on your house or building will change. You can get a quick idea of your open areas and colors.

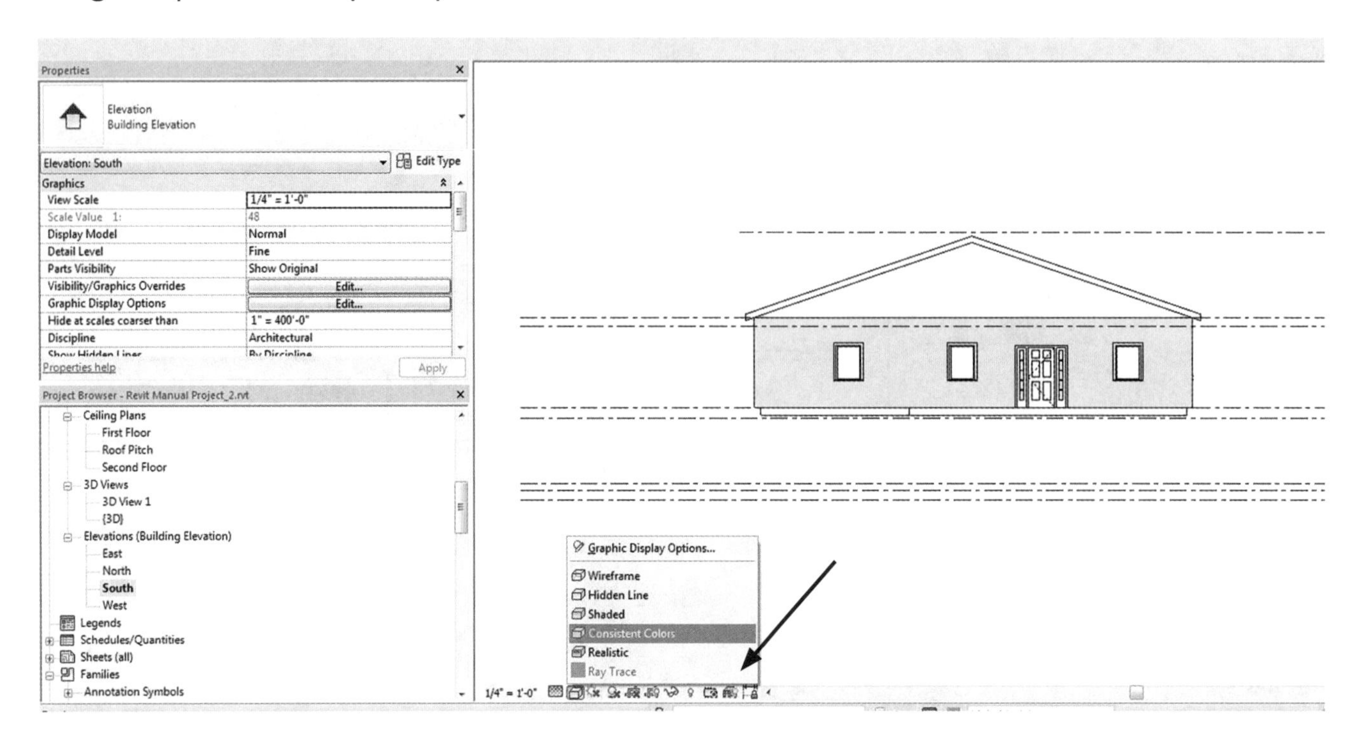

The colors will appear like the following screen view, or similar depending on the graphics of your computer.

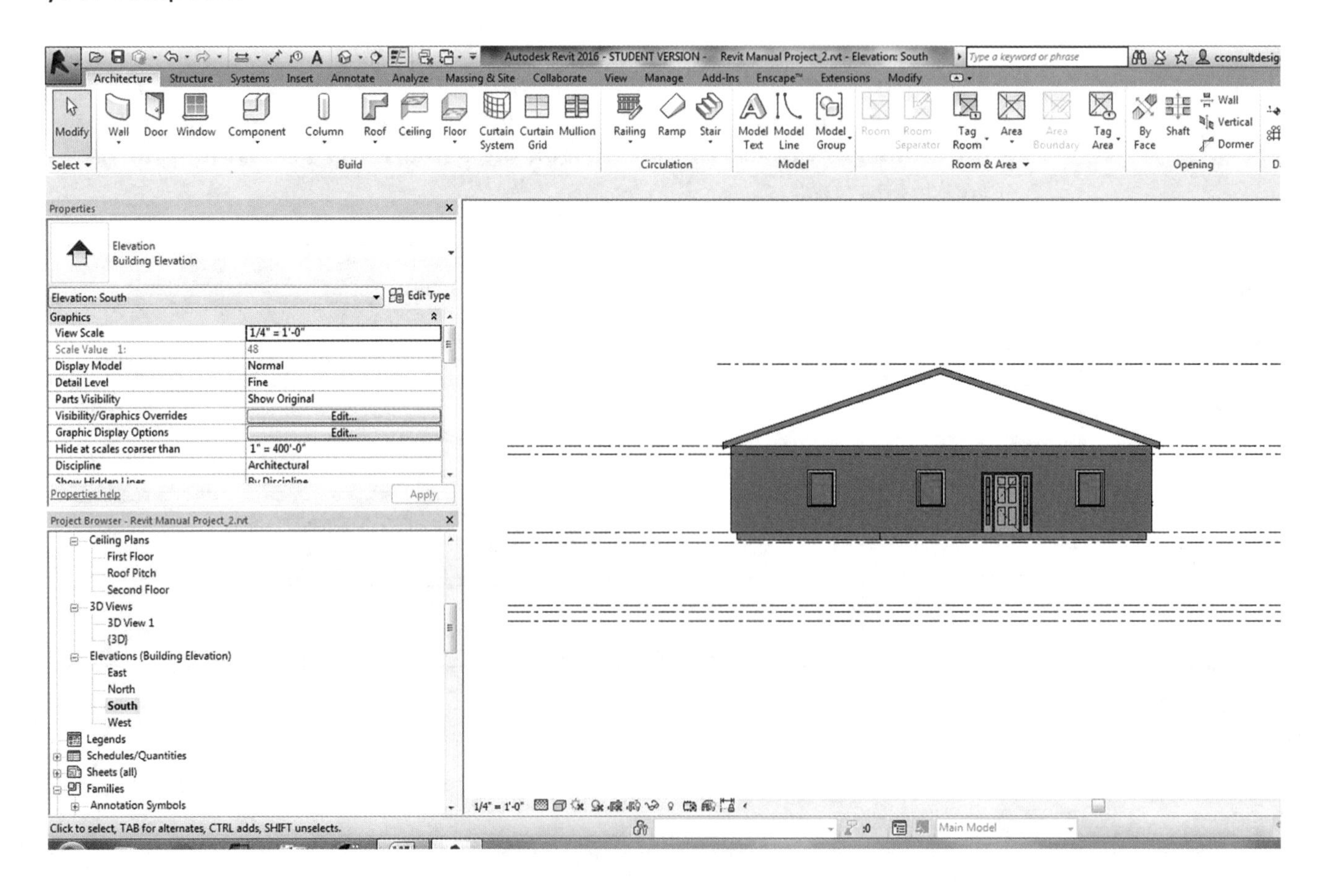

Here is the other elevation view with consistent colors turned using the **VISUAL STYLE** Button:

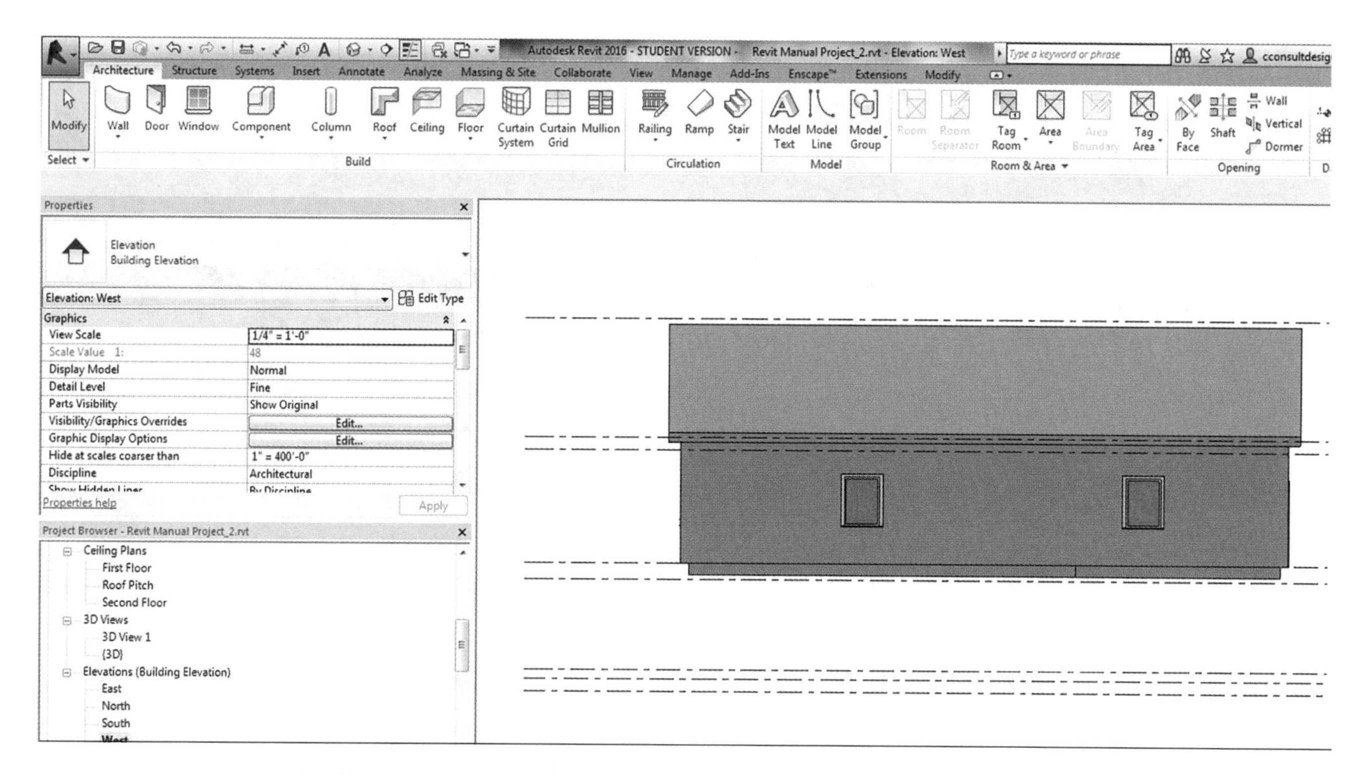

I have shown only two elevations, but you can choose to show all four elevations, if you wish. Either way, you will be able to see the difference between an opening in your house or building compared to a closed connection of wall and roof.

Once you have selected the view that has an opening between the wall and the roof, you are still in that Elevation view. In this case, it will be the South Elevation view, with the wall and roof opening.

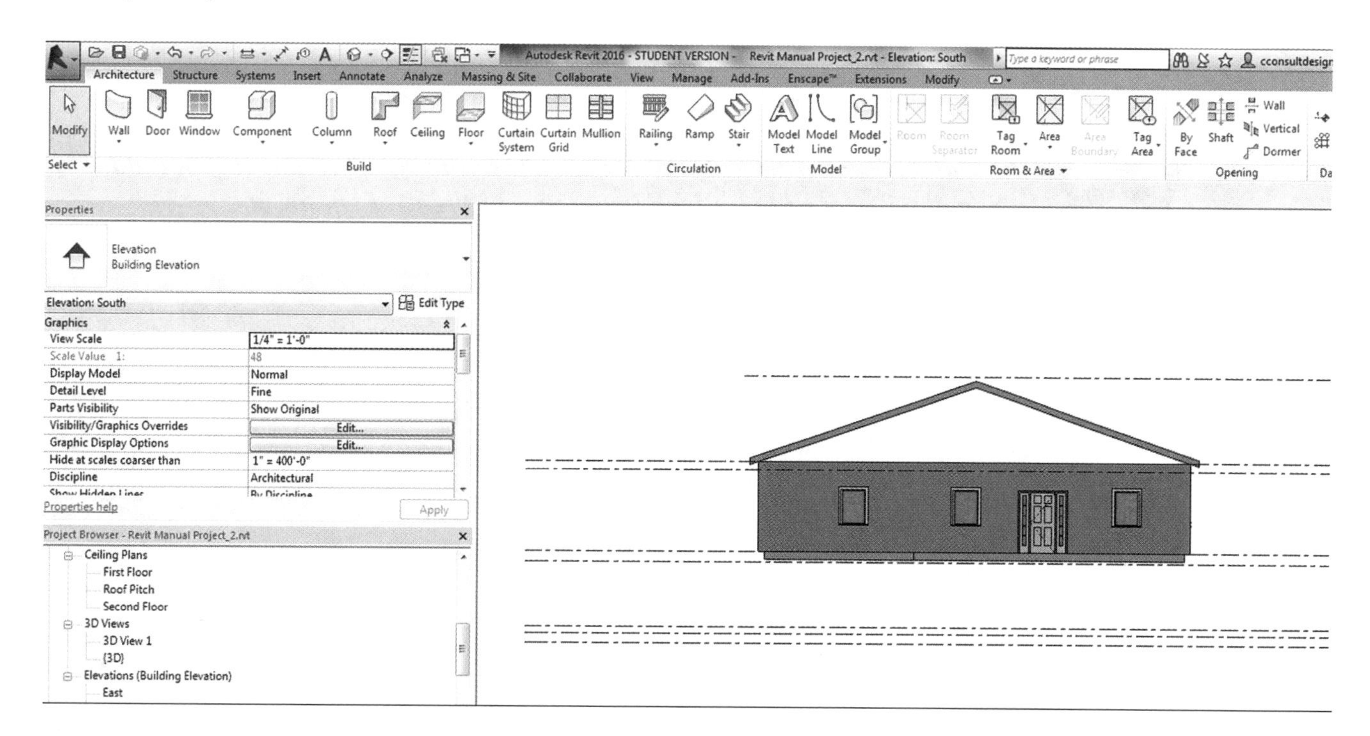

Once in the Elevation View, click on the wall itself.

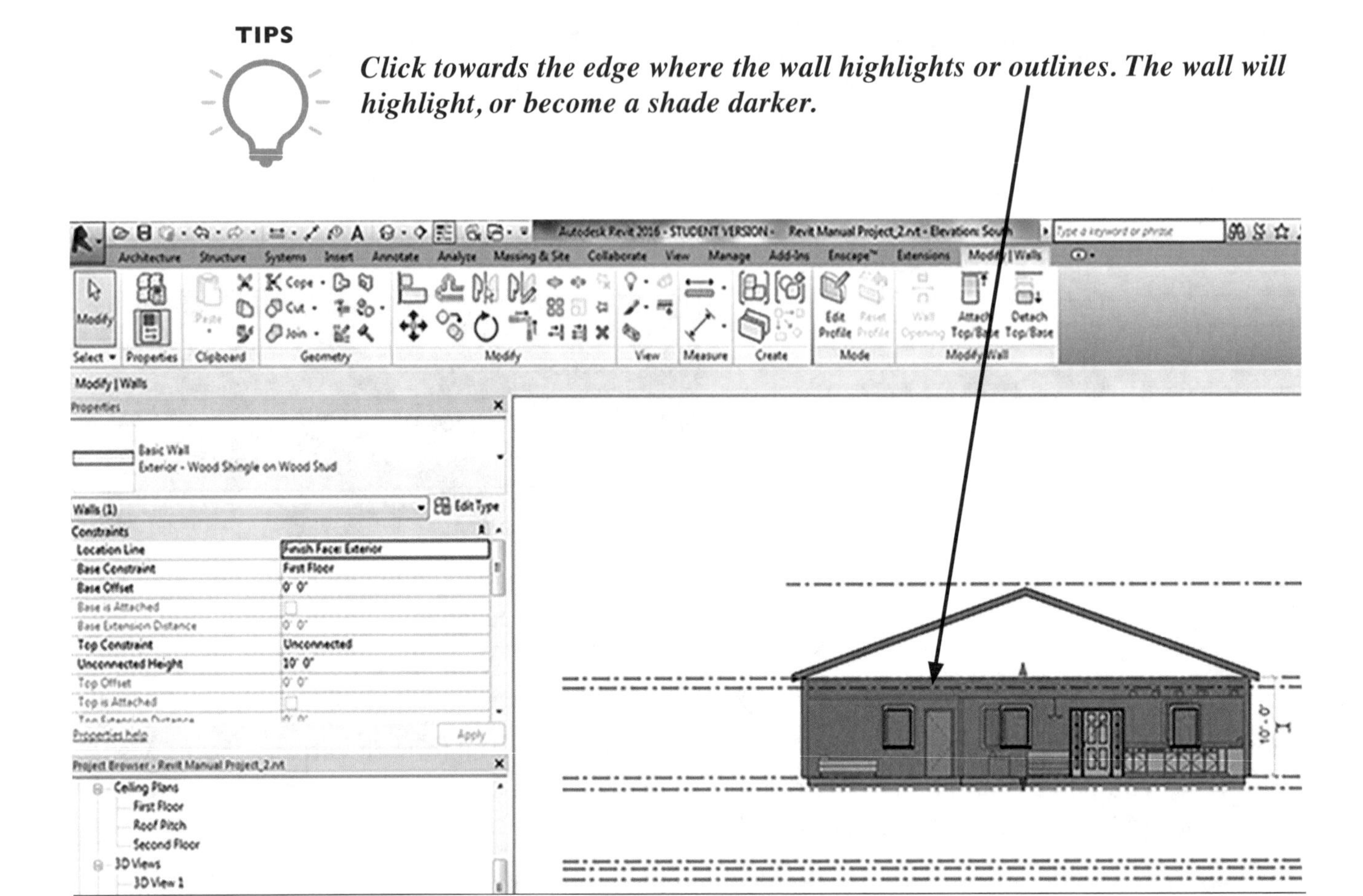

This now has activated the **MODIFY/WALLS** Ribbon command.

WARNING *There are two grips on the wall. Do* not *touch those. The grips at the bottom and the top of the wall allow you to manually lengthen the height of the wall either below ground or above into the roof area. But if you do this, the wall* will not *automatically conform to the roof shape.*

In the Ribbon, click the **ATTACH TOP/BASE** tile on the Ribbon.

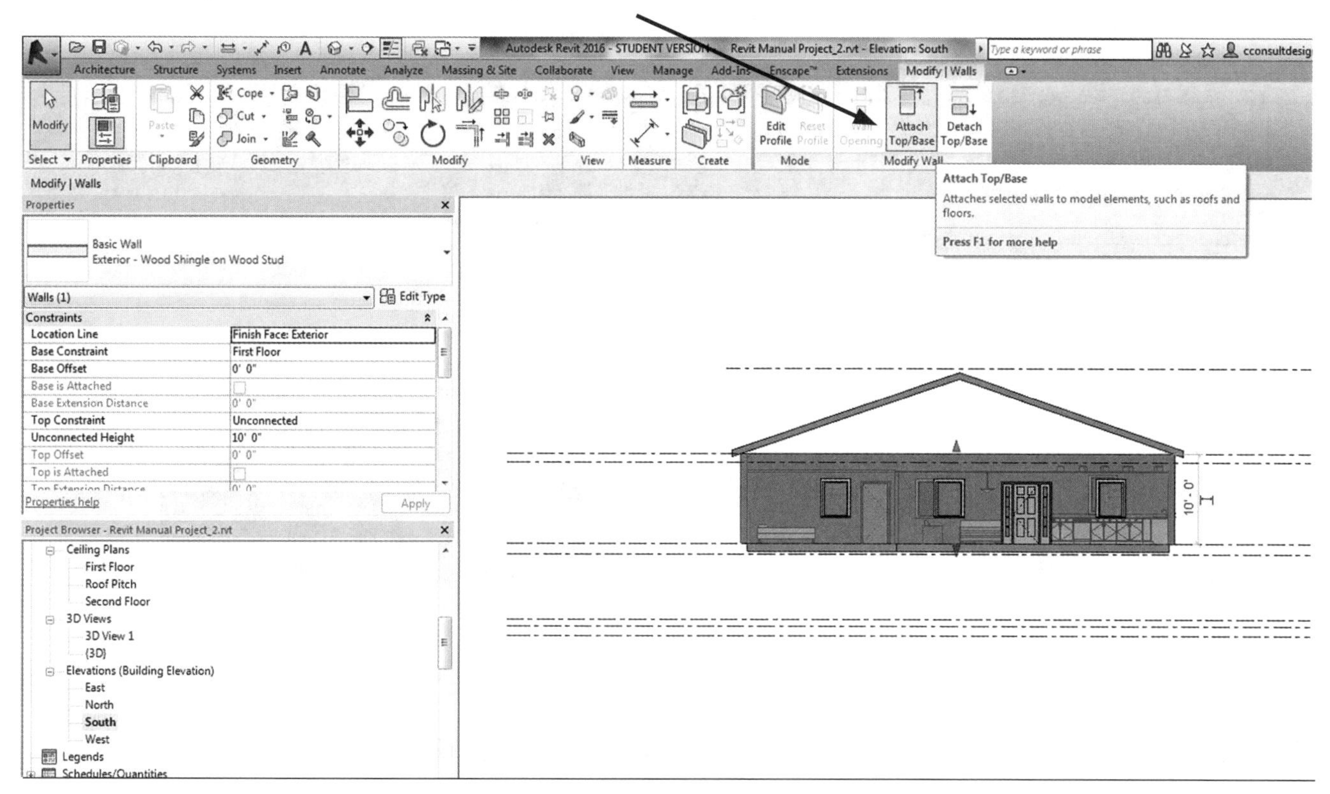

Since you selected the wall in the beginning and then you selected the **ATTACHED TOP/BASE** command, the next step is to just click on the roof.

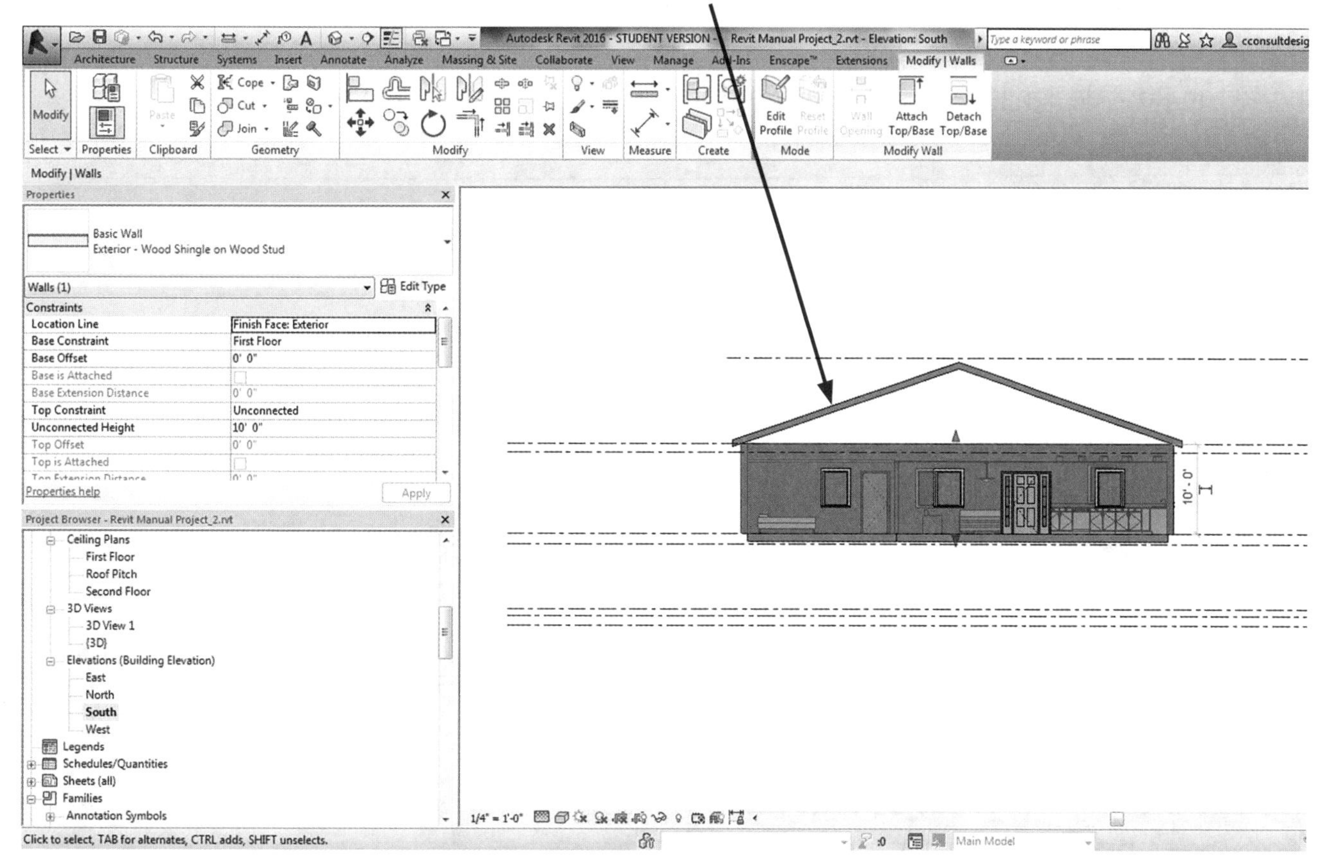

The wall will automatically connect to the roof and the designed space. Then click the **MODIFY** Button to end the command.

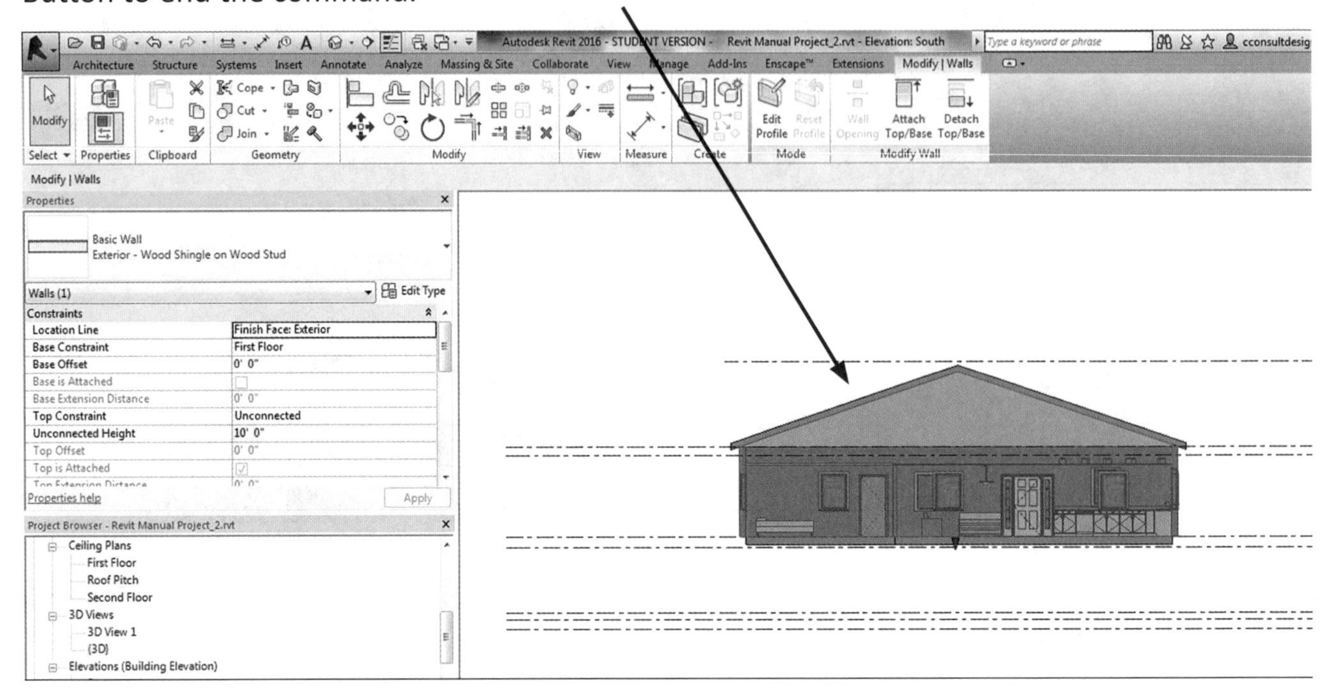

REPEAT THESE STEPS FOR THE OTHER SIDE

Then you can click on **3D** View and check out your work around your house/building. When you are performing the Attach Top/Base function for the Hip Roof, follow the same steps. Click on the wall edge and then click on the roof (regardless if you can't see the roof opening or vacant space).

In the Hip Roof design layout, select any of the elevation views in the Project Browser.

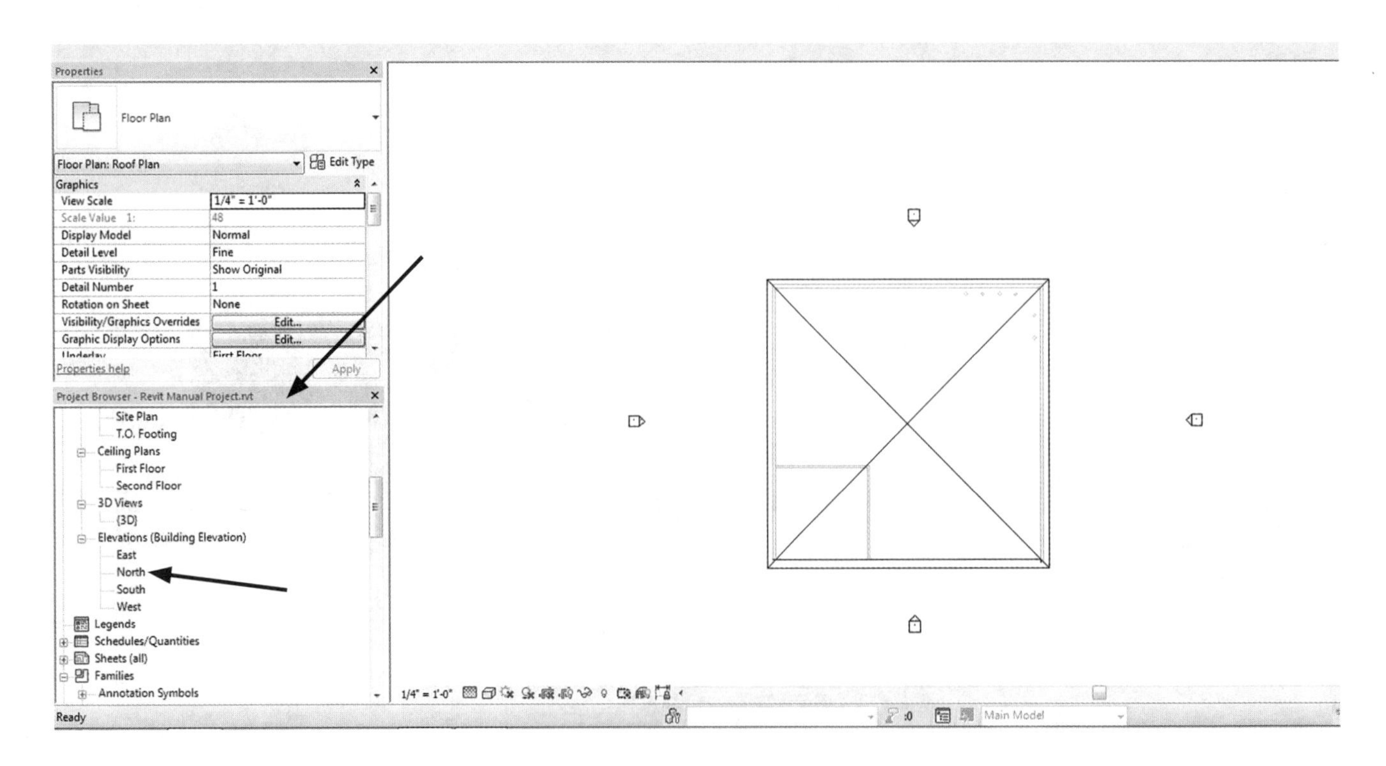

NOTES

Once the elevation view is selected, you can't see that there is an opening between the walls and the roof, but the walls are not securely fastened to the roof. Just like the previous roof design for the Gable Roof, you can click on the VISUAL STYLE ***Button for the elevation views.***

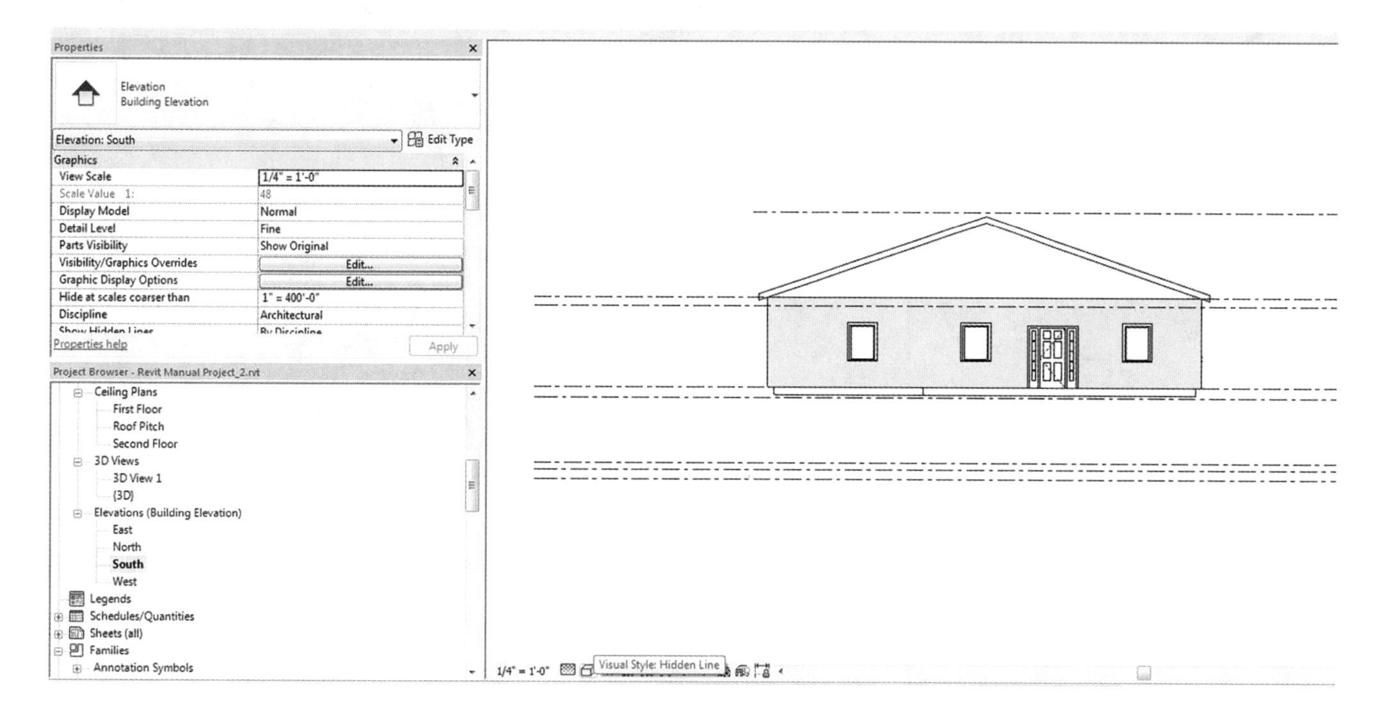

Then click on **CONSISTENT COLORS.** The colors on your house or building will change, and you can get a quick idea of your open areas and colors.

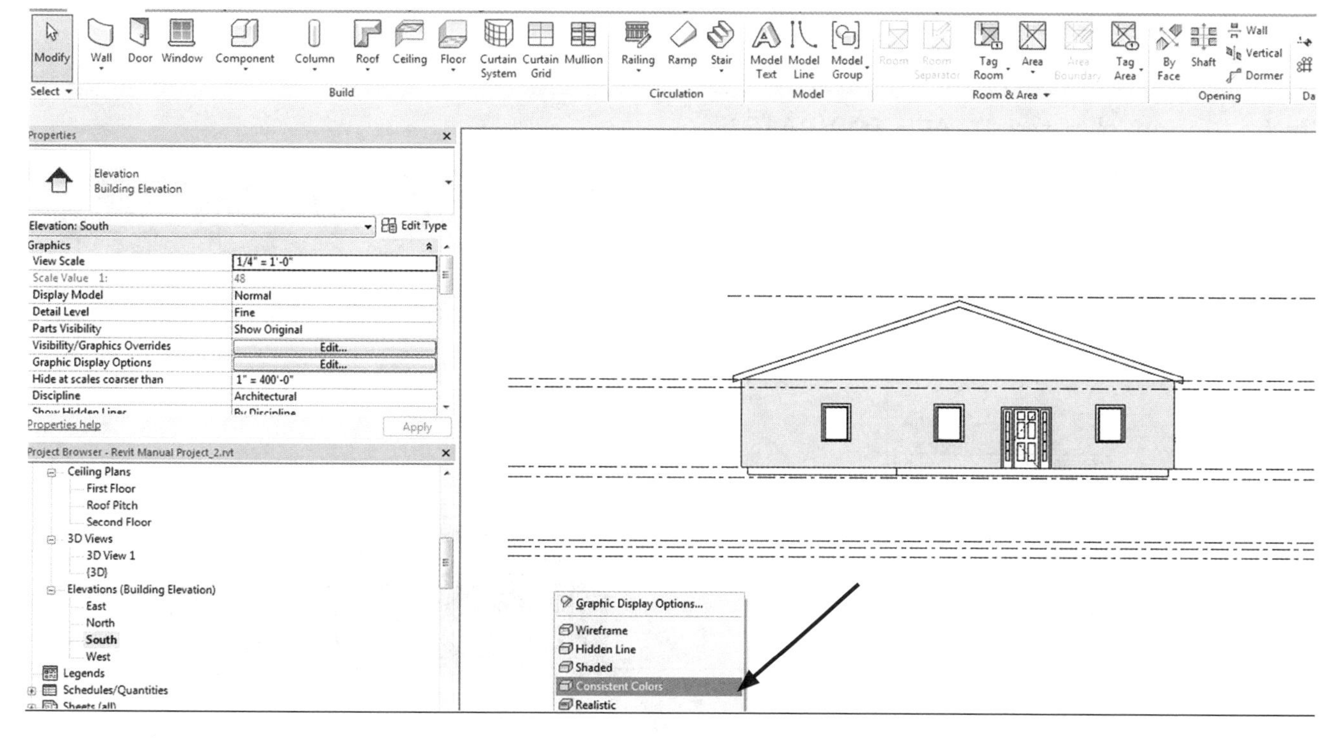

Once in this view, click on the wall itself with the mouse.

TIPS

Click towards the edge, where the wall highlights or outlines. The wall will highlight or become a shade darker.

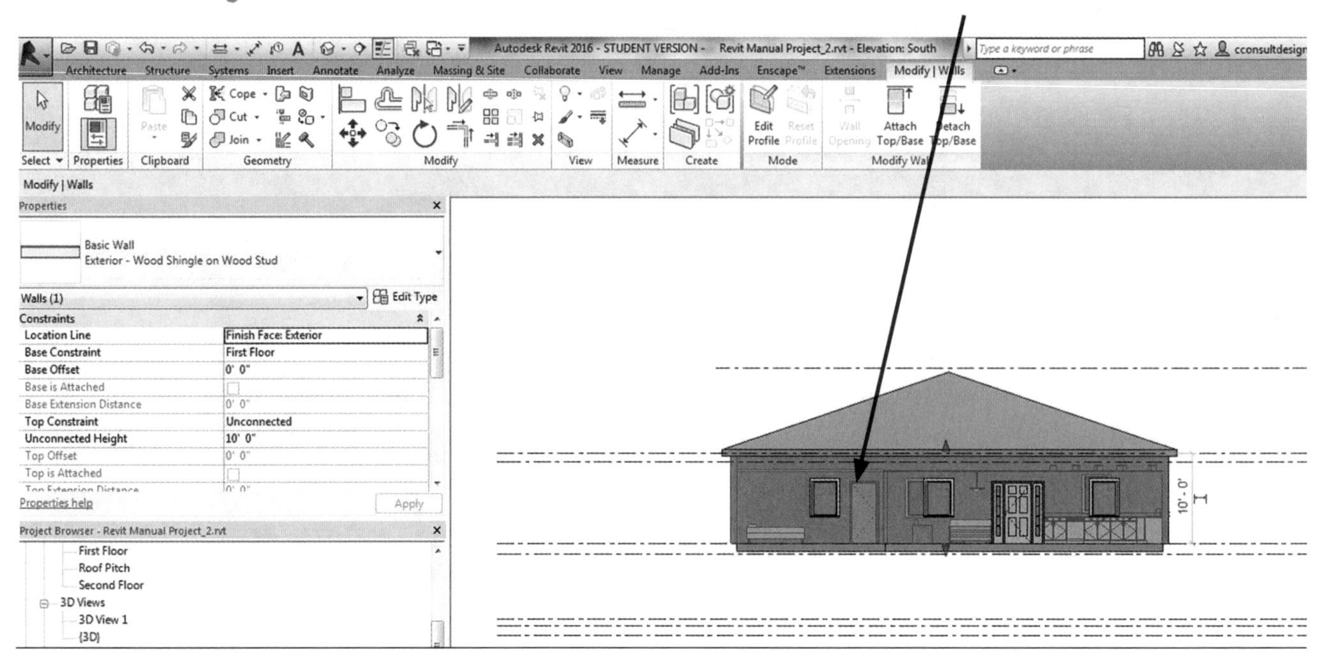

This has now activated the **MODIFY/WALLS** Ribbon command.

REMINDER

There are two grips on the wall. (You may only see one grip because the other is under the roof.) Don't touch those.

In the Ribbon, click the **ATTACH TOP/BASE** tile.

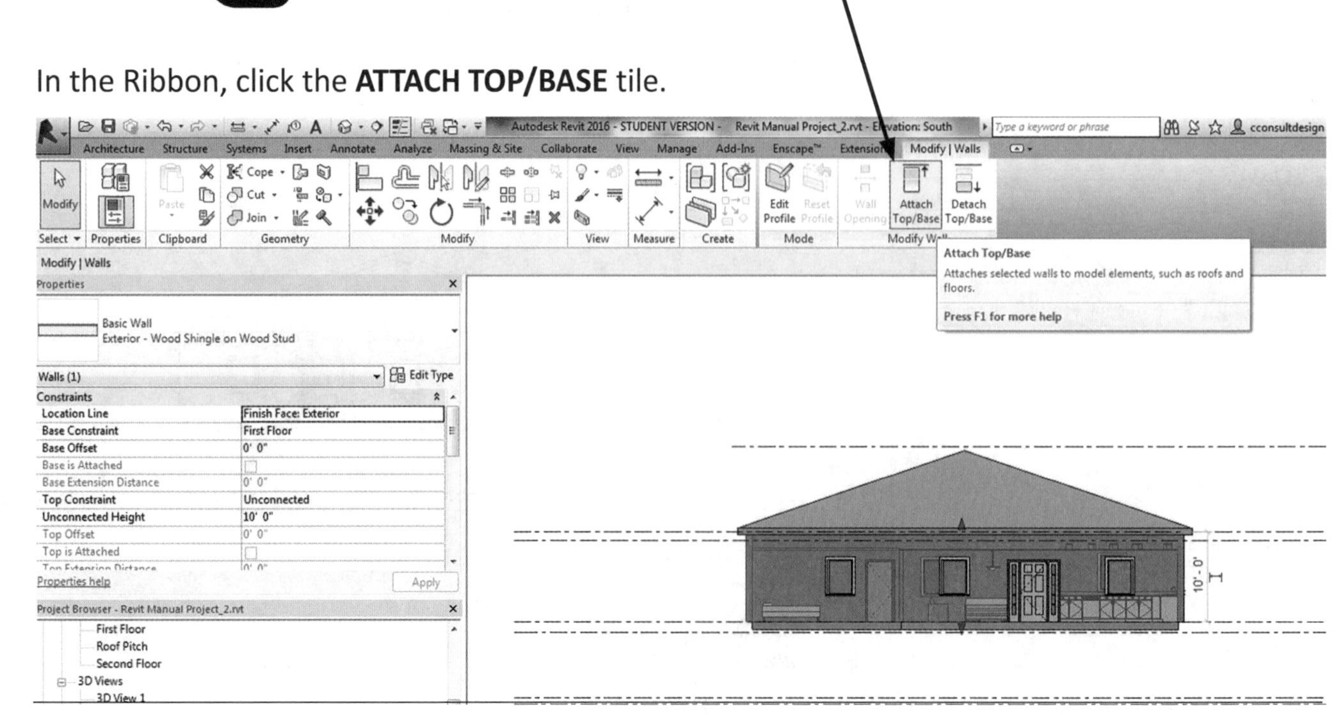

Since you selected the wall in the beginning and then you selected the **ATTACHED TOP/BASE** command, the next step is to just click on the roof.

Then click on each elevation, and repeat the steps to attach the walls to the roof.

NOTES

You won't see the walls attached to the roof like you will with the Gable Roof.

If you are attaching the walls to the Slope Roof, the steps are the same except three of the four walls are *much* higher than the one the roof is sitting on. Follow the same steps in selecting an elevation in the Project Browser.

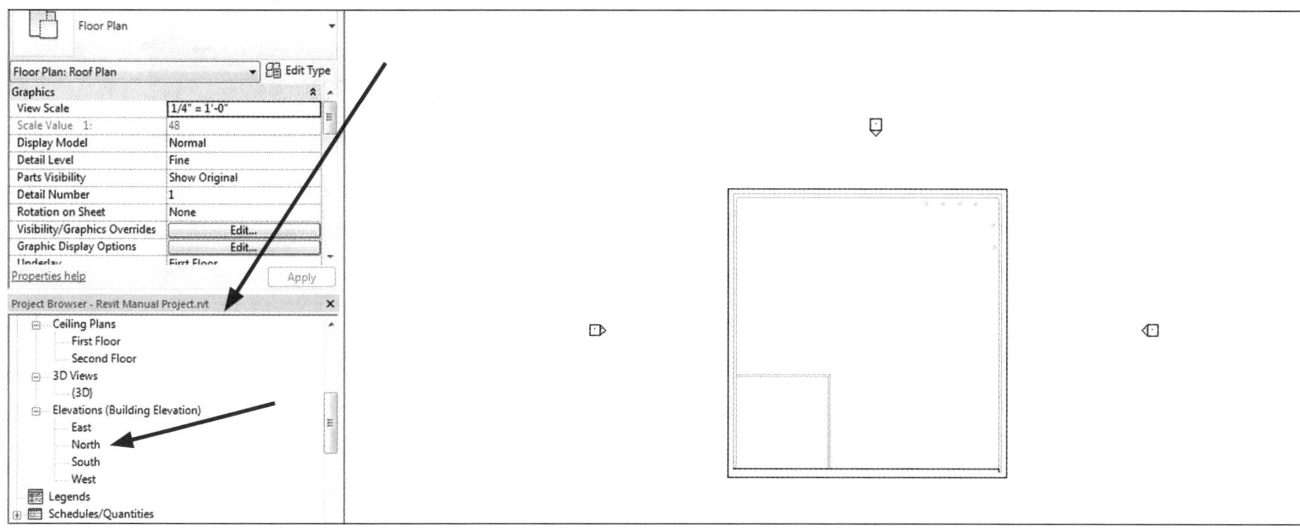

Just like the previous roof design for the Gable Roof and Hip Roof, you can click on the **VISUAL STYLE** Button for the elevation views. One elevation will be all you need, because the other side is connected to the house or building and other sides of the house you will see between the roof and the walls.

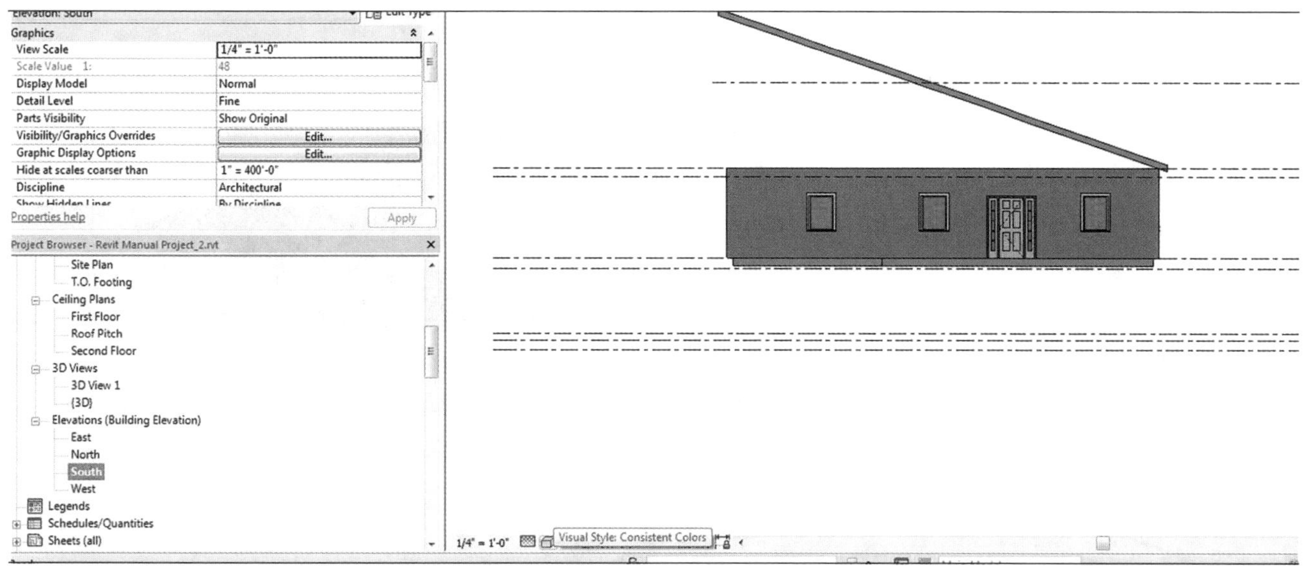

Once you have selected the wall, click on the **ATTACH TOP/BASE** command on the ribbon, and you can select the roof.

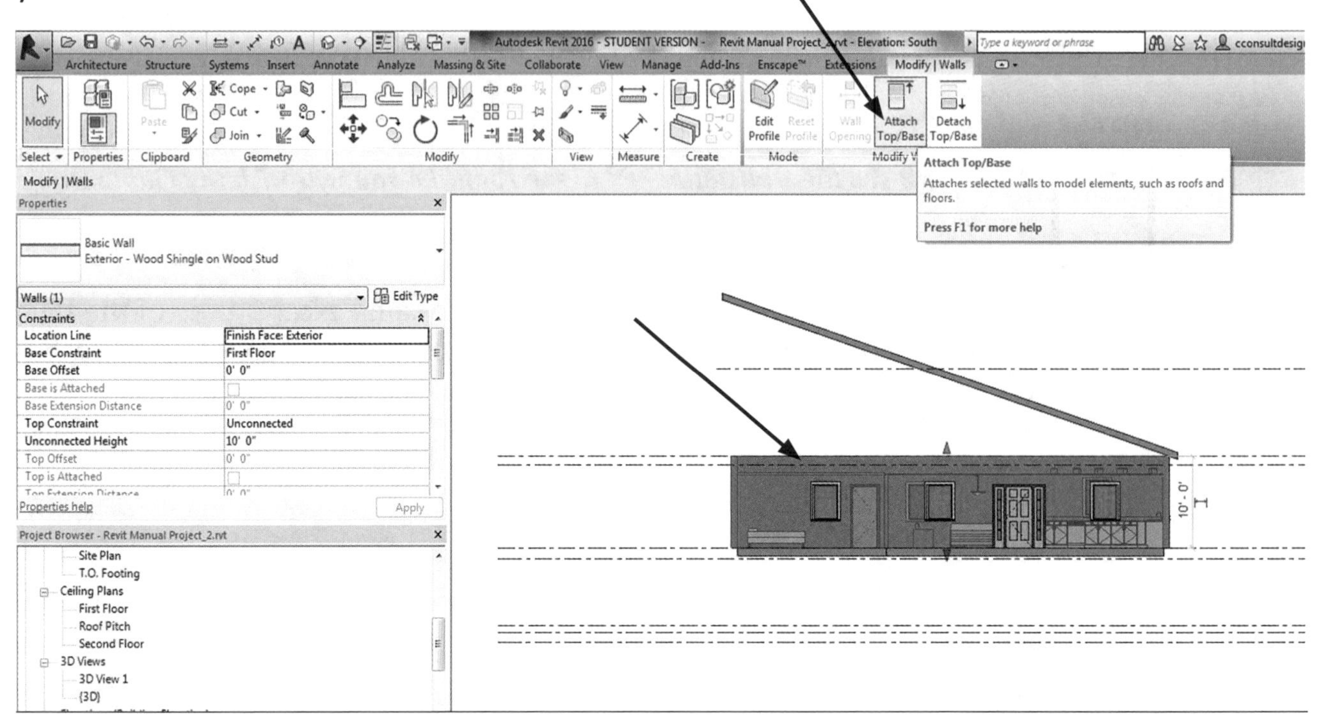

Now the wall and roof are connected in that elevation view. You can switch to a new elevation in the Project Browser where the walls and roof are not connected. (Or it maybe easier to switch to a 3D view to see the opening in the house or building. You can switch to the 3D symbol in your Quick Access Toolbar, at the top of the ribbon by clicking with your mouse).

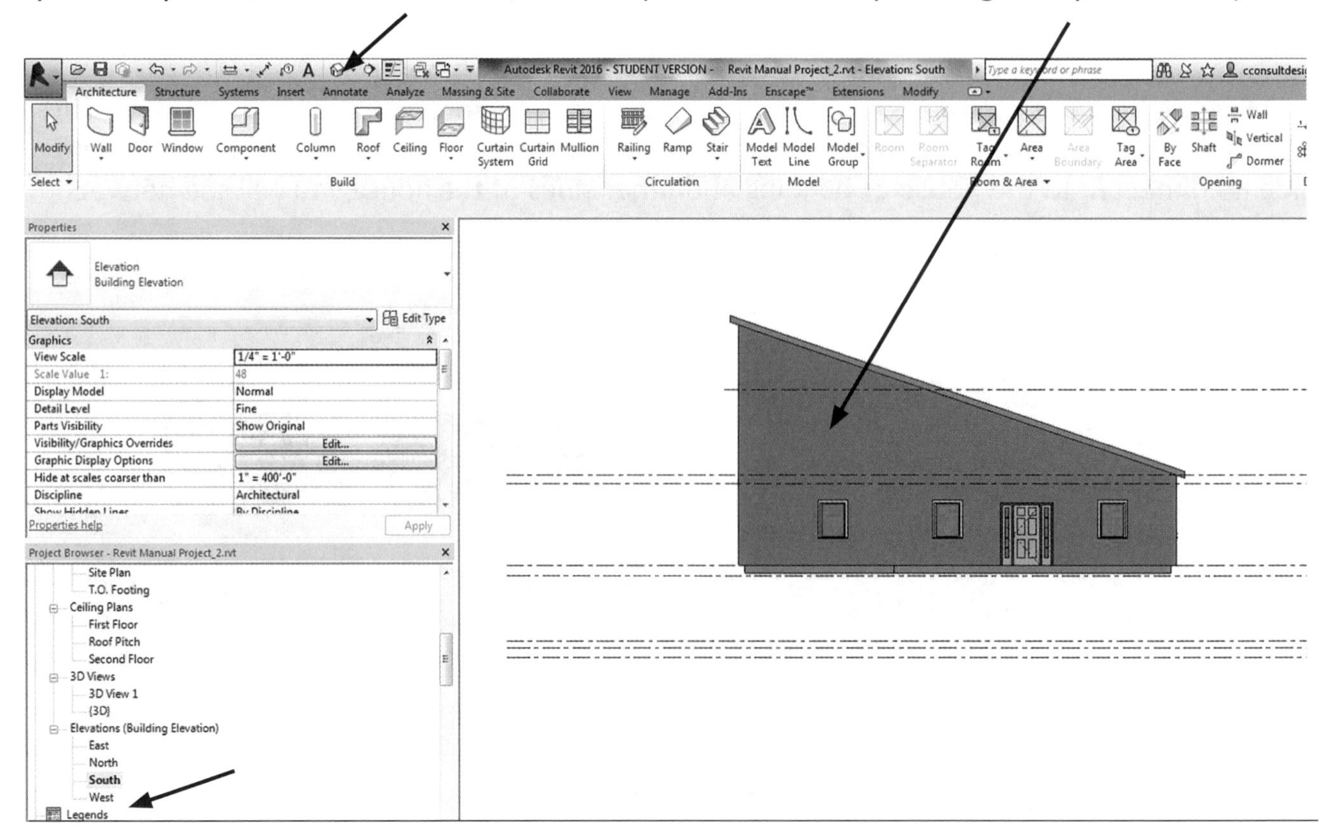

Following is a 3D view where you can see the opening better:

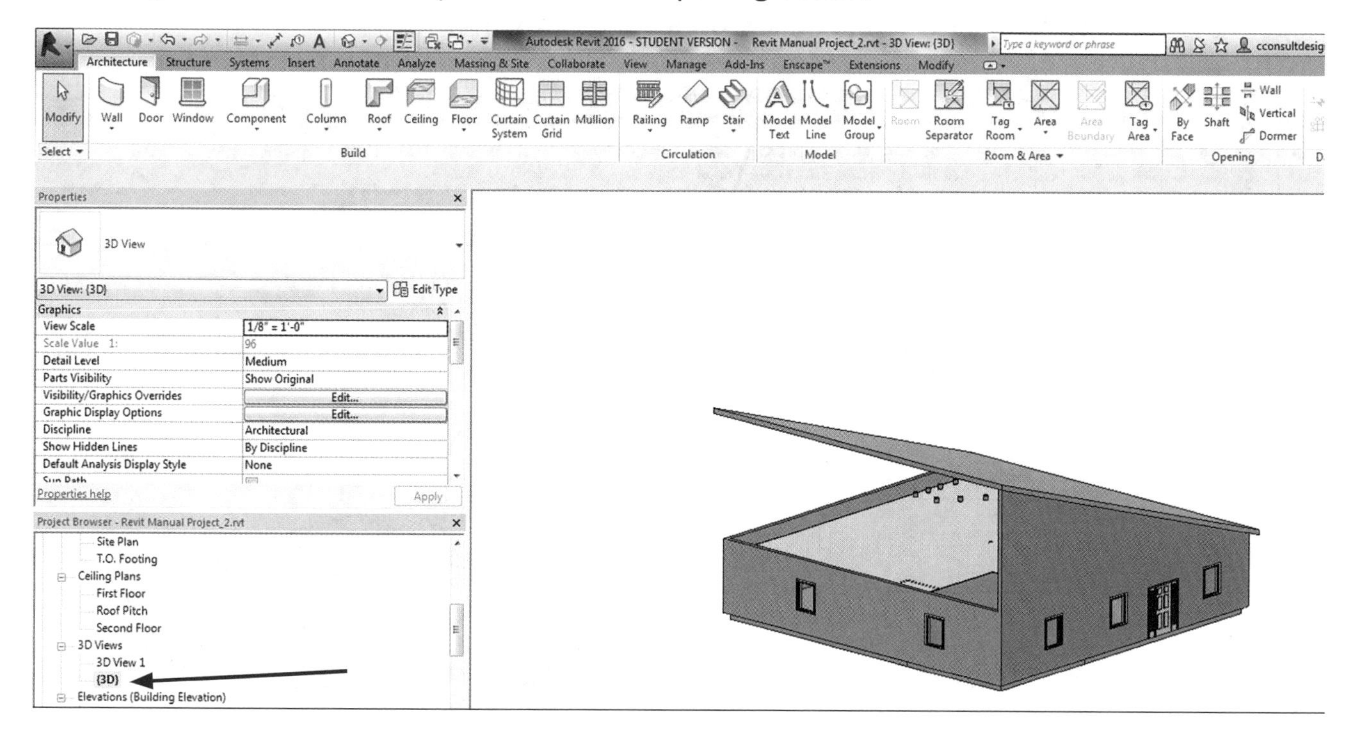

You can select the 3D view by clicking the **3D** command in the Project Browser.

In the 3D view, it's also easier to click on the wall and select the **ATTACH TOP/BASE** command. Click the wall first and then the **ATTACH TOP/BASE** command. Then you can click on the roof.

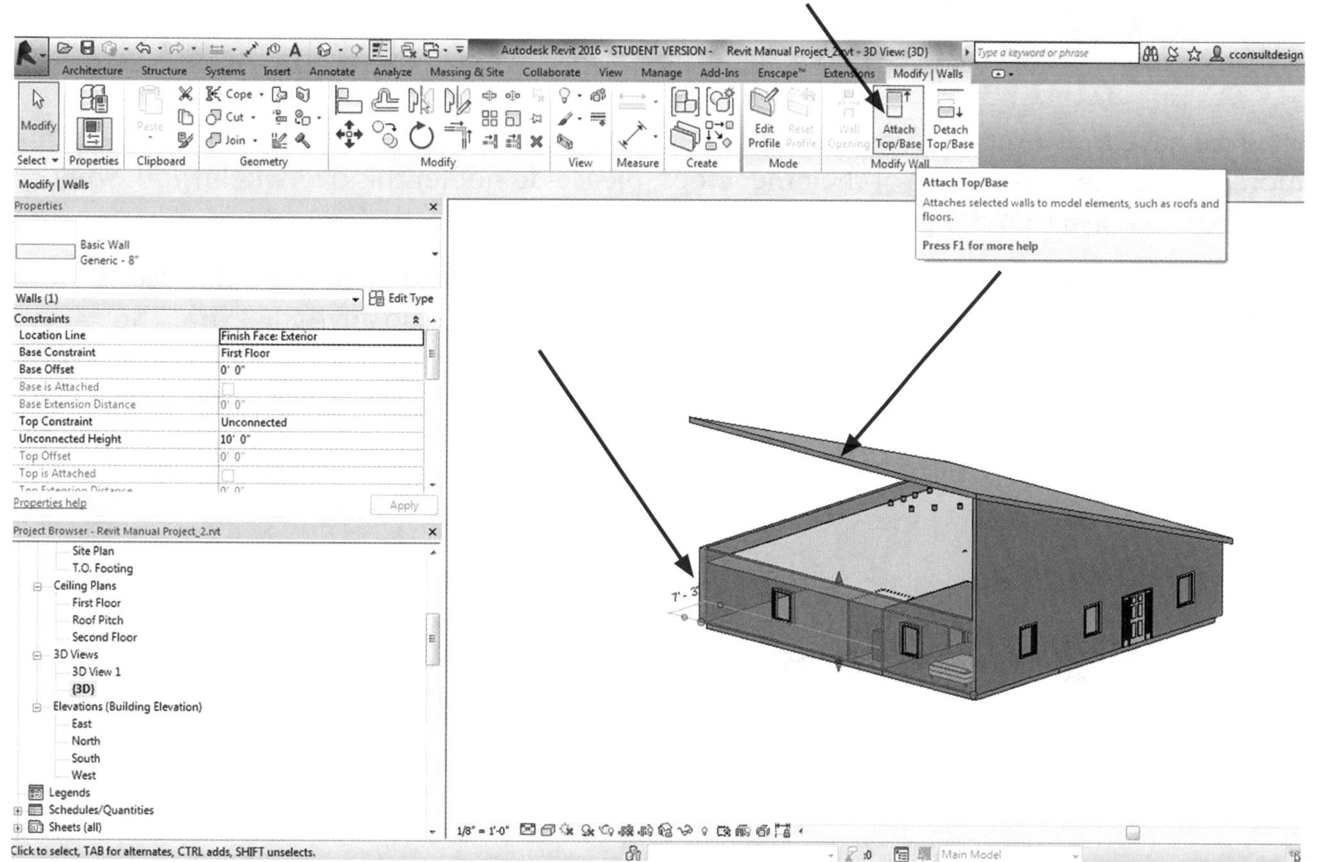

Once you have attached the wall to the roof the connection will look like the following view. Repeat the steps to connect the wall on the other side of the house.

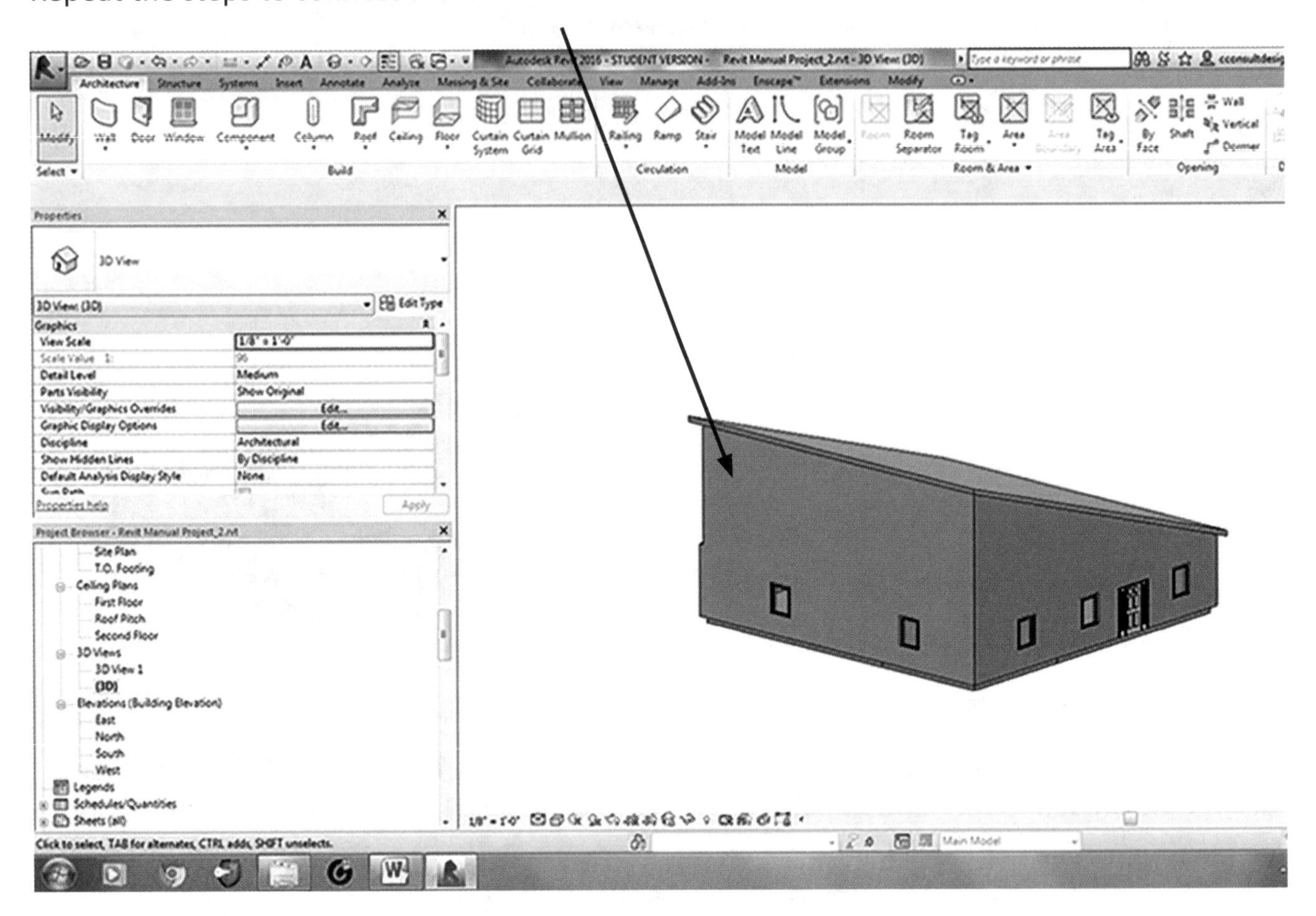

Now we have *truly* experienced how to use Revit to layout different roof designs. While a bit more complicated than some of the other steps, please do not let this overwhelm you when you are creating a roof design. Take each one step-by-step.

A Hip Roof is sloped on all four sides, and is the easiest for Revit to do because it defaults to placing slopes automatically, where the designer doesn't have to do anything extra. The Gable Roof is the same as a Hip Roof, but you need to simply remove two of the slope angles. Revit will do the rest as far as placing the overhang at a default of 1'-0" (which you can *always* change).

REMINDER

You have the ability to change all this information. Revit just quickly defaults to these as they are the industry standards.

TIPS

If you have a complex roof design (or what looks like a complex design), I would suggest that you first draw it in Revit, the way it looks from the sketch with all the slopes and see how Revit creates it. It's a very good possibility that it will work out where you only need to do minor changes. If not, then create the roof in sections, step-by-step.

This manual will assist you in giving you the ability to combine a Hip Roof and a Gable Roof into your house or building, and making either (or both!) look fantastic.

Let's continue to Chapter 15, *Interior and Exterior Camera Views*, so we can see our designs.

CHAPTER 15

Interior and Exterior Camera Views [Perspective Views]

INTERIOR CAMERA VIEW

When setting up a drawing to use camera views, make sure you are in the Floor Plan – First Floor or whatever plan you want to take a picture of.

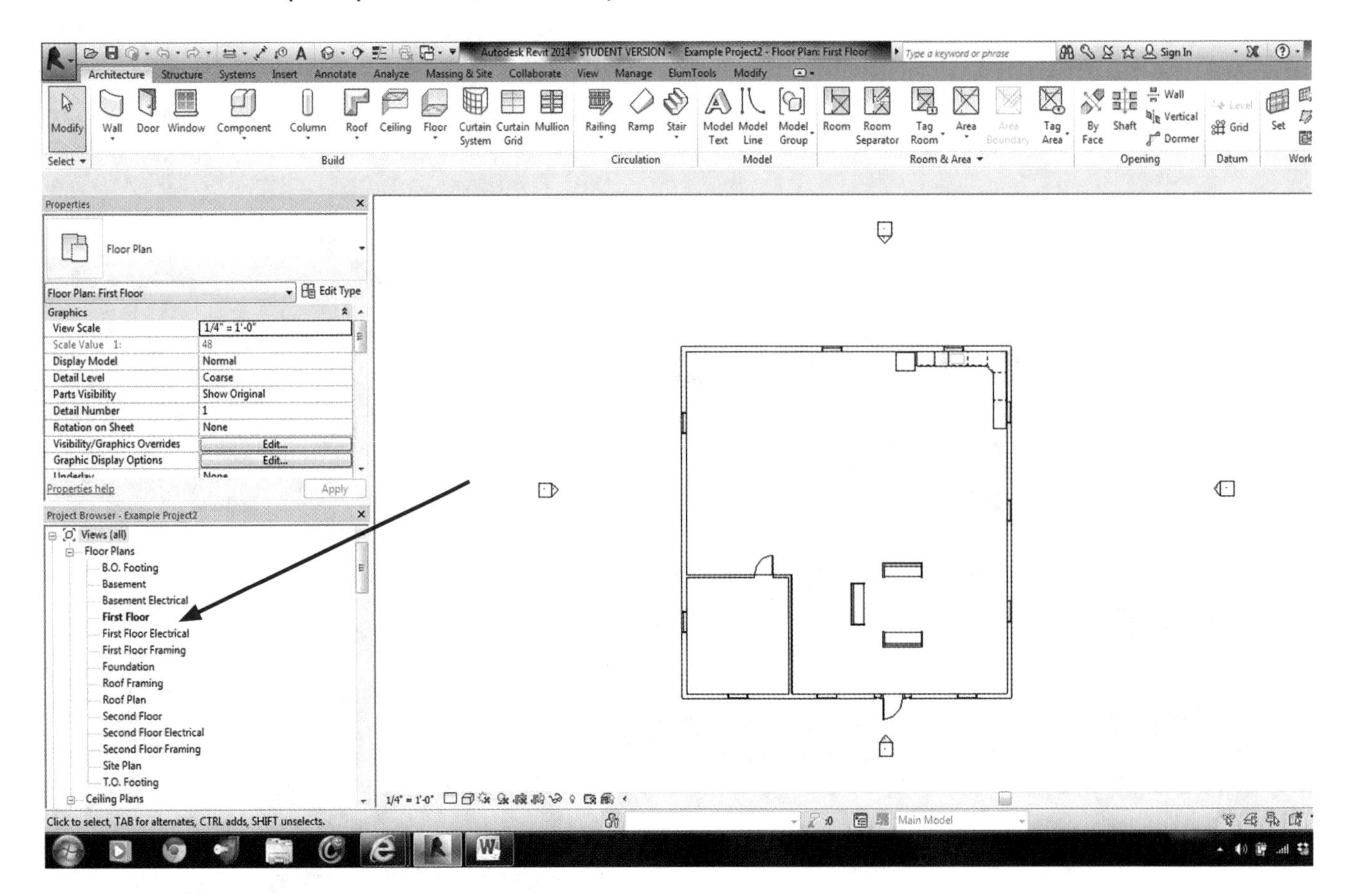

Make sure you are not in the Ceiling Plan. Revit will be placing the camera at 5'-6" from the Floor Plan height. If you are on the Ceiling Plan, then 5'-6" higher will be in the attic.

Click on the **DROP-DOWN ARROW** next to the House symbol at the top of the screen. Then click on the **CAMERA** symbol.

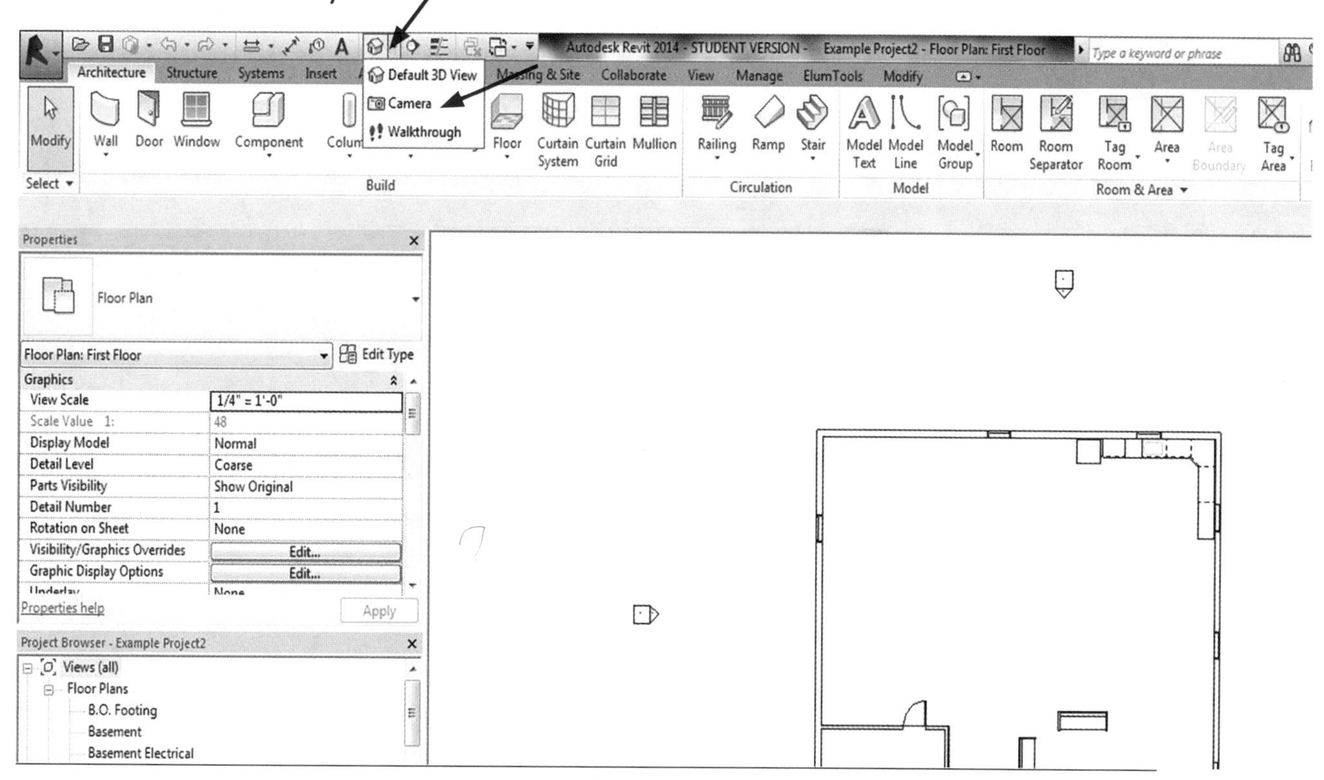

Once you select it, a small camera will appear on the screen, and you can place it wherever you want in your design.

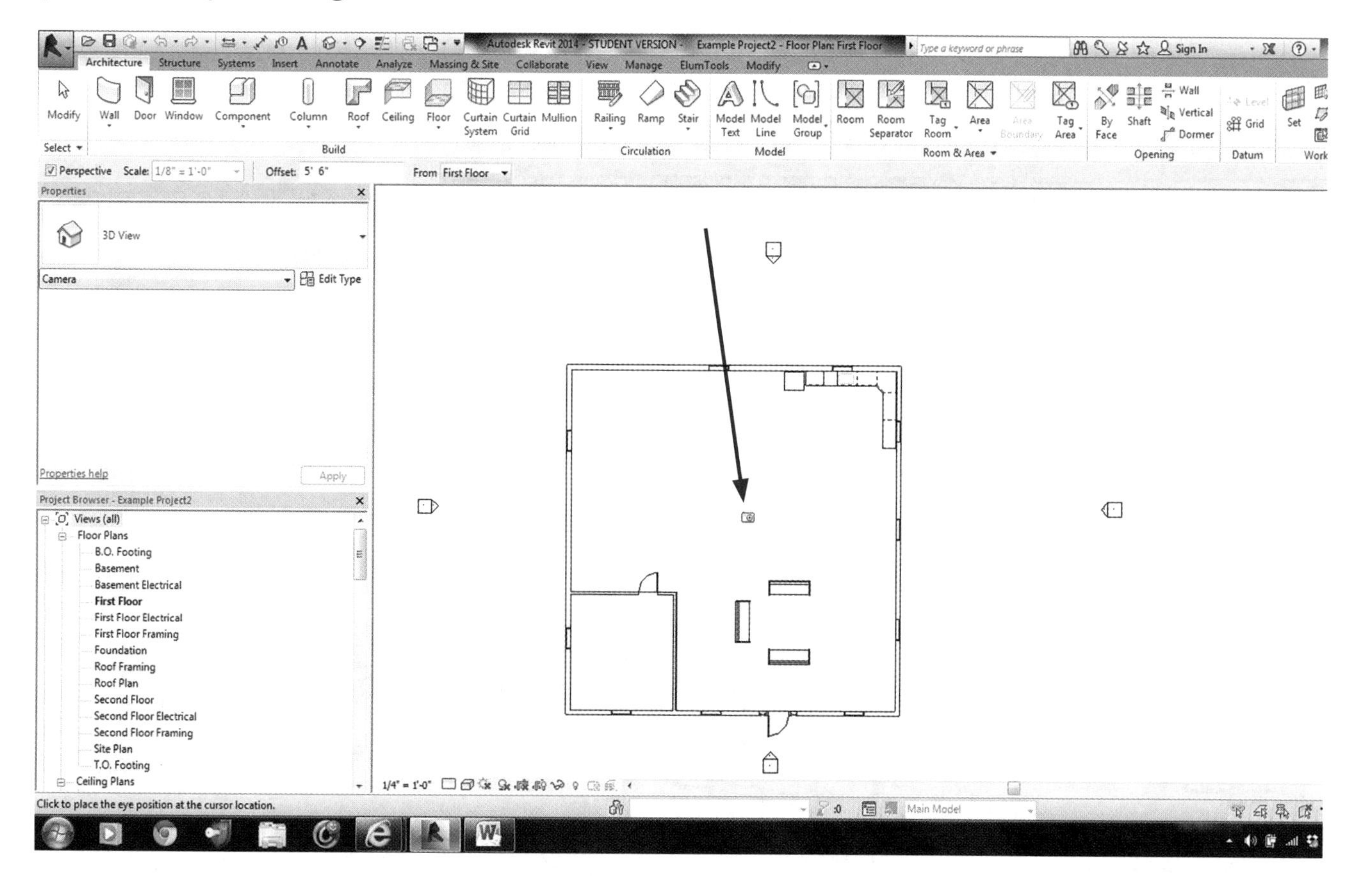

When placed in your design, click the mouse and you will get 3 lines which form your view angle. The middle line is your *sight line*. Pull this line all the way to the outside of your design.

WARNING

If you don't pull your middle line out, you will only get to view your design from a limited distance in the Perspective View.

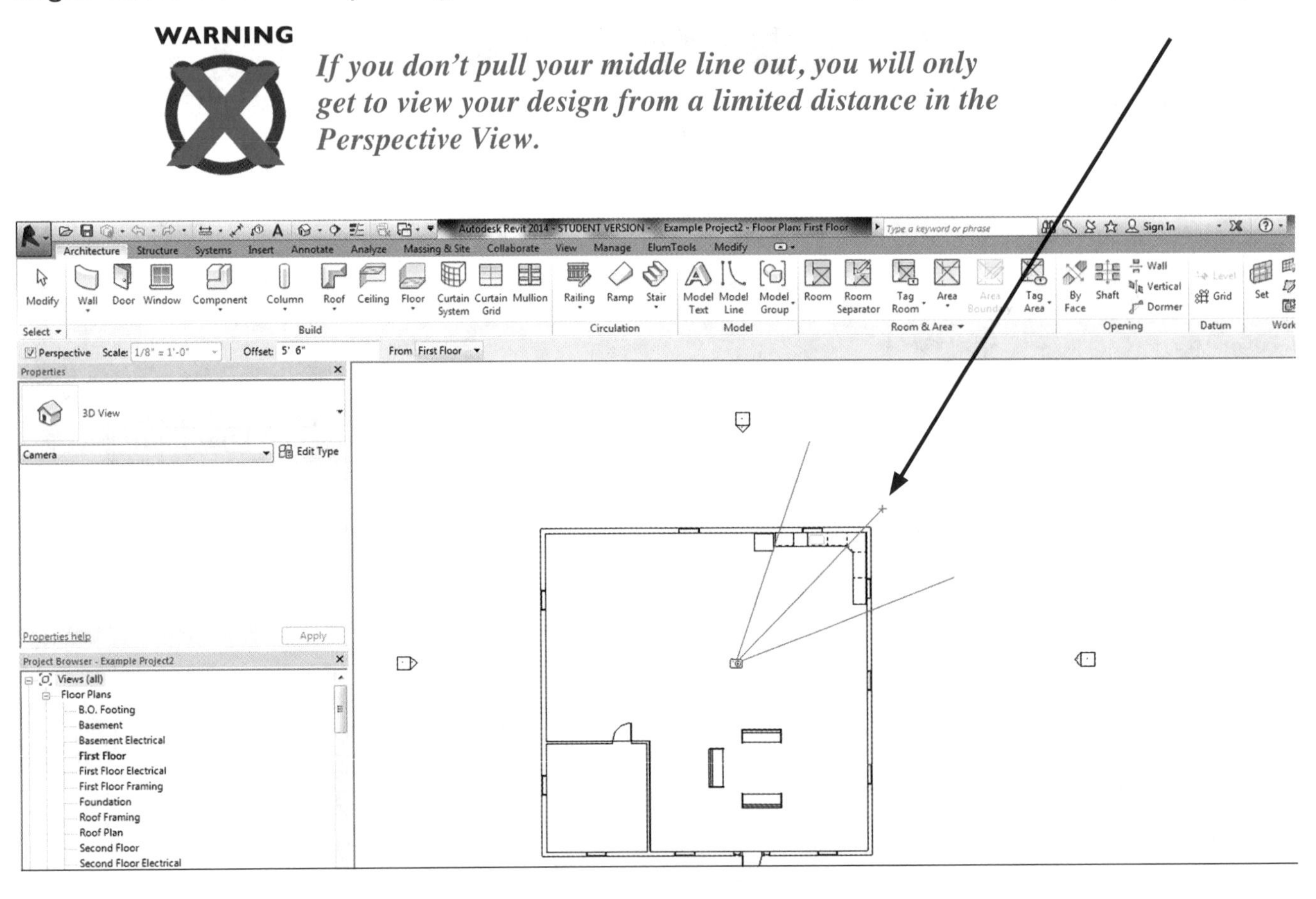

Click the mouse, and your drawing will then switch to a Perspective View. The new view is now in your Project Browser, **3D VIEW1** (Revit Default name).

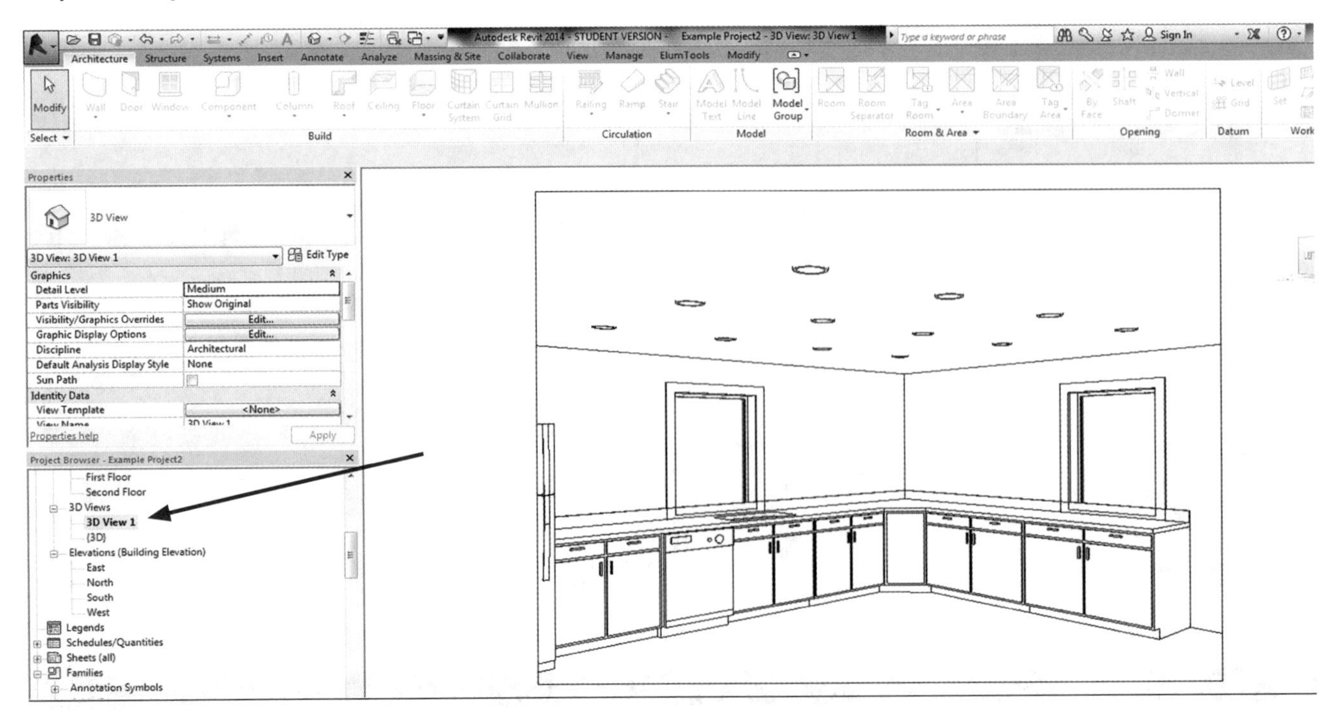

You can rename the view to whatever you want by right-clicking on the name and clicking on **RENAME**.

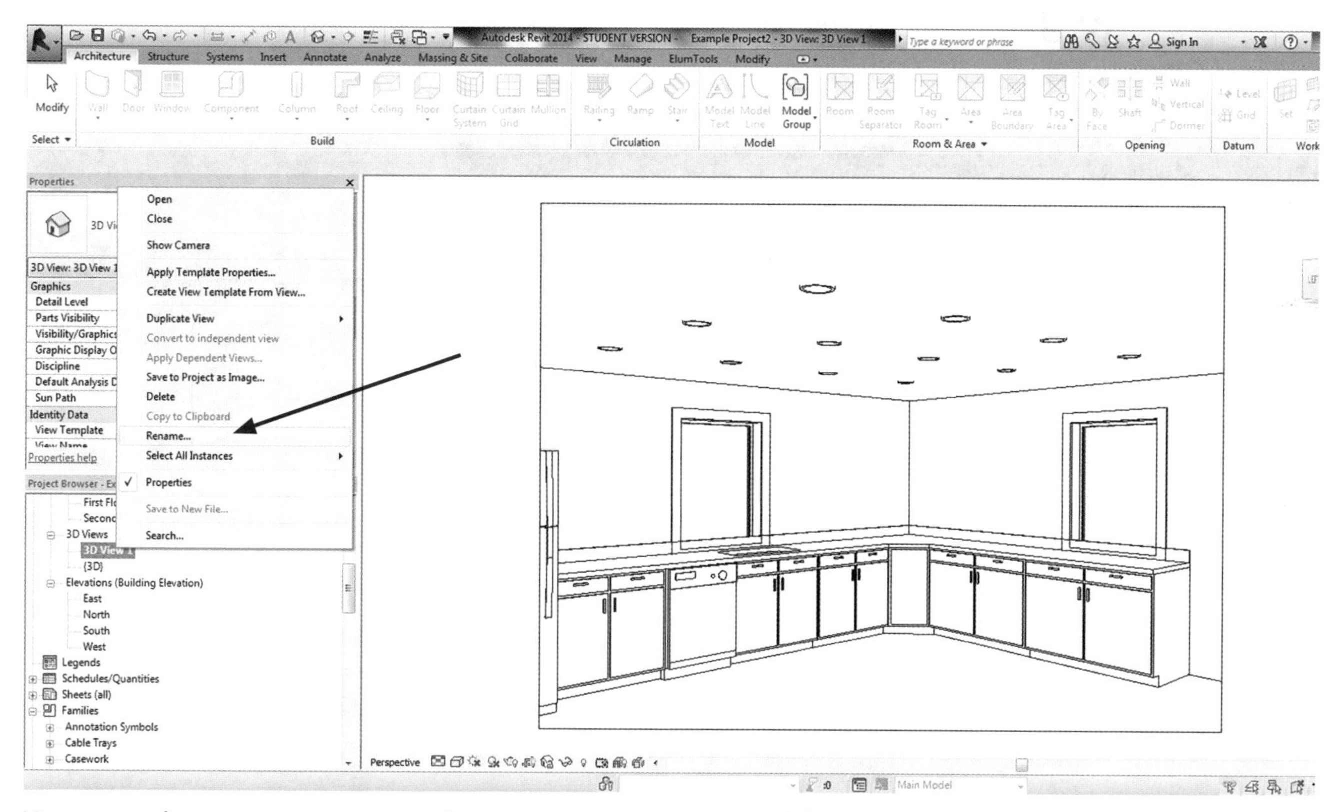

You can take as many camera views as you want or need for your project. Go to the Properties Box, scroll to the floor plan you want, and repeat the steps in the beginning of the chapter.

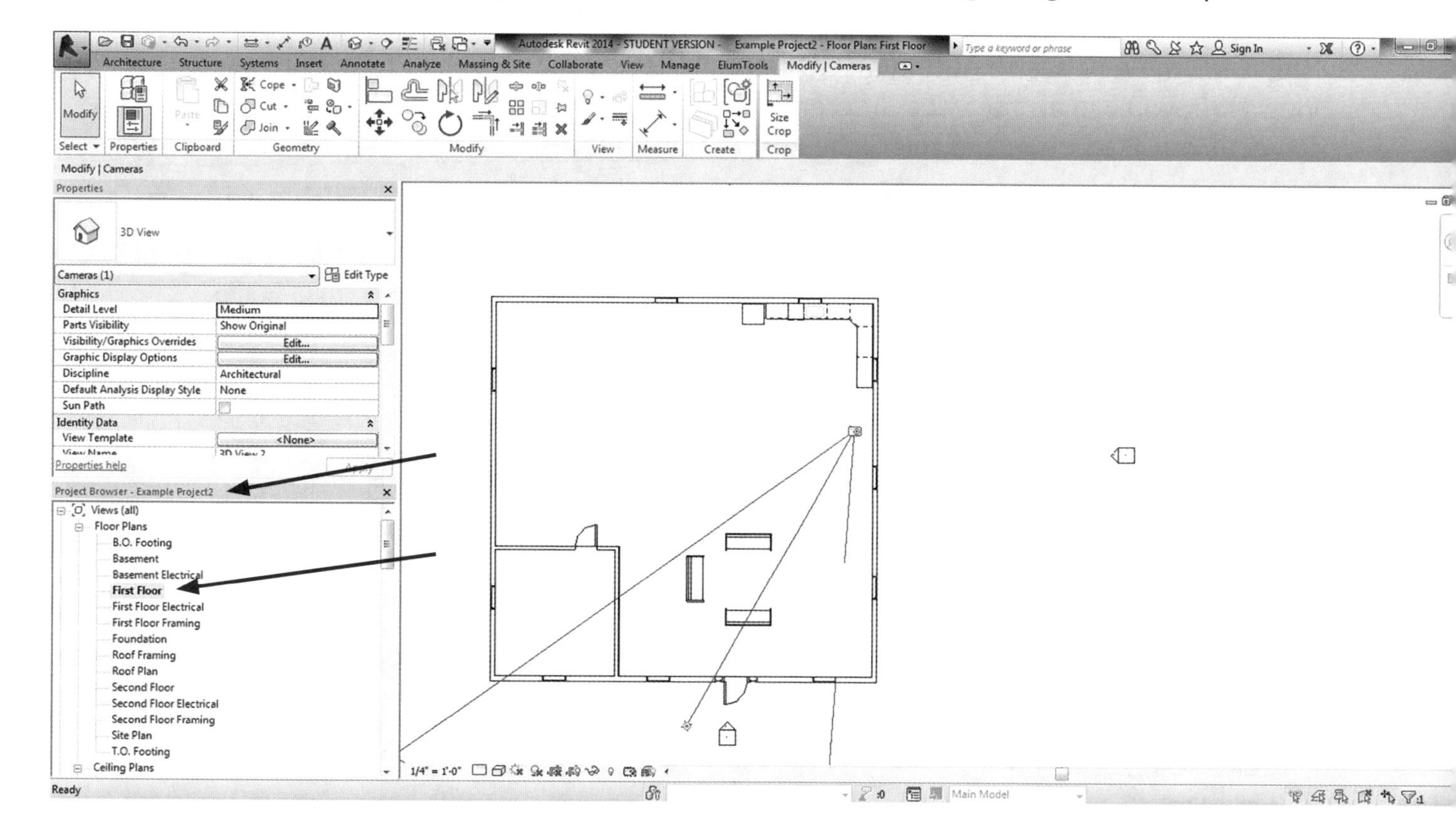

EXAMPLE

e.g.

The following is an example of a 3-point perspective view. (It's pretty cool, huh?)

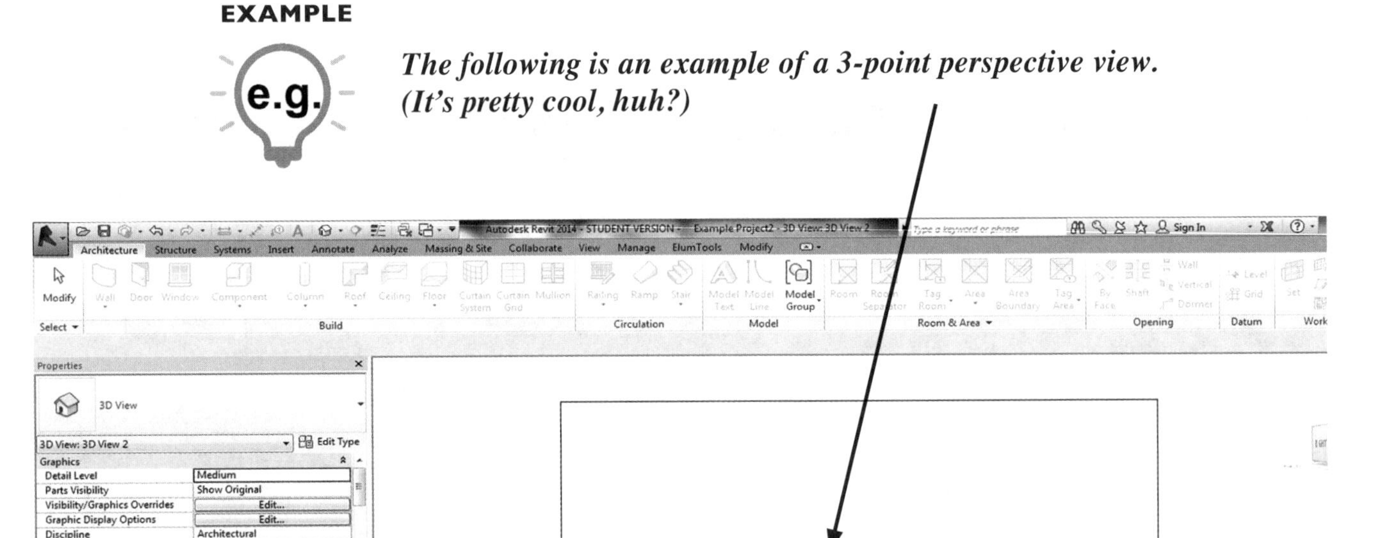

EXTERIOR CAMERA VIEW

Even though Revit has an Automatic 3D Default View, located in the Project Browser, under 3D Views, 3D, you can also select a different **EXTERIOR CAMERA** view. Make sure you are on the correct Floor Plan View, and select the **FIRST FLOOR PLAN.**

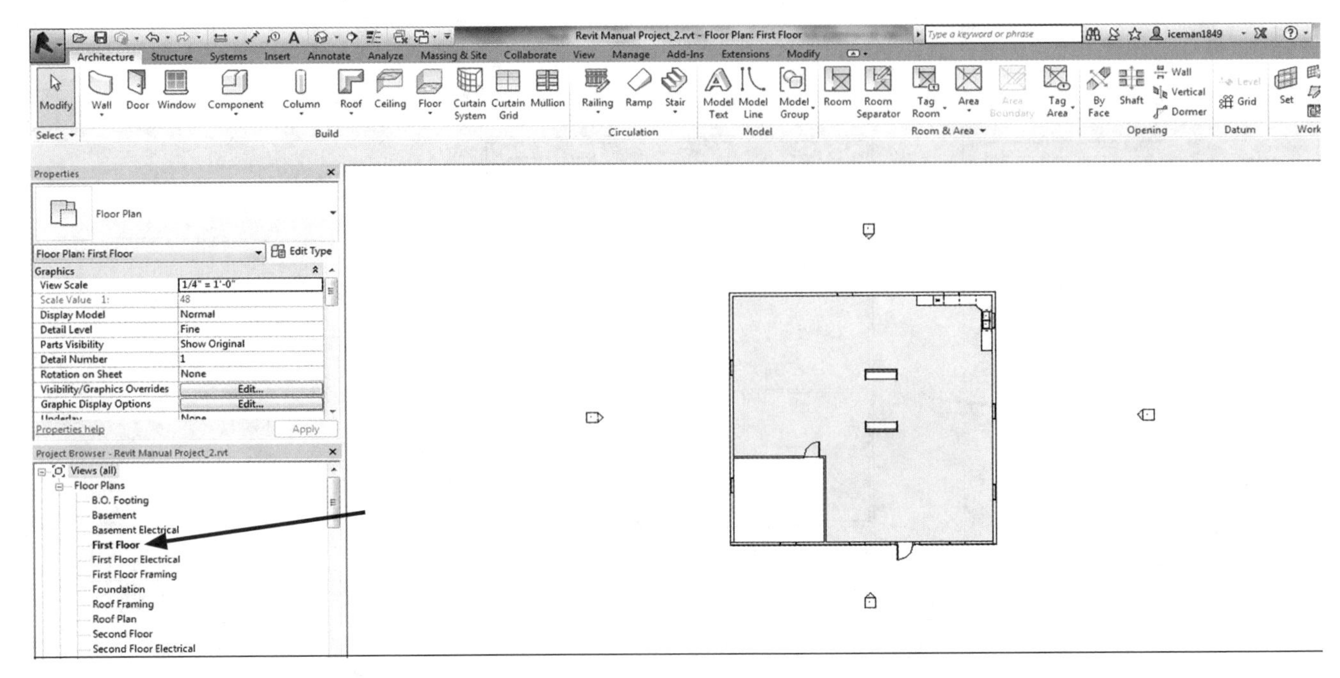

Then click on the **DROP-DOWN ARROW** next to the house in the *QAT* (*QUICK ACCESS TOOLBAR*) Toolbar.

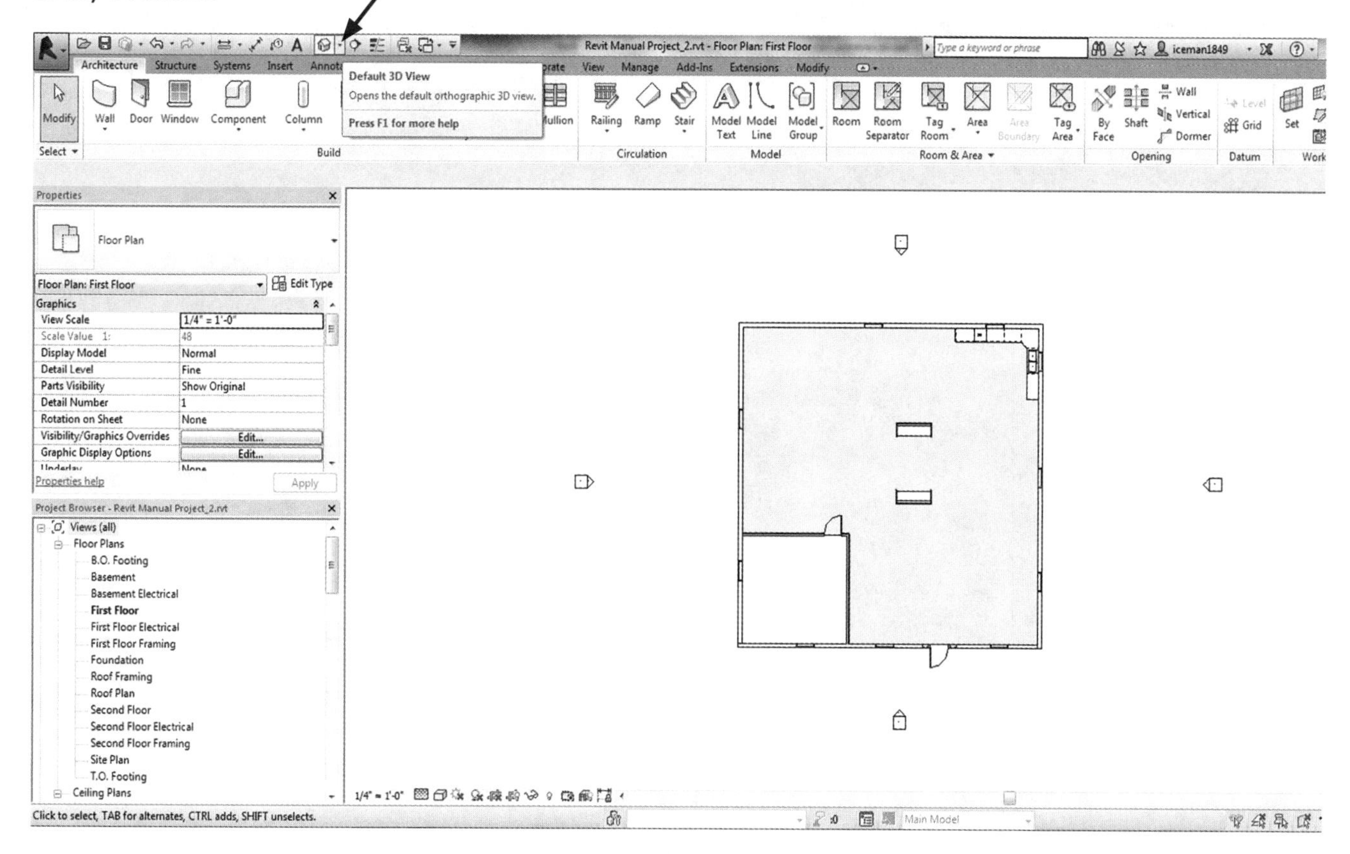

Here you will see three different options. Select the **CAMERA**.

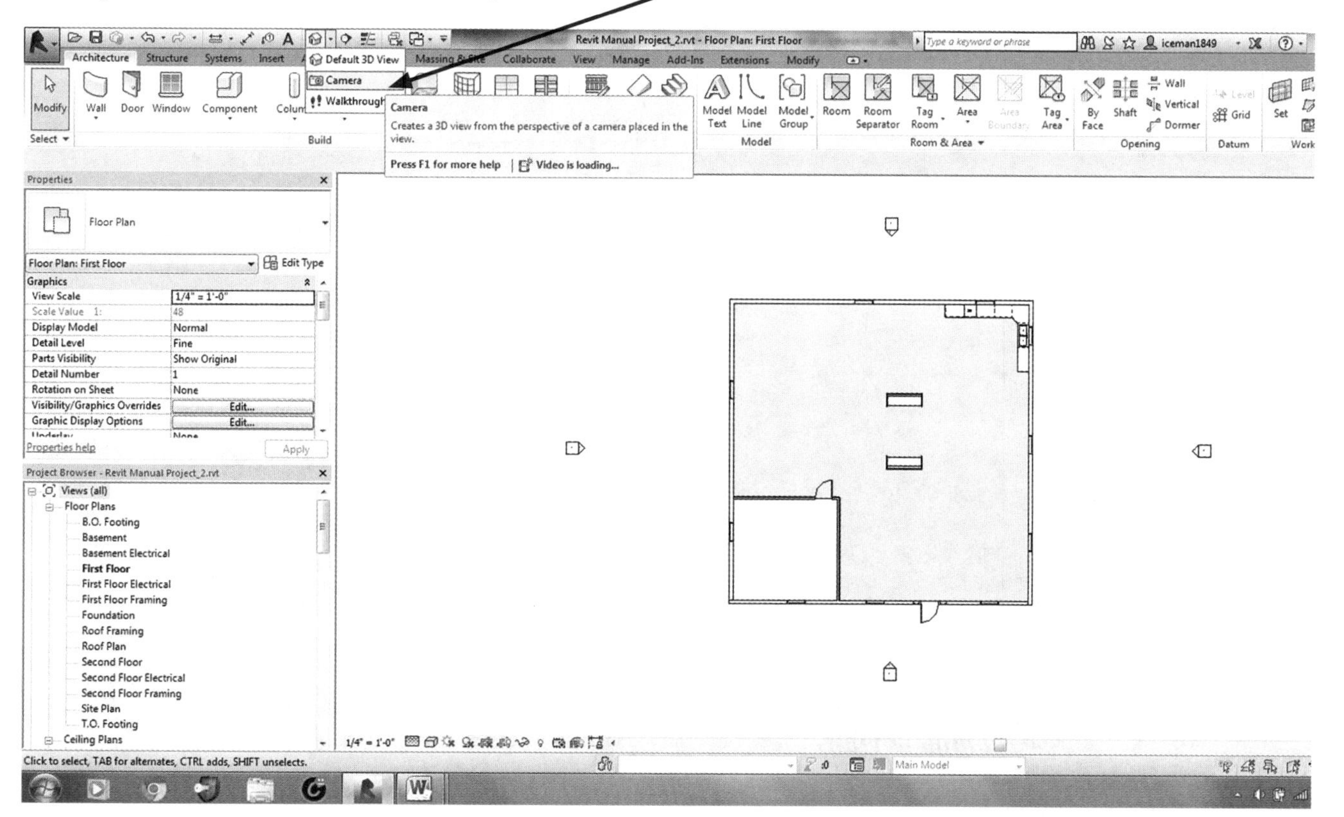

This will allow you to place a camera around the *outside* of the house or building.

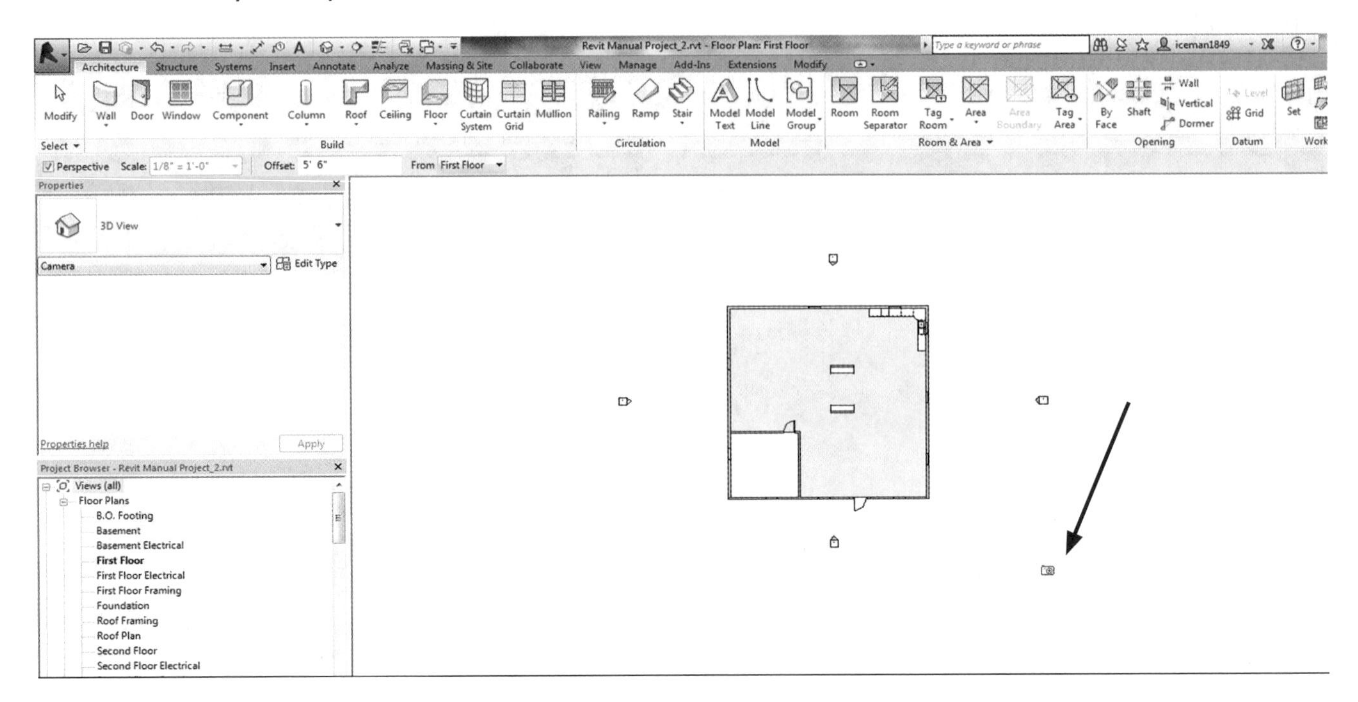

Once you decide where you want to place the camera to get an outside view, click with the mouse and expand the view target lens by pulling the middle line or target line to your desired sight view, by pulling the middle sight line through the other side of your house or building.

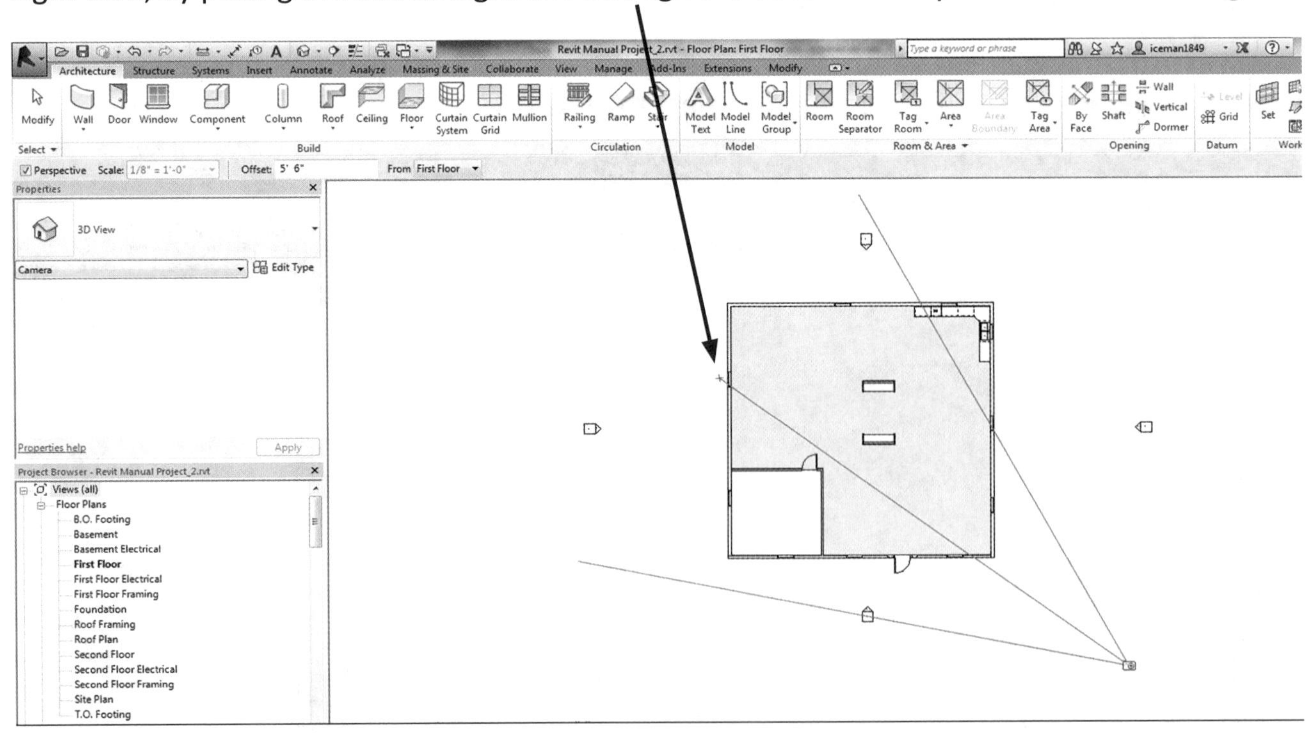

TIPS

Pull the target lens completely through the house or building to see the complete view.

The new 3D view will automatically be displayed.

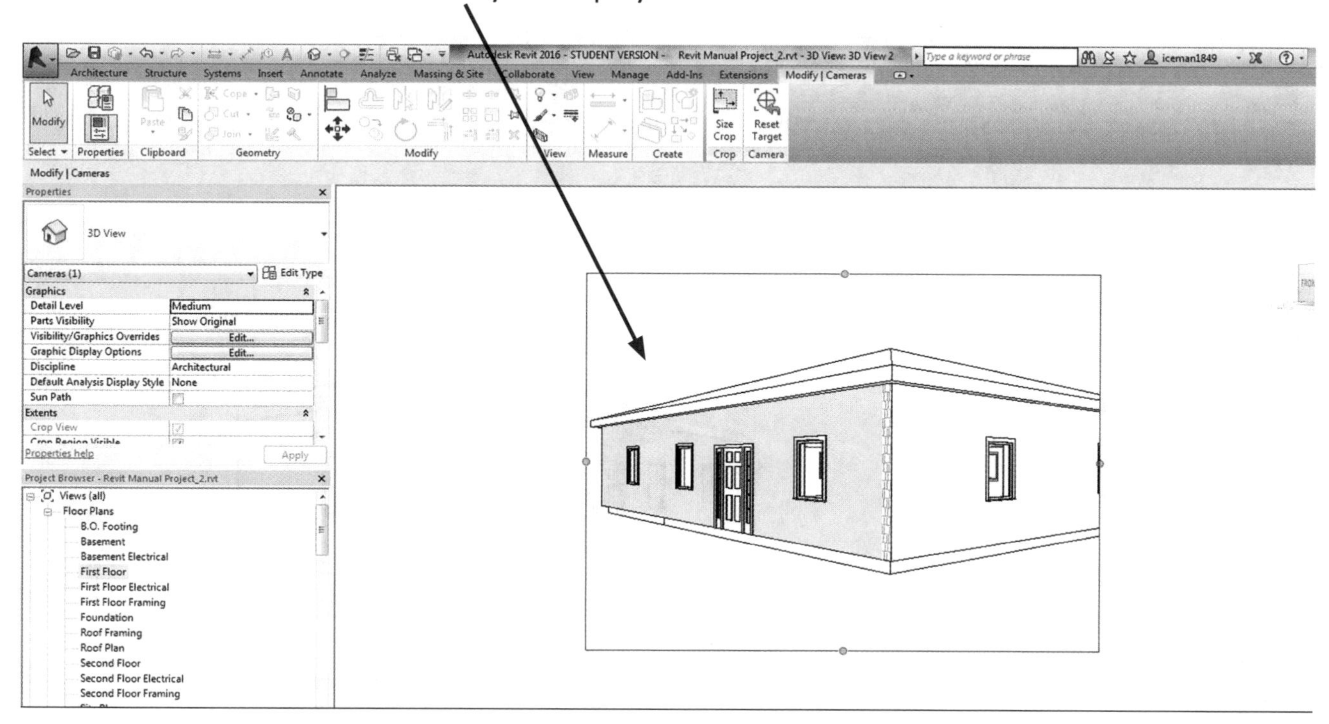

If you scroll down in the PROJECT BROWSER in the 3D View category, the new view will be listed or recorded, and you can right-click on the name and rename the view to keep track of all the different views you create.

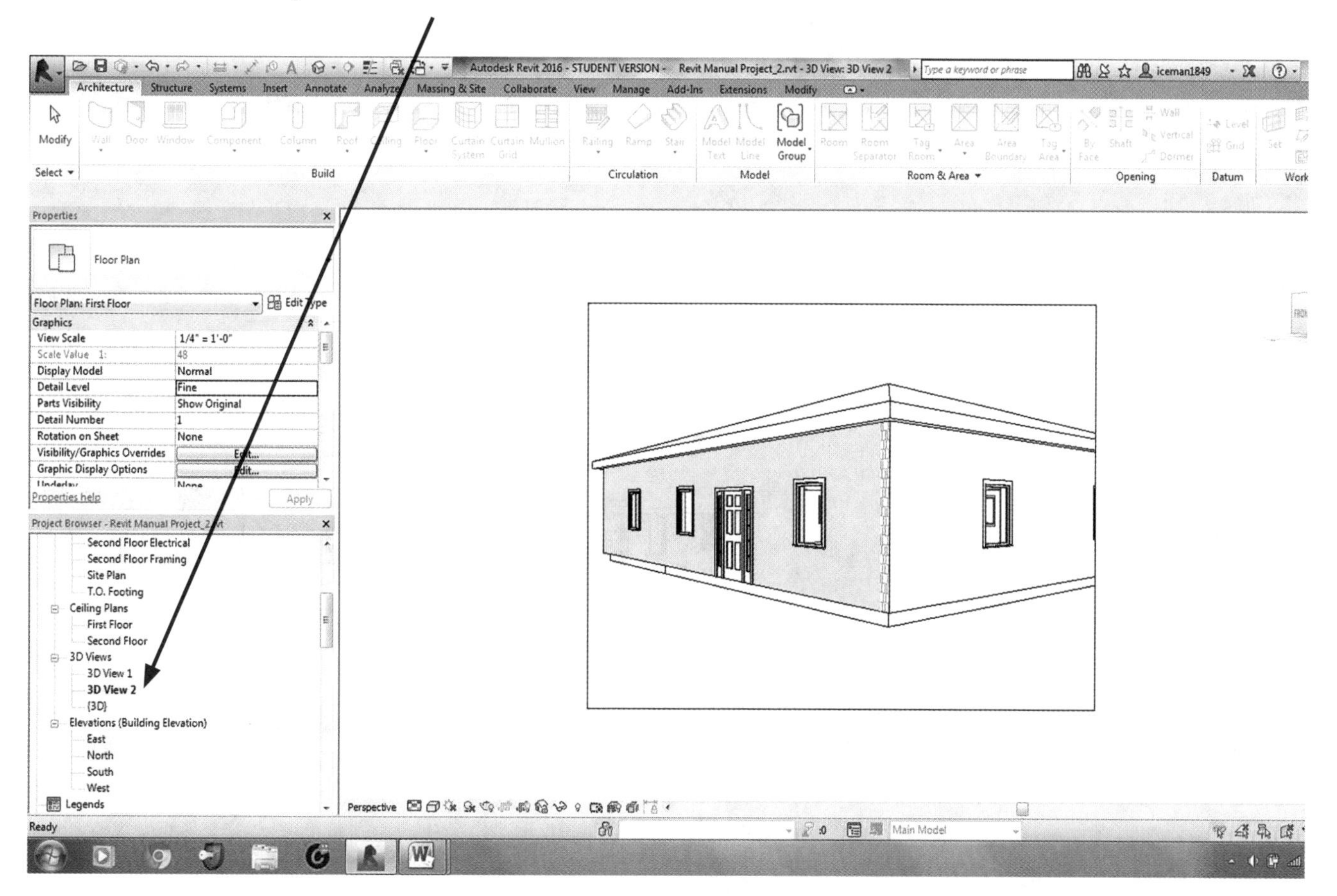

Rename the new 3D view.

Autodesk Revit 2016 - STUDENT VERSION - Revit Manual Project_2.rvt - 3D View: 3D View 2
Architecture Structure Systems Insert Annotate Analyze Massing & Site Collaborate View Manage Add-Ins Extensions Modify
Modify Wall Door Window Component Column Roof Ceiling Floor Curtain System Curtain Grid Mullion Railing Ramp Stair Model Text Model Line Model Group Room Room Separator Tag Room Area Area Boundary Tag Area By Face Shaft Wall Vertical Dormer Level Grid Set
Select Build Circulation Model Room & Area Opening Datum Work
Properties
3D View
3D View: 3D View 2
Graphics
Detail Level
Parts Visibility
Visibility/Graphics
Graphic Display
Discipline
Default Analysis
Sun Path
Extents
Crop View
Properties help
Open
Open Sheet
Close
Show Camera
Apply Template Properties...
Create View Template From View...
Duplicate View
Convert to independent view
Apply Dependent Views...
Save to Project as Image...
Delete
Copy to Clipboard
Rename...
Select All Instances
Properties
Save to New File...
Search...
Ceiling Plans
3D Views
{3D}
Elevations (Building Elevation)
East
North
South
West
Legends
Perspective
Main Model

Rename the view in the dialog box and click **ENTER**.

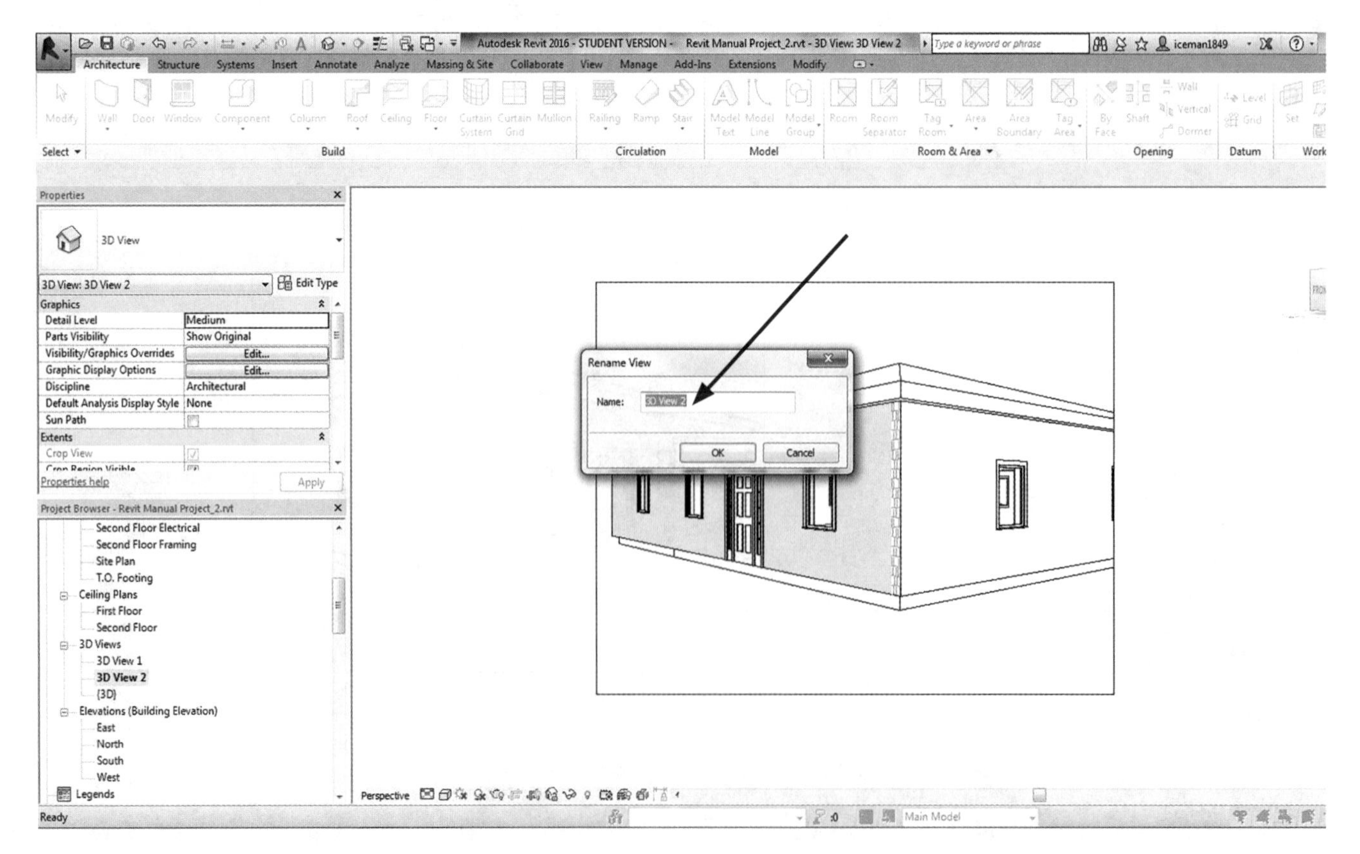

This is truly where the Revit software starts to shine and show its real power in the world of real-life design. Not only can you now design your house the conventional way in 2D or flat on paper, but with the camera views and the 3D capabilities, you can instantly see how your house or building looks or is even beginning to take shape. This really sells a project to your clients (not to mention yourself).

Now is not the time to stop, it's just getting interesting! Let's continue and add colors and shades, by forging ahead to Chapter 16, *Rendering Interior Views*.

CHAPTER 16

Rendering Interior Views

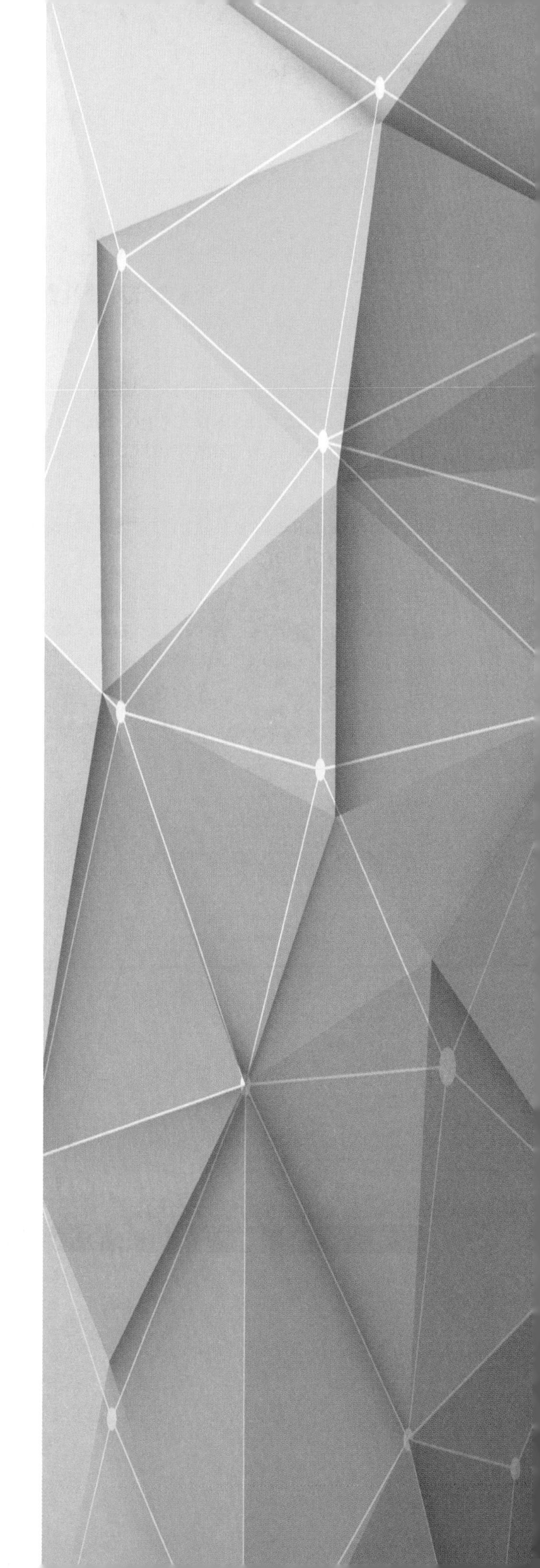

RENDERING INTERIOR VIEWS

After creating the 3D Interior View, you should render the view for presentation. This shows the client what the finished product or design will look like and brings life to the design. You are able to see the lights and colors the client picked out for their house or building.

First, make sure you are in the 3D view you want to render then click on the **TEA POT** in the View Control Bar.

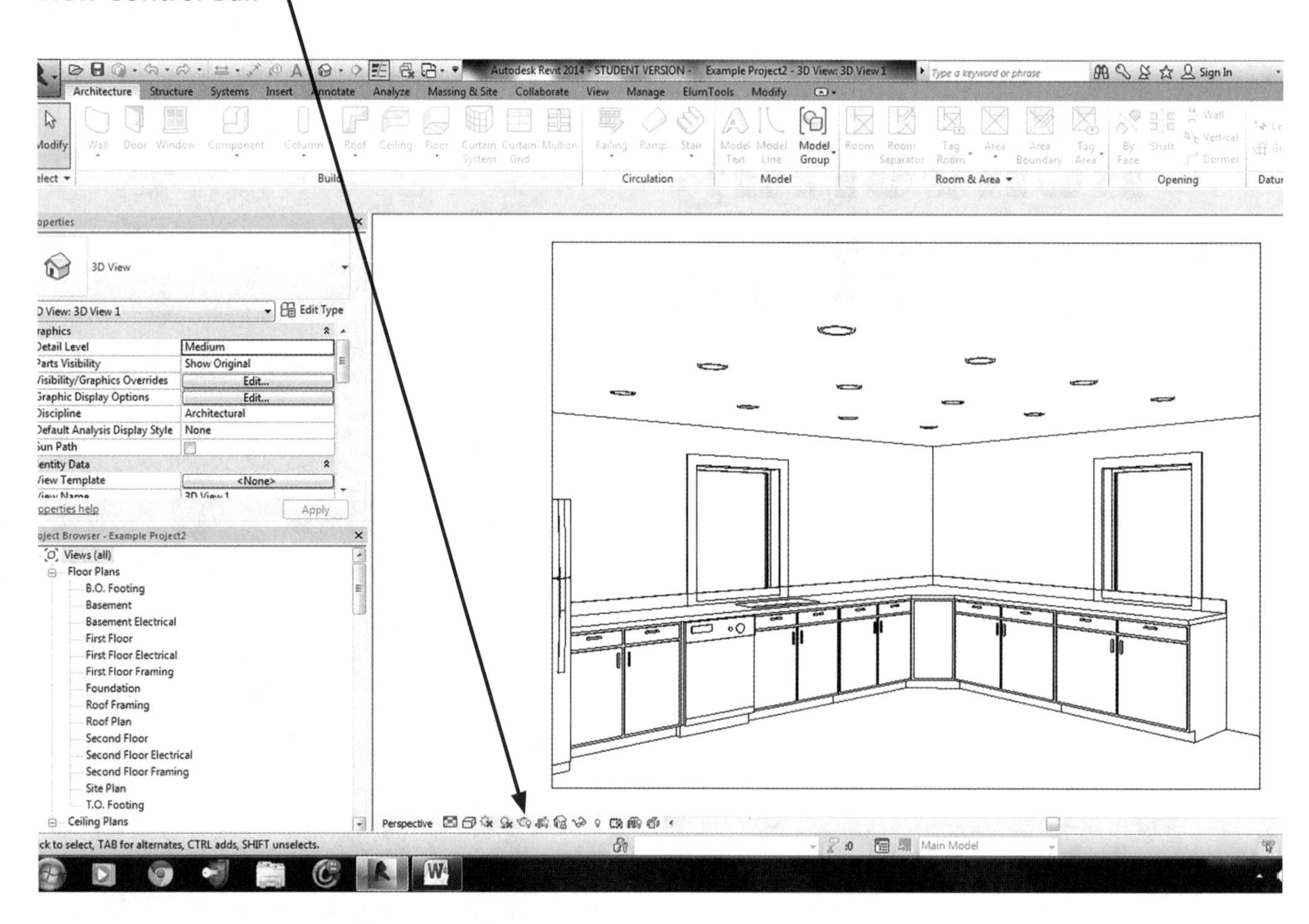

Once you do this, you will see a Render Dialog Box.

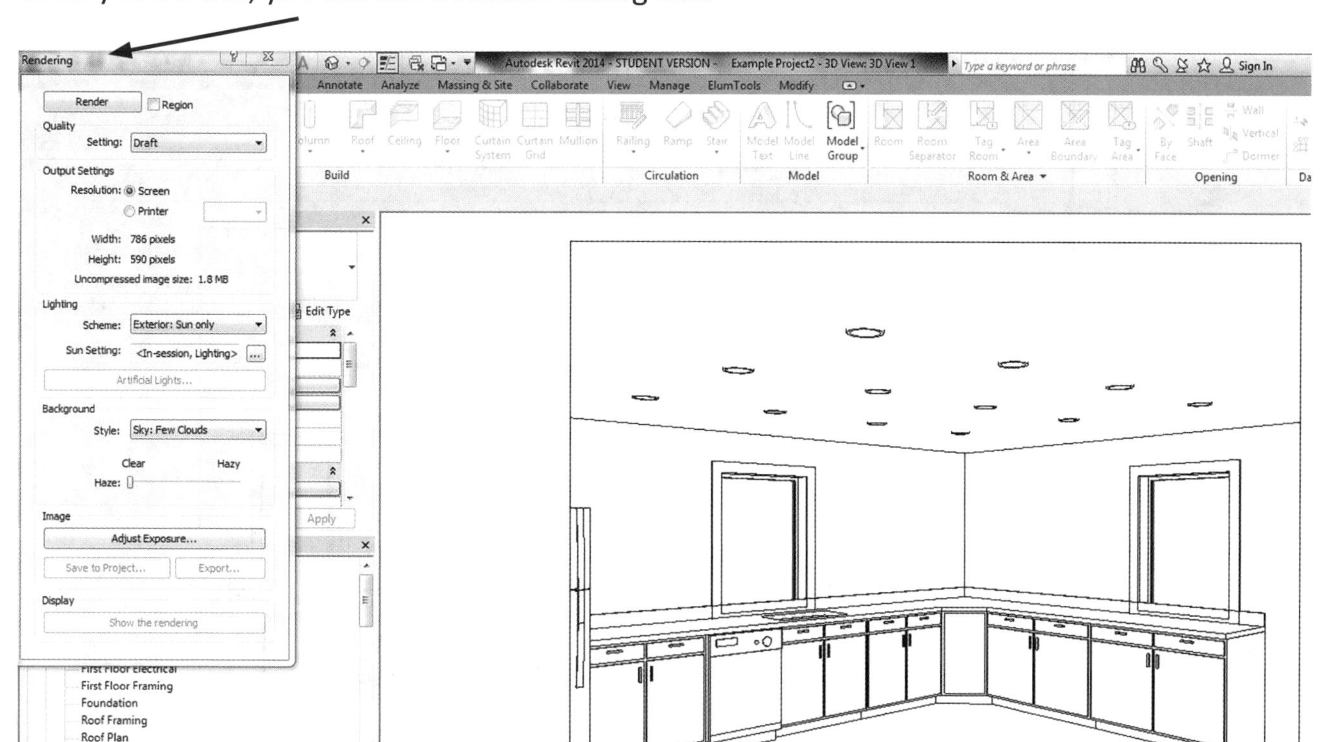

This Render Dialog Box includes several settings the operator needs to understand. These setting options are:

Draft: A quick render of how the drawing will appear that doesn't take much time. You can make changes to these easily.

Low: A little slower than the draft setting, and the pixels are smaller, so the image is beginning to become clearer.

Medium: Slower than the Low setting, and the pixels are much smaller and closer together. The image is getting more defined.

High: Takes longer to develop, and the pixels are *much* smaller. The image is even further defined.

Best: This setting takes a long period of time, such as 12-24 hours to develop, with the pixels becoming much tighter and much more defined, resulting in a photo-realistic picture of the rendering.

In the Settings Box, select **LOW** or **MEDIUM.**

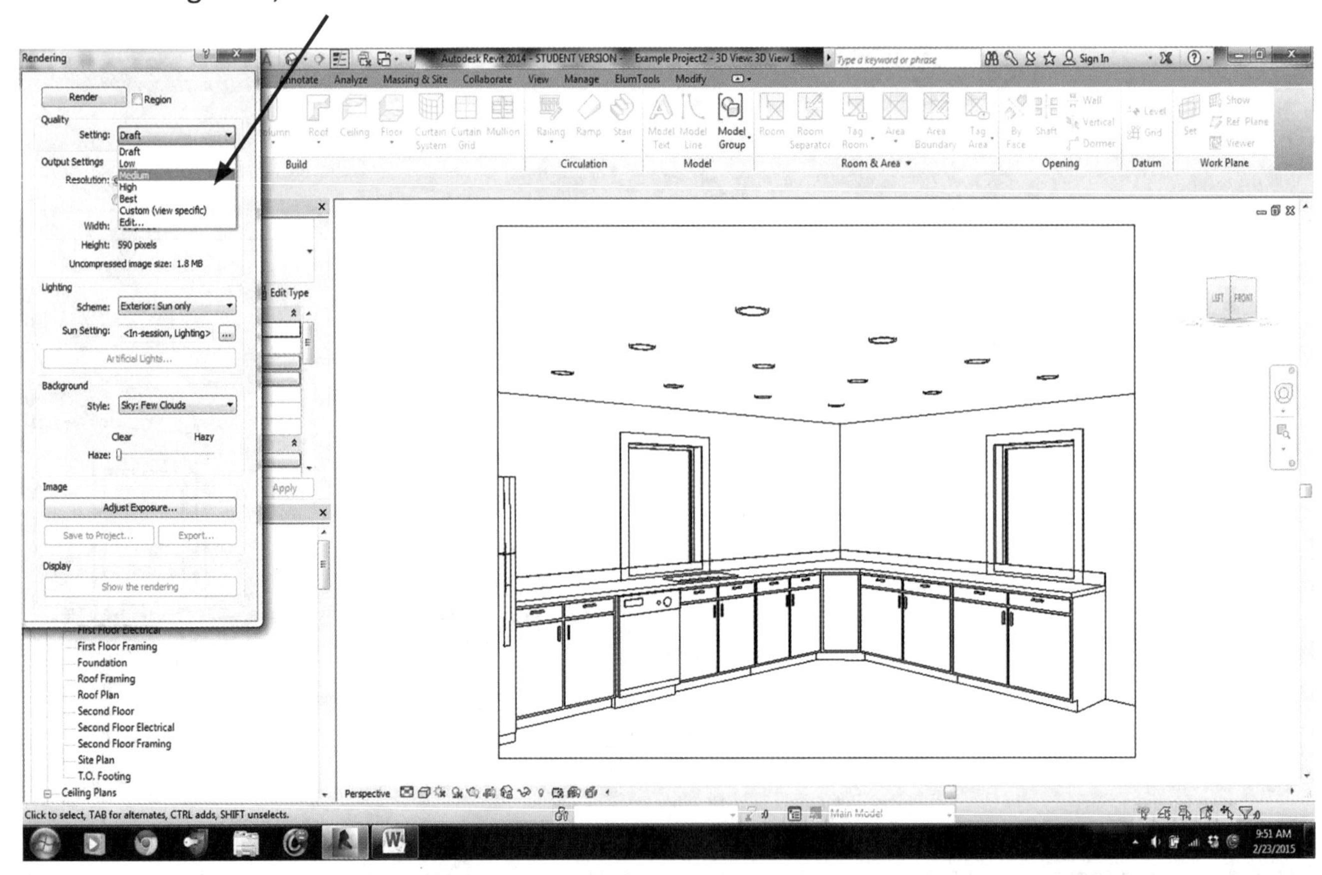

Once you have selected the quality setting for your rendering, go to the Lighting Tab. Then select either **INTERIOR: SUN AND ARTIFICAL LIGHTS** or whichever type of lighting you will use.

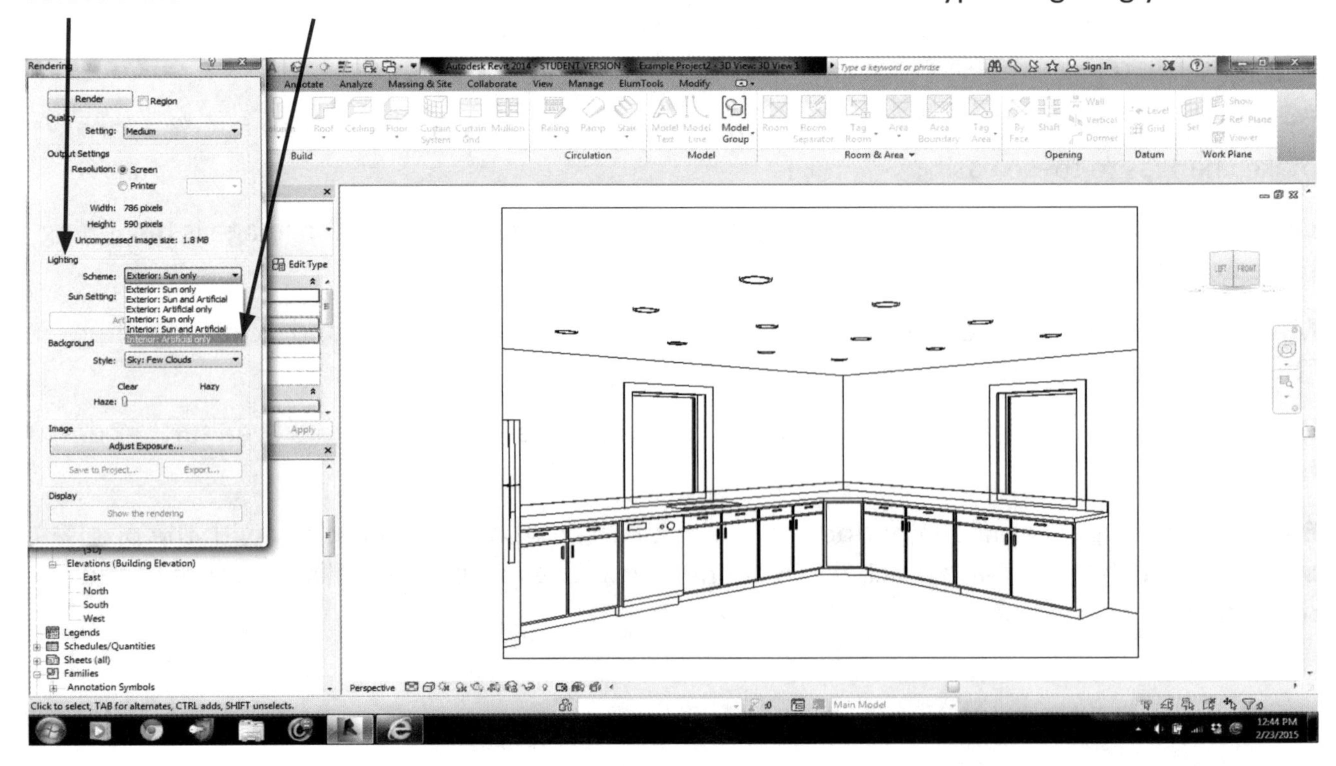

Once that is completed, click on **RENDER.**

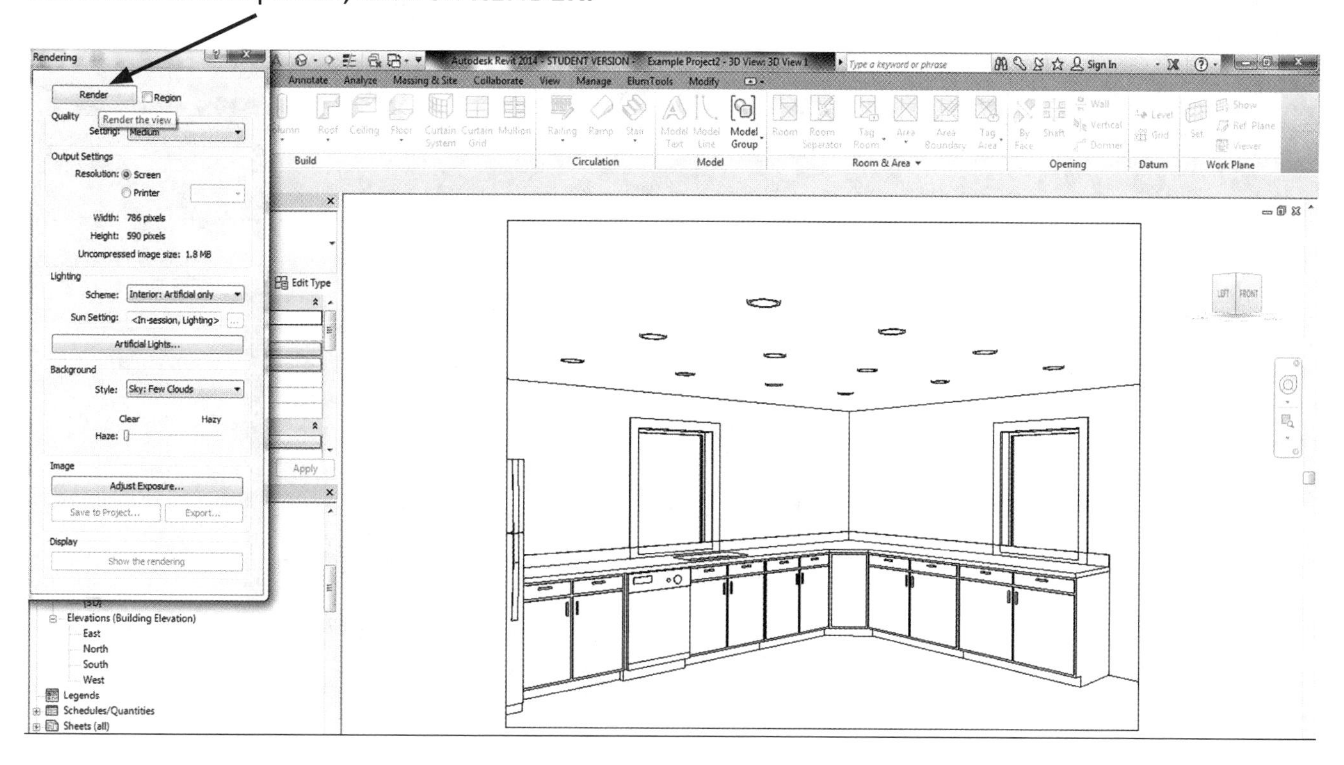

You will see a progress bar appear to let you know that your drawing is in the process of rendering.

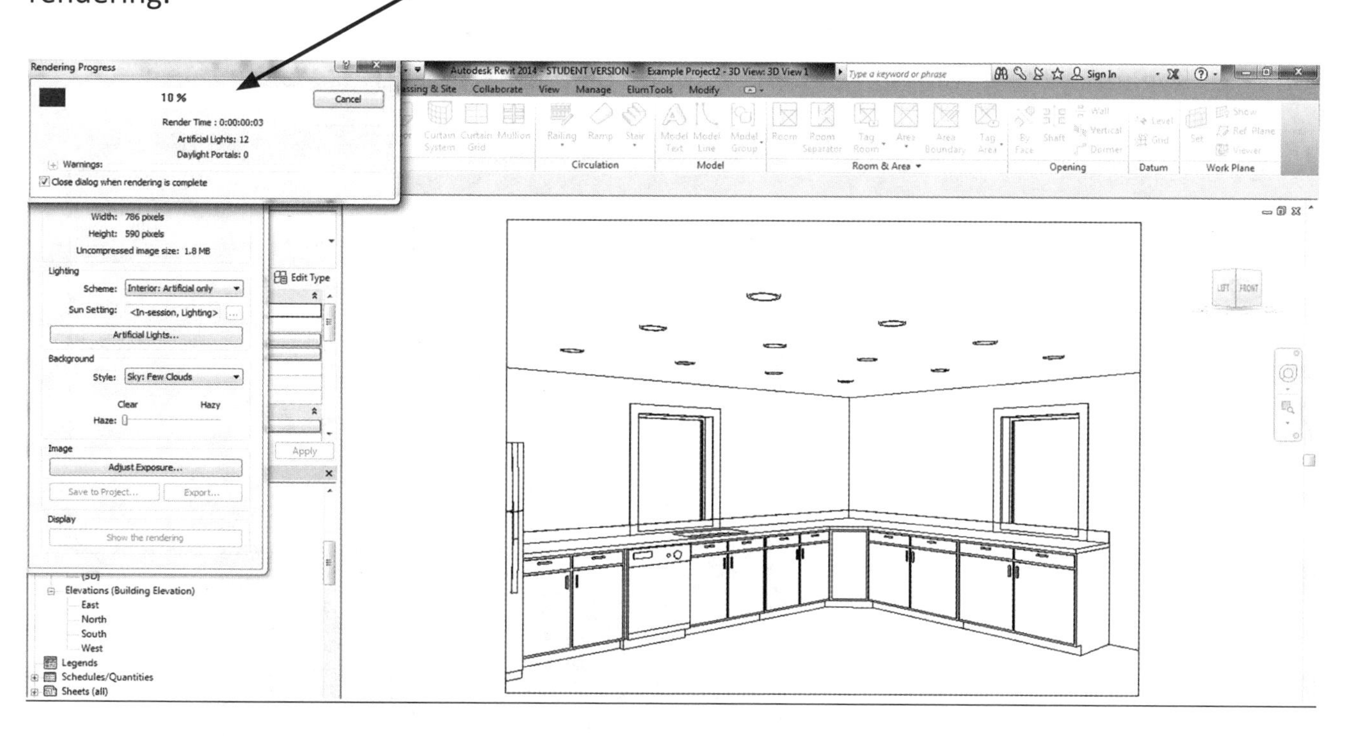

NOTES

Your rendering is complete when the progress bar reads 100%. Then the progress bar disappears.

Once the render is complete, you can then go to the Adjust Exposure Tab and begin adjusting the exposure to enhance the lighting.

Click on the **EXPOSURE VALUES, SHADOWS AND WHITE POINT** to see how final rendering will look. You can make things brighter or darker, cooler or warmer, greyer or more intense.

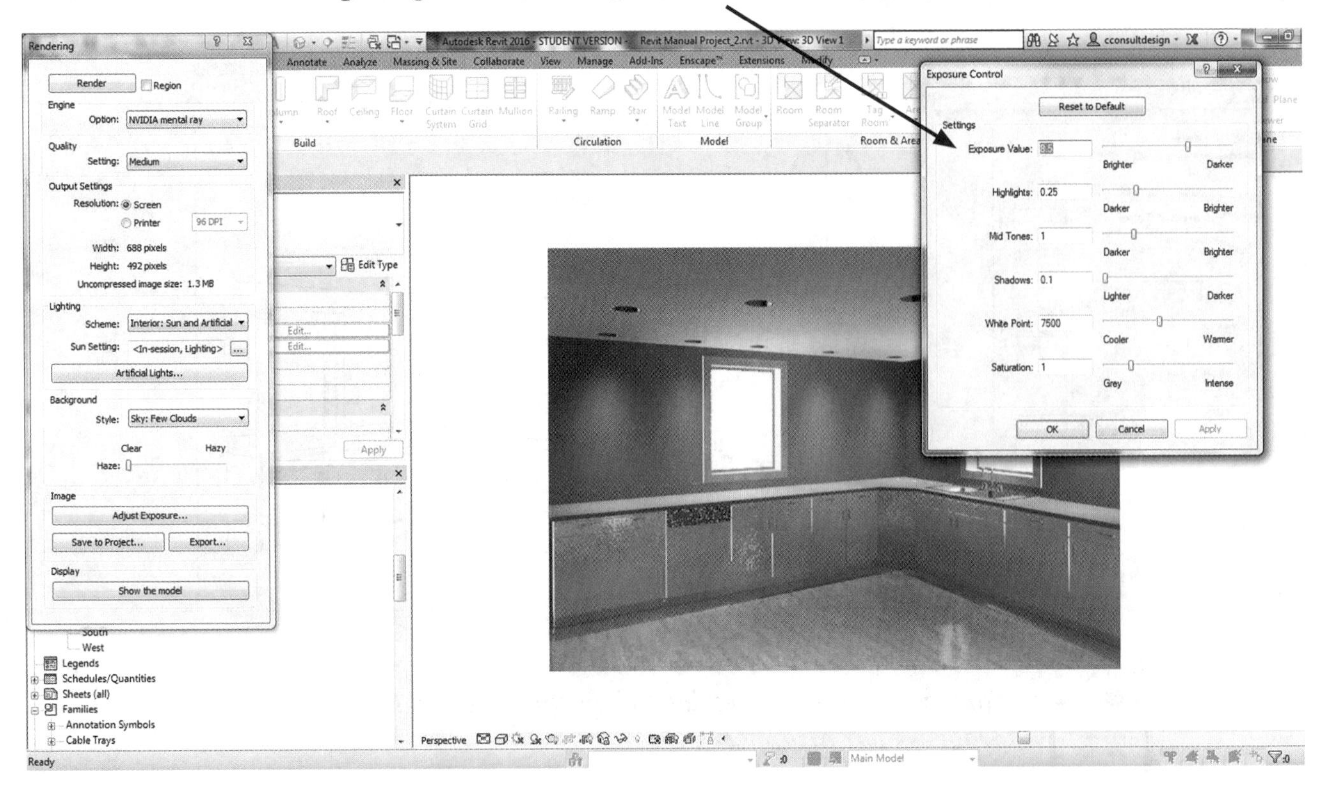

TIPS

Just click the* APPLY *button and see how things look before you click* OK*, so you don't have to repeat the steps again and keep opening up the Exposure Control Box again and again.

If you are happy with your rendering, then click **OK** and you can save the rendering to your project.

You can also save it as a .JPEG to an outside file on your desktop with the Export Button, or to any file location of your preference.

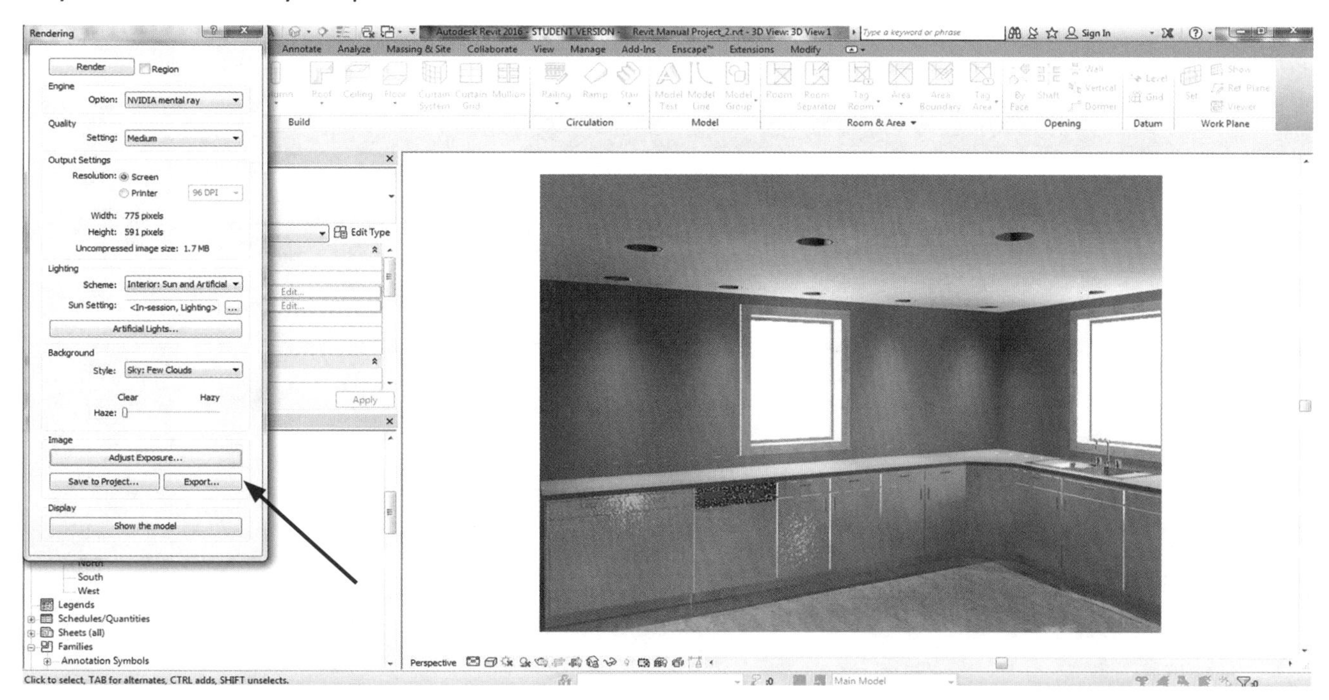

The rendering phase of Revit really brings out the power of the software and allows the designer to show off what the inside of the project will look like when completed or even what a renovation will look like compared to the existing house or building. This has become a major feature that clients have been requesting in designs, and some people don't offer this in their layouts. This can be a huge selling point for you or your firm.

Now let's look at Chapter 17, *Inserting Objects from the Revit Library and the Internet*. This really expands your design possibilities.

CHAPTER 17

Inserting Objects from the Revit Library and the Internet

INSERTING OBJECTS FROM THE REVIT LIBRARY

When there are objects or items you need that Revit doesn't already have, you need to either insert them from the Revit Family or from the Internet.

To insert items from the Revit Family, these are items or families that are already in the Revit software, but you have to load from a different file location, so you click on the **INSERT** Tab.

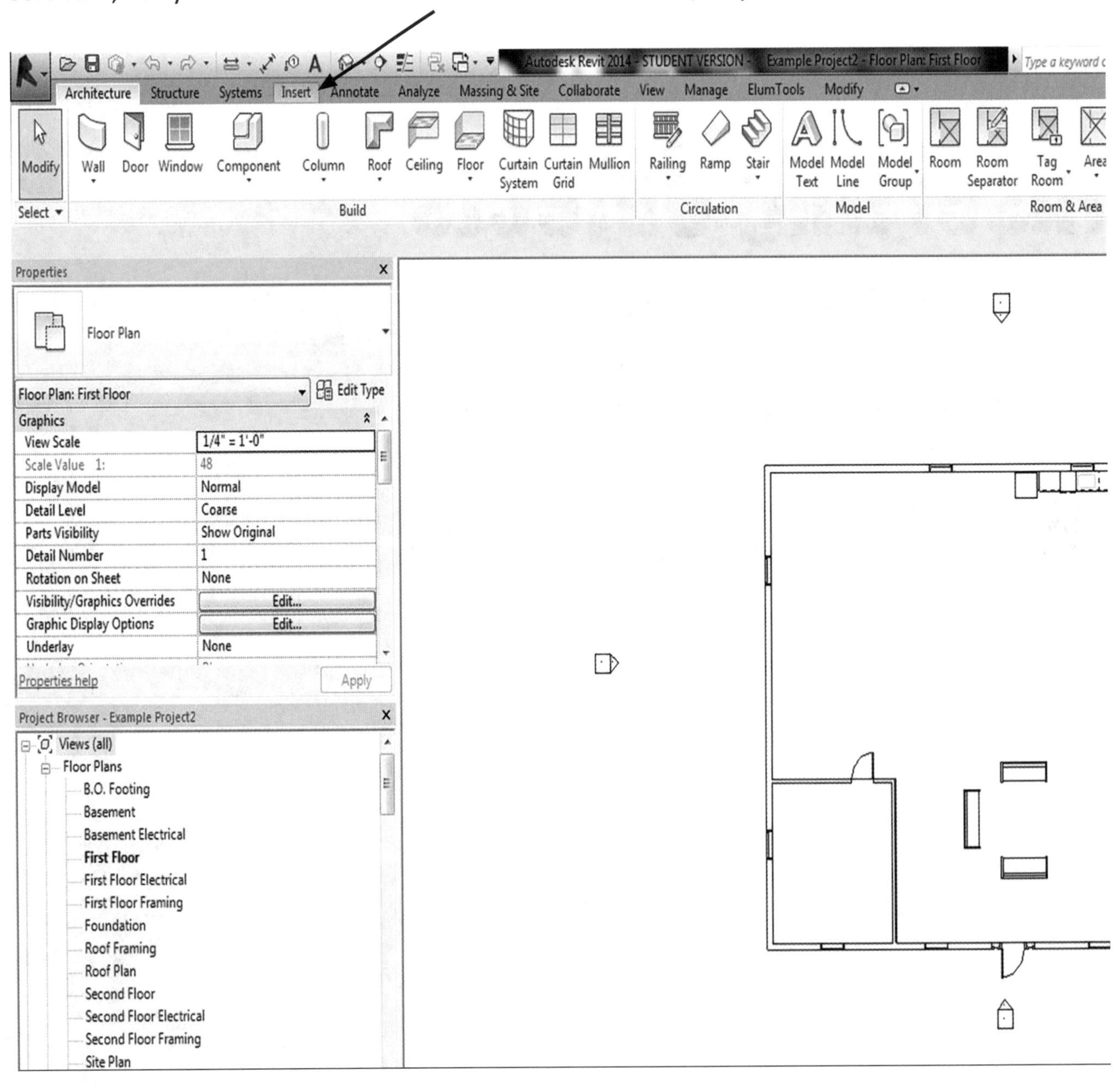

Then click on the **LOAD FAMILY** Tab.

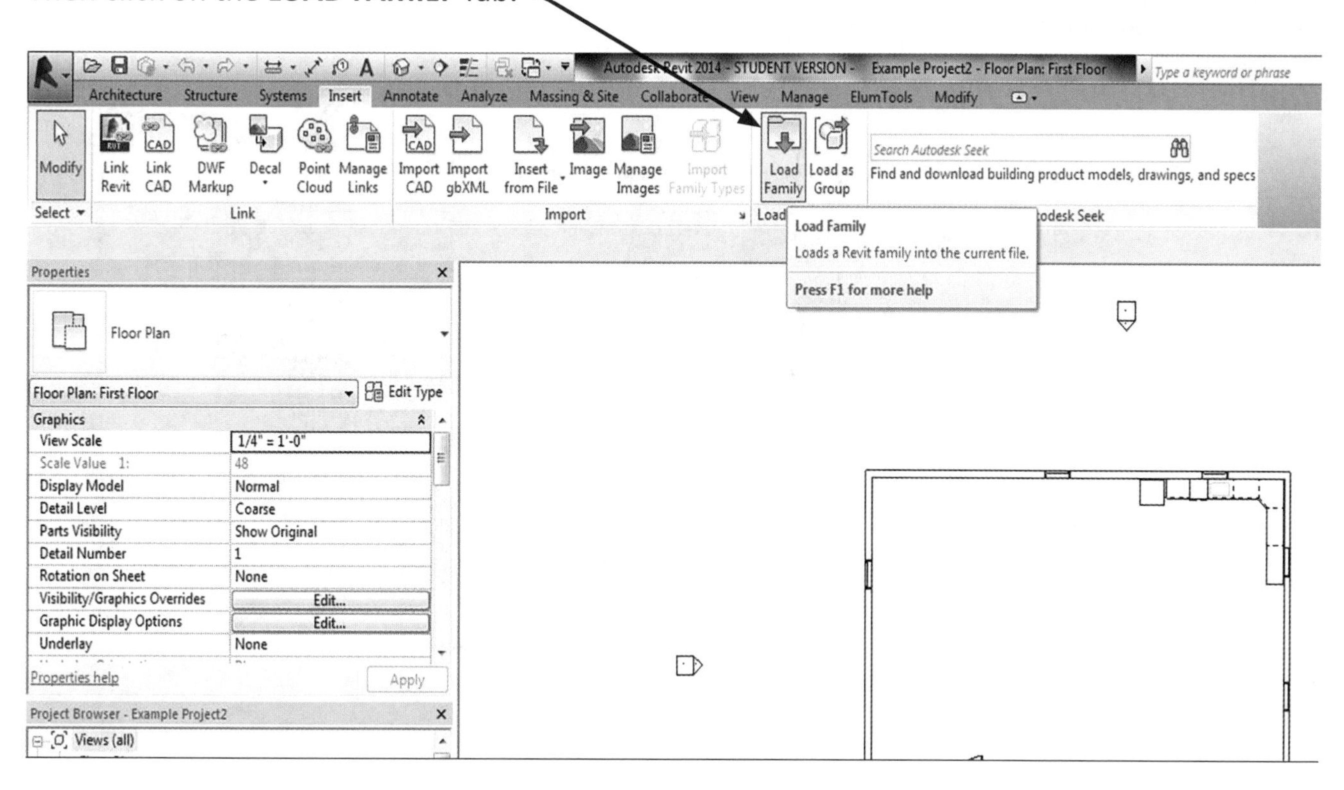

You will then select the folder of the category of the items you want to load in your design.

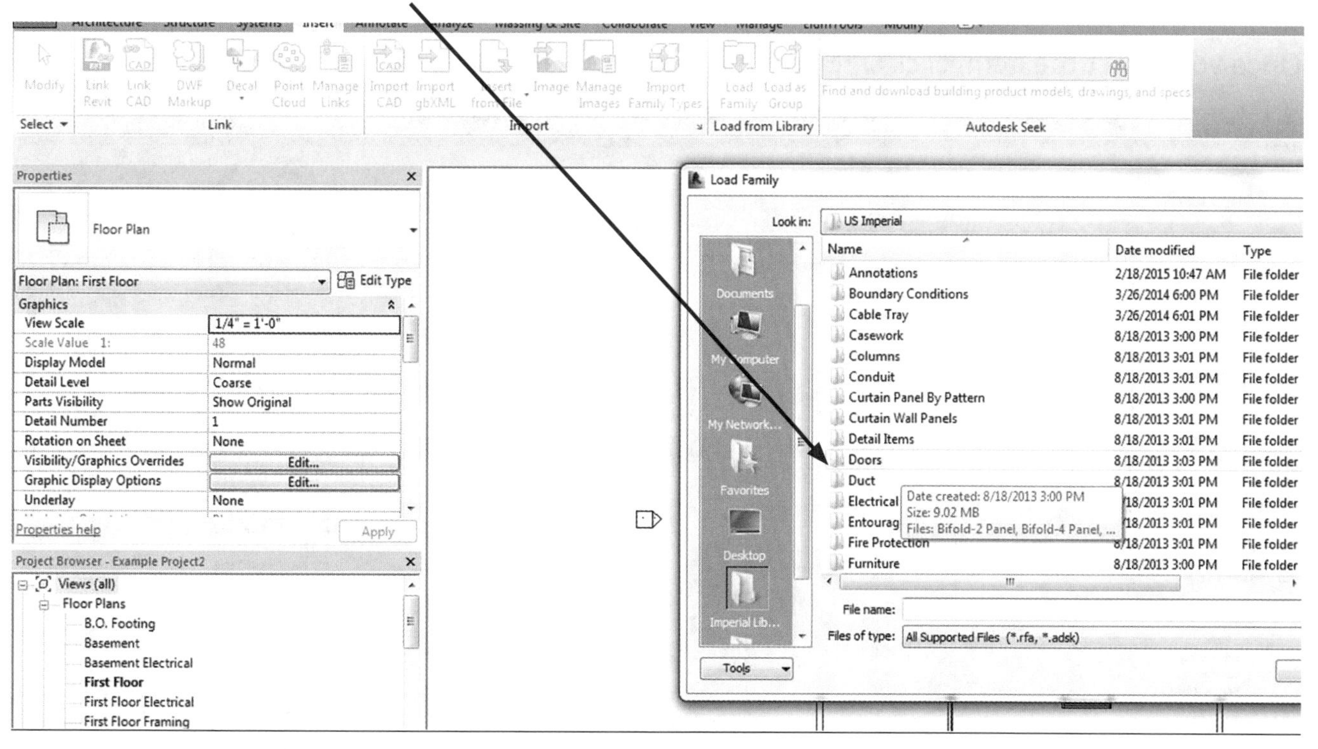

EXAMPLE

In this scenario, we're going to add to the door options.

Open the folder, and select the specific item you want. You can see the view of the item of your choice on the right side of the screen.

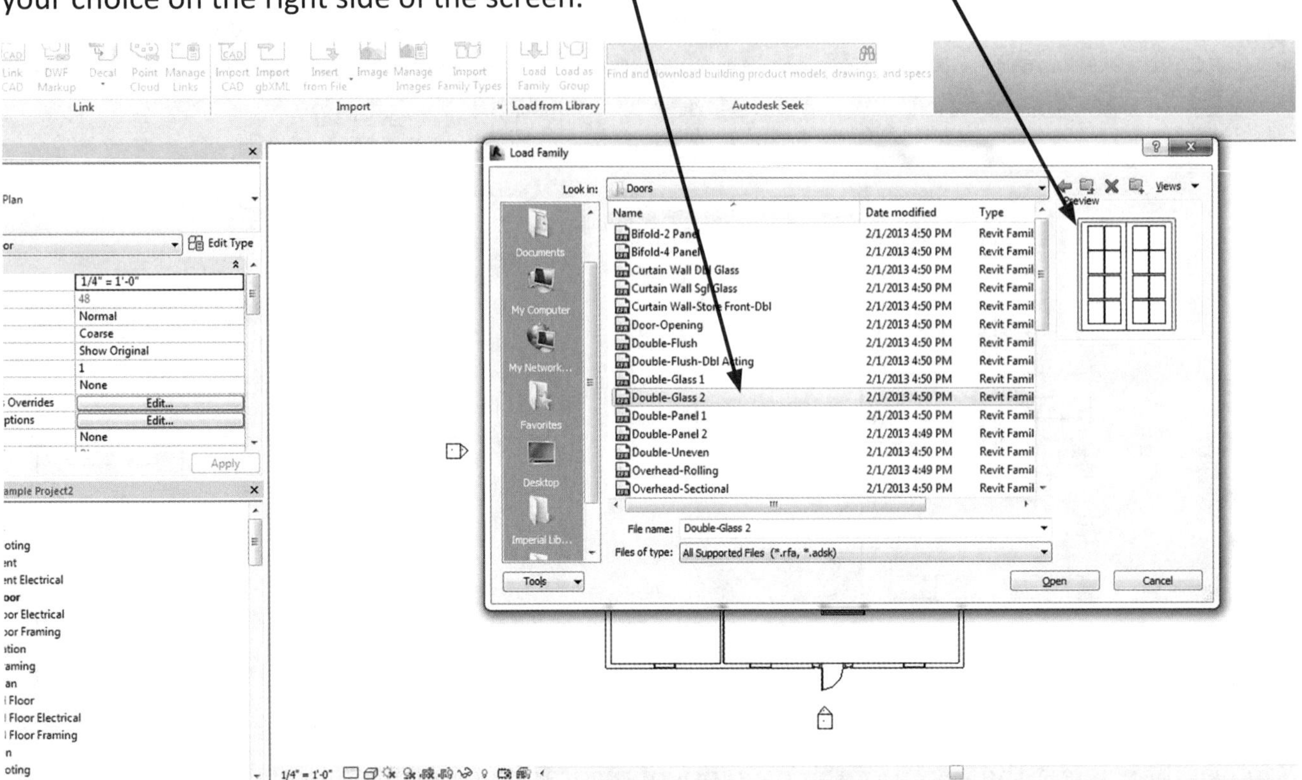

Once you select the item, click **OPEN,** and it will automatically load into your project.

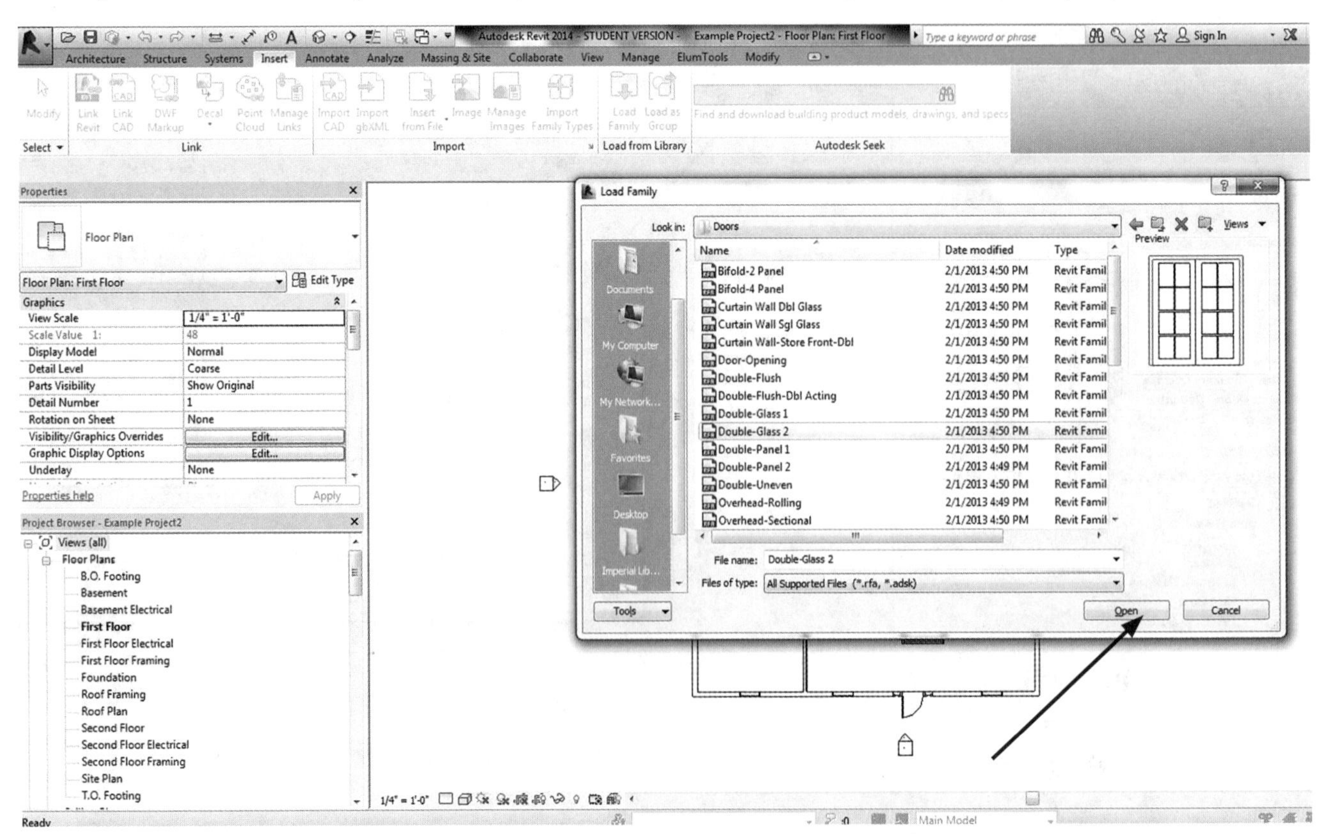

If you can't find a specific object or item you want that's not listed in the Revit library, you need to download it from the Internet. The easiest and most efficient way to get more items and objects is to go to the Internet to the Revit Content Distribution Center.

First go to your search engine.

TIPS

Google Chrome works really well for this application.

In the search bar, type in: **REVIT 2009 CONTENT DISTRIBUTION CENTER.**

TIPS

The reason I recommend the 2009 version is that the 2010 and newer versions of the families are in French. It connects you to the BIM OBJECT ***website, which is*** *not* ***user friendly, in my opinion, through its slowly working out the kinks.***

Once in the Revit Content Distribution Center, the screen will look like the following:

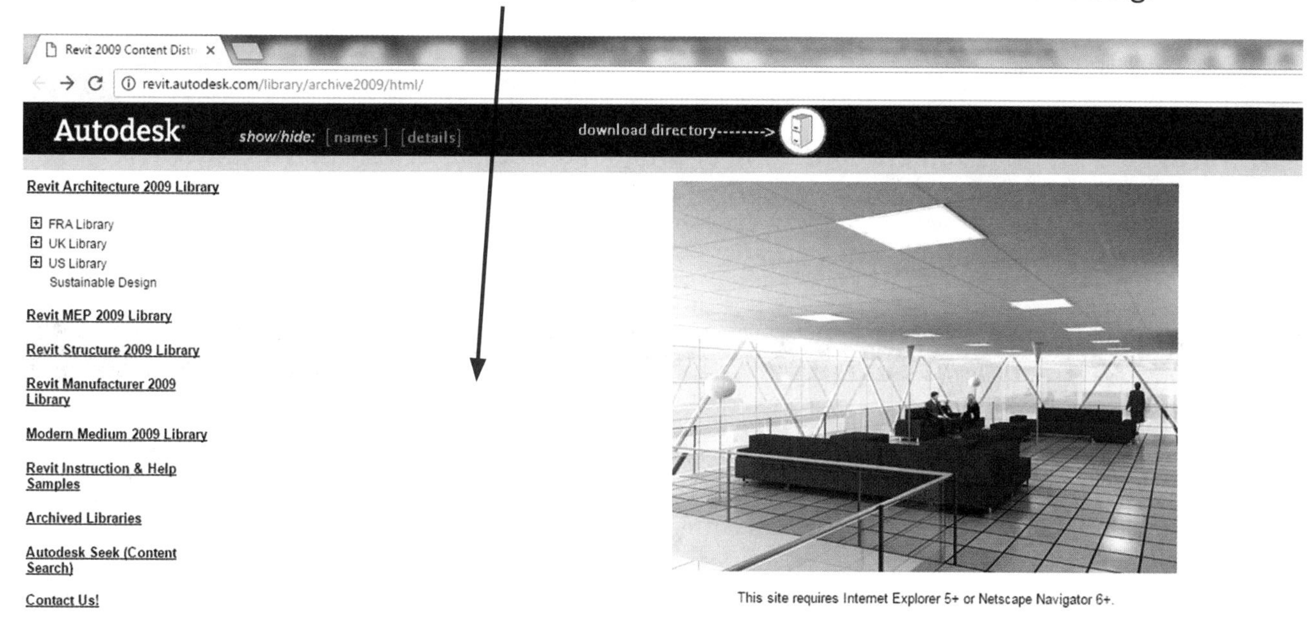

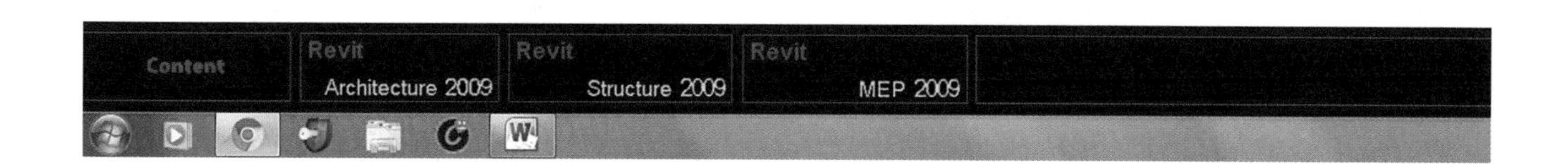

The site defaults to the Architectural Library and you will see the following tabs: FRA Library, UK Library and US Library. Select the **US LIBRARY** drop-down symbol. You can press the **+** symbol to expand the tree.

This will open a lot more folders consisting of other Family Categories. You can pick more objects that the Revit software doesn't already offer.

When you click on one of the categories such as **FURNITURE**, a set of furniture pictures will open up on the right side, so you can see the specific type of furniture in the library.

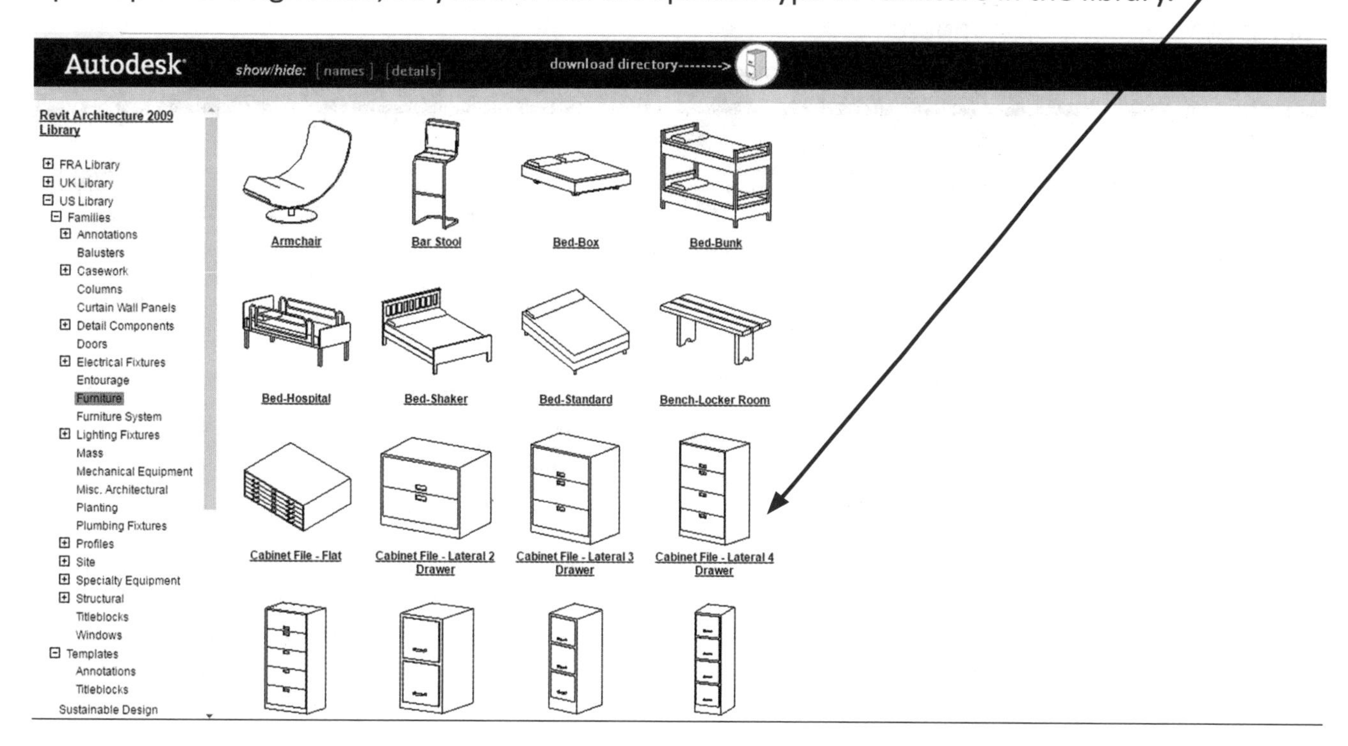

Once you decide on the type of furniture family you want to bring into your Revit project, you *must* click on the **FAMILY NAME**, and not the object itself.

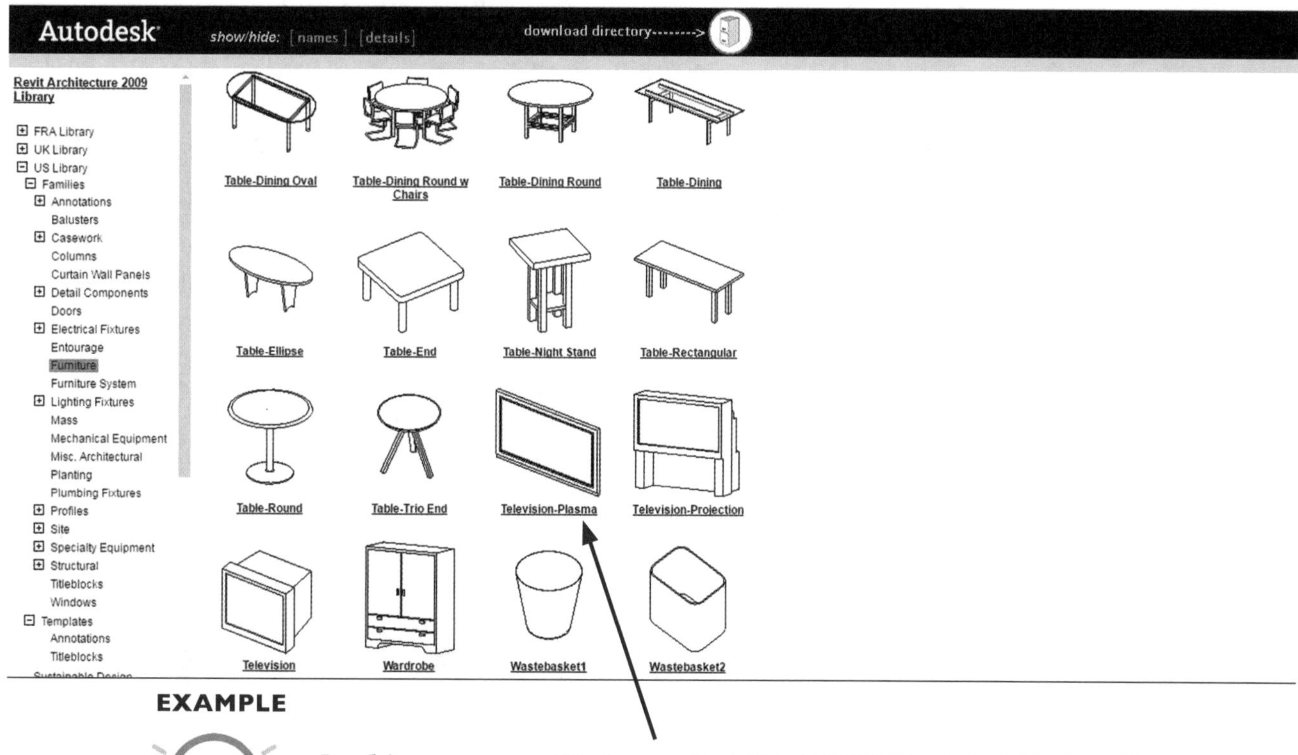

EXAMPLE

e.g.

In this case, we will choose the TELEVISION-PLASMA ***Family.***

Once you click on the **FAMILY NAME**, a dialog box will open and you will be able to select where you want to save the Family. You can save it to either a flash drive or your C: drive on your laptop or desktop computer.

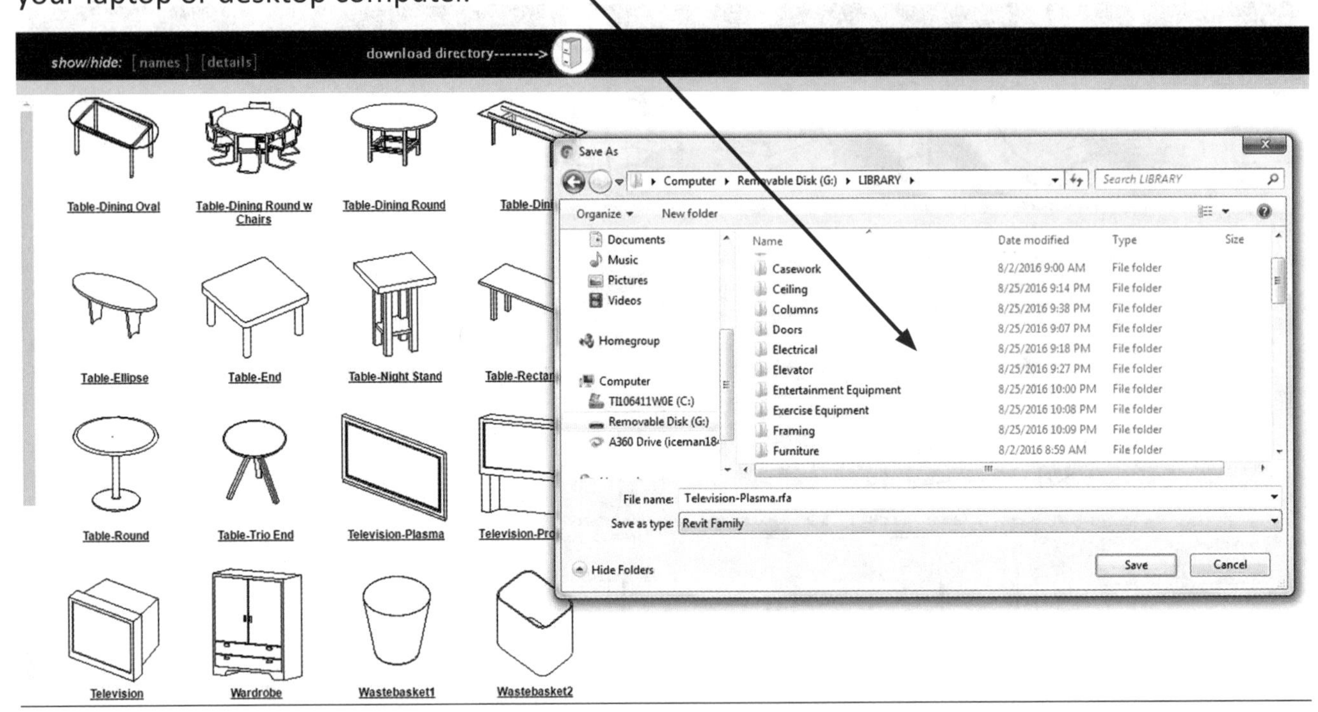

In the Revit 2009 Content Distribution Center, there are numerous Family Categories, and so many different Revit categories such as architecture, furniture, mechanical, electrical and specialty equipment, restaurant, etc. Please visit the different categories to see all that Revit offers.

Once you have downloaded the selected families to either your flash drive, laptop, or desktop, you can then follow the steps in Revit to insert the Family into your project.

Go to the Insert Tab, Load Family, and then select the location where you have saved the new Families you downloaded. The Families will appear in the Revit Families in your project, and will be available to drag into your design.

After you have downloaded the family you selected to your design you can take a 3D perspective camera shot (as explained in Chapter 15, *Interior and Exterior Camera Views*), and see the new family you have inserted.

In the following screen shot you will see the plasma TV inserted in our design in a 3D perspective view:

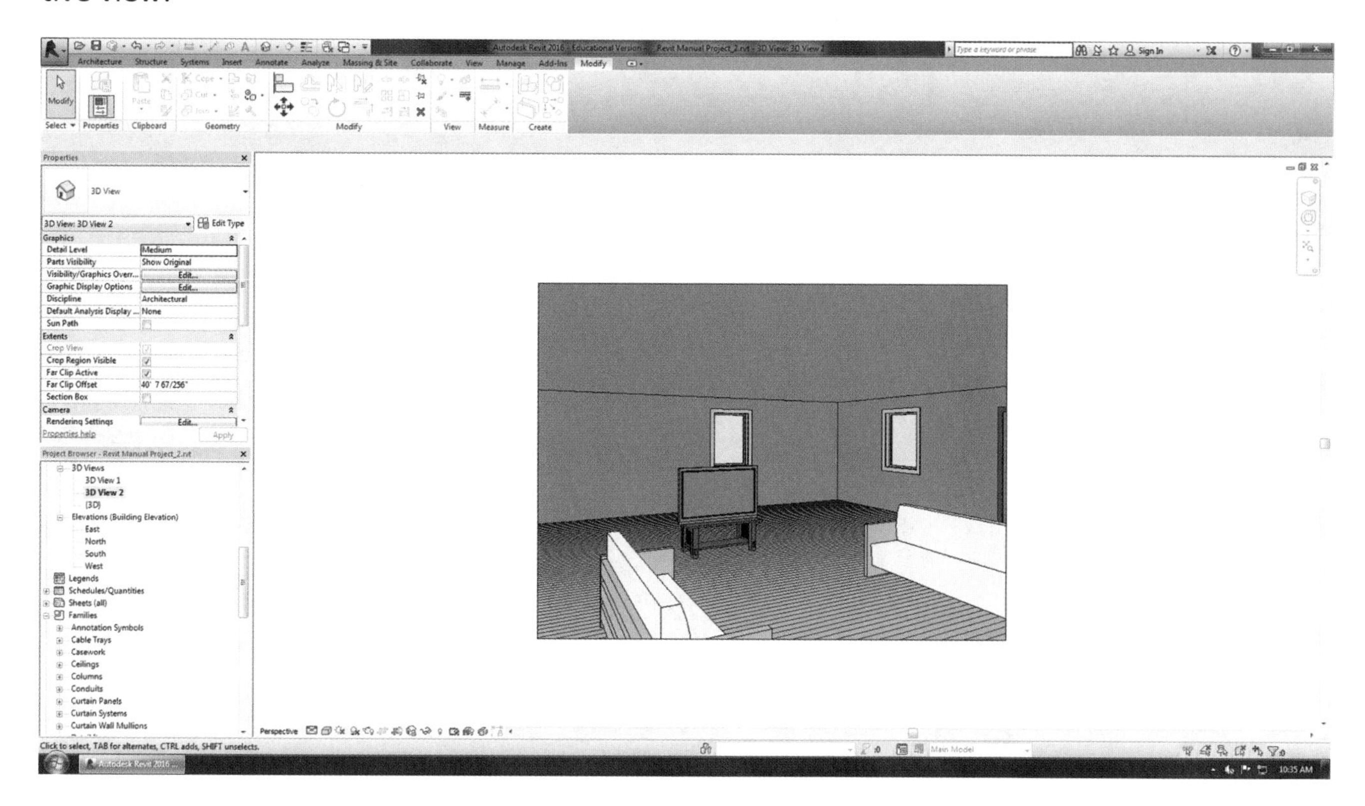

With the ability to bring in more family categories and detailed objects, you are now stretching not only Revit's design and drawing capabilities to make the project stand out, but also your design flair. You can show clients all the amazing features you can add to a house and make it stand above your competitors.

Now it's time to take it a step above in Chapter 18, *Modifying Doors and Windows*. Let's go!

CHAPTER

18

Modifying Doors and Windows

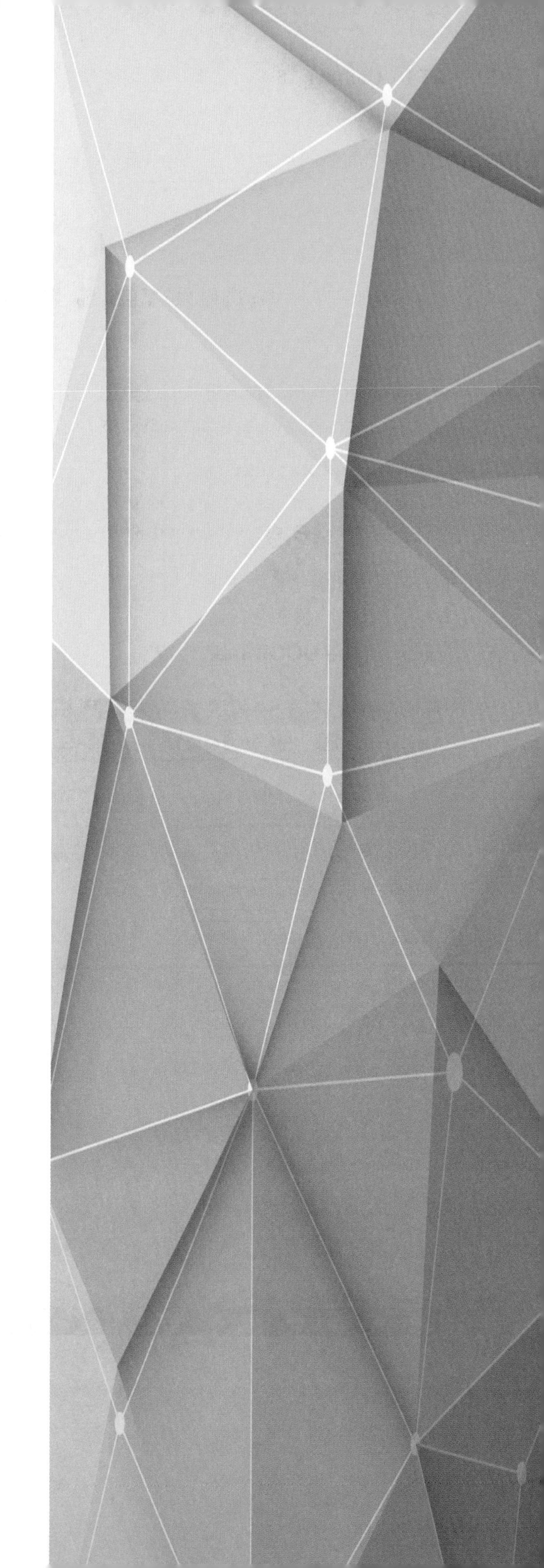

MODIFYING DOORS AND WINDOWS

When modifying a door, window or any object**,** make sure you always *duplicate* the object. Do *not* change what Revit already has in the Family.

EXAMPLE

We are using a door for this example (the steps are still the same, regardless of what object you're modifying).

Click on the **DOOR** Tab.

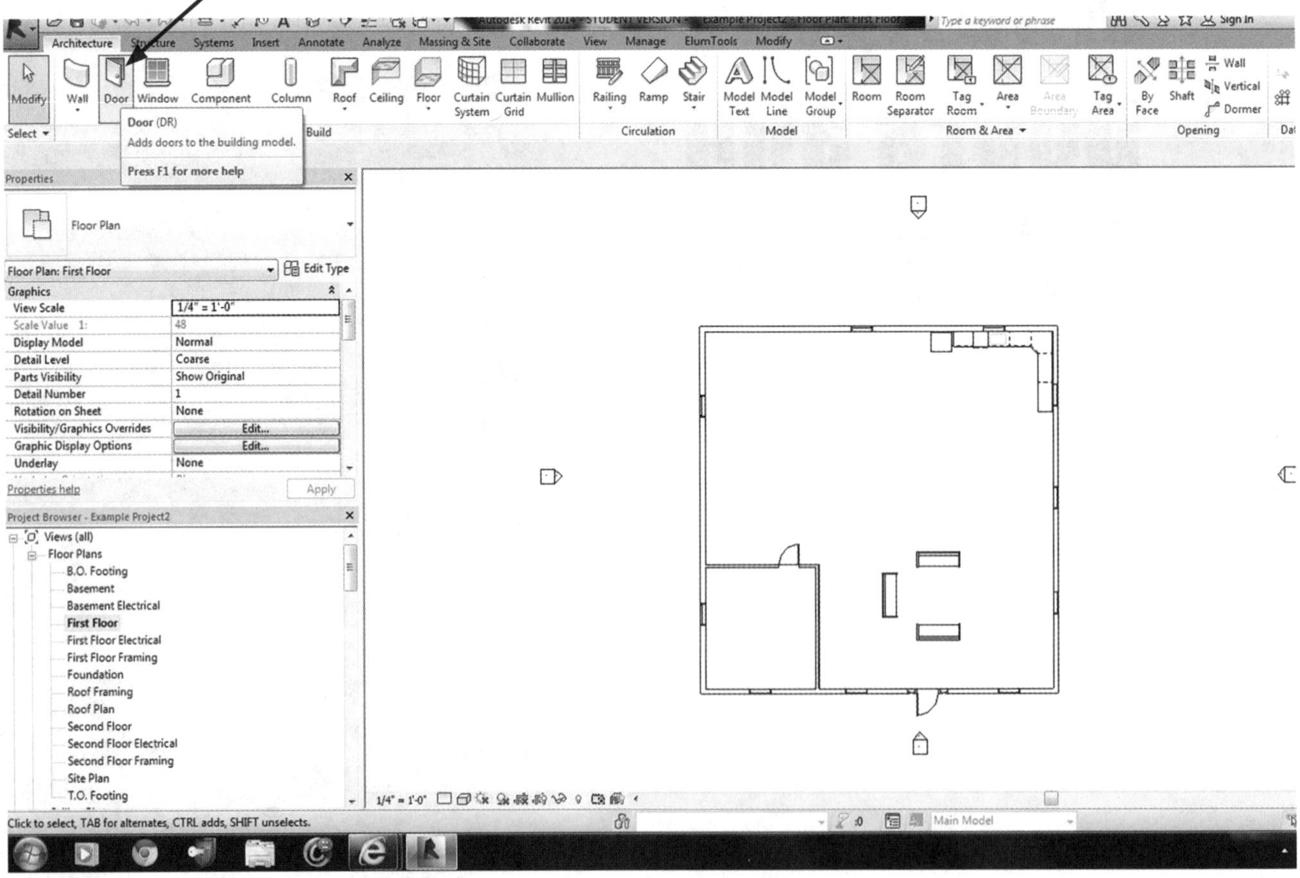

Pick the door you want to modify.

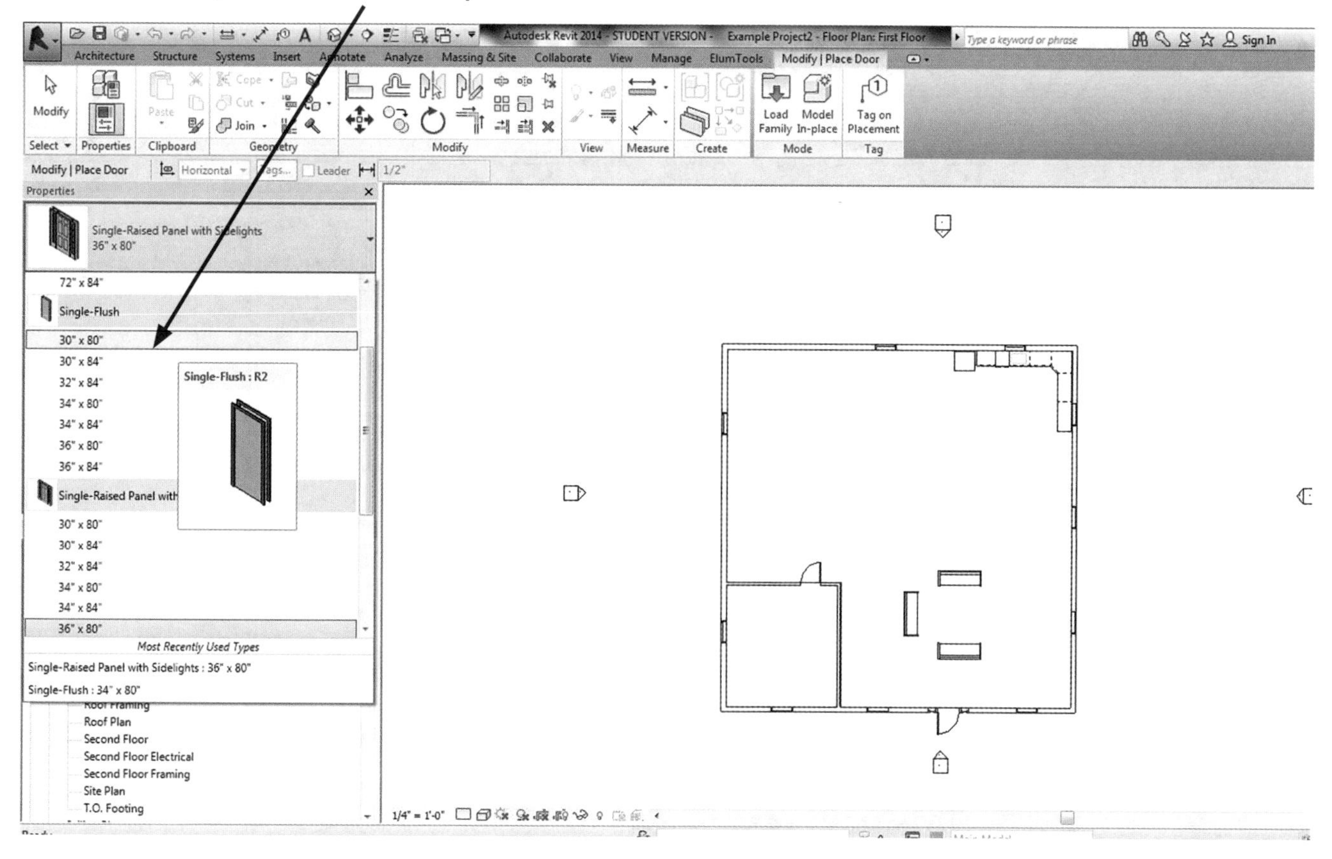

Once you select the door, click the **EDIT TYPE** Button.

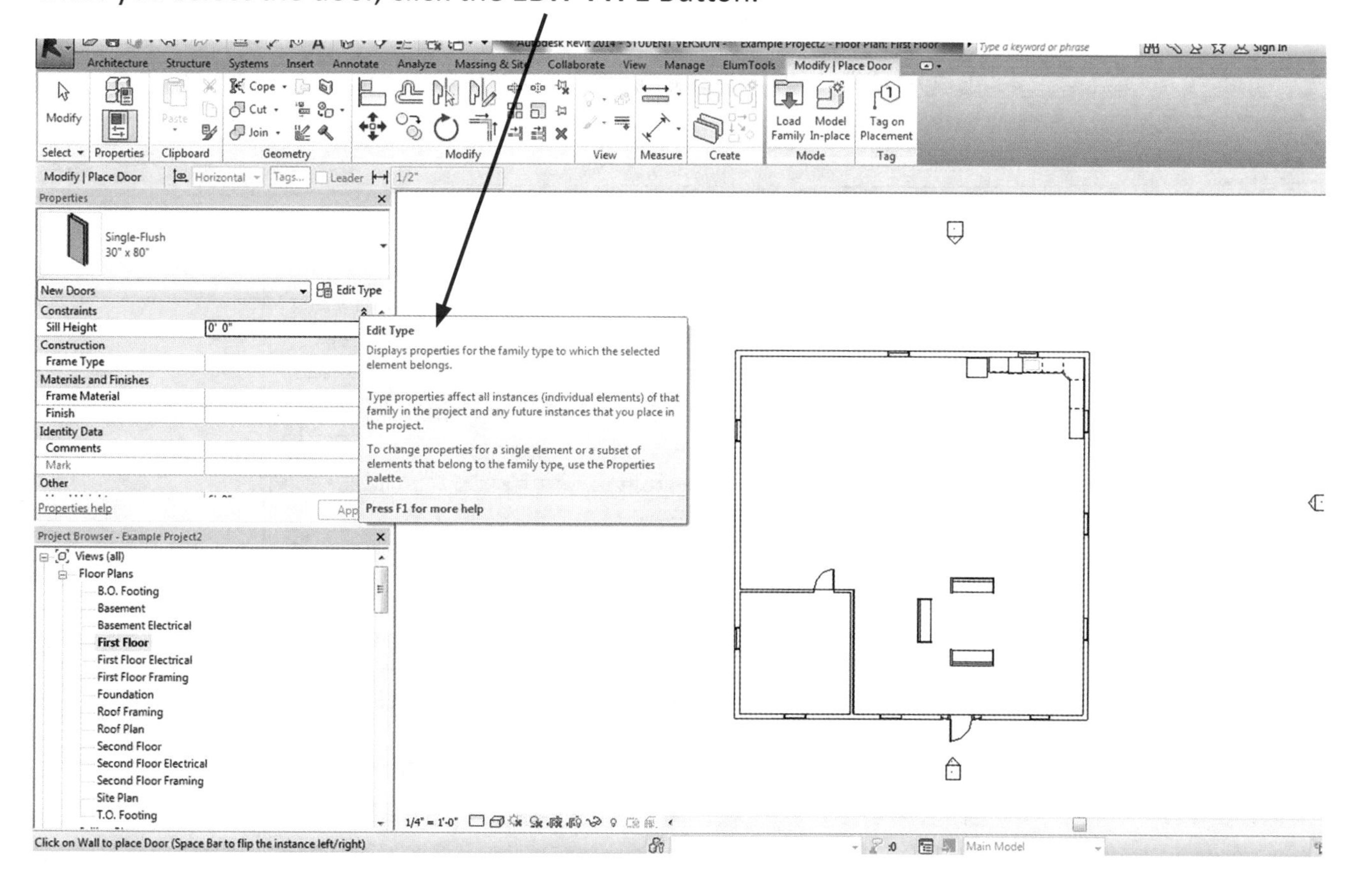

Click the **DUPLICATE** Button.

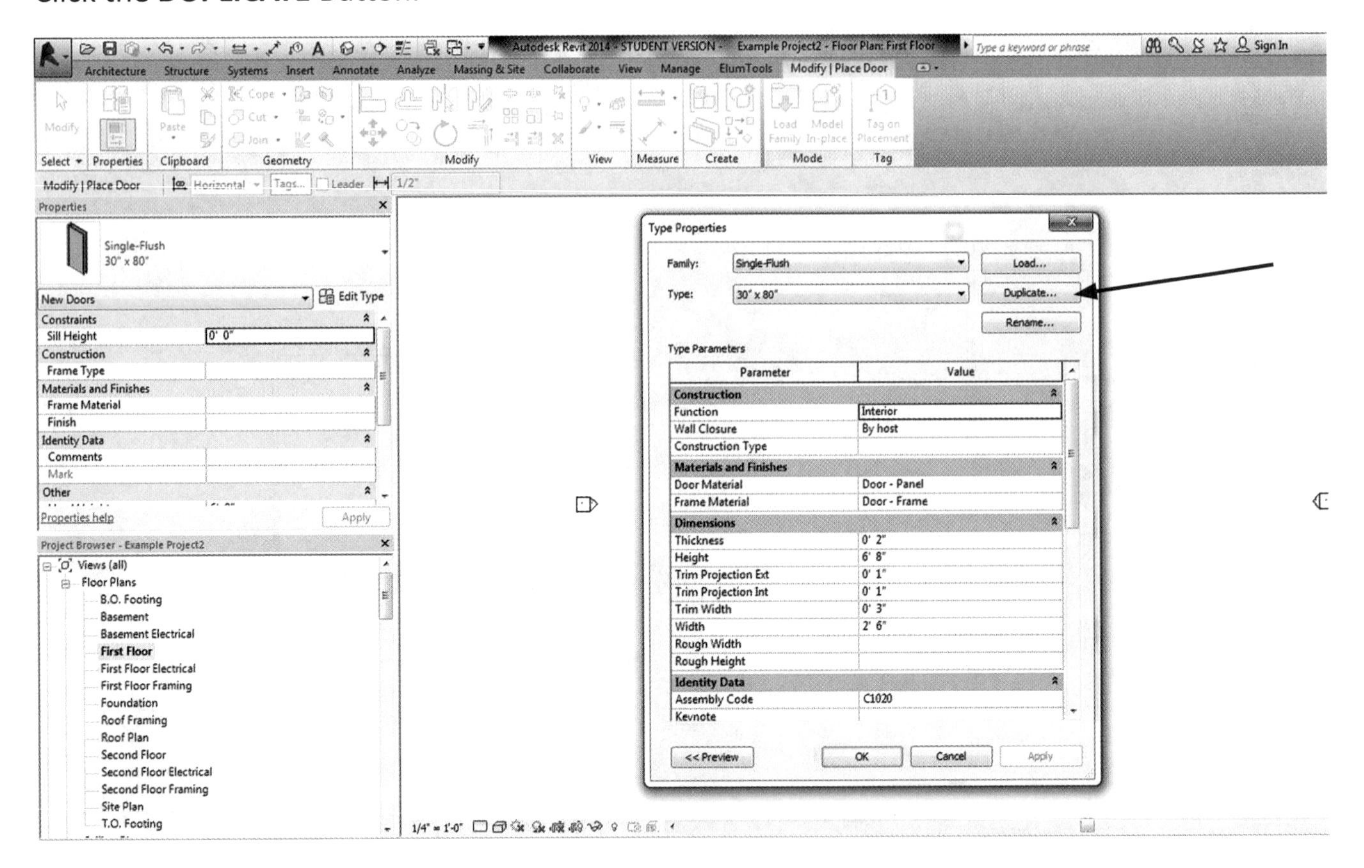

Then rename the object (usually using the new size of the door), and click **OK.**

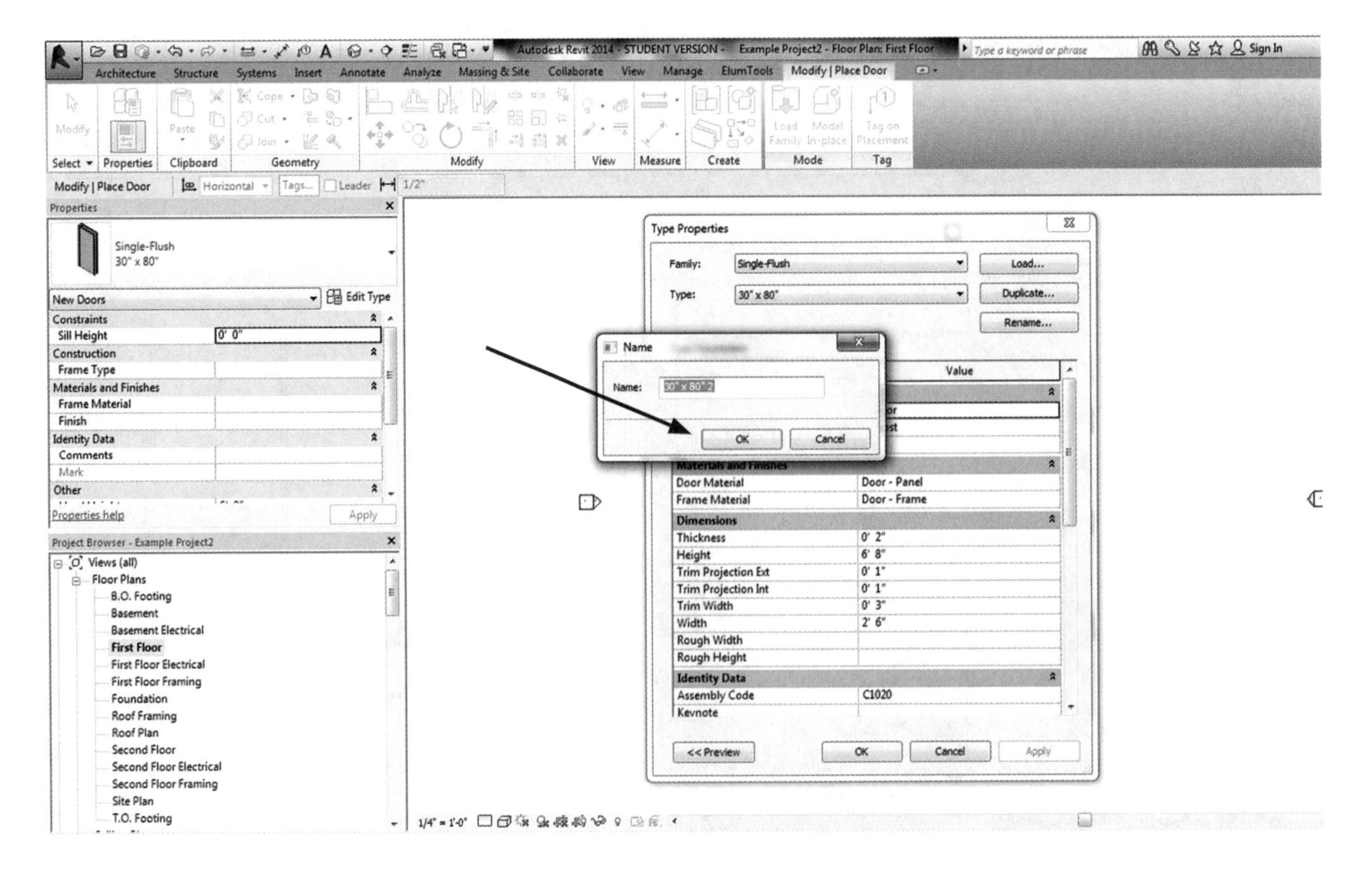

Once you have renamed the object, then you have to change the size of your object. Change the width and the height in the areas provided.

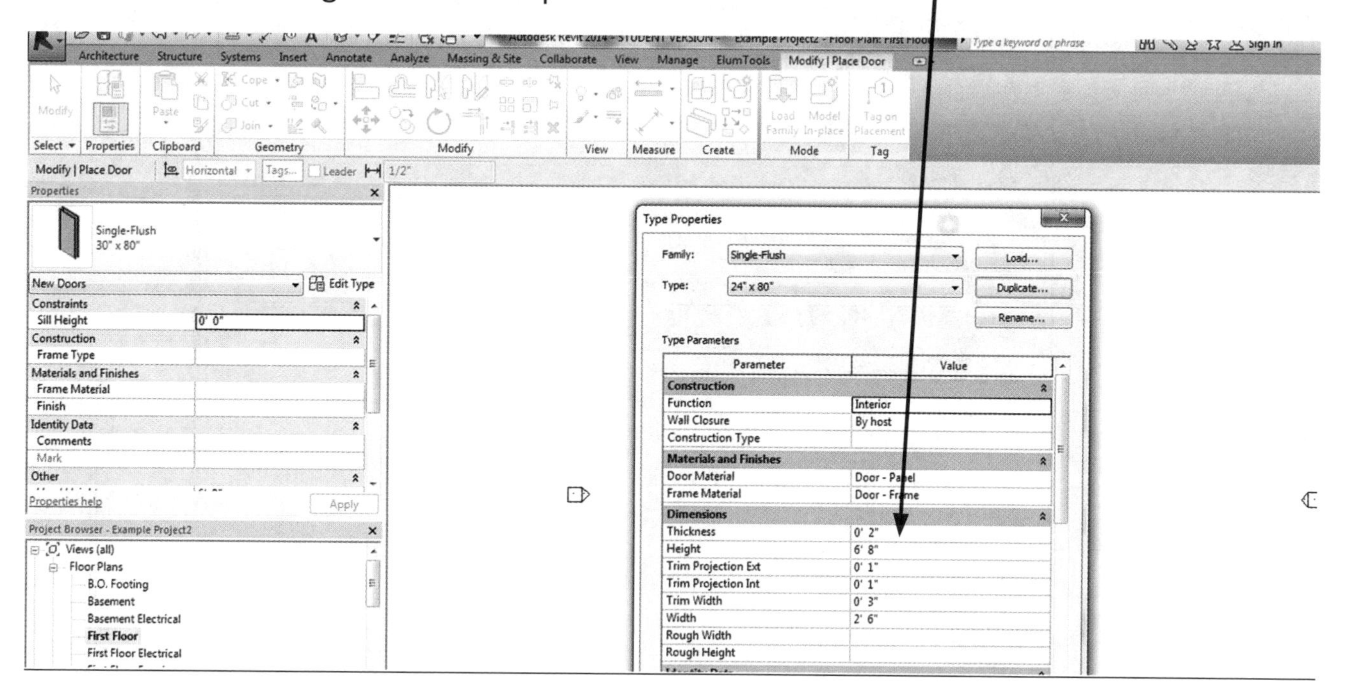

Click **OK** and then the **MODIFY** Button. Congratulations! You have created a new sized door.

MODIFYING A WINDOW

When modifying a new window size, you will follow the same steps as you did when modifying a door. First select the window and click the **EDIT TYPE** Button.

REMINDER

You only want to** duplicate **the windows, not change anything Revit already has in its Library. You will end up needing it later, eventually.

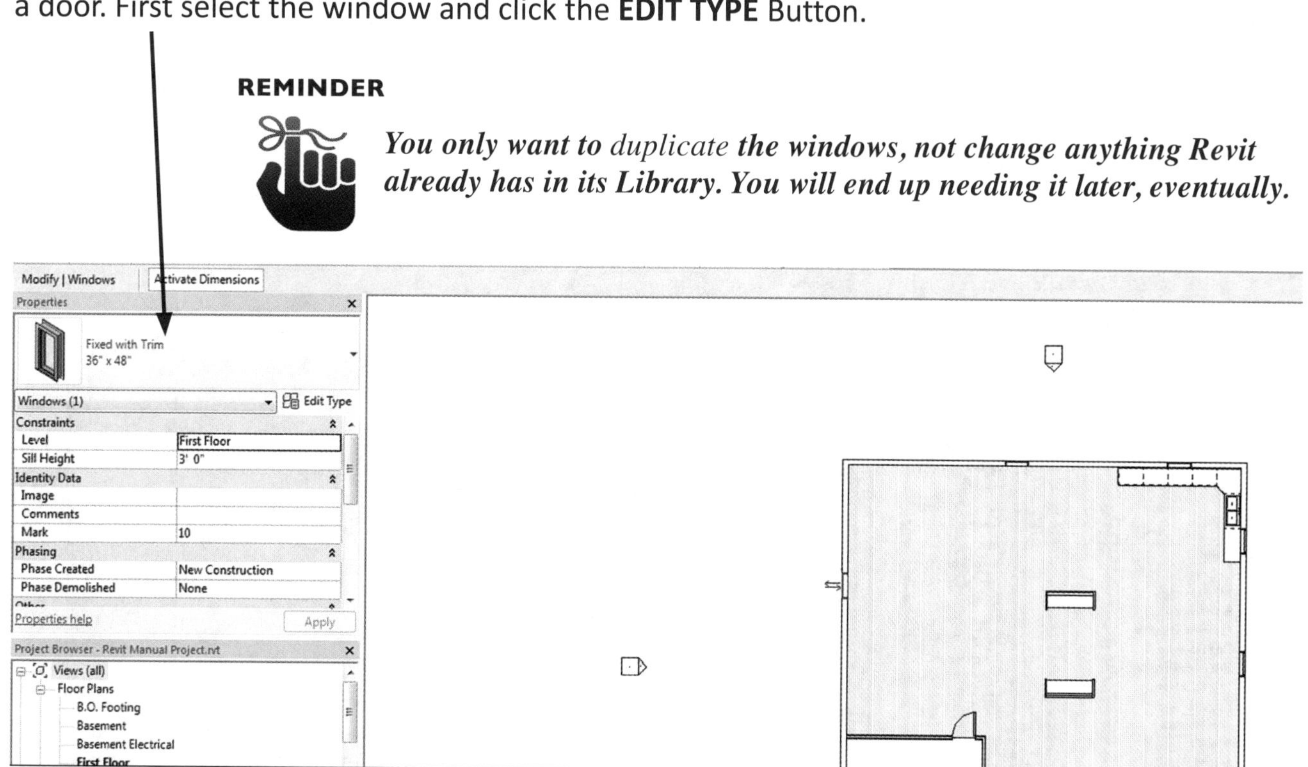

Click **EDIT TYPE,** and a dialog box will appear.

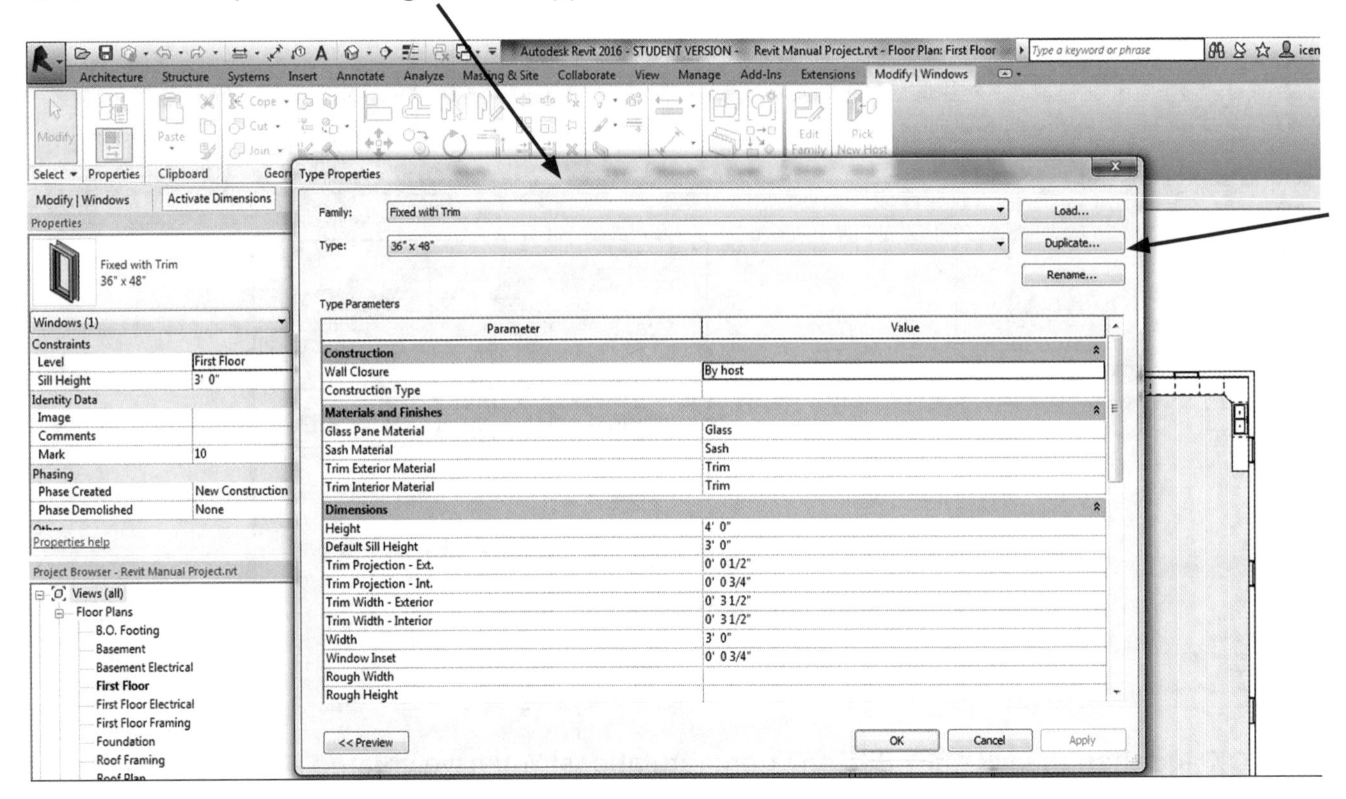

Click on the **DUPLICATE** button.

Then rename the window size name, and click **OK.**

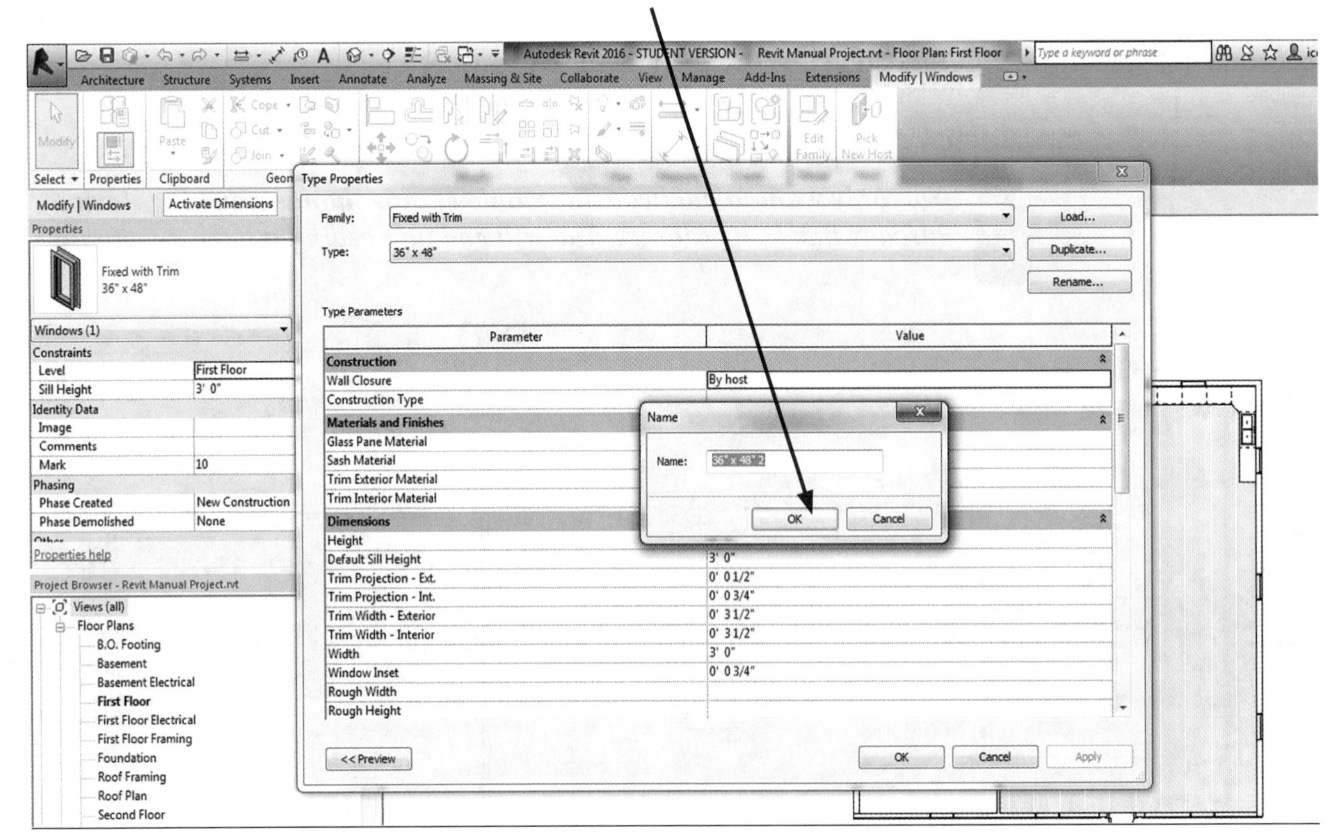

After clicking **OK,** you will need to change the new window sizes to match the new window name, altering the width and the height measurements.

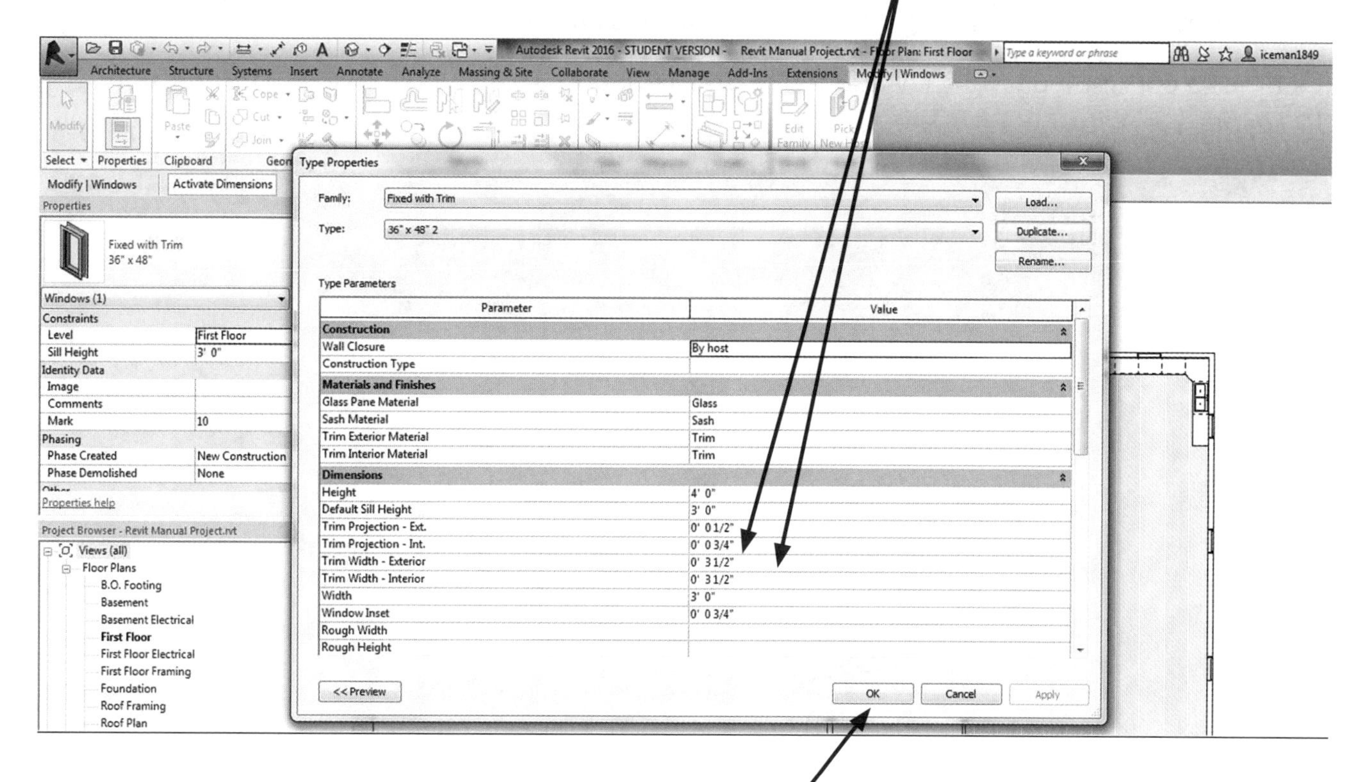

After the sizes have been changed, click **OK,** and the dialog box will disappear. You can place the new window in your project and click the **MODIFY** Arrow.

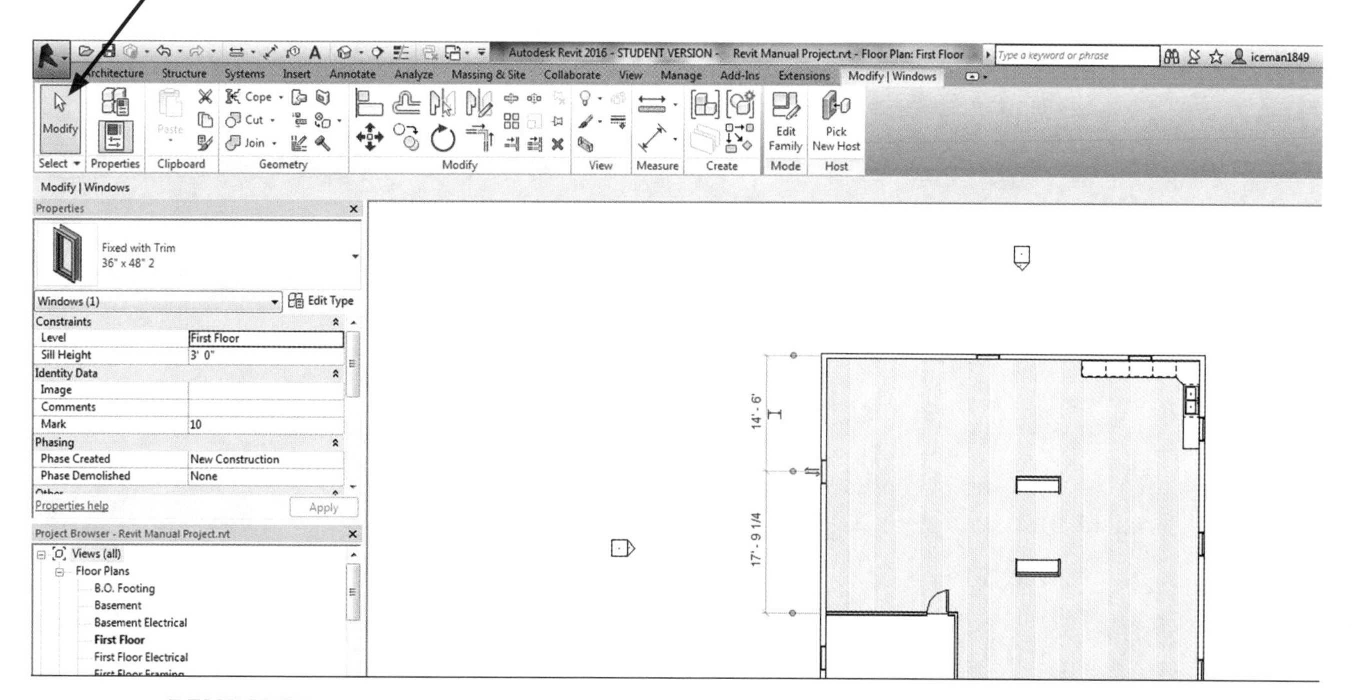

REMINDER

Do not hit the* ESC *Key, or you will undo work that you probably don't want to lose.

You have now created your own doors and windows, and even learned to modify the sizes to accommodate your design for yourself, or perhaps more importantly, the client. With Revit, there is nothing that is stopping you from a great design.

It's time to continue our journey and move on to Chapter 19, *Modifying Wall Materials and 3D Layout Visibility.*

CHAPTER

19

Modifying Wall Materials and 3D Layout Visibility

MODIFYING WALL MATERIAL

When changing material on walls such as tile, first you will determine if all the walls are going to be the same. If only a select few are going to be different, you will need to go to the Wall Tab and select the wall, and follow the duplicate instructions you learned in Chapter 18.

Once you have duplicated the wall and renamed it for your selected wall type, follow these steps:

Select the wall you want to modify.

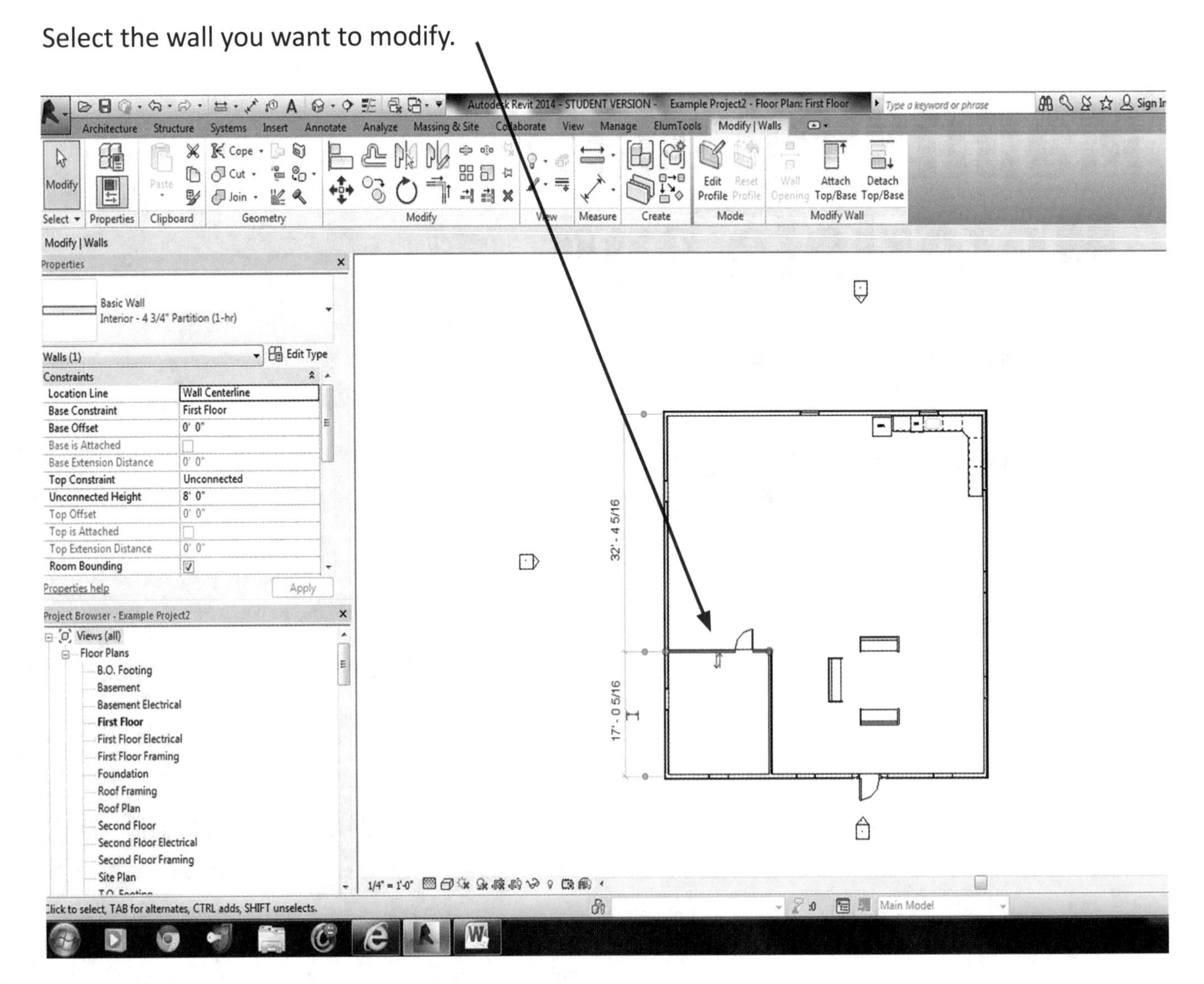

After you have selected the wall, in the Properties Box, select the **EDIT TYPE** Button.

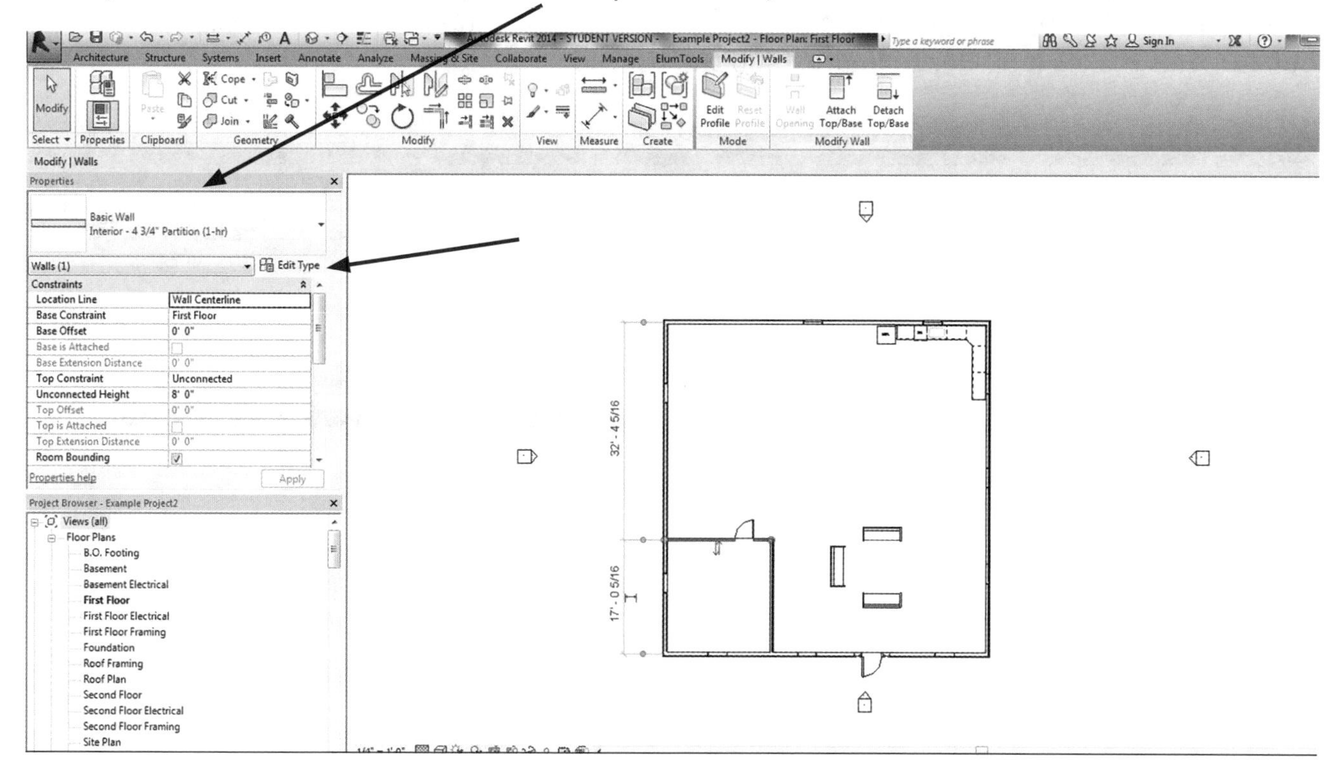

Once the edit dialog box opens up, go to the Edit Button on the Structure Category.

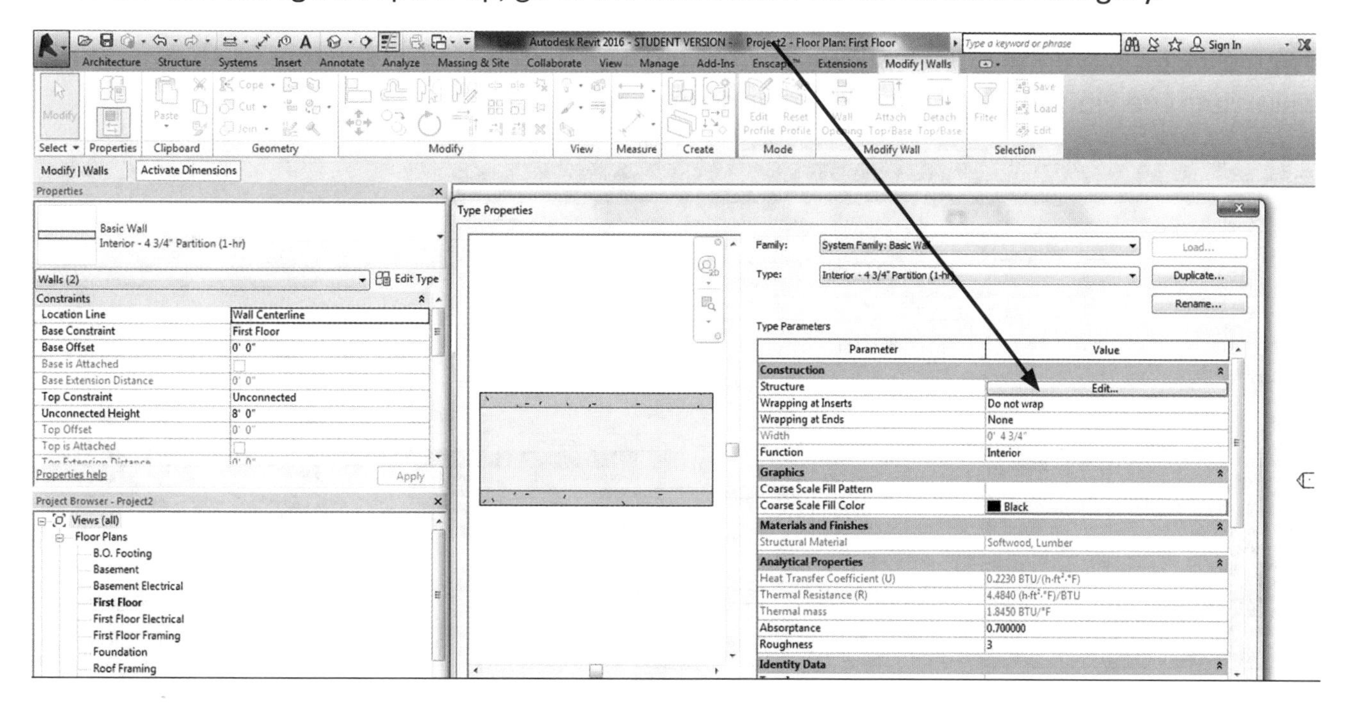

Another dialog box will open so you can edit the wall.

REMINDER

You have to remember whether you're adding material to the exterior or interior of the wall.

After you click on the **EDIT** Button in the structure category, when the edit dialog box first opens the preview of the wall on the left defaults to a horizontal view of the wall. In my opinion, this makes it harder to understand the wall construction.

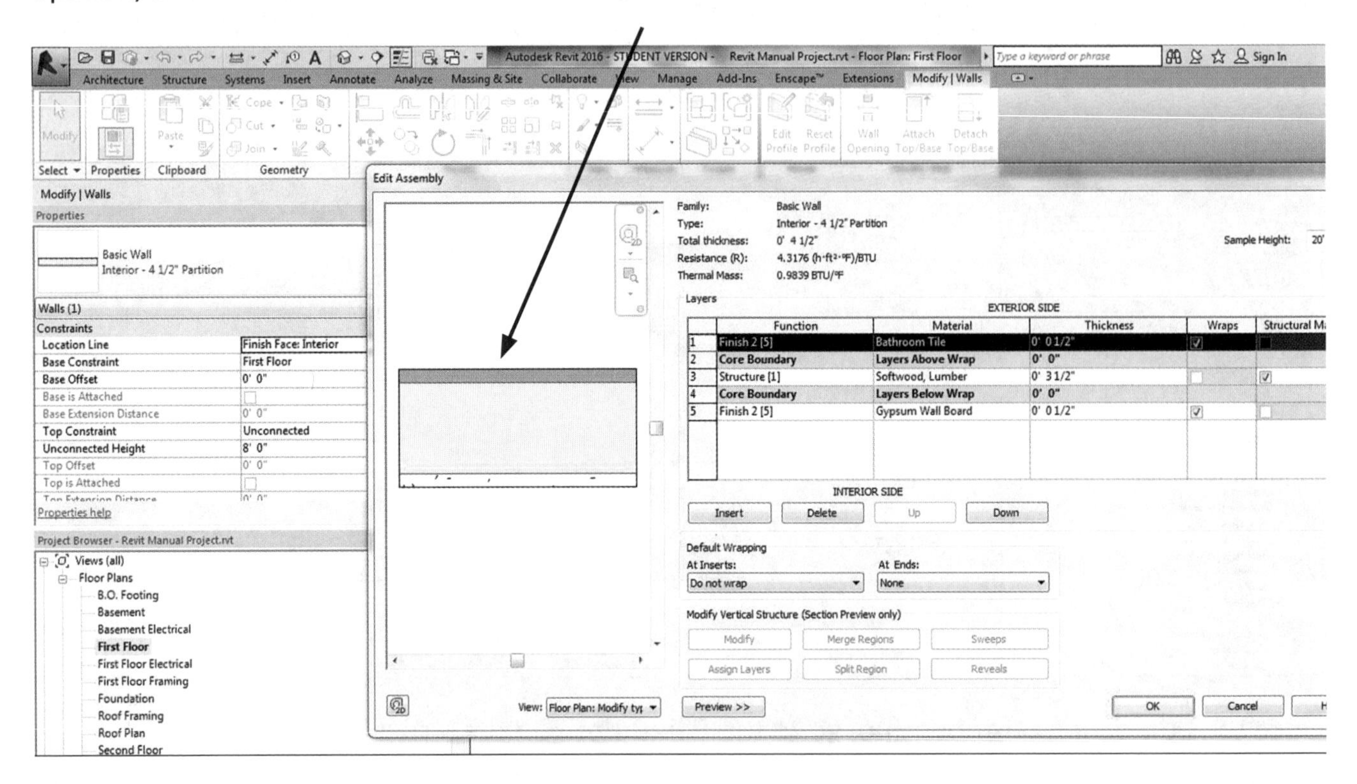

If you click on the **VIEW** Button on the bottom left and switch from Floor plan to Section, this will switch the view in the window from horizontal to vertical.

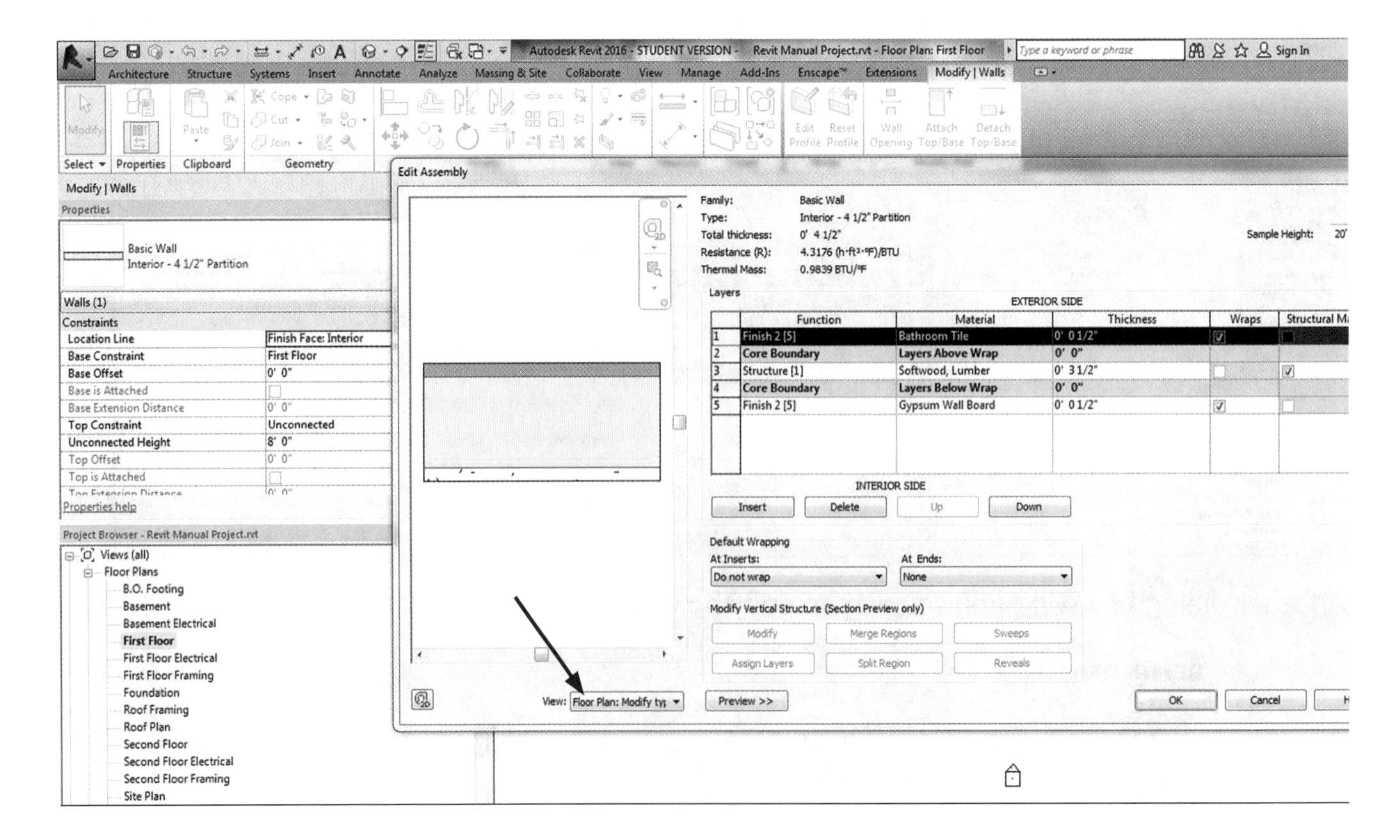

The view in the window will switch from horizontal to vertical, and will help you understand more about how the wall materials are put together.

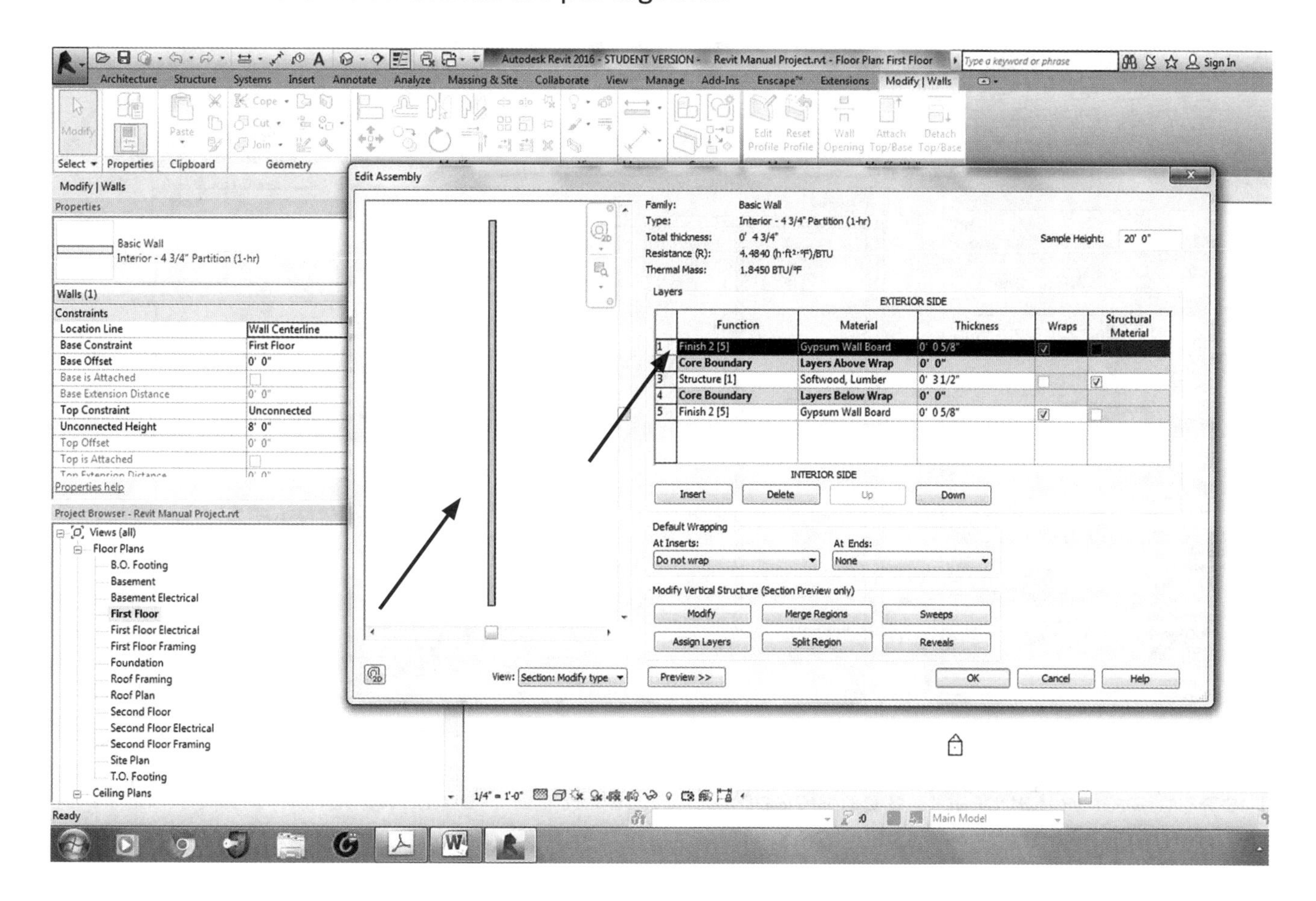

TIPS

If you click in the screen where the wall sample is located and zoom in, this will bring the wall closer. Then click on the numbers to the right. As the rows highlight, you will see the materials highlight and see where the materials are in the wall. This will help you better understand how the wall was put together.

If you want to split the wall and have tile above and below, you can insert another layer by clicking the **INSERT** Button.

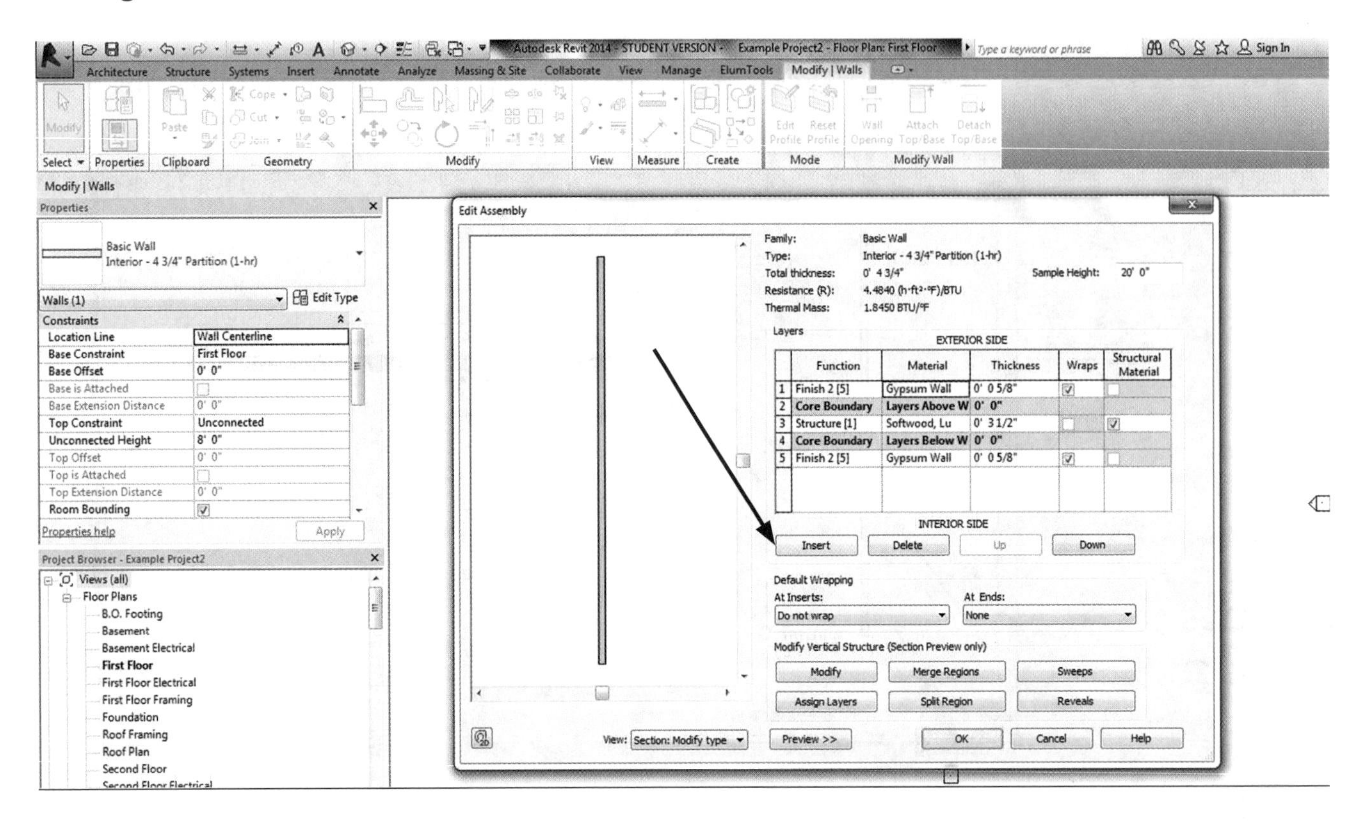

The new layer will appear in the dialog box. Then select the drop-down arrow, and you can change the Finish Category to **FINISH 1 (4)**.

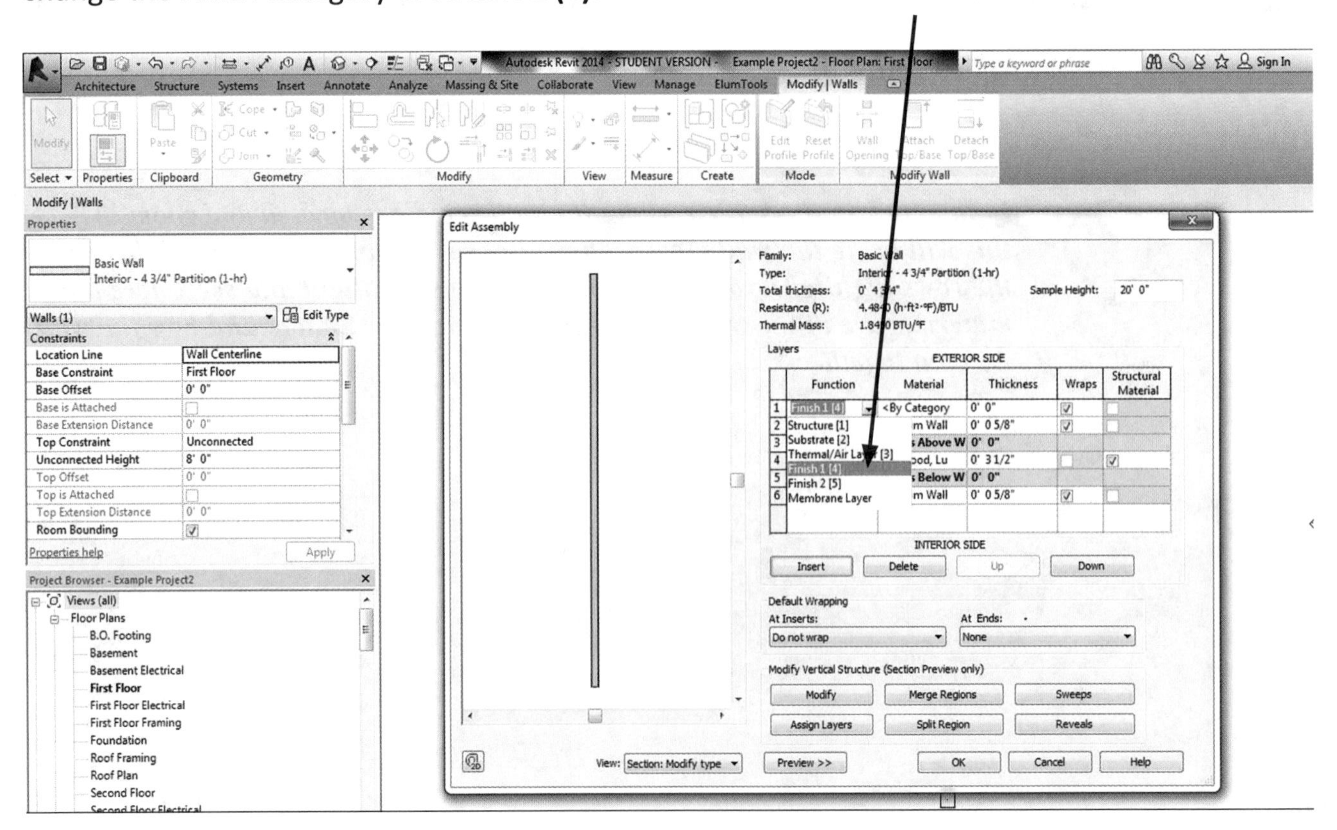

Once the Finish category has changed, click on the material and just to the right, a little dialog symbol will appear.

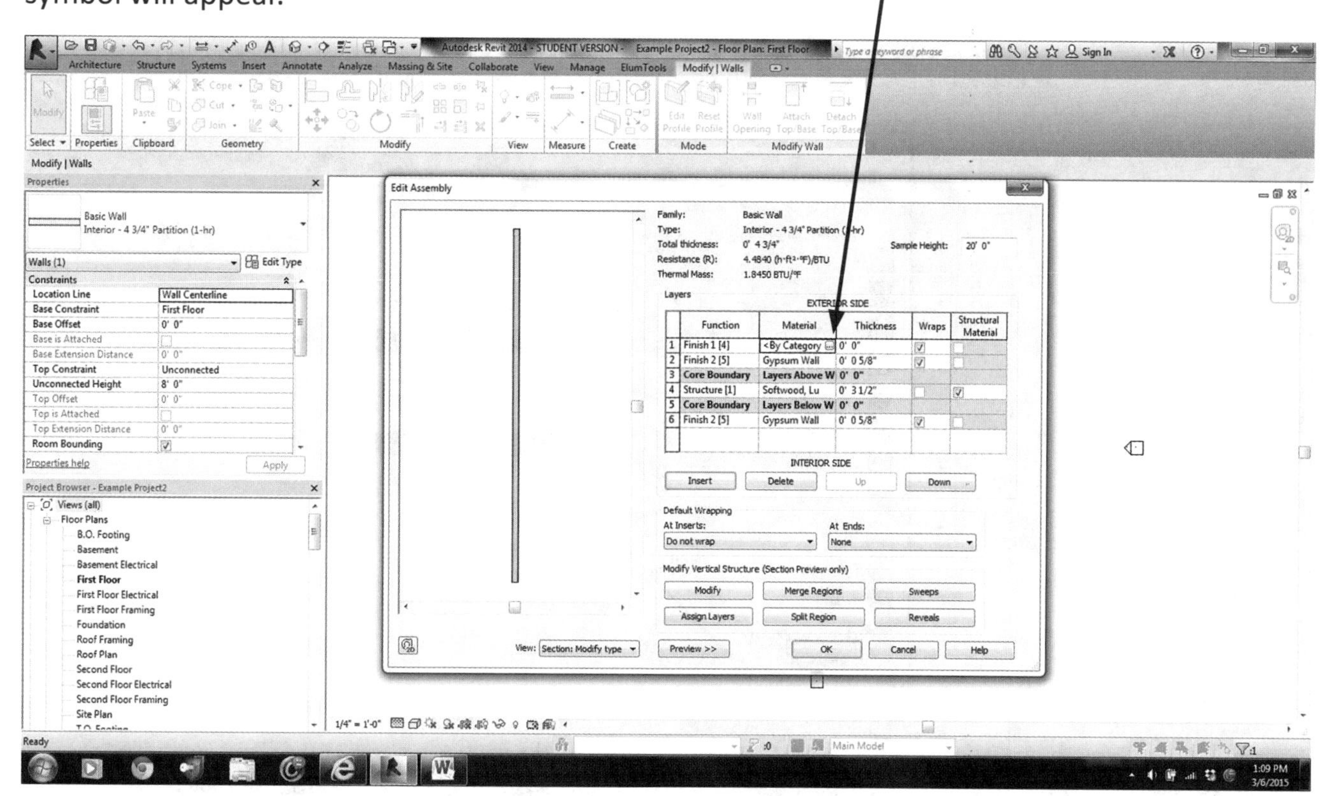

This will open the **MATERIALS BROWSER.**

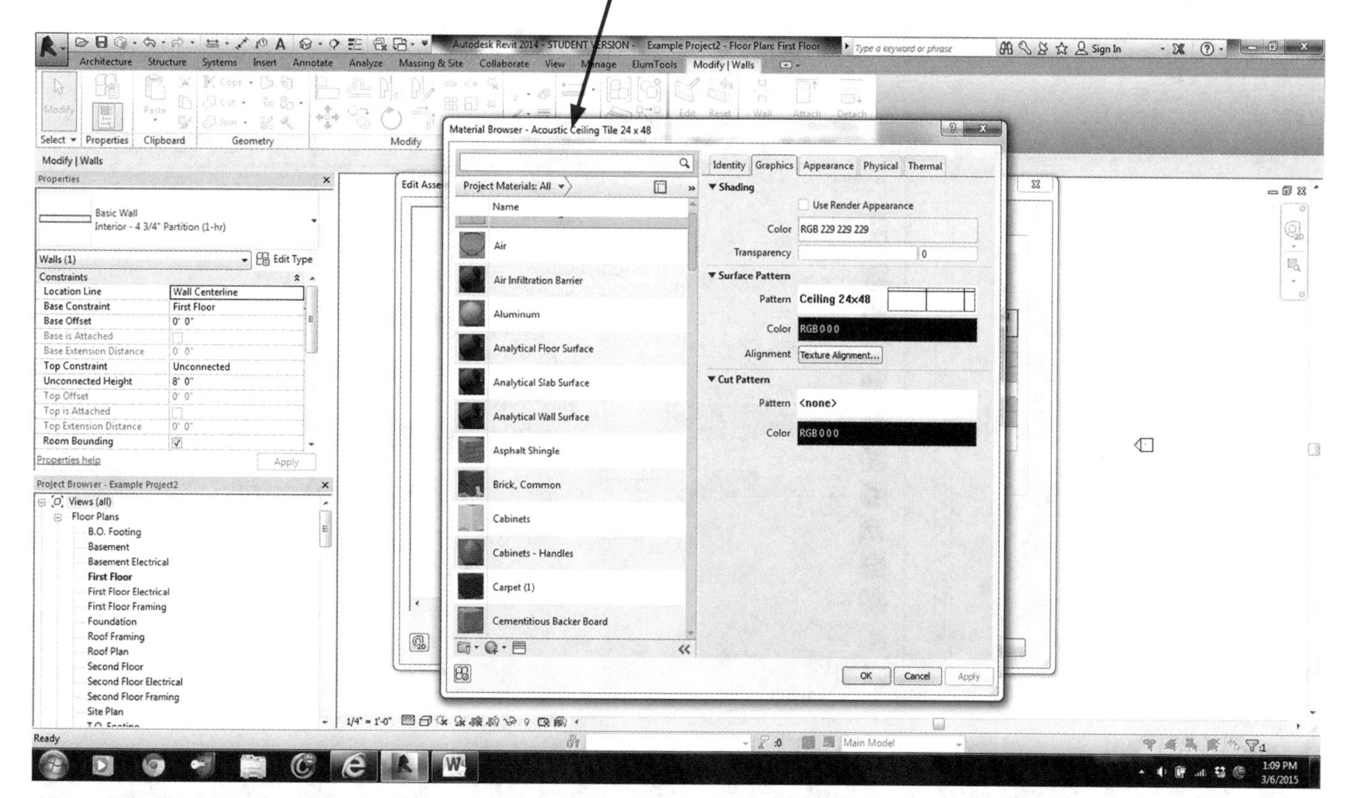

Then click the drop-down arrow beside the ball at the bottom of the screen, and select **CREATE NEW MATERIAL.**

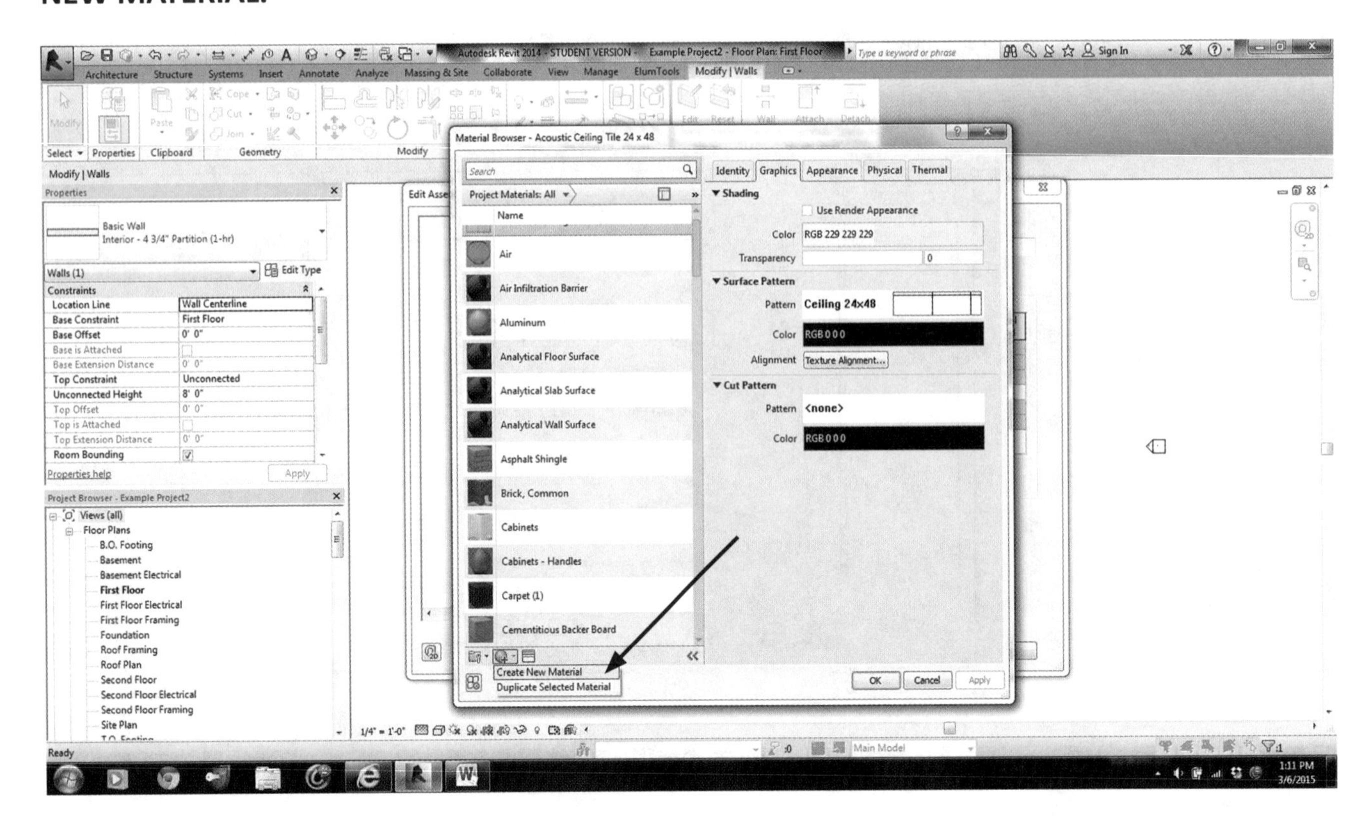

When you have selected that, you will have new material in the list.

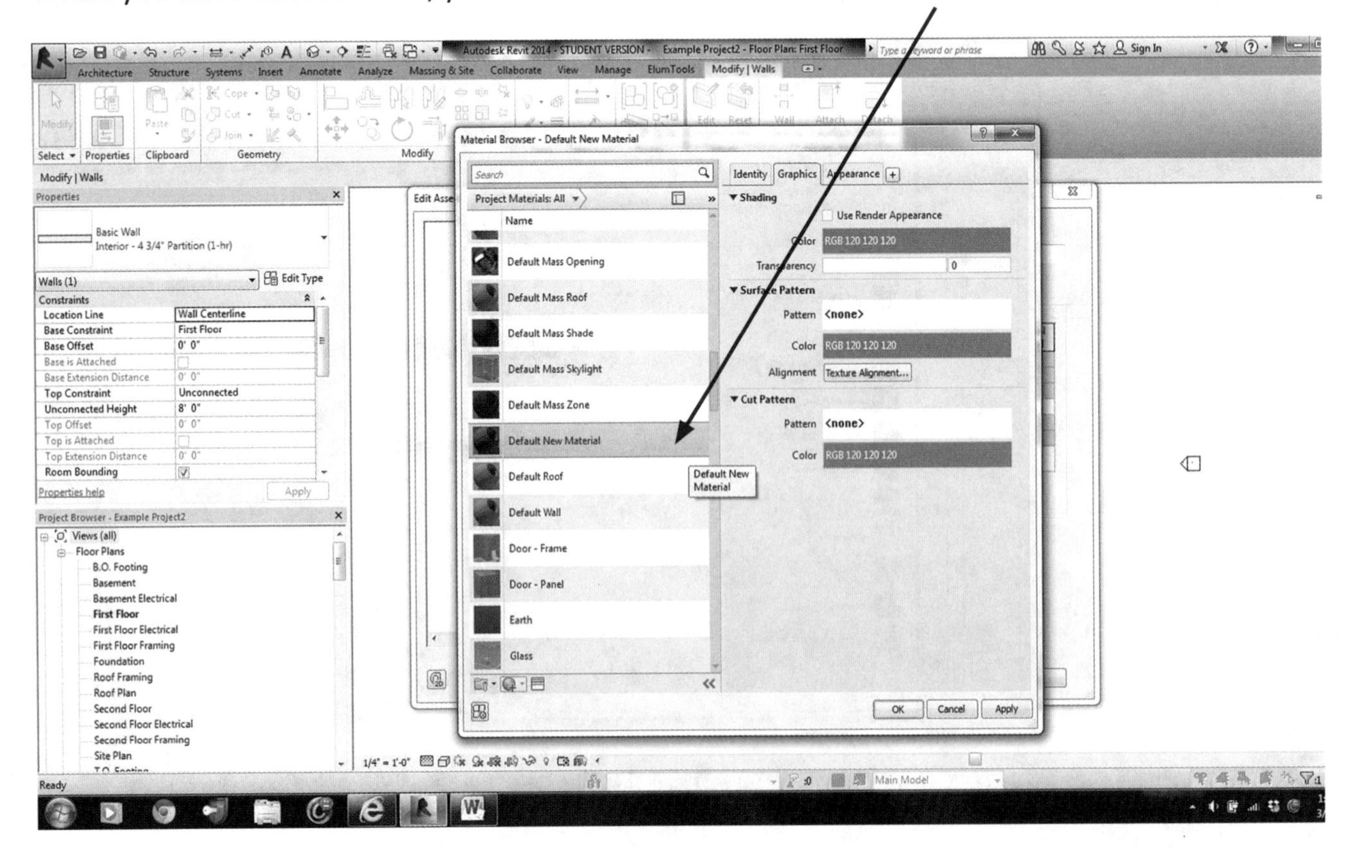

Right-click on the new material and rename it so you don't confuse it with any other wall material when you change it later.

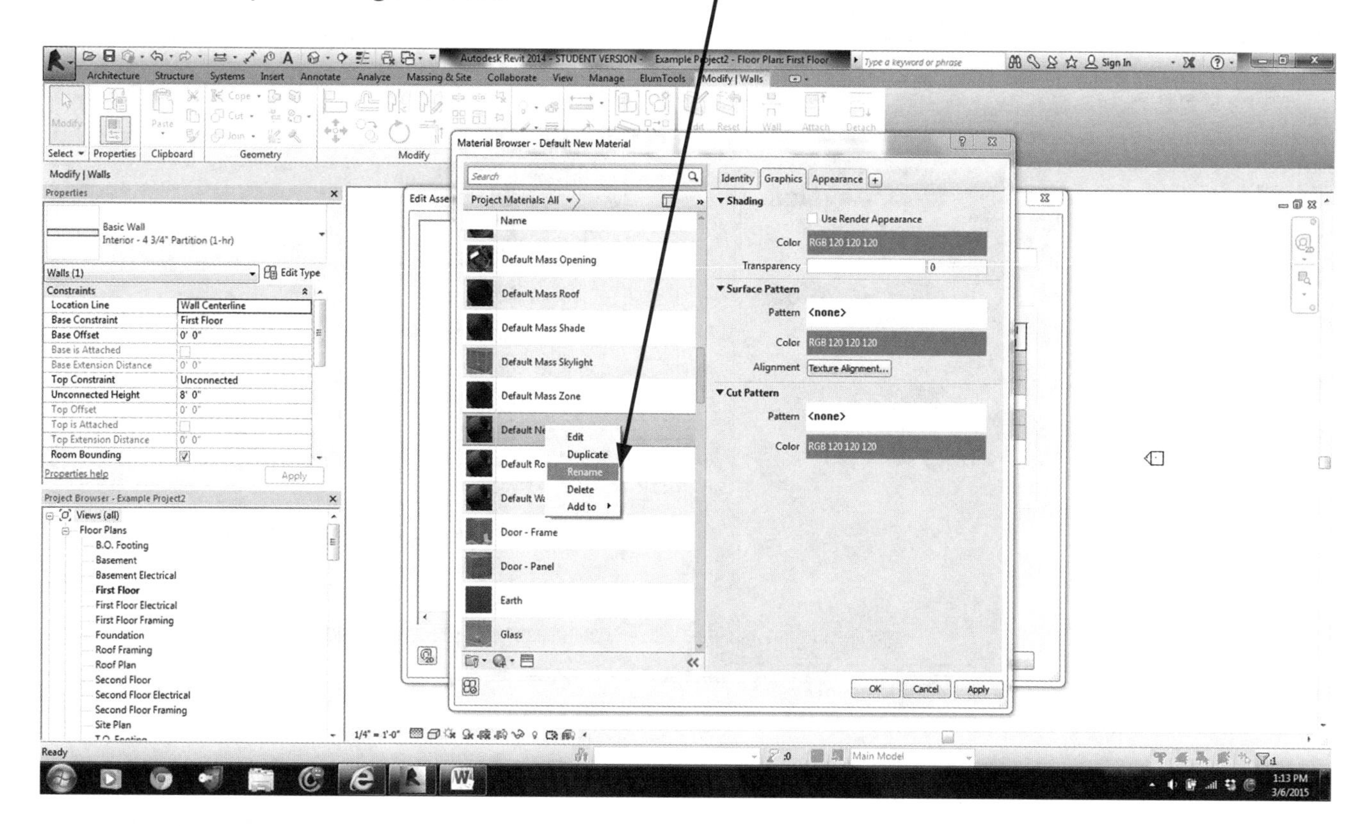

Once you have renamed your new material, click on the box to the right of the New Material Button.

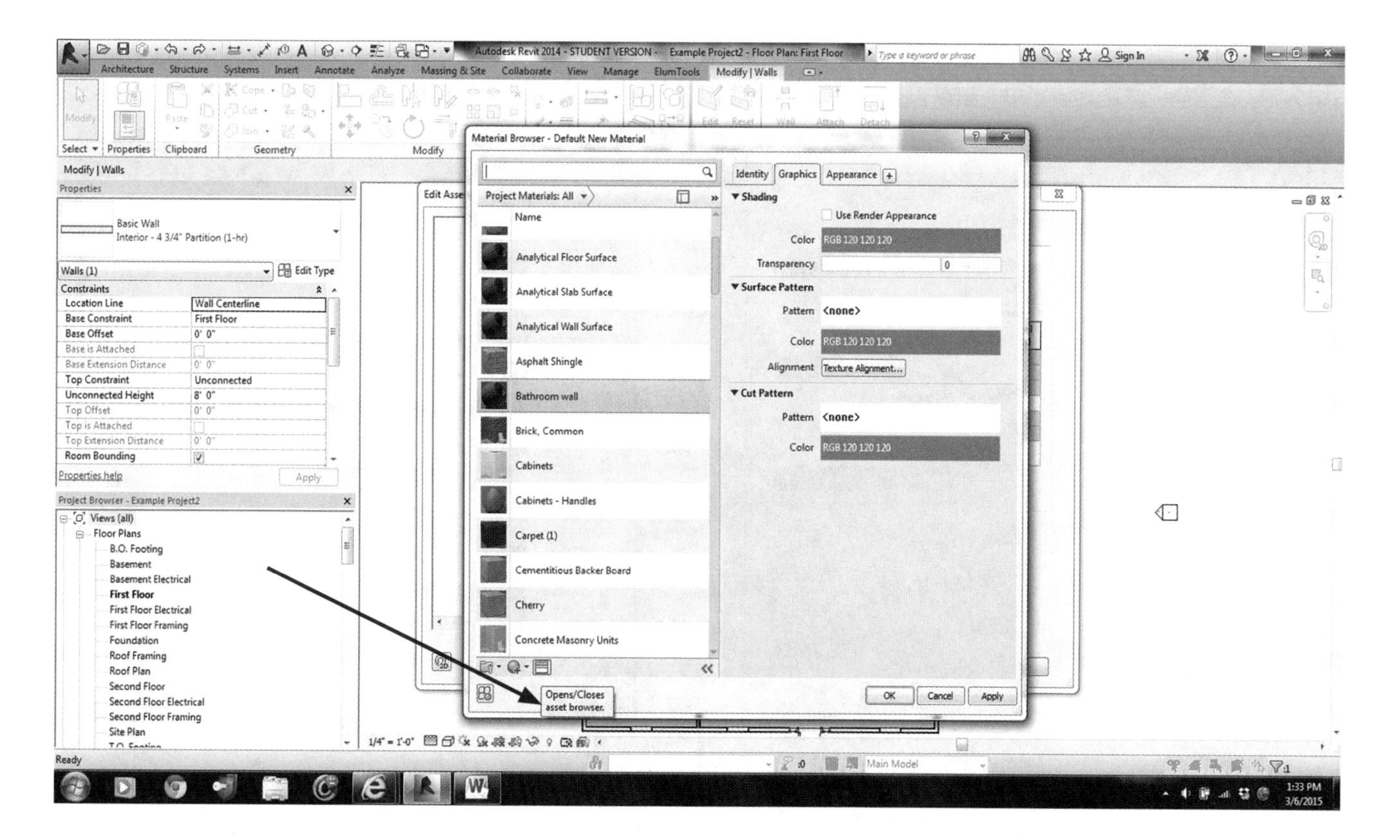

This now opens the **ASSET BROWSER** for you to pick the type of material you want to use.

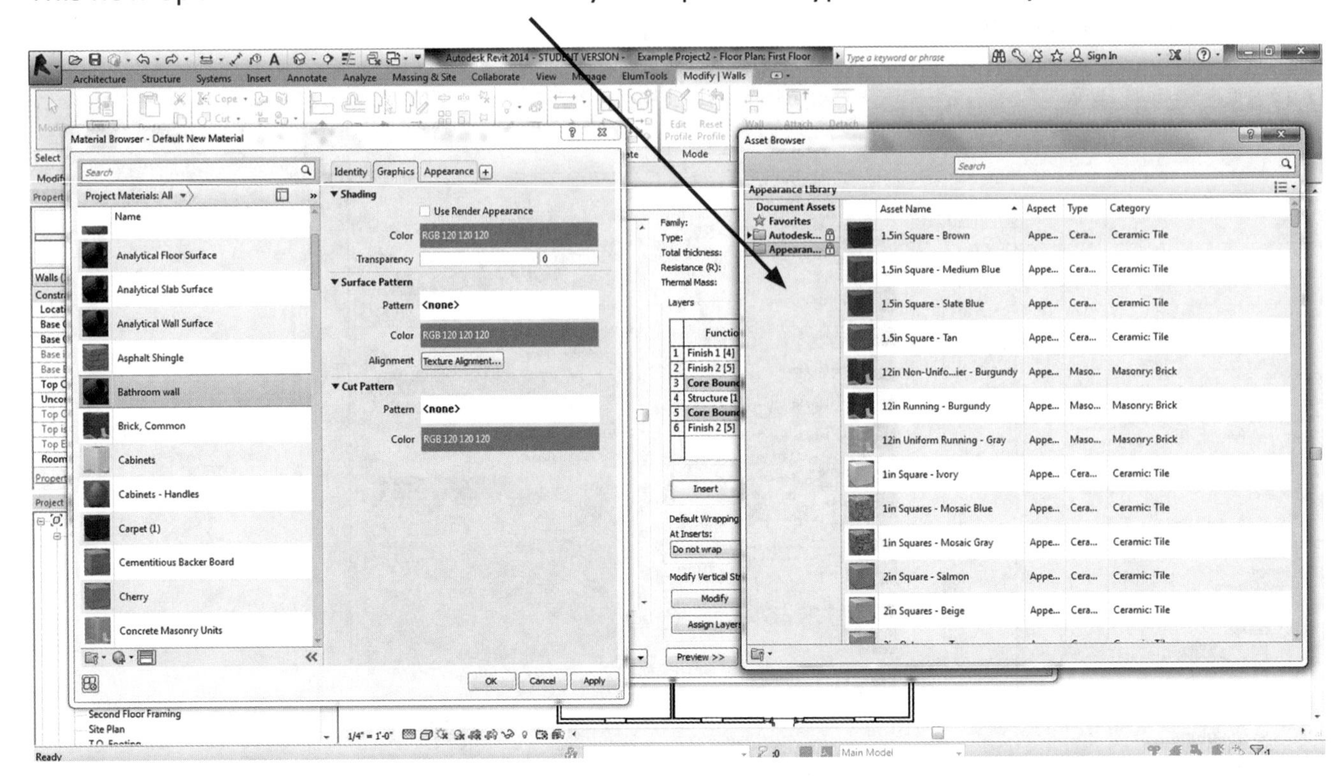

TIPS

I always use the Appearance ***Folder to see a more detailed list of flooring colors, stains, wall paints and covering and metals and coatings, etc.***

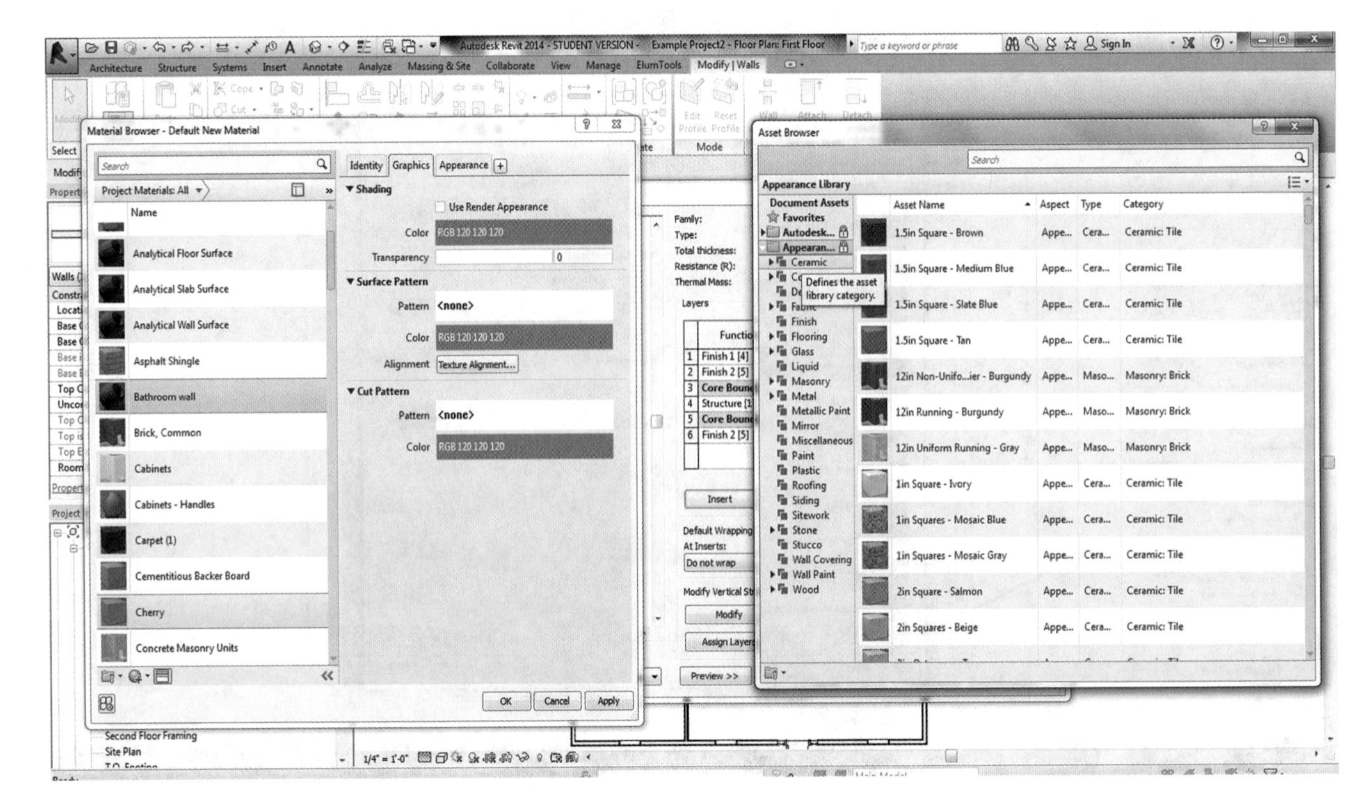

Once you see the materials you want to use, if you hover over to the far right, you will see two arrows. Click on them and it will transfer the material to the new wall material you created.

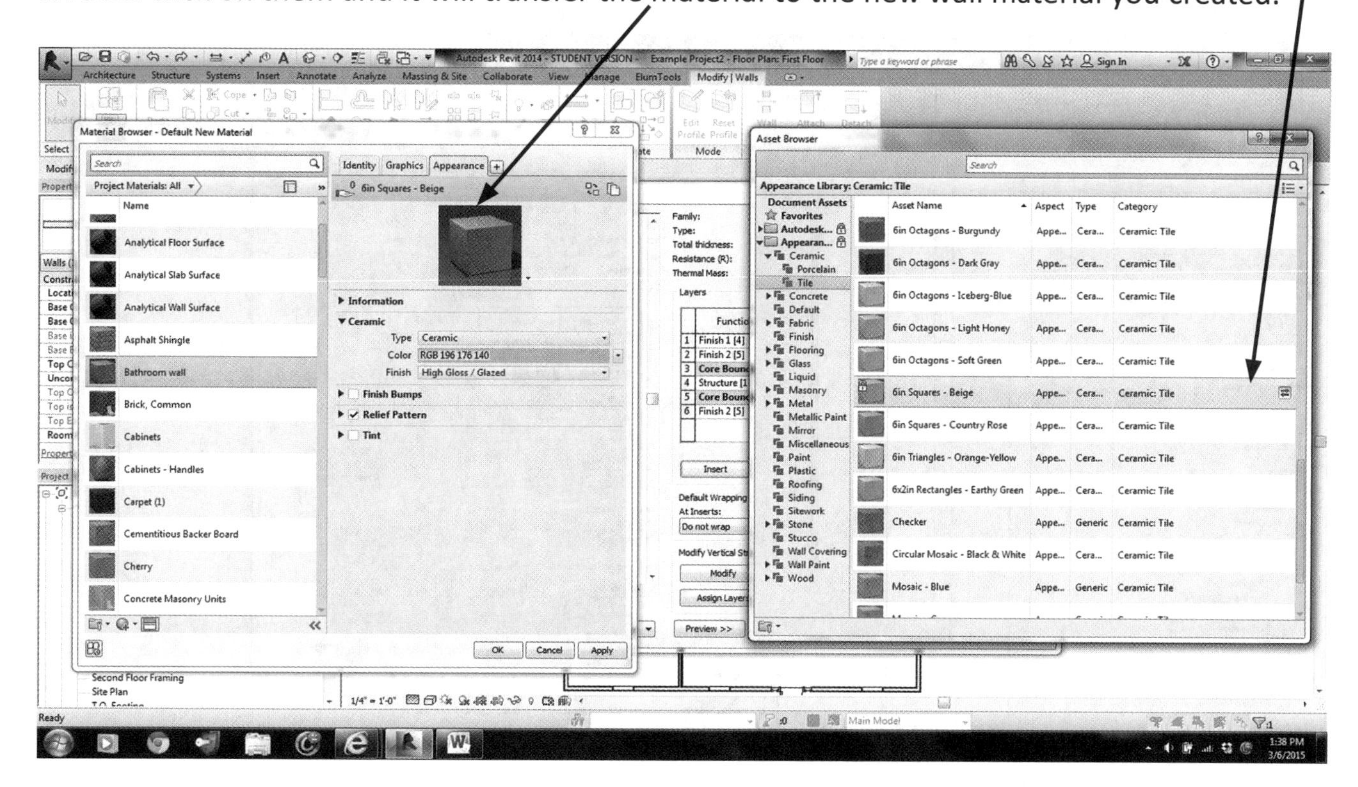

Then click on the **GRAPHICS** Tab and check the **USE RENDER APPEARANCE** Box.

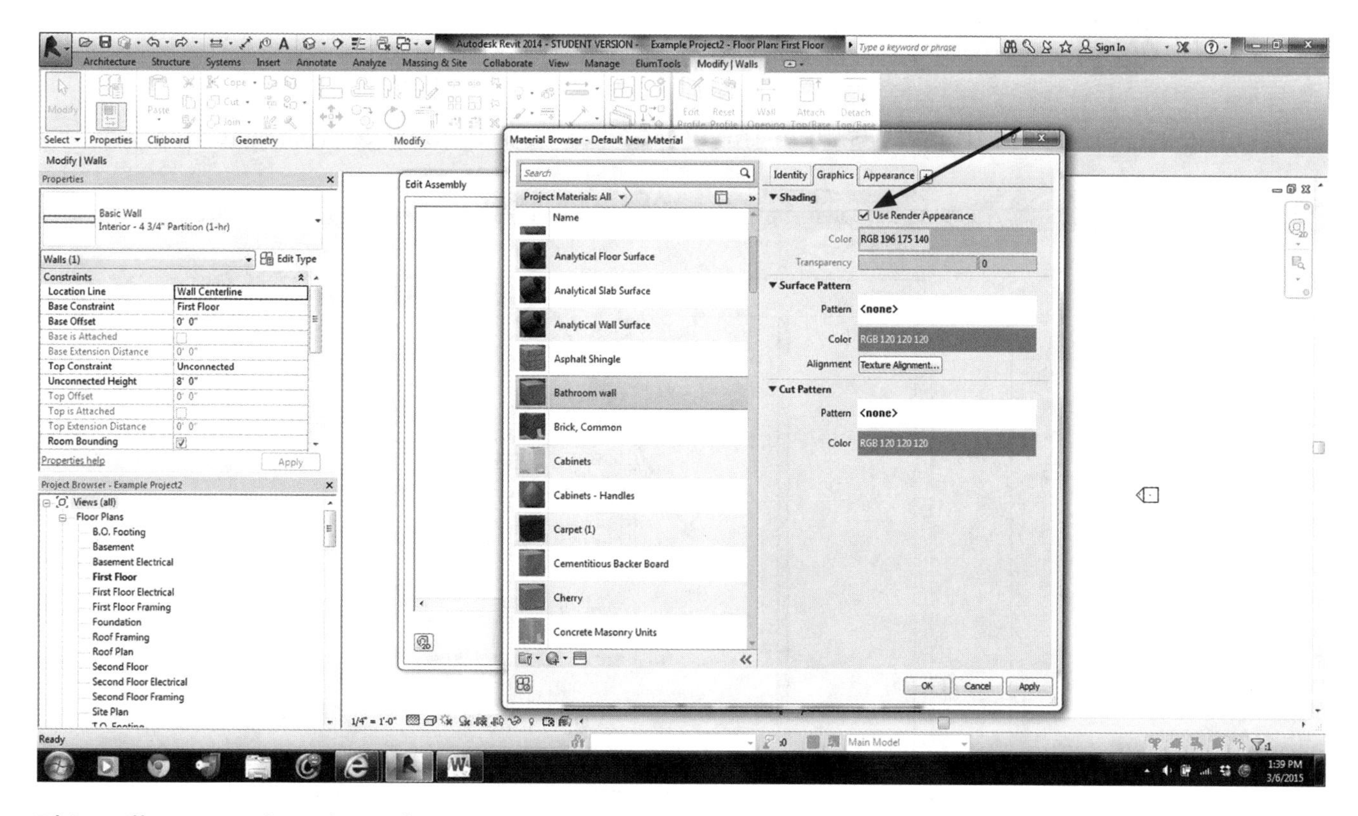

This will ensure that the colors appear correctly in your drawing, and are suitable for presentation.

Once this is complete, it is time for you to customize your wall and place the tile or material in the correct location.

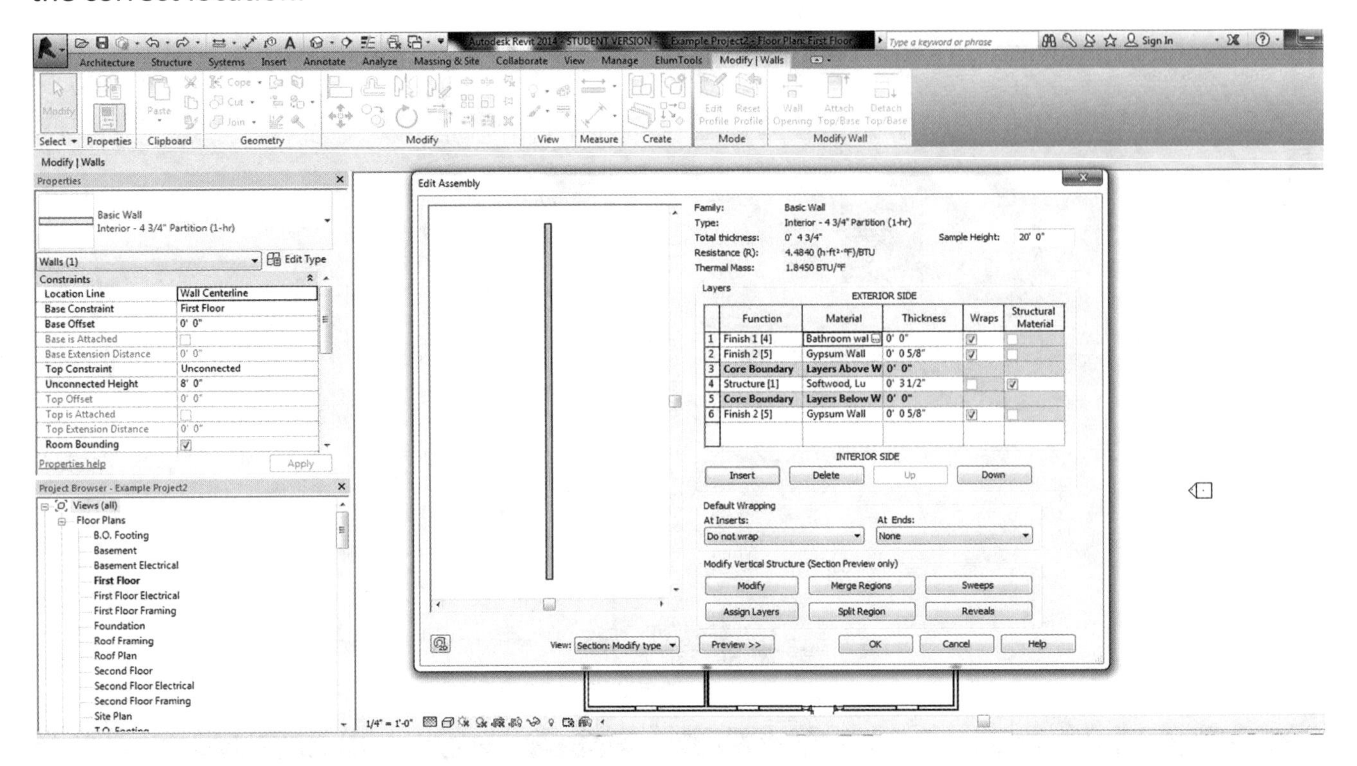

REMINDER

At this point in particular, you do not want to hit* ESC*. You will lose everything you just created!

Click in the screen where the wall is, near the bottom, and zoom in to get closer to the wall.

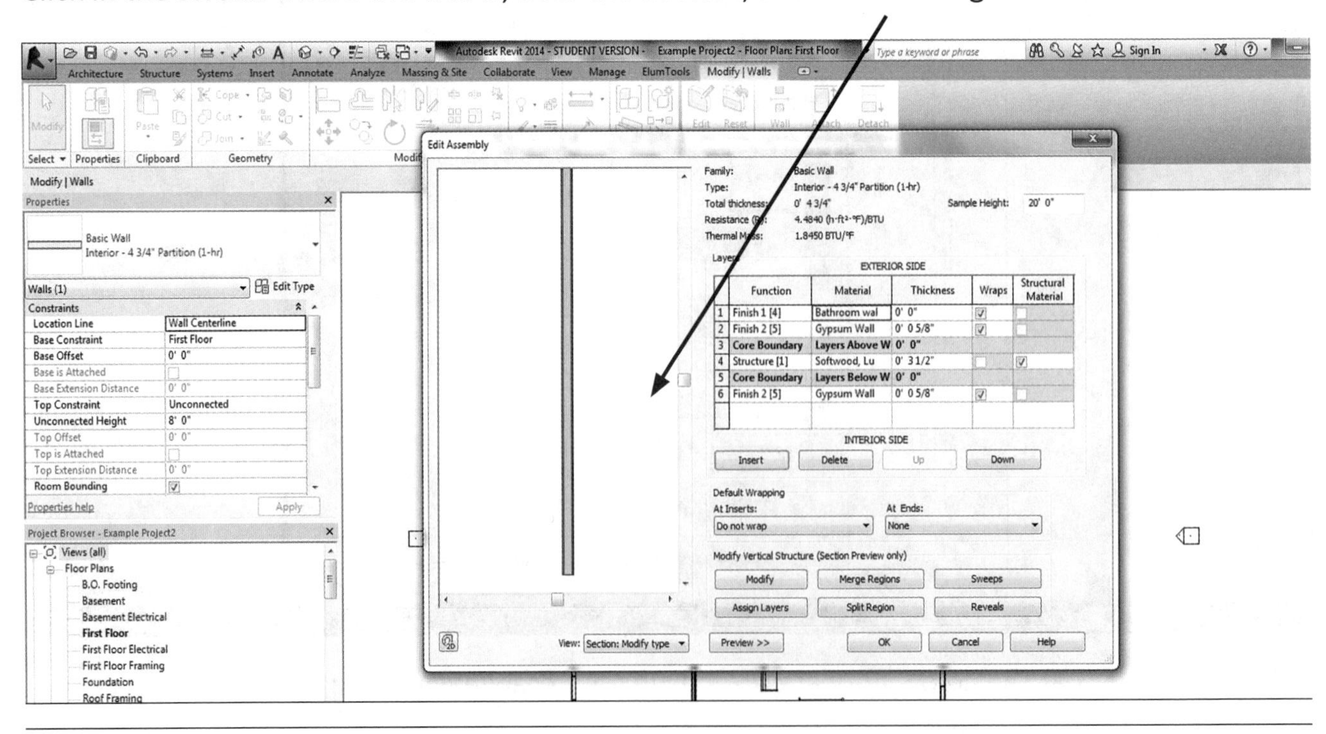

Once the wall is closer, click on **SPLIT REGION**.

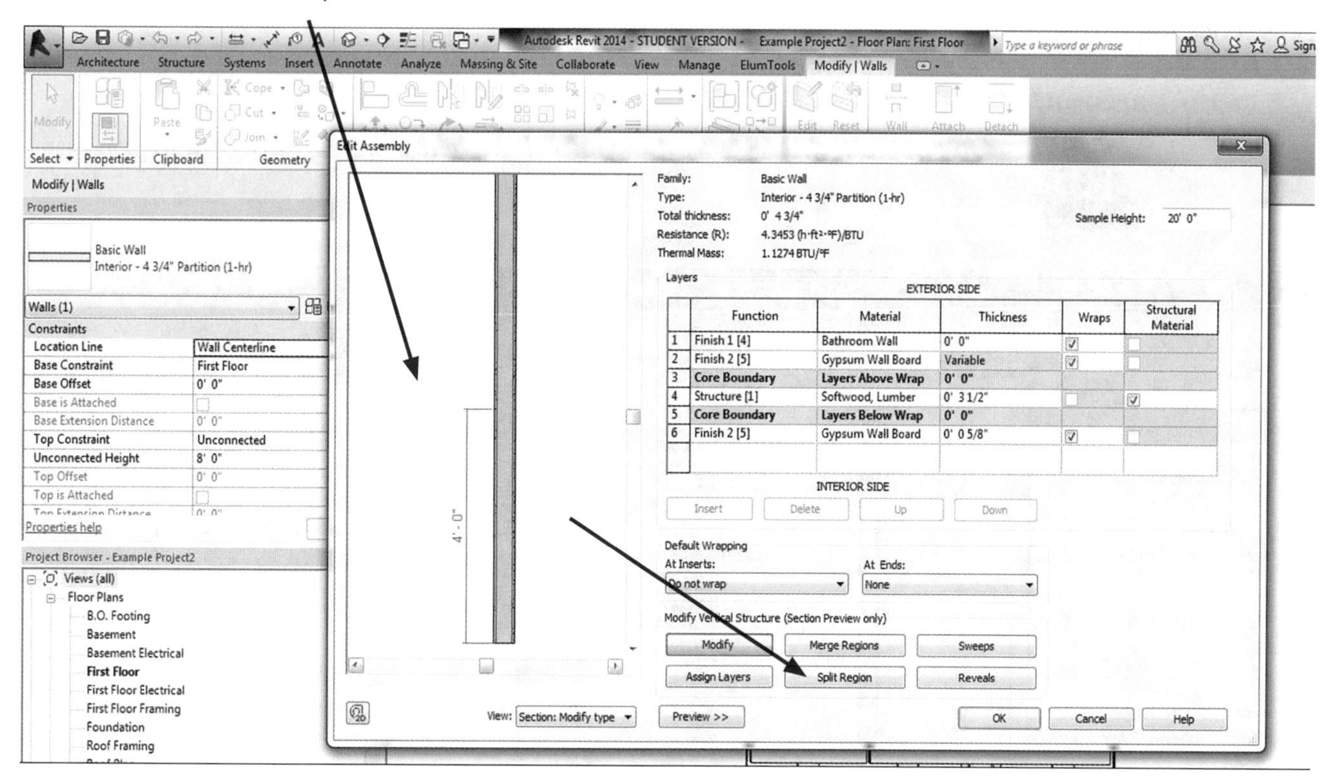

This will allow you to determine how high you want to make the layer material up the wall.

Once you select **SPLIT REGION**, a knife will appear, and you can measure the height up the wall.

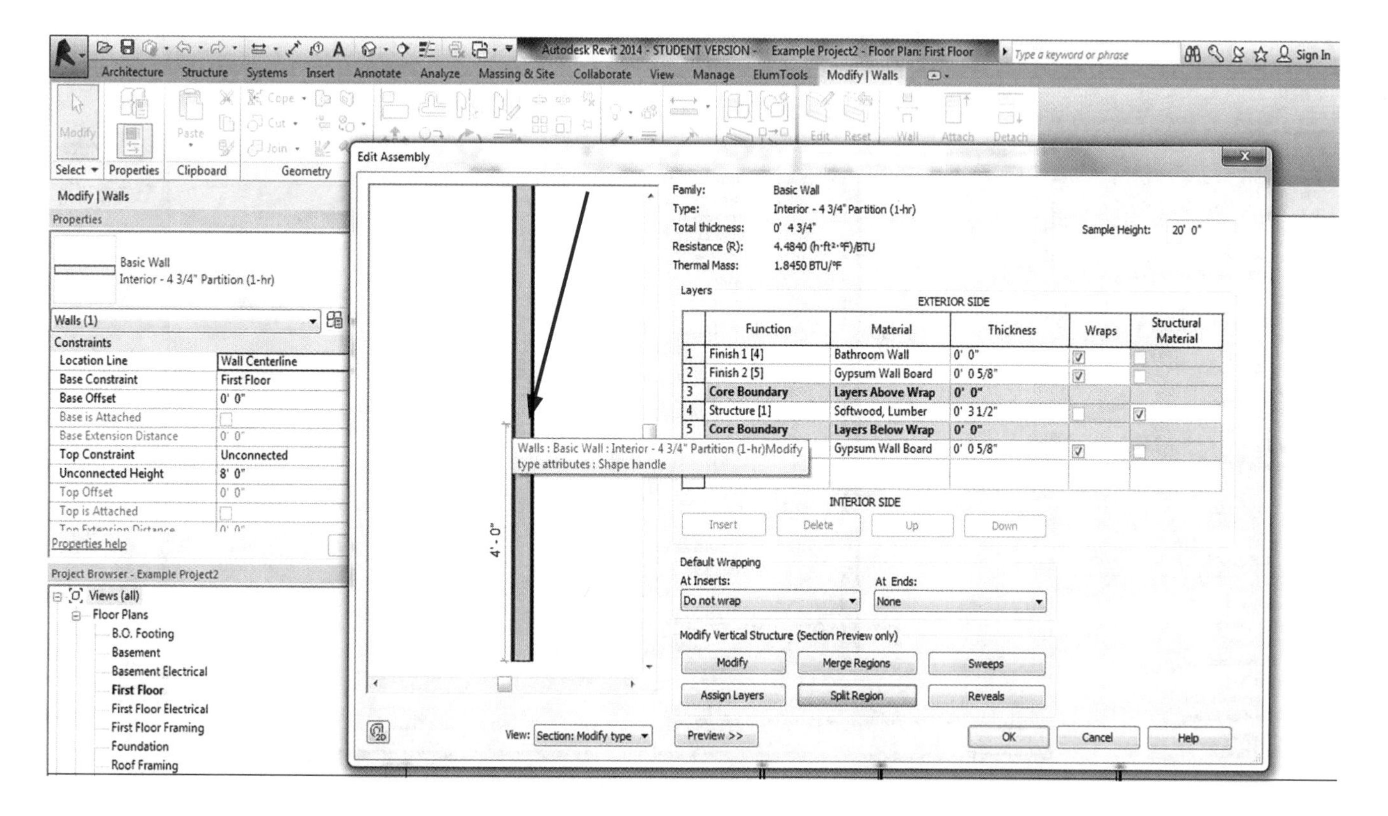

When you pick the height you want click, and that will be your dividing line between the materials. Click the **MODIFY** Button.

REMINDER

Do** not **click the** ESC **Key or you will have to do everything all over again.

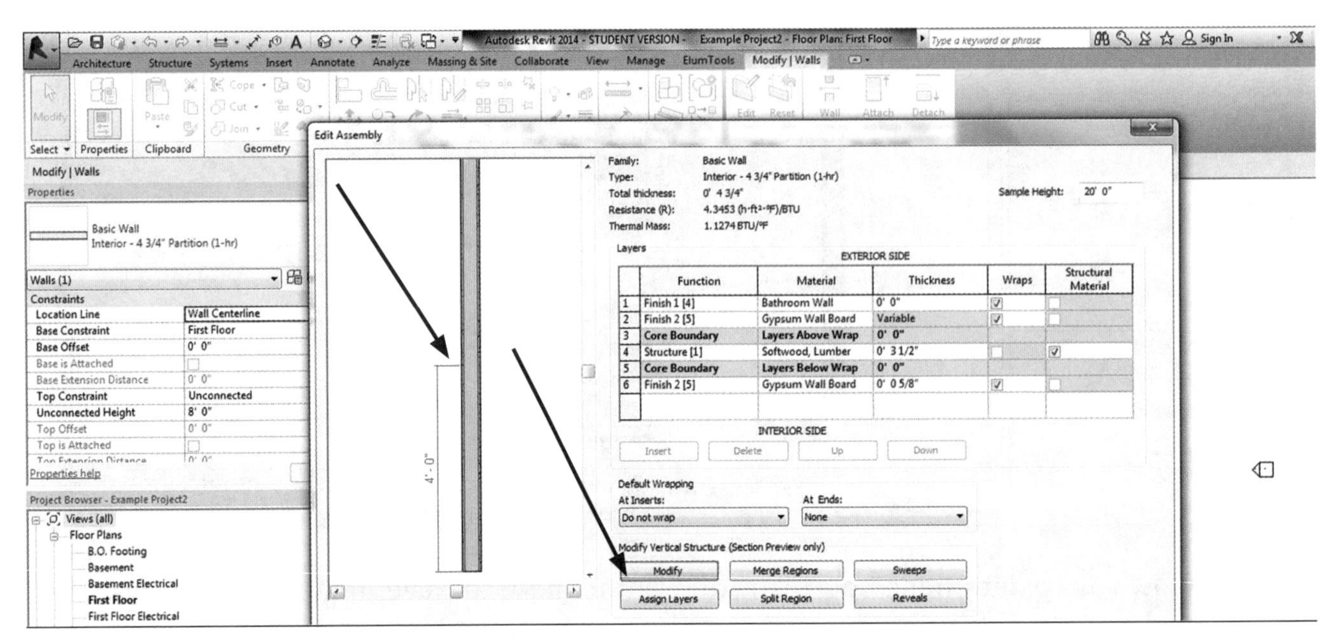

After clicking the **MODIFY** Button, place your cursor close to the line you split on the wall.

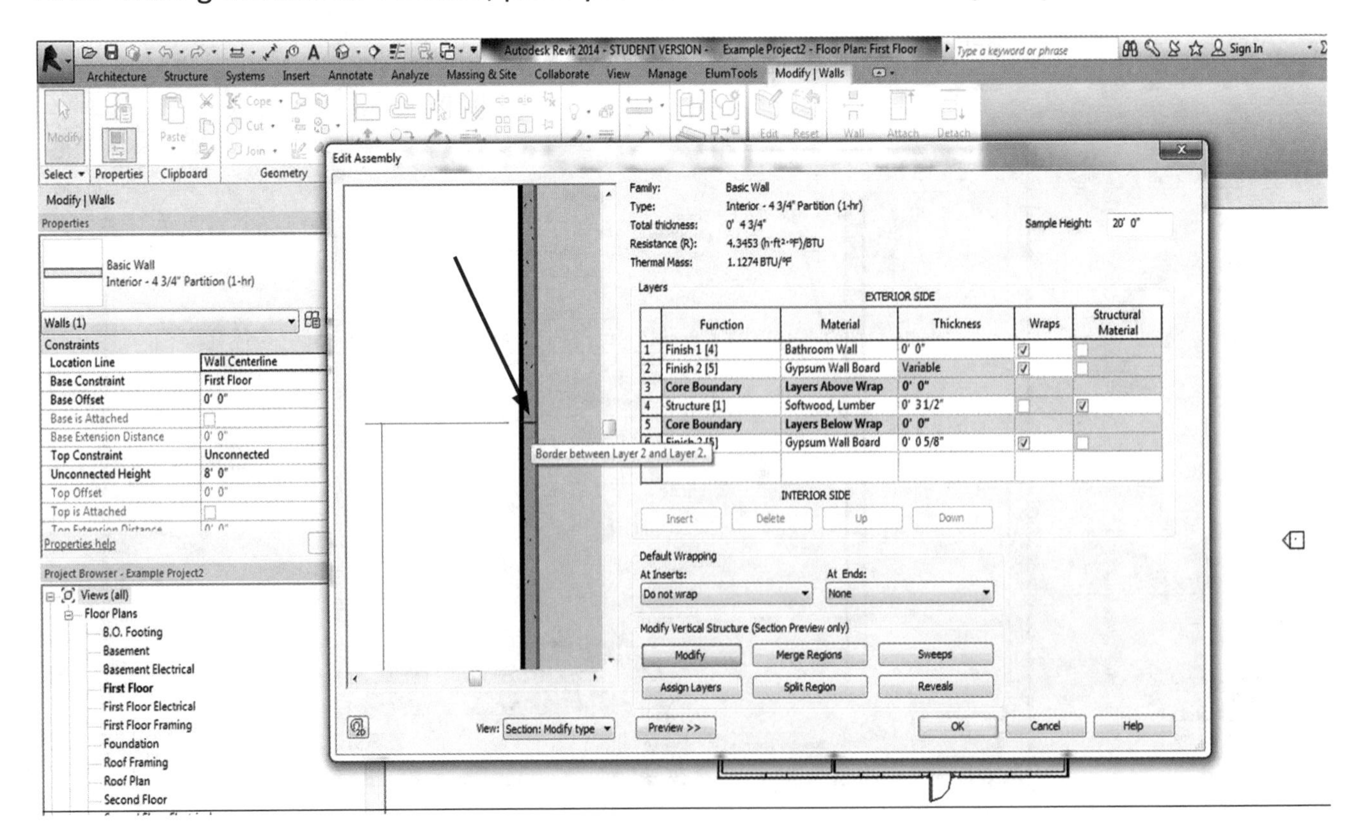

Once you click the line you will get an arrow pointing either up or down. This will determine which direction the new material will be placed, either above or below the line.

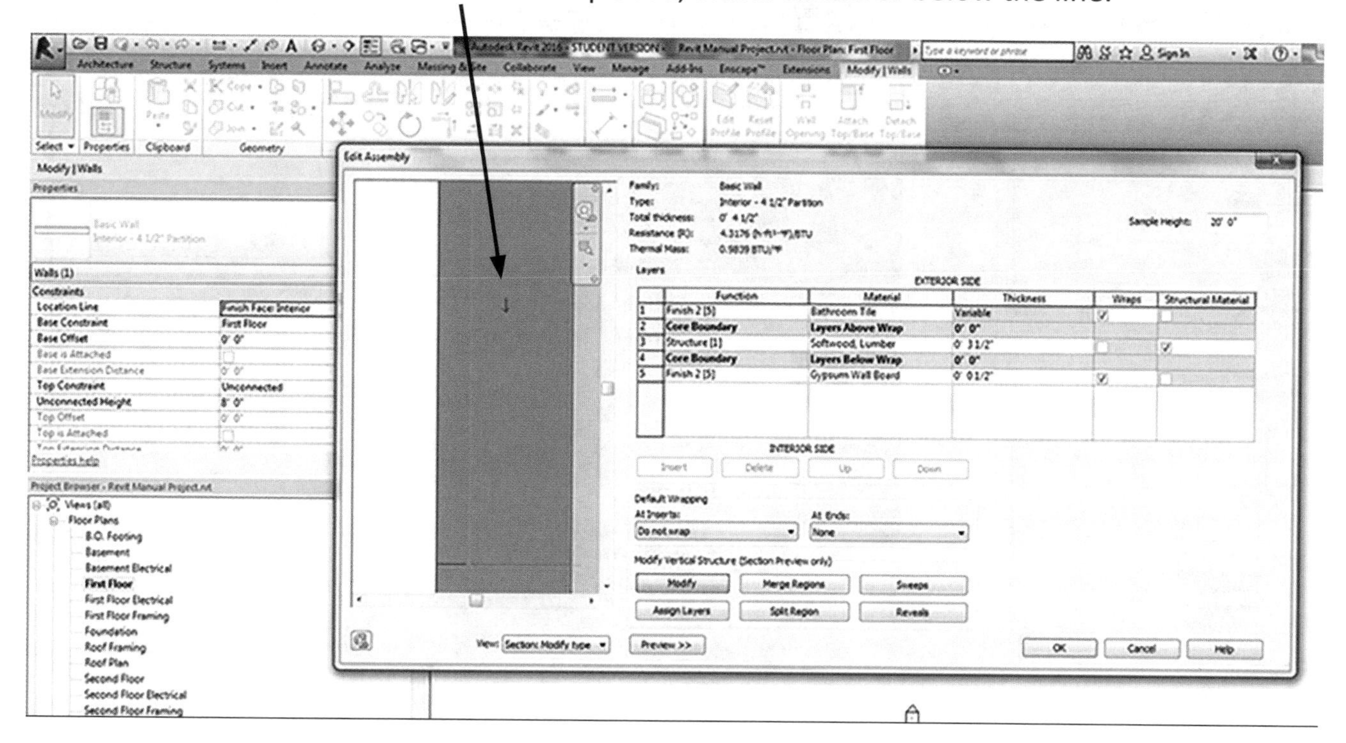

Once you have selected the direction you want to place the new material, scroll down to bottom of the wall in the View Window and select the bottom of the wall material to activate a padlock.

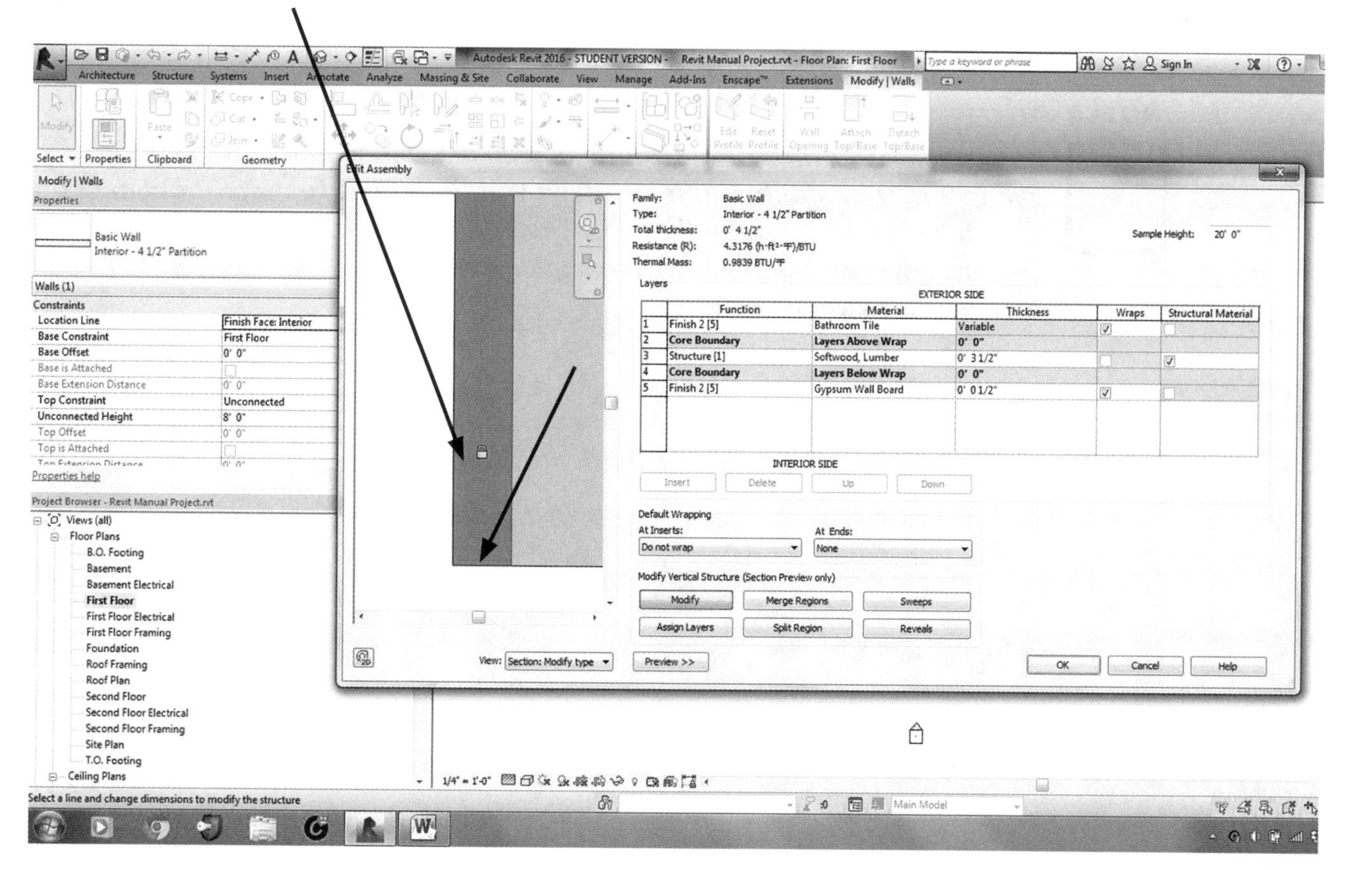

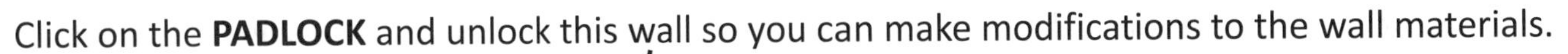
Click on the **PADLOCK** and unlock this wall so you can make modifications to the wall materials.

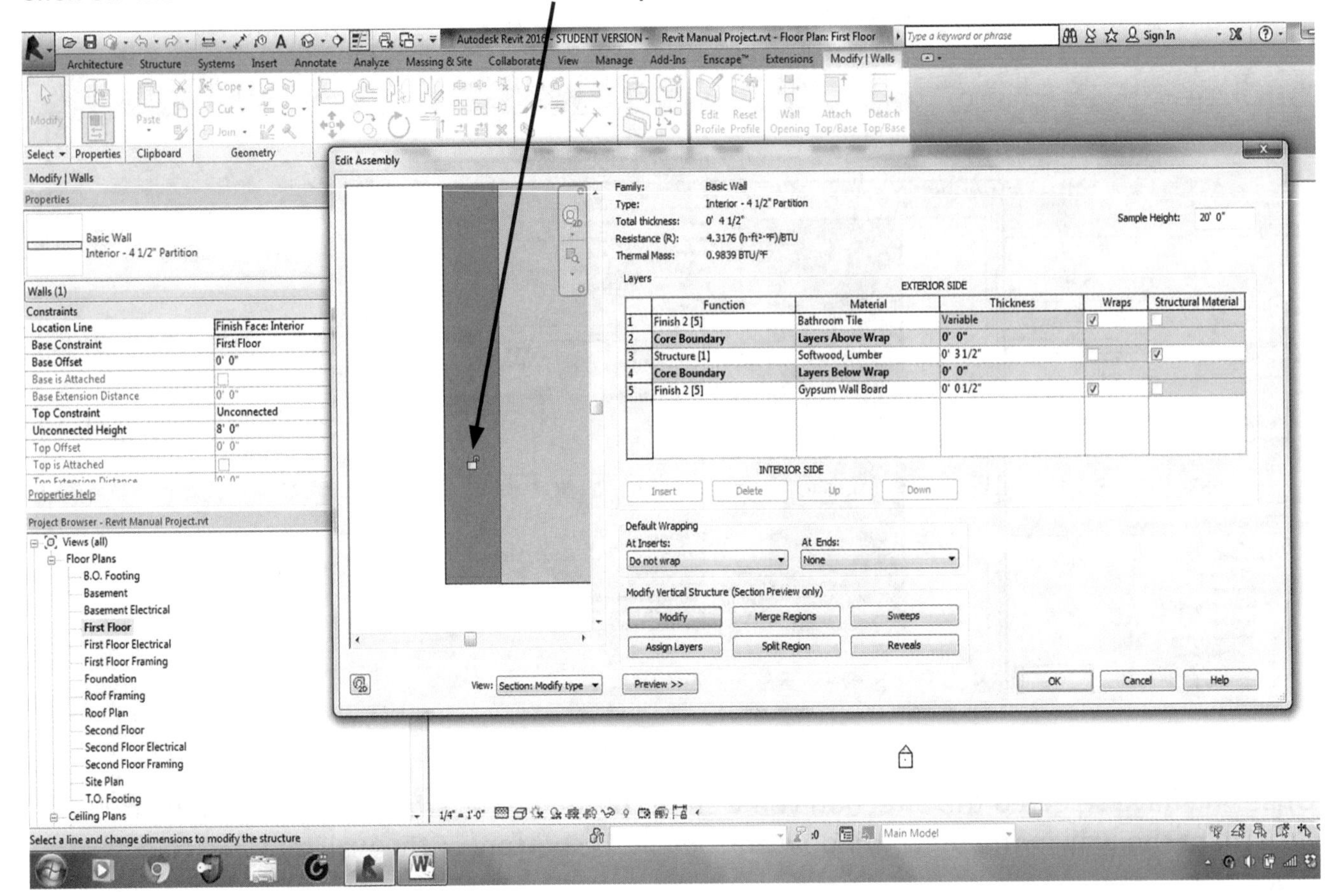

NOTES

The padlock is a safeguard feature done to the wall materials so nothing changes in the default settings of Revit Wall Families. In order for changes to be made and material to be added, you must unlock the padlock of the Default Family after you have duplicated the wall and renamed for it to become your custom wall.

To change the material on the upper section of the wall, select the number of the material you want, click **ASSIGN LAYER,** and the other wall material will change.

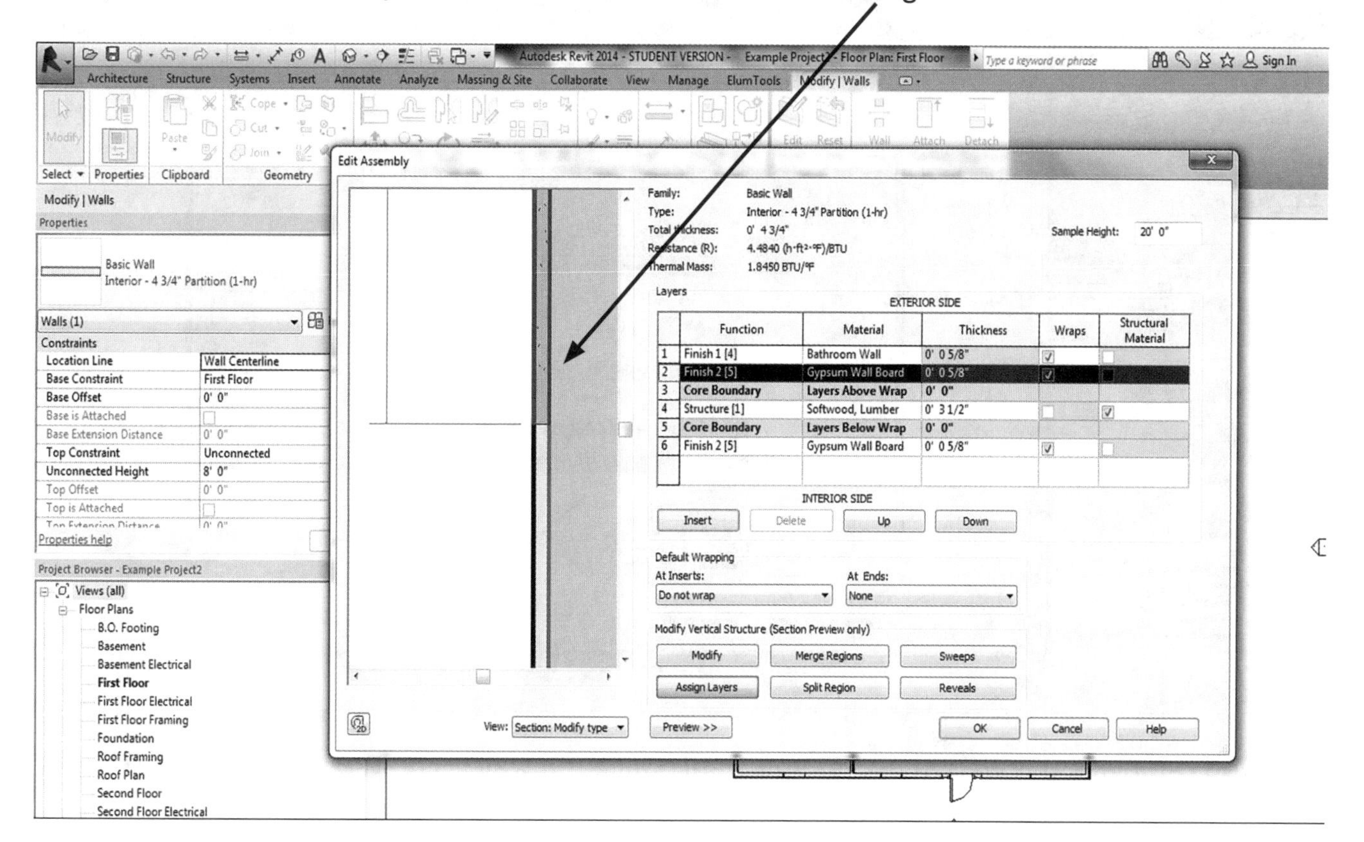

When you are finished with changing the wall material, click the **OK** Button to clear this screen.

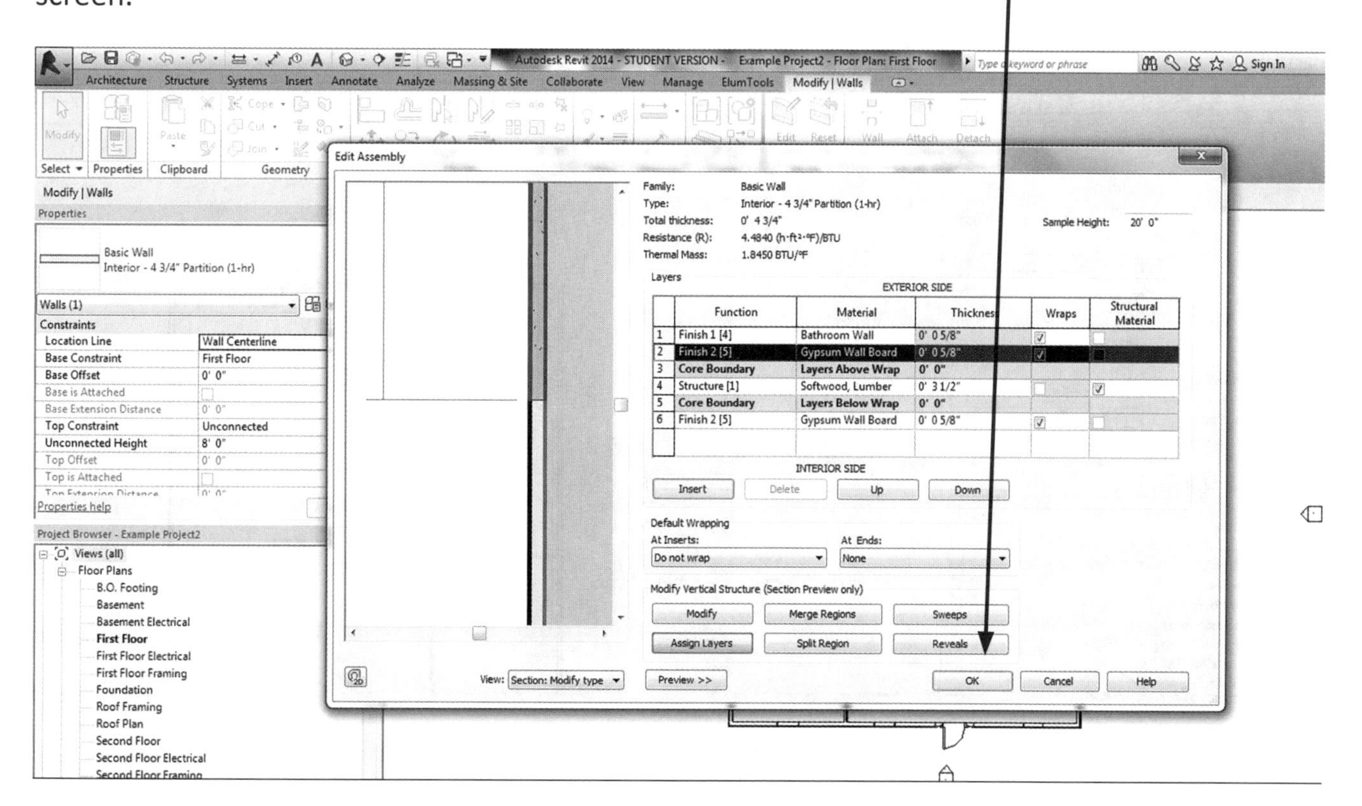

Then click the **APPLY** and **OK** Buttons to lock all the changes to the wall.

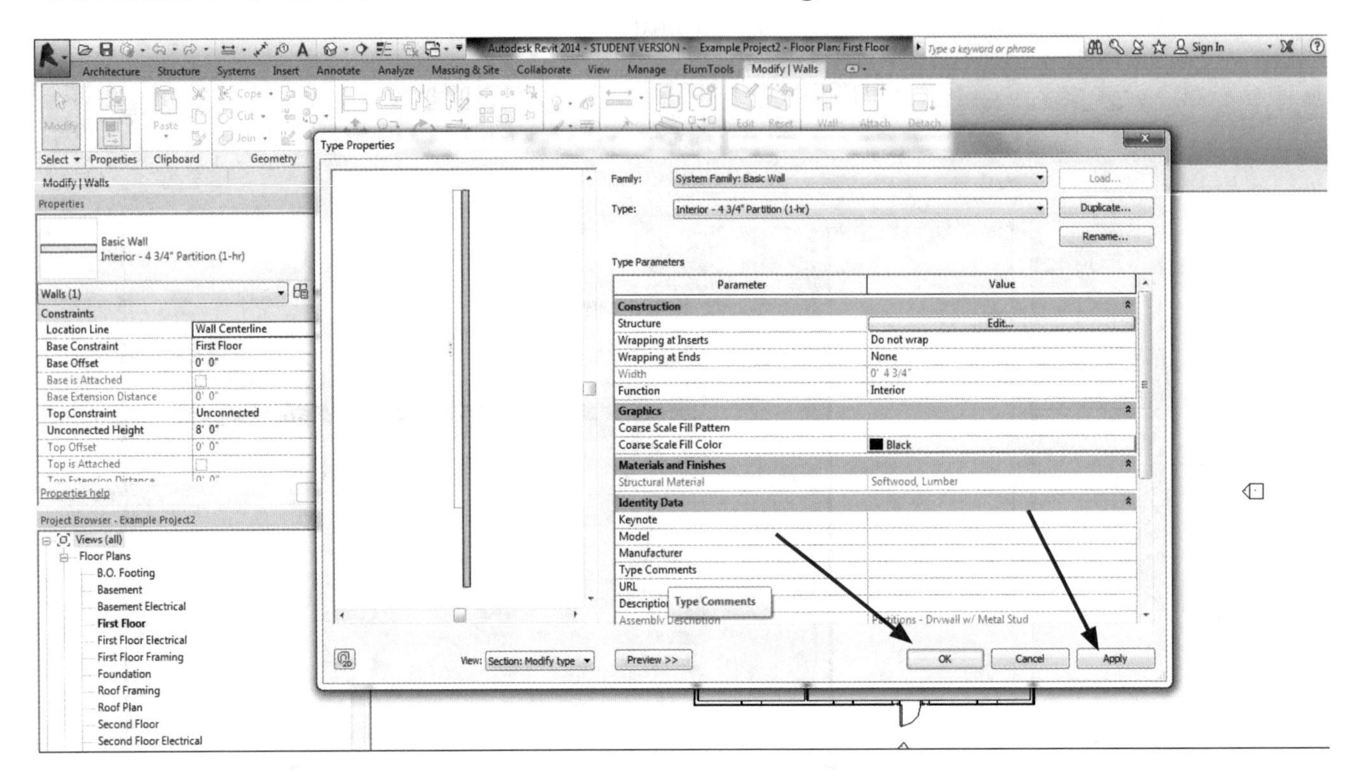

Now you can place this wall wherever you want. Once the wall is created, and you want to show it to a client in 3D view, go to the top of the screen and click the **HOUSE** Button.

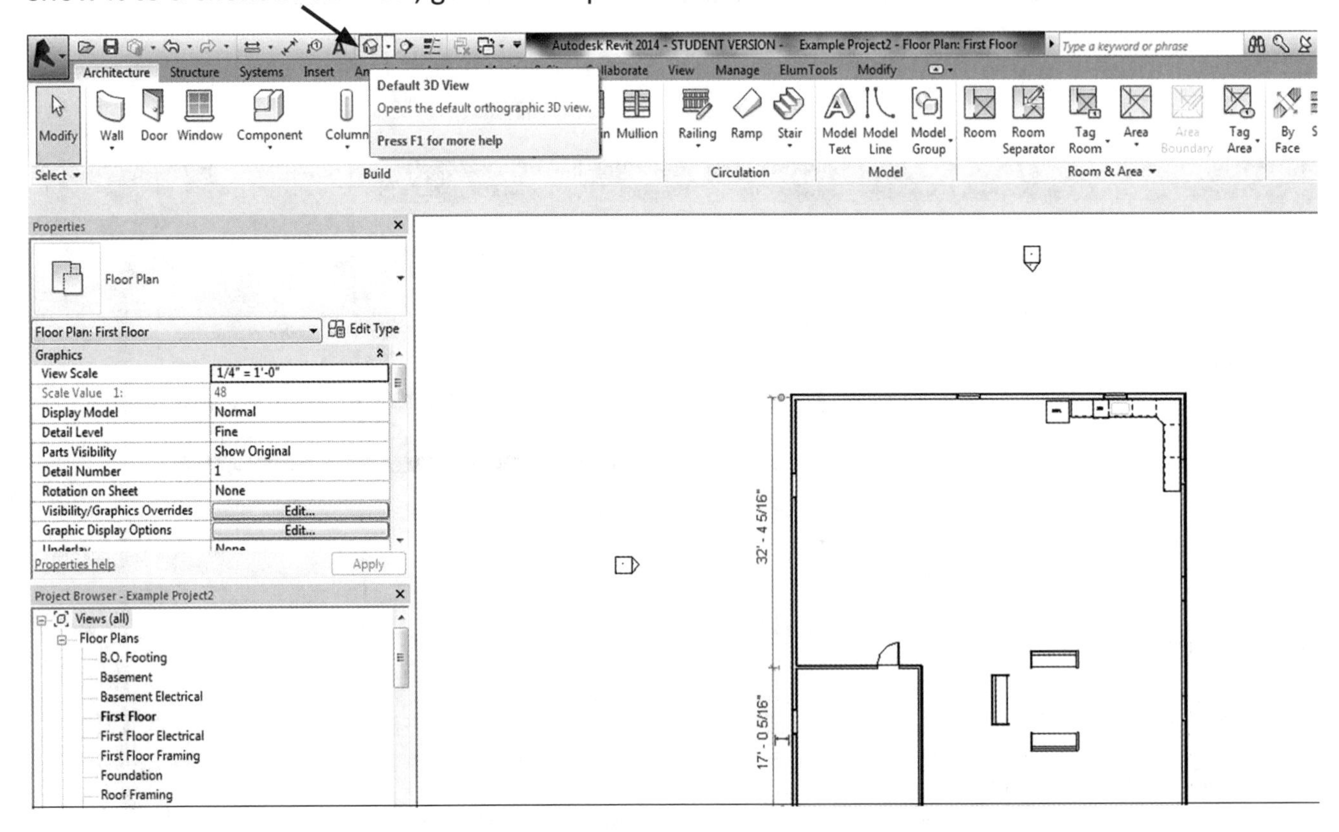

The view will change to 3D, and you can show the client the rendering view.

If you have tile in a basket-weave pattern or a custom tile, I would use Consistent Colors instead of the Rendering View. It will show the design pattern better.

To change the design to Consistent Colors, go to the bottom in the View Style.

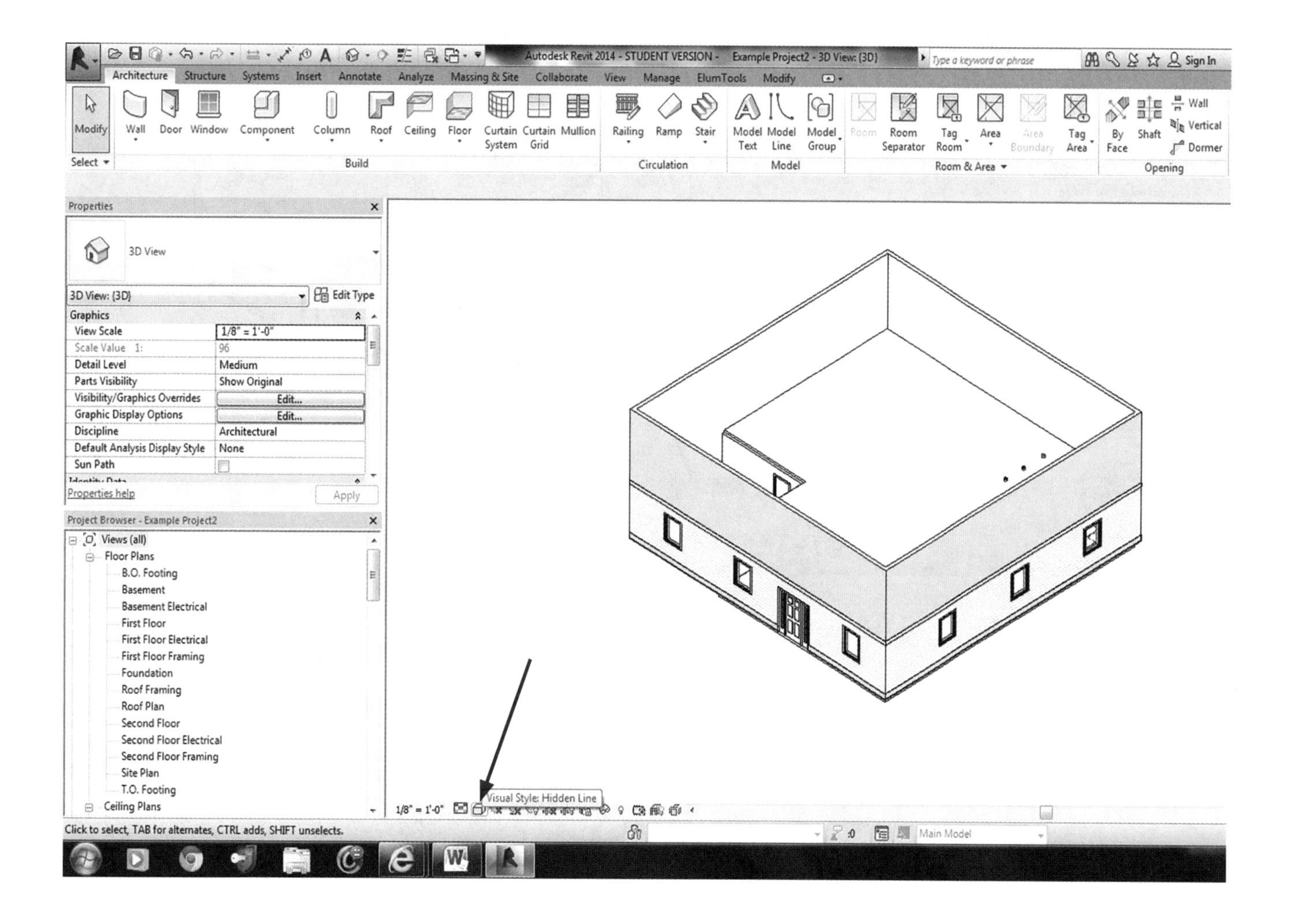

The Options selection will open and you can change it to **CONSISTENT COLORS**.

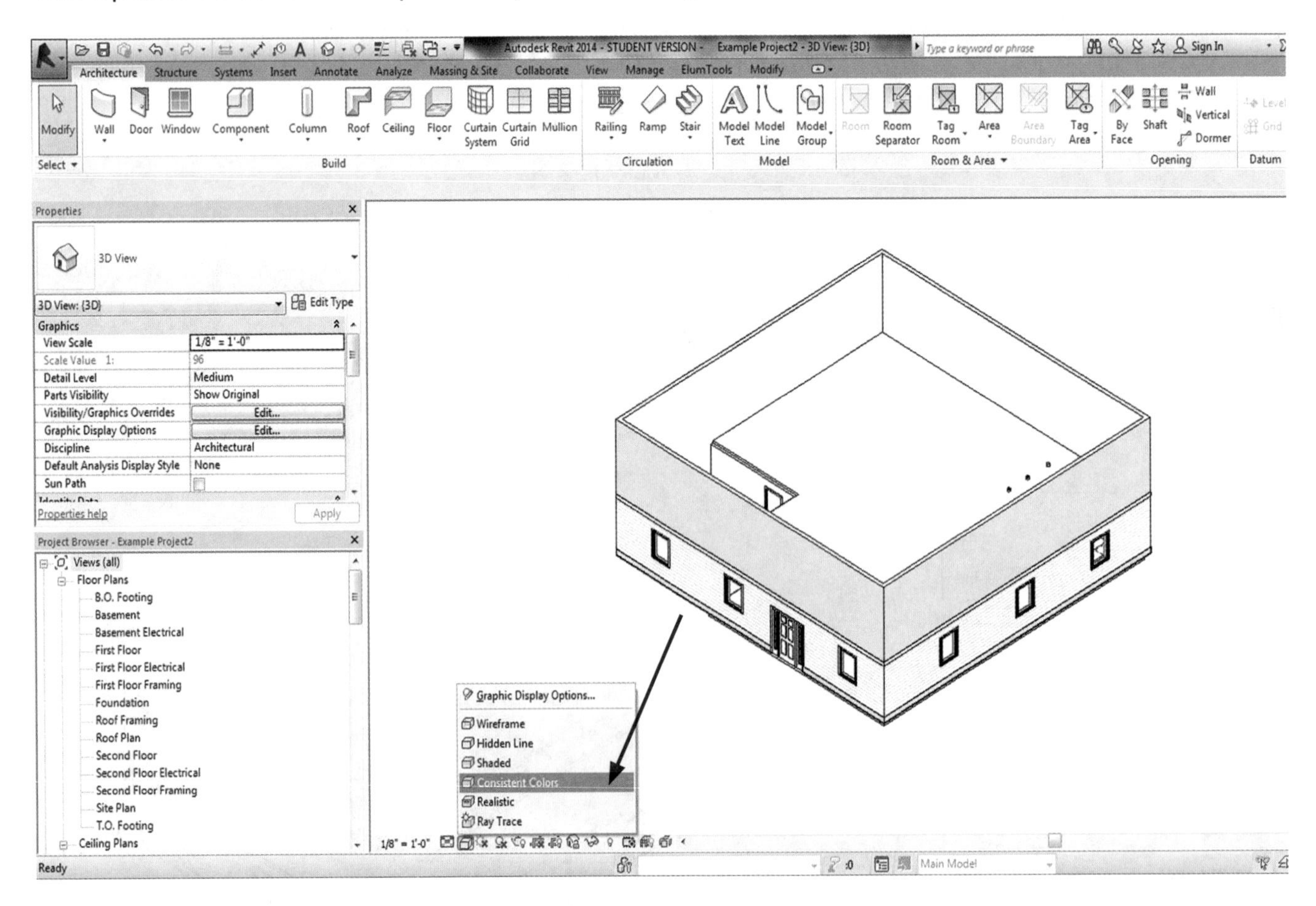

This will change your 3D to show color without having to render, and show the pattern detail more clearly.

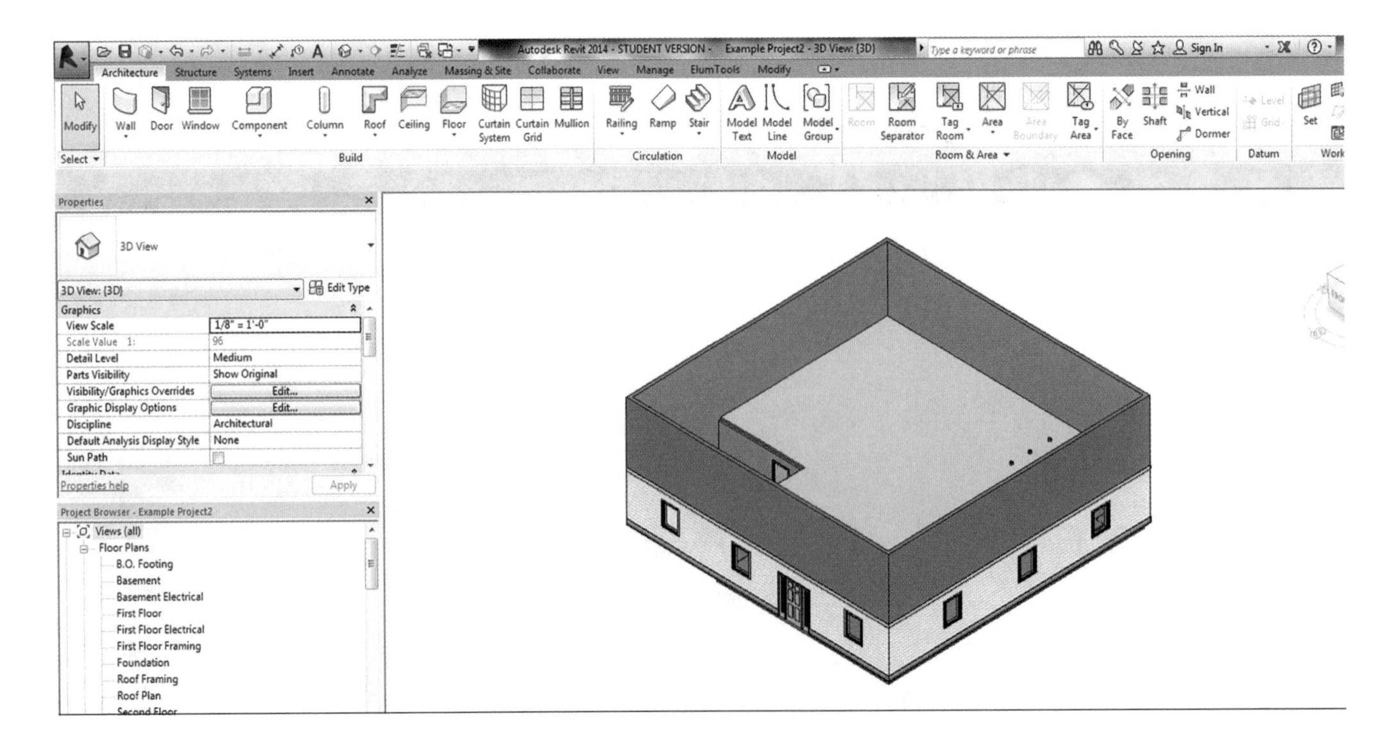

EXAMPLE

Following all the steps we learned in this chapter, I have added a split wall in the kitchen, including a tile back splash around the counters. See the results:

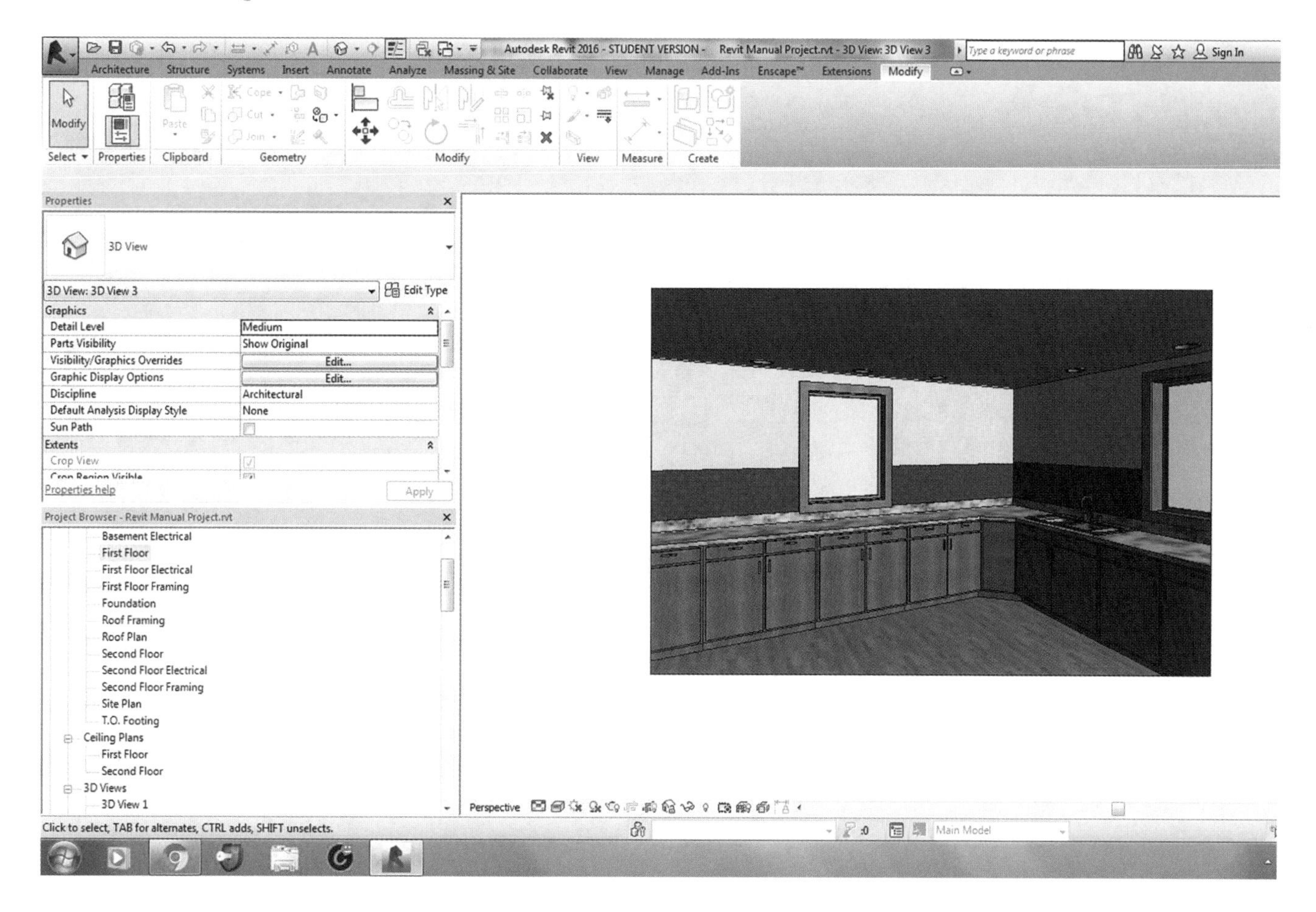

Revit provides pre-designed walls in its own software from the beginning when you open it and start a project. With the chapter you just completed, you can create your "own" wall to meet any special needs, your own or your client's.

Chapter 20, *Revit Modifying Commands*, will explain how to use all the modify commands for the Revit software to make your design more efficient. Soon you'll be able to begin creating your own spectacular designs.

CHAPTER

20

Revit Modify Commands – Copy, Offset, Trim, Mirror, Move, and Rotate

REVIT BASIC COMMANDS

THE MOVE COMMAND

When you begin placing a wall or any object, the Modify Box automatically opens for you to begin modifying objects, walls or doors, etc.

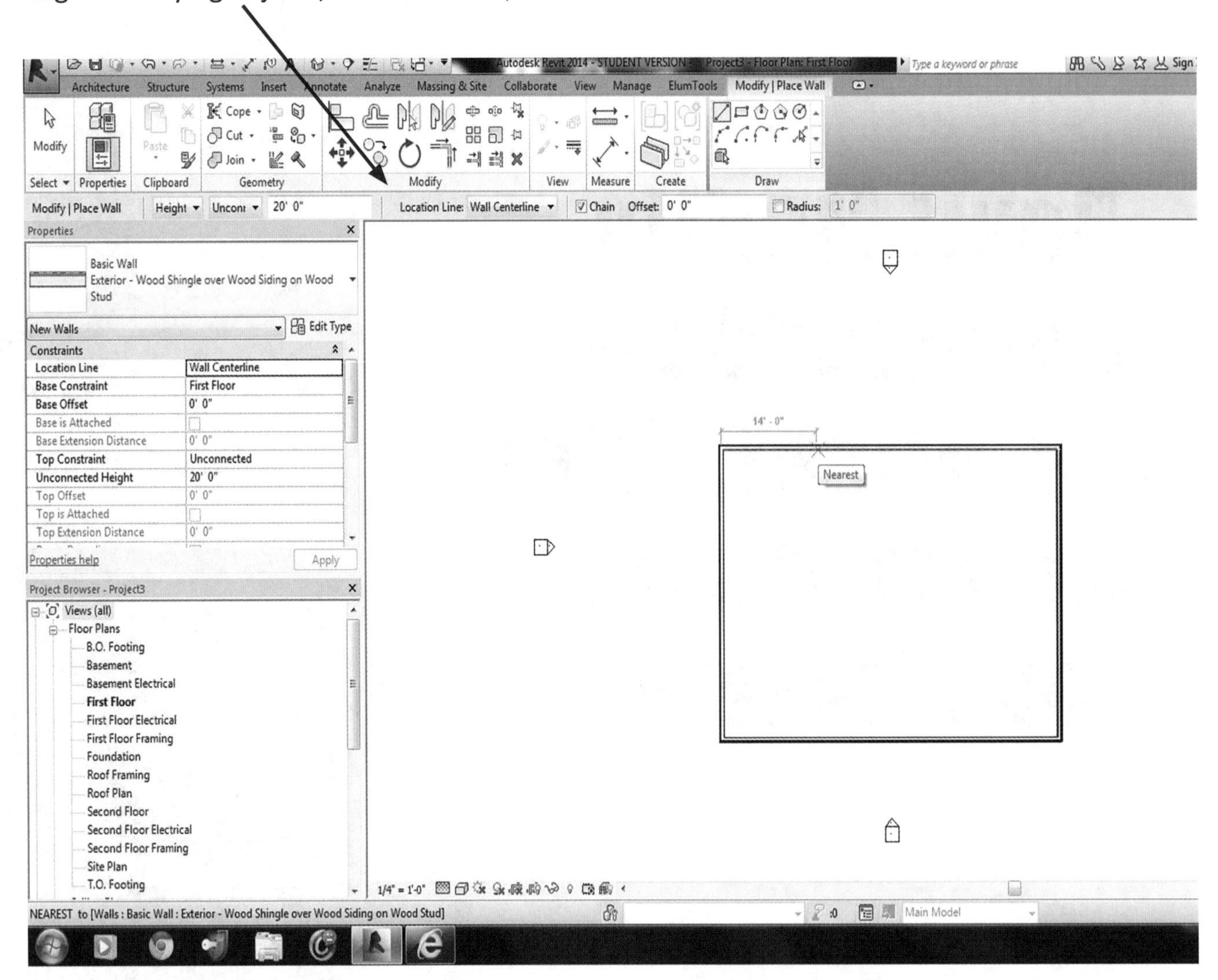

To move an object, select it, then click on the **MOVE** Button.

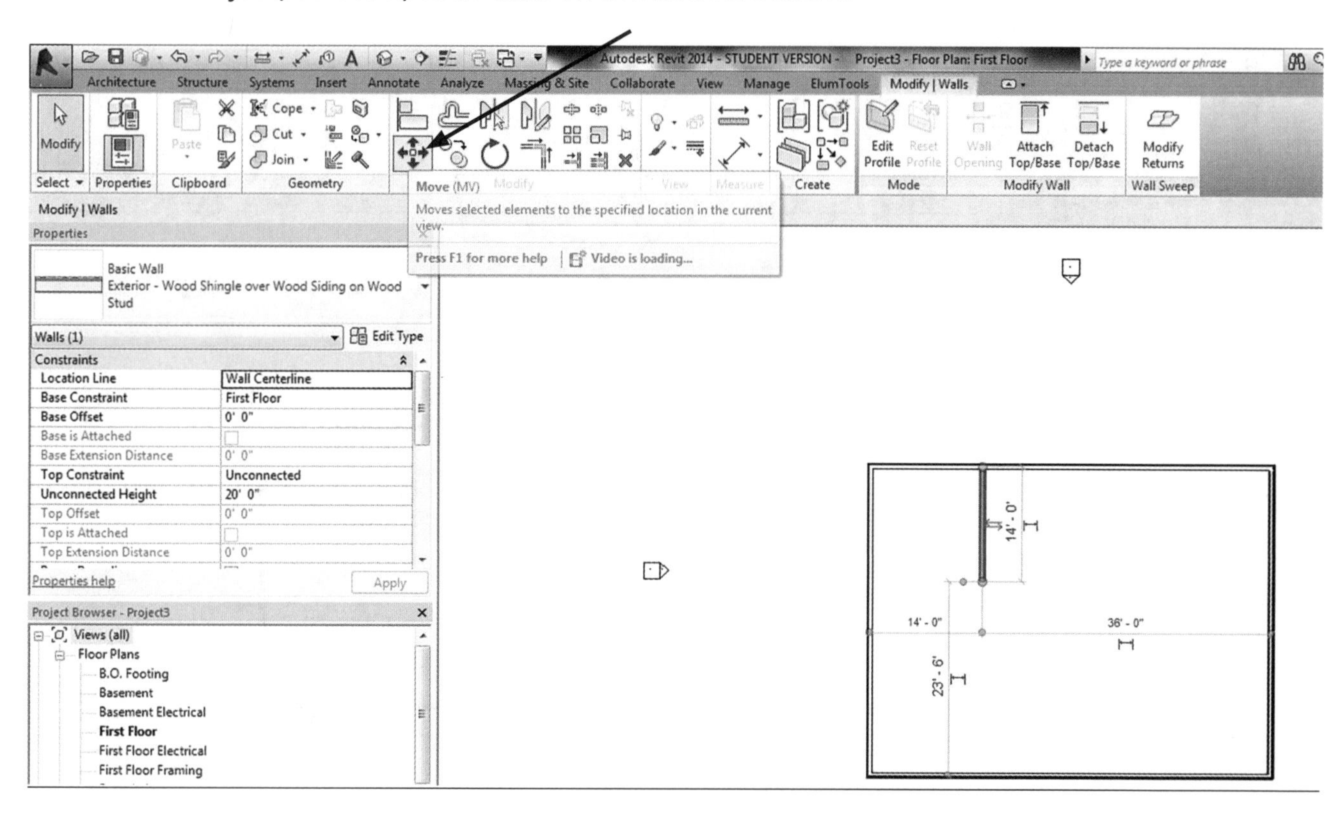

This allows you to manually drag the object where you want it to be. Simply click on the object, and drag it into position.

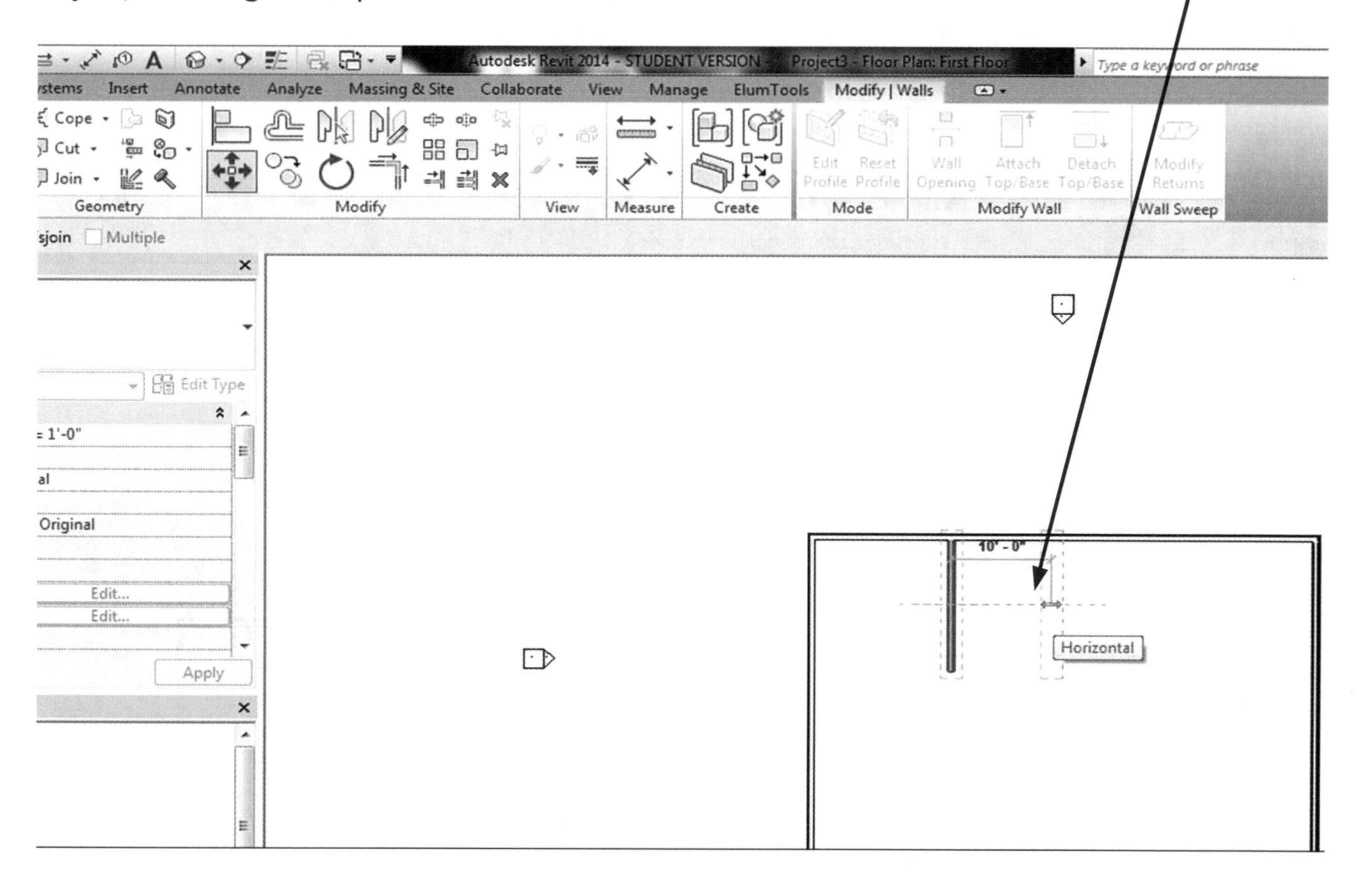

Or, if you prefer, you can type in the specific distance to move the object.

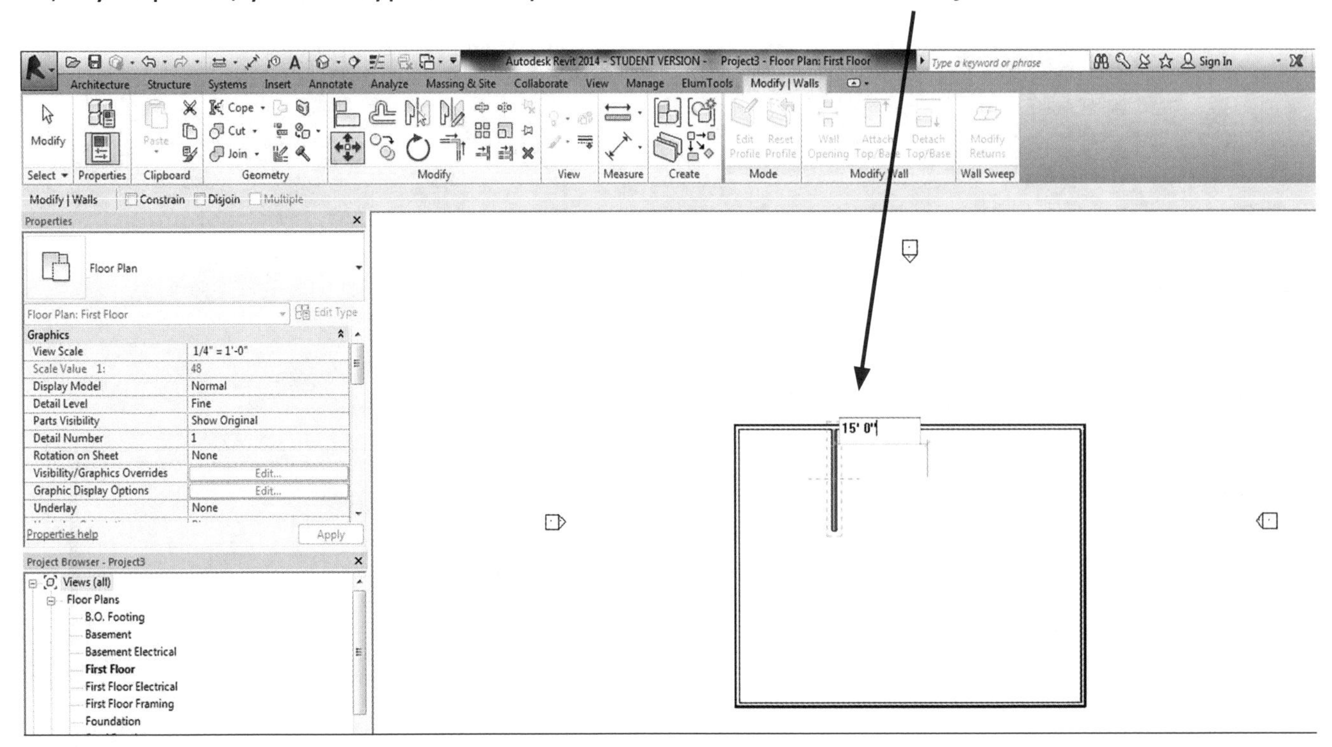

THE COPY COMMAND

Copying an object works almost the same as the Move Command except you can choose to copy once, or you can do it multiple times. Select the object and then click on the **COPY** Button.

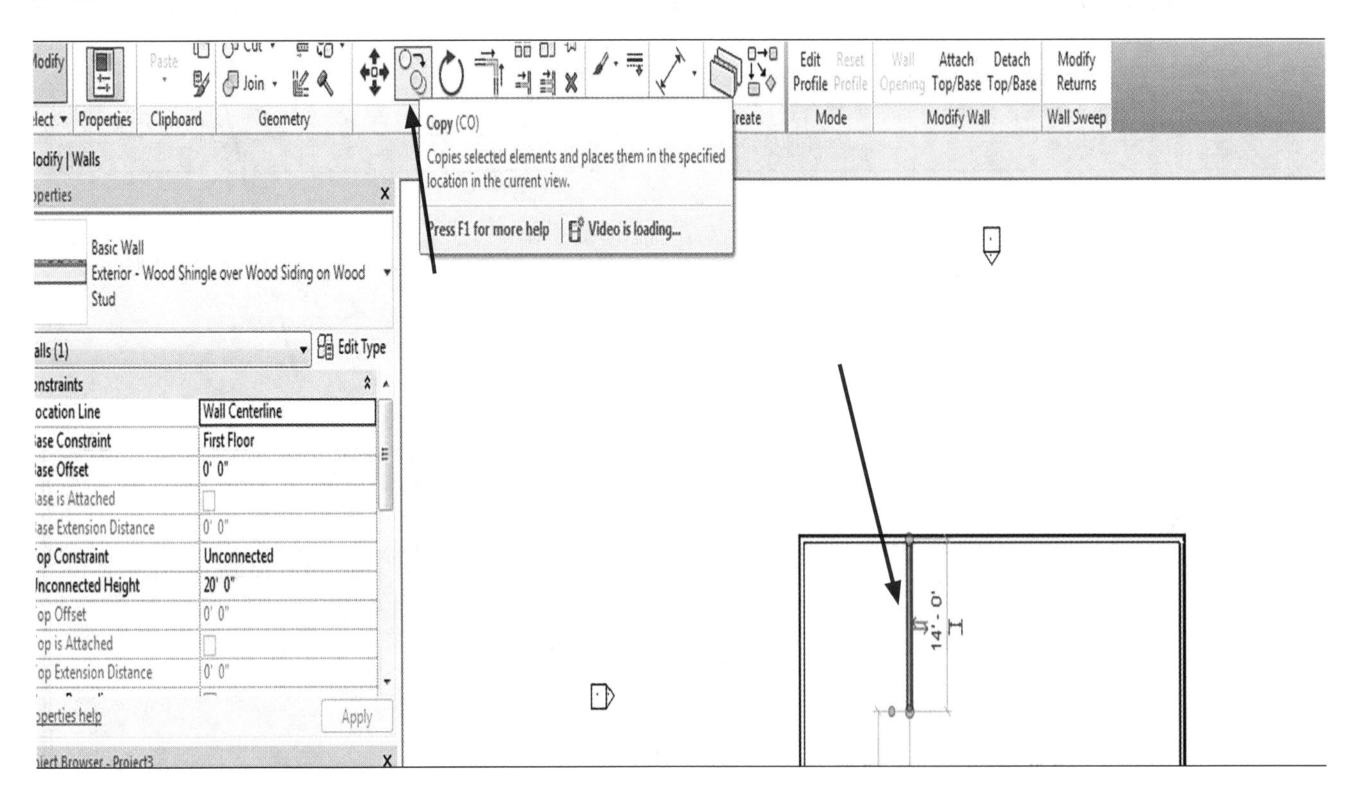

Once you have selected the **COPY** Button, your Options Bar appears.

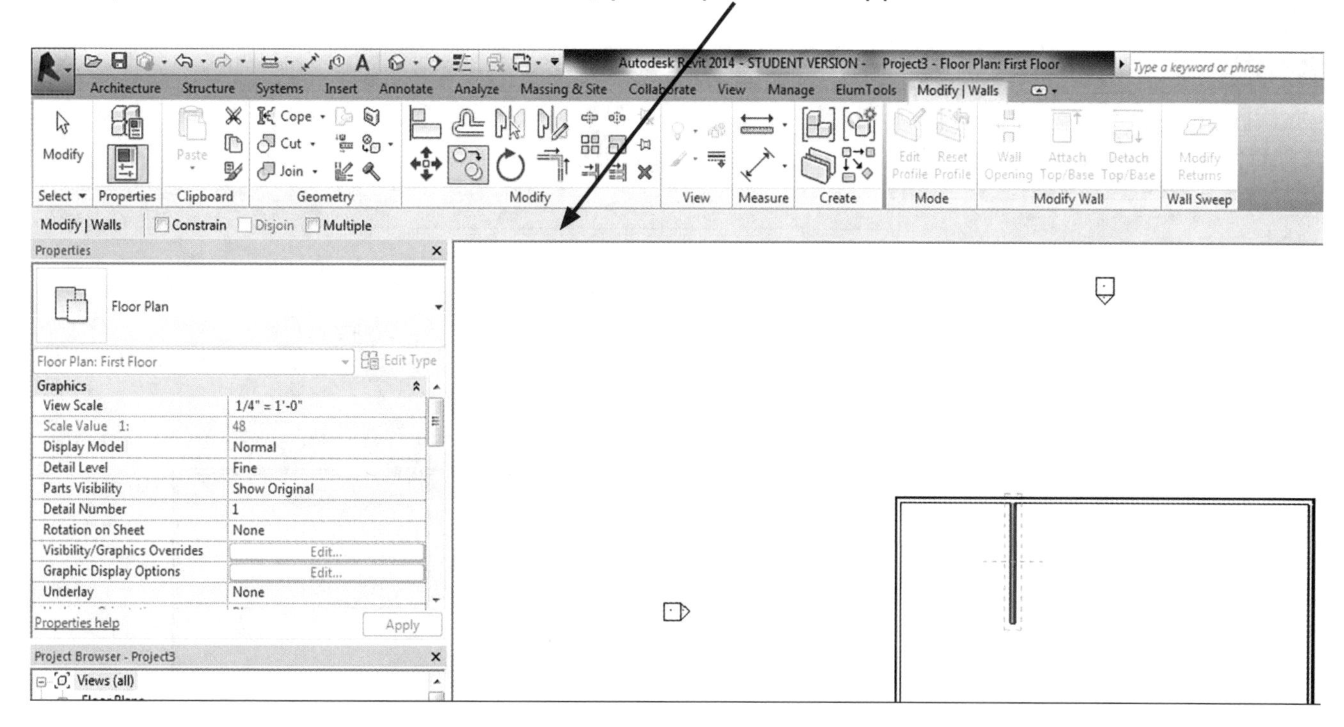

Here you can select **CONSTRAIN** and **MULTIPLE.** Constrain keeps your objects in a straight line, up and down or right to left. Multiple allows you to copy objects multiple times. Click on the object, and you can copy it from one place to another by dragging it.

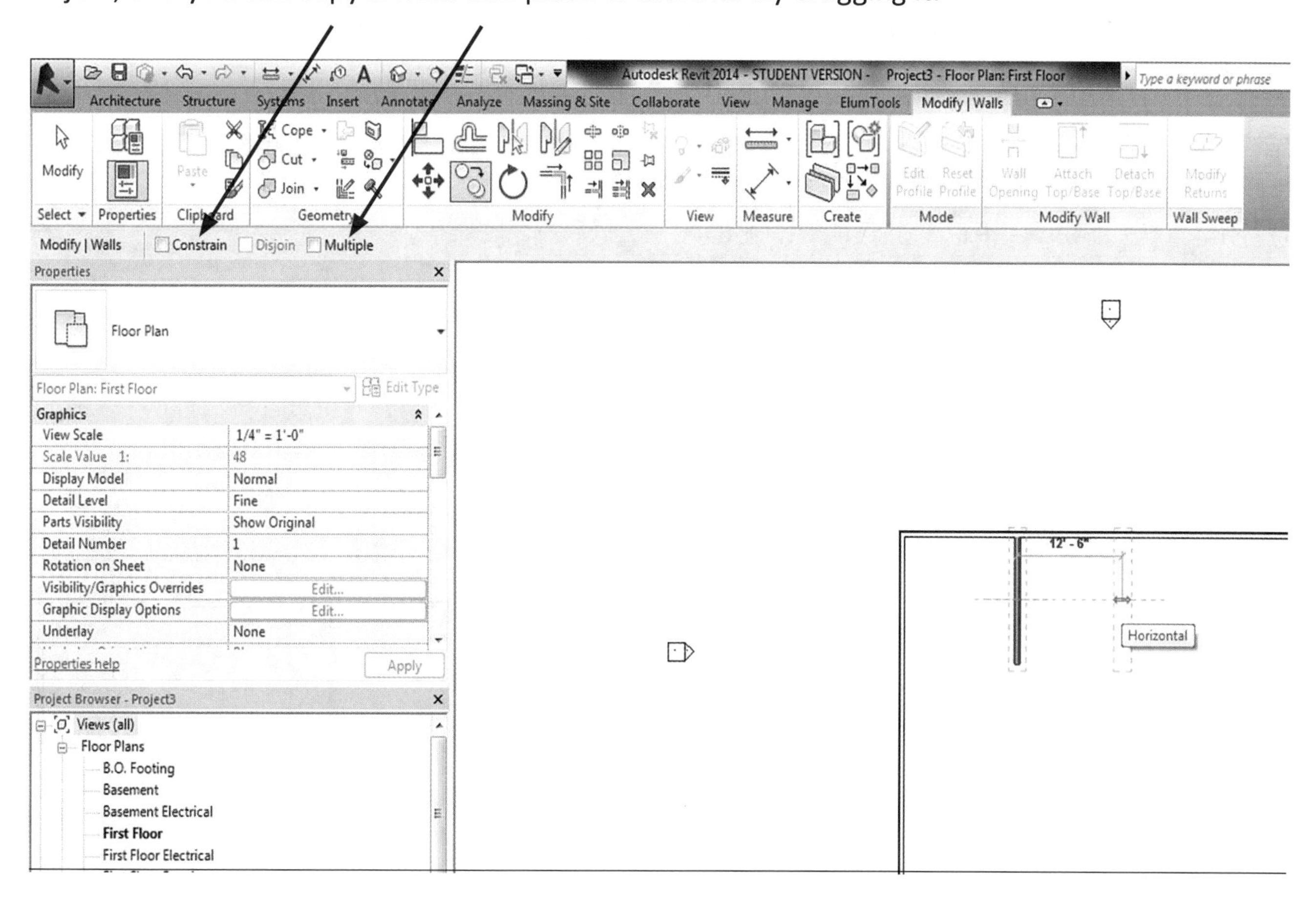

Or, just like with the Move Command, you can copy the object a specified distance.

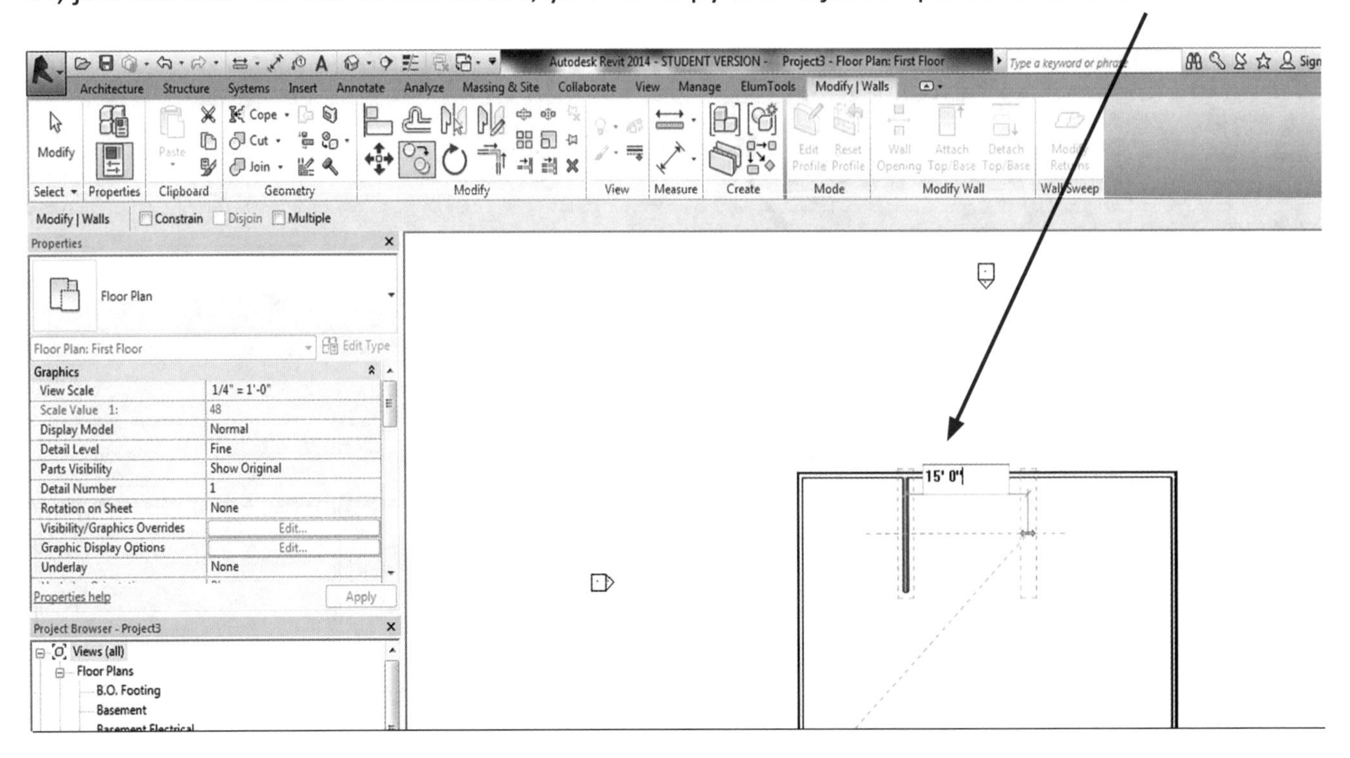

THE OFFSET COMMAND

The Offset Command allows you to select an object and offset it a certain distance, like a roof overhang, or something that will be traced.

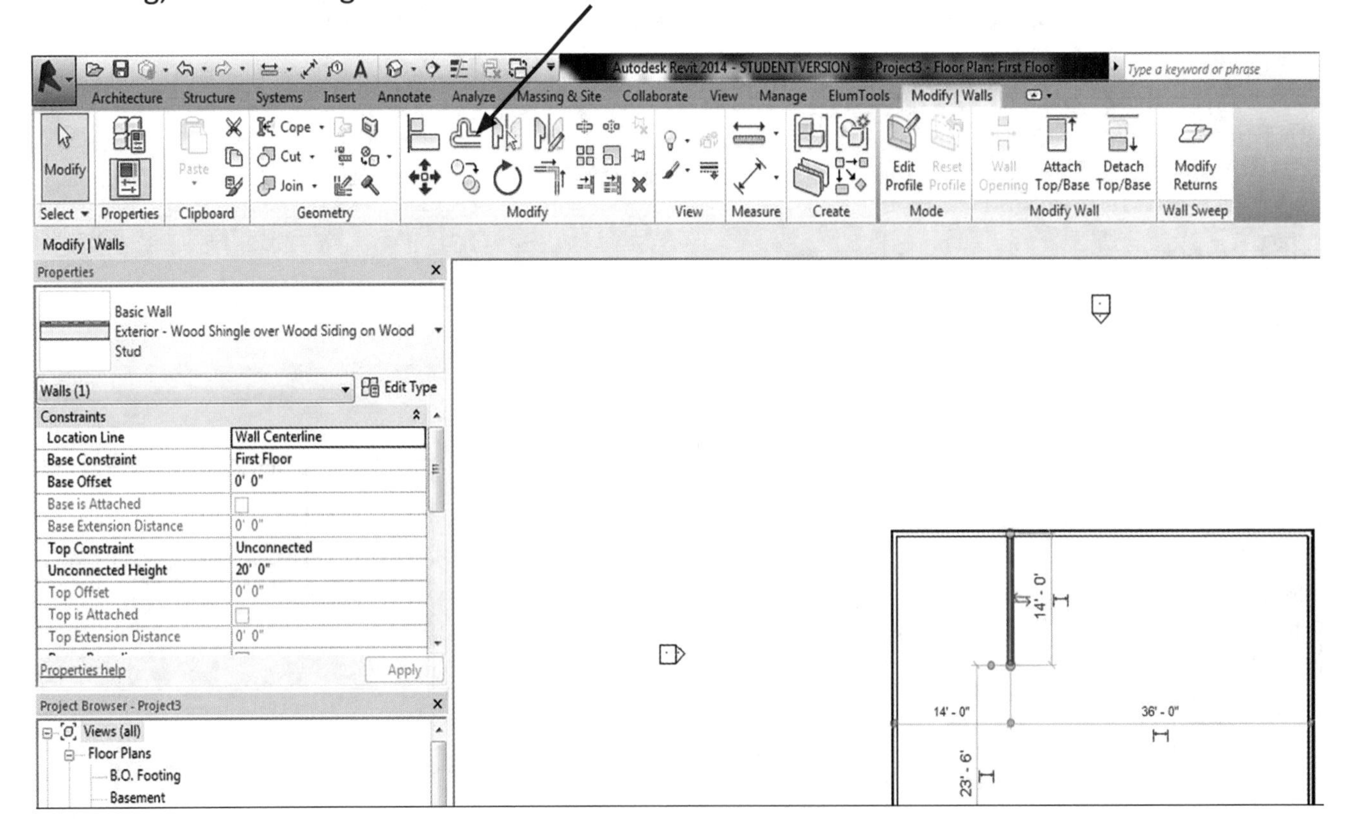

EXAMPLE

As you may recall from Chapter 14,* Creating Roof Designs*, when you click on the* PICK WALLS *command, the roof automatically offsets for the 1'-0" overhang. Or, if you choose the line tool in the roof command, you can trace the roof outline and the overhang will automatically offset the 1'-0" overhang.

Select the object and then click on the **OFFSET** Tab. The Options Bar will open, allowing you to type in the distance to offset the object.

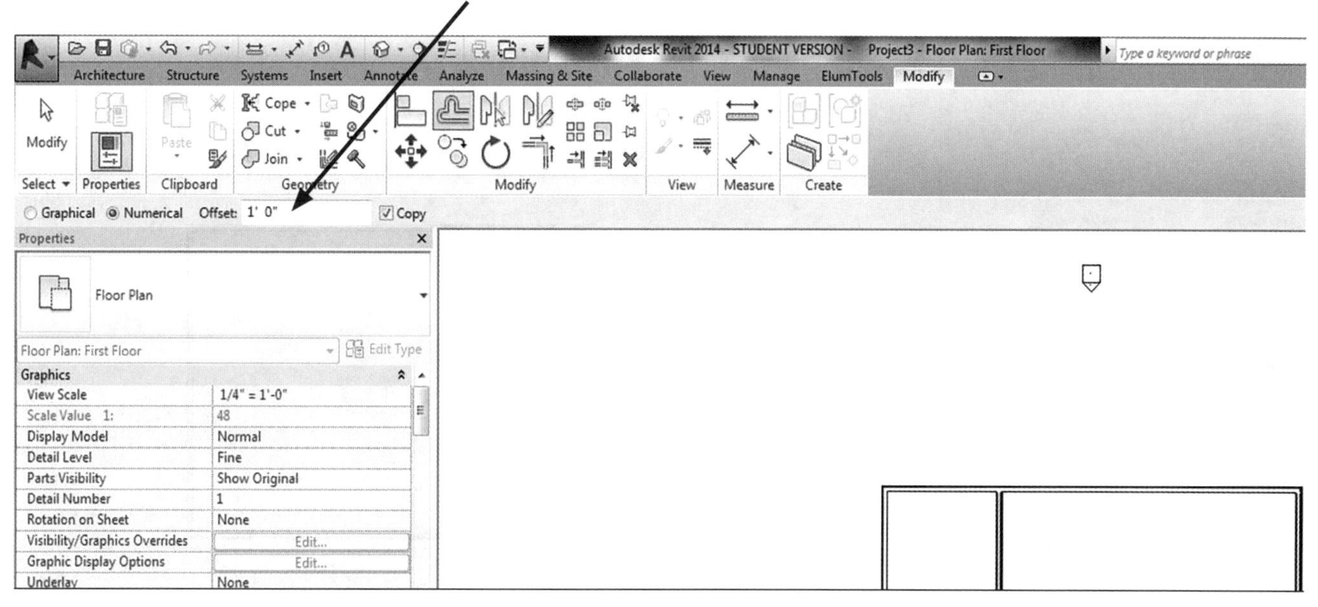

Once that is completed, when you hover over the object you will get a dashed line showing where the new object will be offset before you place it.

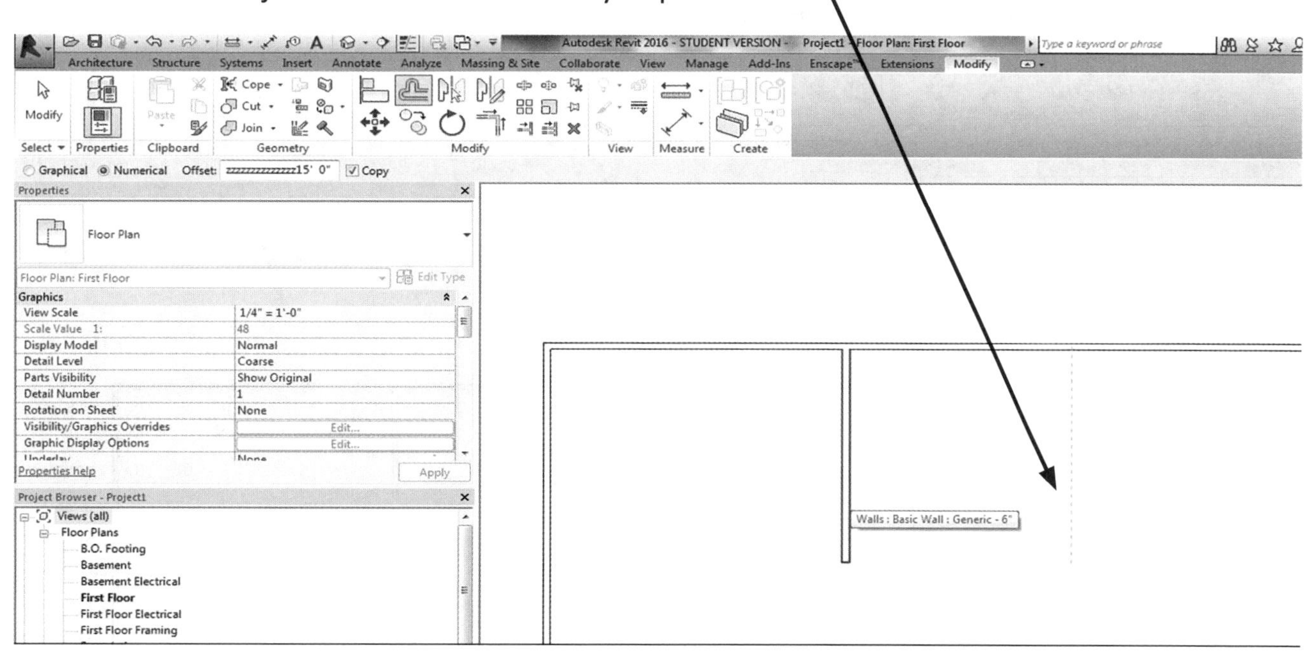

If you are satisfied, then click the location, and you have placed the object you offset.

THE ROTATE COMMAND

You can rotate objects in Revit just as easily as you do any other function. Select the object you want to rotate, and then click on the **ROTATE** Tab.

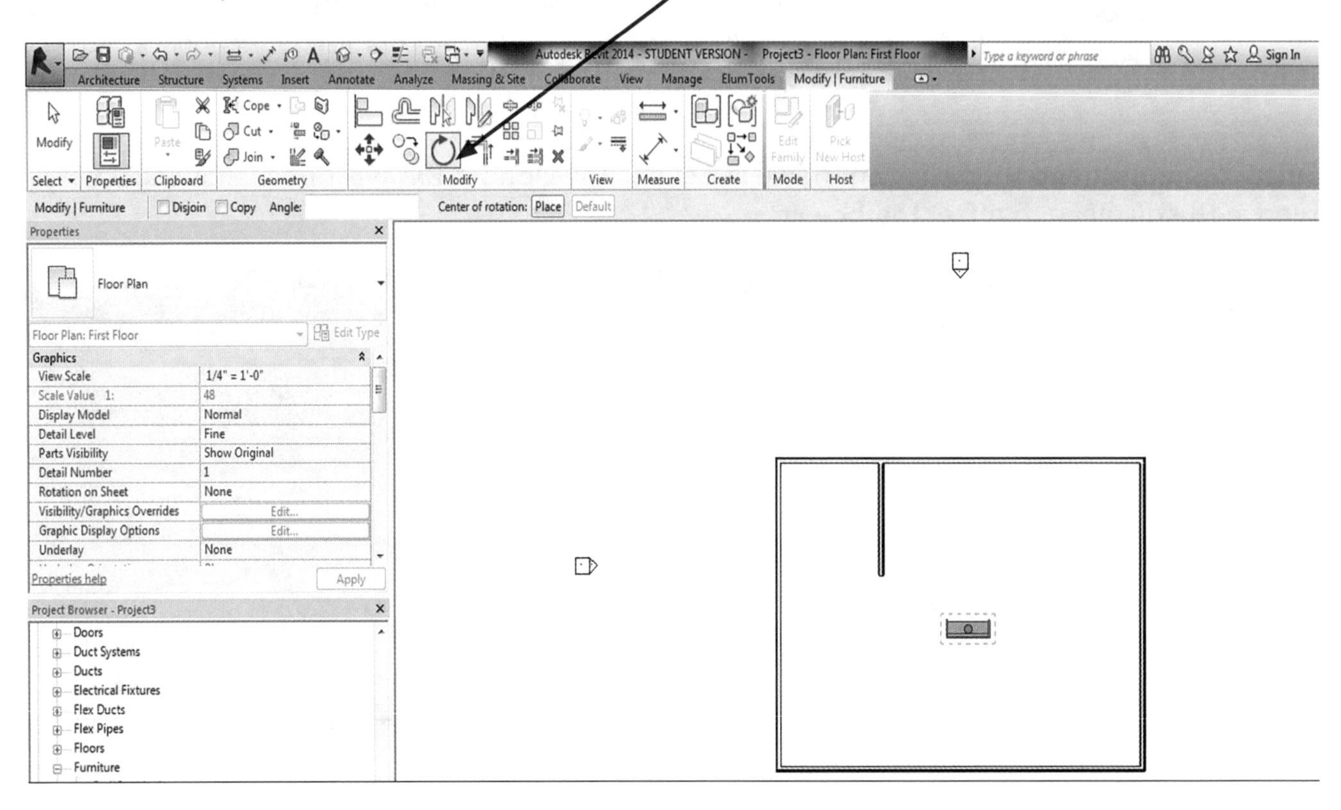

The object will highlight, you can then grab it, and the reference line will appear.

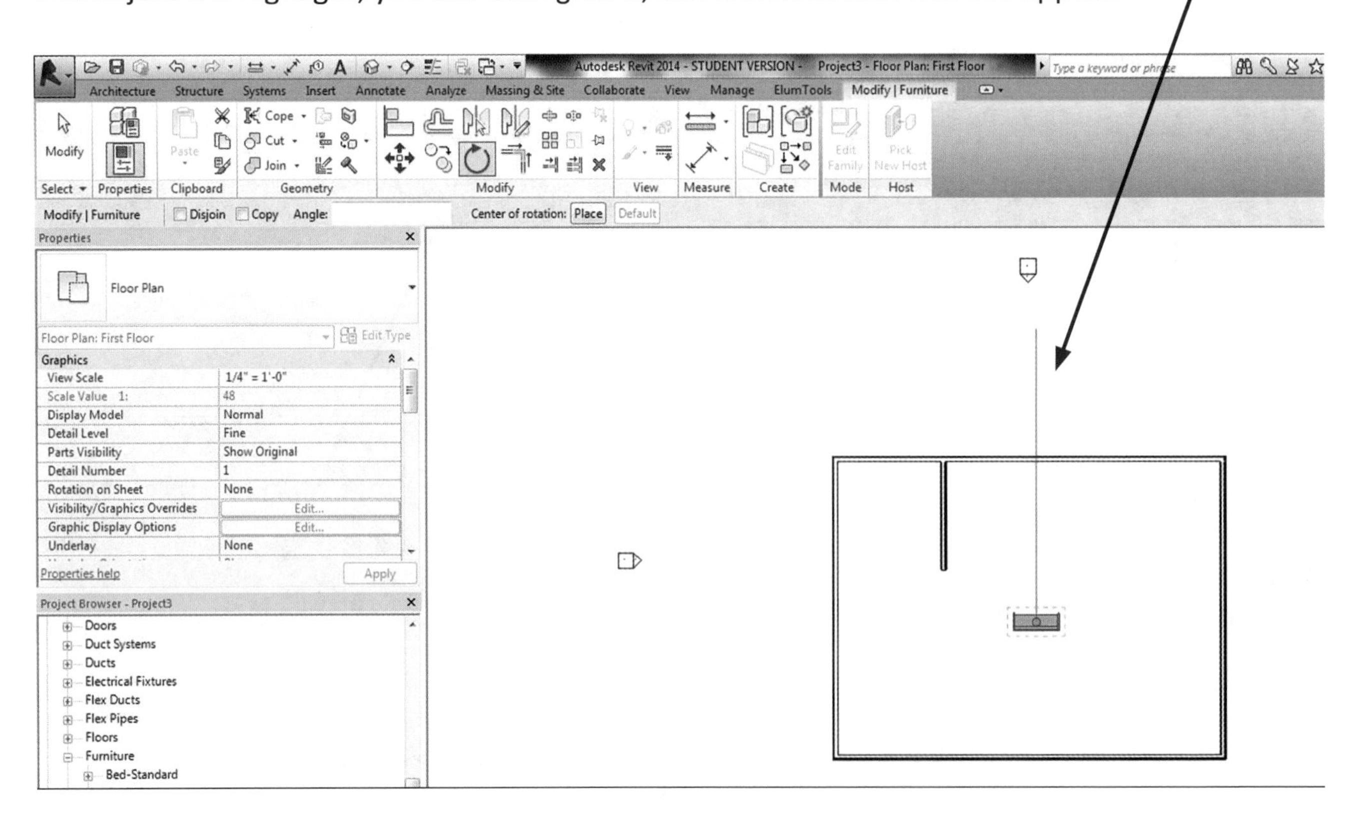

TIPS

I always make my reference line point straight up, for easier drawing.

ESSENTIAL REVIT FACT

Another way to rotate an object is to click on it and click your SPACE BAR. ***The object will rotate.***

Once the reference line is placed, you can move your cursor either left or right to change the angle of your object.

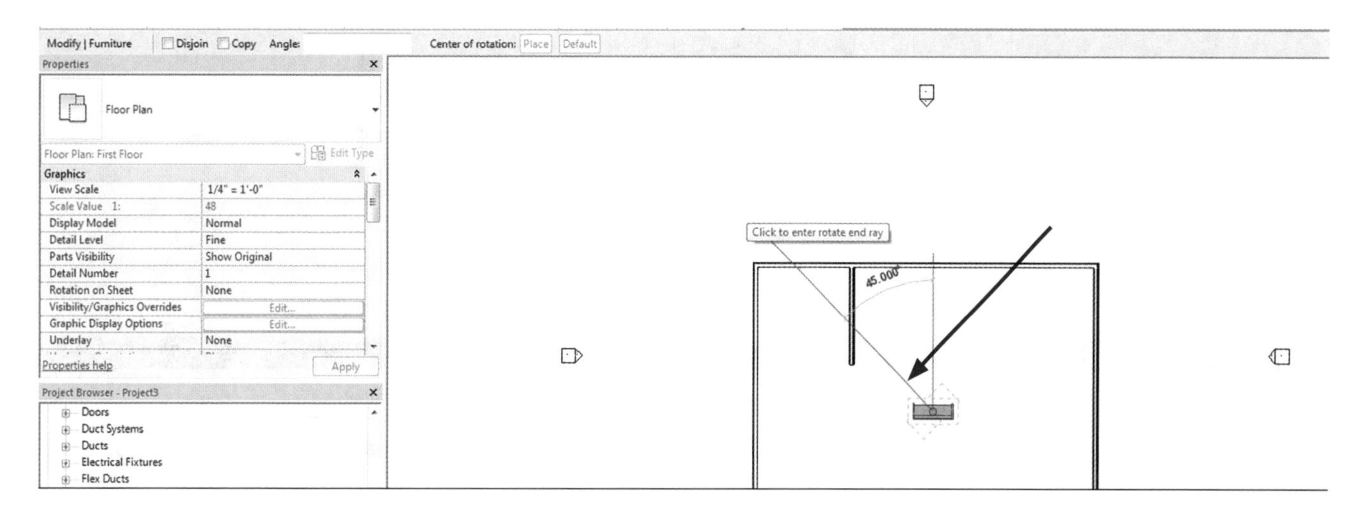

Once you have rotated it to the angle you want, click the mouse and you are complete in your rotation.

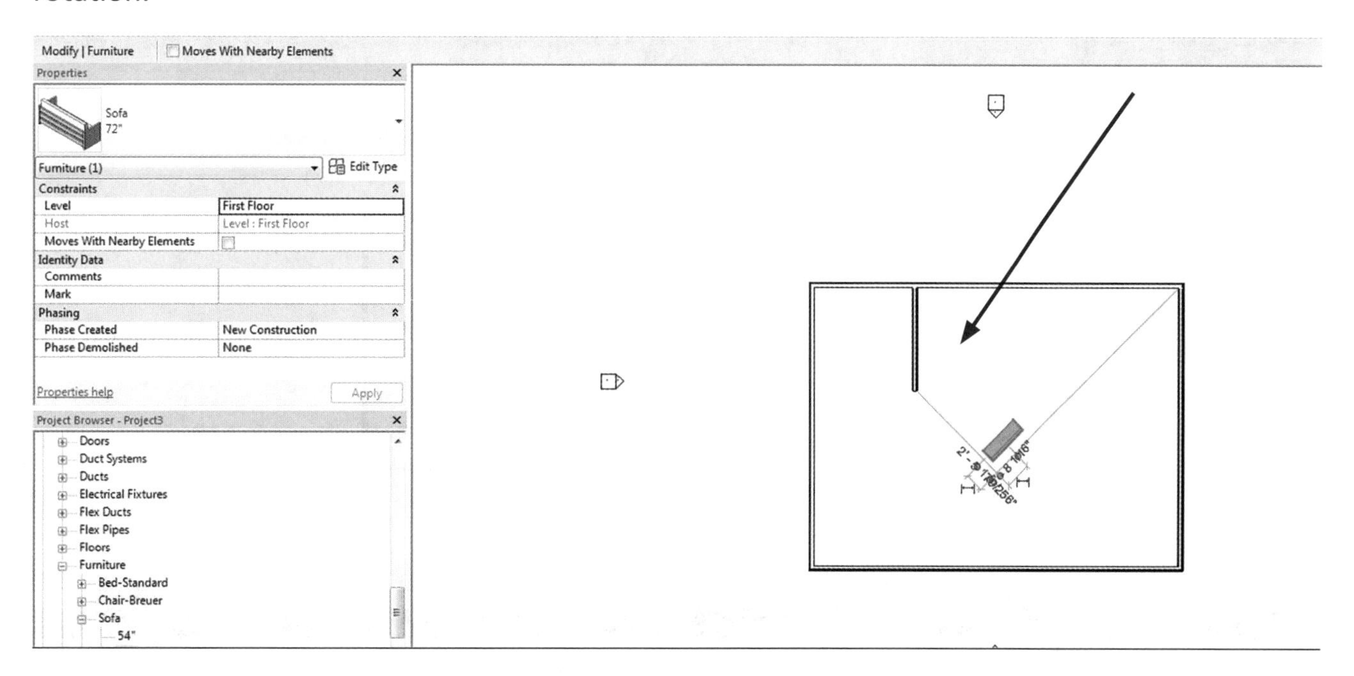

THE MIRROR COMMAND

You can easily create symmetry in your design. To mirror an object to another side of your design, you can select the object and click on the **MIRROR–DRAW AXIS** Tab.

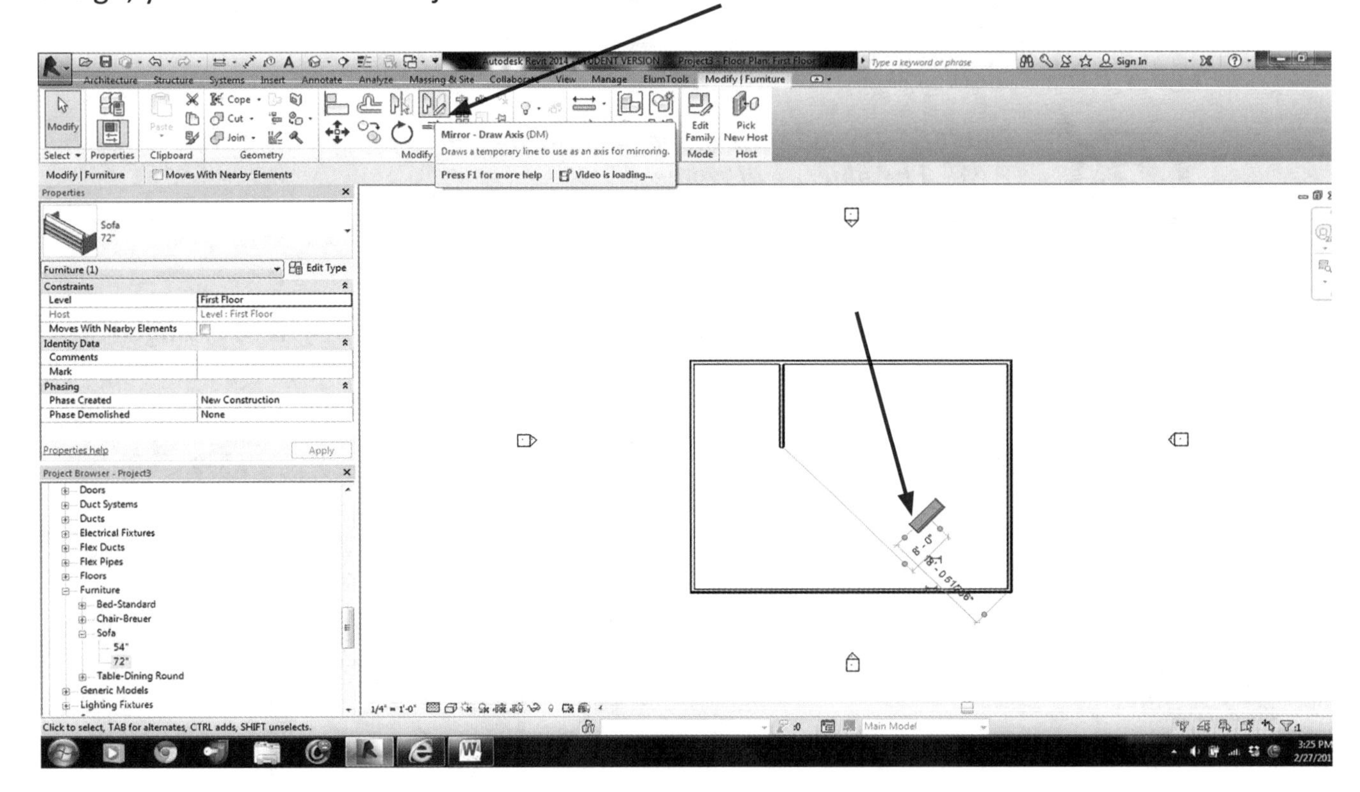

This allows you to pick an axis as your reference line. Draw your reference line through your design.

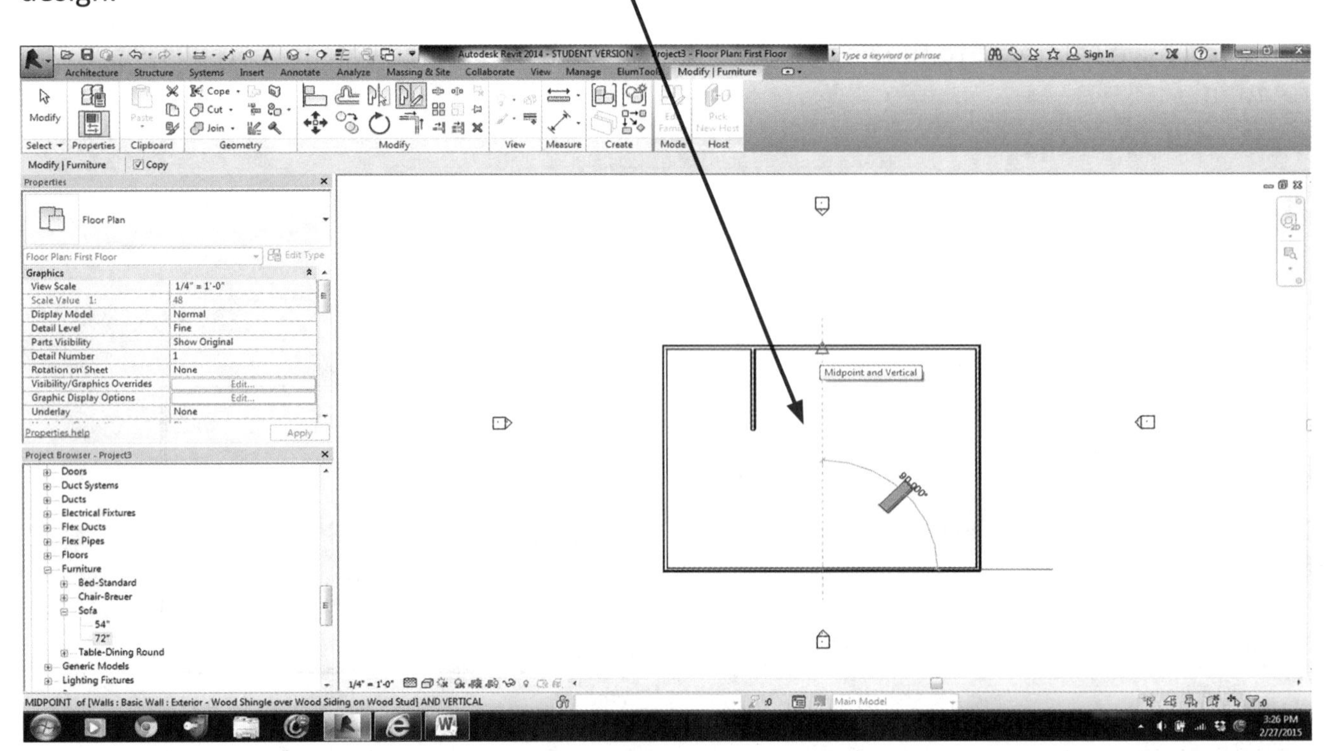

Once this is completed, the object you selected will then automatically mirror to the other side of your axis reference line.

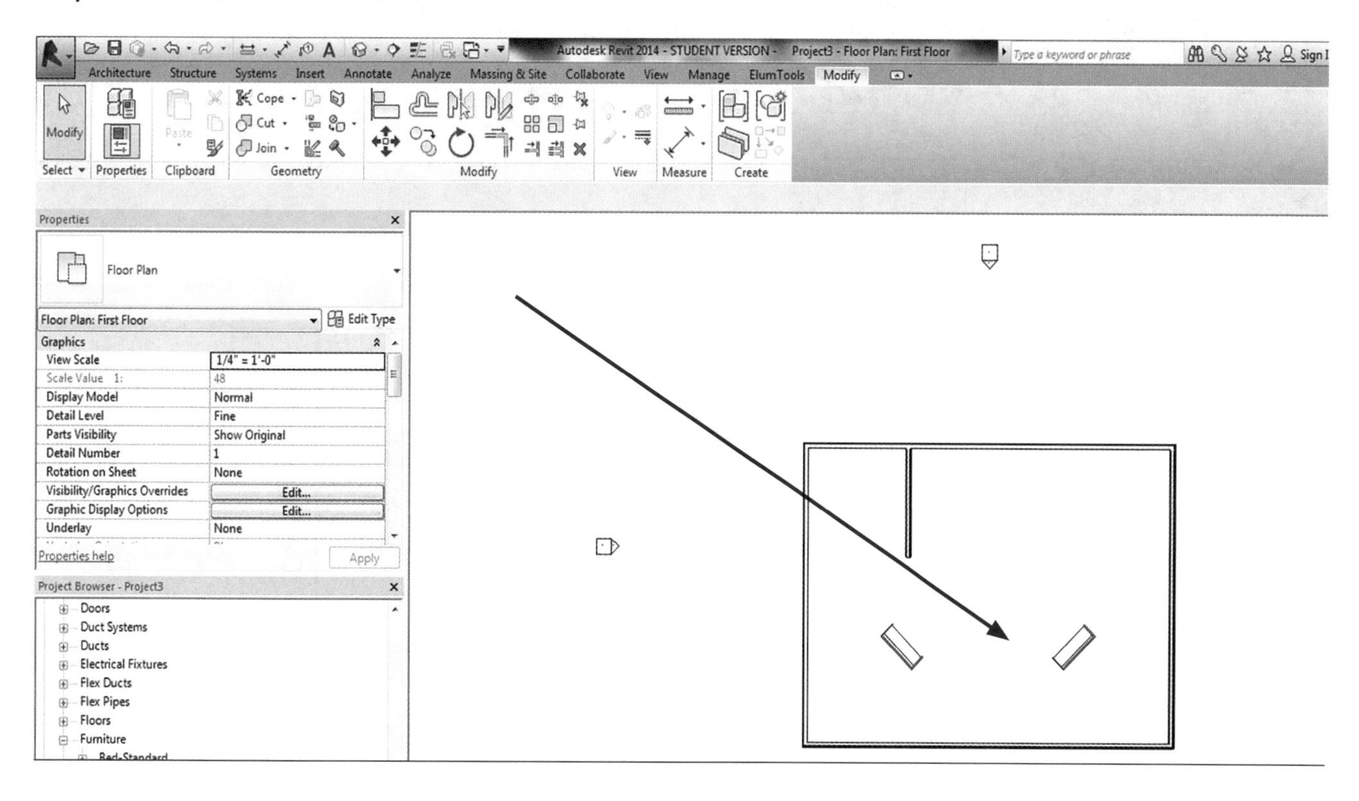

THE TRIM COMMAND

To trim lines or walls in Revit, you will select the object you want to alter. Then click on the **MODIFY** Tab to open up the Modify Tools.

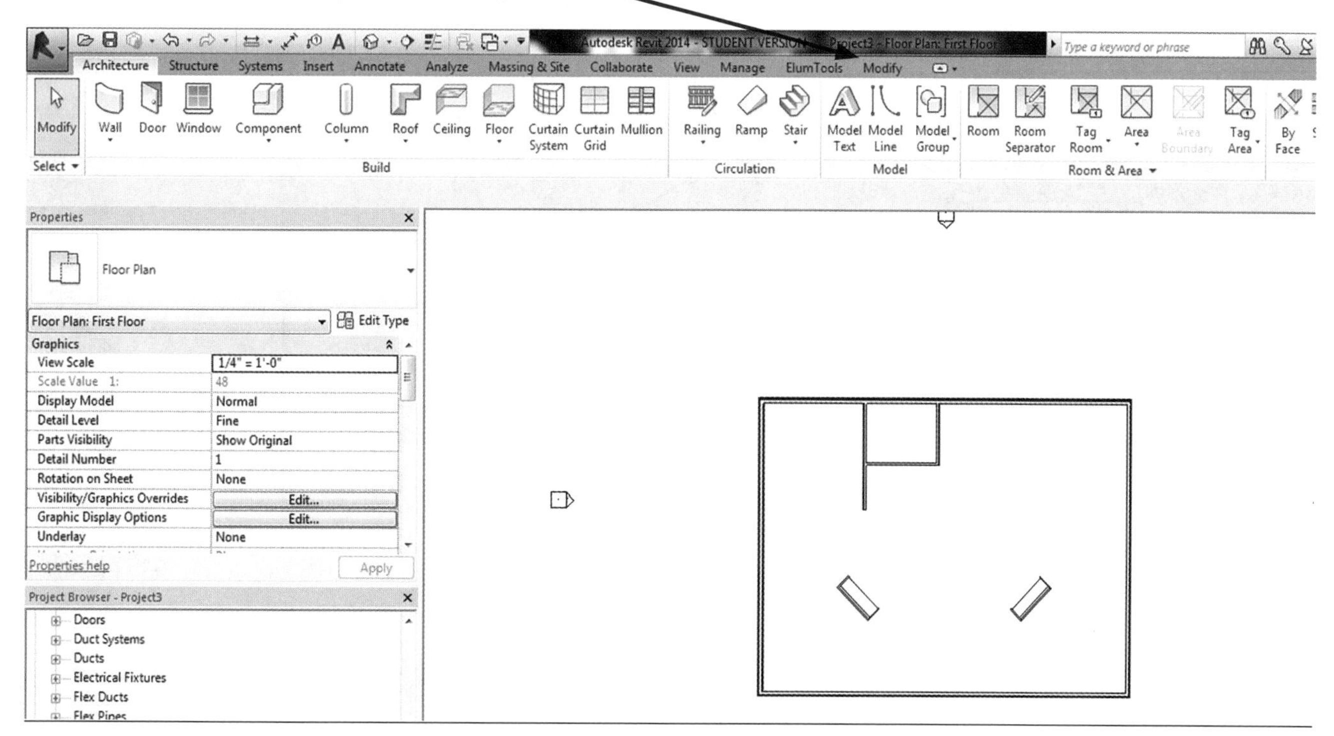

Open the **MODIFY TOOLS** Box, and then select the **TRIM** Command.

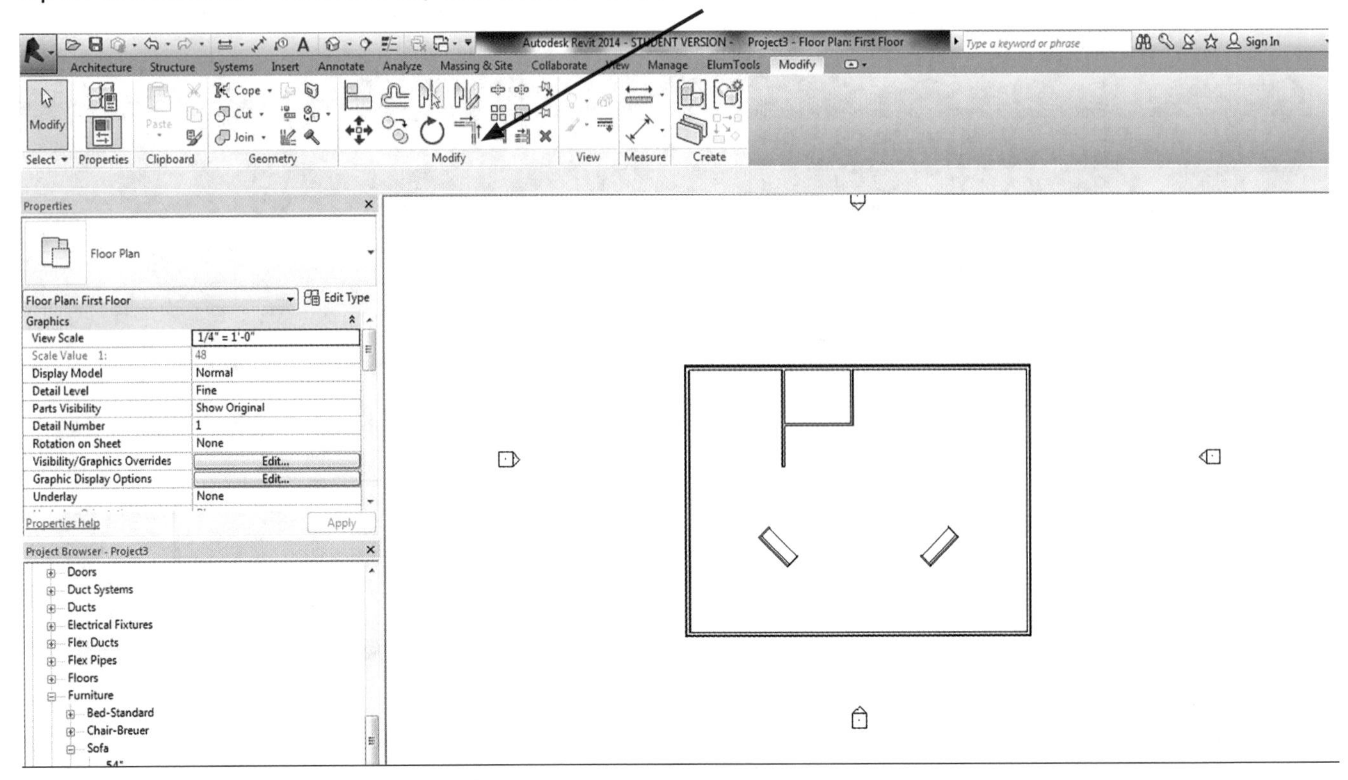

You will then click on the walls or lines you want to *keep*, and *not* the ones you want to get rid of.

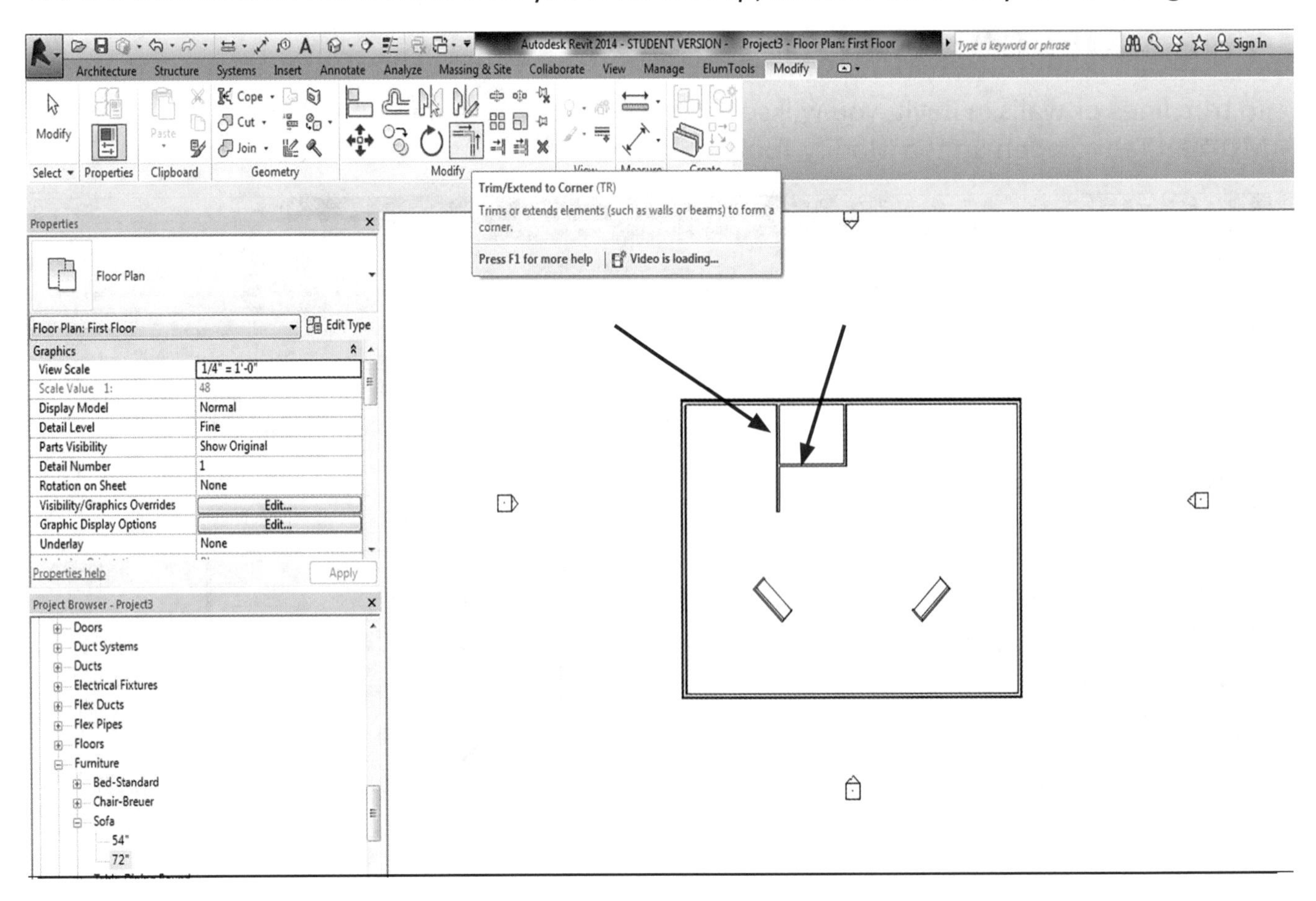

Your unwanted walls or lines will disappear.

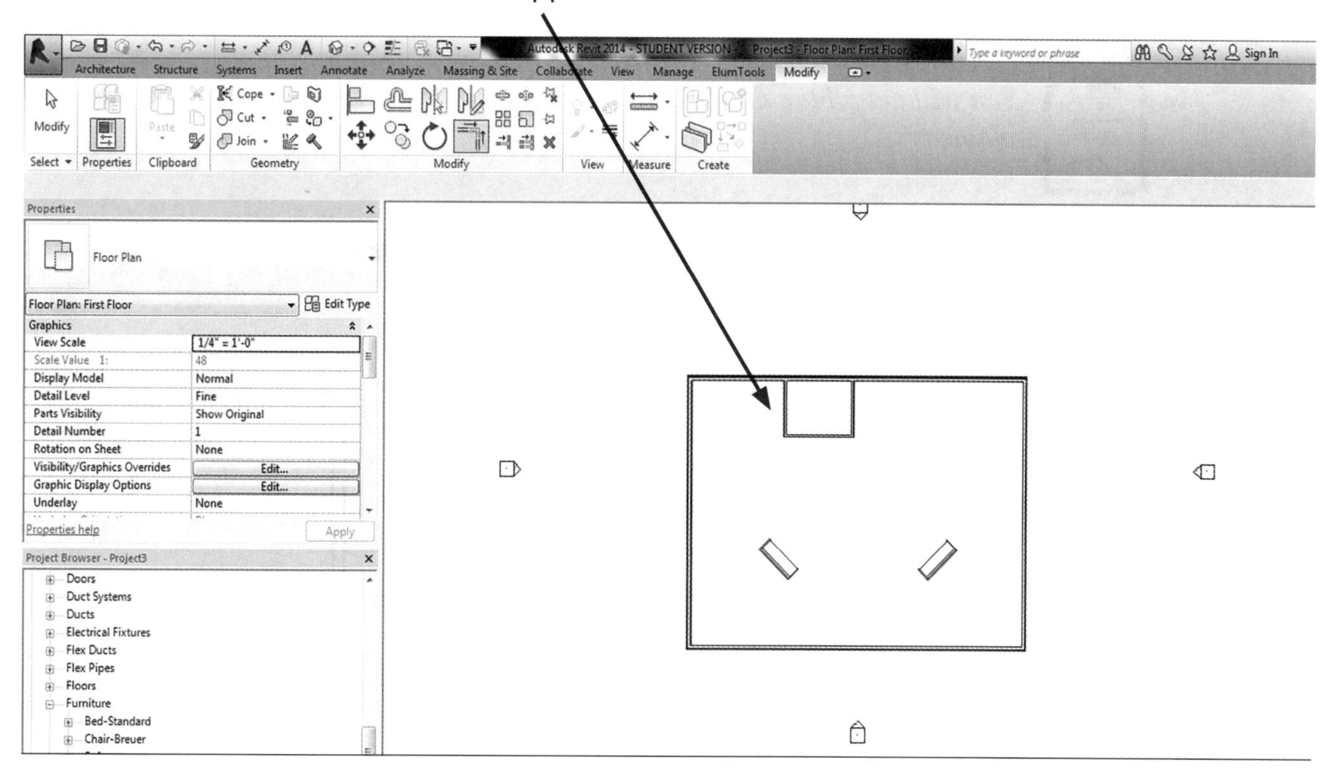

THE ALIGN COMMAND

The Align Tool will allow you to line up objects to make your drawing look more presentable. Click the **MODIFY** Tab and open the **MODIFY TOOLS** Box, then select the **ALIGN** Tab.

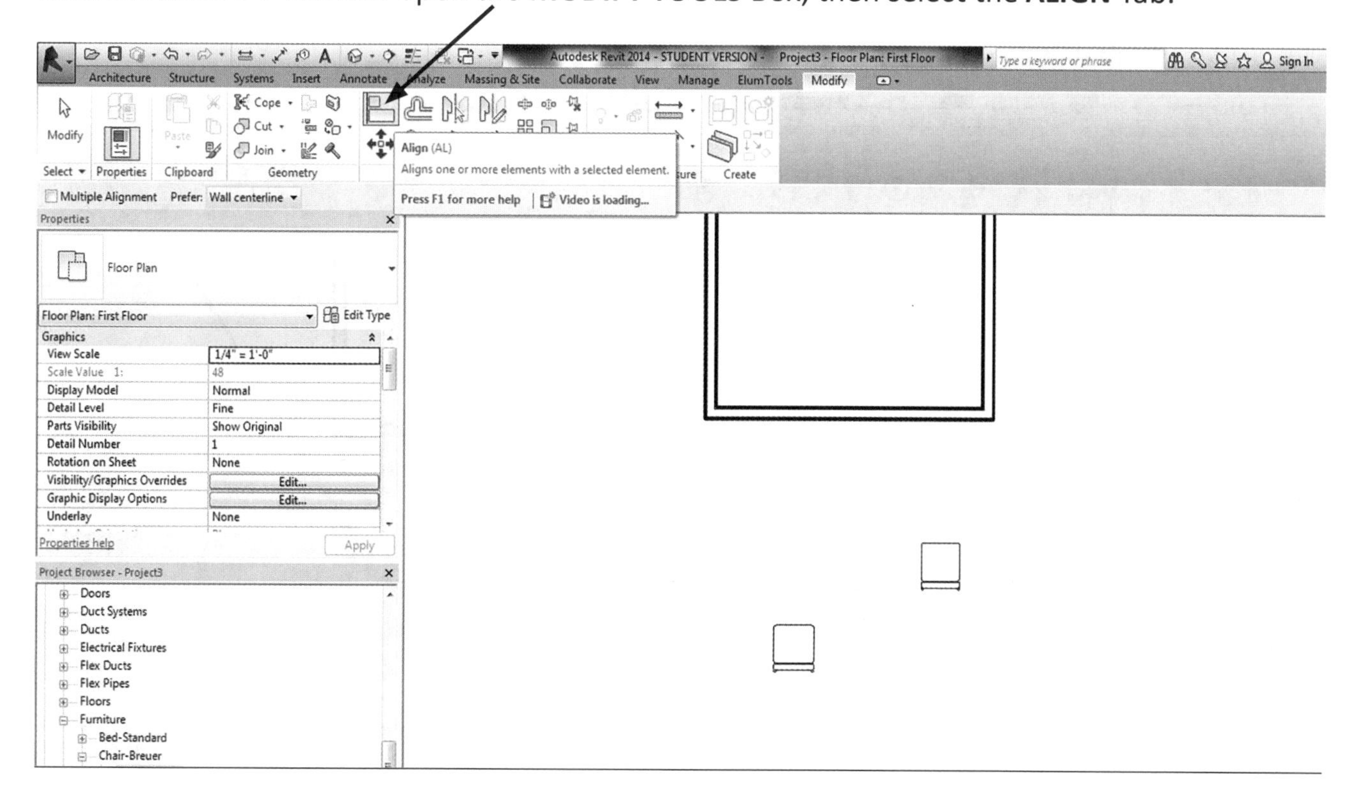

Select the object you want to use as your reference.

NOTES

You will get a blue dotted line or a light shaded dotted line as your reference plane.

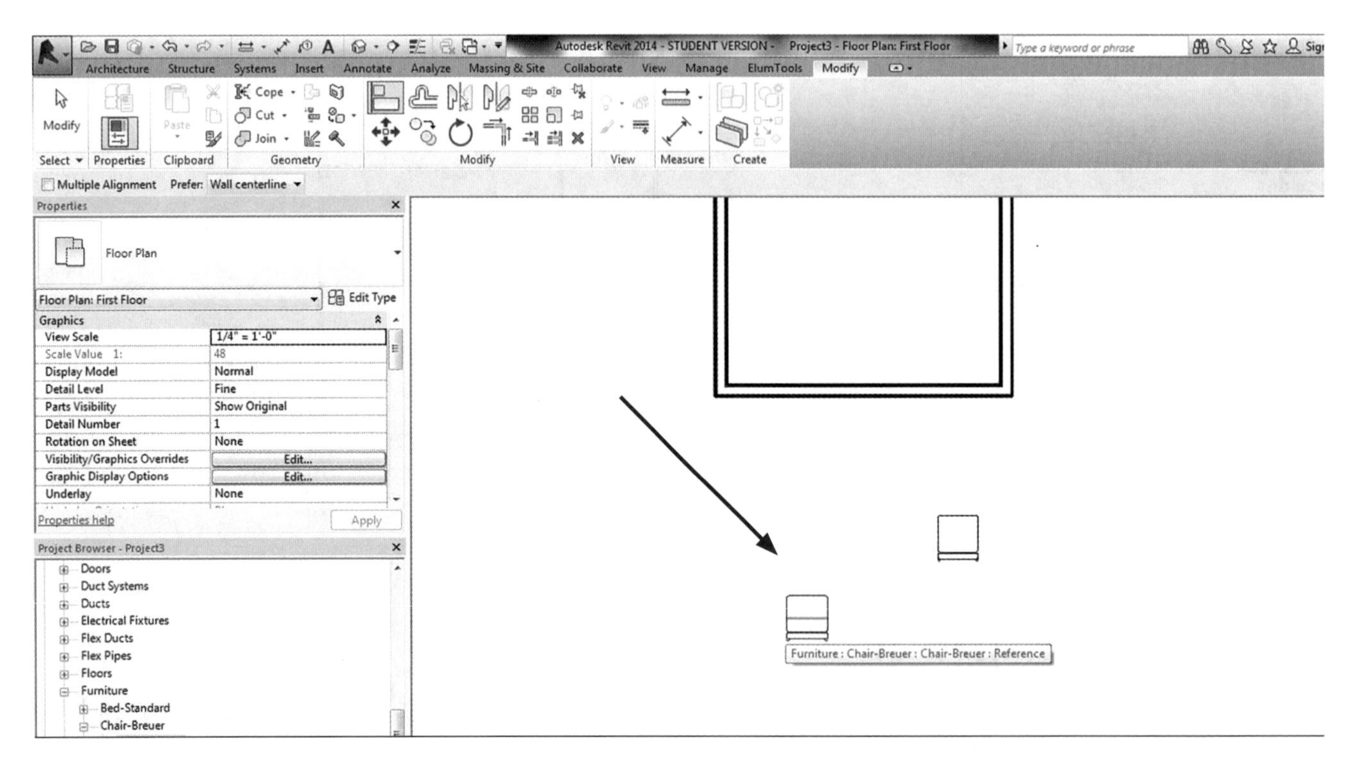

Once you have selected the object as your reference plane, select the object you want to align with the reference object.

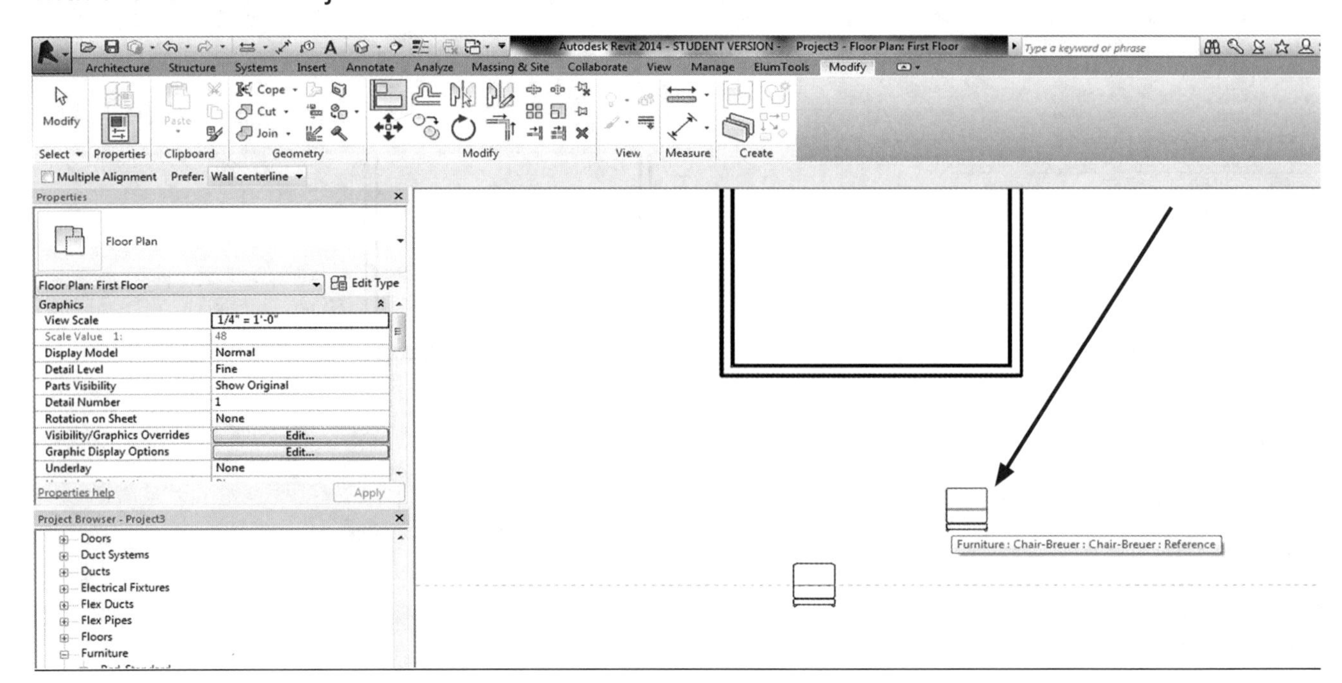

Your objects will align automatically, and you are finished.

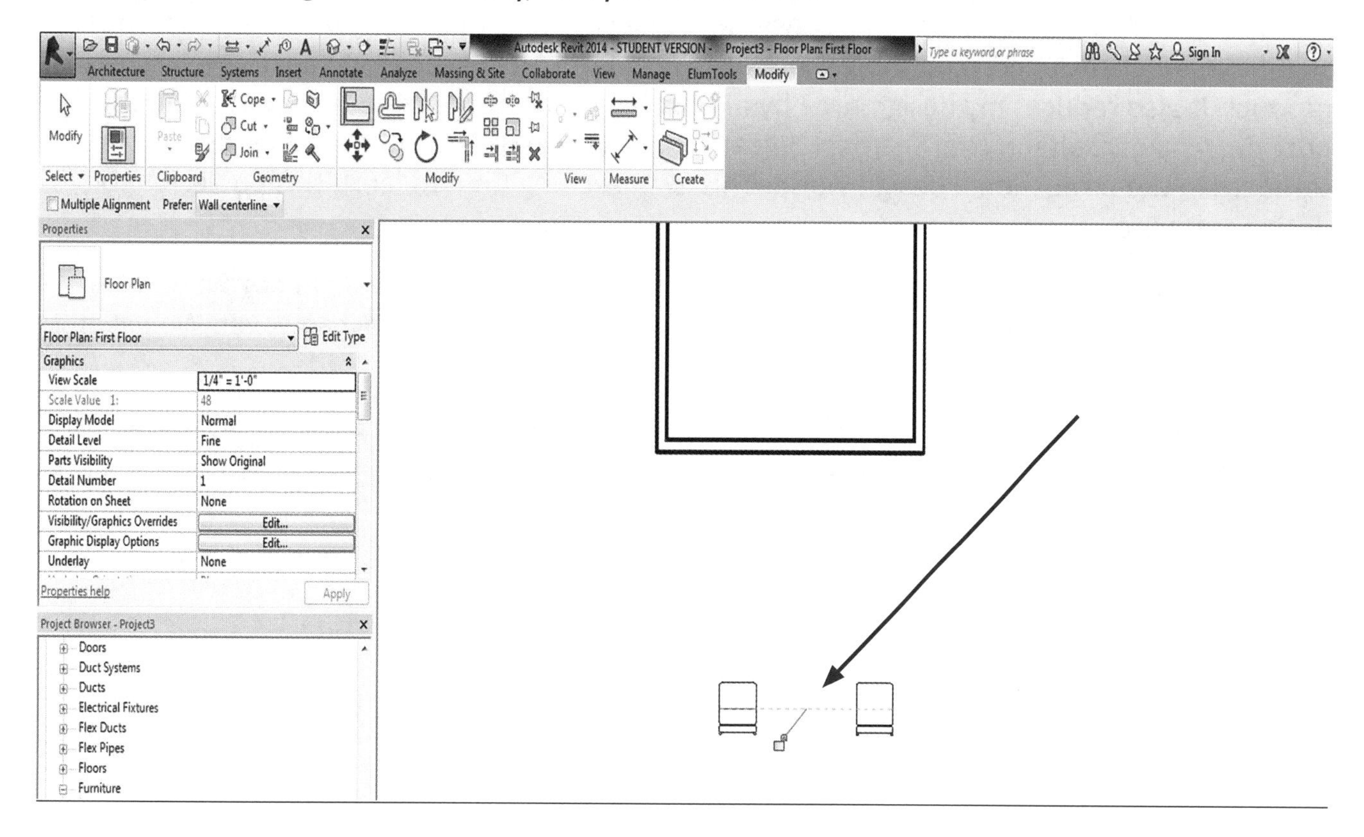

All the commands in Chapter 20 are ones I use on a daily basis because they make my job and designs more efficient. That is why I focused mainly on these six. As you get more familiar with Revit and your designs, and begin to implement your commands, you will see how it all comes together and improves your productivity.

You have taken an exciting step in learning the most powerful 3D architectural/design software which has become today's industry standard. As you have gone through this training manual, chapter by chapter, I hope you didn't get too overwhelmed, and want to be an expert overnight. Be patient.

There's a lot you have learned about the basics of putting together a house or building, from drawing the beginning walls, to the interior walls and kitchen layout, to even the bathroom design. There are chapters that showed how to place lights in the house which can be used in *any* design (not just for the kitchen area as the manual showed).

You were shown how to place furniture in the design, even finding out about the extended offerings in the Revit Library/Revit Distribution Content Library. And you even learned about adding complex mechanical equipment, such as an air conditioner and furnace.

Finally, you learned about the most powerful and exciting part of the Revit software—presenting your design in 3D and rendering with the proper colors to show yourself, friends, and clients.

Over the past 8 years, Revit has quickly become my favorite software from both a professional standpoint and for personal use as a consultant. The software makes my designs more understandable for clients, and I can show them *exactly* what they are going to see in the finished product. It also reduces my design time by almost half of what the other software used to provide. As a consultant, over the past 3 years I have witnessed Revit becoming the industry standard in the professional design world. This is an exciting time to learn this powerful 3D design software.

Please remember, this Revit Training Manual provides the *basics* of the Revit Software. There is so much more to learn about with this software than just one manual can cover. We haven't begun to scratch the surface. This is a great start for the 3D beginner.

Enjoy this manual and the best of luck on your journey to learning and using Revit. Again, be very patient. Most importantly, enjoy!!!

Notes

Notes

Notes

Notes